Magnetism, Magnetic Materials and their Applications

Magnetism, Magnetic Materials and their Applications

Proceedings of the Fourth Latin American Workshop on Magnetism, Magnetic Materials and their Applications, held in São Paulo, Brazil, June 1998

Editor:

F.P. Missell

ttp TRANS TECH PUBLICATIONS LTD
Switzerland • Germany • UK • USA

ISBN 0-87849-827-3

Volumes 302-303 of
Materials Science Forum
ISSN 0255-5476

***Distributed** in the Americas by*

Trans Tech Publications Inc
PO Box 699, May Street
Enfield, New Hampshire 03748
USA
Phone: (603) 632-7377
Fax: (603) 632-5611
e-mail: ttp@ttp.net
Web: http://www.ttp.net

and worldwide by

Trans Tech Publications Ltd
Brandrain 6
CH-8707 Uetikon-Zuerich
Switzerland
Fax: +41 (1) 922 10 33
e-mail: ttp@ttp.net
Web: http://www.ttp.net

Printed in the United Kingdom
by Hobbs the Printers Ltd,
Totton, Hampshire SO40 3WX

IV Latin American Workshop on Magnetism, Magnetic Materials and Their Applications

International Advisory Committee

D. Givord	*A. Hernando*	*F. Leccabue*
R. Grössinger	*H.A. Davies*	*J.L. Morán-López*
J.L. Sánchez	*F.P. Missell*	*F.H. Sánchez*

National Committee

F.P. Missell	*J. d'Albuquerque e Castro*	*A.P. Guimarães*
M.N. Baibich	*S. Gama*	*F.A. Machado*

Local Organizing Committee

F.P. Missell	*H.R. Rechenberg*	*A.D. Santos*
C.S.M. Partiti	*V. Villas-Boas*	*A.C. Neiva*

Contributors to the

IV Latin American Workshop on Magnetism, Magnetic Materials and Their Applications,

São Paulo, Brazil

Fundação de Amparo à Pesquisa do Estado de São Paulo - FAPESP
Conselho Nacional de Desenvolvimento Científico e Tecnológico – CNPq
Coordenadoria de Aperfeiçoamento de Pessoal de Nível Superior – CAPES
Pró-Reitoria de Pesquisa -Universidade de São Paulo
Centro Latino-Americano de Física - CLAF
International Centre for Theoretical Physics – ICTP
International Union of Crystallography - IUCr
CGA Analítica

PREFACE

The **IV Latin American Workshop on Magnetism, Magnetic Materials and Their Applications** was held in São Paulo, Brazil during the week of June 7-11, 1998. This was the fourth of a series of meetings initiated in La Habana, Cuba in 1991 and followed by Workshops in Guanajuato, México (1993) and Mérida, Venezuela (1995). A total of 111 researchers in magnetism participated in the **IV Workshop**, the largest part from Brazilian universities and laboratories. The **Workshop** fulfilled its primary goal of fostering Latin American integration with the participation of researchers from Argentina, Colombia, México, Cuba, Peru, and Venezuela.

Following the format of previous meetings, a number of invited talks were presented by leading specialists from Europe and North America, on topics such as thin films, giant magnetoresistance and magnetoimpedance, nanocrystalline materials, superconducting oxides, magneto-optics and other topics. An almost equal number of invited papers were presented by top-notch young Brazilian researchers, making a total of 20 invited papers. Another 24 papers were presented in oral form and 62 were presented as posters. A prize was given to the best poster: "EXAFS Characterization of Cobalt Nanoparticles Embedded in a Copper Matrix", presented by the student Júlio C. Cezar with collaborators H. Tolentino, M. C. Martins Alves, D. Z. Cruz, F. C. S. Silva, and M. Knobel.

The articles presented here were refereed for their scientific and technical content, in addition to editing of format and style. Although the authors are solely responsible for the veracity and accuracy of the material presented here, a small group of dedicated referees read the articles not only to check the format but also to try to help the authors communicate their ideas more clearly. This process was most successful in those cases where the author respected the conference deadlines for submission of papers. Special thanks for reading papers go to H. R. Rechenberg, V. Villas Boas, A. C. Neiva, M. V. P. Altoé, G. Goya, D. R. Cornejo, A. D. Santos, M. Knobel, F. J. G. Landgraf, W.A. Ortiz, M. S. Lancarotte, F. L. A. Machado, J. d'Albuquerque e Castro, and L. Sampaio.

A special person deserves thanks for his contributions to the **Workshop**. Iran Mamedes de Amorim proved to be an exceptionally able secretary, taking care of manuscripts, finances and other problems as they arose. Both before and after the **Workshop** his dedication was notable.

One of the most difficult problems in the organization of these meetings is clearly that of obtaining the necessary financial support. We have used the traditional sources of registration fees and sponsorship to meet our obligations. Although the list is short, the contributors on the following page were an essential ingredient in the success of this meeting.

Frank Missell　　　　　　　　October 1998

Table of Contents

Fundamental Properties

Thin Films and Magneto-Optics

Manganites and Oxides

Hard Magnets

Low Temperature Magnetism

Ferrites and Steels

Fluids and Particles

Materials Science Forum Vols. 302-303 (1999) pp. 1-10

Neutron Diffraction Studies of Magnetic Materials

W.B. Yelon
University of Missouri, Research Reactor, Columbia, MO 65211, USA

Keywords: Magnetic Materials, Neutron Diffraction, Substitutional Atoms, Interstitial Atoms

Abstract
Neutron scattering is a central tool in the study of magnetic materials. The neutron can probe the magnitude and direction of the site-specific moments in a multi-sublattice system as well as determine the distribution of substitutional atoms on different lattice sites, even for nearby elements in the periodic chart, for which the x-ray contrast is close to zero. The high sensitivity of the neutron to light elements has also allowed the determination of interstitial atom location and concentration in many high performance magnetic systems. Powder diffraction, coupled with Rietveld analysis has proven exceedingly powerful in examining novel materials, and in following the changes in magnetic ordering and crystal bonding as the parent compound is modified either substitutionally, interstitially, or both. Theorists have used the neutron results as the basis for detailed calculation of the changing band structures in these materials, and the results have been used to constrain other analyses (such as Mossbauer) to extract more reliable information. Work at the University of Missouri Research Reactor (MURR) has been carried out over many years, to determine site preferences, crystal structures ($Nd_2Fe_{14}B$, $Nd_3Fe_{29-x}Ti_x$), and interstitial occupancies. Both heuristic models and recent theoretical calculations are being used to predict these effects in new systems.

Introduction
Since the earliest days of neutron scattering in the late 1940s, application of neutron diffraction to magnetic materials has been central. The magnetic moment of the neutron allows for coherent diffraction from the ordered spins in a crystalline lattice. When the periodicity of the magnetic lattice is different from that of the structural lattice, as is often the case for antiferromagnets, new Bragg reflections of purely magnetic character will be seen in the diffraction diagram. When the two periodicities are the same the magnetic intensity will be superposed on the nuclear. In that case the magnetic intensity is often small and difficult to measure with accuracy. Polarized neutron diffraction, which improves the contrast between nuclear and magnetic intensity, can be used to improve the precision of the determination of the magnetic contribution. Dynamical (inelastic) neutron scattering was developed later and is used to measure the spin wave spectrum from which the magnetic exchange parameters and gap energy can be determined. However, both the polarized beam technique and inelastic scattering are intensity limited techniques that usually require single crystal specimens. Their use is, thus, restricted to a small number of systems for which appropriate samples are available and for which the time requirements are compatible with the scientific content.

The development of the Rietveld method [1] in the late 1960s, however, has allowed for reasonable precision in magnetic structure determination even when the magnetic contributions to the Bragg intensities are small. The refinements use the entire diffraction diagram and are able to extract magnetic moments even when the magnetic intensity is only a few percent of the total intensity. For the study of permanent magnet phases we have found that other aspects of neutron scattering are at least as important as the magnetic contribution. In particular, detailed structural knowledge is essential to any thorough understanding of these materials, and conventional neutron diffraction coupled with Rietveld refinement provides this information better than any other technique.

The neutron cross sections for a number of important elements are shown in Table I [2]. The atomic numbers (z) for those are also listed since the x-ray cross section is proportional to z. There are several important features shown. With the exception of a few elements (V, Li) the scattering length b ranges from 0.25 to 1.25 x 10^{-12} cm, and is generally between 0.35 and 0.95 x 10^{-12} cm. Thus, all elements contribute significantly to the scattering, while for x-rays the light elements essentially disappear in any compound with heavy elements such as rare-earths. While the x-ray cross section changes only slowly with z, nearby elements may have significantly different neutron cross sections, such as Fe, Mn and Co, and are thus easily distinguished. Since the neutron scattering cross sections are through neutron-nuclear interactions the scattering for different isotopes (such as H and D) may be quite different. The neutron is a neutral probe and interacts weakly except with those isotopes showing large absorption cross sections (unfortunately including Sm), but even for these elements separated isotopes are often available which can be used for the scattering study. The cross section for nuclear scattering is essentially independent of angle, while the magnetic scattering is characterized by a form factor similar to that for x-ray scattering. This allows the structural parameters to be unambiguously determined from the high angle portion of the scattering diagram allowing the intensities at low angle to be refined for their magnetic contribution. These features combine to make neutron diffraction the ideal probe for investigation of the structure of magnetic materials and especially for changes in the structure and magnetism when interstitial atoms are introduced or when the structure is modified by the partial replacement of one element by another.

Table I: Neutron cross sections for selected elements[1]

	z	σ_{coh}	b	σ_s	σ_a		z	σ_{coh}	b	σ_s	σ_a
H	1	1.76	-0.374	81.67	0.3326	Mn	25	1.75	-0.373	2.1	13.3
D	1	5.6	0.667	7.63	0.0005	Fe	26	11.44	0.954	11.66	2.56
C	6	5.55	0.665	5.56	0.0035	Co	27	0.8	0.253	6.15	37.18
N	7	10.07	0.93	11.5	1.9	Pr	59	2.49	0.445	2.58	11.5
Al	13	1.49	0.345	1.51	0.231	Nd	60	7.43	0.769	16	51
Si	14	2.16	0.415	2.17	0.171	Sm	62	0	0	0	5670

Until the introduction of the Rietveld method, the only reliable way to extract such information, however, was through structure refinement of single crystal data in which each structure factor was independently measured. Powder data, could be employed using only those reflections whose intensities could be cleanly calculated, from regions of the diagram with little overlap of Bragg reflections, which was often insufficient for reliable refinement. The Rietveld method uses the entire powder diffraction diagram, calculating the intensity as a function of angle from a structural model

[1]Cross sections in units of 10^{-24}cm^2, b in units of 10^{-12}cm. The coherent cross section, σ_{coh}, is equal to $4\pi b^2$.

and information about the diffractometer function. The model is parameterized in terms of atomic coordinates, magnetic moments and direction, temperature factors, site occupancy parameters, etc, while the instrument function includes background and peak shape terms. This last may include the effects of particle size and strain broadening. The parameters are then adjusted in a least square minimization. The Rietveld method has now been incorporated into a number of different computer codes, GSAS [3], RIETAN [4] and FULLPROF [5]. While none of these does everything in the most convenient way, they provide considerable power to the user for refinement of powder data, including joint x-ray/neutron refinement and refinement of either steady state (reactor) or pulsed source neutron data. The refinement codes have proven to be quite robust, rarely converging to false minima and often usable in structure solution, from a partial structure model, an aspect not anticipated in the conception of the method.

Instrumentation

Until a few years ago x-ray and neutron instruments for powder diffraction looked quite different, even neglecting the obvious size differences. X-ray diffractometers employed small samples, wide-open beams and focusing optics. Neutron instruments used multiple (Soller) slit arrays to control the beam divergences and large samples (typically 10 gm) to compensate for the intensity losses and the weak sources. In addition, multiple individual detectors (up to 64) are used to collect data simultaneously at from each detector element. Recently, however, the University of Missouri Research Reactor, MURR, has taken the lead in the development of a new type of instrument more closely resembling the x-ray designs. Figure 1 shows the general layout of the MURR powder diffractometer [6]. Rather than controlling the divergence of the beam, the instrument uses a bent, nearly perfect monochromator and small sample which "back-selects" the neutrons which can be scattered at a particular angle. The small sample size matches the resolution element of a position sensitive detector, such that some 400 data points are measured simultaneously using one detector bank. Multiple detector banks could be employed to increase the data rates further. In this way major gains in data acquisition rates have been achieved while the sample dimensions have been reduced to 3 mm diameter and mass of order 0.2-1 gm. This reduction, of course, is useful for those preparing materials, but also allows for measurement of moderately absorbing samples something that has not hitherto been possible. A similar system has been proposed for construction at the Brazilian reactor center IPEN and for a number of other facilities.

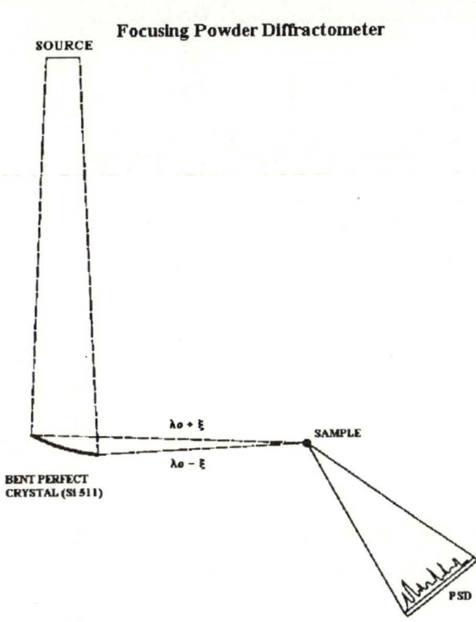

Fig. 1: The general layout of the MURR powder diffractometer.

Results

The first studies of permanent magnet related materials at MURR investigated the substitution of Fe by Mn in the RE_6Fe_{23} phase [7]. This resolved two controversies. The first concerned the magnetic ordering of the pure Mn end member. This was shown to be ferrimagnetic, with two of the four sublattices aligned antiparallel to the others, leading to a small net moment. More important in the future development of the field was the observation (Fig. 2) that the substitution was not random, i.e. equal percentages of Mn on each of the four sublattices. Instead, strong site preferences were exhibited, which can be related to the relative size of the different sites. The Mn atoms preferred to fill the larger sites first. Similar studies with similar conclusions were subsequently made of the $Nd_2Fe_{17-x}Co_x$ system [8]. These and later studies have established that preferential site occupancy is ubiquitous, although size considerations only explain some of the observed effects when the substituent is electronically dissimilar.

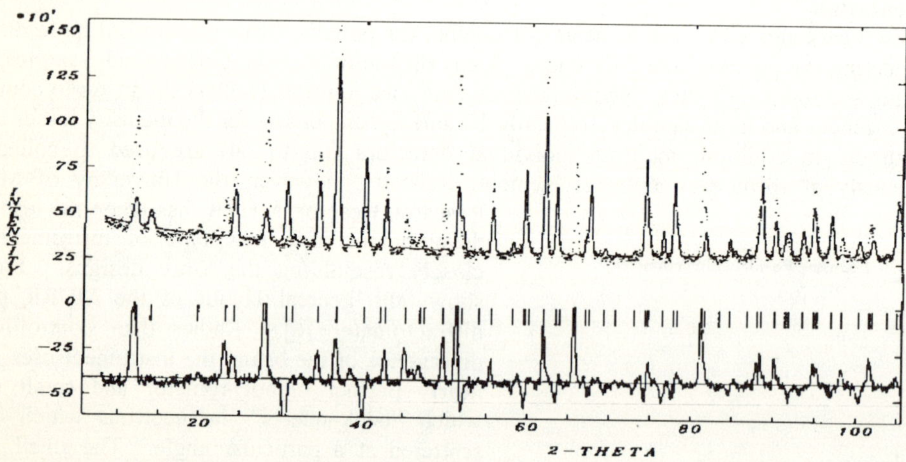

Fig. 2a: Best fit to neutron data for $Y_6(Fe_{0.4}Mn_{0.6})_{23}$ assuming random occupation of transition metal sites by Fe and Mn.

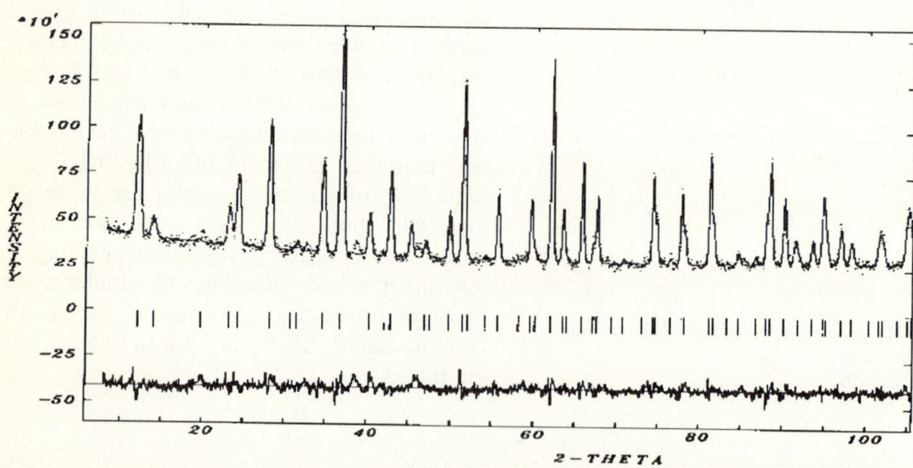

Fig. 2b: Best fit to neutron data for $Y_6(Fe_{0.4}Mn_{0.6})_{23}$ allowing preferential occupation of transition metal sites.

When the $Nd_2Fe_{14}B$ phase was discovered single crystals were not available for structural analysis, although the space group and lattice parameters had been determined from powder x-ray data. The Rietveld method was employed at MURR to construct a model, by assuming that at least a few of the high multiplicity sites were filled with Fe atoms. These converged rapidly into stable positions and fitted much of the observed spectrum. Filling in the rest of the atoms was a simple task, and the solution of the structure, was, at that time, the largest ever directly from powder data. The magnetic refinement showed a particularly interesting result. The j_2 site, which is analogous to the dumbbell site in the RE_2Fe_{17} phase, has a magnetic moment substantially larger than those on the other five Fe sites [9]. The structure of $Nd_3Fe_{29-x}Ti_x$ was also solved from neutron powder data [10]. The close relationship of the monoclinic structure to the $CaCu_5$ structure simplified the analysis, but the initial choice of space group, reported from x-ray measurement [11] proved to be incorrect. Subsequent analysis using the correct space group resulted in substantially similar coordinates, but refinement of Cr and Mn substitution was stable and reasonable, which was not the case with the wrong space group [12].

Neutron diffraction was the perfect tool to investigate systems in which the magnetic properties are strongly enhanced by the incorporation of interstitial atoms such as N or C. These include the RE_2Fe_{17} and $REFe_{12}$ phases. Early studies by thermogravimetric analysis suggested that more than three N atoms in the former phase or more than one in the latter, could be incorporated into the lattice. Neutron diffraction indicated that only the octahedral interstitial sites are occupied (except, perhaps at low N concentration) and that the high concentrations reported were the result of decomposition, which was easily observed in the diffraction diagrams. The data were analyzed to determine the changes in Fe-Fe bonding and theory was used to attempt to separate the effects of lattice expansion from those due to electronic effects on the magnetic properties, especially Curie temperature [13].

When similar effects were observed in compounds in which Fe is partially substituted by Si, Al or Ga, neutron diffraction again played a central role in clarifying the phenomenon [14-16]. It was found that these substituents, at low concentration, prefer the 18h site in the RE_2Fe_{17} phase, which is not the largest (volume) site. Instead, this site has the largest number of RE near-neighbors (3). The enhancement in Curie temperature cannot be attributed to magnetovolume effects since all three species give similar enhancements, but Al and Ga cause the lattice to expand, while Si causes it to contract. Figure 3 shows the occupancy of the four Fe sites in the RE_2Fe_{17} phase as a function of Al concentration. These results have been used as constraints in fitting Mössbauer spectra on the same samples [17] and as input information (along with atomic coordinates and cell parameters) for theoretical modeling [18]. Although increases in T_c can be inferred from the modeling, the details are still not

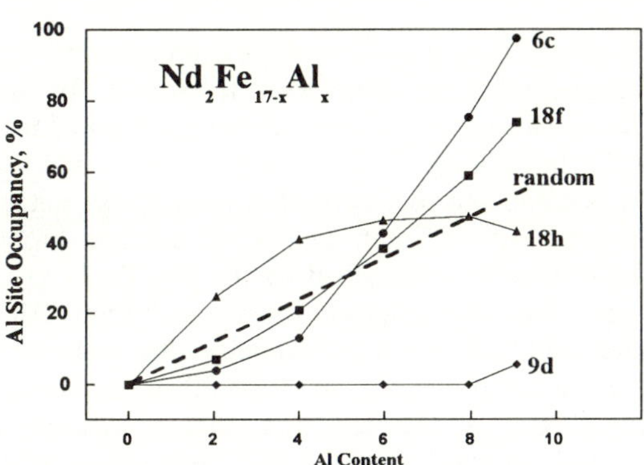

Fig. 3: The occupancy of the four Fe sites as a function of Al concentration.

confirmed. The neutron data collected at MURR and elsewhere has stimulated the development of a model for site preference based on size and chemical effects, which appears to correctly predict most of the observations [19,20].

The highest Curie temperatures in these phases, which are derived from the $CaCu_5$ structure, are achieved for nitrogenated samples without any substitutional atoms. These materials, which are prepared by gas phase insertion of the N atoms, unfortunately are thermodynamically unstable and decompose well below the temperatures required to process dense, sintered magnets. It has been found that Si and Ga substitution allows relatively high C interstitial concentration from melt prepared alloys. An obvious question is whether the substituent occupancies are the same as for samples prepared without C. For Si and Ga this is found to be the case, but transition metal substituents show pronounced changes in occupancy patterns with C incorporation (Fig 4). This is attributed to the high affinity of some of those (V, Nb, Mo, Ti) for bonding with C [21].

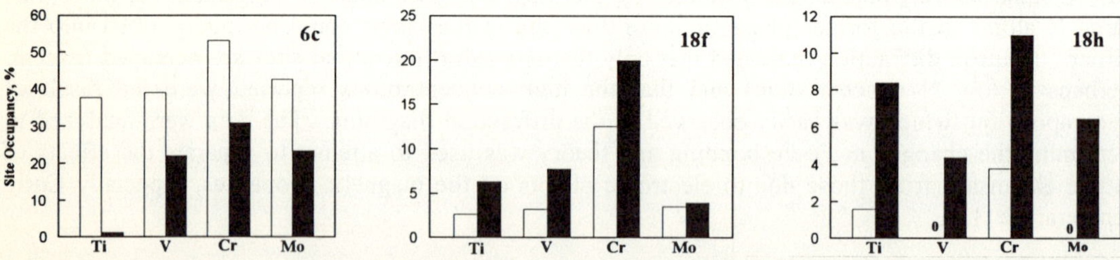

Fig. 4: A comparison of the site occupancy of Ti, V, Cr and Mo in $Nd_2Fe_{17-x}TM_x(C)$ before (open columns) and after (solid columns) C incorporation.

Despite intensive study, none of the $CaCu_5$ derived phases appear to be serious candidates for high energy-product magnet application. Thus, strategies must be developed to search for new candidates. The RE_6Fe_{23} phase and the RE_5Fe_{17} phase may be considered. Although both of these phases are rare-earth rich, they could be economically interesting if high energy-products and high Curie temperatures are achieved. While the former is cubic and therefore does not possess suitable anisotropy, it can be made tetragonal through the incorporation of interstitial H and perhaps through substitution or other interstitial occupation. The latter was first described by the Sao Paulo group [22] (RE = Nd) and has been shown to become uniaxial when Nd is partially replaced by Sm. Thus, we have initiated a collaborative program to understand and attempt to modify this material to develop the required properties. The first step has been a comprehensive neutron diffraction study of the parent compound [23]. In spite of the complexity of the structure, which contains 14 independent Fe sites and 7 independent RE sites, the neutron powder refinement gives greater precision for the atomic coordinates than did the single crystal x-ray analysis [Table II].

Table II: Neutron and x-ray coordinates for Nd_5Fe_{17}[2]

Atom	site	neutron refinement			x-ray refinement			near neighbors	
		x	y	z	x	y	z	Fe	Nd
Nd-1	2a	0	0	0	0	0	0	0	12
Nd-2	4d	0.3333	0.6667	0	0.3333	0.6666	0	12	6
Nd-3	$6g_1$	0.1218(6)	0	0.25	0.1202(4)	0	0.25	4	9
Nd-4	$6g_2$	0.5042(6)	0	0.25	0.5029(4)	0	0.25	13	4
Nd-5	$6g_3$	0.7988(6)	0	0.25	0.7999(4)	0	0.25	9	6
Nd-6	12i	0.1117(4)	0.2234(4)	0	0.1123(1)	0.2246(4)	0	10	7
Nd-7	24l	0.5398(3)	0.1588(3)	0.1151(4)	0.5403(2)	0.1587(2)	0.1153(3)	13	4
Fe-1	6f	0	0.5	0	0	0.5	0	6	6
Fe-2	6g	0.6550(5)	0	0.25	0.653(1)	0	0.25	12	2
Fe-3	$12k_1$	0.7089(3)	0	0.0455(6)	0.71(7)	0	0.049(1)	9	3
Fe-4	$12k_2$	0.3834(4)	0	0.0874(6)	0.3848(7)	0	0.089(1)	9	3
Fe-5	$12k_3$	0.2622(3)	0	0.1435(5)	0.2622(6)	0	0.148(1)	9	3
Fe-6	$12j_1$	0.2953(4)	0.1151(3)	0.25	0.2944(7)	0.1158(7)	0.25	8	4
Fe-7	$12j_2$	0.3905(4)	0.0690(4)	0.25	0.3913(7)	0.0701(7)	0.25	9	3
Fe-8	$12j_3$	0.4419(4)	0.2029(4)	0.25	0.4423(7)	0.2026(6)	0.25	10	2
Fe-9	$12j_4$	0.6610(4)	0.1364(4)	0.25	0.6612(7)	0.1367(7)	0.25	9	3
Fe-10	$12j_5$	0.6909(4)	0.2746(4)	0.25	0.6902(7)	0.275(7)	0.25	6	6
Fe-11	$24l_1$	0.3802(3)	0.1320(3)	0.0633(4)	0.3792(5)	0.1311(5)	0.062(7)	12	2
Fe-12	$24l_2$	0.6326(3)	0.0704(3)	0.0789(4)	0.6333(5)	0.0695(5)	0.0779(7)	9	3
Fe-13	$24l_3$	0.7216(3)	0.2139(3)	0.0808(4)	0.7208(5)	0.2129(5)	0.0817(7)	8	4
Fe-14	$24l_4$	0.7649(2)	0.1184(3)	0.1451(4)	0.7648(4)	0.1194(4)	0.1477(7)	10	2

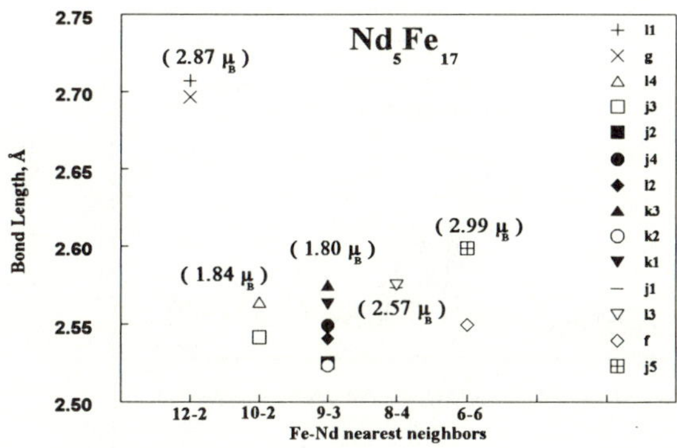

Fig. 5: The average Fe-Fe bond lengths vs. the number of Fe-Nd nearest neighbors. The refined moments for each group are shown in parenthesis.

Unfortunately, the complexity, and the small magnetic scattering contribution does not permit independent refinement of all of the magnetic moments. This problem is solved by grouping the sites according to their coordination. The average Fe-Fe bond lengths within each group are reasonably similar, justifying this grouping, and the refined moments for each group appear to be reasonable (Fig. 5). The results are also consistent with the decomposition of the Mössbauer spectrum into distinct subspectra with different hyperfine splittings and areas [24]. From these results predictions about site substitution can be made which will direct attempts to modify the structure in order to enhance the Curie temperature.

[2] We suspect an error in the reported errors for some x-ray entries with few significant digits shown.

The high performance of the diffractometer has also allowed studies of melt spun ribbon to be carried out using the quantities of materials normally available from single preparation runs. This has the advantage of improving the uniformity of material under examination. Figure 6 shows the results of one such study carried out in collaboration with the Sao Paulo group. Samples were prepared from the same starting material but subject to different post-spinning anneals. The lower temperature treatment resulted in the formation of $Pr_2Fe_{14}B$, Fe_3B and $Pr_2Fe_{23}B_3$ while at the higher temperature the metastable $Pr_2Fe_{23}B_3$ phase was absent and α-Fe was observed. The refinement allowed the relative fractions of each phase to be quantified, and showed dramatic peak broadening for the lower temperature diagram. This is attributed primarily to strain, rather than particle size effects.

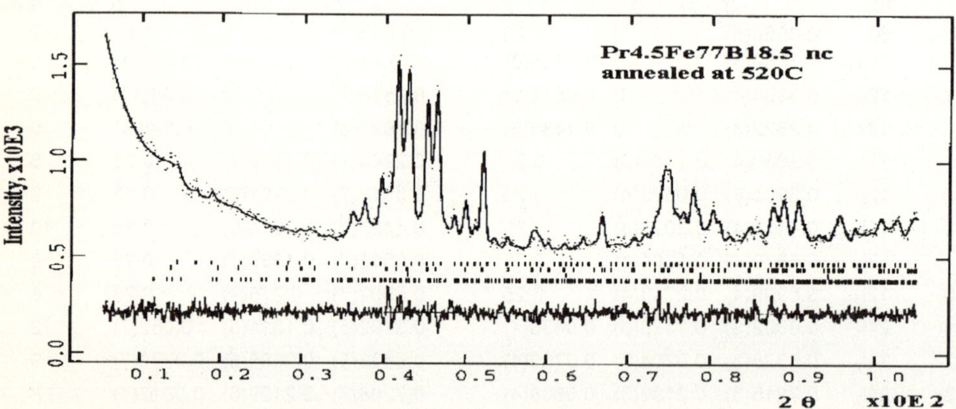

Fig. 6a: Melt Spun $Pr_{4.5}Fe_{77}B_{18.5}$ ribbon flash annealed at 520 °C.

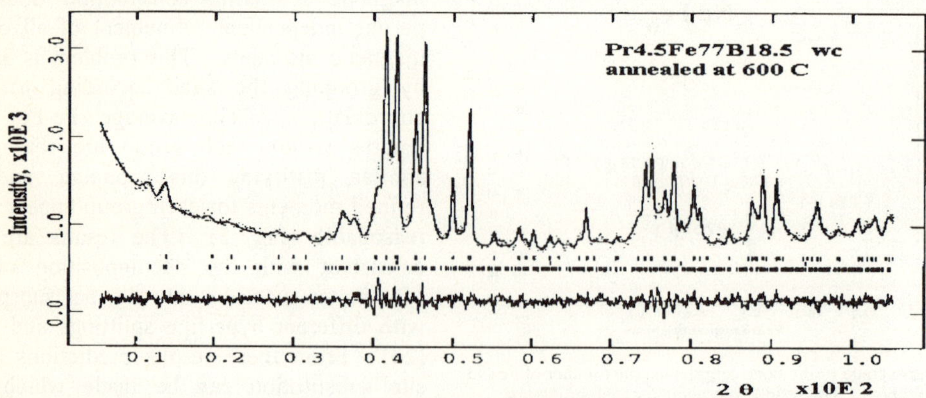

Fig. 6b: The same ribbon as in 6a, flash annealed at 600 °C.

Conclusions:

Neutron diffraction has been, and continues to be an essential tool for the study of magnetic materials, due both to the magnetic scattering cross section and the ability of the neutron to probe many structural details normally not accessible to x-ray diffraction. New, high performance instruments have reduced the requirements for sample volume, and multiphase refinement codes have allowed study of mixed phase materials. The neutron data has provided insight into the processes controlling the properties of permanent magnet phases as well as providing essential data for theory and for the analysis of other experiments such as Mössbauer spectroscopy. The new generation of instruments has also allowed this type of research to be carried out at smaller reactors, where such work had not previously been undertaken.

Acknowledgements:

The author would like to thank the many collaborators and students participating in this work over many years, including W.J. James, G.K. Marasinghe, Z. Hu, H. Luo, M. Chen, J. Herbst, G. Long, F.P. Missell and V. Villas-Boas. Support of the United Staes National Science Foundation through grants DMR-9305782 and DMR-9614596 is gratefully acknowledged.

References:

[1] H.M. Rietveld, J. Appl. Cryst. **2** (1969), p. 65.

[2] V. F. Sears, "Methods of Experimental Physics, **23**, Part A, Neutron Scattering", Academic Press (1986), p. 521., V.F. Sears, Neutron News, **3**, (1992) p25

[3] A.C. Larsen and R.B. Von Dreele, GSAS-General Structure Analysis System, Los Alamos National Laboratory, (1985-1994).

[4] F. Izumi, in The Rietveld Method, R.A. Young, ed. Oxford University Press, Oxford (1993) Chap. 13.

[5] J. Rodriguez-Carvajal, FULLPROF, version 2.4 Institute Laue-Langevin, Grenoble (1993).

[6] W.B. Yelon, R. Berliner and M. Popovici, Proceedings of the International Conference on Neutron Scattering, Toronto, Canada, Aug. 17-21, (1997), Physica B. (in press).

[7] K. Hardman, W. J. James and W. B. Yelon, J. Phys. Chem. of Solids **41** (1980), p. 1105.

[8] J. Herbst, J. Croat and W. B. Yelon, J. Appl. Phys. **53** (1982), p. 250.

[9] J. Herbst, J. Croat, F. Pinkerton and W. B. Yelon, Phys. Rev. **B29** (1984), p. 4176.

[10] Z. Hu, and W.B. Yelon, J. Appl. Phys. **76** (1994), p. 6147.

[11] C.D. Fuerst, F.E. Pinkerton and J.F. Herbst, J. Magn Magn. Mater, **129**, (1994), L115.

[12] W.B. Yelon and Z. Hu, J. Appl. Phys. **79** (1996), p. 1330.

[13] S. S. Jaswal, W. B. Yelon, G. C. Hadjipanayis, Y. Z. Wang and D. J. Sellmyer, Phys. Rev. Letters **67** (1991), p. 644.

[14] W. B. Yelon, H. Xie, G. J. Long, O. A. Pringle, F. Grandjean and K.H.J. Buschow, J. Appl. Phys. **73** (1993), p. 6029.

[15] W.B. Yelon, Z. Hu, S. Mishra, G.J. Long, O.A. Pringle, D.B. Middleton, K.H.J. Buschow, and F. Grandjean,J. Appl. Phys. **76** (1994), p. 443.

[16] G. J. Long, G. K. Marasinghe, S. Mishra, O. A. Pringle, F. Grandjean, K.H.J. Buschow, D. P. Middleton, W. B. Yelon, F. Pourarian and O. Isnard, Solid State Comm. **88** (1993), p. 761.

[17] G.K. Marasinghe, S. Mishra, O.A. Pringle, G.J. Long, Z. Hu, W. B. Yelon, F. Grandjean, D.P. Middleton and K.H.J. Buschow, J. Appl. Phys. **76** (1994), p. 6731.

[18] W.Y. Ching, M.Z. Huang, Z. Hu and W.B. Yelon, J. Appl. Phys. **81** (1997), p. 5618.
[19] E. Girt and Z. Altounian, Phys. Rev. **B57** (1998), p. 5711.
[20] E. Girt, Z. Altounian and I.P. Swainson, Physica **B234** (1997), p. 2438
[21] H. Luo, Z. Hu, M. Chen, W.B. Yelon, G.K. Marasinghe, P.C. Ezekwenna, W.J. James, W.C. Chang and S.H. Tsai, J. Appl. Phys. **81** (1997), p. 4542.
[22] J.M. Moreau, L. Paccard, J.P. Nozieres, F.P. Missell, G. Schneider and V. Villas-Boas, J. Less-Common Met. **163** (1990), p. 245.
[23] W.B. Yelon, H. Luo, F.P. Missell and V. Villas-Boas, (to be submitted)
[24] J.P. Nozieres and H. Rechenberg, Solid State Commun. **79** (1991), p. 21.

Materials Science Forum Vols. 302-303 (1999) pp. 11-20

X-Ray and Neutron Analysis of Magnetism in Reduced Dimension Systems

D. Givord[1], M. Bonfim[1], S. David[1], A. Fontaine[1], W. Neubeck[1,2], J. Rothman[1], C. Vettier[2] and J. Vogel[1,2]

[1] Laboratoire Louis Néel, C.N.R.S., B.P. 166, FR-38042 Grenoble cedex 9, France

[2] E.S.R.F., avenue des Martyrs, B.P. 220, FR-38043 Grenoble cedex, France

Keywords: X-Ray Magnetic Scattering (XRMS), X-Ray Magnetic Dichroism (XMCD), Neutron Reflectometry, Thin Film Magnetism, Magnetic Multilayers

ABSTRACT

Various experimental techniques, involving neutrons or X-ray, exist for the characterisation of magnetism in reduced dimensions. They are element selective and their domain of application is wide. X-ray techniques may offer the additional advantage of being electronic shell selective.
More specifically, neutron reflectometry, X-ray magnetic diffraction in the resonant mode and X-ray Magnetic Dichroism (XMCD) may be used as monolayer-sensitive magnetometry techniques which do not probe the substrate parasitic signal. Neutron reflectometry may allow complex moment configurations to be determined close to a surface or to an interface. X-ray magnetic diffraction in the non-resonant mode and XMCD separate spin and orbital moments. XMCD offers the specific advantage that it may be a relatively simple experimental technique.

I - INTRODUCTION

The evolution of magnetism during the last ten years has been towards the study of objects with reduced dimensions. This has been stimulated by the discovery of original new phenomena such as giant magnetoresistance, oscillatory coupling through metallic non magnetic layers or perpendicular magnetic anisotropy [1]. The quantitative analysis of these phenomena at the atomic scale has required the use of highly sensitive experimental tools. Amongst these, the experimental techniques which use neutrons and X-ray have a wide range of applications. The most important of these techniques are reviewed in the present paper. Their specific advantages are discussed on the basis of illustrative examples. The essential specific features, which explain the impressive development of these techniques, such as ultra-high sensitivity and element or magnetic-shell selectivity, are illustrated by selected examples.
The first part of this article deals with neutron techniques, in particular neutron reflectometry. The second part concerns X-ray magnetic diffraction and XMCD.

II - NEUTRON SCATTERING ANALYSIS OF MAGNETISM IN REDUCED DIMENSIONS

Magnetic neutron scattering originates from the fact that the neutron spin interacts with magnetic electrons in matter. Almost any type of magnetic system may in principle be examined, whatever the atomic arrangement and the nature of the magnetic elements. Neutron scattering cross-sections are small however and neutron diffraction in reduced dimension systems is restricted to a few specific systems, in particular multilayers as shown in the first part of this section.
Grazing incidence neutron diffraction could possibly be another technique to study surface magnetism. First results were obtained recently [2]. The weakness of the measured intensities constitutes for the moment a real obstacle to the development of this technique which is not further discussed in the present article.
Polarised neutron reflectometry is another experimental technique which allows depth sensitive magnetic properties to be probed close to a surface or an interface. As it is general for reflectivity measurements, the experimental intensities are large. This technique offers some unique possibilities.

It is likely that it will develop considerably in the coming years. Some illustrative applications are presented in the second part of this section.

II-1 - MAGNETIC ORDERING IN MULTILAYERS AND HETEROSTRUCTURES

The magnetic structure of almost any type of materials may in principle be derived from neutron scattering studies. The weakness of neutron cross-sections is however an obstacle for the study of reduced-dimension systems. Multilayers, where the same building block is repeated provide the possibility to overcome this experimental limitation. The analysis of the magnetic structures in these systems then involve basically classical neutron scattering. Reviews may be found in [3,4]. Amongst the most spectacular results obtained, one may mention rare-earth based heterostructures, such as Er/Y or Dy/Y. Magnetic satellites of nuclear harmonics peaks were found which demonstrate that helimagnetic structures may propagate over many superlattice periods through the Y non-magnetic layer [4]. In transition metal based multilayers, such as Co/Cu , antiparallel coupling between Co layers was found. It indicates that antiferromagnetic interactions are mediated by the non magnetic Cu layers[5].

II-2 - NEUTRON REFLECTOMETRY

II-2-1 NEUTRON REFLECTIVITY

The interaction energy between a neutron and a magnetic layer, α, may be expressed as :

$$V_\alpha(\mathbf{R}) = \frac{2\pi\hbar^2}{m}\frac{\hbar}{2}\,[\,b_\alpha\rho_\alpha + C\rho_\alpha\,\sigma\,\mathbf{M}_{//}\,] \qquad (1)$$

where m and σ are the neutron mass and spin respectively, b_α and $M_{//}$ are the material Fermi length and in-plane magnetic moment respectively, and ρ_α is the atomic density integrated over the whole layer. The constant $C = \gamma r_o = 0.2696\ 10^{-12}$ cm. Eq. 1 shows that the neutrons interact only with the in-plane component of the moments.

Let us consider the case of a neutron which is **specularly** reflected, i.e. the diffusion vector is perpendicular to the sample plane, and neglect for the moment the magnetic part in the interaction. By analogy with light reflectivity, the neutron refractive index is defined as :

$$n_\alpha = 1 - \frac{\lambda_o^2}{2\pi}\,\rho_\alpha\, b_\alpha$$

where λ_o is the neutron wavelength. The deduced critical angle for total reflection is :

$$\sin(\theta_c) = \lambda_o\sqrt{\frac{\rho_\alpha\, b_\alpha}{\pi}} \qquad (2)$$

The reflectivity coefficient R obtained is :

$$R^2 \approx \frac{\pi^2}{q_o^4}\,[\,b_\alpha^2 + (b_s-b_\alpha)^2 + 2\,b_\alpha\,(b_s-b_\alpha)\cos(2\,q_o d)\,] \qquad (3)$$

The $1/q_o^4$ coefficient in front of Eq. 3 describes the global decrease of reflectivity with increasing the wave vector q. The last term in the part between brackets is ocillatory with period $\Delta q_o = \pi/d$, it represents the well known Kiessing fringes which are due to the interferences occurring between the neutron waves which are successively reflected on the film interfaces.

For a magnetic layer, Eq. 3 applies to first approximation, with b_α being replaced by $b_\alpha + C\,\sigma\,\mathbf{M}_{//}$ (see relation (1)). The scattering is formed of two contributions. The scattering length for non-spin-flip scattering includes a nuclear and a magnetic term, it is equal to b_α +/- C $M_{//}^z$ where $M_{//}^z$ is the projection of $M_{//}$ along the neutron polarisation direction, z, and the sign + or - represents the neutron spin state. The scattering length for spin-flip scattering is purely magnetic, it is proportional to $M_{//}^{\perp}$, the projection of $M_{//}$ perpendicularly to z.

As shown by Fermon [6], the above expressions are only approximate due to the energy separation which exists between neutrons of different spins placed in a given magnetic field. The essential features characterising neutron reflectivity by a magnetic system remain however the same.

II-2-2 : ULTRA-HIGH SENSITIVE DETERMINATION OF SURFACE MAGNETIC MOMENTS : Fe (110) EPITAXIAL FILMS

At the surface of a layer or at the interface between two layers, the electronic properties, including the magnetic properties, are expected to differ from those in the bulk. This can be illustrated by Fe (110) epitaxial films grown on W(110). As shown by Elmers and Gradmann [7], the Fe moments are confined in the film plane, their easy direction is [1$\bar{1}$ 0] at small thicknesses up to 50 Å, and [001] at larger thicknesses. The Fe moment for such ultra-thin films cannot be extracted from bulk magnetic measurements because the magnetic signal from the substrate is in general orders of magnitude larger than the layer magnetic signal. Polarised neutron reflectometry may offer an extremely sensitive determination of the Fe absolute moment in such a ferromagnetic system. Such a study was undertaken on the SPN spectrometer at the IBR-2 reactor in Dubna [8]. The studied sample was a 6Å thick Fe (110) film grown on a W (110) buffer layer, itself grown on sapphire (11$\bar{2}$ 0). A 500 Oe in-plane magnetic field was applied in order to ensure that the Fe moment direction and the z-axis are parallel. The non-spin-flip reflectivities R^+ and R^- were measured at 300 K for the two possible neutron spin states (see Eq. 3) (Figure 1). One can concentrate on magnetic effects by plotting R^+/R^-, as shown in Figure 1 - c. The profile shown in this Figure is the best fit to the data ; the associated Fe moment is 1.80 μ_B. To deduce the Fe moment at low temperatures, it was assumed that thediamagnetic sapphire signal which dominates bulk magnetic measurements is temperature independent. Hence, the temperature dependent signal was attributed to the thermal variation of the Fe moments. From this and the 300K neutron data, the Fe moment obtained at 0K is 2.1 μ_B. Similar experiments were performed on Fe thicker films and the Fe moment obtained was identical to the bulk value, 2.2 μ_B.

According to ab-initio calculations for a seven layers free standing Fe(110) film [9] the ground state magnetic moment at the surface is expected to be 2.65 μ_B, which corresponds to a 19% increase with respect to the bulk. However, calculation for a monolayer of Fe (110) deposited on W [10], shows that the ground state moment of Fe is 2.18 μ_B, and indicates that this reduction with respect to the free standing film is due to hybridisation effects. The experimental data confirm this theoretical result.

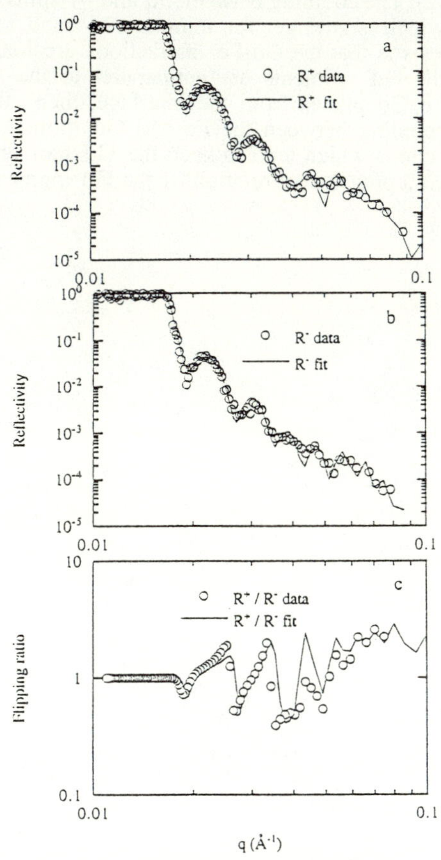

Figure 1 : Specular neutron reflectivity for a 6 Å Fe (110) thin film

II-2-2 : MAGNETIC MOMENT CONFIGURATION CLOSE TO SURFACES : Gd (0001) / Fe(110)

Neutron reflectivity may be used also to reveal the existence of non-collinear magnetic configurations close to film surfaces. This can be illustrated by Gd/Fe bilayers [11]. A 133 Å thick Fe(110) film was grown on a W buffer as above. A 63 Å thick Gd (0001) film was then grown on top of the Fe film. Bulk magnetic measurements in this system revealed complex behaviour. A polarised neutron reflectometry analysis was undertaken on the PADA spectrometer at the Orphée reactor in Saclay. During the measurements, a 300 Oe field was applied in plane to define the quantification axis, z, of the neutrons. The neutron wavelength in this study was 4.14 Å. Non spin-flip reflectivities R^{++} and R^{--} and spin-flip reflectivities R^{+-} and R^{-+} were collected.

The reflectivity profiles obtained at 300 K are shown in Figure 2. A large splitting between the $R^{++}(q)$ and the $R^{--}(q)$ profiles is observed which indicates the existence of a large component of the magnetisation along the direction of the applied field. The most striking feature however is the presence of a distinctive spin-flip reflectivity $R^{+-}(q)$ which shows that the magnetisation is not completely parallel nor antiparallel to the applied field.
The moment configuration was analysed by fitting the experimental reflectivities. The best fit was found for a complex configuration where the Fe moments stay parallel amongst themselves and where the Gd moments nearest the interface are antiparallel and then exhibits a twisted state further from the interface as shown in Figure 3.
This twisted state was interpreted in the light of the generally accepted understanding of magnetic interactions in rare-earth-transition-metal intermetallics [12]. The coupling between 3d and 4f spins is mediated by 5d electrons through 3d-5d hybridisation, which is stronger for minority 3d band and thus antiferromagnetic. The necessary additional hypothesis is that the Gd-Fe interactions are long-ranged and oscillatory. Close to the Gd-Fe interface, the Gd moments are antiparallel to the Fe moments as in bulk GdFe alloys. After approximately 8 Gd planes from the interface, the GdFe interactions change sign and they then favour parallel coupling between the Fe and Gd moments. They are therefore in competition with the Gd-Gd interactions which tend to keep the Gd moments ferromagnetically coupled. The result of this competition is a progressive rotation of the Gd moments similar to that observed in a helimagnetic structure.

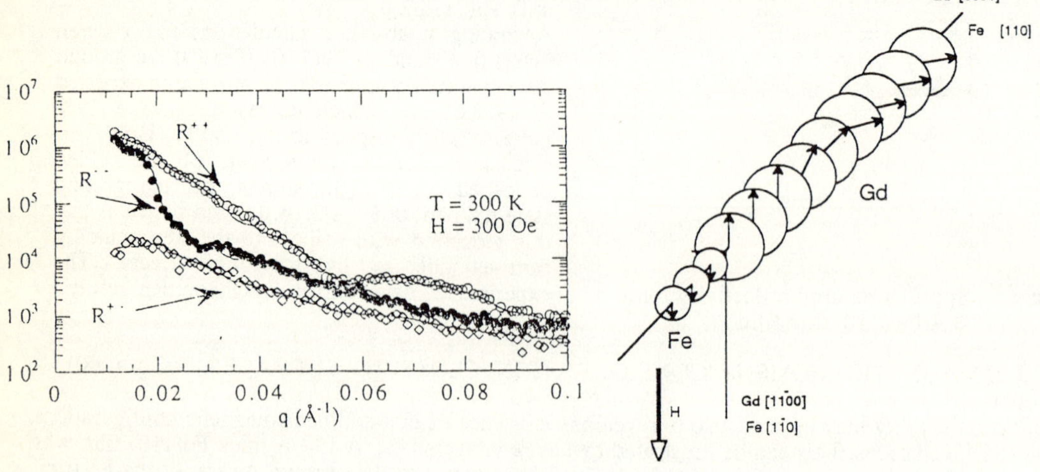

Figure 2 : Polarised neutron reflectivity measurements in Gd/Fe bilayer.

Figure 3 : Schematic representation of the moment configuration in the Gd/Fe bilayer.

II-2-3 REFLECTOMETRY IN La/Fe MULTILAYERS : IMPRINTING MAGNETIC STRUCTURE

When neutron reflectometry is applied to multilayer systems, Bragg peaks are observed which can be viewed as harmonics of the reciprocal space origin. In this case, the q vector is generally large compared to the critical vector q_c for complete reflection. Thus, the kinematic theory of diffraction (Born approximation) may be used to evaluate the intensities. The nuclear satellites due to the chemical periodicity occur at :

$$q = 2\pi\, n / d$$

where d is the chemical periodicity in Å and n is an integer. In a magnetic system, additional peaks occur which can be viewed as magnetic satellites of the nuclear peaks. They occur at :

$$q = 2p\, [(n/d) \pm (1/\Lambda)]$$

where Λ is the magnetic propagation vector (for a ferromagnet $\Lambda = 0$). Of special interest is the case where the magnetic structure is helical ($\Lambda \neq 0$). There are two satellites with intensities $I_m^{\pm}$. These intensities depend on the polarisation of the neutrons and the chirality of the helix through :

$$I_m^{\pm} \approx 1 \pm (\hat{\mathbf{P}} \cdot \hat{\mathbf{q}})(\hat{\mathbf{q}} \cdot \hat{\Lambda}) \qquad (4)$$

where $\hat{\mathbf{P}}$, $\hat{\mathbf{q}}$ and $\hat{\Lambda}$ are collinear unit vectors. $\hat{\mathbf{P}}$ represents the neutron polarisation and $\hat{\mathbf{q}}$ the chirality of the helix. Eq. 4 indicates that for neutron polarised in the state +1(respectively -1), the I_m^+ satellite is only observed (respectively I_m^-). In general however, there is an equal number of domains with + and - chiralities and both satellites have the same intensity. The existence of only one satellite was however recently demonstrated by Felcher et al. [13] in La/Fe multilayers. During the preparation of the sample, the substrate was actually rotated in a small magnetic field ($\approx$ 3 Oe). The chirality was imprinted in the sample as a result of this. In particular, the chirality and the pitch were dictated by the direction and speed of rotation of the substrates.

III - X-RAY ANALYSIS OF MAGNETISM IN REDUCED DIMENSIONS

It has been known for a long time that X-rays couple to magnetisation densities in solids. When the X-ray wavelength is far away from any absorption edge, the only observed effect is the weak non-resonant magnetic scattering. When the photon energy is tuned near an absorption edge of a magnetic element in the material, resonant effects take place which lead to magnetic X-ray resonant scattering (MXRS) and magnetic dichroism. These phenomena arise from the promotion of a core electron from a core level into an available electronic state above E_F. The magnetic sensitivity of such an electrical process originates in the spin-orbit coupling within the core level and the final states. Resonant enhancements of MXRS can be spectacular (up to 6 orders of magnitude at the $M_{4,5}$ edges of actinides); they make the MXRS technique appropriate to investigate magnetic order of small magnetic volumes. The dichroism methods provide information on the average value of the magnetisation.

III-1 X-RAY MAGNETIC SCATTERING

Magnetic Bragg scattering was observed for the first time in antiferromagnetic NiO by de Bergevin and Brunel [14]. The weakness of the magnetic amplitudes, which amount to typically 10^{-3} the classical Thomson amplitudes, has however originally limited the development of X-ray magnetic scattering. It is only recently, with the construction of synchrotron light sources, that reliable measurements have been obtained with this technique.

Non resonant X-ray magnetic scattering offers the unique possibility to separate out the orbital and spin moments by exploiting the polarisation dependence of the diffraction. This has in particular been applied to NiO [15]. The corresponding theoretical magnetic amplitudes are :

$$M_{\sigma\sigma} = \sin 2\theta \sin(\phi)\, S(\mathbf{q}) \qquad (5\text{ a})$$

$$M_{\pi\sigma} = \sin 2\theta \sin\theta\ [(\cos(\phi)\, S(\mathbf{q}) + \cos(\phi+\phi_0) S(\mathbf{q})] \qquad (5\text{b})$$

σ and π in $M_{\sigma\sigma}$ and $M_{\pi\sigma}$ denote a linear polarisation state respectively perpendicular and parallel to the scattering plane, the first index refers to the polarisation of the incoming beam and the second to the polarisation of the outgoing beam. $S(\mathbf{q})$ and $L(\mathbf{q})$ are the spin and orbital form factors respectively, ϕ denotes the direction of the spin moment and ϕ_0 the angular shifts between the spin and angular moments.

The propagation vector characterising the antiferromagnetic structure of NiO is (1/2, 1/2, 1/2) and it is known that the magnetic moments lie along directions such as $(1,\ 1,\ \bar{2})$. The dependence of the measured intensities is shown in Figure 4. There is a $\pi/2$ phase shift between $I_{\sigma\sigma}$ and $I_{\pi\sigma}$, which directly shows that the spin and orbital moments are collinear. However, the simple variations predicted by Eq. 5 are not observed. This is due to the fact that there are 3 equivalent easy directions and thus 3 possible magnetic domains, denoted as S-domains. The fit to the curves allowed the respective volume of each domain to be obtained. The ratio of the orbital to spin moment L/2S was then derived from the ratio between averaged intensities. It amounts to 17 %. This value is significantly larger than it was expected for such 3d system where the orbital moments was essentially thought to be quenched.

As mentioned above, resonant effects were discovered when tuning the photon energy close to absorption edges. The first observations were made by Namikawa et al. [16] and Gibbs et al. [17]. The resonant process effect is maximum when the absorption corresponds to a dipole transition into a strongly spin-polarised excited state. The most appropriate systems for such studies are the actinides because the X-ray wavelength corresponding to the $M_{4,5}$ absorption edge is of approximately 1.5 - 2 Å, i.e. ideally suited for diffraction studies. In the case of micro-crystal samples, XMRS is better suited than neutron diffraction to observe magnetic long range order: the intensity/volume ratio with x-rays exceeds by several orders of magnitude the values observed with neutrons. Furthermore, as observed in the $(Np_{0.5}\ U_{0.5})URu_2Si_2$ compound, this method is species selective and allows a clear separation of long range ordered moment belonging to different chemical species. Moreover, electronic shell selectivity is obtained by studying different absorption edges such as the M and L edges.

All aspects of magnetic x-ray resonant scattering are not fully understood yet and the technique has only been used on a qualitative basis. It remains that the extremely high intensities obtained give to this technique a decisive advantage for the study of very small crystals and thin films.

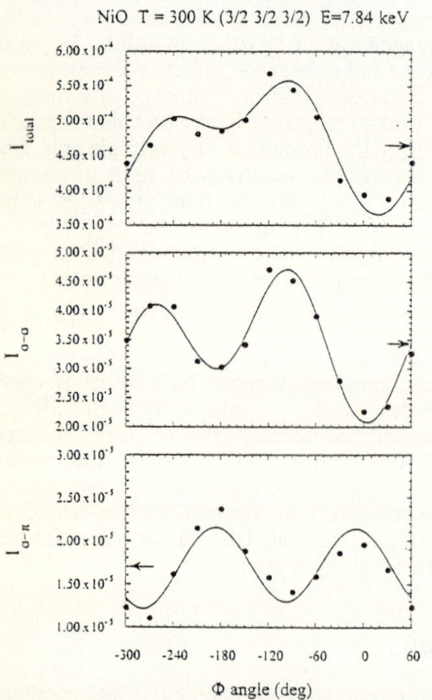

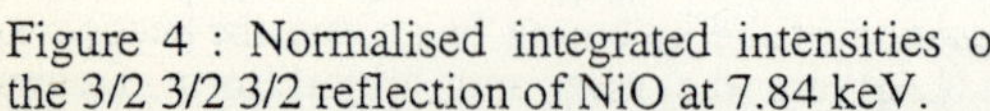

Figure 4 : Normalised integrated intensities of the 3/2 3/2 3/2 reflection of NiO at 7.84 keV.

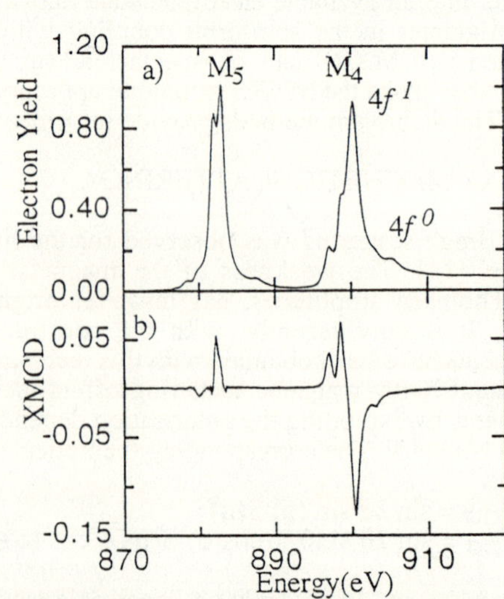

Figure 5 : X-ray absorption and XMCD at the $L_{2,3}$ edges of a 180 Å Ce (111) epitaxial film.

III-2 X-RAY MAGNETIC CIRCULAR DICHROISM (XMCD)

In X-ray absorption experiments, an electron is excited from a core level into an empty final state. When the incoming energy, Eo, reaches the energy of an absorption edge, the absorption cross-section is considerably enhanced. The absorption cross-section is given by the matrix element which couples the initial and the final state. The value of the matrix element depends on the energy and on the polarisation of the light. The interaction of the light with the orbital moment and, through the spin-orbit coupling, with the spin moment, leads to the selection rules. For left (right)-polarised light, the selection rule is Dm = +1 (-1). For a magnetic element, the absorption depends on the polarisation of the light because, in the final state, the electrons occupy a polarised band. The difference in absorption is called XMCD.
The analysis of XMCD experiments is notably different whether the energy of the absorption edge is in the soft X-ray range (0.5-4 keV) or in the hard X-ray range (5-15 keV) [18, 19].

- The soft X-ray range deals with the 2p-3d transitions ($L_{2,3}$ edges) of the 3d elements and with the 3d-4f transitions ($M_{4,5}$ edges) of the rare-earth elements. In this case, the significance of XMCD has been considerably reinforced by the establishment of the so-called sum rules [20]. These allow quantitative values for the orbital and spin moments, projected along the X-ray propagation direction, to be deduced from the experimental spectra.
The relatively low energies of soft X-ray implies that the first atomic layers only contribute and sample preparation and measurements must in most cases be realised under UHV conditions. This is a constraint of the technique. It may turn as beneficial because it means that the technique is surface sensitive ; combined with the element selectivity, even buried interfaces may be studied.

- The hard X-ray range deals with the K edge of the 3d transition elements and with the L edges of the rare-earth elements. The sum rule does not apply because the radial extension of the shell differs for each spin state, due to the interactions between these shells and those containing the magnetic electrons. It results that the experiments are of a more qualitative nature. The higher photon energy, however, tends to make the realisation of experiments easier and measurements at these edges offer the possibility to reveal possible magnetic polarisation of the electrons in the final states, 4p for transition elements and 5d for rare-earth elements.

III-2-1 : ULTRA-HIGH SENSITIVE MAGNETOMETRY : Ce (111) FILMS

XMCD at the L edges of the transition elements or at the M edges of the rare-earth elements may be used as a ultra-sensitive technique of quantitative magnetometry with 1/100 of a monolayer sensitivity. This may be illustrated by the study of Ce ultra-thin films.
Ce metal may present 5 different crystalline phases in the bulk form. In particular, the magnetic bcc γ-phase is stable at room temperature and ambient pressure. It transforms to the β phase just below room temperature and this one in turn transforms to the non-magnetic α-phase of valence 3.3 at approximately 150 K. The α-phase has the same fcc structure as the γ-phase, but the lattice parameter is contracted by approximately 6 %. It is the delocalisation of Ce's 4f electron which is responsible for the properties of Ce, but the actual process by which this occurs is not yet clarified opposing a Kondo-lattice picture to a Mott picture.
Epitaxial Ce films were grown on W(110) [21]. The films had the fcc structure with (111) orientation and a lattice parameter close to that of the γ-phase. No transition from γ to β and to α was observed by Grazing-incidence X-ray diffraction performed between 20 K and 300 K. This result suggests that the γ-phase is clamped by a high density of misfit dislocations down to the lowest temperature.
The valence of Ce in the films was examined by X-Ray Absorption Spectroscopy at the $M_{4,5}$ edges on the beam line ID12B at ESRF, as a function of temperature and film thickness, d_{Ce}=20-200Å. The XAS was obtained in the sample current mode, yielding a surface sensitive probing. Figure 5a presents the XAS spectra for a 180Å thick film. Contrary to bulk γ-Ce, two features are present at each edge, indicating mixing in the ground state between a $4f^0$ and $4f^1$ configuration, i.e. in the probed region the mean valence is bigger than 3.

From macroscopic SQUID magnetometry, is was only possible to extract qualitative information on the Ce susceptibility. This was due to the large contribution of the substrate to the total magnetic signal. In view of obtaining accurate values, we undertook XMCD measurements together with the above described XAS measurements (Figure 5b). The measurements were performed between T=5 K and 300K and with an applied magnetic field between 0 T and 5T. The mean orbital moment

polarisation $<L_z>$ of the probed atoms was evaluated using the first XMCD-sum rule [20]. In principle the Ce spin moment can be obtained using a second sum-rule. However, its application in our case is difficult due to the overlap of the M_4 and M_5 edges.Thus, the total magnetic polarisation $m=<L_z>+2<S_z>$ was instead obtained using the relation $<L_z>/<S_z>=-8$; expected for a 4f 1-configuration (J=5/2), independently of crystalline fields.
$1/(\chi_T)$ deduced from these data is shown for a series of Ce films in Figure 6. At low T, $1/(\chi_T)$ is linear, indicating Curie-Weiss behaviour. The Ce effective moment is independent of the valence and

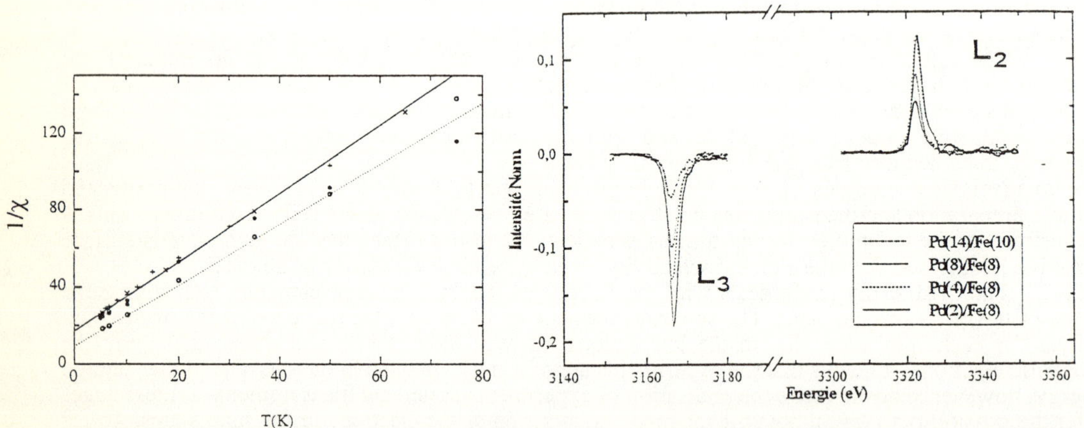

Figure 6 : $1/(\chi_T)$ for Ce epitaxial films

.Figure 7 : XMCD signal at the Pd $L_{2,3}$ edges in Fe/Pd mùultilayers.

intermediate between that for Ce+3 free ion (μ_{eff}=2.54 μ_B) and for a Γ7 doublet (μ_{eff}=1.24 μ_B). Thus the possible influence of the Kondo effect on the moment seems to be small.
These results can be discussed within the Kondo-lattice model. In this description [22], the valence is related to the strength of the Kondo interaction, a high Kondo temperature gives a high valence. In a-Ce, the valence is 3.3 for an estimated Kondo temperature T_K =950K and in γ-Ce the valence is close to 3 for an estimated Kondo temperature T_K=95K. Hence, in this picture the W(110)/Ce(111) interface is characterised by a higher T_K than bulk γ-Ce. According to Doniach [23] this will lead to a decrease in the ordering temperature, which might explain why no ordering is observed in these samples, in opposite to bulk β-Ce, which orders antiferromagnetically at T=12K.

III-2-2 : SPIN VERSUS ORBITAL MOMENT : Fe/Pd MULTILAYERS

The discovery of large magnetic anisotropy in transition metal based multilayers is a subject of considerable interest. The analysis of this phenomenon requires in particular the value of the orbital moment to be linked to the value of the magnetocrystalline anisotropy. It has been noted in particular that, an anisotropy of the orbital moment as small as 0.01 μ_B results in a magnetic anisotropy field of approximately 1T.
The XMCD signal at the L_2 and L_3 Pd edges is shown in Figure 7. The signal does not have a simple Lorentzian shape, a shoulder is visible on the high energy side which may be associated with $4d^8$ states in the Pd ground state [24]. The sign of the dichroism signal indicates that the Pd moment is parallel to the Fe moment in the next layers. The coupling is of the expected sign, since both Fe and Pd are more than half-filled bands. The increase of the integrated signal when the thickness of the Pd layers is decreased shows directly that the Pd atoms at the interface bear a larger moment than in the bulk of the Pd layers. The application of the sum rules allowed the spin and orbital moments to be obtained. The average spin moment varies from 0.14 μ_B in the Pd 14-monolayer sample to 0.34 μ_B in the Pd 2-monolayer sample. The orbital moment deduced is small : from 0.01 μ_B in the Pd 14-

monolayer sample to 0.04 μ_B in the Pd 2-monolayer sample. This result may be related to the relatively small value of the spin-orbit parameter in the Pd layers, as obtained from XAS measurements [25].
A similar analysis was performed for Fe in the FePd multilayers [26] and it was found in this case that the Fe orbital moment is enhanced. It appears therefore that a transfer of the spin-orbit parameter takes place between Pd and Fe this system.

III-2-3 : ELEMENT SENSITIVE HYSTERESIS CYCLES : NdFeB SPRING MAGNETS

As already mentionned, XMCD with hard X-ray offers a method for qualitative magnetometry. Due to the high X-ray energy, the penetration depth is large and the sample preparation is easier than for soft X-rays. The possibility of obtaining element-selective results by this technique is illustrated in this section by the study of NdFeB spring-magnets [27].
Since the discovery of NdFeB magnets in 1983, the experimental search is for magnets with higher performance. The knowledge of the R-M phase diagrams is such that the hope of finding new compounds with significantly improved magnetisation, as required for higher remanence, is low. A very promising possibility has emerged with the concept of spring-magnets [28]. In these materials, a hard medium-magnetisation phase and a soft high-magnetisation phase are coupled at the nanoscopic scale. Reversal in the soft phase is impeded by this coupling and it results that medium coercivity is obtained.
Schematically, it is thought that in medium applied reversed field, reversal in the soft phase leads to the formation of a domain wall at the interface between hard and soft phases. Wall propagation into the hard phase is impeded by the fact that its energy in the hard phase is higher. When the applied field is increased to a value large enough, a second step is observed, at which the wall propagates into the hard phase.
There is no simple experimental means to check the validity of this description with usual macroscopic magnetisation measurements. The element selectivity of XMCD opens up this possibility. To check this idea, alloys with typical composition $Nd_5Fe_{77}B_{18}$ were prepared, in which the hard phase is $Nd_2Fe_{14}B$ and two soft phases are present, α-Fe and Fe_3B[29]. Because Nd atoms are present in the hard phase only, XMCD was measured at the L_2 Nd edges. The hysteresis cycle then obtained (Figure 8) is expected to be representative of reversal of the hard $Nd_2Fe_{14}B$ phase. This cycle is compared to the measured global hysteresis loop in Figure 9. From the difference between these 2 loops, the soft phase loop was deduced. It is also shown in Figure 9. This figure reveals that reversal of the soft phases α-Fe and Fe_3B occurs in lower fields than that of the $Nd_2Fe_{14}B$ phase. This result thus confirms the proposed picture of magnetisation reversal in spring-magnets.

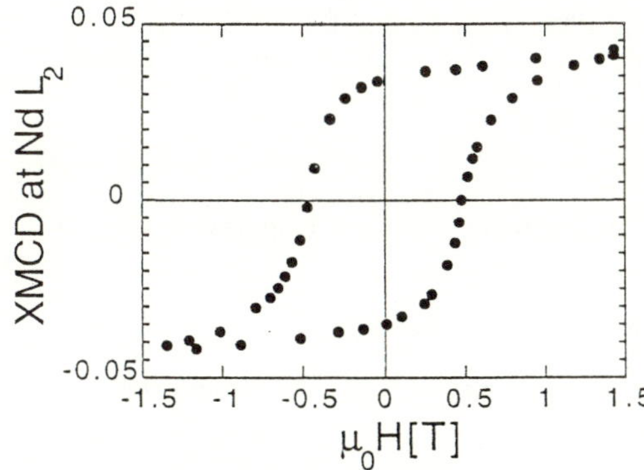

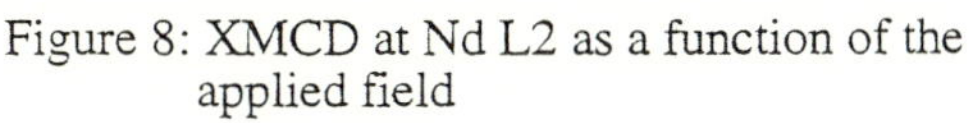
Figure 8: XMCD at Nd L2 as a function of the applied field

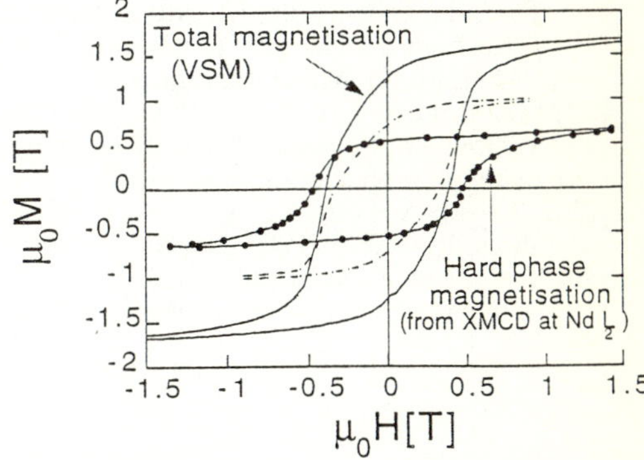

Figure 9: Hysteresis loops:Mtotal, Mhard, Mtotal (classical magnetometry) Msoft (dashed line) (=Mtotal-Mhard.)

IV - CONCLUSIONS

The possibilities offered by neutrons and X-ray to analyse magnetism in reduced dimensions were illustrated by a few selected examples. A priori, for such systems, the challenge is to obtain reliable data despite the small amount of matter which interacts with the neutron or X-ray beam. Most of the result presented have a significance which largely encompass the field of thin film magnetism. This illustrates the outstanding progress which has occurred in the neutron and X-ray experimental techniques during the last years.

REFERENCES

[1]B. Heinrich and J.A.C. Bland, Ultrathin Magnetic Structures, (Springer Verlag, Berlin, 1994)
[2] R. Günther, S. Odenbach, O. Schärpf, H. Dosch, Physica B, 234-236 (1997) p. 508
[3] J.J. Rhyne, R.W. Erwin, J.A. Borchers, M. B. Salamon, T. Tsui, R. Du and C.P. Flynn in Science and Technology of Nanostructured Materials, eds. G. Hadjipanayis and G.A. Prinz (Plenum Press, New York, 1991)
[4] R.A. Cowley, D.F. McMorrow, P.P. Swaddding, R.C.C. Ward and M.R. Wells, Ind. J. Pure and Appl. Phys., 33 (1995) p. 509
[5] J.L. Martinez, A. Cebollada, J.M. Gallego, J.J. de Miguel, C. Ocal, R. Miranda, S. Ferrer, G. Fillion and J.P. Rebouillat, J. Magn. Magn. Mat., 93 (1991) 89
[6] C. Fermon, Physica B 213-214 (1995) 910
[7] H.J. Elmers and U. Gradmann, Appl. Phys., A51 (1990) 255
[8] V. Pasyuk, O.F.K. Mc Grath, H.J. Lauter, A. Petrenko, A. Liénard and D. Givord, J. Magn. Magn. Mat. 148 (1995) 38
[9] C.L. Fu and A.J. Freeman, J. Magn. Magn. Mat., 69 (1987) L1
[10] S.C. Hong, A.J. Freeman and C.L. Fu, Phys. Rev. B38 (1988) 12156
[11] O.F.K. Mc Grath, N. Ryzhanova, C. Lacroix, D. Givord, C. Fermon, C. Miramond, G. Saux, S. Young and A. Vedyayev, Phys. Rev. B54 (1996) 6088
[12] M.S.S. Brooks, L. Nordström and B. Johansson, J. Appl. Phys. 69 (1991) 5683
[13] G.P. Felcher, W. Lohstroh, H. Fritzche, M. Münzenberg, H. Maletta and W. Felsch, Appl. Phys. Lett. (to appear)
[14] F. de Bergevin and M. Brunel, Phys. Lett., 39A2 (1972) 141
[15] V. Fernandez, C. Vettier, F. de Bergevin, C. Giles, W. Neubeck, Phys. Rev. 57 (1998) 7870
[16] K. Namikawa, M. Ando, T. Nakajima and H. Kawata, J. Phys. Soc. Jpn;, 54 (1985) 4099
[17] D. Gibbs et al. Phys. Rev. Lett. 61 (1988) 1241
[18] in Influenced Spectroscopies of Magnetic Solids, 1996, ed. H. Ebert and G. Schütz (Spinger, Berlin)
[19] in Magnetism and Synchrotron Radiation , 1996, ed. E. Beaurepaire, B. Carrière, J.P. Kappler (Les Editions de physique, Les Ulis, France)
[20] B.T. Thole, P. Carra, F. Sette and G. Van der Laan, Phys. Rev. Lett., 68 (1992) 1943
P. Carra, B.T. Thole, M. Altarelli and X. Wang, Phys. Rev. Lett., 70 (1993) 694
[21] J. Rothman, C. Meyer, D. Givord, J. Vogel, M. Finazzi, J.P. Kappler and N. B. Brooks, J. Mag. Magn. Magn. (submitted)
[22] L.Z. Liu and J.W. Allen, Phys. Rev., B45 (1992) 8934
[23] S. Doniach, Physica, B91 (1977) 231
[24] J. Vogel, private communication
[25]J. Vogel, A. Fontaine, V. Cros, F. Petroff, J.P. Kappler, G. Krill, A. Rogalev, J. Goulon, Phys. Rev. B55 (1997) 3663
[26] V. Cros et al. (to be published)
[27] S. David, A. Fontaine, K. Mackay, J. Vogel, S. Pizzini (unpublished)
[28] E.F. Kneller , I.E.E.E. Trans. Mag., 27 (1991) 3588
[29] R. Coehoorn, D.B. de Mooij and C. de Waard, J. Magn. Magn. Mat., 80 (1989) 101

Materials Science Forum Vols. 302-303 (1999) pp. 21-32

X-Ray Magnetic Scattering Studies at the LNLS

C. Giles

Instituto de Física, Universidade Estadual de Campinas, C.P. 6165, 13083-970 Campinas, Brazil

Keywords: Synchrotron Radiation, Magnetism, X-Ray Magnetic Scattering

Abstract. The potential applications of X-ray magnetic scattering techniques to probe microscopic magnetic properties at the Brazilian National Synchrotron Light Source (LNLS) are revisited and discussed. The source, optics and experimental station of a magnetic scattering beamline at the LNLS are described. First preliminary experiments at LNLS bending magnet sources demonstrate the importance of introducing focusing optical elements to make feasible X-ray magnetic scattering experiments at LNLS. Spectral flux calculations of insertion devices demonstrate the potential perspectives of a wiggler beamline for soft X-ray resonant magnetic scattering at the LNLS compared to other synchrotron radiation facilities.

Introduction

New developments in magnetic materials as nanocrystalline structures, thin layers and multilayers are becoming very popular and important for the future technological applications of magnetism [1]. At the same time applications of new experimental techniques to probe microscopic properties of magnetic materials have being essential to understand and control these developments. Along with bulk measurements such as magnetisation and susceptibility measurements, specific heat and electrical resisitivity, which provide the magnetic character of these materials and local probes such as NMR, Mössbauer spectra and high energy spectroscopies, scattering experiments provide information about spatial correlations as well as the dynamics of correlated magnetic moments. Neutron scattering has been since late 40's the only technique available to study microscopic magnetism and has been widely used to obtain a variety of important informations as magnetic structures, critical phenomena, dynamics of correlation functions and magnetic fluctuations [2-5].
Although X-rays do not have a magnetic moment, it was predicted back in the 50's that X-rays are sensitive to magnetisation densities [6]. X-ray magnetic scattering was predicted by Platzman and Tzoar [7] and was first experimentally evidenced by De Bergevin and Brunel [8] in a difficult experiment using a conventional X-ray tube equipment. The advent of dedicated synchrotron radiation facilities has been determinant for the use of X-ray scattering techniques to probe magnetic properties [9-12].
It is the purpose of this paper to describe the project and potential applications of a dedicated experimental station for X-ray magnetic scattering techniques at the Brazilian National Synchrotron Light Source (LNLS).
After a short review of the basic properties of X-rays as a probe for magnetism and the basis for X-ray magnetic scattering and its current applications, we will present the project of an optimised beamline for magnetic scattering studies at the LNLS. First preliminary experiments and new perspectives in the field of magnetism and synchrotron radiation at the LNLS will also be briefly described and discussed.

X-rays and magnetism

Since microscopic magnetism arises from the unpairing of electrons on their electronic band structure, X-rays as an electronic probe is also a probe to magnetic properties, specially in a microscopic scale with small samples and high Q resolution and with chemical and electronic site specificity [13].
Interactions of photons with magnetisation densities were first discovered by Faraday [14] and Kerr [15] demonstrating the influence on the polarisation of visible light respectively transmitted and reflected through magnetic materials. These studies along with the work of Zeeman [16] constitute the basis for the magneto-optics in the visible range. Polarisation effects are difficult to observe in the X-ray range because the complex refractive index is close to unity. Nevertheless they could be observed close to the absorption edges where they are enhanced by resonant effects [17-18]. These enhanced effects could be observed in scattering and absorption experiments giving rise to new techniques as X-ray resonant magnetic scattering [19-20] and X-ray magnetic dichroism [21]. X-ray magnetic linear and circular dichroism has recently become a powerful technique to obtain local magnetic properties with chemical and site specificity [22]. The description of these techniques as well as useful information can be found elsewhere [23-25].
In addition to X-ray scattering and spectroscopy techniques it is worth mentioning reflectivity measurements in resonant conditions [26-27], i.e. the observation of the specular reflected beam at energies close to an absorption edge of the material. In this photon in and photon out process the outcome signal has a contribution both from the real and imaginary parts of the refractive index. Also known as resonant exchange scattering this experimental technique gives spectroscopic information with the unique feature of being a coherent elastic-scattering process with no complicated final-state effects. This technique is especially attractive for the study of surface, interface and thin-film magnetism.

X-ray Magnetic Scattering

There are two regimes for X-ray magnetic scattering. They differ by the presence or not of anomalous scattering. The non-resonant regime, also called the high-energy limit regime, when the photon energy is far from any characteristic energy of the material is a purely magnetic interaction between the incoming photon and the electron magnetic moments.
Relativistic interactions are responsible for these magnetic interactions [9]. De Bergevin and Brunel have given a simple classical picture on the interaction of the electron spin with the electromagnetic field of the photon: the electron magnetic moment is accelerated and reradiates electromagnetic components of the scattered wave [9].
The magnitude of this interaction is given by a correction of hK/mc to the classical Thomsom term on the scattering cross section, where K is the modulus of the scattering vector $\mathbf{K} = \mathbf{k}_i - \mathbf{k}_f$. That means that in a typical diffraction experiment the relative amplitude of magnetic scattering and charge scattering is of the order of 10^{-3}. This figure is further reduced by the fact that only the unpaired electrons take part of the process, which takes the effective intensity of purely magnetic peaks to be 7 to 8 orders of magnitude lower than the Thomson charge scattering.
The second regime is the resonant regime when the photon energy is tuned close to an absorption edge of the magnetic atom. An increase in the amplitude of the magnetic scattering that can be as large as 10^3 is characteristic of this regime. In this case the X-ray scattering amplitude is related to the absorption cross-section through the optical theorem (Kramers-Kronig relation) and is not described by the normal magnetic cross-section. This effect was predicted by Blume [19], first evidenced by Gibbs et al. [10] and theoretically accounted by Hannon et al. [11].
In the resonant regime the enhancement of the magnetic scattering amplitude is the effect of the promotion of an inner shell electron to a magnetisation sensitive unoccupied electronic state,

typically d and f final states. So this addition to the scattering cross-section arising from electric transitions is a probe of the local magnetic moments of the final states. In that sense the resonant magnetic scattering as well as the magnetic dichroism can be understood in terms of electric multipole transitions between an atomic core level and an unfilled atomic shell or a narrow electronic band [20].
The magnitude of the enhancements is predicted to be higher for low-order electric multipole transitions and even higher when the final states correspond to the shell of unpaired electrons. The resonance is further increased when the inner shell corresponds to a well-resolved doublet by spin-orbit coupling like the L_2-L_3 or M_4-M_5 edges.

Applications of X-ray scattering techniques

Powder diffraction is a useful technique for the determination of the order parameter of crystalline structures. Concerning magnetic structures, as the magnetic scattering signal is very weak relative to charge scattering, it has not been possible to use X-ray magnetic powder diffraction. Neutrons remain for the moment the only available tool to solve magnetic structures from powder samples. This fact restrict X-ray magnetic scattering to be used for monocrystalline samples or for structured multilayers systems. In spite of this limitation still several and unique information can be obtained with X-ray magnetic scattering as we will briefly outline below reinforcing the value of this technique.
The non-resonant magnetic scattering regime is characterised to be very weak, nevertheless it has a particular polarisation and scattering geometry dependence that allow the independent determination of the orbital magnetic moment and of the spin magnetic moment in a simple way. If the incoming beam is linearly polarised in a so-called σ polarisation scattering geometry (the electric field is parallel to the scattering plane) the orbital magnetisation density does not contribute to the σ-σ polarisation channel of the magnetic scattered beam (corresponding to a σ polarisation of the scattered beam). In the σ-π channel both the orbital and spin contribute and from these two intensities it is possible to measure the orbital (L) to spin (S) magnetisation ratio (L/S) [29-31].
Different polarisation states of the incoming beam allow different types of experiments as it is the case for circular polarisation that is sensitive to the chirality of magnetic domains. Recent experiments have shown the possibility to distinguish the domain helicity of spiral antiferromagnetic structures as in the case of single crystals of holmium [32].
These examples as well as a number of others put into evidence the importance of having sources of X-rays tuneable in polarisation and also optical elements capable to analyse the polarisation state of the diffracted beam.
Synchrotron radiation is a source of highly polarised X-rays. Very high rates of linear polarisation in the horizontal plane can be obtained from the radiation produced by the conventional bending magnets of the storage ring as well as from insertion devices as wigglers and undulators. Recent advances in synchrotron radiation technology have led to the production of X-ray sources based on special insertion devices as helical undulators or asymmetric wigglers capable to produce X-rays with circular polarisation or linear polarisation in the vertical plane [33]. Also, recent developments in optical elements for polarisation transformation and analysis based on crystal plates are able to produce a variety of polarisation states from a linearly polarised beam in the horizontal plane [34]. These optical elements which are X-ray analogues to compensators, quarter-wave plates or half-wave plates from the visible optics are permanently operated in several beamlines in the more recent third generation sources of synchrotron radiation around the globe [35-38].
Generally speaking antiferromagnetic structures are easily studied with X-ray magnetic scattering since the purely magnetic peaks are separated in the reciprocal space from the charge scattering reflections. This is also the case for helimagnetic systems or other incommensurate magnetic systems as spin density waves [39-40]. Nevertheless in the case of ferro or ferrimagnetic structures

the magnetic peak can be separated from the charge signal using circularly polarised X-rays and a technique called flipping ratio [9]. In this case the magnetic signal can be obtained from the ratio of the signal obtained with the sample magnetisation saturated in two opposite directions.
In addition to these techniques it is possible to considerably enhance the magnetic signal tuning the incoming photon energy through a characteristic and well selected absorption edge of the magnetic atom. This resonance enhancement opens a new variety of experiments as the measurement of weak magnetic signals or the study of different magnetic species in magnetic materials.
Weak magnetic signals may come from systems with small magnetic moments or from small magnetic volumes. As examples of scattering from weak magnetic moments are an extremely weak 0.02 μ_B magnetic moment observed in URu_2Si_2 heavy fermion system [41], or magnetic critical fluctuations on Ho [42] and NpAs [43] revealing two length scales due to surface or near-surface effects and nucleation of magnetic fluctuation on crystallographic defects.
Small magnetic scattering volumes are represented by scattering from surfaces and interfaces. Resonant X-ray scattering can be used to probe important surface phenomena in magnetism. Recently, induced magnetic moment at Pt atoms by Co at the surface of a Co_3Pt crystal has been determined to be a half of its bulk value and interpreted by the lower concentration of Co atoms at the surface [44]. At the same time first order antiferromagnetic to paramagnetic transition at the surface of a UO_2 crystal has been observed with resonant X-ray scattering at the M_4 edge of U, whilst the bulk transition is of second order [45]. These represent the first experiments of a variety of new experiments that remain to be explored.
Finally it is worth mentioning the spectroscopic tool inherent to the resonant X-ray scattering technique. From the analysis of the line shape of the magnetic resonant scattering, i.e. observing the maximum of the resonance enhancement in the energy scale it is possible to analyse the exchange splitting of selected electronic shells [46] and on electronic configurations [47-48] allowing the determination of density of states near E_F, dispersion of valence states and new trends in electronic structures.
The spectroscopic applications of this technique complement powerful information provided by spectroscopic tools as photoemission, spin polarised absorption and magnetic dichroism.

A Magnetic Scattering beamline

The advent of a synchrotron radiation facility in Brazil as LNLS [49] opens a unique opportunity to implement new techniques for the study of magnetism with X-rays in Latin America. The LNLS is already delivering X-rays and VUV photons from a 1.37 GeV storage ring with operation parameters that already exceed the target ones, as the 130 mA current in the storage ring and a lifetime of the beam better than 10 hours. At present seven beamlines are operating and available for the users community, two new beamlines are under commissioning and four new beamlines are being prepared for mid 1999. From the operating beamlines four of them allow spectroscopic experiments, two in the VUV range (TGM and SGM), one in the soft X-ray energy range (SXS) and one in the firm X-ray range (XAS). In most of them first test experiments using X-ray Magnetic Circular Dichroism (XMCD) have already been done and some beamlines continue to implement and upgrade their instrumentation for these new magnetic and spectroscopic techniques.
One X-ray diffraction beamline (XRD) is already operating and among the four new beamlines two will be devoted to scattering techniques. One of these beamlines will be fully equipped to perform X-ray magnetic scattering (XMS) experiments [50].

Beamline requirements for XMS experiments
A beamline dedicated to X-ray magnetic scattering should fulfil a number of requirements concerning the properties of the X-ray beam that is impinging on the sample. We will briefly discuss the main aspects here as well as the more adapted solutions for the XMS beamline at LNLS.

As is evident from the characteristics of XMS, the photon intensity on the sample is a very important factor to the implementation of this technique. Usually non-resonant XMS peaks are 10^6 to 10^8 times smaller than charge scattering peaks. In top of that for the determination of L/S ratio a polarisation analysis of the diffracted beam by the magnetic sample requires another factor of 10 to 10^2. These requirements are less stringent in the case of resonant X-ray scattering and mainly in the soft X-ray energy range where the enhancements are very large (it can reach 10^6 for the M edges of the actinides) the same is not true in the hard X-ray range where an intensity enhancement factor of 2 to 10 is observed in transition metals K edges [17,51].
Most of the X-ray scattering beamlines operating nowadays are optimised for experiments in the X-ray range. This is a natural choice since the wavelengths of electromagnetic waves diffracted by crystals are in the X-ray range. Nevertheless in the case of resonant scattering the larger enhancements on the magnetic peaks are expected for absorption edges in the soft X-ray region. In order to use resonant scattering in the soft X-ray region, artificial multilayered structures have to be used to satisfy Bragg diffraction condition. As we will see this is a very promising perspective for magnetism studies at the LNLS.
The storage ring parameters (energy, current and magnetic field of the bending magnets) are mainly responsible for the spectral flux produced by a synchrotron radiation facility. In some cases insertion devices can be used instead of the bending magnets as the synchrotron source for a specific beamline. Along with the source the measured signal is determined by the optics between the sample and the source and detectors. As an end-station of the beamline an experimental setup with appropriate sample environment is required.

i. Source

All presently installed beamlines at LNLS have bending magnet sources. Insertion devices can be installed in the four available straight sections of the storage ring but they will be more efficient after the installation of a pre-injector booster synchrotron that will allow to operate them with poles gap as small as 20 mm to increase the magnetic field. Projects to install future insertion devices are being prepared and different scientific communities are already preparing their specific projects to implement new beamlines with insertion device sources. The perspective to increase the photon flux at the sample with an insertion device is obviously very important to improve the possibilities to perform X-ray magnetic scattering experiments at the LNLS.
A natural question to be answered at our present stage is: how well suited are the LNLS bending magnet sources for an X-ray magnetic scattering beamline? And furthermore, how can XMS at the LNLS be competitive with other beamlines at other facilities around the world?
This issue can be treated in two steps, first we will analyse the present possibility to perform XMS experiments with a bending magnet source, and second we will show an important use of the LNLS facility for a XMS beamline internationally competitive.

a. Bending magnet source for XMS at LNLS

X-ray magnetic scattering is currently performed at several beamlines around the world and good advances have been done mainly at the National Synchrotron Light Source (NSLS) of Brookhaven, Upton, NY and at the European Synchrotron Radiation Facility (ESRF) in Grenoble, France. New beamlines are being prepared at the Argonne Photon Source (APS, Chicago), the Advanced Light Source (ALS, Berkeley) and at new facilities in Japan and Europe.
Calculations of the spectral flux of the beamlines dedicated to XMS are compared to the spectral flux expected for a beamline at the LNLS (Fig. 1).

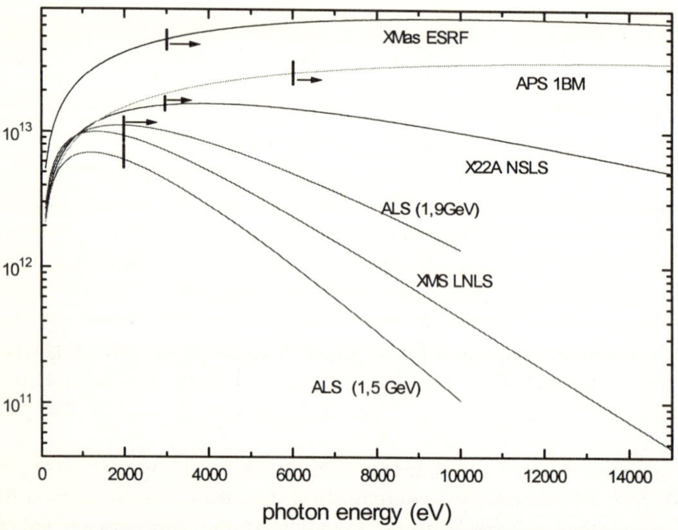

Figure 1. Spectral flux at sample position for different beamlines dedicated to X-ray magnetic Scattering. Arrows indicate the minimum energy of the beamline restricted by optical elements. Calculations take into account nominal values for the energy and current of the respective storage ring and available focusing elements as described in Table 1.

These calculations consider source characteristics and optical elements with focusing capabilities. The main parameters of each beamline are described in Table 1. It is worth noticing that the ratio of flux from a bending magnet ESRF beamline to a LNLS beamline goes from 10^3 at 15 keV to 3 at 2 keV demonstrating the potentiality of the LNLS at the soft X-ray range. At 8keV (typical energy value for X-ray scattering) the intensity ratio of the LNLS to NSLS beamline flux is of the order of 10^{-1}. These figures clearly state that at firm X-rays a XMS beamline at LNLS is feasible but limited to resonant scattering experiments or non-resonant scattering signals with strong magnetic moments. A real improvement should be obtained with a different source.

Table 1 Parameters of X-ray scattering beamlines shown in Fig. 1. (E, I and ε_c correspond to the energy, current and critical energy of the storage ring, mradV and mradH to the maximum angular apertures of the focusing elements and Emin and Emax correspond to the minimum and maximum X-ray energy accepted by the optical elements.)

	E(GeV)	I (mA)	ε_c(keV)	mradV	mradH	Emin (keV)	Emax (keV)
1BM APS	7,0	200	19,2	0.1	3.7	3	15
XMaS ESRF	6,0	200	19	0.2	3.1	3	15
X22A NSLS	2,58	250	4,4	0.25	2.1	3	12
9.3.2 ALS	1,9-1,5	400	2,4-1,5	0.2	2.0	2.2	6
XMS LNLS	1,37	100	2	0.6	5.0	0.8	15

b. Source for a competitive XMS beamline at LNLS

Previous calculations have led us to study the feasibility of using a wiggler as the X-ray source for the X-ray magnetic scattering beamline at LNLS. The maximum magnetic field and the number of poles are the two main parameters of such devices. We have considered two projects one with 37 poles and 1.3 T and another with 52 poles and 1.13 T. The spectral flux obtained from these devices at the LNLS are compared to existing insertion devices at NSLS and ESRF used for magnetic scattering (Fig. 2). Realistic (for the LNLS insertion devices) and current (for the NSLS and ESRF) focusing capabilities of these beamlines were used in these calculations (Table 2).

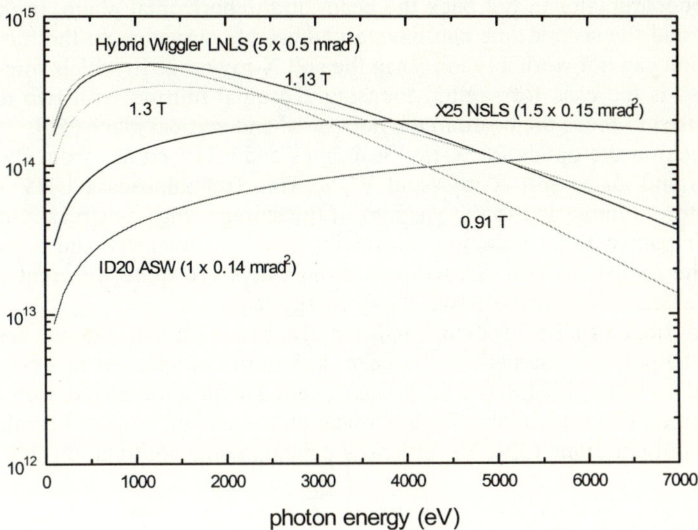

Figure 2. Spectral flux of insertion device beamlines for X-ray Magnetic Scattering. Two projects of hybrid multipole wigglers for the LNLS are compared to existing devices at NSLS (X25) and ESRF (ID20 ASW) considering their focusing optical elements.

It is clear from the calculations in Fig. 2 that very good perspectives exist in the soft to firm X-ray range to have a unique facility for X-ray magnetic scattering. Many potential applications to apply resonant magnetic scattering in small volumes in sufaces and thin films exist and are of technological interest.

Table 2: Insertion device parameters for XMS beamlines and focusing optical elements used in calculations in Fig. 2. (mradV and mradH correspond to the vertical and horizontal angular apertures accepted by the focusing elements.)

	I (mA)	No poles	B(T)	mradV	mradH
X25 NSLS	250	27	1.1	0.15	1.5
ASW ID20	200	8	0.9	0.14	1.0
HW LNLS	100	37	1.3	0.5	5
HW LNLS	100	52	1.13/0.9	0.5	5

Tests experiments to measure magnetic intensities to confirm the potential improvement of a XMS beamline with a wiggler source are being performed and will be reported elsewhere [52].

ii. Optics

The ideal optics on the beamline, is the one with the minimum number of optical elements. In our case essential elements are mirror, monochromator and slits. Double crystal and four crystal monochromators have been developed at LNLS [53] to match the requirements of a fixed exit or a high-resolution monochromatic beam. Silicon or germanium crystals are used in the X-ray energy range and other crystals can be used for soft X-rays. Mirrors are used to clean the beam from the harmonic energies at the sample. Both monochromator and mirror can also have a focusing capability that is an essential factor to increase the intensity at the sample in the case of bending magnets or wigglers.

Indeed our calculations shown in Fig. 1 and Fig. 2 assume a meridional focusing mirror and a double crystal monochromator with sagittal focusing. Another interesting configuration is to use a second mirror after the monochromator to put back the beam on the horizontal plane. In that case the first mirror can be flat and the second one can have an adjustable curvature in the meridional plane [54]. This configuration can not work any longer in the soft X-ray range in which silicon does not diffract anymore and so is the case for sagittal focusing. Toroidal mirrors with two different radius of curvature can be used to focus the beam in the horizontal and vertical plane [55].

Another important difference on the optics for X-ray beamlines and VUV comes from the strong absorption of Be windows and air to soft X-rays and VUV. This fact imposes a UHV vacuum beamline without any windows (connected to the vacuum of the storage ring), whereas X-rays can afford Be windows and air path without reducing drastically the beam intensity and allowing to work without UHV. As the optics for soft X-rays and firm X-rays are quite different is more convenient to build separate beamlines for these two X-ray energy ranges.

Additional optical elements that can be used to condition the beam on the sample are X-ray polarimeters to transform the polarisation state of the beam before the sample. At this point these optical elements developed at ESRF [56], have only been applied with good success on sources with small angular divergences, like undulators. Their application to bending magnet beamlines has been restricted to special configurations [57]. Nevertheless developments continue on this subject [58].

iii. Experimental Station

At the end station of the beamline where the beam is already conditioned in energy resolution, divergence and polarisation an X-ray diffractometer is necessary to measure diffraction profiles. For single crystals, a diffractometer with four-circle geometry is very convenient. This instrument allow to orient the crystal and go for magnetic peaks at any point of the reciprocal space as well as to perform reciprocal space scans or mappings and look for satellite peaks as those created in antiferromagnetic structures. Also, it is very convenient to control the temperature of the crystal in order to study critical phenomena as phase transitions, etc. This is possible with cryostats that have to fit into the diffractometer head. A helium closed circuit refrigerator (displex) with temperature control range of 8 – 350 K is planned for this experimental station. To be able to go to lower temperatures and to apply high magnetic fields as some magnetic systems require, cryomagnets should be used carefully planning how to install it into the diffractometer without degrading its resolution. Furthermore, energy analysers and polarisation analysers will be available in the diffractometer two-theta arm when this will be required as well as scintillators detectors and fluorescence screens as monitoring devices.

First measurements at LNLS

This experimental station is planned to be commissioned by the end of 1998 in the instrumentation beamline (XRI) at the LNLS, whilst the optics described above will be prepared in an X-ray diffraction beamline optimised for XMS. In the meanwhile test experiments have been planned and performed at the XRD beamline. The first tests have demonstrated the need for focusing elements to increase the monochromatic flux at the sample. As an example of that we have performed experiments with single crystals of pure chromium that present Charge Density Waves (CDW) and Spin Density Waves (SDW) phenomena [39]. The same samples were measured at the Magnetic Scattering Beamline (ID20) at the ESRF that is an undulator source beamline (Fig.3) and at the XRD beamline at the LNLS without any focusing device (Fig. 4). As is shown in Fig. 3 CDW peaks are 10^4 times smaller than the Bragg peak and the SDW magnetic peaks are another factor 10^4 times smaller than the CDW peaks. At the ESRF, SDW peaks can be measured with peak intensities of the order of 200 counts/sec. At the LNLS only the CDW peaks are visible with 100 counts/sec showing a 10^4 factor difference as compared to the ESRF beamline. The experiment at LNLS was performed with a narrow solid angle spot of the beam on the sample. Sagittal focusing of the monochromator and meridional focusing of the mirror can improve the flux on a sample spot of 0.5 mm^2 in more than two orders of magnitude. Another factor 10 or 50 could be gained with an insertion device only then becoming comparable to the more brilliant beamlines dedicated to X-ray magnetic scattering.

Spin density wave peaks are indeed very weak and difficult to be observed in the present configuration at the LNLS, nevertheless resonant magnetic scattering and other more intense signals could already be studied at the LNLS. More realistic limits of the capabilities to perform XMS at LNLS will become possible with the first experimental results with the new instrumentation.

More experiments in the soft X-ray range are planned to test the feasibility of working with XMS at this energy ranges [52].

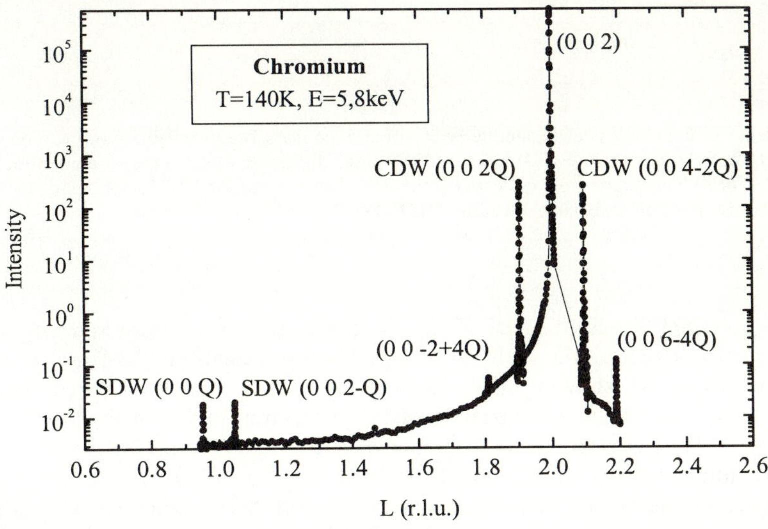

Fig. 3 Relative intensities of reciprocal lattice coordinate L of a Cr single crystal around the Bragg reflection (0 0 2) at 140 K and just below the Cr K edge. Charge density wave peaks (0 0 2Q) and (0 0 4-2Q), its harmonics (0 0 –2+4Q) and (0 0 6-4Q) and Spin Density Wave peaks (0 0 Q) and (0 0 2-Q) are clearly distinguished and analysed showing the capabilities to study very weak magnetic peaks at synchrotron radiation facilities. This experiment was performed at ID20 Magnetic Scattering beamline of the ESRF [59].

Other feasibility studies of synchrotron radiation for the study of magnetism were performed at XAS beamline. X ray Circular Magnetic Dichroism on rare-earth phase Laves were measured through their ordering temperature [60]. Also XMCD measurements with nanosecond time resolution were performed showing the feasibility of this new technique at the LNLS first implemented at the ID24 Energy Dispersive beamline of the ESRF [61].

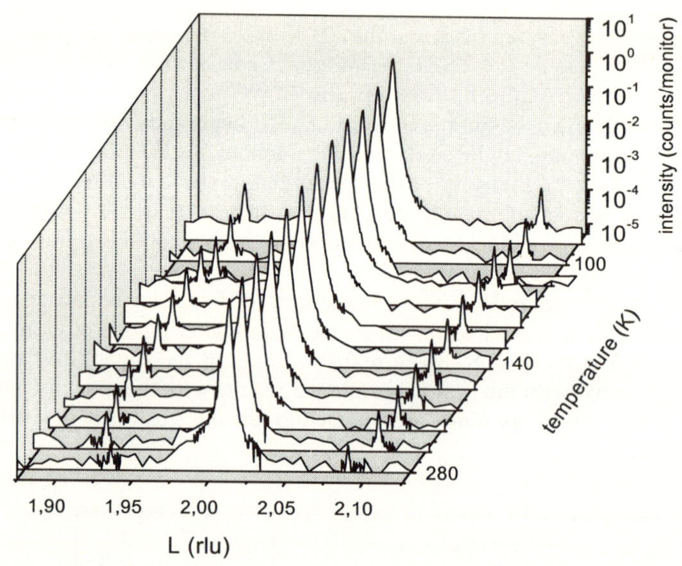

Fig. 4 Temperature dependence of the CDW peaks (satellite peaks around the main Bragg reflection (0 0 2)) on a pure chromium single crystal. At the Néel temperature of 310 K the CDW peaks disappear whereas they are not affected by the spin flip transition at T_{sf}= 90 K. These data were collected at the XRD beamline of the LNLS, not being possible to detect SDW peaks as it was done in a similar experiment at the ESRF (Fig. 3).

Conclusions

We have revised the potential capabilities of X-ray magnetic scattering techniques to study microscopic properties of magnetic materials with synchrotron radiation. A large class of applications have been presented showing that XMS is already a well stablished tool for experiments in microscopic magnetism. New beamlines in synchrotron radiation facilities around the world are being projected and constructed. A Brazilian synchrotron radiation facility like LNLS offers the possibility of implementing this technique, both in the firm X-ray range where more conventional scattering experiments are performed and in the soft X-ray range. In this spectral domain that corresponds to the maximum flux of the LNLS facility, interesting phenomena can be studied by resonant X-ray magnetic scattering at the M edges (transitions from 3d core electrons to 4f or 5f empty states) of rare-earths and actinides and the L edges (2p core electrons to 3d empty states) of transition metals. New possibilities like studying magnetic properties of surfaces, thin films and multilayers are very promising at this energy range for new magnetic materials. Preliminary experiments at bending magnet sources of the LNLS compared to experiments at the ESRF demonstrate the need of a suited focusing optics in order to allow acomparable flux to existing beamlines dedicated to magnetic scattering (like X22 at NSLS). Spectral flux calculations demonstrate the importance of an alternative source (multipolar wiggler) to enable the application of the unique features of magnetic scattering, like the polarisation analysis of the scattered beam in order to determine the orbital to spin (L/S) magnetic moment ratio. It is worth reminding that spectroscopic techniques are also very promising and scattering techniques should complement information coming from photo-emission, spin dependent absorption and magnetic dichroism as well as X-ray reflectivity. Also, in general X-ray techniques for the study of magnetism presents complementary characteristics to neutron experiments, having some advantages as the high Q-resolution and capabilities to probe small magnetic volumes but nevertheless not being competitive in other experiments where neutrons are better suited.

Acknowledgements

The author acknowledges the financial support from FAPESP under contract 1996/5586-6, CNPq and from Ministério de Ciência e Tecnologia (PRONEX). Fruitful discussions with C. Vettier are also acknowledged.

References

[1] 7th Joint Magnetism and Magnetic Materials (MMM)-Intermag Conference Proceedings, San Francisco, California, January 6-9, 1998
[2] J.D. Axe, J.B. Hayter and R. Pynn, Mat. Res. Soc. Bull. XV-**11**, (1990) p. 42
[3] R.J.Bigeneau, E.Bucher, J.P.Maita, L.Passell and K.C.Tuberfield, Phys. Rev. B**8** (1973) p.5345
[4] A.D. Taylor, R. Osborn, K. A. McEwen, W. G. Stirling, Z. A. Bowder, W. G. Williams, E. Balcar and S. W. Lovesey, Phys. Rev. Lett. **61** (1988) p. 1309
[5] S.K. Sinha, in: Handbook on the Physics and Chemistry of Rare Earth Metals, ed. R.J. Elliott (Plenum Press, New York, 1972) p. 187
[6] M. Gell-Mann, M.L. Golberger, Phys. Rev. **96** (1954) p. 1433
[7] P.M. Platzman, N. Tzoar, Phys. Rev. B**2** (1970) p. 3536
[8] F. de Bergevin, M. Brunel, Phys. Lett. **A39**, (1972) p. 141-142
[9] F. de Bergevin, M. Brunel, Acta Cryst. **A37** (1981) p.314
[10] D. Gibbs, D.R. Harshman, E. D. Isaacs,D.B. McWhan, D. Mills, C. Vettier, Phys. Rev. Lett. **61**, (1988) p. 1241-1244
[11] J.P. Hannon, G. T. Trammell, M. Blume, D. Gibbs, Phys. Rev. Lett. **61**, (1988) p. 1245-1248
[12] E. D. Isaacs, D. B. McWhan, C. Peters, G.E. Ice, D.P. Siddons, J.B. Hastings, C. Vettier, O. Vogt, Phys. Rev. Lett. **62**, (1989) p. 1671
[13] C. Vettier, Physica B **192** (1993) p. 1
[14] M. Faraday, Trans. Roy. Soc. (London) **5** (1846) p. 592
[15] J. Kerr, Phil. Mag. **3** (1877) p. 339
[16] P. Zeeman, in: Research in Magneto-Optics (MacMillan, London, 1913)
[17] K. Namikawa, M. Ando, T. Nakajima, H. Kawata, J. Phys. Soc. Japan **54**, (1985) p. 4099-4102
[18] G. van der Laan, B.T. Thole, G.A. Sawatsky, J.B. Goedkoop, J.C. Fuggle, J.M. Esteva, R. Karnatak, J.P. Remeika, H.A. Dabkowska, Phys. Rev. **B34** (1986) p. 6529
[19] M. Blume, J. Appl. Phys. **57** (1985) p. 3615
[20] C. Vettier, J. Magn. Magn. Mater. **129** (1994) p. 59
[21] G.Schütz, W.Wagner, W.Wilhelm, P. Kienle, R.Zeller, R. Frahm, G. Materlik, Phys. Rev. Lett. **58** (1987) p. 737
[22] F. Sette, C.T. Chen, Y. Ma, S. Modesti, N.V.Smith, in: X-ray Absorption Fine Structure, ed.5.5 Hainin (Ellis Horwood Limited Chichester, 1991)
[23] B.T. Thole, G. van der Laan, Phys. Rev. Lett. **70** (1993) p. 2499
[24] R. Wu, D. Wang, A.J. Freeman, J. Magn. Magn. Mater. **132** (1994) p. 103
[25] G. van der Laan, J. Magn. Magn. Mater. **156** (1996) p. 99
[26] C.C. Kao, C.T.Chen, E.D. Johnson, J.B. Hastings, H.J. Lin, G.H. Ho, G. Meigs, J.M. Brot, S.L. Hubert, Y.U. Idzerda, C. Vettier, Phys. Rev. **B50** (1994) p. 9599
[27] M. Sacchi, J. Vogel, S. Iacobucci, J. Magn. Magn. Mater. **147** (1995) p. L11
[28] O. Klein, Y. Nishina, Z. Phys. **52** (1929) p. 853-868
[29] M. Blume, D. Gibbs, Phys. Rev. **B37** (1988) p. 1779

[30] S. Landgridge, G.H. Lander, N. Bernhoeft, A. Stunault, C. Vettier, G. Grübel, C. Sutter, F. de Bergevin, W.J. Nuttal, W. G. Stirling, K. Mattenberger, O. Vogt, Phys. Rev. **B55** (1997) p. 6392
[31] V. Fernandez, C. Vettier, F. de Bergevin, C. Giles, W. Neubeck, Phys. Rev **B57** (1998) p. 7870
[32] C. Sutter, G. Gruebel, C. Vettier, F. de Bergevin, A. Stunault, D. Gibbs, C. Giles, Phys. Rev. B**55**(2), (1997) p. 954-959
[33] P. Elleaume, J. Chavanne, Nucl. Instrum. Meth. In Phys. Res. **A308** (1991) p. 363-366
[34] C. Giles, C. Vettier, F. de Bergevin, C. Malgrange, G. Gruebel, F. Grossi, Rev. Sci. Instrum. **66**(2), (1995) p. 1518-1521
[35] C. Giles, C. Malgrange, J. Goulon, F. de Bergevin, C. Vettier, E. Dartyge, A. Fontaine, C. Giorgetti, S. Pizzini, J. Appl. Cryst. **27** (1994) p. 232-240
[36] S. Pizzini et al, to be published
[37] L. Varga, C. Giles, C. Neumann, A. Rogalev, C. Malgrange, J. Goulon and F. de Bergevin, J. Phys. IV France **7** (1997) p. C2-309-313
[38] C. T. Venkataraman, J. C. Lang, C.S. Nelson, G. Srajer, D.R. Haelfner, S. D. Shastri, Rev. Sci. Instrum. **69** (1998) p. 1970
[39] E. Fawcett, Rev. Modern Phys. **60** (1988) p. 209
[40] J.P. Hill, G. Hegelsen, D. Gibbs, Phys. Rev. **B51** (1995) p. 10336
[41] E.D. Isaacs, D.B. McWhan, R.N. Leiman, D. J. Bishop, G.E. Ice, P. Zschak, B.D. Gaulin, T.E. Mason, J.D. Garett, W. L. Buyers, Phys. Rev. Lett. **65** (1990) p. 3185
[42] T.R.Thurston, G.Hegelsen, D.Gibbs, B.D.Gaulin, G.Shirane, Phys.Rev. Lett. **70** (1993) p. 3151
[43] S. Landgridge, W.G. Stirling, G.H. Lander, J. Rebizant, J.C. Spirlet, D. Gibbs, O. Vogt, Europhysics Lett. **25** (1994) p.137
[44] S. Ferrer, P. Fajardo, F. de Bergevin, J. Alvarez, et al.,Phys. Rev. Lett. **77**, (1996) p. 747-750
[45] G. M. Watson, D. Gibbs, et al., Phys. Rev. Lett. **77**, (1996) p. 751-754
[46] D. Gibbs, G. Grübel, D.R. Harshman, E.D. Isaacs, D.B. McWhan, D. Mills, C. Vettier, Phys. Rev. **B42** (1992) p. 10772
[47] C.C Tang, W.G. Stirling, G.H. Lander, D. Gibbs, W. Herzog, P. Carra, K. Mattenberger, O. Vogt, Phys. Rev. **B46** (1992) p. 5287
[48] F. de Bergevin, M. Brunel, R.M. Galera, C. Vettier, E. Elkaïm, M. Bessière, S. Lefèbvre, Phys. Rev. **B42** (1992) p. 10772
[49] A.R.D. Rodrigues, Liu Lin, Synchr. Rad. News **10**(2) (1997) p. 27
[50]C. Giles, Projeto FAPESP n 1996/5586-6 (1997)
[51] J.P. Hill, C.C. Kao, D.F. McMorrow, Phys. Rev. **B55** (1997) p. R8662
[52] C. Giles, C. Vettier, LNLS project 293/97 (1998) unpublished
[53] H. Tolentino, J. Durr, I. Mazzaro, D. Udron, C. Cusatis, Rev. Sci. Instrum. **66** (1995) p. 1806
[54] A. Stunault et al, J. Synchrotron Radiat. (to be published).
[55] ESRF Beamline Handbook, XmaS beamline BM28 Magnetic Scattering CRG (1995)
[56] C. Giles, C. Malgrange, J. Goulon, C. Vettier, F. de Bergevin, A. Freund, P. Elleaume, E. Dartyge, A. Fontaine, C. Giorgetti, S. Pizzini, **SPIE** Vol. **2010** (1993) p. 136-149
[57] C. Giles, C. Malgrange, J. Goulon, F. de Bergevin, C. Vettier, A. Fontaine, E. Dartyge, S. Pizzini, F. Baudelet, A. Freund, Physica B **208&209** (1995) p. 784-786
[58] F. de Bergevin, C. Malgrange, C. Giles, C. Vettier, Y. Flora to be published
[59] C. Giles, F. Yokaichiya, P.C. de Camargo, A. J.A. de Oliveira, C. Vettier, to be published
[60] F. Garcia, L. Sampaio, H. Tolentino, A. Fontaine, to be published
[61] M. Bonfim, K. Mackay, S. Pizzini, A. SanMiguel, H. Tolentino, C. Giles, T. Neisius, M. Hagelstein, F. Baudelet, C. Malgrange, A. Fontaine, J. Synchrotron Rad. **5** (1998) p. 750

Materials Science Forum Vols. 302-303 (1999) pp. 33-37

Investigation of the Spin-Density-Wave and Charge-Density-Wave in Antiferromagnetic Cr and Cr-0.2 at. %V using X-Ray Diffraction

P.C. de Camargo[1], A.J.A. de Oliveira[2], C. Giles[3], F. Yokaichiya[3] and C. Vettier[4]

[1] Laboratório de Materiais, DF, Universidade Federal do Paraná, CP. 19081, 81531-990, Curitiba, PR Brazil

[2] Grupo de Supercondutividade e Magnetismo, Departamento de Física, Universidade Federal de São Carlos, CP 676, 13565-905, São Carlos, SP Brazil

[3] Instituto de Física, Universidade Estadual de Campinas, 13083-970 Campinas, SP Brazil

[4] European Synchrotron Radiation Facility, BP 220, FR-38043 Grenoble Cedex, France

Keywords: Itinerant Antiferromagnetism, Spin-Density-Wave, Charge-Density-Wave, Magnetic Scattering

Abstract

Chromium and Chromium-0.2at%Vanadium single crystals were investigated using synchrotron radiation at energies of 5.8 KeV and 13.04 KeV. A description of the spin-density-wave (SDW) and charge-density-wave (CDW) peaks and their temperature dependence for Cr-0.2at%V is presented. A single Q state was observed for all measurements on a (100) surface. Results for pure chromium are compared with those for the alloy.

Introduction

The antiferromagnetic (AFM) phase in chromium and chromium-vanadium dilute alloys is characterized by a transverse spin-density-wave (TSDW), where the spin polarization (**S**) is orthogonal to the magnetic periodicity wave vector (**Q**) below T_N , until a spin-flip transition occurs at temperature (T_{SF}). Below T_{SF} a longitudinal spin-density-wave (LSDW) is formed [1]. Pairs ($\mathbf{Q}_i,\mathbf{S}_j$) with $i \neq j$ for (TSDW) characterize magnetic domains corresponding to orthorombic symmetry above T_{SF} and to tetragonal symmetry for (LSDW) where i=j, i and j referring to the crystal axis. Associated to the SDW there is a charge-density-wave CDW with periodicity 2**Q**. In this work, results of CDW and SDW positions, width and temperature dependence for a single crystal of Cr-0.2at%V which was previously investigated by many other techniques [2] including neutron elastic and inelastic scattering [3,4], are presented and compared to results for chromium. Due to the strong interaction with eletronic charges, x-rays scattering from metallic samples come from only a few micrometers depth, for the energies used, and can be limited to micrometer width spots, therefore, regions of better crystal quality can be selected. A pure CDW satellite would not be detected by neutron scattering, being seen only due to the strain wave (SW) associated to the CDW and SDW.

X-ray-scattering in chromium, due to spin-density-wave (SDW) and charge-density-wave (CDW), has been investigated by Hill et al [4] mapping out several points of the reciprocal space. The Cr-0.2at%V alloy is paramagnetic (P) above the Néel temperature T_N= 288.7 K, as determined by the temperature where the line width increases above the resolution limit for a neutron coherent

Bragg peak [3]. The SDW and CDW satellites are described by their position in reciprocal space. The representation (0,0,**Q**) for SDW and (0,0,2**Q**) for CDW, where $\mathbf{Q} = 2\pi/a(1-\delta)$ and δ measures the deviation from commensurability, indicates that the CDW is the second harmonic of SDW. The incommensurability δ is given by the size difference of electrons and holes Fermi surfaces, where nesting of electrons occur. Antiferromagnetic (AFM) crystals have a charge periodicity superposed by the magnetic periodicity, which for the case of a BCC commensurate systems correspond to a structure where the body central atoms have opposite spin in respect to the corner ones. The expected neutron diffraction would have two distinct contributions, the chemical at reciprocal space points (0,0,2) and the magnetic contribution at (0,0,1).

Experimental Procedure.

Chromium and chromium-0.2at%vanadium alloy single crystals were prepared for X-ray measurements at (0,0,1) face following standard procedure to prepare samples for metallographic analysis. This alloy is the same sample investigated by ultrasound [6], thermal expansion [7], AC magnetic susceptibility [8], and neutron scattering [3,4], all previous properties have shown a sharp Neél transition occuring within a few degrees. In order to remove a few hundreds of damaged surface layers the samples were attacked electrochemically with oxalic acid at current density of $1 mA/cm^2$, during a few minutes. After mounting the system into a cryostat at the beam line, the energy of 13.08 keV was selected and Bragg peaks corresponding to (200), (013) and (004) directions found. The experiments were performed at the Magnetic Scattering beam line ID-20 of the European Synchrotron Radiation Facility, using energies of 5.8 keV and 13.08 keV with a double crystal Si monochromator, aluminum attenuators, direct beam monitor, 13 circle difractometer with $(\omega,\chi,\varphi,2\theta)$ with analysing crystal. The detection system had direct detection after the sample, detector monitoring the scattered beam on a capton scatterer and a detector after the analysing crystal (Al or pyrolitic graphite) with diffraction in the vertical plane. The beam size was limited to about one milimeter square to allow the selection of the best spot as far as crystal quality was concerned. All final scans were performed for regions were the Bragg peak had a FWHM of less than 0.03 degrees.

Following a procedure similar to that of Hill at al. [5] the samples were cooled down to 140 K and the CDW satellites positions (0,0,2Q) and (0,0, 4-2Q) found. Once the reciprocal space directions are identified the energy could be changed to 5.8 keV and new scans of the satellites performed. For large temperature scans a manual adjustment was necessary. As one goes through Bragg, CDW and SDW peaks, the intensity varies six orders of magnitude ($1:10^{-4}:10^{-6}$ respectively), so that filters were used and removed, both automatically and manually, either to protect the detectors or to improve resolution.

Results and Discussion.

A diagram of reciprocal space is quite useful in order to discuss many aspects of this work. Therefore, in Figure 1 the reciprocal space plane [H-L] is shown indicating points for nuclear, CDW and SDW reciprocal lattice, at respective positions. For incommensurate systems, satellites at the scanning axis will appear at points that can be represented in general by, (H,0,L) with H or L equal to R, and R = N +/- n(1-Q) where N = 0, 1, 2, 3, 4, ... corresponding to the reciprocal nuclear lattice points and n = 1, 2, ... referring to the order of the harmonic of interest, with $n \geq N$. N being odd refers to SDW satellites and even numbers correspond to CDW. The first harmonic of the SDW around (0,0,1) is located at R = Q and R = 2 – Q, and those for CDW appears around (0,0,2) happening at R = 2Q and R = 4 – 2Q. Similarly for neutrons, the Q state of each satellite is characterized scanning reciprocal space. Satellites along the scanning axis correspond to domains with Q parallel to the diffracting vector, meaning that in the present case a single Q state perpendicular to (0,0,1) plane, is characterized if no scattering is found for the off-axis satellites.

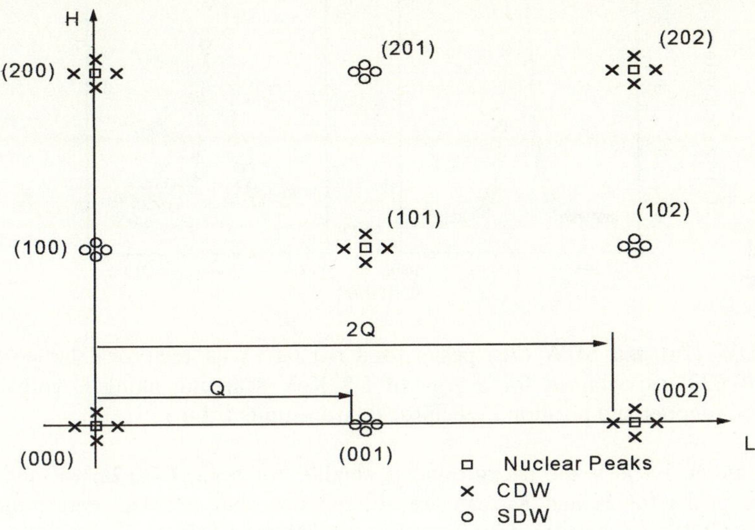

Figure 1.- Reciprocal space points for the H-L plane for nuclear (□), CDW (x) and SDW (0).

Reciprocal space, CDW (0,+/- 2δ, 2), (0,0,2Q), (0,0,4-2Q) and SDW (0,0,Q) and (0,0,2-Q) satellites positions for Cr and Cr-0.2at%V samples were scanned for both energies 5.8 keV and 13.04 keV. In principle, the higher energy is more convenient because it leads to larger magnetic scattering cross section [5] and allows access to more points of reciprocal space. However, the scattered radiation for 13.04 keV has presented a large background, which is about the same magnitude as the CDW peaks, therefore not allowing the detection of harmonics of the CDW. The background detected at this energy is not due to fluorescence and appears to be caused by an anomalous specular reflection, which form ridges detected as large broad peaks when scanning along H and K directions. The same effect was observed for pure Cr and mentioned by Hill et all [5]. A CDW peak at (0,0,2Q) centered at 2Q = 1.8988 r.l.u. and a SDW peak at (0,0,Q) centered at Q = 0,94985 r.l.u., obtained for Cr-0.2at%V, at 5.8 keV are shown at Figures 2 (a) and 2 (b) respectively. The CDW is displaced from (0,0,2) a distance 2δ = 0.101 r.l.u. consistently with the center position of SDW. The area under the CDW peak, which is proportional to the scattered integrated intensity, is 0.1336 a.u. (arbitrary units) while that of the SDW is 1.546×10^{-5} a.u.

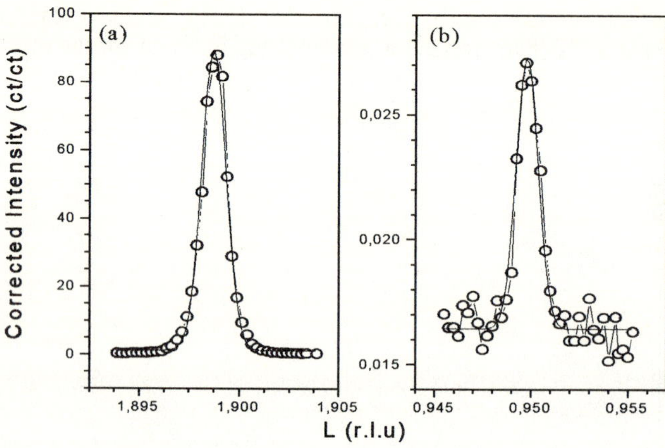

Figure 2. - CDW (2a) and SDW (2b) peaks for Cr-0.2at%V at reciprocal lattice positions (0,0,2Q) and (0,0,Q) as obtained for energy of 5.8 KeV scanning along L with corrected intensities in count/count and position in reciprocal lattice units (r.l.u.).

At energy of 5.8 keV the background is smaller for both, Cr-0.2at%V and Cr. The ridges showing peaks for H and K scan are still present, but one can detect the second harmonic of the CDW. As suggested by Tsunoda et al[8] the CDW peaks are symmetrically located around the Bragg peak, however, their intensity is asymmetric. This asymmetry is likely due to the superposition of a phase modulation (SW) and an amplitude modulation (CDW) as described by Tsunoda et al [8].

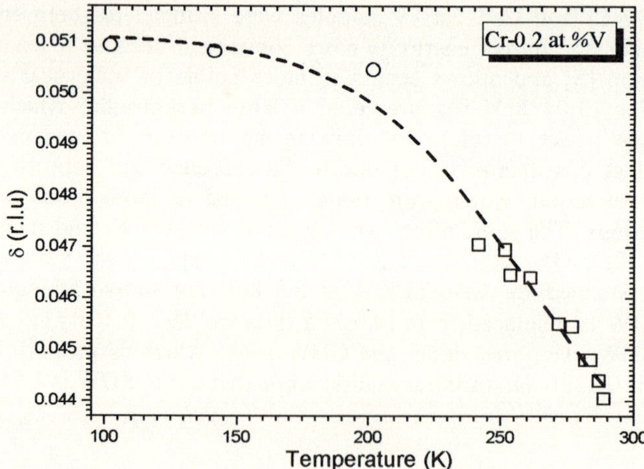

Figure 3 – Evolution of δ in function of temperature for Cr-0.2 at.%V. The dots are a guide to the eye and circles and squares correspond to SDW and CDW respectively.

The evolution of the incomensurability δ in reciprocal lattice units (r.l.u.) as a function of temperature is shown at Figure 3 with points obtained from CDW and SDW peaks. The temperature dependence follows a second order transition as it is expected comparing to neutron diffraction [5], however the temperature resolution for this experiment was not good enough to allow a detailed critical behavior analysis. The transition temperature found was 288.7 K, which is exactly the same as that found by neutron scattering [5]. Resistivity data from the same sample have shown a Néel temperature 0.8 K above this. Actually, it was shown by de Oliveira et al. [8] from AC magnetic susceptibility investigation that two transitions may be defined for Cr-0.2at%V, the usual Néel transition at $T_N = 287.9$ K and a $T_{CW} = 300.4$ K, where a Curie Weiss temperature dependence starts. More detailed investigation of the temperature dependence is necessary to clear up the origin of this difference, however, it is clear that the bulk Neél transition and the single Q a few micrometers from the surface lead to the same T_N. The broadening, represented by the FWHM, for the alloy increases at constant rate until the intensity is negligible, within the resolution of the experiment, contrasting with the case for chromium where the broadening sharply increase close to T_N [5].

Acknowledgments

This work was supported by the European Synchrotron Research Facility and brazilian agencies CNPq, Ministério de Ciência e Tecnologia (PRONEX) and FAPESP.

References

[1] - E. Fawcett. Rev. Modern Phys. 60 (1988) 209.
[2] - E. Fawcett, H.L. Alberts, V. Yu. Galkin, D. R. Noakes and J. V. Yakhmi. Rev. Mod. Phys. 66 (1994) 25.
[3] - D. R. Noakes, T. M. Holden, E. Fawcett and P. C. de Camargo. Phys. Rev. Letters 65, (1990) 369.
[4] - D. R. Noakes, T. M. Holden and E. Fawcett . J. Appl. Phys. 67 (1990) 5262.
[5] - J.P. Hill, G. Helgensen and D. Gibbs. Phys. Rev. B. 51, (1995) 10336.
[6] - E. P. Castro, P. C. de Camargo and F. R. Brotzen. Solid State Commun. 57, (1986) 37.
[7] - P.C. de Camargo, E. P. Castro and E. Fawcett. J. Phys. F: Met. Phys. 18 (1988) L219.
[8] - A .J. A.de Oliveira, W. A. Ortiz, P. C. de Camargo and V. Yu. Galkin. J. Magn. Magn. Matter 152 (1996) 86.
[9] Y. Tsunoda, M. Mori, N. Kunitomi, Y. Teraoka and J. Kanamori. Solid State Commun. 14, (1974) 287.

*Corresponding author. e-mail: camargo@fisica.ufpr.br

Materials Science Forum Vols. 302-303 (1999) pp. 38-42

EXAFS Characterization of Cobalt Nanoparticles Embedded in a Copper Matrix

J.C. Cezar[1,2], M.C. Martins Alves[1], D.Z. Cruz[1,2], F.C.S. da Silva[2], M. Knobel[2] and H. Tolentino[1]

[1] National Synchrotron Light Laboratory (LNLS), C.P. 6192, 13083-970 Campinas (SP), Brazil

[2] Universidade Estadual de Campinas (UNICAMP), Instituto de Física Gleb Wataghin, Brazil

Keywords: EXAFS, Granular Alloys, Giant Magnetoresistance

Abstract In this work samples with different microstructures, departing from as-quenched $Cu_{100-x}Co_x$ (x = 3, 10 e 25) melt-spun ribbons, were studied using the X ray absorption spectroscopy (XAS) technique. The spectra (XANES, EXAFS) of these samples, as well as of pure fcc Cu and hcp Co used as reference, were measured in the XAS beam-line of LNLS. As a function of heat treatment, the spectra in the Co K edge region shift, in terms of average Co-metal distances, from fcc Cu towards pure hcp Co. A detailed analysis of this evolution could, in principle, give indications of the Co clusters size.

I. Introduction

A great effort has been concentrated in materials that exhibit magnetoresistive properties [1], because besides the search for a physical model for these characteristics, there is also the technological interest of the magnetic media and sensors industries.

Granular alloys formed by a magnetic element (Co, Ni or Fe) in a metallic non magnetic matrix (Cu or Ag) have been extensively studied in the last years, mainly because they display giant magnetoresistance (GMR) effect similar to the one observed previously in magnetic multilayers. Although many theoretical and experimental studies have helped to elucidate many points, there are still open questions concerning the role of the local atomic order in the GMR, mainly at the interface between magnetic and non-magnetic regions. However, owing to the inherent difficulty to structurally characterize these materials (low contrast, small grain sizes), different experimental techniques, provided by synchrotron radiation, are needed.

We have studied the structure of nanoparticles of cobalt embedded in a copper substrate, using the EXAFS spectroscopy, as a function of cobalt concentration and the thermal treatment done on the samples.

II. Samples and Technique description

The samples were produced by melt-spinning method [4,5,6]. As result, one obtains metallic ribbons with thickness of order of 50 μm and 5 mm width, where cobalt atoms form a metastable solution in the copper matrix[7]. This solution is metastable due the fact that the solubility of cobalt in copper, for temperatures below 700 °C, is less than 1.1% [8].

These samples were submitted to a thermal treatment, that enables the coalescence of the cobalt atoms. This treatment consisted in a flow of electric current through the sample during one minute (Joule effect treatment). According to the thermal treatment (i.e., different currents through the sample), one obtains different average particle sizes and distance between particles. These microstructural characteristics are closely related to the GMR phenomena.

Several structural characterization techniques have been used in this kind of materials, namely transmission electron microscopy (TEM) [9,10,11], X ray diffraction (XRD) [7,12] and X-ray absorption spectroscopy (XAS) [13], among others. Due the low atomic number difference between cobalt and copper (Z_{Co}=27 and Z_{Cu}=29), we choose a element sensitive structural technique, extended X-ray absorption fine structure (EXAFS) spectroscopy, that can give us the average interatomic distances, average coordination number and the projected Debye-Waller factor of each coordination shell [14,15,16].

To get a EXAFS spectrum, one must measure the ratio between the incident and transmitted X-ray beam through the sample, in a interval from some tens of eV before the absorption edge of the element studied, to some hundreds of eV above this edge. The spectrum will exhibit some oscillations that one can relate to the structural and electronic properties of the sample. One can extract these electronic properties from model compounds or first principles calculations, and use them to determine the structural parameters of the samples [14,15,16].

The measurements have been performed at the XAS beam-line [17] at Laboratório Nacional de Luz Síncrotron (Synchrotron Light National Laboratory) at Campinas, São Paulo. We have measured the spectra around the cobalt K edge (7709 eV), in the transmission mode, with two ionization chambers as incident and transmitted beam intensities detectors. With these spectra, we realized the standard EXAFS data analysis, that consists in the extraction of the EXAFS signal ($\chi(E)$) from the raw data and transformation of this signal from energy dependence ($\chi(E)$) to wave number modulus dependence ($\chi(k)$). After that we can calculate the Fourier transform (FT) of this signal, and we obtain a pseudo-radial distribution function (RDF) in the real space. In this RDF we isolate the contribution of each coordination shell to the EXAFS signal, and then, with data from model compounds, we fit the structural parameters. As model compound we used a bulk cobalt standard. In figure 1 we present one example of this procedure, for the as quenched sample of 10 at. % cobalt.

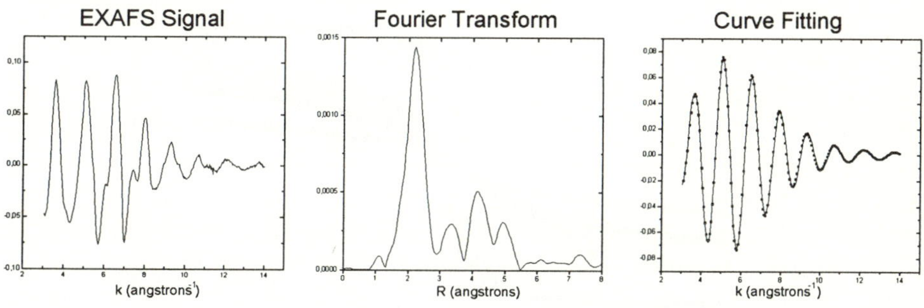

Figure 1: EXAFS signal, the Fourier transform and fitting for the first coordination shell, of the sample with 10 at. % cobalt as quenched.

III. Results and Discussion

We have done the experiments on samples with 3, 10 and 25 at. % cobalt. The results of the data analysis are shown in the figure 2. In this figure the dotted lines are plotted just to guide throughout the experimental points. The two horizontal lines correspond to the Cu-Cu and the Co-Co distances in the model compounds.

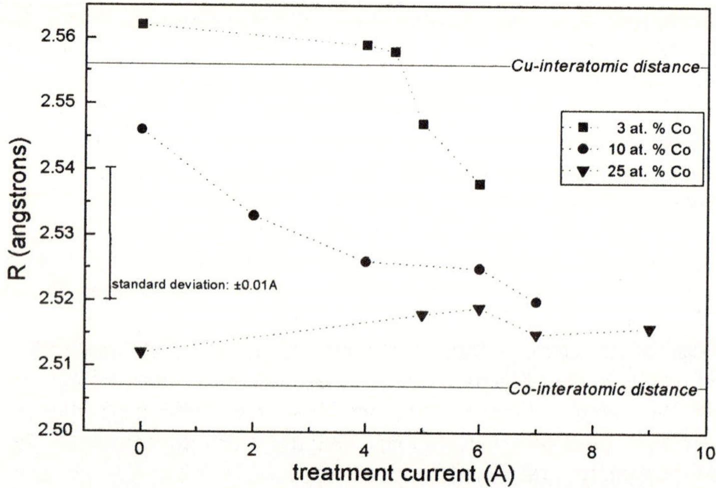

figure 2: Results for the cobalt first neighbor distance, from the curve fitting for three set of samples (3, 10 and 25 at.%) as a function of the thermal treatment, indicated as the current (in Ampères) used to heat each sample

We can note from the figure 2 that as the cobalt percentage and the treatment are increased, the Co-first neighbor distance approaches that of bulk cobalt. For the 3 at. % samples, we need a current greater than 5 A to produce the coalescence of the cobalt atoms, whereas for the 25 at. % practically the final distance is already reached in the as quenched sample. For the 10 at. %, we note a roughly linear relation between the current used in the treatment and the cobalt-first neighbor distance.

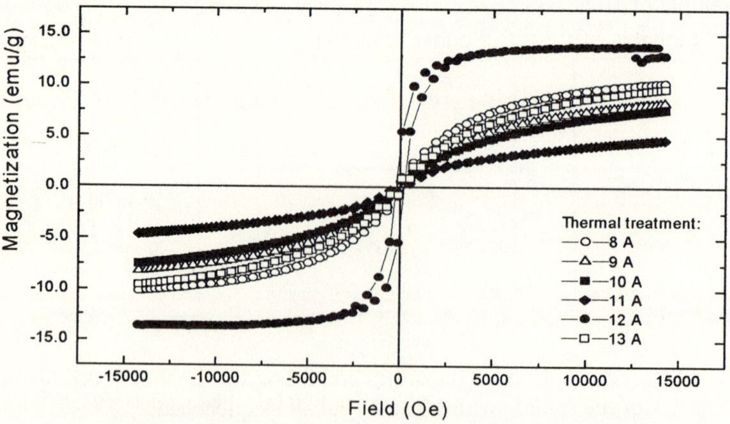

Figure 3: Magnetization curves of the 10 at. % sample.

Magnetic measurements have been performed on these samples, and in figure 3 we present the magnetization curves of the 10 at. % set. The growth of the Co particles is clearly visible in this figure. The as-quenched ribbon and those treated at low currents display a typical superparamagnetic behavior, with very weak hysteresis and no tendency to saturation. When the samples are further treated, one can observe the appearance of the hysteresis, and the increase of the saturation magnetization, indicating the formation of larger Co clusters. From the magnetization measurements it is possible to infer the structural parameters [18]. Analysis of these data, together with magnetoresistance measurements, are currently being performed and they can result in a independent way to obtain the clusters size.

The Co-first neighbor distance depends on the size of the cobalt cluster. As the surface to volume ratio of the cluster decreases (the size of the nanoparticle is increased) the tendency is for the cobalt to turn from the fcc copper interatomic distances to the hcp cobalt interatomic distances. This result means that the cobalt clusters are growing with the percentage and thermal treatment. With these datum it is possible, in principle, to determine the average cobalt cluster size. There are some works [19] that do it for copper particles embedded in a non-metallic substrate (alumina in that case). For the cobalt-copper system this is intrinsically more complex, due to the metallic nature of both components and the similar structural properties of them.

IV. Acknowledgments

The authors wish to express their gratitude to the group of Prof. J.E. Schmidt (Porto Alegre), where some magnetic measurements were performed and to E.F.Ferrari, who had allowed us to use the magnetization curves

. We thank A. Stantero, P. Tiberto and F. Vinai from the IENGF, Torino, Italy, for making the as-quenched samples available to us. This research was supported by CNPq and FAPESP.

V. Bibliography

[1] J.A.Brug, L.Tran, M.Bhattacharyya, J.H.Nickel, T.C.Anthony, J.Jander, *J. Appl. Phys.* **79** (8), p. 4491, 1996

[2] M.N.Baibich, J.M.Broto, A.Fert, F.Nguyen Van Dau, F.Petroff, P.Eitenne, G. Creuzet, A.Friederich, J.Chazelas, *PRL* **61**, p. 2472, 1988

[3] T.L.Hylton, K.R.Coffey, M.A.Parker, J.K.Howard, *Science* **261**, p. 1021, 1993

[4] A.E.Berkowitz, J.R.Mitchell, M.J.Carey, A.P.Young, S.Zhang, F.E.Spada, F.T.Parker, A.Hutten, G. Thomas, *PRL* **68** (25), p. 3745, 1992

[5] J.Q.Xiao, J.S.Jiang, C.L.Chien, *PRL* **68** (25), p. 3749, 1992

[6] P.Allia, M.Baricco, M.Knobel, P.Tiberto, F.Vinai, *J. Mag. Mag. Mat.* **140-144**, p. 617-618, 1995

[7] R.H.Yu, X.X.Zhang, J.Tejada, M.Knobel, P.Tiberto, P.Allia, *J. Appl. Phys.* **78** (1), p. 392, 1995

[8] J.Wecker, R.von Helmolt, L.Schultz, K.Samwer, *Appl. Phys. Lett.* **62** (16), p. 1985, 1993

[9] MA.Howson, et al., *J. Appl. Phys.* **75** (10), p. 6546,1994

[10] R.Busch, F.Gärtner, C.Borchers, P.Haasen, R.Bormann., *Acta Metall. Mater.* **43** (9), p. 3467, 1995

[11] R.Busch, F.Gärtner, C.Borchers, P.Haasen, R.Bormann., *Acta Metall. Mater.* **44** (6), p. 2567, 1996

[12] C.L.Chien, J.Q.Xiao, J.S.Jiang, *J. Appl. Phys.* **73** (10), p. 5309, 1993

[13] T.Haulbold, F.Boscherini, S.Pascarelli, S.Mobilio, H.Gleiter, *Phylos. Magaz. A* **66** (4), p. 591, 1992

[14] E.A.Stern, et al, *Basic Principles and Applications of Exafs*, in *Handbook on Synchrotron Radiation*, Vol. I, ed. E.E.Koch, North-Holland Publishing Company, 1993

[15] D.E.Sayers, B.A.Bunker, in *Data Analysis*, D.C.Koningsberger, R.Prins, *X-Ray Absorption: Principles, Applications, Techniques of EXAFS, SEXAFS and XANES*, John Wiley & Sons, 1988

[16] B.K.Teo, *EXAFS: Basic Principles and Data Analysis*, Springer-Verlag, 1986

[17] H.Tolentino, J.C.Cezar, D.Z.Cruz, V.Compagnon-Cailhol, E.Tamura, M.C.Martins Alves, to be published in the *Proceedings of SRI 97, J. Synchrotron Rad.* **5**, 1998

[18] E.F. Ferrari, F.C.S.da Silva, M.Knobel, *Phys. Rev. B* **56**, p. 6086 (1997).

[19] M.Borowski, *J. Phys. IV, Colloque C2, Supplemént au Journal de Physique III*, avril 1997, p. 259

Materials Science Forum Vols. 302-303 (1999) pp. 43-52

Application of the Preisach Model of Hysteresis

G. Bertotti

IEN Galileo Ferraris and INFM, Corso M. d'Azeglio 42, IT-10125 Torino, Italy

Keywords: Hysteresis Modeling, Preisach Model, Metastability, Magnetic Viscosity

Abstract The article analyzes the connection between bistability, Preisach-type hysteresis and magnetic viscosity. Expressions for the magnetization of a Preisach system are derived, whereby one can separate the role of past history from the influence of microstructural features. Thermal effects are then added to the picture. On this basis, some aspects of magnetic viscosity are discussed.

INTRODUCTION

The simplest system displaying hysteresis is probably a bistable object, characterized by two locally stable states separated by an energy barrier [1]. A well-known example in magnetism is a magnetic particle with uniaxial anisotropy. The next phenomenological step of increasing complexity is to consider not one, but a collection of many independent bistable units characterized by different state energies and energy barriers. This approach is at the basis of at least two widely employed classes of models: thermal-activation-type models of magnetic viscosity [2] and Preisach-type models of hysteresis [1,3]. A direct justification of these models, by showing that the physical constituents of the system have indeed the character of bistable units, in most cases is hardly possible. Even a collection of magnetic particles, which at first sight would seem to be a good candidate for this description, contains substantial complications – the vector nature of the particle state, the existence of particle-particle interactions – that make by no means obvious the validity of the decomposition into bistable contributions. In magnetic viscosity studies, the focus is on the distribution of energy barriers separating higher form lower energy states. One assumes that this distribution can be unambiguously defined in the neighborhood of the initial global state of the system, and that it is not substantially altered during the small field and magnetization variations taking place in a typical relaxation experiment. In Preisach-type models, on the other hand, one makes the much more drastic assumption that the decomposition of the system into bistable units can be carried out a priori, independently of the particular state considered, and that the same decomposition holds under whatever field history may be applied to the system.

These considerations might give the impression that the applicability of the Preisach model as a physical model can be hardly justified. This is not the case, however. A remarkable property of the Preisach model is that one does not need to identify real bistable contributions in a system in order to justify the Preisach approach. The theorem discussed in [4] shows that there exist two properties, known as wiping-out (or return-point-memory) and congruency which are necessary and sufficient conditions for a system to be represented by the Preisach model. These properties have to do with general aspects of measured magnetization curves. Therefore, it is possible to test these properties by experiments on the system of interest. If the test is positive and the properties are verified, one knows for sure that the hysteresis properties of the system can be described by the Preisach model, without going into the troubles of a direct microscopic justification of the decomposition into bistable units. In this respect, wiping-out and congruency appear as the macroscopic fingerprint of Preisach-type hysteresis. This conclusion, however, is valid only at zero temperature. It says nothing about the relaxation effects induced by thermal fluctuations. In order to study this aspect, one needs to go back to the decomposition of the system into bistable units, and study their relaxation by thermally-activated hopping over energy barriers. We will discuss the application of this approach to the study of magnetic viscosity and its dependence on past history.

BISTABLE UNITS AND THE PREISACH APPROACH

As mentioned in the introduction, in the Preisach model one accepts the hypothesis that the hysteresis properties originating from a complex free energy profile with multiple metastable states can be described in terms of an equivalent collection of elementary bistable units (Preisach units) [1,5]. The energy profile associated with an individual unit is shown in Fig. 1. The energy is a function of the magnetic moment m and is confined to the interval $-\Delta m \leq m \leq +\Delta m$, where Δm is a characteristic magnetic moment of the description. The magnetic moments involved are scalar, which means that one assumes that the field is applied along a fixed direction and that one looks at the magnetic response of the system along that direction only. The free energy represented in Fig. 1 is to be interpreted as a Gibbs energy of the form $G(m\,;H) = F(m) - \mu_0 Hm$. $F(m)$ describes the internal properties of the unit. It is identified by its values at the points $-\Delta m$, 0, $+\Delta m$, which we shall express in the form $-\mu_0 h_u \Delta m$, $\mu_0 h_c \Delta m$, $\mu_0 h_u \Delta m$, where h_u and h_c ($h_c \geq 0$) are appropriate equivalent fields. The term $-\mu_0 Hm$ of $G(m\,;H)$ describes the linear coupling of m to the magnetic field. $G(m\,;H)$ controls the stability of the unit. Depending on the relative values of h_u, h_c, and H, G will display one or two minima. The various possibilities can be graphically represented by introducing the so-called Preisach plane, with axes h_c and h_u (Fig. 2). A given Preisach unit is identified by the representative point of coordinates (h_c, h_u) in this plane. The lines $h_u = H \pm h_c$ divide the Preisach plane into three regions. A unit with $h_u > H + h_c$ has a G profile with only one minimum in correspondence of the (-) state of Fig. 1. Analogously, a unit with $h_u < H - h_c$ is certainly in the (+) state. The units lying inside the cone delimited by $h_u = H \pm h_c$ (broken lines in Fig. 2) have two accessible energy minima, and which minimum they occupy will depend on past history. When H changes in time, the G profile is altered in shape by the $-\mu_0 Hm$ term and one of the two minima may be destroyed. When this occurs, any unit occupying the state being destroyed will jump to the other remaining state, giving rise to a magnetic moment change from $-\Delta m$ to $+\Delta m$ or viceversa.

Let us consider now a collection of Preisach units, randomly distributed in the Preisach plane. The state of the units inside the metastability cone of Fig. 2 is undetermined. In principle, any partition of the cone into (+) and (-) subregions is acceptable, and represents a possible metastable state of the system. The analysis that follows will be restricted to the particular class of metastable states represented by a partition of the Preisach plane in only one (+) and one (-) region. This class of states is fully described by the shape of the boundary line $b(h_c)$ between the (+) and (-) region (Fig. 2). It will be shown shortly that any state generated by an arbitrary field history starting from positive or negative saturation belongs to this class. The average magnetization of the system, M, results from the sum of the $\pm\Delta m$ contributions coming from all the units. Let $p(h_c,h_u)$ be the probability density to have a unit characterized by given (h_c, h_u) values, and M_s the saturation magnetization realized when all units are in their (+) state. Under the assumption, actually not restrictive, that all units are characterized by the same value of Δm, the average magnetization of the system in the state specified by a certain $b(h_c)$ line will be

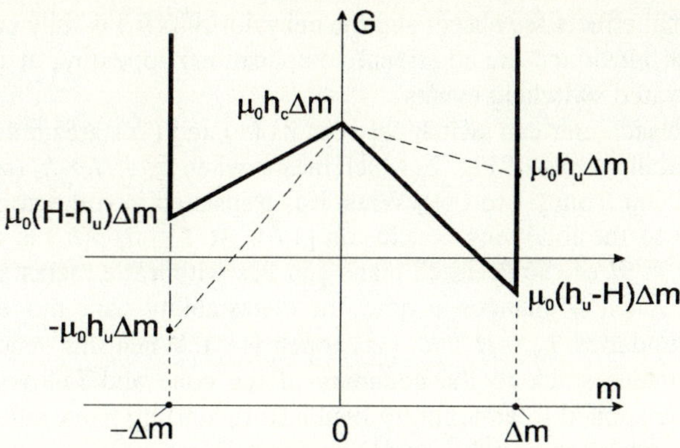

Fig. 1. Free energy profile of a single Preisach unit. The broken line represents the profile under zero field.

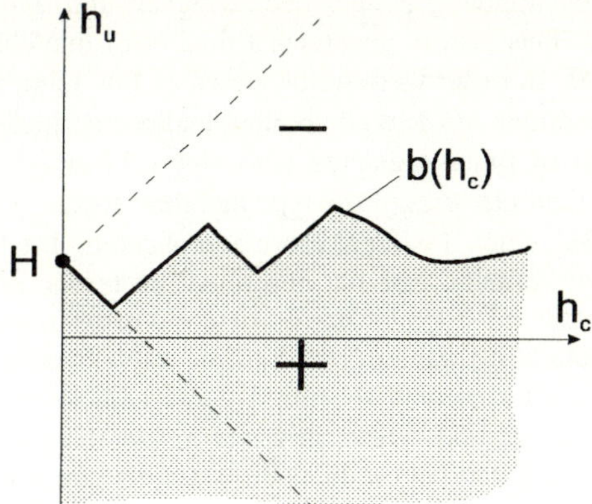

Fig. 2. Preisach plane with $b(h_c)$ state line. Units belonging to the shaded (white) region occupy the (+) ((-)) state of Fig. 1.

$$M = 2M_s \int_0^\infty dh_c \int_0^{b(h_c)} p(h_c, h_u)\, dh_u \tag{1}$$

provided $p(h_c,h_u)$ obeys the symmetry property $p(h_c,h_u) = p(h_c,-h_u)$. Equation (1) may be interpreted as the history-dependent constitutive law of the system [6]. The law involves two functions, the boundary line $b(h_c)$, which identifies the state of the system and carries the information about past history, and the Preisach distribution $p(h_c,h_u)$, which summarizes all the microstructural and magnetic features affecting the magnetization process. The central problems of this description are to clarify the history-dependent rules obeyed by $b(h_c)$ and the main aspects of the connection between $p(h_c,h_u)$ and given microstructural properties.

THE STATE LINE $b(h_c)$

The behavior of $b(h_c)$ is the direct consequence of the properties of the energy profile of the individual Preisach units, previously described. We shall first consider the limit case of zero temperature, where thermal effects are absent and the behavior of $b(h_c)$ is fully controlled by the field history, and then we shall introduce the additional complications appearing at nonzero temperature because of thermally activated switching events.

At T = 0, a given Preisach unit can switch between its (+) and (-) states only when it crosses the boundaries of the metastability cone of Fig. 2, which means when $h_u = H + h_c$ (switching from (+) to (-)) or $h_u = H - h_c$ (switching from (-) to (+)). When it is translated into the corresponding behavior of $b(h_c)$, this result leads to the following conclusion [1,6]. At T = 0, $b(h_c)$ is a line that originates from the point $h_c = 0$, $h_u = H$ of the Preisach plane and lies within the metastability cone of Fig. 2 under all circumstances. When H changes in time, the metastability cone moves accordingly in the Preisach plane and its boundaries, $h_u = H \pm h_c$, may touch $b(h_c)$. When this occurs, $b(h_c)$ is modified in such a way that it remains stuck to the boundary of the cone and follows it until the field is reversed. When this rule is applied to an arbitrary field history starting from saturation (say, all units in the (+) state), a state consisting of only one (+) and one (-) region is generated, whose $b(h_c)$ line is made up of a chain of straight segments of alternating slope $db/dh_c = +1$ and $db/dh_c = -1$, one for each past field reversal. We can generate $b(h_c)$ lines that contain arbitrarily fine mixtures of segments of alternating slope by applying oscillating field histories, in which the mean value and the amplitude of the oscillation change slowly from one period to the next. An important example is the demagnetized state, achieved by applying a field oscillating around zero, with amplitude slowly decreasing from infinity to zero. This history generates a final $b(h_c)$ line characterized by a fine saw-tooth structure around $b(h_c) = 0$. In order to describe states of this kind, it is convenient to extend the space of admissible state functions $b(h_c)$ to all continuous lines with slope $-1 \leq db/dh_c \leq +1$. All the lines that are not made up of finite segments with slope +1 or -1 should be interpreted as appropriate limits of oscillating field histories of the type just described.

The evolution rules for $b(h_c)$ when T ≠ 0 are more complicated, due to the presence of thermal fluctuations. Thermal relaxation effects can be described in terms of the thermally-activated switching of the individual Preisach units. Under not particularly restrictive conditions, one can show that this approach can be translated into the motion of a time-dependent state line, $b(h_c,t)$, governed by an evolution equation that will be derived shortly. To this end, let us analyze the approach to equilibrium of an individual Preisach unit. Only the units inside the metastability cone of Fig. 2 exhibit two energy minima and participate in the relaxation. Let us consider the unit associated with the Preisach point (h_c,h_u) under the action of the magnetic field H. The transition rates w_+, for a thermally activated jump from $-\Delta m$ to $+\Delta m$, and w_-, for a jump from $+\Delta m$ to $-\Delta m$ are given by

$$w_- = \frac{1}{\tau_0}\exp\left(-\frac{h_c + h_u - H}{H_f}\right)$$
$$w_+ = \frac{1}{\tau_0}\exp\left(-\frac{h_c - h_u + H}{H_f}\right) \tag{2}$$

where τ_0 is a typical attempt time and we have introduced the fluctuation field $H_f = k_B T/\mu_0 \Delta m$. To keep the discussion simple, we shall assume that Δm is just a constant, the same for all units. The master equation for the unit is

$$\frac{dP_+}{dt} = w_- - (w_+ + w_-)P_+ \tag{3}$$

where $P_+(t; h_c, h_u)$ is the probability to find the unit in the (+) state at time t. As shown in [7], the study of the relaxation of a Preisach system amounts, in rigorous terms, to the study of infinite evolution equations, one for each Preisach unit of the Preisach plane. However, when the temperature is not too high $P_+ \approx 0$ or $P_+ \approx 1$ for most of the units and most of the time. The relaxation is controlled by the units for which $P_+ \approx 1/2$ at a given time, and these units occupy a limited portion of the Preisach plane. To illustrate this point, let us consider the case where the system is initially in the state described by the state line $b(h_c)$. This means that the units with $h_u < b(h_c)$ are in the (+) state and the ones with $h_u > b(h_c)$ are in the (-) state. We can describe this initial condition by the formula

$$P_+(0;h_c,h_u) = \theta[b(h_c) - h_u] \tag{4}$$

where $\theta(x)$ is the Heaviside step function. The units will evolve according to Eq. 3 till the final equilibrium state is reached, described by the stationary solution of Eq. 3:

$$P_{+,eq}(h_c,h_u) = \frac{1}{2} - \frac{1}{2}\mathrm{th}\left(\frac{h_u - H}{H_f}\right) \tag{5}$$

In this state, the Preisach units are in the (+) state ($P_{+,eq} \approx 1$) or in the (-) state ($P_{+,eq} \approx 0$) everywhere except for a transition layer of width of the order of H_f, in which $P_{+,eq}$ passes from 0 to 1, and the average magnetic moment of the unit passes from $-\Delta m$ to $+\Delta m$. As the relaxation proceeds, the probability $P_+(t;h_c,h_u)$ behaves like a front propagating from $P_+(0\,;h_c,h_u)$ toward $P_{+,eq}(h_c,h_u\,)$. The width of the front is zero at the beginning, but soon acquires a width of the order of H_f. The magnetization is given by the Preisach integral of the average magnetic moment $\Delta m[2P_+(t\,;h_c,h_u) - 1]$ of the individual Preisach units. In analogy to Eq. 1, one has

$$M(t) = M_s \int_0^{\infty} dh_c \int_{-\infty}^{+\infty} [2P_+(t;h_c,h_u) - 1]\, p(h_c,h_u)\, dh_u \tag{6}$$

If the Preisach distribution $p(h_c,h_u)$ is approximately constant over h_u intervals of the order of H_f, the term $2P_+(t\,;h_c,h_u) - 1$ can be safely approximated by a step function, from +1 to -1, located at $h_u = b(h_c,t)$, where $b(h_c,t)$ describes the position of the propagating front. $b(h_c,t)$ may be defined, for example, as the point where $P_+(t\,;h_c,h_u) = 1/2$ This approximation makes Eq. 6 identical to Eq. 1, with $b(h_c,t)$ playing the role of time-dependent state line. The assumption that the Preisach distribution is constant over field intervals of the order of H_f is well justified in most cases and permits one to apply reliably the $b(h_c,t)$ approximation.

The central problem is now to derive the equation governing the time evolution of $b(h_c,t)$. This information can be obtained by making a specific assumption about the shape of the probability front, and by inserting it into Eq. 3. Following Eq. 5, we choose

$$P_+(t;h_c,h_u) = \frac{1}{2} - \frac{1}{2}\mathrm{th}\left(\frac{h_u - b(h_c,t)}{H_f}\right) + R(t;h_c,h_u) \tag{7}$$

where R is a residual correction, not important in the present context. By inserting Eq. 7 into Eq. 3 one finds

$$\frac{\partial b(h_c,t)}{\partial t} = \frac{2H_f}{\tau_0}\mathrm{ch}\left(\frac{b(h_c,t) - h_u}{H_f}\right)\mathrm{sh}\left(\frac{H - b(h_c,t)}{H_f}\right)\exp\left(-\frac{h_c}{H_f}\right) + \ldots \tag{8}$$

where the sum continues with unimportant correction terms. Equation 8 shows that the point of the probability front at which $h_u = b(h_c,t)$ propagates at the velocity

$$\frac{\partial b(h_c,t)}{\partial t} = \frac{2H_f}{\tau_0}\,\mathrm{sh}\left(\frac{H(t)-b(h_c,t)}{H_f}\right)\exp\left(-\frac{h_c}{H_f}\right) \tag{9}$$

This is the fundamental equation describing the evolution of the state line of a given system. Note that no specific assumption about H was made in the derivation of Eq. 9. Therefore, the equation holds not only under constant field, but also when the field changes in time.

The evolution of $b(h_c)$ at T = 0 can be obtained from Eq. 9 by taking appropriate limits. In fact, when T $\rightarrow$ 0, $H_f = k_BT/\mu_0\Delta m \rightarrow 0$ as well. Therefore, the hyperbolic sine of Eq. 9 can be approximated as $\mathrm{sh}(x) \approx \pm\exp(|x|)/2$, depending on the sign of x, and Eq. 9 can be written in the form $\partial b/\partial t \propto \pm\exp[\pm(H - b \mp h_c)/H_f]$. Given the fact that $H_f \rightarrow 0$, this approximate form shows that $\partial b/\partial t \approx 0$ for any point of the state line lying inside the metastability cone of Fig. 2. One can have $\partial b/\partial t \neq 0$ only when the argument of the exponential vanishes, that is, when $b = H \pm h_c$. This is the moment where the state line touches the boundary of the cone, in agreement with our previous conclusions.

THE PREISACH DISTRIBUTION $p(h_c,h_u)$

The Preisach distribution $p(h_c,h_u)$ appearing in Eq. 1 contains the information on the material microstructure, as it is probed by domains during the magnetization process. In principle, the function $p(h_c,h_u)$ should be derived from some model of the particular mechanisms that are active in a given material. In this respect, the real breakthrough would be to clarify under what approximations one may expect the Preisach model to be an acceptable approximation to the general equations of micromagnetics, when the free energy of the system acquires a complicated multi-valley structure, as a consequence of the presence of structural disorder. This program is extremely difficult to carry out in detail for realistic situations, due to the mathematical difficulties inherent to the micromagnetic approach [8]. A less ambitious but nonetheless fruitful possibility is to derive information on $p(h_c,h_u)$ from a suitable set of experimental magnetization curves, and to use this information to compare different materials and to monitor the influence of given microstructural conditions on magnetization. In rigorous terms, this approach amounts to solving the so-called identification problem for the description, whereby one gives a precise description of the set of experimental data required in order to reconstruct the functions and the parameters of the model. In particular, as regards the Preisach model, it can be proven [3,4] that the Preisach distribution $p(h_c,h_u)$ can be unambiguously determined from the knowledge of first-order return branches, for all possible values of the reversal field.

Solving the identification problem is of conceptual importance, but it has two practical shortcomings. First, the identification relies on the assumption that the system considered can be exactly described by the model, which in general is only approximately true, and, second, the reconstruction procedure usually involves a large amount of experimental data, which may not be easily accessible. The first difficulty may be partly circumvented by considering, instead of first-order curves, appropriate sets of minor loops [9] or averages of several return branches associated with the same field interval [10]. As to the second problem, it can often be acceptable a parametric approach, where one makes some a priori assumption about the analytical form of the Preisach distribution, focusing then attention on the behavior of the parameters of the chosen representation. For example, it has been observed that Si-Fe electrical steels are described well by lorentzian distributions [11], and gaussian distributions are often assumed for the description of magnetic recording media [12]. Whatever the approach followed, once the Preisach distribution is approximately known, one has a sort of representation of the material microstructure, as it is probed by magnetic domains. In this way, one has the possibility of monitoring microstructural changes and their influence on the magnetization process. This idea has been applied to the study of the onset of partial crystallization in soft amorphous alloys [13] and the coexistence of soft and hard magnetic phases in spring magnets [14,15].

MAGNETIC VISCOSITY

The interest of Eq. 9 lies in the fact that it permits one to study the evolution of a Preisach system under the joint action of field and temperature under quite general conditions. In this respect, a conceptually important point is that in a system with hysteresis thermal relaxation is expected to depend on past history, because a system with hysteresis occupies states that depend on past history and thermal relaxation is certainly affected by the initial state of the system. Equation (9) is the appropriate tool for the study of this problem. In fact, the information on past history is carried by the initial state line $b(h_c,0)$, and Eq. 9 predicts the evolution of the system starting from this initial condition. To illustrate the physical consequences of this point, let us consider a system which is initially in the state of positive saturation, and then it is brought to the so-called remanence coercivity, that is to the field such that the system is left at zero magnetization when the field is removed. The state generated in this way has zero magnetization under zero field, but it is not the demagnetized state of minimum free energy. Therefore, it will relax toward equilibrium in a certain way. Both the initial and the final equilibrium state have zero magnetization, and it is not immediately obvious what path the system will follow to approach equilibrium. We shall see shortly that, according to Eq. 9, the system will relax by first increasing its magnetization and then by decreasing it back to zero, although this second step may set in after exceedingly large times. This example illustrates well the fact that the magnetization M is not always a good parameter to characterize the state of a system with hysteresis. Many different states may be associated with the same values of H and M, and each of them will follow a different path of approach to equilibrium. In particular, M can move away from the final equilibrium value even if the system is approaching equilibrium. In addition, the example makes clear that the law of approach to equilibrium can be non-monotone and quite complicated, depending on past history. The common statement that the relaxation of magnetic systems approximately obeys a logarithmic law cannot be of general validity, but only applies to the particular history (i.e., decreasing the field from positive saturation to the measurement point) typically considered in magnetic viscosity experiments.

The main physical consequences of Eq. 9 can be appreciated by studying the solution of Eq. 9 under constant field. When H is independent of time, Eq. 9 can be integrated, giving the result

$$b(h_c,t) = H - 2H_f \,\mathrm{Arth}\left[\mathrm{th}\left(\frac{H-b(h_c,0)}{2H_f}\right)\exp\left(-\frac{2t}{\tau_0}\exp\left(-\frac{h_c}{H_f}\right)\right)\right] \tag{10}$$

At large times, when the relaxation is complete, the state line approaches the stationary solution $b_{eq}(h_c) = H$. Note that, for $H = 0$, this final equilibrium state coincides with the demagnetized state generated by an alternating field of decreasing amplitude. To study the path followed by the system to reach equilibrium, let us introduce the time-dependent field

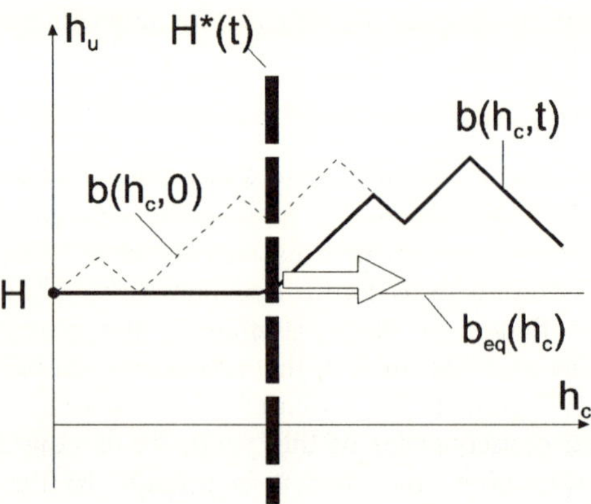

Fig. 3. Relaxation front propagating according to Eq. 11. $b_{eq}(h_c)$ is the final equilibrium line at the ened of the relaxation.

$$H^*(t) = H_f \ln\left(1 + \frac{t}{\tau_0}\right) \tag{11}$$

The argument of the exponential of Eq. 10 is itself an exponential function of h_c, which takes values of the order of unity when $h_c \sim H^*(t)$. In the portion of the Preisach plane where $h_c << H^*(t)$, the argument of the exponential is large and negative, the exponential vanishes and $b(h_c,t) \cong H$. In this region, the relaxation is complete and the state line has already reached the final equilibrium condition. The relaxation proceeds by a relaxation front, located at $h_c \sim H^*(t)$, which progresses toward higher h_c values at a speed proportional to the logarithm of time (see Fig. 3). In the region $h_c > H^*(t)$, the argument of the exponential of Eq. 10 is small, the exponential is only slightly different from unity and the properties of the solution are controlled by the nontrivial interplay of the exponential and logarithmic dependencies hidden in the Arth and th functions. A more accurate analysis is needed, which can be carried out by introducing a suitable approximation to Eq. 9 itself. As a first step, let us decompose the state line as

$$b(h_c,t) = b(h_c,0) + \Delta b(h_c,t) \tag{12}$$

It can be verified that the sign of Δb is always that of $H - b(h_c,0)$, and that $|\Delta b|$ always increases in time. Accordingly, one has

$$|H - b(h_c,t)| = |H - b(h_c,0)| - |\Delta b(h_c,t)| \tag{13}$$

In the region $h_c > H^*(t)$, where $b(h_c,t)$ has not yet reached the asymptotic value H, one can use the approximation sh$(x) \approx \pm$ exp$(|x|)/2$, where the sign to choose is that of Δb. Accordingly, Eq. 9 can be transformed into the following approximate equation for $|\Delta b|$:

$$\frac{\partial|\Delta b|}{\partial t} \cong \frac{H_f}{\tau_0} \exp\left(\frac{|H - b(h_c,0)| - h_c}{H_f}\right) \exp\left(\frac{-|\Delta b|}{H_f}\right) \tag{14}$$

Equation (14) can be integrated. The result is

$$\exp\left(\frac{|\Delta b|}{H_f}\right) - 1 = \frac{t}{\tau_0}\exp\left(\frac{|H - b(h_c,0)| - h_c}{H_f}\right) \tag{15}$$

that is,

$$|\Delta b(h_c,t)| = H_f \ln\left(1 + \frac{t}{\tau(h_c)}\right) \tag{16}$$

with

$$\tau(h_c) = \tau_0 \exp\left(\frac{h_c - |H - b(h_c,0)|}{H_f}\right) \tag{17}$$

For given h_c, Eq. 14 applies up to the time at which the state line reaches the equilibrium value $b(h_c,t) \approx H$. From this time on, no further relaxation takes place at that h_c value.

As an example, we shall apply the results just obtained to the problem of thermal relaxation after a double field reversal, mentioned at the beginning of this section. Let us assume that the system is initially at positive saturation. The field is first decreased from $+\infty$ down to the value H_1, and then increased from H_1 up to the final value H_2, which is then kept constant in time (Fig. 4). The initial state line is

$$b(h_c,0) = \begin{cases} H_2 - h_c & \text{for}\, h_c < (H_2 - H_1)/2 \\ H_1 + h_c & \text{for}\, h_c > (H_2 - H_1)/2 \end{cases} \tag{18}$$

According to our previous results on $b(h_c,t)$, the two portions of the state line indicated by the black arrows in Fig. 4 relax in opposite directions, and give opposite contributions to the magnetization. The two processes have however completely different time constants. By inserting Eq. 18 into Eq. 17, one obtains

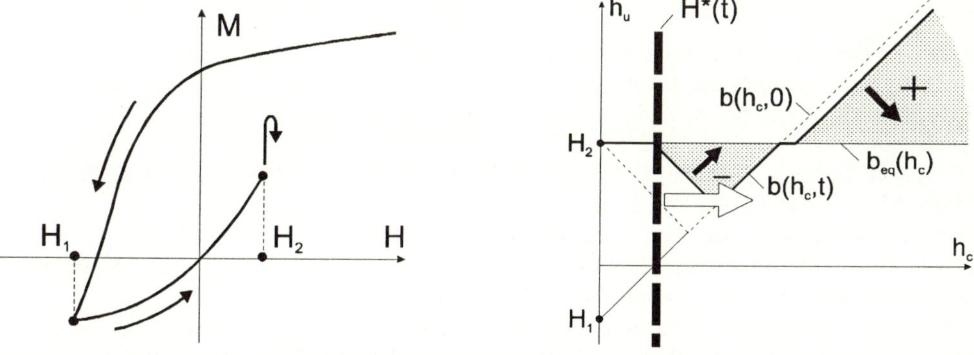

Fig. 4. Non-monotone relaxation consequent to double field reversal and corresponding representation in the Preisach plane. The shaded Preisach areas give opposite contributions to the relaxation.

$$\tau(h_c) = \begin{cases} \tau_0 & \textit{for } h_c < (H_2 - H_1)/2 \\ \tau_0 \exp[(H_2 - H_1)/H_f] & \textit{for } h_c > (H_2 - H_1) \end{cases} \tag{19}$$

A non-monotone relaxation takes place, in which the magnetization first increases, while the left region of Fig. 4 is relaxing, and then decreases, when this contribution is exhausted and the right region remains the only active one. By inserting the expression for $b(h_c,t)$ obtained from Eqs. 12, 16, 19 into Eq. 1, one obtains a non-monotone relaxation law of the type

$$\Delta M(t) = \begin{cases} +S_1 \ln\left(1 + \dfrac{t}{\tau_1}\right) & \tau_1 = \tau_0 & \text{for small times} \\ -S_2 \ln\left(1 + \dfrac{t}{\tau_2}\right) & \tau_2 = \tau_0 \exp[(H_2 - H_1)/H_f] & \text{for large times} \end{cases} \tag{20}$$

where S_1 and S_2 are positive coefficients. The time constant τ_2 can be very large, because it depends exponentially on the value of the two reversal fields. Therefore, it may happen that $t/\tau_2 << 1$ during the whole relaxation experiment, which means that one should observe a *logarithmic* increase of magnetization at small times and a *linear* decrease of magnetization at large times.

Careful experiments on the influence of past history on magnetic viscosity are presented and discussed in [16]. They show indeed the type of phenomenology here anticipated. The application of the Preisach approach to the interpretation of those results is discussed in [14].

REFERENCES

[1] G. Bertotti, "Hysteresis in Magnetism", Academic Press, Boston, 1998.
[2] R. Street and J.C. Woolley, Proc. Phys. Soc. A62 (1949), p. 562.
[3] I.D. Mayergoyz, "Mathematical Models of Hysteresis", Springer-Verlag, New York, 1991.
[4] I.D. Mayergoyz, Phys. Rev. Lett. 56 (1986), p. 1518.
[5] J. Souletie, J. Physique 44 (1983), p. 1095.
[6] G. Bertotti, Phys. Rev. Lett. 76 (1996), p. 1739.
[7] C.E. Korman and I.D. Mayergoyz, IEEE Trans. Magn. 32 (1996), p. 4204.
[8] A. Aharoni, "Introduction to the Theory of Ferromagnetism", Oxford University Press, Oxford, 1996.
[9] V. Basso, G. Bertotti, A. Infortuna, and M. Pasquale, IEEE Trans. Magn. 31 (1995), p. 4000.
[10] D.A. Philips, L.R. Dupré, and J.A.A. Melkebeek, IEEE Trans. Magn. 30 (1994), p. 4377.
[11] L.L. Rouve, Th. Waekerlé, A. Kedous-Lebouc, IEEE Trans. Magn. 31 (1995), p. 3557.
[12] E. Della Torre and F. Vajda, IEEE Trans. Magn. 30 (1994), p. 4987.
[13] V. Basso, G. Bertotti, P. Duhaj, E. Ferrara, V. Haslar, L. Kraus, J. Pokorny, and K. Zaveta, J. Magn. Magn. Mat. 157/158 (1996), p. 217.
[14] M. Lo Bue, V. Basso, G. Bertotti, and K.-H. Muller, IEEE Trans. Magn. 33 (1997), p. 3862.
[15] M. Lo Bue, D.R. Cornejo, V. Basso, F.P. Missell, and G. Bertotti, MMM-Intermag Conference, San Francisco 1998, to appear on IEEE Trans. Magn. (1998).
[16] K.-H. Muller, M. Wolf, A. Handstein, D. Eckert, C. Andrikis, and S. Collocott, in "Magnetic Hysteresis in Novel Magnetic Materials", G. Hadjipanayis ed. (Kluwer, Dordrecht, 1997), p. 243.

Materials Science Forum Vols. 302-303 (1999) pp. 53-58

Study of Reversible and Irreversible Magnetization Processes using the Moving Preisach Model

D.R. Cornejo and F.P. Missell

Instituto de Física, Universidade de São Paulo, São Paulo, S.P., Brazil

Keywords: Hysteresis Modeling, Preisach Model, Reversible and Irreversible Magnetization

Abstract

Recent studies of demagnetization processes in permanent magnets have shown that the reversible magnetization can be written as $M_{rev}(H_i,M_{irr})$ where H_i is the internal field and M_{irr} the irreversible magnetization. The dependence of M_{rev} on M_{irr} is given in terms of the so-called inter-relation function $\eta = (\partial M_{rev} / \partial M_{irr})_{H_i}$. Previous discussion of η showed that it is possible to find a hysteretic behavior in the reversible magnetization. In the present work, η is studied using the moving Preisach model, which facilitates a separation of the magnetization into components M_{rev} and M_{irr}. In that model, the effective magnetic field acting on the system is $H = H_i + k_m M$, where $k_m M$ is a mean field term associated with the total magnetization M. It is found that the function η is naturally associated with long range interactions as expressed by the moving parameter k_m, through the relation $\eta = k_m \chi^i_{rev}$ where χ^i_{rev} is the intrinsic reversible susceptibility. Then, for $k_m \neq 0$, M_{rev} is actually hysteretic and $\eta \neq 0$. Also, we show that M_{rev} is a reversible quantity (non-hysteretic) when considered to be a function of the effective field $H = H_i + k_m M$ acting on each elementary loop of the Preisach distribution.

I. Introduction

The description of magnetization processes in ferromagnetic materials is still far from a well-established theoretical structure, especially as regards the relation between physical processes and the details of the magnetization curves. It is well know that the elementary processes that contribute to changes in magnetization are both reversible and irreversible in nature. In a thermodynamic sense, irreversible magnetization changes can be characterized by the sudden passage of the system between two metastable configurations. In these processes, part of the free energy involved will be dissipated as heat. When the change is reversible, on other hand, there is no production of heat. In this case, after application and removal of a magnetic field, the magnetization returns to its original value. It is common, therefore, to separate the total magnetization M in reversible, M_{rev}, and irreversible M_{irr} components. Implicitly, it has been assumed that M_{irr} is hysteretic and that M_{ver} is a single-valued function of the internal field H_i. However, recent studies of magnetization processes have shown that there is an inter-relation between reversible and irreversible magnetization components, and that M_{rev} can show hysteretic behavior [1,2]. Thus, M_{rev} is not a reversible function of the internal field as is generally assumed. This was expressed phenomenologically by Cammarano *et al.* [1] using the equation

$$dM_{rev} = \chi^i_{rev}\, dH_i + \eta\, dM_{irr} \tag{1}$$

where χ^i_{rev} and η are given by

$$\chi^i_{rev} = (\partial M_{rev}/\partial H_i)_{M_{irr}} \qquad \eta = (\partial M_{rev}/\partial M_{irr})_{H_i} \tag{2}$$

The inter-relation function η measures the change in M_{rev} when the domain configuration of the system, represented by M_{irr}, is changed. The internal field H_i is $H_i = H_{ap} - D\,M$, where H_{ap} is the applied magnetic field and D is the demagnetization factor of the sample.

In this work, we use the moving Preisach model in order to simulate the behavior of reversible and irreversible magnetization components of a nanocrystalline magnetic material. We show that the inter-relation function η is proportional to the intrinsic reversible susceptibility χ^i_{rev}. Also, we show that M_{rev} is a reversible quantity (non-hysteretic) when considered to be a function of the effective field $H = H_i + k_m M$ acting on each elementary loop of the Preisach distribution.

II. Relation between η and χ^i_{rev}

Previously [4], we showed that experimental hysteresis curves in nanocrystalline SmFeCo could be well represented by a Preisach distribution function with a Dirac delta function along the diagonal $h_u = 0$ contributing to the magnetization, but not to the losses

$$p(h_c, h_u) = p_{irr}(h_c, h_u) + f(h_c)\,\delta(h_u) \tag{3}$$

The second term describes the reversible magnetization processes, which are responsible for the nonzero slope generally observed at the turning point of measured minor loops. In this expression h_c is the elementary loop half-width and $-h_u$ is the interaction field. The magnetization of the system can be written as

$$M=2M_s\left\{\int_0^{\infty} dh_c \int_0^{b(h_c)} dh_u p_{irr}(h_c,h_u) + \int_0^{\infty} dh_c \delta(h_c) \int_0^{b(h_c)} dh_u f(h_u)\right\} \tag{4}$$

The function $b(h_c)$ is a line in the Preisach plane which describes the distribution of switching units over their +1 and -1 states and is determined by the field history $\{H(t)\}$ [5]. It has the property $b(0) = H$. Using the delta function to evaluate the second term on the right hand side, the final expression for the magnetization is given by

$$M = 2M_s \int_0^{\infty} dh_c \int_0^{b(h_c)} dh_u p_{irr}(h_c,h_u) + M_s \int_0^{H} dh_u f(h_u) \tag{5}$$

where the first term represents the irreversible magnetization M_{irr} and the second term the reversible magnetization M_{rev}. We have

$$M_{irr} = 2M_s \int_0^{\infty} dh_c \int_0^{b(h_c)} dh_u p_{irr}(h_c,h_u) \quad \text{and} \quad M_{rev}(H) = M_s \int_0^{H} dh_u f(h_u) \tag{6}$$

where the magnetic field $H = H_i + k_m M$ is the effective field acting on each elementary loop of the moving Preisach model [5]. The moving parameter is k_m and is associated with the presence of long range order in the system. It is important to notice that, in Eq. (6), M_{rev} is a reversible quantity (non-hysteretic) when considered to be a function of the effective field H. This is not true for H_i, and, in the MPM, M_{rev} exhibits hysteresis when taken to be a function of H_i.

Calculating the differential of M_{rev} in Eq. (6), we find

$$dM_{rev} = \frac{M_s f(H) dH_i + k_m M_s f(H) dM_{irr}}{1 - k_m M_s f(H)} \tag{7}$$

Therefore, comparing the above relation with Eq. (1), we obtain

$$\chi^i_{rev} = M_s \frac{f(H)}{1 - k_m M_s f(H)} \quad \text{and} \quad \eta = M_s \frac{k_m f(H)}{1 - k_m M_s f(H)} \tag{8}$$

and the final result is

$$\eta = k_m \chi^i_{rev} \tag{9}$$

This relation shows that the dependence of the reversible magnetization on the irreversible magnetization is a consequence of the presence of long range order in the system.

The relationship between η and χ^i_{rev} was also previously obtained by integrating the identity $(\partial\eta/\partial H_i)M_{irr} = (\partial\chi^i_{rev}/\partial M_{irr})H_i$ [6]. Using Eq. (2) the above identity becomes

$$\chi^i_{rev}\,(\partial\eta/\partial M_{rev})\,M_{irr} = \eta\,(\partial\chi^i_{rev}/\partial M_{rev})\,H_i \tag{10}$$

One can show that $(\delta M_{rev})M_{irr} = (\delta M_{rev})H_i$ in the limit that $\delta H_i \to 0$. Then, for $\delta H_i \to 0$, we have the relation $d\eta/\eta = d\chi^i_{rev}/\chi^i_{rev}$, which can be integrated to give

$$\eta = (\eta_o/\chi^i_{rev_o})\,\chi^i_{rev} \tag{11}$$

The initial value of η in Eq. (11) is taking as $\eta_o = \eta(H_i = 0) = M_{R_{rev}}/(M_R - M_{R_{rev}})$, where $M_{R_{rev}}$ is the reversible magnetization at remanence ($H_i = 0$) and M_R is the total magnetization at remanence. M_{Rrev} was determined by integrating χ^i_{rev} between $H_i = 0$ and the value of H_i corresponding to the maximum in χ^i_{rev} along the demagnetization curve ($M_{rev} = 0$). χ_{rev_o} is the value of χ^i_{rev} corresponding to $H_i = 0$. Referring to Eqs. (9) and (11) we see that determining $\eta_o/\chi^i_{rev_o}$, the constant of proportionality between η and χ^i_{rev}, thus provides a way to estimate k_m for the system.

Choosing suitable analytic expressions for p_{irr} and f in Eq. (5), it is possible to evaluate the behavior of reversible and irreversible magnetization components in a hysteresis loop. On other hand, we can combine Eq. (1) and (9), and, using $dM_{irr} = \chi_{irr}\,dH_i$, we have

$$dM_{rev} = \chi^i_{rev}(1 + k_m\chi_{irr})dH_i \tag{12}$$

were χ_{irr} is the irreversible susceptibility. Then, using experimental values of χ^i_{rev} and χ_{irr} along the initial magnetization curve and demagnetization curve, it is possible to integrate Eq. (12) and thereby determine $M_{rev}(H_i)$

$$M_{rev}(H_i) = \int_0^{H_i} \chi^i_{rev}(H_i')\left[1 + k_m\chi_{irr}(H_i')\right] dH_i' \tag{13}$$

The irreversible magnetization can be evaluated from

$$M_{irr}(H_i) = M(H_i) - M_{rev}(H_i) \tag{14}$$

were $M(H_i)$ are the experimental magnetization data.

III. Experiment

Alloy of final composition $Sm_{18.1}Fe_{11.0}Co_{70.9}$ (determined by energy dispersive x-ray analysis) was produced by mechanical alloying of elementary Sm, Fe, and Co powders for 30 hours with hardened steel balls in a hardened steel vial using a Fritsch Pulverisette-5 mill. As-milled powders were cold-pressed into 3 mm diameter cylinders and heated in vacuum for 25 minutes at 963 K. A Rietveld analysis of the x-ray diffraction pattern indicated that the main hard magnetic phase formed was $Sm(Fe,Co)_7$ (70-80%), followed by about 10% of α-(Fe,Co) and $SmCo_5$, and about 5% of Sm_2Co_7 and Sm_2Co_{17}. The alloy presents remanence enhancement ($M_r/M_S = 0.62$ at 300 K), with saturation magnetization $\mu_0 M_S = 0.66$ T and coercivity $\mu_0 H_C = 1.96$ T. All magnetization measurements were carried-out at room temperature using a vibrating sample magnetometer (VSM) mounted in a 9 T superconducting solenoid.

IV. Results

In Ref. [4] we show that analytic expressions of the form

$$p(h_c,h_u)=\frac{A}{4\pi h_c\sigma_c\sigma_u}\exp\left[-\frac{\log^2(h_c/H_0)}{2\sigma_c^2}\right]\exp\left(-\frac{h_u^2}{2\sigma_u^2}\right)+f(h_u)\delta(h_c) \quad (15)$$

where

$$f(h_u)=f_0+\frac{A_w}{\pi\, w}\;\frac{1}{1+(h_u/w)^2} \quad (16)$$

give a reasonable description of the magnetic behavior of nanocrystalline SmFeCo. This form was used to calculate the minor hysteresis loops, Henkel plots and recoil curves corresponding to an ac demagnetization in several alloys of $Sm_{18}(Fe,Co)_{82}$ [4,7].

In this work we use Eq. (6) with (15) and (16) to evaluate M_{rev} and M_{irr} for $Sm_{18.1}Fe_{11.0}Co_{70.9}$. In Figure 1*a* we show the results. We can see that, effectively, the reversible magnetization shows hysteretic behavior as a function of H_i.

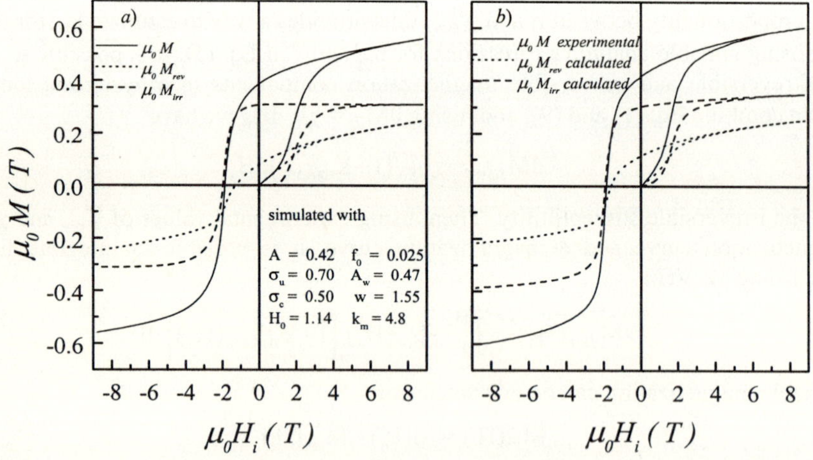

Fig 1: a) Total magnetization and reversible and irreversible components calculated from the MPM
b) Experimental magnetization and its components calculated from Eqs. (13) and (14).

In Figure 2 we show the experimental values of χ^i_{rev} and χ_{irr} for the sample. Calculated values of M_{rev} and M_{irr} from Eqs. (13) and (14) are show in Figure 1*b*. The agreement between this determination and that shown in figure 1*a* is excellent.

Finally, we can comment on the origin of the hysteresis observed in M_{rev}. Using the relation $\eta = k_m\,\chi^i_{rev}$ in Eq. (1) above, we obtain

$$dM_{rev}\,(1+k_m\,\chi^i_{rev}) = \chi^i_{rev}\,(dH_i + k_m dM) \quad (17)$$

We define the total internal field as $H_{it} = H_i + k_m\,M$ and the total reversible susceptibility as $\chi_{rev_t} =$

$dM_{rev}/dH_{it} = \chi^{i}_{rev}/(1 + k_m \chi^{i}_{rev})$. This total internal field H_{it} is analogous to the effective field acting in the moving Preisach model [5]. We note that χ_{rev_t} is a function of both H_i and M_{irr} and, therefore, is not reversible in regard to variations in H_i. On the other hand, χ_{rev_t} is a single-valued function of H_{it} and is completely reversible with regard to that field [7]. In Figure 3 we show the calculated data for the magnetization (Figure 1a) as function of the effective field $H_i + k_m M$. It is shown that the reversible component of magnetization is non-hysteretic when considered to be a function of the effective field H.

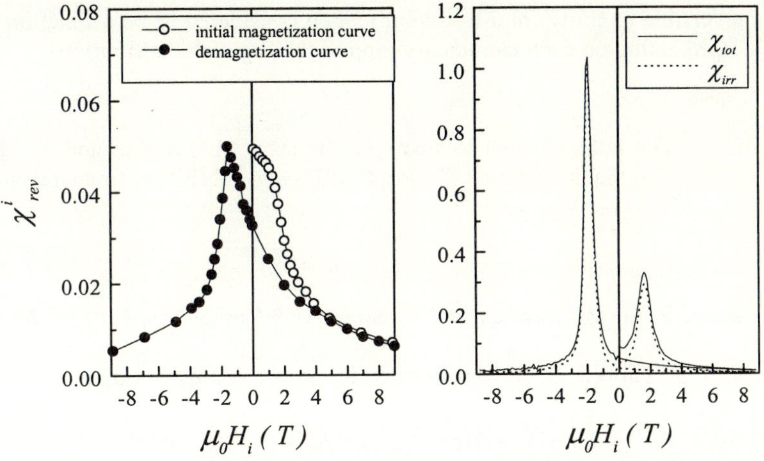

Fig. 2: Reversible and irreversible susceptibilities of $Sm_{18.1}Fe_{11.0}Co_{70.9}$.

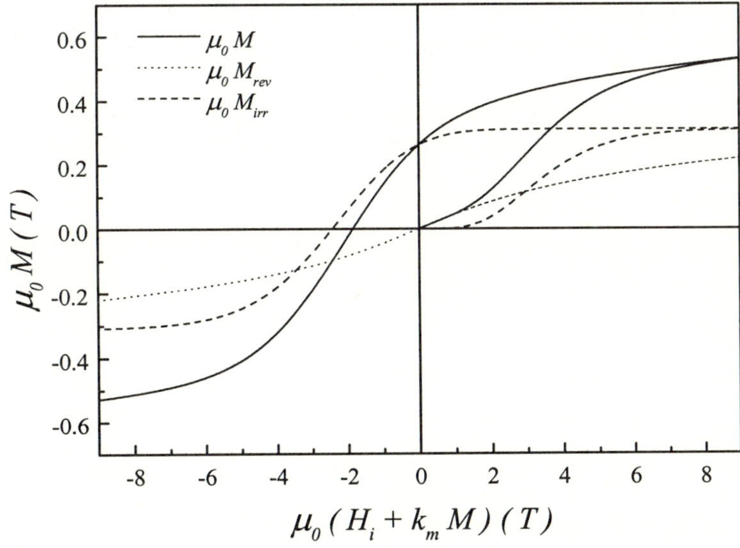

Fig 3: Total magnetization and its components, calculated from the MPM as a function of the effective field $H_i + k_m M$.

V. Conclusions

We have used the moving Preisach model to describe the behavior of reversible and irreversible magnetization components in a $Sm_{18.1}Fe_{11.0}Co_{70.9}$ alloy, prepared by mechanical alloying and subsequent annealing. We show that the inter-relation function η between reversible and irreversible magnetization components, is proportional to the intrinsic reversible susceptibility χ^{i}_{rev}. Determining the constant of proportionality between η and χ^{i}_{rev} thus provides a way to estimate the moving parameter k_m for the system. The hysteresis observed in the reversible magnetization is found to be a manifestation of the fact that $k_m \neq 0$ for this material. Also, we show that M_{rev} is a reversible quantity (non-hysteretic) when considered to be a function of the effective field $H = H_i + k_m M$ acting on each elementary loop of the Preisach distribution.

Acknowledgment - The authors wish to thank G. Bertotti, M. Lo Bue and V. Basso for useful conversations. They also thank FAPESP, CNPq, CAPES and FINEP for financial support.

References

[1] R. Cammarano, P. G. McCormick and R. Street, *J. Phys. D: Appl. Phys.* **29**, 2327 (1996).
[2] D. C. Crew, S. H. Farrant, P. G. McCormick, R. Street, *J. Magn. Magn. Mater.* **163**, 299 (1996).
[3] M. Lo Bue, D. R. Cornejo, V. Basso, F. P. Missell and G. Bertotti: *IEEE Trans. Magn.* **34** (1998).
[4] D. R. Cornejo, M. Lo Bue, V. Basso, G. Bertotti, and F. P. Missell: *J. Appl. Phys.* **81** (1997) 5588-5590.
[5] G. Bertotti: *Phys. Rev. Letters* **76** (1996) 1739.
[6] D. R. Cornejo, V. Villas-Boas and F. P. Missell: *J. Appl. Phys.* **83** (1998).
[7] D. R. Cornejo, Ph.D. thesis, Instituto de Física, Universidade de São Paulo, 1998.

Materials Science Forum Vols. 302-303 (1999) pp. 59-63

Study of Thermally Activated Phenomena in Nanocrystalline Soft Magnetic Materials

V. Basso, M. LoBue, C. Beatrice, P. Tiberto and G. Bertotti

IEN Galileo Ferraris and INFM, C.M. D'Azeglio 42, IT-10125 Torino, Italy

Keywords: Thermal Activation, Dynamic Hysteresis Loops, Nanocrystalline Soft Materials, Magnetic Hardening

Abstract - **We present experiments on nanocrystalline materials where thermal activation gives rise to dynamic effects on hysteresis curves. A scheme of interpretation is presented and the typical activation volume is found to be about 100 nm.**

INTRODUCTION

Multi-phase nanostructured materials are actively studied because of their innovative properties. Remanence enhancement in hard materials and vanishing magnetostriction and anisotropy in soft materials are examples of effects coming from the coexistence of different magnetic phases and from their mutual coupling [1]. In particular, the FeSiBNbCu (Finemet) material discovered in 1988 [2], consisting of Fe-Si crystallites of size around 10 nm imbedded in an amorphous matrix, has outstanding soft magnetic properties that make it of great technological interest for a number of applications. At the same time, Finemet materials offer unique opportunities for the fundamental study of magnetization processes. In fact, dramatic changes in the magnetic behavior can be induced by a change in the measurement temperature, as a consequence of the fact that the Curie temperature of the crystalline and amorphous phase are quite different. Above the Curie temperature of the amorphous phase, this class of materials generally shows the so-called magnetic hardening, that is, a strong increase of coercivity due to the fact that the nanograins are no longer exchange-coupled and the material becomes a system of weakly interacting magnetic particles. Another interesting feature of Finemet materials is the important role played by thermal activation. Thermal-activated phenomena are relevant to many hard magnets, but are often difficult to detect in soft materials, where activation volumes are generally large. However, when a Finemet material is brought beyond the Curie temperature of the amorphous matrix, activation volumes become small because of grain decoupling and thermally activated processes become important.

We investigated nanocrystalline Finemet, by measuring hysteresis curves at various temperatures in the range 20°C-500°C and at various exciting frequencies. The magnetic hardening effect was observed, with a maximum around 370 °C. Moreover we found that at elevated temperature hysteresis curves exhibit dynamic effects that can only be ascribed to thermal activation, the measurement frequencies being so low ($f < 100$ Hz) that eddy currents are fully negligible. In particular, we observed that the coercive field increases with the field rate dH/dt according to the law $H_c \approx H_f \ln(dH/dt)$, where H_f is a suitable constant and that when the field is reversed at $H=H_p$, the differential susceptibility after the reversal point is initially negative and follows the law $dI/dH \propto -\exp(-|H-H_p|/H_f)$, where H_f is the same constant as before. We present a scheme of interpretation based on the Preisach model of hysteresis and described in a recent paper [3], where the shape of hysteresis loops and thermal relaxation effects can be jointly predicted. With simple hypotheses on the activation energy distribution, we were able to reproduce the laws observed experimentally with good accuracy. The model shows that the constant H_f coincides with the fluctuation field $H_f=k_BT/I_s v$ where v is the activation volume of the process.

EXPERIMENTAL

The measurements were performed on a single strip (30 cm long, 10 mm wide and 20 μm thick) of nanocrystalline Finemet of composition $Fe_{73.5}Cu_1Nb_3Si_{13.5}B_9$, submitted to a conventional annealing in furnace at 550°C for 1h, which induces the growth of Fe-Si nanograins of about 10nm size. The sample was placed inside an induction furnace. The temperature in the oven ranged from 20°C to 500°C, and never exceeded the original annealing temperature. The sample, the solenoid to generate the field and the compensated pick-up coils were inserted in a tube kept under controlled Ar atmosphere. The temperature, measured by a thermocouple, was checked to be constant along the sample. The large thermal inertia of the furnace permitted us to perform measurements with the heater off in order to reduce electrical disturbances. Experiments were performed up to a maximum temperature of 500°C and it was checked that no structural changes were induced by the measurement at the highest temperature. All experiments were performed under a triangular field waveform, in order to have a well defined rate dH/dt in the range 10^3-10^6 A/ms. From 20°C to 400°C, we observed the increase of coercivity due to magnetic hardening (see Fig.1a). At each temperature, a set of twenty minor loops was measured, in order to reconstruct the Preisach distribution by the method discussed in the next section, and the major loop was measured at different magnetization rates, to study the dependence of the coercivity on dH/dt (see Fig. 1b). We found that above $T \cong 350$°C the coercive field follows the law:

$$H_c \approx H_f \ln(dH/dt) \qquad (1)$$

where H_f is a suitable constant (see Fig.1b). In addition, we observed that when the field is reversed at the turning point $H=H_p$, the differential susceptibility after the reversal point is initially negative and follows a law independent of dH/dt, of the type:

$$dI/dH=-\chi \exp(-|H-H_p|/H_f) \qquad (2)$$

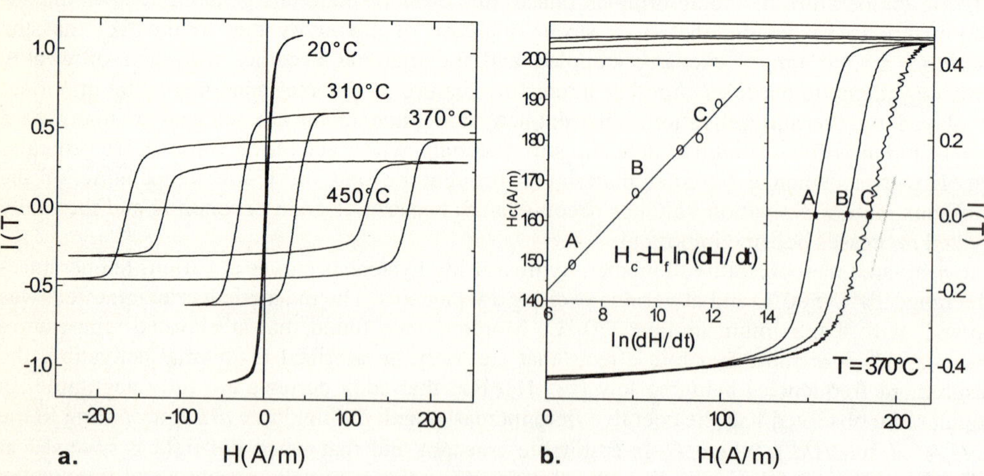

Fig. 1. a. Major hysteresis loops measured at different temperatures on nanocrystalline Finemet. **b.** Rate dependence of major hysteresis loops and coercivity (inset). Major loops correspond to field rate (A) 10^3; (B)10^4; (C)10^5 A/ms. H_f=7A/m.

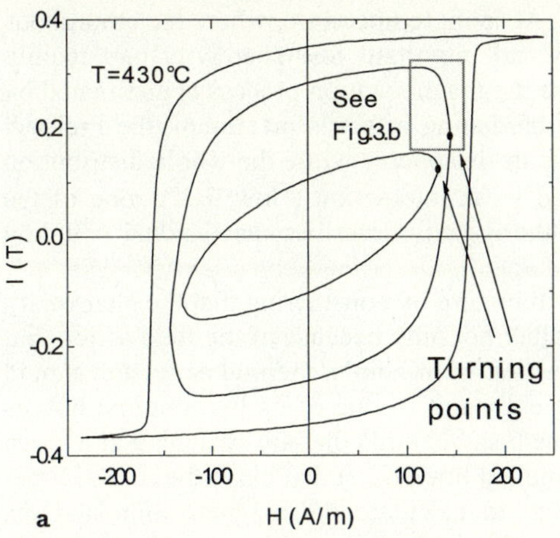

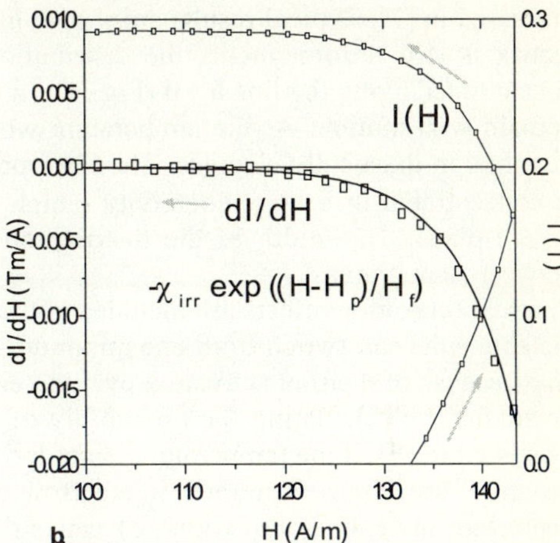

Fig.2. a. Hysteresis loops measured on Finemet at 430°C. The effect of thermal activation can be seen in the negative slope after the turning points. **b.** Detail of hysteresis loop. The susceptibility after the reversal point is compared with (2). The fluctuation field is found to be H_f=6A/m.

where χ is approximately equal to the susceptibility just before the turning point and H_f is a constant (see Fig.2). This effect was observed for several temperatures and peak field amplitudes. The model discussed in the next section shows how the experimental laws (1) and (2) can be justified and derived theoretically.

THEORY AND DISCUSSION

Our scheme of interpretation is based on the Preisach model, where a given magnetic system is described by the superposition of elementary bistable units with two energy minima labeled (+) and (-). Each elementary unit can be characterized by two energy values, the activation energy $\mu_0 \Delta m h_c$ and the energy difference between the two minima $2\mu_0 \Delta m h_u$ where h_c represents a local coercive field, h_u represents a local interaction field, and Δm is the magnetic moment associated with the unit. The contribution of each elementary unit to the magnetization process is weighted by a function $p(h_c,h_u)$ defined on the (h_c,h_u) plane and known as the Preisach distribution. The collective state of the bistable units is described by the borderline $b(h_c)$, defined in the plane (h_c,h_u), separating units in the (+) or (-) state. Accordingly, the magnetization I of the system is given by the integral:

$$I = 2I_s \int_0^{\infty} dh_c \int_0^{b(h_c)} dh_u \, p(h_c,h_u) \tag{3}$$

Equation (3) plays the role of history-dependent constitutive law. The state of the system is described by the $b(h_c)$ state line. The history-dependent behavior of $b(h_c)$ under varying field is controlled by definite switching rules for the elementary units, described in [4]. Conversely, the information on the microstructural properties of the material is carried by the Preisach distribution $p(h_c,h_u)$. If one is able to determine this function for a given system, one obtains a representation of the material microstructure as probed by magnetic domains. One possibility is to reconstruct the Preisach distribution from an appropriate set of measured hysteresis loops of variable amplitude. The reconstruction was carried out at different temperatures by the method discussed in [7]. Typical results are shown in Fig. 3. At room temperature, where the amorphous matrix is still ferromagnetic, the distribution shows an important low coercivity part mainly concentrated along the line h_c=0 (Fig.3a). In this case the magnetization process is dominated by domain wall motion. At the temperature where the hardening effect is maximum, the Preisach distribution drastically changes. The low coercivity part disappears while the whole distribution is concentrated in a large-coercivity (high h_c) and weak-interaction (low $|h_u|$) zone of the (h_c,h_u) plane. The width of the distribution along the h_u axis describes the residual coupling between nanograins.

Thermal relaxation effects are included in the Preisach picture by considering that the elementary bistable units can switch from one minimum to the other not only because of the field action, but also because of thermal activation over the energy barrier. In principle, thermal activation should be studied by calculating the probability of finding each unit in one of its two minima but, as discussed in [5], if the temperature is not too high, the bistable units that are switching at a given instant of time are concentrated in a narrow band around a line $b(h_c,t)$ that plays the role of time-dependent state line. Equation (3) can still be used to calculate the magnetization and the problem is reduced to deriving the evolution equation obeyed by $b(h_c,t)$. This equation turns out to be [5]:

$$\frac{\partial b(h_c,t)}{\partial t} = 2\frac{H_f}{\tau_0}\,\mathrm{sh}\left[\frac{H-b(h_c,t)}{H_f}\right]\exp\left[-\frac{h_c}{H_f}\right] \tag{4}$$

where $H_f=k_BT/\mu_0\Delta m$ is the fluctuation field, k_B is the Boltzmann constant, T is the absolute temperature and $\tau_0{\sim}10^{-10}$s is a typical attempt time. Equation (4) can be applied equally well to a typical magnetic viscosity experiment, where the magnetic field H is constant and $b(h_c,t)$ evolves till the final state $b(h_c,\infty)=H$ is reached [6], as well as to the case considered in this work, where the field $H(t)$ changes in time and the behavior of the system results from the competition between field action and thermal activation. In particular, (4) can be used to predict the two experimental quantities shown in Figs. 1 and 2: the field rate dependence of the coercive field and the differential susceptibility after a turning point. We shall present the results of this analysis and discuss their physical meaning. For the details of the calculations, which cannot be

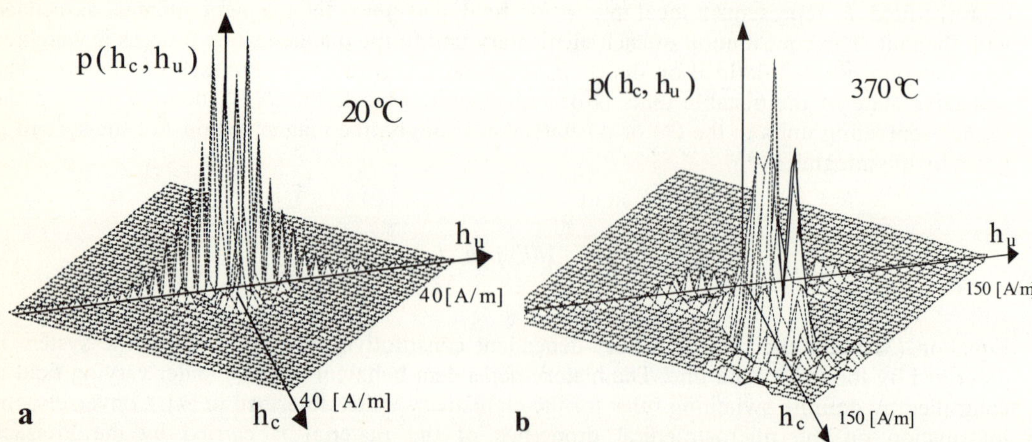

Fig.3. a. Reconstructed Preisach distribution at room temperature and, **b.** at 370°C, where the magnetic hardening is maximum.

given here for lack of space, see [3] and [5]. As to the coercive field dependence on dH/dt, when thermal activation is present one finds from (4) that

$$H_c = H_c^i + H_f \ln\left[\frac{\tau_0}{H_f}\left|\frac{dH}{dt}\right|\right] \tag{5}$$

where H_c^i is the coercive field in the limit of high field rates. This result coincides with the experimental law (1). In our experiments $H_f \cong 7$A/m and $H_c^i \sim 250$A/m. As to the susceptibility, by deriving (3) with respect to H one finds

$$\frac{dI}{dH} = 2I_s \int_0^\infty dh_c\, p[h_c, b(h_c,t)] \frac{\partial b/\partial t}{dH/dt} \tag{6}$$

The term $\partial b/\partial t$ is given by (4) and is in general a complicated function of time. However, for the present purposes, it is enough to calculate the derivative at the time t_0 of field inversion, where $H = H_p$ and $b(h_c,t_0) = H_p - H_f \ln[(\tau_0/H_f)|dH/dt|] - h_c$. One obtains

$$\frac{dI}{dH} = -2I_s \int_0^\infty dh_c\, p[h_c, b(h_c,t_0)] \exp\left[\frac{H - H_p}{H_f}\right] \tag{7}$$

The integral term in front of the exponential turns out to be equal to the susceptibility χ before the turning point. Therefore, one precisely obtains (2), which shows that H_f actually represents the fluctuation field. The theoretical predictions are in agreement with the experimental susceptibility curves in the range of temperatures T=370~450°C — where the system consists of nanograins (size ≈ 10nm) dispersed in a paramagnetic amorphous matrix — if one assumes H_f=6±1A/m, in agreement with the coercive field behavior. From this value, we were able to give an estimate of the activation volume involved in the process. We found $v^{1/3} \cong 100$ nm, which corresponds to clusters of about 10^3 nanograins coherently reversing their magnetization.

REFERENCES

[1] A.Hernando and T. Kulik, *Phys.Rev. B* **49** (1994) pp.7064-7067.
[2] Y. Yoshizawa et al. , *J. Appl. Phys.* **64** (1988), pp.6044-6046.
[3] V. Basso, M. LoBue, C. Beatrice, P. Tiberto, G. Bertotti, Joint MMM-Intermag Conference, San Francisco 1998, to appear in IEEE Trans. Magn.
[4] I.D.Mayergoyz, *Mathematical models of hysteresis,* Springer, New York, 1991.
[5] G.Bertotti , *Hysteresis in magnetism*, Academic Press, Boston, 1998.
[6] M.LoBue, V.Basso, G,Bertotti and K.H.Muller, *IEEE Trans. Magn.* 33 (1997) pp. 3862-3864.
[7] M. Lo Bue, V. Basso, C. Beatrice, P. Tiberto, SMM Conference, Grenoble 1997, to appear in Journal de Physique.

Materials Science Forum Vols. 302-303 (1999) pp. 64-75

Magnetic Multilayers: Interlayer Coupling in Fe/Cr/Fe

S.M. Rezende, C. Chesman, M.A. Lucena, M.C. de Moura, A. Azevedo and F.M. de Aguiar

Departamento de Física, Universidade Federal de Pernambuco, 50670-901 Recife, Brazil

Keywords: Magnetic Multilayers, Ferromagnetic Resonance, Biquadratic Exchange

The exchange coupling between neighboring magnetic layers in multilayer systems consisting of stacks of ferromagnetic layers separated by nonmagnetic metallic layers plays a central role in the properties of these novel artificially structured materials. Here we review the most common techniques for measuring this coupling, namely magneto-optical Kerr effect magnetometry, Brillouin light scattering and ferromagnetic resonance. The theoretical background for interpreting experimental data in trilayers formed by two magnetic layers separated by a nonmagnetic layer is presented, based on a phenomenological model energy including bilinear and biquadratic exchange couplings, as well as surface, in-plane uniaxial and crystalline cubic anisotropy contributions. Accurate quantitative values for the coupling constants and the other magnetic parameters are measured in the prototype system (100) Fe/Cr/Fe grown by sputtering for several Cr spacer thickness. Consistent values are obtained with all three techniques for both the bilinear (J_1) and biquadratic (J_2) exchange coupling constants. In most of the Cr thickness range corresponding to the first antiferromagnetic peak, J_2 follows J_1 with a ratio $J_2 / |J_1| \cong 0.1$. In the range corresponding to the second antiferromagnetic peak J_2 also follows J_1, but with a much larger ratio $J_2 / |J_1| \cong 1.0$, indicating that the origin of the biquadratic coupling in the two ranges resides in different mechanisms.

I. INTRODUCTION

Magnetic multilayers, consisting of stacks of ferromagnetic layers separated by nonmagnetic metallic layers, as shown in Fig.1(a), have attracted considerable attention due to their unique physical properties and potential for technological applications. Many multilayer systems exhibit a coupling between the magnetic layers mediated by the nonmagnetic spacers, which oscillates periodically between ferromagnetic (FM) and antiferromagnetic (AF) as the spacer-layer thickness varies in the range 5-50 Å. The existence of this coupling, reported in 1986 by Grünberg and collaborators [1], was the first of a series of important discoveries that aroused interest from the scientific community on magnetic multilayers. The next important finding was the giant magnetoresistance, reported in 1988 by Baibich et al. [2]. This is observed in multilayers with AF coupling, where the resistance measured across the structure (in-plane or perpendicular to the plane) decreases considerably with increasing applied external field. This is due to the change in electron mean free-path as the magnetizations rotate from the AF alignment in zero-field to parallel alignment at sufficiently strong fields. This phenomenon immediately found application in magneto-resistive reading reads in magnetic recording [3]. The oscillation in the interlayer coupling was first reported by Parkin et al. in 1990 [4].

Due to its central role in the properties of magnetic multilayers, the coupling between the magnetic layers through the nonmagnetic metallic spacer has been the subject of extensive investigations for nearly ten years. Experimentally this coupling is more conveniently studied in a trilayer structure, formed by two magnetic thin films separated by a nonmagnetic layer, as shown in Fig.1(b). The buffer layer has the purpose of matching the magnetic material structure to that of the substrate, whereas the cap layer provides symmetry and protection for the top magnetic layer. The most common elements used to make the ferromagnetic layers are Fe, Ni and Co, pure or in alloys, while the metallic spacers are usually Cr, Cu, Au, Ag, Pd and Ru. The most studied system is Fe/Cr/Fe, which has been used in all three mentioned important discoveries [1,2,4]. Extensive reviews of the work on this and other systems can be found in the literature [5].

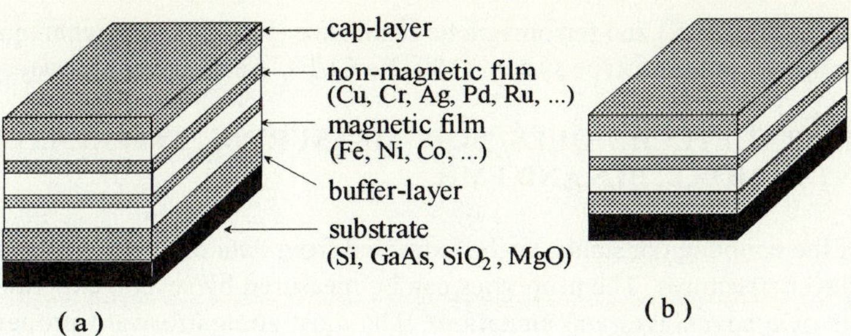

FIG. 1. Schematics of a magnetic multilayer (a) and a trilayer structure (b).

The coupling between the magnetic layers is usually dominated by a mechanism which can be modeled by an interaction energy of the form $-J_1\mathbf{m}_1\cdot\mathbf{m}_2$, where $\mathbf{m}_1$ and $\mathbf{m}_2$ are the unit magnetizations of the two magnetic layers and J_1 is called the bilinear exchange coupling constant. The origin of this coupling lies in the interaction between the s-electrons in the metallic spacer and the d-electrons in the magnetic layers [5,6,7]. Its magnitude can be calculated by RKKY treatments, resulting in a periodic oscillation with period inversely proportional to the length of the spanning vectors joining extremal points of the Fermi surface, normal to the spacer layer crystal growth direction [8]. As a result, in systems with two or more quite different extremal points, several periods of oscillation may occur. Indeed, a short period corresponding to two Cr monolayers (ML) in addition to the known long period of 12.5 ML (18 Å) was observed [9,10] in Fe/Cr structures grown on Fe (001) whiskers, which provide extremely flat substrate surfaces. However, modest interface roughness washes out the short-period oscillation and only the long-period one is usually observed.

Later it was observed [11-13] that under certain conditions the magnetic moments of the two layers tend to align at 90° with respect to each other. This alignment may be accounted for through an interaction energy described by a phenomenological biquadratic exchange coupling (BEC) $-J_2(\mathbf{m}_1\cdot\mathbf{m}_2)^2$, where J_2 is called the biquadratic coupling constant. This coupling has been investigated in several systems, such as (100) Fe/Cr/Fe wedges and trilayers [11,13-16], trilayers of (100) Fe/Al/Fe [12], (001) Co/Cu/Co [17], (100) Fe/Ag/Fe [18,19] and (211) Fe/Cr/Fe [20] among others. Over the last few years several mechanisms have been proposed for the origin of the biquadratic coupling [21-24], some attributed to intrinsic properties of the spacer layer and others to extrinsic factors, such as the presence of impurities or interface roughness. Intrinsic mechanisms seem not to account for the observed coupling strengths and signs, nor for its temperature dependence [21]. On the other hand, if the mechanisms based on extrinsic factors prevails in the BEC, one expects the value of the coupling constant J_2 to be quite sensitive to details of the sample preparation conditions. Indeed, there is a considerable spread in the values of J_2 measured by different groups for nominally the same system, indicating the dominance of extrinsic mechanisms. However, there are conflicting reports on the temperature and spacer layer thickness dependences of J_2 and only in very few cases there seems to be a reasonable connection between data and some specific extrinsic mechanism [19]. The fact is that the whole question of the origin of the BEC is still quite controversial and deserves further investigation.

In order to understand the origin of the interlayer coupling in detail it is important to be able to measure the values of J_1 and J_2 reliably in trilayers structures. Here we review three of the best techniques for such measurements, namely, magneto-optical Kerr effect (MOKE) magnetometry,

Brillouin light scattering (BLS) and ferromagnetic resonance (FMR). These techniques are used for a detailed investigation of the prototype system (100) Fe/Cr/Fe, with varying Cr spacer layer thickness.

II. EXPERIMENTAL TECHNIQUES FOR MEASURING INTERLAYER COUPLING CONSTANTS – MOKE, BLS AND FMR.

The values of the coupling constants can be extracted from data on various magnetic and dynamic properties of trilayer structures. The properties can be measured by several experimental techniques, each one with its own advantages and limitations. The most straightforward property to measure is the magnetization versus applied field curve, $M \times H$. Due to the very small volumes and consequently small moments of the magnetic films, the vibrating sample magnetometer is hardly useful for trilayer structures. The SQUID magnetometry, on the other hand, has sufficient sensitivity but is difficult to interpret because of the additional contributions from the sample holder, substrate and other films of the structure. It turns out that the most useful technique to measure $M \times H$ in trilayers is the MOKE magnetometry. The MOKE magnetometry is also the simplest of the three techniques that will be considered here.

A MOKE magnetometer consists basically of a modulated laser beam incident on the sample, two crossed polarizers, an electromagnet to apply the field on the sample and the peripheral electronic equipment. As the field across the sample is swept, the changes in the magnetization produce changes in the polarization and/or amplitude of the reflected light. Under certain conditions, the output signal is proportional to the change in the magnetization, resulting in a $M \times H$ hysteresis curve. Our MOKE set-up employs a He-Ne laser modulated in amplitude at 50 kHz by an elasto-optic modulator placed between crossed polarizers. Magnetic hysteresis loops are measured in the longitudinal Kerr effect configuration, with the light polarized along the dc magnetic field in the film plane.

In order to interpret the MOKE measurements one needs to compare the $M \times H$ data with curves calculated with an adequate model and realistic parameters. We consider two magnetic thin single-crystal films, 1 and 2, having cubic lattice structure, thickness d_1 and d_2 and separated by a nonmagnetic spacer layer with thickness s. The coordinate system is chosen so that the xz plane is parallel to the film surface, with the x and z axes along [100] and [001] crystal directions, respectively. We study only the situation where the external static magnetic field $\mathbf{H}_0$ is applied in the plane of the films, at an arbitrary angle θ_H with respect to the [001] direction. In this case, the equilibrium directions of the magnetizations of the two films, $\mathbf{M}_1$ and $\mathbf{M}_2$ are also in the xz plane, characterized by the polar angles θ_1 and θ_2. In general, each interaction tries to align the magnetizations of the two magnetic films along different directions. The equilibrium directions of $\mathbf{M}_1$ and $\mathbf{M}_2$ are determined by the minima of the total free energy. We consider a free energy per unit area with four basic contributions,

$$E = E_z + E_a + E_{ex} + E_{dip} \quad , \tag{1}$$

where the subscripts denote, in order, Zeeman, anisotropy, exchange and dipolar terms. Assuming that the system is in a static situation, this expression yields,

$$E = \sum_{i=1}^{2} d_i \left[-M_i H_0 \cos(\theta_i - \theta_H) + \frac{1}{4} K_1^{(i)} sin^2 2\theta_i - K_u^{(i)} \cos^2\left(\theta_i - \theta_u^{(i)}\right) \right]$$
$$- J_1 \cos(\theta_1 - \theta_2) + J_2 \cos^2(\theta_1 - \theta_2) \quad , \tag{2}$$

where $K_1^{(i)}$ is the first-order cubic anisotropy constant of film i and $K_u^{(i)}$ is the uniaxial in-plane anisotropy constant in a direction defined by a polar angle $\theta_u^{(i)}$. Note that the anisotropy constants $K_1^{(i)}$ and $K_u^{(i)}$ have units of erg/cm^3 and each one is associated with an effective anisotropy field $H_a^{(i)} = 2K^{(i)} / M_i$ with units of Oe. Usually the anisotropy parameters are the same for both magnetic films, so that the superscript (i) can be dropped for simplicity. It is also common to associate effective exchange fields to the bilinear and biquadratic coupling $H_{ex1}^{(i)} = J_1 / M_i d_i$ and $H_{ex2}^{(i)} = J_2 / M_i d_i$ with units of Oe.

In simple situations the equilibrium configuration can be obtained analytically by equating to zero the derivatives of the energy in (2) with respect to θ_1 and θ_2. However, in more general situations this leads to transcendental equations which cannot be solved analytically. In the cases studied in this paper, the equilibrium configuration was determined numerically by varying θ_1 and θ_2, firstly in steps of 5° , to locate the range of minimum energy, and then in steps of 0.5° to obtain accurate minima values. Once θ_1 and θ_2 are found, we obtain normalized values for the magnetization component in the field direction from

$$\frac{M(H_0)}{M_S} = \frac{M_1 \cos(\theta_1 - \theta_H) + M_2 \cos(\theta_2 - \theta_H)}{M_1 + M_2}, \tag{3}$$

where M_S is the total saturation magnetization. This equation will be used to fit MOKE data as presented in Section III. Although the MOKE technique is simple and very convenient to study multilayers, in some cases it does not yield accurate values for the coupling constants J_1 and J_2 separately. For this reason it is important to have available other techniques to measure the coupling constants and to compare the values obtained with all of them independently.

BLS and FMR are among the best experimental techniques to determine the interlayer exchange coupling constants. The basic idea behind the use of these techniques to measure the coupling is that they serve to determine the frequencies of the spin-wave excitations of the system and these depend on the values of the coupling constants. It is easy to see that two identical excitations of the individual uncoupled films with the same frequency will give rise to two modes with different frequencies when the coupling is turned on, having frequency separation that depends on the strength of the coupling constants.

In the BLS technique [25], a laser light incident on the film plane produces a reflected beam which contains a small component due to inelastic scattering from the spin-wave excitations of the system. Due to energy and momentum conservation, the frequency shift of the light equals the frequency of the excitation with wavenumber $q = 2k_L \sin\alpha$, where $k_L = 2\pi/\lambda_L$ is the laser wavenumber. BLS measurements were carried out in the backscattering geometry, using a Sandercock-model tandem Fabry-Perot interferometer with active stabilization in a (2×3)-pass configuration. The light source was a single-mode stabilized argon ion laser operating at 5145 Å, with incident power in the range 60-100 mW. Light detection was made by photon counting using a cooled EG&G photodetector with 40% quantum efficiency and an average dark noise of 1.5 cps. The spectra were stored in 256 channels, each with gate duration of 1 ms. Several hundred to a few thousand interferometer scans were used to record the spectra, depending on the signal intensity. All measurements were done at room temperature with the sample between the poles of an electromagnet, with the field in the film plane and perpendicular to the incidence plane of the light beam. The sample was mounted on a goniometer to allow for two independent rotations: rotation around the normal as to vary the field direction in the film plane; rotation around the field direction to vary the incident angle α and thus the spin-wave wavenumber. Magnons with wavenumbers in the range $0.4 \times 10^5 \text{ cm}^{-1} < q < 2.1 \times 10^5 \text{ cm}^{-1}$ could then be probed in a magnetic field varying from 0 to 6 kOe.

In the FMR technique one excites directly the $q = 0$ spin-wave modes with microwave radiation. In order to enhance the signal the sample is placed in a resonant structure, such as a microwave cavity. Hence the frequency has to be kept constant. In this case, one sweeps the field and records the absorption spectrum with the signal reflected from the cavity. In this work FMR measurements were carried out with a home-made microwave spectrometer employing a sweep oscillator with frequency stabilized at the cavity resonance. The spectra were obtained by sweeping the field at fixed frequency and monitoring the derivative of the absorption lines provided by field modulation at 1 kHz with Helmholtz coils mounted on the cavity walls. In order to change the frequency several retangular cavities were constructed to operate in the TE_{102} mode in the X and K microwave bands. All measurements were made with the field applied in the film plane. In order to investigate the in-plane magnetic anisotropy the sample was rotated about the normal to its plane, maintaining the field in the plane.

In order to interpret the BLS and FMR experiments it is necessary to have the spin-wave dispersions for a trilayer structure including all interaction energies present in Eq. (1). The calculation of the spin-wave frequencies takes into account an applied static magnetic field in the film plane, surface and in-plane uniaxial and cubic crystalline anisotropies, bilinear and biquadratic exchange interactions with arbitrary signs and magnitudes. The approach used is based on the equations of motion for the small-signal magnetization deviations from the equilibrium directions, similar to that of Hillebrands [26], and Cochran and Dutcher [27]. However, besides introducing the biquadratic coupling, our calculation leads to an analytical expression obtained with an improved treatment on the dipolar interaction, which makes it suitable for reasonably thick magnetic layers [28]. The equation of motion for the magnetization in film i is written as

$$\frac{\partial \mathbf{M}_i}{\partial t} = \gamma_i \mathbf{M}_i \times \mathbf{H}_{eff}^{(i)} , \tag{4}$$

where $\gamma_i = g_i \mu_B / \hbar$ is the gyromagnetic ratio (2.8 GHz/kOe for $g_i = 2$), and $\mathbf{H}_{\text{eff}}^{(i)}$ is the effective field acting on $\mathbf{M}_i$. All fields and magnetizations are decomposed in a static part and a small-signal dynamic component [28]. The effective fields corresponding to the anisotropy, surface demagnetizing, bilinear and biquadratic exchange energy contributions E_λ are given by $\mathbf{H}_\lambda = -\nabla_{\text{M}}(E_\lambda / d)$. Assuming the exp($i\omega t$) time variation and retaining only terms to first order in small quantities, we obtain a set of four coupled equations, the solutions of which are found by requiring that the secular determinant vanishes. This leads to magnetic excitation frequencies which are given by the zeroes of the equation

$$\left(\frac{\omega}{\gamma}\right)^4 + a\left(\frac{\omega}{\gamma}\right)^2 + b\left(\frac{\omega}{\gamma}\right)^2 + c = 0 , \tag{5}$$

where a, b and c are given by lengthy expressions involving the trilayer parameters, the equilibrium angles, and the external field [28]. For any given applied field, Eq.(5) has two real solutions, corresponding to the acoustic and optic modes. This expression has quite general applicability for the in-plane configuration and can be readily used by experimentalists. In the BLS technique one measures the frequencies as a function of H_0 for a given q determined by the experimental geometry. Fitting the theoretical curves to the data yields accurate values for all magnetic and coupling parameters. However, one disadvantage of the BLS technique is that due to the small signal from the inelastic scattering, several minutes of photon counting for each field value are needed to record a good spectrum, so that several hours are necessary to obtain reliable data. In the FMR technique one measures the spectrum of the $q = 0$ mode at fixed frequency as the field is swept. Usually this is done repeatedly as the sample plane is rotated in the field. Fitting the resonance fields versus the in-plane field angle with the theoretical curves allows the determination of the parameters. The FMR technique is the best one to determine the anisotropy parameters. However, the coupling constants can only be determined accurately when both acoustic and optic modes resonances are observed.

III. APPLICATION TO (100) Fe/Cr/Fe

The techniques described in the previous section have been used to study in detail trilayers of the prototype system (100) Fe/Cr/Fe. The samples are single-crystal structures prepared at the IBM Almaden Research Center by magnetron sputter deposition in an ultrahigh vacuum chamber equipped with six 2-in dc magnetron sputtering sources. The base pressure prior to deposition was typically 2×10^{-9} Torr and the sputter pressure was usually 3×10^{-3} Torr Ar. The Fe/Cr/Fe trilayers were grown on MgO (100) substrates. Initially a Cr seed layer, at least 100 Å thick, was deposited on the substrate at temperatures ranging up to 525 °C in order to establish epitaxy. The subsequent Fe/Cr/Fe layers were grown at 150-180 °C, forming (100) planes with the Fe [100] along the MgO [110] direction. All samples have the same Fe layer thickness, $d = 40$ Å, and a thin Cr cap layer. Initial characterization of the coupling was made by MOKE with a Cr wedged sample, with $0 < s < 70$ Å. Then a series of samples was prepared with uniform Cr thickness varying from 5 Å to 35 Å, a range that corresponds to the first two antiferromagnetic peaks.

Due to space limitation, we show below only representative results for a few samples of the series. Consider first the sample with Cr thickness $s = 11$ Å, a value close to the first AF maximum. The initial characterization is made with MOKE magnetometry. Figure 2(a) shows data (open circles) measured with the field along the hard [101] axis. The solid line represents a theoretical fit with the calculation described in Section II, using the following parameters: $4\pi M = 19.5$ kG, $H_{ca} = 2K_1/M = 0.57$ kOe, $H_u = 0$, $H_{ex1} = J_1/d_1M = -0.89$ kOe ($J_1 = -0.55$ erg/cm^2), $H_{ex2} = J_2/d_1M = 0.07$ kOe ($J_2 = 0.044$ erg/cm^2, $J_2/J_1 = -0.08$). Note that the behavior of the $M \times H$ curves is characterized by three distinct phases: The magnetizations remain close to an antiferromagnetic (AF) alignment along the [001] axis at fields below a certain critical value of 0.5 kOe. At this field there is a sudden transition to a spin-flop (SF) state. As the field increases above this value, the spins rotate towards the field and at 2.7 kOe the system attains saturation, with $\mathbf{M}_1$ and $\mathbf{M}_2$ aligned with $\mathbf{H}_0$. Using the energy in equation (2), it is easy to show that the saturation field is given by

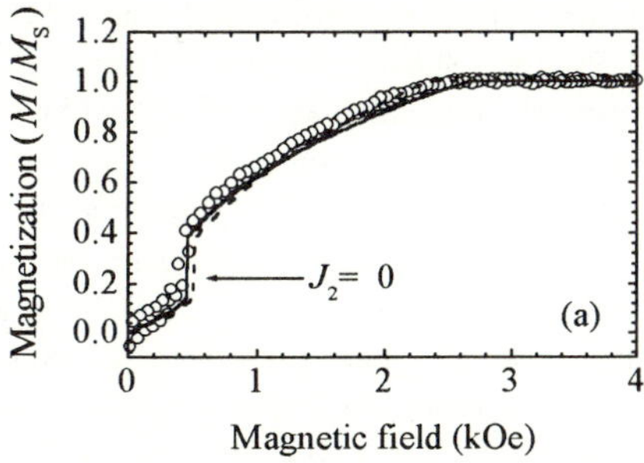

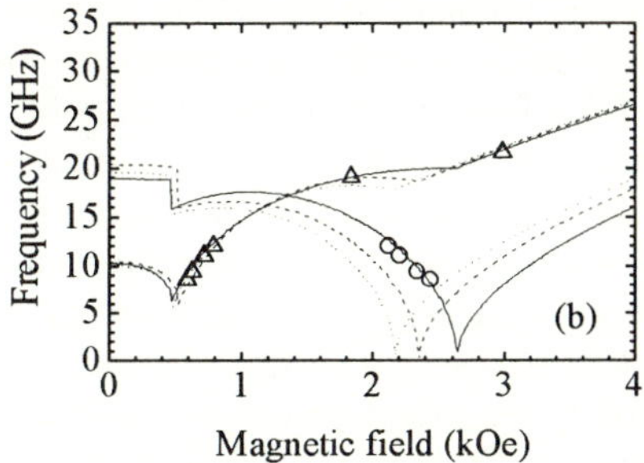

FIG. 2. Data (symbols) and theoretical fits (lines) in (100) Fe(40Å)/Cr(11Å)/Fe(40Å) with $\mathbf{H}_0$ applied along [101]: (a) Normalized magnetization measured by MOKE, and (b) variable-frequency FMR, as described in the text.

$$H_{SAT} = 2|H_{ex1}| + 4H_{ex2} \pm H_{ac} \quad , \tag{6}$$

where the sign + (–) occurs when the field is applied along a hard (easy) magnetization axis. The parameters above lead to a critical field H_{SAT} = 2.7 kOe for the field applied along the hard [101] axis. Note that the presence of a small biquadratic exchange does not change the nature of the equilibrium states and of the phase transitions, it only affects the curvature of the magnetization curves and changes the values of the critical fields. This is clearly seen in Eq. (6) and in the curve obtained with J_2 = 0, shown by dashed line in Fig.2(a). For this reason, when only the MOKE technique is used to measure the coupling, the uncertainty in the value of H_{ex2} is considerable if its magnitude is much smaller than H_{ex1}. Better accuracy in the determination of all magnetic parameters is achieved with the BLS and FMR techniques. In order to obtain reliable fits of the theory to experimental data it is necessary to measure the variation of some spin wave quantity as a function of a convenient parameter. In the case of FMR one can measure the field for resonance at constant microwave frequency as a function of the angle as the sample is rotated in its plane [5,17]. Alternatively, it is possible to employ several microwave cavities to vary the frequency discretely and obtain the q = 0 spin-wave dispersion relations [29]. However, in both cases it is necessary to observe both acoustic and optic modes in order to extract accurate values for the exchange coupling constants.

Figure 2(b) shows the dispersion relations for the q = 0 modes with the magnetic field $\mathbf{H_0}$ applied along the hard [101] axis. The symbols represent the FMR data and the lines are theoretical results obtained with Eq.(5) with the same parameters used to fit the MOKE data. Like the equilibrium state, the dispersion relations are characterized by three distinct phases: AF for $0 < H_0 < 0.5$ kOe; spin-flop (SF) for 0.5 kOe $< H_0 <$ 2.7 kOe; and saturated or FM for $H_0 >$ 2.7 kOe. As seen in Fig.2(b), the behavior of the frequency vs field for both acoustic and optic modes is different in each phase. This results from the fact that the frequencies of the magnetic excitations depend directly on the configuration of the magnetizations in both films. The dashed lines in the figure are obtained with H_{ex2} = 0, keeping the same values for the other parameters. As expected, this produces a large departure of the optic mode line from the data because the critical field for the SF-FM transition is very sensitive to the value of H_{ex2}. One may ask what happens if we keep H_{ex2} = 0 and vary the other parameters to obtain a least square fit to the data. The result is shown by the dotted lines in Fig.2(b), obtained with $4\pi M$ = 19.5 kOe, H_{ca} = 0.53 kOe, H_{ex1} = – 0.82 kOe and g = 2.1. This shows that although H_{ex2} is small, it is not possible to fit the data to the theory without inclusion of the biquadratic exchange energy in the calculation. The estimated maximum error in the values of the parameters extracted from the fits was 5% for H_{ex1} and 10% for H_{ex2}.

Another way to obtain data with the FMR technique is by keeping the frequency fixed and measuring the fields for resonance for the acoustic and optic modes as the sample plane is rotated about its normal, maintaining the field in the plane. The measured fields variations with angle exhibit four-fold symmetry due to the cubic anisotropy of Fe (100). Figure 3 shows the data obtained with a frequency of 9.5 GHz in one quadrant together with theoretical lines calculated with Eq.(5). The solid lines are obtained with parameters identical to the previous values, namely: $4\pi M$ = 19.5 kOe, H_{ca} = 0.57 kOe, H_{ex1} = – 0.89 kOe, H_{ex2} = 0.07 kOe and g = 2.1. In order to demonstrate the accuracy of the fit we also show in (a) the results obtained with the same H_{ex2} = 0.07 kOe, but with H_{ex1} = – 1.0 kOe (dashed), – 0.80 kOe (dotted) and in (b) the results with H_{ex1} = – 0.89 kOe and H_{ex2} = 0.14 kOe (dashed), 0 (dotted). These results show that in the SF phase the effect of changing J_1 is to shift both resonance fields upwards or downwards. On the other hand, changes in the value of J_2 move the two fields in opposite directions, resulting in larger changes in the separation between the two modes. It turns out that the fact that changes in J_1 and J_2 have distinct effects in the frequencies of the acoustic and optic modes in the AF and SF phases is what allows an accurate determination of these parameters from the FMR and BLS data. Finally we mention that the dispersion relations obtained with the first set of parameters for $q \neq 0$ are also in excellent agreement with the BLS data for this sample.

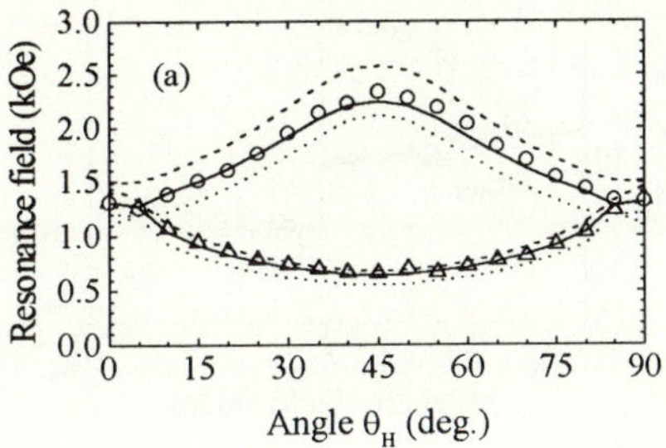

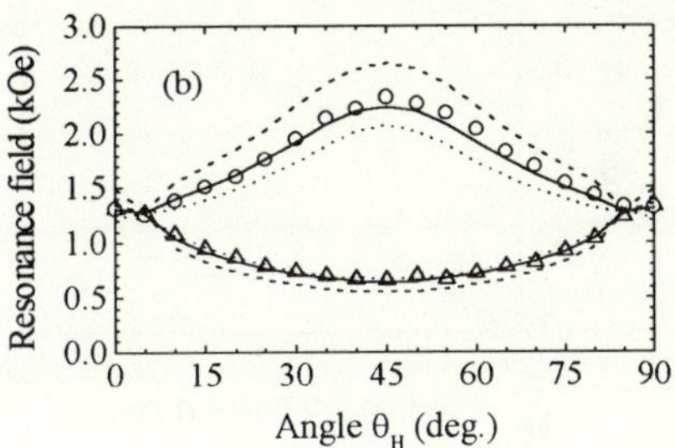

FIG. 3. Comparison between FMR data and theory for the resonance fields vs in-plane angle θ_H in (100) Fe(40Å)/Cr(11Å)/Fe(40Å), as described in the text.

When the sample with Cr thickness $s = 15$ Å was first investigated it was expected to have a small antiferromagnetic coupling and correspondingly small critical fields for all magnetic phase transitions, because this thickness is close to the value where the coupling crosses from AF to FM. However, besides the expected low critical fields, the data furnished by the various techniques displayed surprising sudden jumps as the field varied. This was soon found to be a result of abrupt phase transitions in the magnetic state produced by the relatively large biquadratic exchange [16]. Figure 4(a) shows MOKE measurements in the field range $0 < H_0 < 0.4$ kOe, applied along the easy [100] axis, and the theoretical lines obtained with the parameters $4\pi M = 19.0$ kG, $H_{ca} = 0.55$ kOe, $H_{ex1} = -0.15$ kOe, $H_{ex2} = 0.05$ kOe. These exchange fields correspond to $J_1 = -0.09$ erg/cm^2 and $J_2 = 0.03$ erg/cm^2. The relatively large biquadratic coupling in this sample is responsible for the two first-order phase transitions observed in the equilibrium state. At $H_0 \cong 0.1$ kOe the alignment of the two magnetizations changes from AF to nearly 90° and at $H_0 \cong 0.22$ kOe it changes from 90° to FM. Steps due to first-order phase transitions are also observed in the spin-wave frequency versus field data obtained with BLS, as shown in Fig.4(b). The lines represent the spin-wave frequencies calculated from Eq.(5), using the same consistent set of parameter as in Fig.4(a). In order to investigate further the effect of the biquadratic coupling in the dynamics, we show in Fig.4(b) by the dotted lines the frequencies calculated with $H_{ex2} = 0$ in Eq.(5), keeping $H_{ex2} = 0.05$ kOe in the energy expression (2), as to preserve the correct ground states. As with the previous sample, we found it was not possible to fit the data with theory without inclusion of the biquadratic coupling in the dynamics. We also note that the calculated resonance fields vs in-plane angle using the same parameter set are in excellent agreement with the FMR data.

It turns out that with the $s = 15$ Å sample all three technique provide good accuracy in the measurement of J_1 and J_2. The main reason for this is the existence of the two first-order phase-transitions arising from the relatively large biquadratic exchange. In fact, it can be shown [30] that if H_0 is applied along the easy [100] axis and if the conditions $H_{ca} > 2\,|H_{ex1}|$ and $H_{ex2} \,/\, |H_{ex1}| < 1$ are satisfied, the critical field for transition between the AF phase and the 90-deg phase is given approximately by

$$H_{c1} \cong |H_{ex1}| - H_{ex2} \ , \tag{7}$$

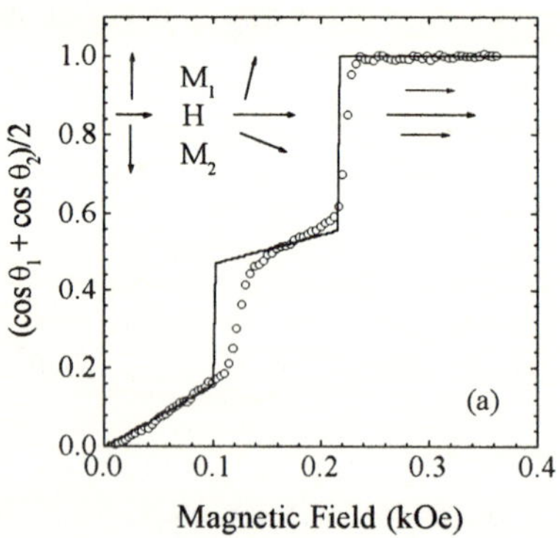

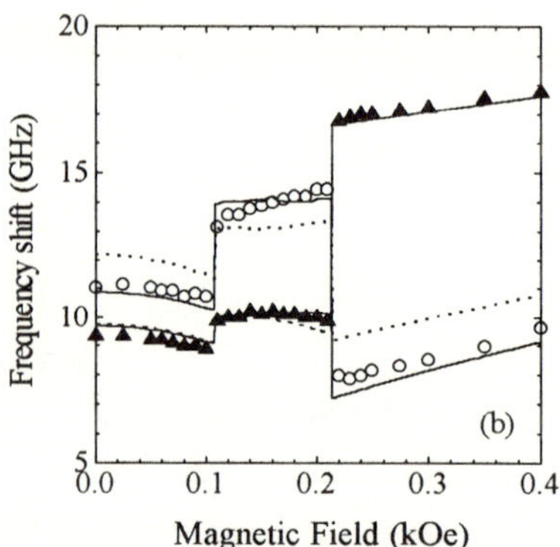

FIG. 4. Data (symbols) and theoretical fits (lines) in (100) Fe(40Å)/Cr(15Å)/Fe(40Å) with $\mathbf{H_0}$ applied along [100]: (a) Normalized magnetization measured by MOKE and (b) spin-wave dispersion relation measured by BLS.

while the transition from the 90-deg phase to the FM phase occurs at

$$H_{c2} \cong |H_{ex1}| + H_{ex2} \quad . \tag{8}$$

Hence, any technique that probes a quantity sensitive to the magnetic state can be used to measure the critical fields H_{c1} and H_{c2}, allowing the two exchange fields to be determined from (7) and (8).

Regarding the samples with Cr layer thickness 25 Å < s < 35 Å, a range corresponding to the second AF peak, the situation is quite different from the previous ones. These samples display only one phase transition, as can be seen in Fig.5(a) for the s = 25 Å sample. This was found to result from a larger biquadratic coupling. In fact, it can be shown [30] that if $J_2/|J_1| > 1$, the AF phase is wiped out, and only the 90-deg and FM phases exist. For $\mathbf{H_0}$ along [100], as is the case of Fig.5(a), the critical field for the transition 90-deg to FM is given by Eq.(8). So, measurement of H_{c2} alone does not allow the determination of J_1 and J_2 separately [30]. In order to obtain both J_1 and J_2 it is necessary to measure another independent quantity that depends on the coupling constants. One may consider measuring $M \times H$ with $\mathbf{H_0}$ applied along the hard [101] axis, where a transition between the spin-flop and the saturated phases occurs at a critical field given by (6). However, this field cannot be measured with good accuracy because of the second-order nature of the transition. Moreover, since the exchange coupling fields are small in samples with Cr layer thickness at the second AF peak, the value of H_{SAT} is dominated by the term $H_{ca} \cong 0.55$ kOe, so the uncertainty in the values of H_{ex1} and H_{ex2} is quite large.

We have found that in samples with $J_2/|J_1| \geq 1$ the first-order transition persists as the field direction is changed between the [100] and [101] axes at quite large angles θ_H , allowing for the precise measurement of a critical field that depends on J_1 and J_2 but is not related to H_{c2}. This is illustrated in Fig.5(b) for the sample with s = 25Å, with the field along $\theta_H = 35^\circ$. So, fitting the theoretical curves to the $M \times H$ data for $\theta_H = 0$ and $\theta_H = 35^\circ$ allows an accurate determination of H_{ex1} and H_{ex2} separately. The best fits shown in Fig.5 were obtained with $H_{ex1} = -0.033$ kOe and $H_{ex2} = 0.0336$ kOe, using the same anisotropy field determined by FMR, namely, $H_{ca} = 0.55$ kOe.

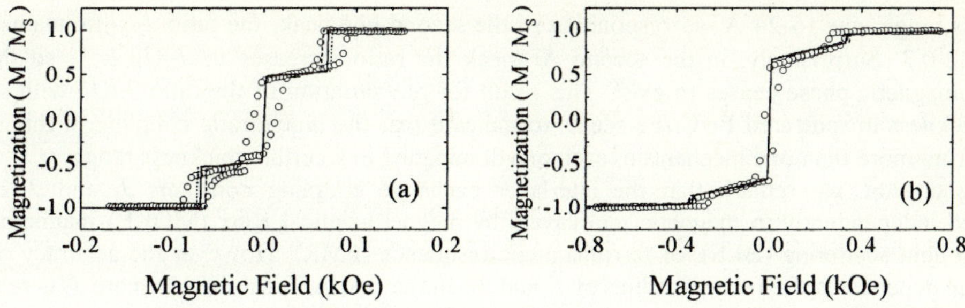

FIG. 5. MOKE data (symbols) and theoretical fits (lines) in (100) Fe(40Å)/Cr(25Å)/Fe(40Å) with (a) $\theta_H = 0$ and (b) $\theta_H = 35°$.

IV. SPACER LAYER THICKNESS DEPENDENCE OF J_1 AND J_2

Figures 6 (a) and (b) summarize the measurements of the exchange coupling constants J_1 and J_2 in sputtered (100) Fe(40Å)/Cr(s)/Fe(40Å) carried out with the three techniques employed in this work, namely, MOKE, BLS and FMR. Note that only the negative values of J_1 are plotted in a log scale. Two AF peaks alternating with one FM peak are observed in J_1 in the thickness range 5Å < s < 35Å, a well known result which has been obtained by many authors. The maximum (absolute) value of J_1 in the first AF peak is 0.59 erg/cm^2, for s = 9.5Å, a value somewhat smaller than $J_1 \cong 1$ erg/cm^2 reported for some MBE grown samples [5,14,29], but similar to those reported for other MBE [13] and sputtered [20] (100) Fe/Cr/Fe trilayers.

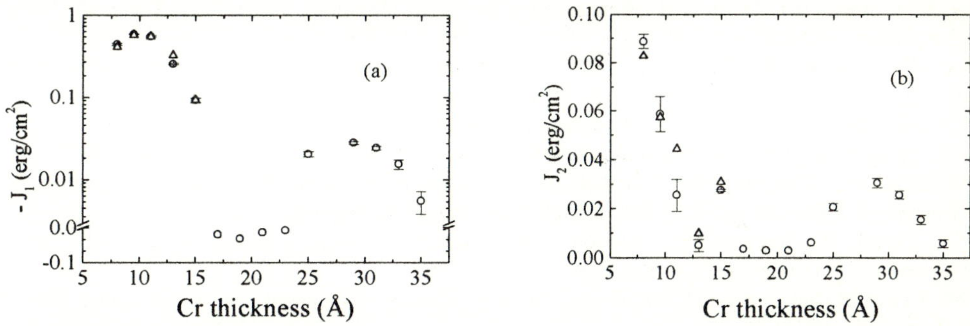

FIG. 6. Exchange coupling constants measured in sputtered (100) Fe(40Å)/Cr(s)/Fe(40Å) by MOKE (circles) and FMR (triangles), as described in the text.

The result for J_2 is not so well known. In fact, to our knowledge, this is the first measurement of J_2 vs spacer-layer thickness in the second FM and AF peaks. The data show that J_2 is positive in the whole range, and that its ratio to $|J_1|$ varies considerably with s. In most of the first AF peak, J_2 follows a dependence with s similar to that of J_1, with $J_2/|J_1| \cong 0.1$. However, near the crossing from AF to FM (s = 15 Å), this ratio increases to $J_2/|J_1| \cong 0.3$, which is similar to that measured in a structure Fe(28Å)/Cr(15.8Å) grown by MBE on a (100) Fe whisker [31]. Throughout the Cr thickness range s = 16-24 Å, corresponding to the second FM peak, the ratio $J_2/|J_1|$ remains in the range 0.2-0.3. Surprisingly, in the second AF peak the ratio increases to $J_2/|J_1| \cong 1$, so that the antiferromagnetic phase ceases to exist. This result for the variation of the ratio $J_2/|J_1|$ with spacer-layer thickness in sputtered Fe/Cr/Fe seems to indicate that the biquadratic coupling in this system results from more than one mechanism, each one dominating in a certain thickness range.

In conclusion, we remark that the interlayer exchange coupling constants J_1 and J_2 can be measured independently in magnetic multilayers by magneto-optical Kerr (MOKE) magnetometry, Brillouin light scattering (BLS), or ferromagnetic resonance (FMR). However the accuracy of each technique depends on the relative values of J_1 and J_2. In the first AF peak of J_1, where J_2 is relatively small, BLS and FMR provide better accuracy than MOKE, provided that both the acoustic and optic modes are detected. For larger ratios, such as in the second AF peak of Fe/Cr/Fe, where $J_2/|J_1| \cong 1$, all three techniques have similar accuracy. In this case the MOKE technique has the advantage of simplicity over the other two.

ACKNOWLEDGMENTS

We thank Dr. S.S.P. Parkin, of the IBM Almanden Research Center, for providing the samples. This work has been supported by the Brazilian federal agencies CNPq, FINEP, PADCT and CAPES and the Pernambuco state agency FACEPE.

[1] P. Grünberg, R. Schreiber, Y. Pang, M.B. Brodsky, and H. Sowers, Phys. Rev. Lett. **57** (1986) 2442.
[2] M.N. Baibich, J.M. Broto, A. Fert, F. Nguyen Van Dau, F. Petroff, P. Etienne, G. Creuzet, A. Friederich and J. Chazelas, Phys. Rev. Lett. **61** (1988) 2472.
[3] See, for example: R.L. White, IEEE Trans. Magn **28** (1992) 2482; G.A. Prinz, Phys. Today **48** (1995) 58.
[4] S.S.P. Parkin, N. More, and K.P. Roche, Phys. Rev. Lett. **64** (1990) 2304.
[5] See, for example: B. Heinrich and J.F. Cochran, Adv. Phys. **42** (1993) 523; *Ultrathin Magnetic Structures*, ed. by B. Heinrich and J.A.C. Bland, Springer-Verlag, Berlin (1994).
[6] Y. Wang, P.M. Levy, and J.L. Fry, Phys. Rev. Lett. **65** (1990) 2732.
[7] D.M. Edwards, J. Mathon, R.B. Muniz, and M.S. Phan, Phys. Rev. Lett. **67** (1991) 493, 1476.
[8] P. Bruno and C. Chappert, Phys. Rev. Lett. **67**, 1602, 2592 (1991); Phys. Rev. **B46** (1992) 261.
[9] J. Unguris, R.J. Celotta, and D.T. Pierce, Phys. Rev. Lett. **67** (1991) 140.
[10] S.T. Purcell, W. Folkerts, M.T. Johnson, N.W.E. McGee, K. Jager, J. aan de Stegge, W.B. Zeper, W. Hoving, and P. Grünberg, Phys. Rev. Lett. **67** (1991) 903.
[11] M. Ruhrig, R. Schäfer, A. Hubert, R. Mosler, J.A. Wolf, S. Demokritov, and P. Grünberg, Phys. Stat. Sol. (a) **125** (1991) 635.
[12] C.J. Gutierrez, J.J. Krebs, M.E. Filipkowski, and G.A. Prinz, J. Magn. Magn. Mater. **116** (1992) L305.
[13] U.Köbler, K. Wagner, R. Wichers, A. Fuss, and W. Zinn, J. Magn. Magn. Mater **103** (1992) 236.
[14] R.J. Hicken, C. Daboo, M. Gester, A.J.R. Ives, S.J. Gray, and J.A.C. Bland, J. Appl. Phys. **78** (1995) 6670.
[15] A.J.R. Ives, J.A.C. Bland, R.J. Hicken, and C. Daboo, Phys. Rev. **B55** (1997) 12428.
[16] A. Azevedo, C. Chesman, S.M. Rezende, F.M. de Aguiar, X. Bian, and S.S.P. Parkin, Phys. Rev. Lett. **76** (1996) 4837.
[17] B. Heinrich, J.F. Cochran, M. Kowalewski, J. Kirschner, Z. Celinski, A.S. Arrot, and K. Myrtle, Phys. Rev. **B44** (1991) 9348.
[18] Z. Celinski, B. Heinrich, and J.F. Cochran, J. Appl. Phys. **73** (1993) 5966.
[19] M. Schäfer, S. Demokritov, S. Müller-Pfeiffer, R. Schäfer, M. Schneider, P. Grünberg, and W. Zinn, J. Appl. Phys. **77** (1995) 6432.
[20] M. Grimsditch, S. Kumar, and E.E. Fullerton, Phys.Rev. **B54** (1996) 3385.

[21] J. Slonczewski, Phys. Rev. Lett. **67** (1991) 3172; J. Magn. Magn. Mater. **150** (1995) 13.

[22] J. Barnás and P. Grünberg, J. Magn. Magn. Mater. **121** (1993) 326.

[23] R.P. Erickson, K.B. Hathaway, and J.R. Cullen, Phys. Rev. B**47** (1993) 2626.

[24] D.M. Edwards, J.M. Ward, and J. Mathon, J. Magn. Magn. Mater. **126** (1993) 380.

[25] See, for example: P. Grünberg, in *Light Scattering in Solids V*, ed. by M. Cardona and G. Güntherodt, *Topics in Applied Physics*, Vol.66, Springer-Verlag, Berlin (1989), Chap.8; M. H. Grimsditch, ibid., Chap.7; J. R. Dutcher, in *Linear and Nonlinear Spin Waves in Magnetic Films and Superlattices*, ed. by M. G. Cottam, World Scientific, Singapore (1994), Chap.6

[26] B. Hillebrands, Phys. Rev. **B37** (1988) 9885; Phys. Rev. **B41** (1990) 530.

[27] J. F. Cochran and J.R. Dutcher, J. Appl. Phys. **64** (1988) 6092.

[28] S.M. Rezende, C. Chesman, M.A.Lucena, A . Azevedo, F. M. de Aguiar, and S.S.P. Parkin, J. Appl Phys, to be published.

[29] J. J. Krebs, P. Lubitz, A. Chaiken, and G. A. Prinz, Phys. Rev. Lett. **63** (1989) 1645; J. Appl. Phys. **67** (1990) 5920.

[30] C. Chesman, M.A. Lucena, A. Azevedo, F.M. de Aguiar, S.M. Rezende, and S.S.P. Parkin, Phys. Rev. **B,** to be published.

[31] M. From, L.X. Liao, J.F. Cochran, and B. Heinrich, J. Appl. Phys. **75** (1994) 6181.

Materials Science Forum Vols. 302-303 (1999) pp. 76-80

Ferromagnetic Resonance in Co hcp and fcc Single Crystal Films on MgO

V.C. Santos[1], A.A.R. Fernandes[1], E.E. Fullerton[2] and C.A. Ramos[3]

[1] Instituto Miliar de Engenharia, Praça Gen. Tiburcio 80, 22290-270 Rio de Janeiro, Brazil

[2] IBM Almaden, 650 Harry Road, San Jose, CA, 95120-6099, USA

[3] Centro Atómico Bariloche and Instituto Balseiro, (8400) S.C. de Bariloche, Ríó Negro, Argentina

Keywords: Magnetic Thin Films, Cobalt, Magnetic Anisotropy

Abstract. Epitaxial Co fcc and hcp thin films were studied by ferromagnetic resonance (FMR) and magnetization (M) measurements. Fcc Co(001) (50 nm thick) was obtained by rf sputtering on MgO(001), having Co [100] || MgO [100] as observed by x-ray diffraction. Using FMR we determined the cubic anisotropy field, K_1^c/M (100K $<$ T $<$ 300K). A variation of over 10% of the effective demagnetization field was observed on going from 100K to 300K. Hcp Co (50 nm thick) was grown by incorporating a 20nm Cr(211) buffer layer on MgO(110). X-ray diffraction and FMR results show that the epitaxial relations are Co($10\bar{1}0$) || MgO(110) and Co [0001] || MgO [001]. From magnetization measurements we obtained the uniaxial anisotropy fields, K_1/M and K_2/M as a function of T.

1 Introduction

Single crystals films of Co can be grown adopting either bcc [1], fcc [2] or hcp [3] structures. The demagnetizing fields, crystalline structure, surface and interface effects determine the characteristics of the magnetization processes and the spin wave excitation spectra [3].

In this work we report magnetization (M(H,T)) and ferromagnetic resonance (FMR) experiments on fcc and hcp Co films. These films were grown under the same conditions as the films used in recent Brillouin [3] (Co hcp) and magneto-optic Kerr effect [2] studies (hcp and fcc films) and we considered of interest to study the anisotropy constants as a function of temperature using FMR and M(H,T) measurements.

2 Experimental

The Co films were prepared using a rf sputtering system. A single crystal fcc Co(100) film, 50 nm thick, was grown on a MgO(100) substrate held at 600°C. A top Cr layer, 1.5 nm thick, was deposited at 300°C to protect the film from oxidation. The Co($10\bar{1}0$) hcp 50 nm film was grown at 300°C over a 20 nm Cr(211) buffer deposited at 400°C on a MgO(110) substrate. A final top layer of Cr, 1.5 nm thick, was deposited.

The samples were characterized by $\theta - 2\theta$ x-ray scans and rocking curves. In Fig. 1a we show the results for the fcc sample. The full width at half maximum (FWHM), $\Delta 2\theta$, gives an estimate of the structural coherence length,ξ, through the Scherrer relation $\xi = 0.9\lambda/(\Delta 2\theta \cos\theta)$, where $\lambda = 0.15406$ nm is the radiation wavelength used. In fcc Co (200) $\Delta 2\theta = 7.2$ mrad, which leads to $\xi_\perp = 25$ nm. No hcp peak was observed. A rocking curve of the Co(200) peak yields a FWHM = 16 mrad, indicating good epitaxial alignment. From transmission x-ray scans we determined that the in-plane epitaxial relation was Co[100] || MgO[100] and $\xi_{||} = 15$ nm. The lattice constant, a, was measured perpendicular and parallel to the film face and we found

$a_\perp$ = 0.35396± 0.00013 nm and $a_{||}$ = 0.35362 ±0.00043 nm. The averege a is slightly smaller than the reported bulk fcc Co value (0.35447 nm).

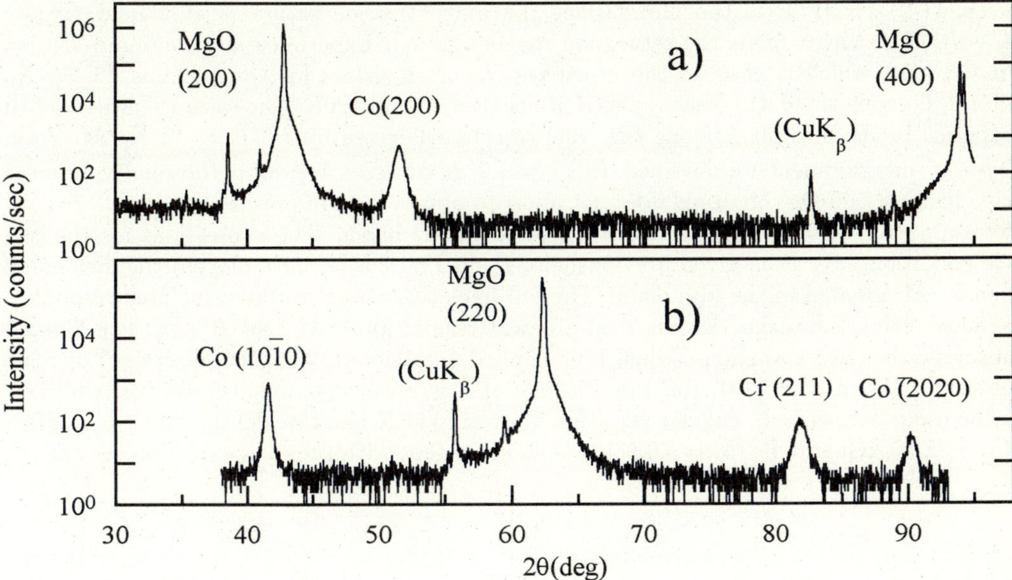

Fig. 1: a) $\theta - 2\theta$ scan of Co fcc thin film (001) on MgO(001). (The small peaks marked CuK_β are from the substrate). b) Idem a) taken on the hcp Co film deposited on MgO(110). The Cr(211) crystalline buffer is also observed.

In Fig. 1b we show the θ–2θ scan on the Cr/Co/Cr/MgO(110) crystal. The Co peak corresponds to the (10$\bar{1}$0) plane. The Cr(211) peak is also clearly observed and confirms recent observations of Lo *et al*[4] showing that the in-plane epitaxial relationship is Co[0001]//MgO[001]. This was also confirmed by magnetization and FMR. From the position of the (10$\bar{1}$0) peak and its width ($\Delta 2\theta$= 5.4 mrad) we obtain a lattice constant of a = 0.2511±0.0002nm, which compares well with the bulk hcp value of a = 0.2507 nm, and $\xi_\perp$ = 27nm. Rocking curve FWHM scan yields $\Delta\theta$ = 30 mrad.
The magnetization measurements were carried out in a SQUID magnetometer. The FMR measurements were carried at X- and Q-band Bruker spectrometer.

3 Magnetic Anisotropy

3.1 fcc Co

The free energy per unit volume that describes the magnetization and FMR measurements can be expressed as:

$$E = -\ (\mathbf{M}.\mathbf{H}) + 2\pi M_s^2 \alpha_3^2 + E_A \tag{1}$$

where the first term is the Zeeman interacction, the second corresponds to the demagnetization energy, and the last term is the crystalline anisotropy given by:

$$E_A(fcc) = K_1^c(\alpha_1^2\ \alpha_2^2 + \alpha_2^2\ \alpha_3^2 + \alpha_3^2\ \alpha_1^2) \tag{2}$$

where K_1^c is the lowest order cubic anisotropy constant and α_i are the direction cosines of **M**. The equilibrium orientation of **M** can be calculated by minimizing eq.(1) with respect to the polar and azimutal angles (θ, ϕ) that define its spatial orientation.
From eq. (1-2) for H $\perp$ to the film surface the magnetization saturates at a field $H_{sat}^{\perp} = 4\pi M_s - K_1^c/M_s$, where M_s is the saturation magnetization. Experimentally we found $H_{sat}^{\perp} = 18.3 \pm 0.5$ kOe, which is close to the previously reported values for Co fcc films [3, 5]. No significant T-dependence was observed in H_{sat} between 10 and 300K. The saturation field when H is applied in the film plane along the [100] direction is given by $H_{sat}^{[100]}$ = - 2 K_1^c/M. From this type of measurement we obtained $H_{sat}^{[100]} = 1.2 \pm 0.3$ kOe. However, the relatively large coercive field ($\approx$ 500 Oe) observed does not allow to obtain a more precise value.
To determine $K_1^c/M(\mathrm{T})$ we performed FMR at X and Q bands. The expressions for the resonance condition have been given previously [6, 7]. In Fig. 2.a we have plotted the dispersion relation for **H** oriented in the film plane. The full lines correspond to the calculated resonance fields values using the magnetization measurement results for $4\pi M_s$ and K_1^c/M_s. For X-band frequencies either two absortions (when **H** is applied very close to the [100] direction) or none are observed. When $H \parallel (010)$ (off the film plane) it is always possible to observe two lines. From the room temperature angular variation, measured at X-band we determined g = 2.14(4), $4\pi M_s$= 17.8(5) kG, and K_1^c/M_s= - 0.53(3) kOe, consistent with the magnetization results.

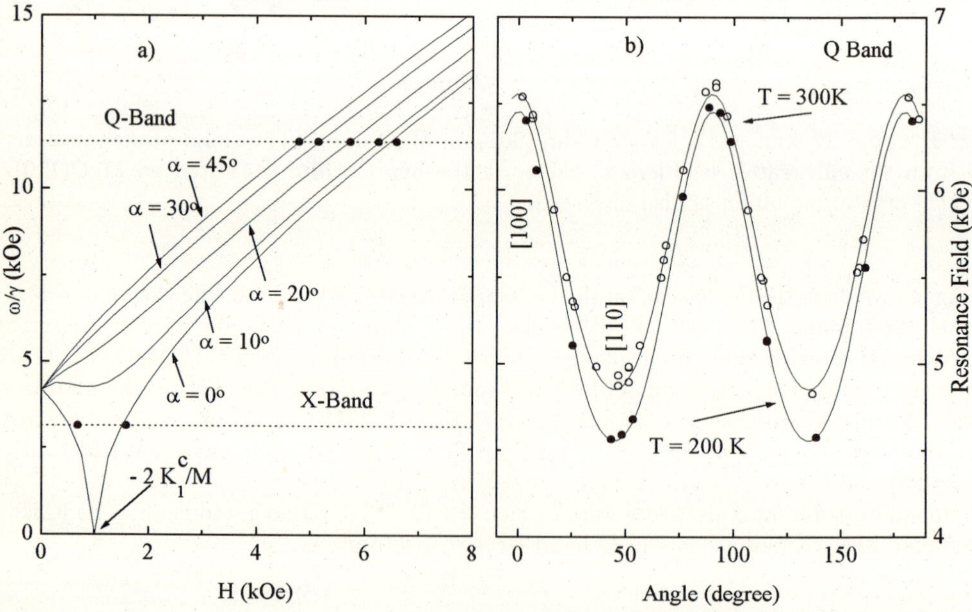

Fig. 2: a) Dispersion relation for Co fcc with H in the film plane (001). Here $\omega = 2\pi\nu$ where ν is the microwave frequency and $\gamma = g\mu_B/\hbar$. b) Resonance field angular variation for **H** in the film at $\nu = 34.2$ GHz.

In order to observe the cubic anisotropy effect on the angular dependence of the resonance field it is convenient to measure with H in the film plane at frequencies higher than the gap (see Fig. 2.a). In Fig. 2.b we present the Q-band results taken at T = 200K and 300K. In Fig. 3.a we have plotted the values obtained for K_1^c/M_s *vs* T. In bulk fcc Co Rodbell [8] found that K_1^c/M_s goes as $[\mathrm{M(T)/M(0)}]^p$ with $p = 8.8$. In Fig. 3 we have fit the T-dependence of K_1^c/M_s assuming

M(T) = M(0) - b $T^{3/2}$ where b was obtained from the M(T) data quoted in Ref. [8]. Within this model (full line in Fig. 3.a) the best fit value of p is $p = 8.7 \pm 1.0$. Yet this agreement may be fortuitous as the T-dependence is poorly reproduced, indicating other mechanisms must be present.

Also a variation of over 10% on going from 100 to 300K is observed in the effective demagnetizing field, $4\pi M_{eff}$ ($\neq 4\pi M_s$ in general, see for example discussion in [7, 9]). An effect of this sort has been observed in Fe/MgO(100) by Goryunov *et al.* [9]. A possible explanation may relay on the appearance of effective in-plane biaxial anisotropy produced by stacking faults. If this biaxial anisotropy is directed along the [100] and [010] directions, the resultant angular variation of the resonant field would still have a four-fold symmetry when H is rotated in the (100) plane but give a larger effective demagnetizing fields. In support of this idea we must mention that Ref. [9] reports Co/MgO(100) grown at T = 300°C in fact grows as a twinned hcp structure, with the c-axes in the film plane.

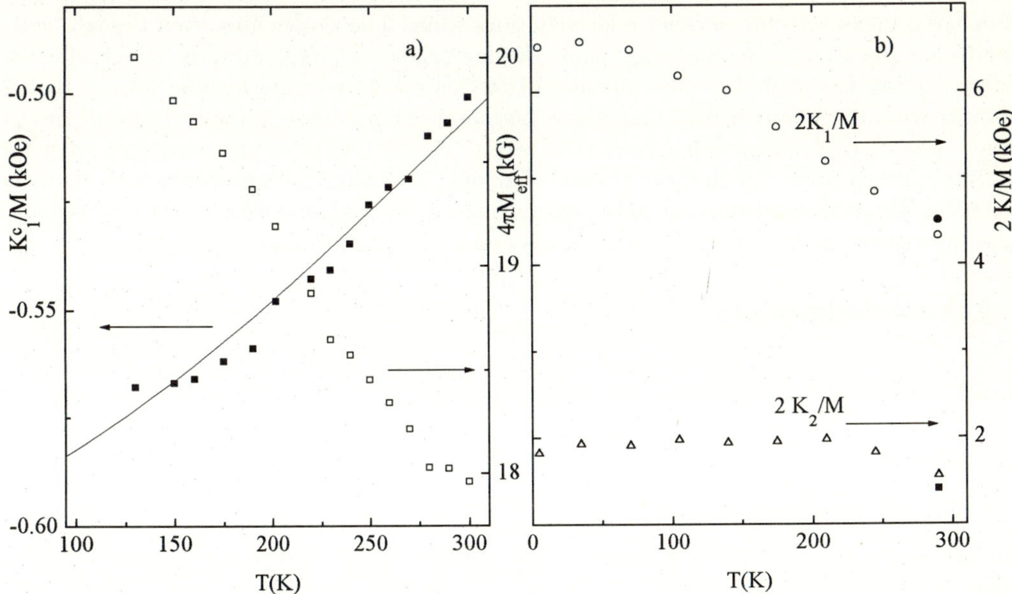

Fig. 3: a) Anisotropy field, K_1^c/M in fcc Co. The full line is explained in the text. We also plotted the effective demagnetizing field, $4\pi M_{eff}$ as extracted from the FMR data. b) Anisotropy fields, $2K_1/M$ and $2K_2/M$ in hcp Co. The empty symbols were obtained from M(H) measurements with H in the film plane and perpendicular to the easy axis. Also shown as full symbols are the data as obtained from FMR measurements.

3.2 hcp Co

The hcp Co sample has the c-axis on the film plane. Considering a set of axes with the z-axis along the c-axis and the y-axis perpendicular to the plane, then the crystalline anisotropy free energy is written as:

$$E_A(hcp) = K_1 \sin^2\theta + K_2 \sin^4\theta \qquad (3)$$

where θ is the angle between **M** and the z-axis and the demagnetization energy is now given by $2\pi M_s^2 \alpha_2^2$. The crystalline anisotropies can be determined by measuring the magnetization along the film plane in the direction perpendicular to the c-axis. Along this direction the magnetization, M, is given by [10]:

$$H = (2K_1/M_s)M/M_s + (4K_2/M_s)(M/M_s)^3 \tag{4}$$

where M_s is the saturation magnetization. Fitting this functional form to M(H,T) we obtained the values plotted in Fig. 3.b. The saturation magnetization measured was $4\pi M_s = 16.6\pm0.8$ kG. In Fig. 3.b we have also included in the uniaxial anisotropy values determined from fitting the FMR. Also from FMR at T = 295K we obtained $4\pi M_{eff} = 16.6\pm0.9$ kG and g = 2.14±0.04 which are in excellent agreement with M(H) and recently published results [3].

4 Conclusions

We have measured the anisotropy constants in 50 nm Co fcc and hcp films in the range 100-300K and 10-300K respectively. The T-dependence of these parameters are, in general, consistent with bulk values o measurements in other relatively thick films. The Co fcc film, even though, looks perfectly homogeneous from the x-ray point of view, shows a large variation of the effective demagnetization field, $4\pi M_{eff}$. We postulate that fcc stacking faults leading locally to hcp structures with the c-axis in the film plane (as observed by other authors [9]) will induce a larger "effective" demagnetizing field. The results for the hcp film are consistent with the recently published data [3]. Here we showed that magnetization measurements with H in the plane of the film and perpendicular to the easy axis leads to the same values for the anisotropy constants as obtained by fitting the angular dependence of the FMR.

5 Acknowledgments

One of us (C.A.R.) would like to thank CONICET (PIA 7102/97), Fundación Balseiro and Compañía de Electricidad Bariloche for finantial support.

References

[1] G.A. Prinz, Phys. Rev. Lett.**54** (1985) 1051.

[2] R.M. Osgood, K. T. Riggs, A. E. Johnson, J. E. Mattson, C. H. Sowers, and S. D. Bader, Phys. Rev. B **56** (1997) 2627.

[3] M. Grimsditch, E. E. Fullerton, and R. L. Stamps Phys.Rev. B **56** (1997) 2617.

[4] C.K. Lo, Y. Liou, C. P. Chang, I. Klik, Y. D. Yao, and J. C. A. Huang, J. Appl. Phys. Lett **68** (1996) 2155.

[5] T. Suzuki, D. Weller, C.-A. Chang, R. Savoy, T. Huang, B. A. Gurney, and V. Speriosu, Appl. Phys. Lett. **64** (1994) 2736.

[6] O. Kohomoto and C. Alexander, Jpn. J. Appl. Phys. **31** (1992) 2101.

[7] R. Naik, C. Kota, J. S. Payson, and G. L. Dunifer, Phys. Rev. B, **48** (1993) 1008.

[8] D. S. Rodbell, J. Phys. Soc. Jpn., **17** (1962) 313.

[9] Yu. V. Goryunov, G. G. Khaliullin, I. A. Garigullin, L. R. Tagirov, F. Schreiber, P. Bödeker, K. Bröhl, Ch. Morawe, Th. Mühge, and H. Zabel, J. Appl. Phys. **76** (1994) 6096.

[10] B. D. Cullity, *Introduction to Magnetic Materials* Addison Wesley, Reading, 1972.

This article was processed using the LaTeX macro package with TTP style

Materials Science Forum Vols. 302-303 (1999) pp. 81-85

Magnetic Properties of Ti/Fe Double Layers Grown on MgO(100) by DC Magnetron Sputtering

J.R. Fermin, A. Azevedo, B. Li, F.M. de Aguiar and S.M. Rezende

Departamento de Física, Universidade Federal de Pernambuco, Recife 50670-901, PE, Brasil

Keywords: Bilayers, FMR Linewidth, Angular Dispersion

ABSTRACT. The purpose of this work is to investigate the magnetic properties of monocrystalline Ti/Fe bilayers deposited by DC magnetron sputtering onto MgO(100) substrates. We have employed in-plane ferromagnetic resonance (FMR) to measure the resonance field, H_R, and linewidth, ΔH, as a function of the azimuthal in-plane angle and the coverage thickness. All measurements were carried out at room temperature and at 12.3 GHz. Ti overlayers with thicknesses, t_{Ti} , in the range 5 Å$<t_{Ti}<120$ Å were deposited onto an single-crystalline Fe film with thickness of about 140 Å. We show that, the deposition of a coverlayer of Ti onto an Fe thin film changes the crystalline and magnetic properties. The deposition of the Ti also provokes angular dispersions on the effective field that broadens the FMR linewidth.

EXPERIMENT. The films were grown by DC magnetron sputtering onto commercially obtained MgO(100) substrates. As is known[1] , magnetron sputtering is a suitable low-cost technique to grow epitaxial Fe films. Prior to deposition the substrates were cleaned in ultrasound baths of acetone and methanol for 10 min each, and then dried in flowing nitrogen. The deposition was carried out in a 3.4×10^{-3} Torr argon atmosphere with a base pressure of 1.0×10^{-7} Torr before sputtering. The purity of the Fe and Ti target was 99.9 % and that of the argon gas was 99.999% . The Fe films were deposited with the substrate at 130 °C, an electrical input power of 22 W for targets with 2 inches diameter. The growth rate was around 1.16 Å/s for Fe and 1.0 Å/s for Ti. Neither buffer layer was deposited onto the substrate nor cover layer onto the Ti overlayer. The thickness of the films was controlled by a crystal oscillator. Both ferromagnetic resonance field (H_R) and the linewidth (ΔH) have been measured as a function of the in-plane azimuthal angle for every sample. We employed a home-made FMR spectrometer in which the microwave power is provided by a synthesized sweep oscillator with the frequency stabilized at the resonance of a TE_{102} rectangular cavity (Q = 2500). The sample was mounted on the tip of an external goniometer that is introduced through a hole in the end of the cavity. Thus, the film can be rotated in the same plane of the back wall of the cavity with respect to an external magnetic field provided by a 9" VARIAN electromagnet. The in-plane field was modulated by Helmholtz coils at 1 kHz. The signal thus detected corresponds to the derivative of the absorbed microwave power. The FMR spectra were obtained at room temperature and at a microwave frequency of 12.3 GHz.

PHONOMENOLOGICAL MODEL. To interpret our data on the resonance field we used a phenomenological model presented previously [2]. The model takes into account a total magnetic free energy density appropriate for a (001)-oriented film, given by

$$E = -HM\sin\theta\cos(\varphi-\varphi_H)+\frac{K_1}{4}\left(\sin^2 2\theta+\sin^2 2\varphi\sin^4\theta\right)$$
$$+K_u\sin^2\theta\sin^2(\varphi-\varphi_u), \qquad (1)$$

where, θ and φ are the polar and azimuthal angles of the magnetization vector, respectively; M is the saturation magnetization; H is the applied magnetic field; K_1 is the first-order cubic magnetocrystalline anisotropy constant; K_u is the in-plane uniaxial anisotropy; φ_H, φ_u are the azimuthal angles of the applied field and uniaxial in-plane axis with respect to the [100] direction, respectively; M_{eff} is an effective magnetization defined by $4\pi M_{eff} = 4\pi M - 2K_n/M$ and K_n is the perpendicular anisotropy.

The ferromagnetic resonance frequency is expressed in terms of the second derivatives of the free-energy density as[3]

$$\left(\frac{\omega}{\gamma}\right)^2 = \frac{1}{M^2 \sin^2\theta}\left[\frac{\partial^2 E}{\partial\theta^2}\frac{\partial^2 E}{\partial\varphi^2} - \left(\frac{\partial^2 E}{\partial\theta\partial\varphi}\right)^2\right]_{\theta_0,\varphi_0} , \qquad (2)$$

where ω is the angular frequency, γ (= 2.98 kOe / kGHz) is the gyromagnetic ratio and φ_0 and θ_0 are the equilibrium positions of the magnetization. For the linewidth we have used the following expression[4]

$$\Delta H = \frac{2}{\sqrt{3}}\frac{1}{|\partial\omega / \partial H|}\frac{G}{M^2}\left(\frac{\partial^2 E}{\partial\theta^2} + \frac{1}{\sin^2\theta}\frac{\partial^2 E}{\partial\varphi^2}\right) + \left|\frac{\partial H_0}{\partial\varphi_H}\right|\Delta\varphi_H + \Delta H_0 \qquad (3)$$

where the first term is the intrinsic contribution or Suhl's term[3], the second term is a contribution from the angular dispersion of the effective field, as defined by Chappert et. al.[5], $H_0 = (\omega/\gamma)_{eq}$, as defined by eq. (1), G is the Gilbert damping parameter, $\Delta\varphi_H$ is assumed to be the angular dispersion of the effective field and ΔH_0 is the frequency independent contribution to the linewidth. Since ΔH_0 is related with the thermal history of the material as well as the preparation process and not with the magnetic properties of the sample itself, we will not attempt to explore this parameter here. The calculations of the frequency and linewidth are based on minimization of the free energy density (1) to determine the equilibrium direction of **M** for each orientation of the applied field, determined by the condition $\partial E/\partial\theta = \partial E/\partial\varphi = 0$. The sample parameters were obtained by least square fittings of the experimental data with Eqs (1) and (3).

RESULTS. The figure 1 shows the in-plane angle dependence of the resonance field for the Fe without Ti overlayer (solid circles) and with Ti overlayers of about 5 Å (solid squares) and 60 Å (solid triangles). The expect four-fold symmetry expect for a (001)-oriented Fe film is observed. A shift of the resonance field to lower values is also observed near the [100] easy axis. The solid lines in figure 1 are theoretical fits to the experimental data. The values of the sample parameters ($2K_1/M_S$; $4\pi M_{eff}$) obtained from our fits are shown in Fig. 2 as a function of the Ti coverage. Capping the Fe with a Ti overlayer induces changes in the crystalline and magnetic properties. The magnetocrystalline anisotropy constant, $2K_1/M_S$, and the effective magnetization, $4\pi M_{eff}$, are increased up to saturation values close to that of the bulk iron. These can be explained as a result of the surface reconstruction favoring the fourfold symmetry and a decrease of the perpendicular anisotropy, $2K_n/M_S$, respectively.

The observed linewidth, ΔH, as a function of the in-plane angle is shown in Fig. 3 for the Fe film with a 5 Å thick Ti coverage. The continous line is a numerical fit with Eq. (3). Figure 4 shows the behavior of the angular dispersion of the effective field, $\Delta\varphi_H$, obtained from these fits, and the average linewidth, $\langle\Delta H\rangle$, as a function of Ti coverage. Both ΔH and $\Delta\varphi_H$ present qualitatively the same in-plane angle behavior, a rapid decrease at low Ti coverages followed by a linear increase at larger Ti thicknesses. This means that the Ti coverage causes an in-plane angular dispersion on the effective field that broadens the FMR linewidth.

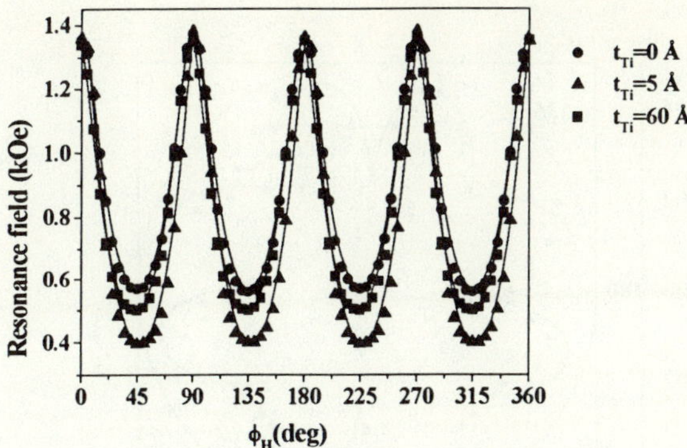

Figure 1. In-plane angle dependence of the resonance field for Fe(140 Å) (solid circles); Ti(5 Å)/Fe(140 Å) (solid squares) and Ti(60 Å)/Fe(140 Å) (solid triangles).

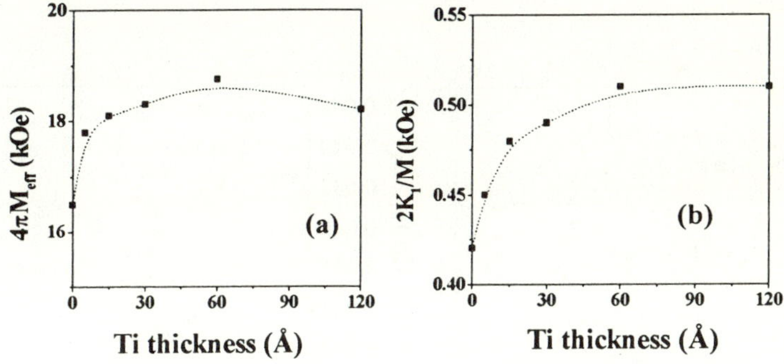

Figure 2. Ti coverage dependence of (a) $4\pi M_{eff}$; and (b) $2K_1/M_S$. The dashed lines are to guide the eye.

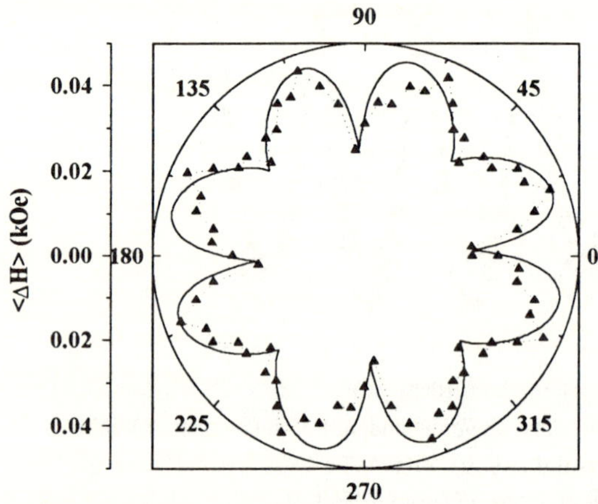

Figure 3. Polar plot of the average in-plane FMR linewidth for Ti(5 Å)/Fe(140 Å). The solid line is a numerical fit using Eq. (3)

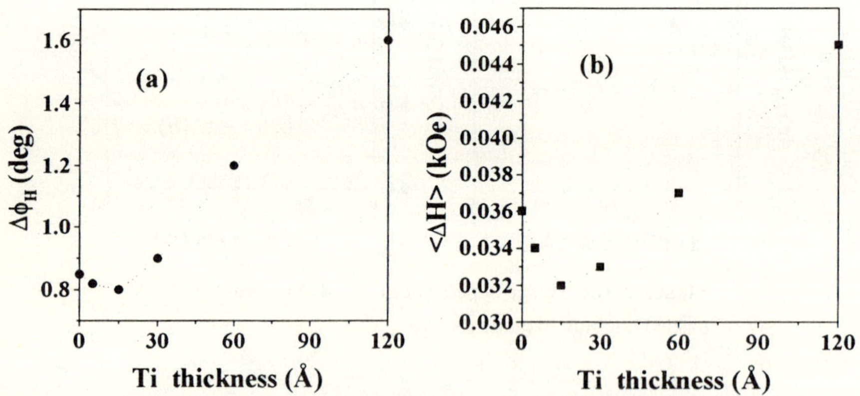

Figure 4. Ti coverage dependence of (a) angular dispersion of the effective field and (b) average FMR linewidth

CONCLUSIONS. In conclusion, we have shown that capping an iron thin film with Ti overlayers can change the crystalline and magnetic properties. The Ti coverage induce surface reconstruction favoring the fourfold symmetry, decreases the perpendicular anisotropy during the Ti deposition. The FMR linewidth as a function of the Ti thickness, can be well explained as resulting from an angular dispersion of the effective field. This dispersion may be associated as an interface effect between the iron film and the titanium coverlayer.

REFERENCES.

[1] Biao Li, J. R. Fermin, Antonio Azevedo, F. M. de Aguiar and Sergio M. Rezende, to be published in Appl. Phys. Lett.

[2] S. M. Rezende, J. A. S. Moura, F. M. de Aguiar and W. H. Schreiner, Phys. Rev. B **49**, 15105 (1994)

[3] H. Suhl, Phys. Rev. **97**, 555 (1955)

[4] Yu. V. Goryunov, N. N. Garif'yanov, G. G. Khallinlin, I. A. Garifullin, L. R. Tagirov, F. Schreiner, Th. Mühge and H. Zabel, Phys. Rev. B **52**, 13450 (1995)

[5] C. Chappert, K. Le Dang, P. Beauvillain, H. Hurdequint and D. Renard, Phys. Rev B, **34**, 3192 (1986)

Materials Science Forum Vols. 302-303 (1999) pp. 86-90

Crystallographic and Magnetic Characterization of CoCr Thin Films

E.F. Chinaglia and I.C. Oppenheim

Instituto de Física, Universidade de São Paulo, C.P. 66318, 05315-970 São Paulo, SP, Brazil

Keywords: Thin Films, CoCr, Crystallographic Characterization, Magnetic Anisotropy

Abstract

Thin films of CoCr were deposited by e-beam evaporation on amorphous substrates (C/Si) and on Ti underlayers (Ti/C/Si). We studied the behavior of the saturation magnetization and the coercive field as a function of the films' composition and the substrate temperature during deposition. We also considered the influence of the Ti underlayer on these properties. The maximum {0002} texture and the best degree of crystallinity were obtained for Ti and CoCr films deposited at 120°C. CoCr films deposited on Ti/C/Si showed a stronger {0002} texture than those deposited on C/Si. The saturation magnetization showed a linear decay with the increase of the Cr at% in the films and this behavior was independent of the substrate temperature during deposition and of the presence of the Ti underlayer. The saturation magnetization values of the CoCr films are larger than those reported for bulk alloys indicating the existence of compositional inhomogeneities in the films. Maximum values of the perpendicular coercive field were obtained for films with 15 to 22 at% of Cr. The texture of the films has a weak influence on the coercive field, indicating that its behavior is more dependent on Cr segregation processes to the grain boundaries. Magnetization curves parallel and perpendicular to the external applied field indicate that the perpendicular magnetic anisotropy is favored for films with compositions around 20 at% of Cr and with a high degree of crystallinity of the (0002) planes.

Introduction

Thin films of CoCr with perpendicular magnetic anisotropy have been studied since the mid 70's because they can be used as media in high density digital magnetic storage devices [1]. The perpendicular magnetic anisotropy of the hcp CoCr thin films are related to the strong crystallographic orientation of the c-axis perpendicular to the surface of the film, to the columnar grain structure of the film and to the degree of Cr segregation to the grain boundaries [2-4]. These films' properties depend on the deposition parameters and on the deposition process itself.

The influence of the c-axis orientation and the compositional inhomogeneities, owing to the Cr segregation, on the magnetic properties of CoCr thin films have been studied by many authors [3-8]. CoCr thin films show a strong {0002} texture when deposited on amorphous surfaces or on Ti {0002} surfaces [5]. Compositional inhomogeneities are generally observed for films with compositions around 18 to 22 at% of Cr and deposited at high temperatures (200 to 300°C).

The goal of this work is to analyze the crystallographic quality and the magnetic behavior of CoCr films deposited by e-beam evaporation as a function of the substrate temperature during evaporation (T_S), the films' composition and the presence of a Ti underlayer.

Experimental Procedure

C, Ti and CoCr films were deposited by e-beam evaporation using an Edwards E-306 deposition chamber with a base pressure of $7x10^{-7}$ Torr. The CoCr films were prepared on Si substrates covered with an amorphous C layer (C/Si), and on C/Si covered with a Ti underlayer (Ti/C/Si). The C deposition rate was less than 0.1 nm/s, and the C layer was thick enough to hide the underlaying Si surface texture. The Ti deposition rate was 0.1 to 0.2 nm/s, and the Ti underlayer thickness was around 40 nm.

$Co_{90}Cr_{10}$ alloys prepared in an arc-melting furnace were used as evaporation sources for the CoCr films. The films were simultaneously deposited on C/Si and Ti/C/Si surfaces. The CoCr deposition rate was 0.3 to 0.4 nm/s and the films' thickness was around 95 nm.

Four series of samples were fabricated. In three of them the Cr content of the films was varied from 0 to 27 at% while T_S was maintained at 120°C (Series S1), 200°C (Series S2), and 280°C (Series S3). In the fourth series the Cr content in the films was maintained around 18 to 22 at% and T_S was varied from 85 to 300°C (Series S4).

The composition and thickness of the films were obtained with Rutherford Backscattering Spectroscopy (RBS) using a $^4He^+$ beam with an energy of 2.0 MeV. θ-2θ X-ray diffraction was performed with Cu-K_α radiation in a Philips - PW1710 diffractometer. The magnetic properties of the films were determined with a Vibrating Sample Magnetometer (VSM) PAR-EGG 4500.

Results and Discussions

The crystallographic analyses of Ti films deposited at different temperatures showed that all films have {0002} preferential texture and that they present a better degree of crystallinity when deposited at 120°C.

Fig. 1 presents the intensity of the {0002} texture of the CoCr films as a function of the Cr content for series S1. The observed {0002} relative intensity of the films is much higher than the expected for powder-Co. Series S1, S2, and S3 showed that films deposited on Ti/C/Si have a stronger {0002} texture than films deposited on C/Si. CoCr (0002) planes are those of minimum surface free energy [5] and Ti (0002) planes favor the epitaxial growth of CoCr (0002) planes since Ti and Co have similar lattice parameters (a_{Ti} = 0.295 nm, and a_{Co} = 0.251 nm).

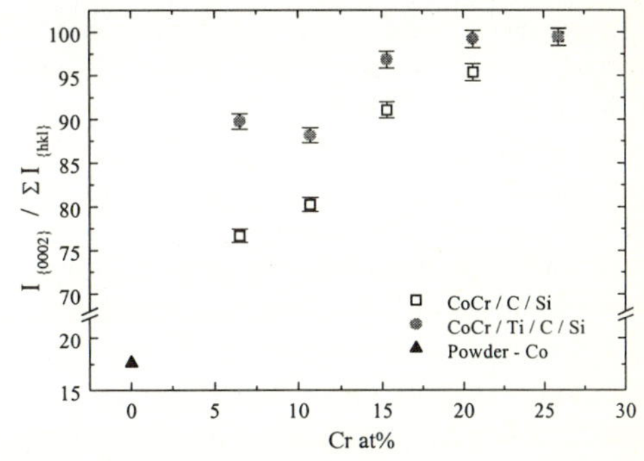

Figure 1: Intensity of the {0002} texture of CoCr films deposited at 120°C as a function of the Cr at% in the films.

In Fig. 2 we show the intensity of the {0002} texture as a function of T_S for samples of series S4. The {0002} texture intensity of films deposited on Ti/C/Si show a weak temperature dependence, and the texture intensity of films deposited onto C/Si decreases as T_S increases. Therefore, the maximum {0002} texture for CoCr on C/Si was observed for T_S = 120°C. The best degree of crystallinity [9] for CoCr on Ti/C/Si was also obtained for T_S = 120°C.

The lattice parameters of the CoCr films (c = 0.44±0.02 nm) and the Ti films (c = 0.50±0.03 nm) calculated from the experimental data are larger than those expected for bulk Co and Ti by 8% and 7%, respectively. In both cases the lattice parameters are temperature independent, and for the CoCr films they do not depend on the films' composition.

Our results indicate that the microstructure formation processes have a stronger temperature dependence at the CoCr/C interface than at the CoCr/Ti interface. The behavior of the {0002} texture intensity with the composition of the films is a function of T_S, and may depend on several factors such as stress states owing to the Cr inclusion, grain interface energies during growth, etc.

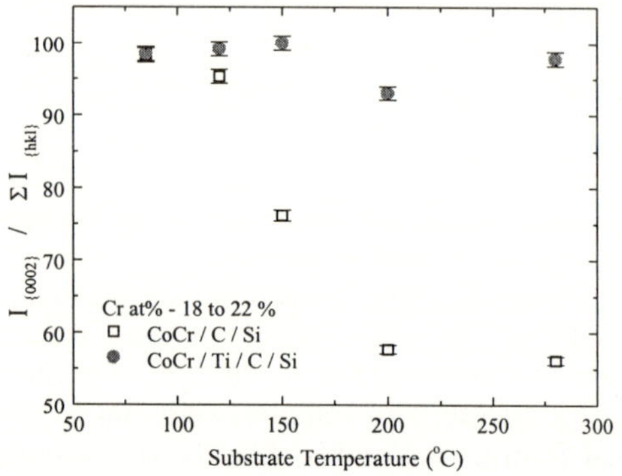

Figure 2: Intensity of the {0002} texture of CoCr films as a function of T_S.

For samples of series S1, S2, and S3 deposited onto Ti/C/Si the saturation magnetization (M_S) decreases linearly with the increase of the Cr content in the films, as shown in curve (1) of Fig. 3. In curve (2) of Fig. 3 we plot M_S for bulk CoCr alloys according to values reported by Farcas [10] and Bozorth [11]. The angular coefficients of curves (1) and (2) are -(42±2) emu/cm^3.Cr at% and -(66.8±0.7) emu/cm^3.Cr at%, respectively.

Bulk CoCr alloys are homogeneous in composition while Cr segregation to grain boundaries is expected for CoCr thin films. Therefore, the number of Co-Cr bonds should be maximum for bulk CoCr alloys. Since it is known that Co-Cr bonds decrease the magnetic moment of Co atoms [12-14], the larger M_S values of curve (1) indicate the presence of Cr compositional inhomogeneities in our CoCr films. The smaller angular coefficient of curve (1) may be the result of an increase of the Cr segregation to the grain boundaries as the Cr content in the films increases.

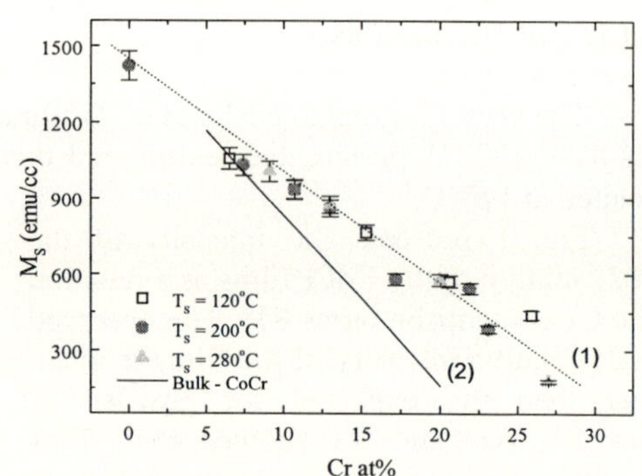

Figure 3: Saturation Magnetization of CoCr films deposited at T_S = 120°C, T_S = 200°C, and T_S = 280°C on Ti/C/Si.

For CoCr films with the same average composition, we did not observe appreciable changes in M_S as a function of T_S and the deposition surface (C/Si or Ti/C/Si). So we may conclude, that these factors do not strongly affect the Cr segregation processes.

In Fig. 4 we show the perpendicular coercive field ($H_{C\perp}$) of the CoCr films as a function of the Cr content in the films for samples of series S1, S2, and S3 deposited on C/Si (Fig. 4a) and Ti/C/Si (Fig. 4b). Maximum $H_{C\perp}$ values were observed for films with 15 to 22 at% of Cr. This result can be explained by considering compositional inhomogeneities of different magnitudes [15]. For films with low Cr at%, there are not enough Cr atoms in the grain boundaries to magnetically isolate the grains and we observe low $H_{C\perp}$ values owing to the granular magnetostatic interactions. As the Cr at% increases, the number of Cr atoms in the grain boundaries also increases leading to a better magnetic isolation of the grains, and therefore to an increase in the $H_{C\perp}$ values. If we further increase the Cr at% we may reach a saturation point of the Cr segregation processes and thus obtain more homogeneous films and stronger intergranular magnetic interactions. We may also be approaching the paramagnetic limit of the material. Both of these factors cause a decrease of the value of $H_{C\perp}$.

Since CoCr films deposited onto Ti/C/Si showed systematically a stronger {0002} texture than films deposited onto C/Si, the deviation of the $H_{C\perp}$ maximum values to lower Cr concentrations, observed in case of Fig. 4b, may be the result of an improved texture and/or the result of an enhancement of the Cr segregation processes owing to the inclusion of the Ti underlayer.

To analyze the influence of the texture on the perpendicular magnetic anisotropy of the films we acquired magnetization curves in the direction parallel (M_X) and perpendicular (M_Y) to the external applied field. In Fig. 5 we show these two components for the external field applied in the direction perpendicular to the surface of the samples.

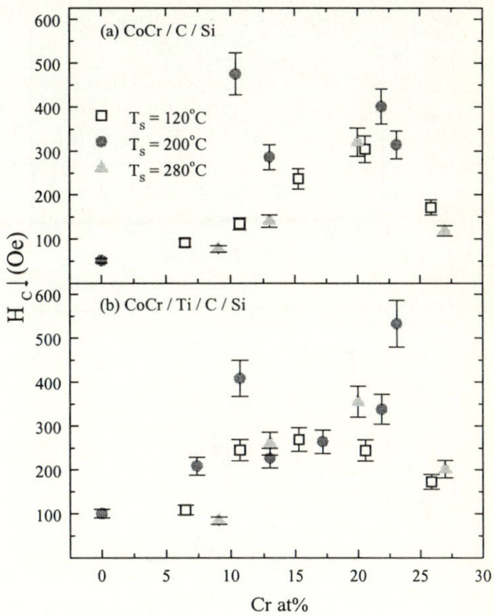

Figure 4: Perpendicular Coercive Field of CoCr films deposited at T_S = 120°C, T_S = 200°C, and T_S = 280°C on (a) C/Si, and (b) Ti/C/Si.

The magnetization component parallel to the plane of the film (M_Y) is strongly dependent on the films' composition. Samples with low Cr at% (Fig. 5a) have high M_Y values that indicate a strong influence of the shape anisotropy of the films on their magnetic behavior. In this case, the perpendicular anisotropy field is not large enough to surpass the demagnetizing field, and the magnetic moments stay parallel to the film's surface. For films with approximately 20 at% of Cr (Figs. 5b, 5c, and 5d) the M_Y component is smaller which indicates that more magnetic moments lay in the direction perpendicular to the surface of the film. The sample deposited with T_S = 120°C (Fig. 5d), has a better degree of crystallinity and a lower M_Y value than the samples deposited with T_S = 200°C (Fig. 5b) and T_S = 280°C (Fig. 5c). Therefore, films with approximately 20 at% of Cr and a good crystallographic quality, have an enhanced perpendicular magnetic anisotropy.

Conclusions

We observed that both the Ti and the CoCr films have a dominant {0002} texture. CoCr films deposited on Ti-{0002} underlayers (Ti/C/Si) are more strongly textured than films deposited on amorphous C (C/Si). Maximum texture and the best degree of crystallinity were obtained for Ti and CoCr films deposited with T_S = 120°C.

M_S decreases linearly with the increase of the Cr at% in the films, and have larger values than those reported for bulk CoCr alloys. The larger M_S values are evidence of the existence of Cr segregation to the grain boundaries in CoCr films. The observed M_S as a function of Cr at% was independent of T_S and the presence of the Ti underlayer.

The behavior of $H_{C\perp}$ shows a stronger dependence on the films' composition than on their texture. This may be an evidence that the intergranular magnetic coupling, directly affected by the Cr segregation processes, is more relevant for determining the $H_{C\perp}$ behavior than the magnetocrystalline anisotropy.

The study of the magnetization curves showed that both the composition and the crystallographic quality of the films are important factors for obtaining films with perpendicular magnetic anisotropy.

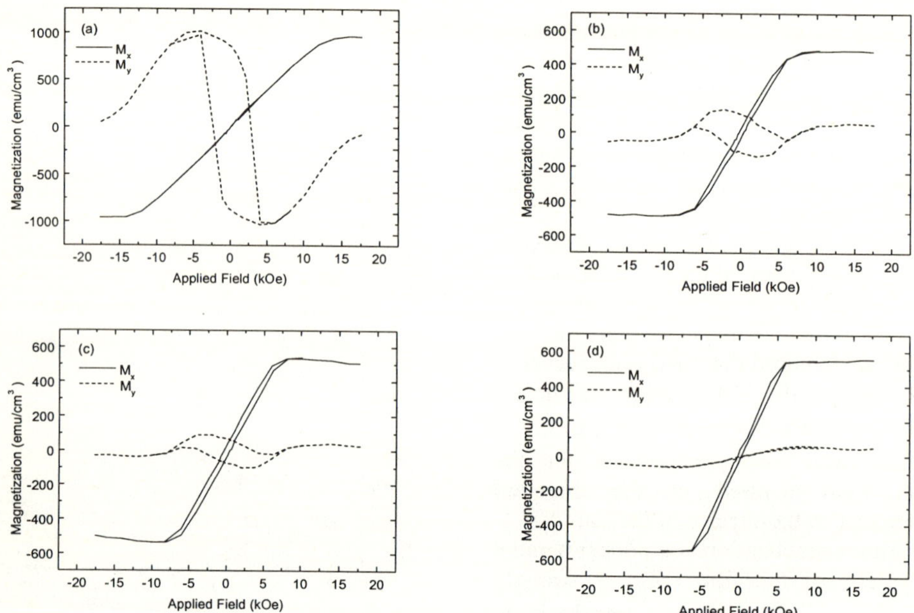

Figure 5: Magnetization curves parallel (M_X) and perpendicular (M_Y) to the applied fields of samples: (a) S2-$Co_{92.6}Cr_{7.4}$ (200 ℃)/Ti/C/Si, (b) S2-$Co_{78.1}Cr_{21.9}$ (200 ℃)/Ti/C/Si, (c) S3-$Co_{80}Cr_{20}$ (280 ℃)/Ti/C/Si and (d) S1-$Co_{79.4}Cr_{20.6}$ (120 ℃)/Ti/C/Si. The applied field was in the direction perpendicular to the samples' surface.

Acknowledgments

We would like to thank Prof. M. C.A. Fantini and Dr. Ernesto Marinero for useful discussions and Prof. A.D. Santos for his help in the laboratory. Financial support was provided by CNPq and FAPESP.

References

[1] S. Iwasaki and Y. Nakamura, *IEEE Trans.Mag., **MAG-13**, (5), (1977), p.1272-1277*
[2] Li Cheng-Zhang, J.C. Lodderaand J.A. Szpunar, *IEEE Trans.Mag., **30**, (4), (1994), p.1373-1379*
[3] Y. Maeda, T. Ohkubo, K.Takei, D.J. Rogers and K.L. Babcock, *J. Mag.Soc.Japan, **19**, (1995), p.706-711*
[4] J.C. Lodder, *J. Magn. Magn. Mater, **159**, (1996), p.238-248*
[5] G. Pan, D.J.Mapps, M.A. Akhter, J.C. Lodder, P. Berge, H.Y. Wong and J.N. Chapman, *J. Magn. Magn. Mater, **113**, (1992), p.21-28*
[6] T. Tanaka and H. Masuya, *Jap.J. Appl. Phys., **26**, (6), (1987), p.897-902*
[7] O. Kitakami, Y. Ogawa and H. Fugiwara, *IEEE Trans.Mag., **24**, (6), (1988), p.2353-2355*
[8] O. Kitakami, K. Ojima, Y. Ogawa, T. Maro and H. Fugiwara, *IEEE Trans.Mag., **MAG-23**, (5), (1987), p.2797-2799*
[9] M.C.A.Fantini, A.Gorenstein, K.Subramanian, N.Mainkar, R.L.Stockbauer and R.L.Kurtz, *Brazilian Journal of Physics, **24**, (3), (1994), p.775-784*
[10] T. Farcas, *Ann.Rev.Phys., **8**, (1946), p.146-151*
[11] R. M. Bozorth, "FERROMAGNETISM", Van Nostrand, *1951*
[12] J.S. Nelson and C.Y. Fong, *J. Appl. Phys., **63**, (8), (1988), p.3049-3051*
[13] F. Bolzoni, F. Leccabue, R. Panizzieri and L. Pareti, *J. Magn. Magn. Mater, **31-34**, (1983), p.845-846*
[14] J.C. Lodder, *Thin Solid Films, **281-282**, (1996), p.474-483*
[15] J.I. Horikawa, M. Kobayashi, K. Moroishi, T. Shimatsu, M. Takahashi, K. Osakabe, S. Nakayama, Y. Yamaguchi, A.J. Ikushima and T. Wakiyama, *IEEE Trans.Journal on Magn. in Japan, **6**, (8), (1991), p.641-647*

Materials Science Forum Vols. 302-303 (1999) pp. 91-95

Structural and Magnetic Properties of Sm/Fe Thin Films

H.J. Santos[1], F.A. Pinheiro[1], L.C. Sampaio[1], A.Y. Takeuchi[1], D.C. Ramos[2], R.A. Simão[2], C.A. Achete[2] and M. Cremona[3]

[1] Centro Brasileiro de Pesquisas Físicas/CNPq, Rua Dr. Xavier Sigaud, 150, Urca, Rio de Janeiro, RJ, 22290-180, Brazil

[2] Programa de Engenharia Metalúrgica e de Materiais, COPPE, UFRJ, Ilha do Fundão, Rio de Janeiro, RJ, Brazil

[3] Dept. de Física, PUC, 22452-970, Rio de Janeiro, RJ, Brazil

Keywords: Magnetic Thin Films, Sm Films, In-Plane Magnetic Anisotropy

Abstract – Thin films of quartz/Sm(250 Å)/Fe(150 Å) have been studied by grazing incidence X-ray diffraction, Atomic Force Microscopy and longitudinal magneto-optic Kerr effect between 78 K and room temperature. Both the Sm- and Fe-layers exhibit growth in a preferential direction and a texture effect was observed. The Fe-layer is composed of small elongated grains, 60nm long and 20nm wide. These grains are monodomains and produce a uniaxial in-plane magnetic anisotropy in the film. The Sm-layer exhibits magnetic order at 140 K and aligns antiparallel to the Fe-layer.

I. Introduction

Over the last years the growth of epitaxial magnetic thin films on single crystal substrates has permitted the study of a wide variety of magnetic properties [1]. Some work has been done in rare-earth/transition metal bilayers, mainly in Fe/Gd [2], in order to investigate the magnetic order in the layers as well as coupling between them. More recently, a growing interest has been directed to the construction of magnetic and magneto-electronic devices using thin films [3]. Despite the fact that the majority of magnetism groups concentrate their research to obtain samples with high quality surfaces and interfaces, it is equally meaningful to direct efforts to obtain suitable properties working with simplest handle samples. In the present paper, we will report structural and magnetic properties of quartz/Sm(250 Å)/Fe(150 Å) where a strong uniaxial in-plane magnetic anisotropy was found, and which makes this system very promising for future technological applications.

II. Structure analyses

Films of quartz/Sm/Fe have been grown in an Ultra High Vacuum (UHV) chamber with a base pressure lower than $2x10^{-9}$ Torr. Both Sm and Fe were evaporated using the commercial Omicrom electron-beam evaporator in low deposition rate (2 ML/min). The quartz substrates were cleaned in acetone before loading to the chamber and by annealing at 600 K for one hour in UHV conditions. Firstly the Sm-layer was evaporated onto quartz substrate at room temperature followed by a fast heating up to 900 K. Afterwards the Fe-layer was evaporated onto the Sm one at room temperature. In this paper we will concentrate our discussions on the system composed of Sm(250 Å)/Fe(150 Å) without loss of generality. The characterization of films was made *ex-situ* through grazing incident (~1 degree) and large angle X-ray diffraction and Atomic Force Microscopy

(AFM). The AFM analyses were done using a commercial system (Topometrix, Discoverer model) in the non-contact mode.

Analyzing the 100 and 110 reflections for Sm- and Fe-layers, respectively, on grazing incidence and rotating the film around an axis perpendicular to the film plane an interesting picture arise – the peak intensity exhibit a periodic behavior. It means the Sm-layer deposition lead to preferential growth along a crystallographic orientation. Furthermore, a texture effect was observed (more details will be published elsewhere). AFM measurements provided the mean size for Sm grain of about 80 nm. The Fe-layer also exhibit a grain structure dominated by the texture effect. AFM analyses show that the Fe grains are elongated and oriented in a preferential direction (see Fig. 1). These grains are approximately 60nm long and 20nm width. It is important to note that the preferential growth and texture for Fe-layer is induced by the textured Sm-layer, therefore inducing an anisotropy for the Fe grain shape. This account is in agreement with X-ray measurements at grazing incident angle.

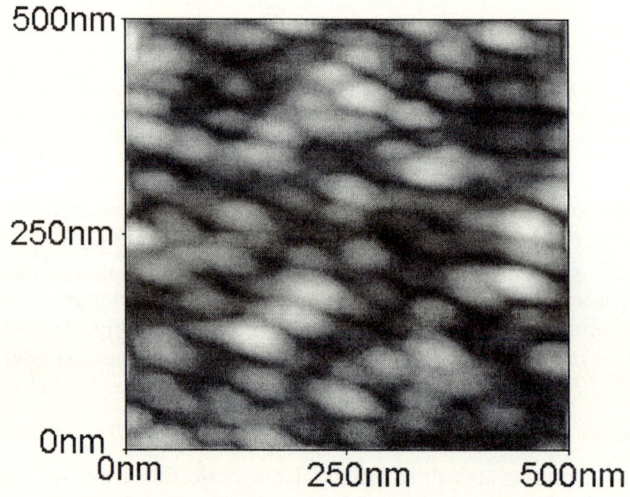

Figure 1 – AFM image for Fe-layer. The grains are 60nm long and 20nm wide.

III. **Magnetic measurements and Discussions**

Magnetization measurements were performed *ex-situ* through longitudinal magneto-optic Kerr (MOKE) effect between 78 K and room temperature. Our experimental setup is composed of a diode laser emitting red light (wavelength of 670 nm), the polarizer and the analyzer were arranged almost crossed and the magnetic field is applied in the plane of incidence. The incident light was linear p-polarized and lock-in detection with light amplitude modulation was used.

Hysteresis loops were performed at room temperature with the magnetic field applied in the film plane and forming an angle (θ) respect to an arbitrary direction in the sample. Figure 2 show typical hysteresis loops for two almost orthogonal orientations - this figure clearly shows the existence of a magnetic anisotropy in the film plane. Considering that the Sm-layer is not magnetically ordered at room temperature, the magnetic measurement is attributed to the Fe-layer only.

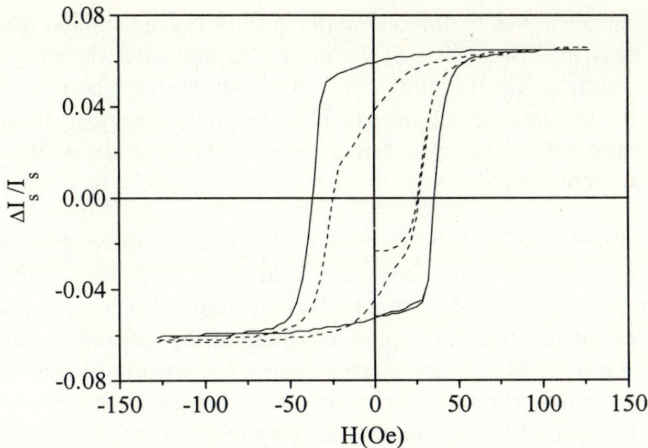

Figure 2 – Hysteresis loops measured at room temperature for the magnetic field applied in the film plane on different orientations, θ=180° (--) and 315 ° (–).

In order to investigate in more details the magnetic anisotropy we show in Figure 3 the coercive field (H_C) at room temperature for the field applied in different orientations. The coercive field follows a sinusoidal dependence on θ, oscillating between 23 and 38 Oe and has the periodicity of 180 degree. In fact, this periodicity agrees with AFM measurements where elongated particles exhibit the same symmetry. Hence, we attribute the observed magnetic anisotropy to the shape anisotropy. In addition, this idea is convincing because the size of Fe grains observed here is very near to the critical value for Fe (~60nm as mean diameter) for which a magnetic particle becomes monodomain. If the Fe grains are greater than the critical value, magnetic domains will be present inside the grains and no asymmetry in magnetic properties should be observed. Concluding, this microstructure produces a uniaxial magnetic anisotropy in the film.

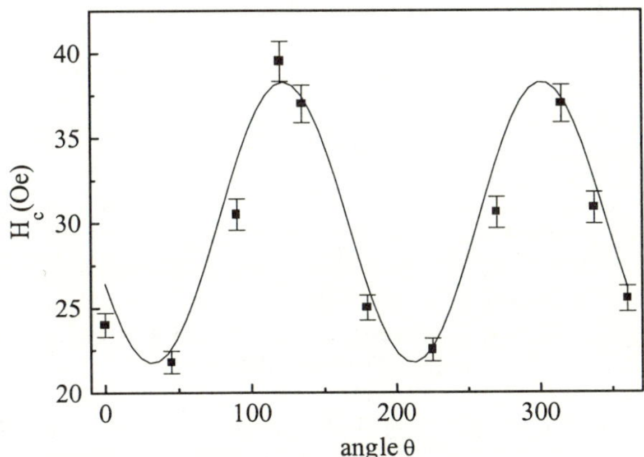

Figure 3 - Coercive field (H_c) at room temperature for magnetic field applied in different directions. The line is a sinusoidal fitting.

In the light of our model we can discuss in more detail the hysteresis loops. The squareness behavior typical of angles where H_C has the maximum value, e.g., in Fig. 2 for θ =315 degree, is characteristic of magnetization curves in monodomain particles for the applied magnetic field in the direction parallel to the easy axis (long axis of the grain). In this case, the magnetization reversal occurs through coherent rotation. On the other hand, for orientations where H_C has the minimum value, e.g., in Fig. 2 for θ =180 degree, the hysteresis curve should correspond to the hard axis. In that case, the magnetization rotates up to a critical value, and just above it, the magnetization suddenly turns to the field direction [4].

Concerning the influence of the Sm-layer on the magnetic properties we have measured the dependence of the magnetization on temperature at the saturation field. Since MOKE measurements do not provide the absolute magnetization value, we have measured the relative variation of light intensity for positive and negative values of the saturation field, $\Delta Is \equiv I(Hs)-I(-Hs)$. In Figure 4, from the plot of $\Delta Is/Is$ versus temperature, one clearly observes a magnetic phase transition (T_{Sm}) at 140 K which should correspond to the magnetic ordering of Sm, although, according to the literature bulk Sm exhibits the magnetic order at 106 K. This apparent disagreement can be attributed to the stress effect on the Sm/Fe interface. Turning now to a second interesting point, the reduction in magnetization below T_{Sm}, we consider that both Sm and Fe contribute to the magneto-optic measurement in spite of the fact that each one contributes with different weights. The rare-earth metals are also sensitive in the magneto-optic measurement, for instance, MOKE measurements carried out in Gd thin films have been reported by several authors [5]. In bulk Sm below the ordering temperature, the spins have a complex antiferromagnetic ordering and it is not straightforward to extend the same behavior to thin films. From our magnetic measurements, we could suggest that spins in the Sm-layer are ferromagnetically ordered and coupled antiparallel to the Fe-layer. Nonetheless, further work is needed to clarify this point.

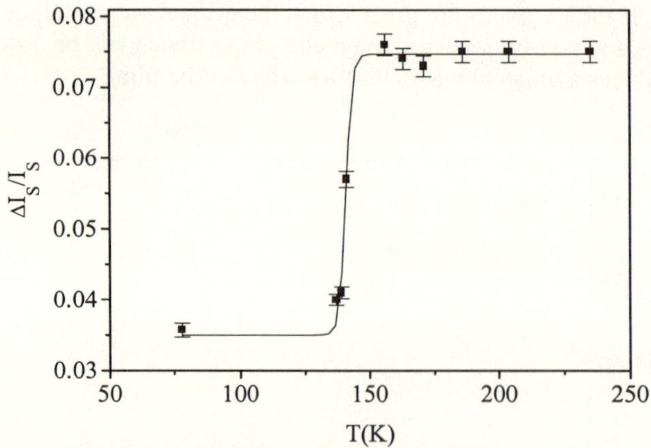

Figure 4 – Temperature dependence of ΔIs ($\equiv I(Hs)-I(-Hs)$). Hs is the saturation field. The line is a guide to the eyes.

Summarizing, structural and magnetic analyses performed on quartz/Sm(250 Å)/Fe(150 Å) thin films have shown unexpected properties. A texture effect for both the Sm- and Fe-layers was detected what produces in the microstructure of the Fe-layer an advantageous component, i.e., the presence of elongated grains (60nm long and 20nm width) which are magnetic monodomain. The magnetic shape anisotropy of the grains gives rise to a uniaxial in-plane magnetic anisotropy in the film. This property seems to be very promising for use in technological applications, mainly considering the relative simplicity of growing samples on quartz instead of single crystal substrates.

Acknowledgments
The authors are grateful to A. Domingues dos Santos and A. Berger for helpful discussions on the MOKE setup. This work is partially supported by CNPq.

References

[1] B. Heinrich and J.A.C. Bland (Eds), Ultrathin Magnetic Structures, Springer-Verlag, 1994, Chap. 5.
[2] L.T. Baczewski, R. Kalinowski and A. Wawro, J. Magn. Magn. Mat. 177, 1305 (1998), Y. Li, C. Polaczyk, F. Mezei, D. Riegel, Physica B 234, 489 (1997), N. Cherief, D. Givord, O. McGrath, Y. Otani and F. Robaut, J. Magn. Magn. Mat. 126, 225 (1993).
[3] M. Zolfl, M. Brackmann, M. Kohler, S. Kreuser, T. Schweinbock, S. Miethaner, F. Bensch, G. Bayreuther, J. Magn. Magn. Mat. 175, 16 (1997) and references therein.
[4] C. Daboo, R.J. Hicken, E. Gu, M. Gester, S.J. Gray, D.E.P. Eley, E.Ahmad, J.A.C. Bland, R. Poessly and J.N. Chapman, Phys. Rev. B 51, 15964 (1995).
[5] M. Gajdzik, T. Trappmann, C. Surgers, H. von Lohneysen, Phys. Rev. B 57, 3525 (1998), A. Berger, A.W. Pang, H. Hopster, Phys. Rev. B 52, 1078 (1995).

Materials Science Forum Vols. 302-303 (1999) pp. 96-100

Hysteresis and Interactions in Nanoparticulate Fe Films

J.M. González[1], M.I. Montero[1], L. Vázquez[1], J.M. Martín Gago[1], D. Givord[2], C. de Julián[2] and K. O'Grady[3]

[1] Instituto de Ciencia de Materiales de Madrid-CSIC, Cantoblanco, ES-28049 Madrid, Spain

[2] Laboratoire de Magnetisme Louis Néel-CNRS, BP 166X, Grenoble cedex, France

[3] School of Electronic and Engineering Systems, UCNW, Bangor, LL57 ITU, UK

Keywords: Pulsed Laser Ablation, Nanoparticulate Fe Films, Coercivity, Remanence, Magnetic Interactions

Abstract- We report on the preparation and properties of granular Fe films. The films were grown by using the pulsed laser ablation technique and characterized by means of atomic force microscopy and alternating gradient field and vibrating sample magnetometries. Our results showed that the average Fe particle diameter varied from 13 nm up to 25 nm when the deposition time was increased. The corresponding room temperature coercive force values took values in the range from 240 Oe down to bulk-like ones. Finally, we evidence, from the results of the field evolution of the demagnetization and isothermal remanences, that the interactions play a crucial role on determining the film properties.

1. Introduction

Since the early 1950s magnetic systems formed by particles having sizes of the order of some nm have generated a great deal of work [1] both from the fundamental point of view, (problems related to the influence on the extrinsic behavior of the reduced size and the surface effects) and from that the technological applications, problems related to the development of magnetic recording media. Regarding the development of high density media it is particularly interesting the growth of nanoparticulate thin films having large magnetization and coercivity. Optimized values of both parameters allow to take advantage of the reduced size of the magnetic entities in order to achieve large signal-to-noise ratios and high media stability against accidental (field or thermally induced) demagnetization. Nanoparticulate thin Fe films, candidate to meet these requirements, have been previously prepared by the so-called gas-deposition method (GDM) [2] in which ultrafine particles are produced through the condensation of a (thermally) evaporated metal around inert gas atoms and the ulterior coalescence process of the initially formed clusters. After their formation, the ultrafine particles, having dimensions in the 10-20 nm range, are injected (by differential pumping) over a substrate. An alternative technique to produce thin film nanoparticulate deposits is the pulsed laser ablation which has already been used to prepare ferrite films [3].

In the present work we will report on the preparation, by pulsed laser ablation in low pressure Ar atmosphere, of thin nanoparticulate Fe films. Data on the structural, morphologic and magnetic properties of these films will also be presented.

2. Preparation of samples and experimental techniques

The deposition chamber we used to grow our nanoparticulate films is presented in Fig. 1. A laser beam of $\lambda = 532$ nm, generated by a $Nd3^{+}$:YAG laser emitting 9 ns pulses at a frequency of 10 Hz, was focussed onto a polished, pure Fe target (20 mm x 20 mm x 10 mm). The laser beam, incident at an angle of 45° with the target normal, was rastered, by means of a vibrating mirror, over different rectangular 2 mm x 3 mm regions. The effective power density of the radiation was estimated in 10^{13} W/m^2. The deposition proceeded in a chamber with ultrapure Ar atmosphere (at a pressure of 40 mTorr), previously the chamber was evacuated down to a pressure of 10^{-6} Torr. In order to avoid any large particle in the sample (which could completely mask the magnetic behavior of the deposit) we used an "off-axis" deposition geometry: the substrates (corning glass, 15 mm diameter, kept at room temperature) were located, at the plane containing the illuminated surface of the target (the distance between the centers of the target and the substrate was of 30 mm). The deposition times ranged from 5 up to 45 minutes. Two series of samples were prepared with (samples on which the magnetic characterization was performed) and without (samples used on the morphology study) a protecting capping Cu layer (in order to grow this capping layer a pure Cu target was located beside the Fe one).

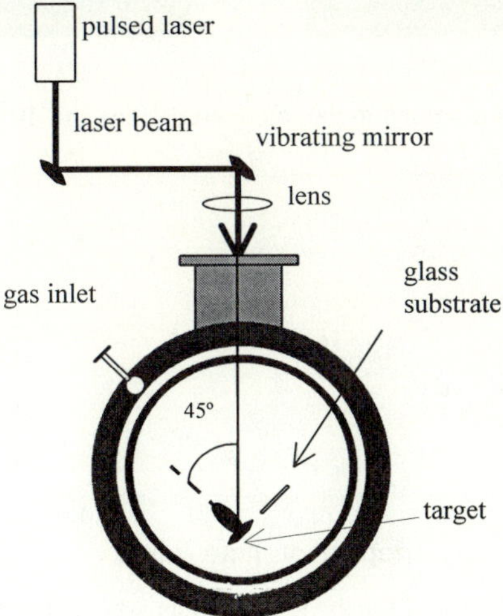

Figure 1. Schematic diagram of the deposition chamber.

The morphologic and structural study of the films was performed by means of an atomic force microscope (AFM). For the magnetic characterization were used an alternating gradient field magnetometer, as well as a vibrating sample one (in which film stacks were measured).

3. Experimental results

We show in Fig. 2 AFM images obtained in the samples deposited without the capping Cu layer for 10 and 40 minutes. The images evidence that even for short deposition times, a quite uniform coverage of the substrate surface was achieved. The deposited particles were quasi-spherical in shape and very homogeneous in size. The average particle diameter, d (measured in regions of the deposits which were located at the same distance from the target center during the deposition process) increased with the deposition time (see Figure 3) from 13 ± 2 nm, in the case of the sample deposited for 10 minutes , up to 25 ± 4 nm in that of the sample deposited for 40 minutes. The increase of the average particle dimensions with the deposition time indicated the occurrence of in-substrate particle growth propitiated by the deposition chamber atmosphere which allowed some degree of thermal contact between the laser-induced plasma and the substrate. Due to their granular nanostructure, the film thickness strongly fluctuated but in all the samples its average value was of the order of the corresponding average particle size.

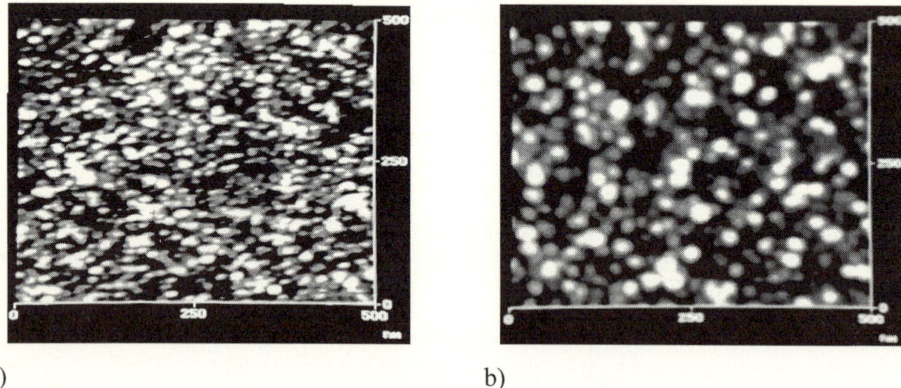

a) b)

Figure 2. AFM images obtained in the uncovered 10 a) and 40 b) minutes deposited samples.

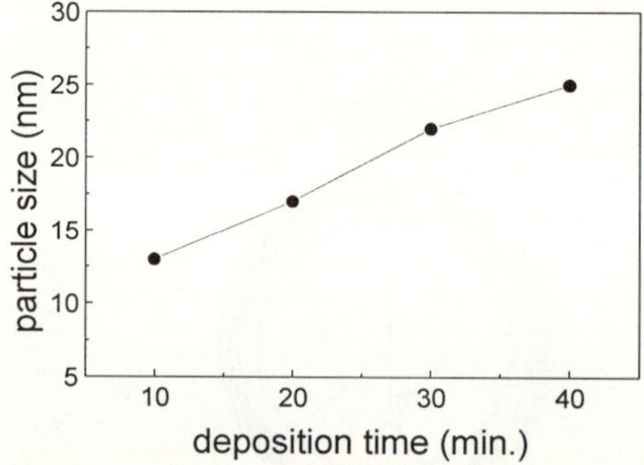

Figure 3. Evolution of the average particle size with the deposition time.

The hysteresis loops measured, at room temperature, in the Cu-layer capped films (deposition times from 5 up to 20 minutes) evidenced the occurrence of coercivity in all the samples (see Figs 4 to 7 where the loops measured in the samples deposited for 5, 10, 15 and 20 minutes, respectively, are shown). Fig. 8 shows the deposition time dependence of the coercivity of the samples which decreased with the increase of the deposition time from 235 Oe down to a Fe bulk-like value of 19 Oe. The coercivity value obtained in the sample deposited for 5 minutes is larger than the maximum coercivity reported in [2] for GDM deposited samples which can be related both to the smaller average particle size of our sample and to the absence on it of large particle agglomerates. The occurrence of giant coercivity (coercive force values not compatible with the bulk magnetocrystalline anisotropy) is linked in our samples to average particle dimensions smaller than 15 nm. Let us point out here that a rough estimation of the average blocking temperatures, T_B, corresponding to the measured average particle diameters ($T_B = (4\pi/3)K(d/2)^3/25k_B$, where we have considered non-interacting particles and a bulk Fe anisotropy constant K given by 4.8 x 10^5 erg/cm^3 [4]) , yields values of 160 K, in the case of the 13 nm particles (film deposited for 10 minutes), and of 1140 K in that of the 25 nm ones (40 minutes deposited film). Since our films exhibited quite a narrow particle size distribution, we should associate the observation of room temperature coercivity in the samples deposited for a short time to the occurrence of magnetic interactions.

The temperature dependence of the coercive force of the 20 minutes deposited film is presented in Fig. 9, where it is possible to observe a steady increase of that quantity with the decrease of the temperature. It is worth to mention that the maximum measured coercivity (460 Oe at 80 K) is larger than those obtained in Fe-Cu or Fe-SiO_2 nanocomposite samples [5].

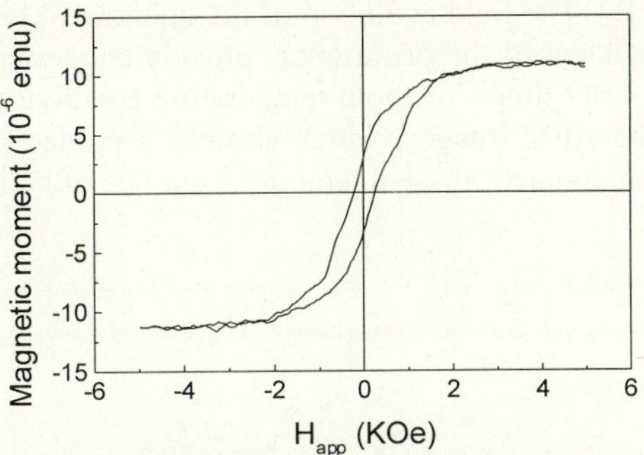

Figure 4. Hysteresis loop measured, at room temperature, in the sample deposited for 5 minutes.

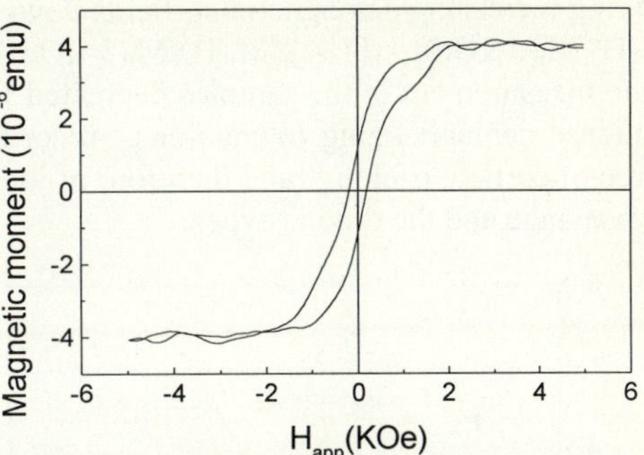

Figure 5. Hysteresis loop measured, at room temperature, in the sample deposited for 10 minutes.

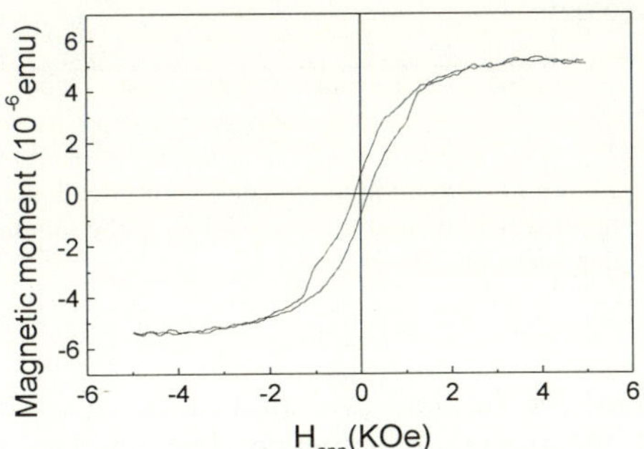

Figure 6. Hysteresis loop measured, at room temperature, in the sample deposited for 15 minutes.

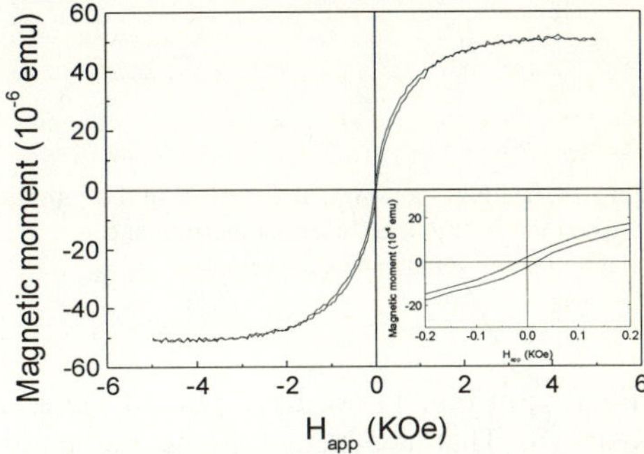

Figure 7. Hysteresis loop measured, at room temperature, in the sample deposited for 20 minutes.

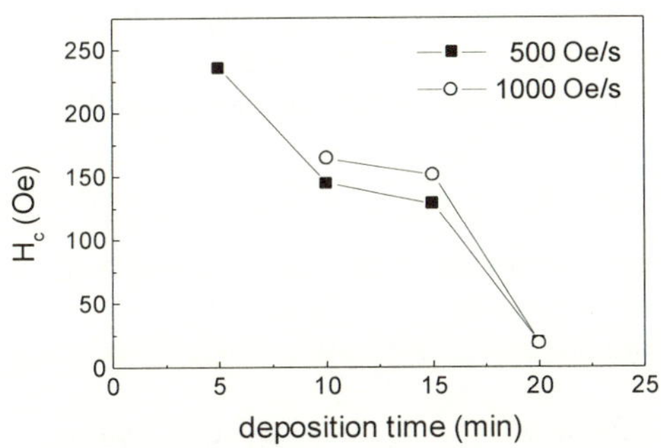

Figure 8. Evolution of the coercive force with the deposition time measured at two field sweep rates.

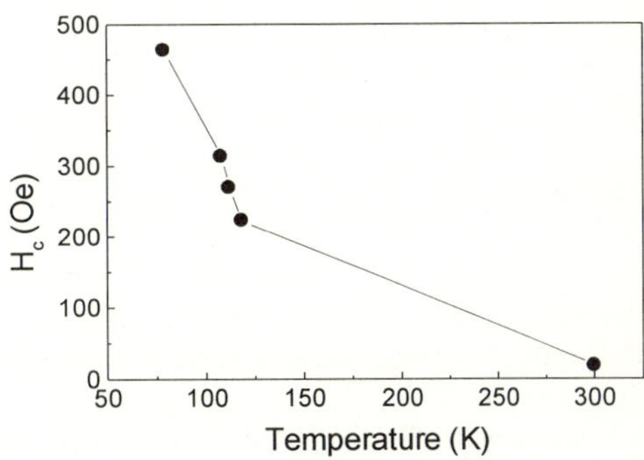

Figure 9. Temperature dependence of the coercive force in the sample deposited for 20 minutes.

In Fig. 10 we present our results for the field evolution of the isothermal, M_i, and demagnetization, M_d, remanences corresponding to the film deposited for 20 minutes (M_i was measured by increasing, up to 5 kOe and from the demagnetized state, the maximum applied field and M_d was obtained by first saturating the sample under a magnetizing 5 kOe field and then applying increasing demagnetizing fields down to -5 kOe). The field evolution of the quantity $\delta M = [M_d(H)/M_i(5\ kOe)] - \{1 - 2[M_i(H)/M_i(5\ kOe)]\}$, [6]) evidenced the occurrence, already suggested by the measurement in the samples deposited for the shorter times, of room temperature coercivity, of intense demagnetizing interactions. According to the AFM images, which showed a moderate degree of particle packing (and therefore of intergranular contact), these interactions may be of both the exchange and the dipolar types.

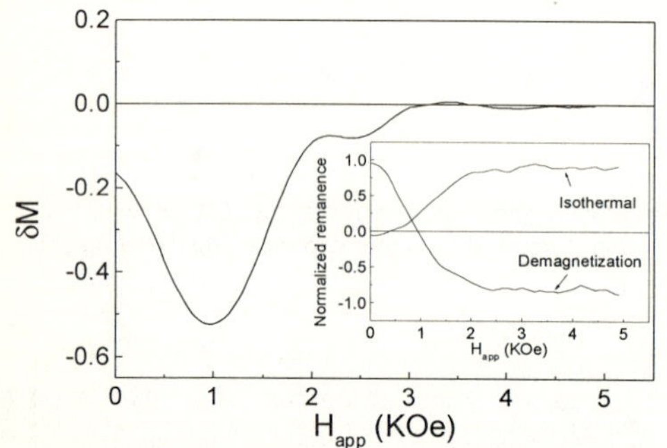

Figure 10. δM plot measured at T = 80 K in the sample deposited for 20 minutes. Inset: isothermal and demagnetization remanences curves.

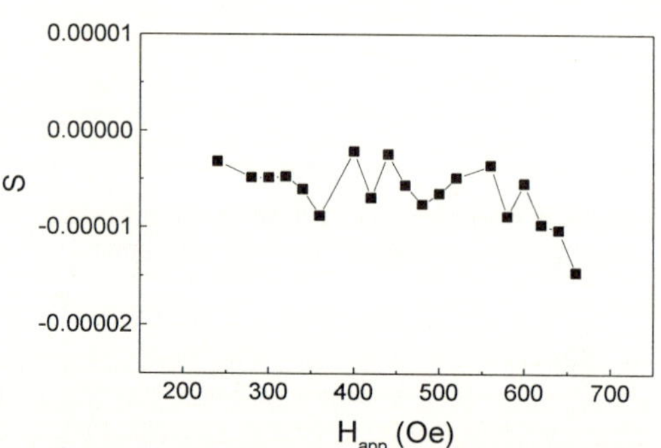

Figure 11. Evolution of the coefficient S with the applied field measured, at T = 80 K, in the sample deposited for 20 minutes.

Finally, in Fig. 11 we have plotted the results obtained for the field evolution of the magnetic viscosity S. That coefficient gives the magnitude of the thermally activated demagnetization observed at T = 80 K and constant applied field. (In all the cases the time evolution of the magnetization was describable as $M(t) = const - S\ln t$). Our results, similar to those found in aligned particulate materials show an steady decrease of S with the increase of the applied field [7].

In summary, we have shown that the pulsed laser ablation deposition technique allows to prepare highly homogeneous films with good control of the average particle size. These films show giant coercivities for average particle sizes below 15 nm. Our remanence measurements evidence that these coercivities are largely influenced by the ocurrence of intense interparticle interactions.

4. References

[1] J.L. Dormann, D. Fiorani and E. Tronc, *Advances in Chemical Physics,* vol. XCVIII, I. Prigogine and S.A. Rice, Eds. John Wiley & Sons (1997) p. 283.

[2] Y. Sasaki, M. Hyakkai, Eiji Kita, A. Tasaki, H. Tanimoto and Y. Iwamoto, *J. Appl. Phys*. vol. 81 (1997) p. 4736.

[3] F.J. Cadieu, R. Rani, W. Mendoza, B. Peng, S.A. Shaheen, M.J. Hurben and C.E. Patton, *J. Appl. Phys*. vol. 81 (1997) p. 4801.

[4] B.D. Cullity, *Introduction to Magnetic Materials,* Addison-Wesley, Reading (1972) p. 234.

[5] N.B. Schevchenko, J.A. Christodoulides, X. Meng-Burany and G.C. Hadjipanayis, *J. Appl. Phys*. vol. 81 (1997) p. 5564.

[6] D. Richards, J.W. Harrel and M.R. Parker, *J. Magn. Magn*. Mater. vol. 120 (1993) p.164.

[7] K.W. Chantrell and K. O'Grady in *"Applied Magnetism"* Eds. R. Gerber et all, Kluwer Academic Publisher, Amsterdam (1994) p. 113.

Materials Science Forum Vols. 302-303 (1999) pp. 101-104

Spin Wave Propagation in Thin Ferromagnetic Films on Vicinal Antiferromagnetic Substrates

A.L. Dantas[1,2] and A.S. Carriço[1†]

[1] Departamento de Física Teórica e Experimental, Universidade Federal do Rio Grande do Norte, Natal-RN, Brazil 59072-970

[2] Faculdade de Ciências Naturais, Universidade Estadual do Rio Grande do Norte, Mossoró-RN, Brazil 59610-210

Keywords: Spin Waves, Domain Wall Modes

Abstract

We consider a thin uniaxial ferromagnetic film on a high Miller indexes two-sublattices antiferromagnetic substrate. The substrate consists of a periodic repetition of terraces separated by steps, leading to a magnetic stripe structure. Neighboring terraces have opposite sublattices of the antiferromagnet and, through a contact interaction, lead to the formation of a periodic structure of Nèel domain walls in the ferromagnet if the F/AF exchange contact interaction strength is beyond a threshold value. We study the domain wall nucleation and spin wave spectrum of the ferromagnet in the limit of wide terraces. We consider long wavelength spin waves propagating along the step edge direction. The interface exchange produces a significant reduction of the domain wall width in low anisotropy ferromagnets, leading to domain wall modes practically localized at the step edges.

1. Introduction

The effects of contact interaction are a topic of recurring interest. The concept was used to explain the observed hysteresis shift on magnetization measurements on samples containing cobalt fine particles many years ago.[1] In recent years it has become widely accepted as a prominent feature of the magnetism of thin films. Thin ferromagnetic films on antiferromagnetic substrates may exhibit shifted hysteresis if the sample is field cooled from the Nèel temperature of the antiferromagnet.[2] The bias field results from the contact interaction with the substrate.

The effective interface effect is strongly affected by interface roughness.[3] For a two-sublattice substrate one may find either one sublattice at the interface. The overall interface effect depends on the relative fraction of the interface area which exhibit each sublattice of the antiferromagnet. The measured values of the hysteresis shift correspond typically to interface exchange fields which are orders of magnitude smaller than the expected values.[4] The nucleation of domain walls at the interface has also been proposed as a mechanism for the small values of the hysteresis shift.[3]

Magnetization measurements either via SQUID or VSM and by optical means (MOKE) informs about the average interface effect on the ferromagnet. These techniques sample macroscopic areas of the interface. Therefore the actual value of the local interface exchange field strength is not available from these measurements. In order to access the local coupling it is necessary to use techniques that probe small areas of the interface.

The effective interaction between two ferromagnetic thin films, coupled trough a nonmagnetic spacer (Fe/Cr/Fe at room temperature, for instance), present similar features. Since the coupling depends on the spacer thickness it is necessary to use techniques that probe small surface areas of the trilayer in order to access the local value of the exchange coupling. The measurement of low energy excitations of coupled domain walls was recently proposed as a possible means to study the local coupling. It was shown that the restoring force of domain wall resonance involves the coupling in microscopic areas of the trilayer.[5]

In order to access the local contact interaction between a ferromagnetic thin film and an antiferromagnetic substrate it is desirable to find local modes pinned to interface defects, whose amplitude is significant in areas of microscopic size. For this purpose the measurement of the dispersion relation of domain wall localized spin wave modes may be helpful.

2. Domain Wall Modes

We consider a thin uniaxial ferromagnetic film on a vicinal two-sublattice antiferromagnetic substrate, as shown in Fig.1. The easy axis of the ferromagnet ($\hat{z}$) is supposed to be parallel to that of the antiferromagnet. The substrate consists of terraces, of width w, separated by steps of odd number of lattice constants. The step edges are along the z-axis. Each terrace contains a single sublattice and neighboring terraces have magnetic moments in opposite directions. The ferromagnet is subjected to an interface field that oscillates in direction according to the orientation of the spins of the antiferromagnetic substrate. The oscillating interface field produces a stripe magnetization pattern in the ferromagnetic film if the interface coupling energy overrides the anisotropy barrier for wall formation.

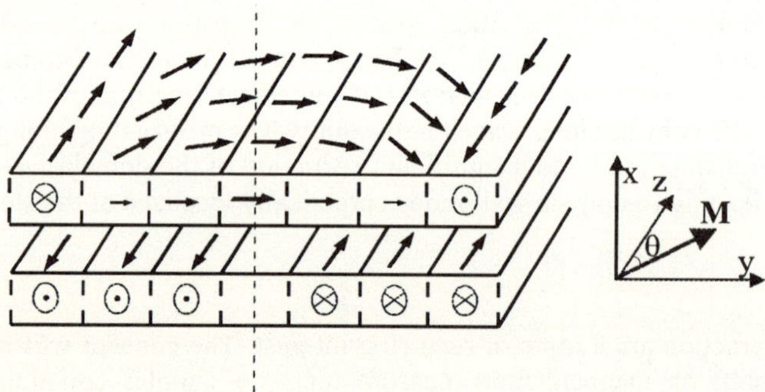

Figure 1 Nèel wall pinned at step defect where the contact interaction changes sign.

In our model the antiferromagnet is frozen and the equilibrium pattern involves only spatial rearrangement of the moments in the ferromagnetic film. Furthermore, we consider a high Miller indexes substrate and the terraces are much wider than the intrinsic domain wall width of the ferromagnet, $w >> \Delta_0$. The interface pinned walls are narrower, $\Delta < \Delta_0$, due to the contact interaction.

The nucleation of walls centered at the step edges depends on the balance between the wall energy and the interface coupling energy $E_C = -\int_{-w}^{w} M.H_e dx$, where H_e is the effective field of the contact interaction and M is the magnetization. While the wall energy, $4\sqrt{AK}$, is independent of the terrace width, provided $w >> \Delta_0$, the interface coupling energy is a strongly decreasing function of w. Therefore the threshold value of H_e which stabilizes the π walls is negligible. Only for w of the order of Δ_0 would the threshold exchange be finite.

We calculate the equilibrium magnetization profile of the ferromagnet, as well as the low energy excitations, neglecting any interaction between the domain walls. Therefore it suffices to consider the structure that forms near one step edge. Assuming a magnetization profile given by $tg(\frac{\theta}{2}) = e^{-\frac{y}{\Delta}}$, where θ is the angle between the magnetization M(y) and the easy axis, we calculate Δ so as to have the total energy minimized. We find that Δ is reduced with respect to the intrinsic value due to the interface exchange field H_e. Δ is given by:

$$\Delta(J)=\frac{\Delta_0}{\sqrt{1+\frac{J}{2K}\ln(2)}} \tag{1}$$

where $J=H_eM$. The interface effect in the domain wall width is controlled by the ratio $\frac{J}{K}$. Therefore a large reduction of the domain wall width is expected, favoring the wall nucleation even if the pinning centers are not too far apart.

We also find that the threshold value of J for wall nucleation is given by

$$J_{th}=\frac{2K}{\ln\left(ch\left(\frac{w}{\Delta}\right)\right)} \tag{2}$$

Therefore if $w >> \Delta_0$ the threshold for wall nucleation, $J_{th} \cong \frac{2K\Delta}{w}$, is negligible.

The equations of motion for spin waves are:

$$\begin{aligned}\frac{dS_x}{dt}&=--J_iSa^2\nabla^2S_{\tilde{y}}+2KS\left(\cos^2\theta-\sin^2\theta\right)S_{\tilde{y}}+2\tilde{K}SS_{\tilde{y}}\\ \frac{dS_{\tilde{y}}}{dt}&=--J_iSa^2\nabla^2S_x-2KS\left(\cos^2\theta-\sin^2\theta\right)S_x-4\pi g\mu_BMS_x\end{aligned} \tag{3}$$

where J_i, is the intrinsic exchange constant of the ferromagnet, $\tilde{K}=\frac{J}{2}$ describe the wall stiffness and a is the lattice parameter. The amplitudes, S_x and $S_{\tilde{y}}$, measure oscillations around the local equilibrium configuration of the magnetization, as determined by the wall profile $\theta(y)$. Spin waves localized in the domain wall and propagating in the z-axis direction are described by:

$$\begin{aligned}S_x&=B\sin\theta e^{1(qz-\Omega t)}\\ S_{\tilde{y}}&=C\sin\theta e^{1(qz-\Omega t)}\end{aligned} \tag{4}$$

The eigenvalue equations, which follow from eqs.(3) and eqs.(4) are

$$\begin{pmatrix}-i\Omega & J_iSa^2q^2+2\tilde{K}S\\ J_iSa^2q^2+4\pi g\mu_BM & -i\Omega\end{pmatrix}\begin{pmatrix}B\\ C\end{pmatrix}=0 \tag{5}$$

Then

$$\Omega=\sqrt{\left(J_iSa^2q^2+2\tilde{K}S\right)\left(J_iSa^2q^2+4\pi g\mu_BM\right)} \tag{6}$$

For $q=0$ eq. (6) gives the frequency of domain wall resonance. In this case it is clear that the domain wall stiffness originates from interface coupling energy $(\tilde{K})$. Also the q-dependence of the spin wave frequency allows a direct comparison between the strength of the interface coupling exchange (J) and the intrinsic ferromagnetic exchange constant (J_i). The frequency of the spin waves propagating along the easy axis direction within interface pinned domain walls, is likely to be higher than that of domain modes (FMR) where the restoring force is due to anisotropy energy variations.

3. References

† Electronic mail: acarrico@dfte.ufrn.br

[1] W. H. Meiklejohn and C. P. Bean, Phys. Rev.105, 204 (1957)

[2] X. Lin, G. C. Hadjipanayis and S. I. Shah, J. Appl. Phys.75, 6676 (1994) and O. Allegranza and M. M. Chen, J. Appl. Phys.73, 6218 (1993)

[3] A. P. Malozemoff, J. Appl. Phys.63, 3874 (1988)

[4] R. Jungblut et al, J. Appl. Phys.75, 6659 (1994)

[5] R. L. Stamps, A. S. Carriço, and R. P. Wigen, Phys. Rev.B55, 6453 (1997)

Materials Science Forum Vols. 302-303 (1999) pp. 105-114

Near Field Optics and Magnetism

Y. Souche[1], M. Schlenker[1], R. Raphel[1] and L. Alvarez-Prado[2]

[1] Laboratoire Louis Néel, CNRS, BP 166, FR-38042 Grenoble Cedex 9, France

[2] Depart. de Fisica, Fac. de Ciencias, C. Calvo Sotelo, S/N ES-33007, Oviedo, Spain

Keywords: Microscopy, Magneto-Optics, Domains, Rayleigh Criterion, Local Probe

Abstract

Near field optical microscopy allows resolution far beyond the half wavelength that corresponds to the Rayleigh criterion. This is achieved by detecting the evanescent waves in the near field, typically some nanometers, from the object. Optical and topographic images are made simultaneously by scanning the surface. High resolution images of magnetic domains or singularities were obtained with a combination of near field optical microscopy and magneto-optical effects. A wide review of results will be presented.

It used to go without saying that optical microscopy relies on waves that have traveled over distances large with respect to their wavelength λ from the sample, hence must operate in the far field. In this classical form, it has for several decades been the most useful, convenient and ubiquitous technique in the laboratory. Advances in optical combinations led the development of objective lenses close to the resolution limit set by diffraction, the Rayleigh criterion. Sophisticated variations, such as the confocal and scanning (far-field) microscopes, or differential interference contrast, further contributed to its flexibility.

In 1984 two teams [1, 2] simultaneously demonstrated experimental setups for near field optical microscopy, based on the diffraction of evanescent waves in the visible range, which pushed back the traditional resolution limit of optical microscopy by an order of magnitude. Synge [3] proposed this concept as early as 1928. He is generally acknowledged as the father of this technique even though he wrote in his paper that the idea was suggested by a friend. However, beyond the fact that his paper had fallen into oblivion, the technology for dealing with the optical near-field, i.e. for probing the vicinity of a surface at distances smaller than λ, with control over the displacements along the surface on the same scale, just did not exist until the scanning tunneling microscope (STM) was invented by Binnig and Rohrer [4].

A few pioneers were able to experimentally demonstrate that resolution beyond the Rayleigh limit was obtained through use of the near field. They got round the technical difficulties by working with long wavelength radiation: microwaves for Ash and Nicholls in 1972 [5], infrared for Massey in 1984 [6].

In this paper we briefly review the Rayleigh criterion in part I, and describe the basis of near field optical microscopy, beginning, in part II, by defining the near field and by connecting the high spatial frequencies in the object with the existence of diffracted evanescent waves. The basic scheme of near-field scanning optical microscopy (SNOM) and some experimental setups are presented in part III. Part IV is devoted to the application of near field optics to magnetism. Some magneto-optical effects are recalled and their relation with the near field is presented. The main results are exposed through a review of literature.

I. Resolution in far field optics: Rayleigh's criterion

Rayleigh's criterion deals with the resolution of an optical system under incoherent illumination when the distances object-objective lens-image are very large compared with the wavelength of light. The object diffracts light into an angular spectrum of plane waves which is filtered by the pupil function of the objective lens. The intensity distribution I(x',y') is expressed by the convolution [7]:

$$I(x', y') = O \otimes D$$

where O is the transmitted or reflected intensity distribution leaving the object and D the distribution of intensity in the image of a point object. This well known expression can also be written as a simple product of Fourier transforms (FT):

$$FT(I) = FT(O) \times FT(D)$$

FT(D), the *frequency response function* [8] of the instrument, is the auto-correlation function of the complex amplitude transmission function of the imaging system. In the case of an instrument with a uniform circular aperture this contrast function vanishes when the spatial frequency is $\pm 2\alpha'/\lambda$, where α' is the numerical aperture of the imaging system in the image space. This means that the lowest transmittable period is $\lambda/2$. It corresponds to an instrument which accepts the light coming from the whole half space, i.e. $\alpha'=1$. This limit, known as *the Rayleigh'criterion* , is close to the value of the radius of the Airy pattern, $1.2\lambda/2\alpha'$.

II. How to go beyond the Rayleigh criterion?

II-1 Far field and near field

The highest detectable spatial frequencies that contribute to the image in the above far-field approach correspond to diffracted light propagating parallel to the plane of the object. That means that all the light available in the far field is used. What about the light diffracted by higher spatial frequencies?

Dürig *et al* [9] and Vigoureux and Courjon [10] analyzed the elegant model based on the relationship between the Rayleigh criterion and the uncertainty principle, or equivalently the inequality between the width of a function and its Fourier transform.

They apply the Heisenberg relations to the position-wave vector pair $\Delta s \Delta k_s$, where s is the transverse coordinate x,y.

$$\Delta s \Delta k_s \geq 2\pi$$

The Rayleigh criterion ($\Delta s=\lambda/2$) corresponds to a total interval for Δk_s of $4\pi/\lambda$ (k_s goes from $-2\pi/\lambda$ to $+2\pi/\lambda$). To increase resolution Δs should become small and Δk_s, corresponding to the interval of detected values of k_s, must increase. As the wave vector **k** has a constant amplitude $2\pi/\lambda$, such that:

$$k^2 = k_x^2 + k_y^2 + k_z^2 = (2\pi/\lambda)^2$$

$k_x^2 + k_y^2 > (2\pi/\lambda)^2$ means that k_z^2 becomes negative. k_z is then imaginary, such that $k_z = j K_z$, where K_z is real. The electric field of the wave must therefore be written as:

$$E=E_0 \exp(-K_z) \exp(j\omega t - k_x x - k_y y)$$

This is an inhomogeneous wave whose equal-amplitude and equal-phase planes are mutually perpendicular. It is an evanescent wave whose propagation is parallel to the plane (x,y) and whose amplitude decreases exponentially along z. This defines the conditions for observing details smaller than $\lambda/2$ in an object. It shows that the detection of far field light in the whole half space cannot take into account all the useful light. *The near field can be defined as the region where non-radiative waves are confined.*

II-2 High spatial frequencies and evanescent waves

Several comments follow from the preceding part.
1- Resolution better than $\lambda/2$ can be obtained if evanescent waves can be detected or generated. This is a necessary condition.
2- Evanescent waves will be diffracted only if high spatial frequencies, typically $1/p>2/\lambda$, are present in the optical structure of the surface object.
Due to the Fourier components of the sample's transmittivity or reflectivity, an incident wave will generate a whole range of plane waves. If one Fourier component only, corresponding to a period p, is taken into account, the diffraction process generates a diffracted beam at a Bragg angle depending on 1/p. When p is large compared with λ, these orders correspond to progressive waves measurable in the far field. When p is small with respect to λ , the Bragg angle is larger than 90° under normal incidence. The corresponding light no longer exists in the far field half space, but is present, and can be detected as an evanescent wave in the near field. Conversely, such near field light is diffracted by the appropriate Fourier component of the surface and, then, propagates as a progressive wave.
A hole on a screen diffracts light. If the dimension of the hole is decreased down to a nanometer scale, it radiates evanescent waves. If this hole is brought very close to the surface of a sample, the condition of detection is fulfilled.
3- A complementary view [11, 12] consists in considering a local oscillating dipole or a local nano-antenna. When the gap width separating the dipole from the surface is decreased, all the terms involved in the fields radiation must be taken into account within the coupling, i. e. even the terms in $1/r^2$ and in $1/r^3$ which are negligible when r becomes large with respect to λ. Oscillations appear in the far field light due to the interference of the direct radiation of the dipole with the one reflected by the surface. One uses the representation by an image dipole. Oscillations are stronger when the surface is a perfect conductor. Moreover, oscillations depend on the direction of the dipole [11]. In the case of a finite conductivity of the surface, there is a specific behaviour for low values of r corresponding to the near field.
Any perturbation in the near field light will affect the far field light.
4- Evanescent waves can also be generated by refraction. Three centuries ago, Newton frustrated the total internal reflection by bringing a spherical dielectric surface near a plane glass-air boundary without contact. He thus probed the evanescent waves existing in the second medium (air). In this case, high spatial frequencies will only pertub the distribution of the evanescent field.
5- To reach a resolution better than the Rayleigh criterion imposes the evanescent waves to be detected. However, this necessary condition is not sufficient as, in addition, the lateral dimensions of the probe have to be small.
6- Actual objects include generally a large Fourier spectrum and both propagative and evanescent waves are generated from an incident beam [13].

III The Near Field Optical Microscope

III-1 The basic scheme

A near field optical microscope is a scanning microscope. Light, generated by a source located in the far field, interacts with the object and with the local probe before going back to the far field where it is detected, typically by a photomultiplier tube. Depending on which scheme is adopted, more or less light is converted into evanescent waves. A large proportion can even be propagating waves. The gap width between the probe and the sample is typically a few nanometers. Therefore, an auxiliary setup will be needed to monitor this gap. The polarization, wavelength and focusing of the incident far-field light can be adjusted. The setup can operate in normal room conditions but also in vacuum and over a large range of temperature.

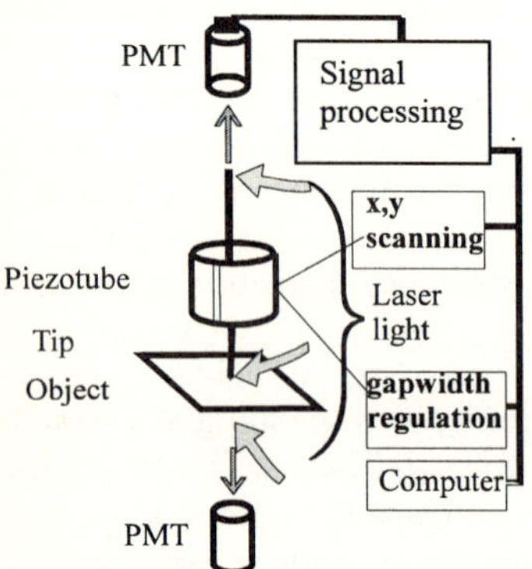

Figure 1: The basic scheme of a near field microscope

III-2 Contribution of previously developed local probe microscopies

The pioneering work of Ash and Nicholls [5] demonstrates how delicate the engineering of a near field optical microscope was. They used microwaves for their experimental setup in order to increase the near field zone and could therefore avoid nanometric displacements. The advent of the scanning probe, e.g. tunnel and atomic force, microscopies provided innovative technologies based on piezoactuators working in the Ångstrom range and, at the same time, improved the acquisition and processing tools. Many authors proposed microscopes associating near field optical microscope with another scanning probe microscope [9, 14, 15, 16]. Besides the imaging processes, the aim of these experimental couplings was to set and regulate the gap width.

III-3 A technical description of near field optical microscopes

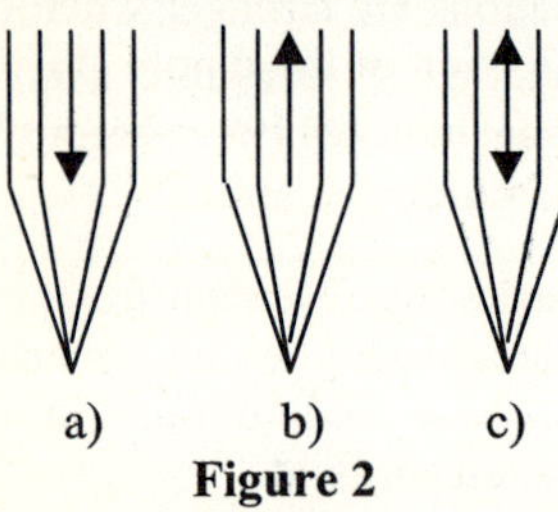

Figure 2

Modes of operation of an optical fiber probe

The local probe is the heart of the microscope. It can bring the light to the near field (fig. 2a) [9], collect light from the near field (fig 2b)[17], achieve both these two tasks (fig. 2c)[18] or perturb the non-radiative waves and convert them into propagating waves [19, 20, 21].

Differents acronyms were coined depending on the way of producing the evanescent waves: SNOM or NSOM (near-field optical microscopy) [22], originating from the general designation Scanning Probe Microscopy (SPM), can work either in transmission, or in reflection [23]. When evanescent waves are generated by total reflection, it is usually named a Photon Scanning Tunnelling Microscopy (PSTM) [24][25] or Scanning Tunneling Optical Microscopy (STOM) [26]. NFMOM could also be proposed for Near Field Magneto-Optical Microscopy.

The local probe can be a nanoaperture. It is very often the tip of a tapered optical fiber coated with an absorbing metal (Al) except at the very end. The beam of light, generated by a gas or diode laser, has to be sharply focused on the core of the fiber, with the numerical aperture of the objective lens adapted to that of the fiber. Scanning is achieved by using piezoactuators. The most commonly used are segmented piezotubes, able to move the tip or the sample along three perpendicular axes, x and y for scanning, z for the gap width regulation. The planarity of the surface must be good enough in order to stay within the range in z accessible to the piezotube. The sensitive detector, located at the end of the fiber or in free space, is usually a photomultiplier. A modulation technique and a lock-in amplifier are very often used for a better amplification of the resulting signal.

An auxiliary setup deals with the gap width regulation. The most commonly used is based on the slight change in the shear force acting on the tip when the gap is changed. The initial setup was introduced by E. Betzig *et al* [27]. The tip is brought into forced vibration at resonance by means of piezoelectric components. The strength of the shear force acting between tip and surface changes progressively between contact, where a tip is locked, and clear non contact where the amplitude of vibration is maximum. There exists a zone of a few nanometers where the amplitude of vibration changes linearly with the gap width, with an associated change in the phase. Many methods of detection of the vibration of the tip can be found in the literature. Optical methods measure the amplitude and phase of the vibration by collecting the light scattered by the tip [27]. Karrai and Grober [28] proposed a method where the tip is attached to a tiny piezoelectric fork: the eigenfrequency changes with the gap width [see also ref. 29]. Another method maintains a constant impedance in a circuit including the small piezoelectric holder [30]. In all cases a feedback loop monitors the voltage V applied on the z piezoactuator. Such a distance regulation provides a topographic image V(x,y) of the surface.

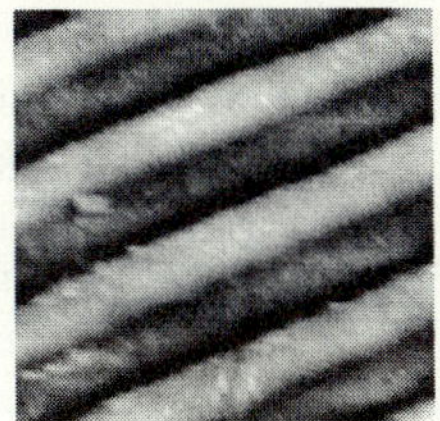 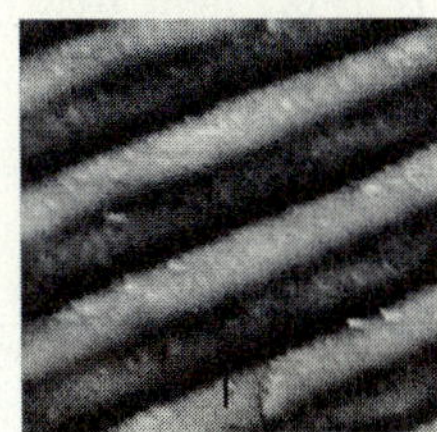

Figure 3
Topographic and optical images of a relief grating
(period: 1.6 μm, height: 80 nm)

IV Near Field Optical Microscopy and Magnetism

Interactions between light and magnetic materials are at the heart of the magneto-optical effects. They are a powerful means of characterization of magnetic materials. However, here also the spatial resolution is subject to the Rayleigh criterion. Nevertheless, these effects are widely used for the reading part of high density magneto-optical recording and the basic scheme is used also for the writing sequence by heating locally above the Curie temperature. However, even though other near field methods, such as Magnetic Force Microscopy (MFM), or Scanning Electron Microscopy with spin Polarization Analysis (SEMPA), are able to provide more accurate images, near field optics could not remain a long time outside this area of interest to the magnetism community.

IV-1 Magneto-optical effects in the far field

Magneto-optical effects are named differently depending on the geometric and optical configuration of light with respect to the sample and to the magnetization vector [31].
The most commonly used magneto-optical effects are:
- the *Faraday effect* including circular dichroism, in the transmission mode. The angles Θ_F and η_F characterize the rotation of the plane of polarization and the ellipticity induced by the effect
- the magneto-optical Kerr effects (MOKE) in the reflection mode. Depending on the relative geometry of the wave vector, the magnetization vector **M** and the polarization, these are the *polar, the longitudinal and the transverse Kerr effects*.

The polar and longitudinal Kerr effect are characterized by the angle Θ_K and the ellipticity η_K. The transverse Kerr effect changes slightly the reflectivity R_p of a p-polarized beam when the magnetization is reversed. The relative change $\Delta R_p/R_p$, typically about some 10^{-3}, is the characteristic quantity of this effect.

The transmitted and reflected fields are expressed as a function of the incident field by the 2x2 matrix of the Fresnel coefficients $\{r_{ij}\}$. These can be deduced from the Maxwell equations and from the boundary conditions.

Magneto-optical effects can be enhanced in several ways: antireflection combinations of dielectric and metallic layers are classical methods widely used for magneto-optical recording. Surface plasmons corresponding to oscillations in the density of electrons at the surface of metal were proposed in 1977 by P.E. Ferguson *et al* [32]. Later, resonance like peaks in the magnitude of the Kerr effect were interpreted as the presence of a plasma edge at this frequency of light [33, 34, 35].

Depending on which magneto-optical effect is considered, experimentalists measure intensities (dichroïsm, TMOKE), polarization rotation Θ_K or Θ_F and/or ellipticities (Faraday, LMOKE, PMOKE). The aim when imaging regions with different magnetization directions is to achieve sufficient contrast as a function of the magnetization **M** over a given area and so be able to "map" **M** over this surface.

IV-2 Magneto-optical effects in the near field

Calculations on magneto-optical effects start from the Maxwell equations, consider the boundary conditions, and conclude with the Fresnel coefficient matrix. These are valid for the far and near fields. However, two specific aspects are particularly delicate.

- The evanescent waves at the investigated surface are perturbed by the high spatial frequencies. A perturbative approach can be applied if the corrugation of the surface is small compared to λ.
- The probe-sample coupling is often neglected by considering the probe as passive. The square modulus of the electric field at the end of the tip is assumed to represent the actual field distribution.

There exist only a few theoretical works about magneto-optical interactions in the near field. A perturbative approach to the Rayleigh theory [36] was applied to the near field study of a relief grating coated with a ferromagnetic metal [37] previously studied in far field conditions [38].

In the case of near field optical effects where the level of the detected signals is low, all methods of enhancement have been taken into consideration. Many authors have studied surface plasmons in the near field [39] so that the application of surface plasmons to the magneto-optical near field was inevitable.

A paper, by V. A. Kosobukin [40], taking into consideration a semi-infinite ferromagnet and a small non-magnetic metal particle located nearby and showing a large plasmon resonance, deals with the conditions of an experiment previously published by T.J. Silva, S. Schultz and D. Weller [41]. The result concerns the longitudinal Kerr effect with s incident polarization and seems not to be easy to apply.

Experimental papers which deal with near field optics and magnetism are more numerous. One can use the same classification as for the description of the magneto-optical effects.

The first paper related to magneto-optics, the Faraday effect, and the near field was published in 1992 by E. Betzig and co-workers of AT&T Bell Laboratories, Murray Hill and of the Data Storage Systems Center of Carnegie Mellon University, Pittsburgh [41]. They obtained contrasted images of domains conventionally written in Co/Pt multilayers. Polarization was well controlled and the inversion of contrast was obtained by the usual rotation of an analyzer. Moreover, they were able to write small domains at a very high density of about 45 Gbits/in^2.

Some substituted yttrium iron garnets are well known for their large Faraday rotation, typically about several thousands deg.cm^{-1}. Consequently, they were tried by some groups for demonstrating the possibilities of near field magneto-optical microscopy.

The basis experimental scheme reproduces the far field's one: contrast between antiparallel domains due to the rotation of the plane of polarization of light after crossing the garnet layer under quasi normal incidence is revealed by means of an analyzer. This setup was proposed by Betzig *et al* [42], Hartmann [43] who obtained also images of domain walls on a thick (2.69 μm) YSmBiGaFe garnet layer, and by G. Eggers *et al* [44].

The magneto-optical garnet layer was also studied in a PSTM type setup by the Safarov and Lampel group [45]. The layer was investigated as a phase object and the microscope could make images of the domain wall motion when the magnetic field was modulated sinusoidally (f=190 Hz) with a small amplitude of .3 Oe. It appears that, due to the large thickness of the garnet layer and to the large angular aperture of the detected beam, light involved in the image formation crosses a large volume of the sample. The resolution is rather poor but differential imaging allows the contrast of the domain wall images to be enhanced.

Circular dichroism coupled with the Faraday effect was also used to obtain images of magnetic samples with perpendicular anisotropy. Two techniques were developed.

The more recent one [46] consists in illuminating a Co/Au multilayer from the far field with circularly polarized light, alternatively modulated left and right, and in collecting the transmitted light with the probe in the near field. The quantity of detected light depends on the modulus and direction of the magnetization vector, at the point under consideration. The tip-sample distance is regulated by maintaining constant the tunneling current between the gold coated tip and the sample surface. Resolution seems to be better than 100 nm.

Another interesting configuration was proposed by Prins *et al* [47] from the University of Nijmegen. This microscope also combines circular dichroism and tunnel microscopy. Modulated circular light transmitted by the sample illuminates a GaAs tip located in the near field of the sample. Light creates photocarriers in the semiconductor. The voltage applied between sample and tip causes a direct current across the gap. This current is proportional to the intensity of light transmitted across the sample. The sample here was a Pt/Co multilayer, exhibiting perpendicular magnetic anisotropy, deposited onto a glass substrate. Thermally written magnetic dots were imaged and a resolution of 6 nm was estimated to be achievable in the better conditions. The Nijmegen group hopes, under different conditions, to succeed in using the same setup as a spin polarized STM.

The production of surface plasmons by total reflection at the surface of a prism or a semi-spherical lens has been known for a long time [48]. This effect was reproduced in order to enhance the magneto-optical effect of a Au/Co/Au multilayer structure deposited on a glass plate under UHV conditions. The authors assume that the physical phenomenon under consideration is the Faraday effect as light travels across the sample and that it manifests itself mainly by a rotation of the plane of polarization. The angle of incidence is changed from 35 to 55 degrees. The plate was optically coupled to the flat face of a half-cylindrical prism with an adapting refractive index liquid [44]. The local probe frustrated the total reflection (PSTM configuration) so that the resulting signal could be detected by two ways:

- by picking up the light in the evanescent waves with the fiber and feeding it into a photomultiplier
- by detecting the perturbed reflected light.

Both ways exhibited resonance like peaks with incident p- and s-polarized light for the magneto-optical response S_{MO} plotted as a function of the angle of incidence. S_{MO} is the figure of merit of the system.

Many magnetic samples are metallic and optically thick so that the previous setups cannot be used. For this reason, near field magneto-optics in reflection is being presently developed.

T.J. Silva and S. Schultz, D. Weller [49] working with the Center for Magnetic Recording Media at the University of California at San Diego and, later, T.J. Silva and A.B. Kos at the National Institute of Standards and Technology in Boulder [50] built a microscope which uses a small silver particle ($\approx$ 40 nm) as local probe. This is optically excited at the plasma edge frequency of silver, generating the surface plasmon resonance ($\lambda \approx$ 440 nm). Thus, the particle scatters light and behaves as an electric dipole radiator. The scattering cross section, hence the electric polarization efficiency, become quite large. The magneto-optical response is therefore enhanced.

The setup devised by T.J. Silva and S. Schultz works in dark field conditions. The particle probe is placed onto the flat face of a hemispherical prism. Light is focused through the glass onto the region with the particle, and then almost totally reflected. The gap width is set at a typical near field

distance, i. e. some nanometers. It can be supposed that the mechanisms of scattering and diffraction which take place between particle and sample occur in several ways: from the particle to the sample and conversely.
The polarization of the emerging light is analyzed by using a phase modulation technique that involves a photoelastic modulator and an analyzer before the photomultiplier tube and the signal acquisition.
The probe/sample spacing regulation is achieved with an additional setup based on the stabilization of a Newton rings pattern while scanning.
Fairly good images of magnetic domains, 0.5 μm in diameter, thermomagnetically written on Co/Pt multilayer films, were obtained. Hysteresis loops were measured for a Permalloy layer (10 nm thick) which was used for tests on the probe/sample spacing.
When scattering is not enhanced through the surface plasmon processes, the light must be collected over a large angular aperture. C. Durkan, I.V. Shvets, J.C. Lodder [51] and G. Eggers et al [43] used a large elliptical mirror, with the sample and the photomultiplier tube located at the foci. The latter group used a more sophisticated phase modulation detection including a photoelastic modulator. Both groups imaged bits thermomagnetically written (0.5 μm wide) on a Pt/Co multilayer.
B.D. Terris, H.J. Mamin and D. Rugar from IBM Research Division, San Jose, proposed an adaptation of the solid immersion lens (SIL) [52] previously described by Kino and Mansfield [53]. There is no tip but a very small spot of light focused by a high index plane-convex lens. Coupling between the SIL and the sample surface is made via the evanescent field through a narrow air gap. B.D. Terris adapted this device to magnetic recording by combining the SIL with an air-bearing slider. Recorded tracks on TbFeCo based multilayers with perpendicular anisotropy could be read with a good signal to noise ratio.
Near field magneto-optical microscopy related to measurement or imaging of surfaces with in-plane magnetization is less frequent. However, it can be imagined that researchers capable of imaging domains magnetized perpendicularly could do the same when the magnetization is in-plane.
We recently proposed [54] a near field differential setup based on the transverse magneto-optical Kerr effect (TMOKE).
In the usual TMOKE experiment, the magnetic field is reversed so that reflectivity of the surface for p polarized light goes from $R_{p+}=R_{p0} + \Delta R$ to $R_{p-}=R_{p0} - \Delta R$. The relative change $\Delta R/R$ is proportional to magnetization M. The typical value is 10^{-2}-10^{-3}.
The same result can also be reached by reversing the sense of propagation of the incident light. In all cases, an observer located on the beam and receiving light can see the reversal of the magnetization vector **M** and, thus, the change of reflectivity.
Let us define a plane of incidence with a p polarized nanosource in the far field and two detectors, placed symmetrically relative to this point, receiving the light reflected by a flat ferromagnetic surface. If this surface is magnetically saturated in the specimen surface but perpendicular (M_+) to the plane of incidence, the difference ΔS of the two photosignals is proportional to $2\Delta R$. If the magnetization is reversed (M_-) it will become $-2\Delta R$. The difference cancels the non magnetic part proportional to R_{p0}. This allows the useful part of the photosignal to be amplified more easily.
This was extrapolated to the near field by decreasing the surface-nanosource gap width and a magneto-optical contrast remained.
Fig. 5 represents the schematic diagram of the configuration and figure 6 is an image of the surface (ðXð μm) of a magnetically saturated soft sample, FeSiAl, whose magnetization was reversed by the application of magnetic field pulses while scanning.

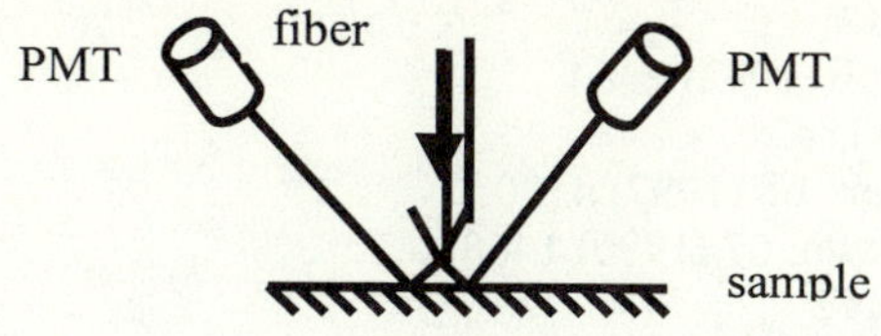

Figure 5: Differential Transverse Kerr effect

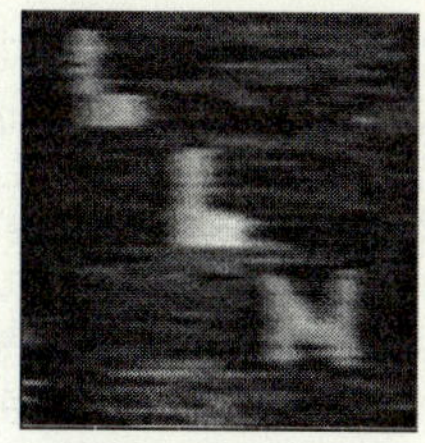

Figure 6: FeSiAl sample (.5 x .5 μm^2)

Conclusion:

Near-field optical microscopy is a recent local probe microscopy the possibilities of which are being explored [56]. In magnetism, several imaging capabilities based on far-field configurations have been demonstrated. Development of the theory might suggest new, typically near-field, configurations. The perspectives of high density magneto-optical recording contribute to the development of these methods and even begin to succeed in an industrial technology [57] which should allow to reach a density limit beyond the superparamagnetic barrier.

References:

[1] D.W. Pohl, W. Denk, F. Rohner, Appl. Phys. Lett. 44 (1984) 651
[2] A. Lewis, M. Isaacson, A. Harootunian, A. Muray, Ultramicroscopy 13 (1984) 227
[3] E.H. Synge, Philos. Mag. 6 (1928) 356
[4] G. Binnig, H. Rohrer, Helv. Phys. Acta 55 (1982) 726
[5] E.A. Ash, G. Nicholls, Nature 237 (1972) 510
[6] G.A. Massey, Appl. Optics 23 (1984) 658
[7] A. Maréchal, M. Françon, "Diffraction, Structures des images", 1960, Ed. La Revue d'Optique théorique et Instrumentale
[8] M. Born, E. Wolf, "Principles of Optics", 6th edition, 1980, Pergamon Press
[9] U. Dürig, D.W. Pohl, F. Rohner, J. Appl. Phys. 59 (1986) 3318
[10] J.M. Vigoureux, D. Courjon, Appl. Opt. 16 (1992) 3170
[11] A. Sommerfeld, F. Renner, Ann. D; Physik 41 (1942) 1
[12] D.W. Pohl, Advances in optical and electron microscopy, 12 (1991) 243
[13] E. Wolf and M. Nieto-Vesperinas, J. Opt. Soc. Am. A2 (1985) 886
[14] E. Betzig, P.L. Finn, J.S. Weiner, Appl. Phys. Lett. 60 (1992) 2484
[15] R. Toledo-Crow, P.C. Yang, Y. Chen, M. Vaez-Iravani, Appl. Phys. Lett. 60 (1992) 2957
[16] N.F. Van Hulst, M.H.P. Moers, O.F.J. Noordman, T. Faulkner, F.B. Segerink, K.O. Van der Werf, B.G. de Groot, B. Bölger, Scanning probe microsc. 1639 (1992) 36
[17] ref. 97
[18] M. Spajer, A. Jalocha, Near Field Optics-1, Arc et Senans 1992, NATO Series E: Applied Sciences (1993) Ed. D.W. Pohl, D. Courjon, Kluwer Academic Publishers
[19] F. Zenhausern, M.P. O'Boyle, H.K. Wickramasinghe, Appl. Phys. Lett. 65 (1994) 1623
[20] F. Zenhausern, Y. Martin, H.K. Wickramasinghe, Science 269 (1995) 1083
[21] L. Aigouy, J.C. Rivoal, S. Gresillon, A. Larech, H. Cory and A.C. Boccara, E-MRS Strasbourg (06/1998)
[22] E. Betzig, J.K. Trautman, T.D. Harris, J.S. Weiner, R.L. Kostelak, Science, 251 (1991) 1468
[23] N. Cerre, F. de Fornel, J.P. Goudonnet, Appl. Optics 31 (1992) 903
[24] R.C. Reddick, R.J. Warmack, D.W. Chilcott, S.L. Sharp, T.L. Ferrell, Rev. Sci. Instrum. 61 (1990) 3669
[25] T.Ferrel, J.P. Goudonnet, R.C. Reddick, S.L. Sharp, R.J. Warmack, J. Vac. Sci. Technol. B9 (1991) 525

[26] C. Girard, D. Courjon, Phys. Rev. B 42 (1990) 9340
[27] E. Betzig, P.L. Finn, J.S. Weiner, Appl. Phys. Lett. 60 (1992) 2484
[28] K. Karrai, R.D. Grober, Appl. Phys. Lett. 66 (1995) 1842
[29] A. Débarre, A. Richard, P. Tchénio, Rev. Sci. Instrum. 68 (1997) 4120
[30] Mark Lee, E.B. McDaniel, J.W.P. Hsu, Rev Sci. Intrum. 67 (1996) 1468
[31] M.J. Freiser, IEEE Trans. on Magn. MAG4 (1968) 152
[32] P.E. Ferguson, O.M. Stafsudd, R.F. Wallis, Physica 89B (1977) 91
[33] W. Reim, O.E. Hüsser, J. Schoenes, E. Kaldis, P. Wachter, J. Appl. Phys. 55 (1984) 2155
[34] J. Schoenes, Phys. Rep. 66 (1980) 187; H. Feil, C; Haas, Phys. Rev. Lett. 58 (1987) 65
[35] T. Katayama, Y. Suzuki, H. Awano, Y. Nishihara, N. Koshizuka, Phys. Rev. Lett. 60 (1988) 1426
[36] D. Van Labeke, A. Vial, V.A. Novosad, Y. Souche, M. Schlenker, A.D. Santos, Optics Comm. 124 (1996) 519
[37] D. van Labeke, A. Vial and D. Barchiesi, Ultramicroscopy 67 (1995) 51-55
[38] Y. Souche, M. Schlenker, A.D. Santos, J. Magn. Magn. Mater. 140-144 (1995) 2179
[39] Surface Plasmon near field microscopy related references by A. Kryukov : http://pubweb.nwu.edu/~kryuk/spom/spomrefs.htm
[40] V. Kosobukin, SPIE, (1995) Vol. 2535, 9
[41] T.J. Silva, S. Schultz and D. Weller, Appl. Phys. Lett., 65 (1994) 658; T.J. Silva, S. Schultz, (1993) in Near Field Optics, Kluwer Academic Publishers, Eds. D.W. Pohl and D. Courjon; T.J. Silva, S. Schultz, (1993) in Scanning Probe Microscopies II, SPIE Proceedings, Vol. 1855, Ed. Clayton C. Williams
[42] E. Betzig, J.K. Trautman, R. Wolfe, E.M. Gyorgy, P.L. Finn, M.H. Kryder, C.H. Chang, Appl. Phys. Lett. 61 (1992) 142
[43) E. Betzig, J.K.Trautman, J.S. Weiner, T.D. Harris, R. Wolfe, Appl. Optics 31 (1992) 4563
[44] U. Hartmann, J. Magn. Magn. Mater., 157-158 (1996) 545
[45] G. Eggers, A. Rosenberg, N. Held, A. Münnemann, G. Güntherodt, P. Fumagalli, Surface and Interface analysis, 25 (1997) 483; see also the beautiful Web page at www.physik.rwth-aachen.de.
[46] V. Safarov, V.A. Kosobukin, C. Hermann, G. Lampel, C. Marlière, J. Peretti, Phys. Rev. Lett. 73 (1994) 3584
[47] V. Kottler, N. Essaidi, N. Ronarch, C. Chappert, Y. Chen, J. Magn. Magn. Mater. 165 (1997) 398
[48] M.W.J. Prins, M.C.M.M. van der Wielen, D.L. Abraham, H. van Kempen, H.W. van Kesteren, IEEE Trans. Magn. **30** (1994) 4491; M.W.J. Prins, R.H.M. Groeneveld, D.L. Abraham, H. van Kempen, H.W. van Kesteren, Appl. Phys. Lett. **66** (1995) 1141
[49] see for example R. Bruns, H. Raether, Z. Phys. 237 (1970) 98; E. Kretshmann, Z. Phys. 241 (1971) 313
[50] S. Schultz, T.J. Silva, Proceedings of "Scanning Probe Microscopies", Los Angeles, SPIE 1639 (1992) 31 ; T.J. Silva, S. Schultz, D. Weller, Appl. Phys. Lett. 65 (1994) 658 ; T.J. Silva, S. Schultz, Rev. Sci. Instrum. 67 (1996) 715
[51] T.J. Silva, A.B. Kos, SPIE 2535 (1995) 2
[52] C. Durkan, I.V. Shvets, J.C. Lodder, Appl. Phys. Lett. 70 (1997) 1323
[53] B.D. Terris, H.J. Mamin and D. Rugar, Appl. Phys. Lett. 68 (1996) 141
[54] S. M. Mansfield, G.S. Kino, Appl. Phys. Lett. 57 (1990) 2615
[55] Y. Souche, Y. Benmati, M. Ramonda, M. Schlenker, S. Manson-Smith, M. Spajer, Conference Digest, 15th International Colloquium on Magnetic Films & Surfaces (1997) 206
[56] D. Courjon, C. Bainier, Rep. Prog. Phys. 57 (1994) 989; D. Courjon, M. Spajer in Handbook of Microscopy, Methods I 1997 Ed. S. Amelinckx, D. Van Dyck, J. van Landuyt, G. van Tendeloo, VCH
[57] TeraStor® in Data Storage (April 1997) 8

Materials Science Forum Vols. 302-303 (1999) pp. 115-119

Blue/Green/Red LED Based TMOKE

A.P.B. Tufaile[1], A.D. Santos[1], Y. Souche[2] and V. Novosad[2]

[1] Instituto de Física da Universidade de São Paulo, CP 66318, CEP 05315-970, São Paulo, SP, Brazil

[2] Laboratoire Louis Néel-CNRS, Univ. J. Fourier, BP 166, FR-38042 Grenoble Cedex 9, France

Keywords: Magneto-Optical Kerr Effect, Thin Films, Rare Earth, Amorphous Alloys, Magnetometry

Abstract

We developed a LED based Transverse Magneto-Optical Kerr Effect (TMOKE) magnetometer that operates at room temperature, in an up to 12 kA/m magnetic field. This equipment works with three wavelengths: blue (λ= 450 nm), green (λ= 563 nm) and red (λ= 660 nm). As a test for the equipment, we obtained the dependence of the saturation TMOKE signal as a function of the wavelength. We measured Co and Fe thin films, and $RE_{1-x}Co_x$ amorphous films (where RE = Dy, Ho, Gd and Y; and x = 0.6, 0.67). Our results showed that the saturation TMOKE signal raises when decreasing the wavelength for RE-Co amorphous films, while it is almost constant for Co. It means that the rare earth contribution for the TMOKE signal increases for blue light. For $x \approx 0.6$, this contribution decreases from Dy-Co to Gd-Co and to Y-Co.

Introduction

The magneto-optical (MO) recording techniques have been strongly improved in the last years. The next step of this evolution could be the substitution of the red or infrared laser light source by a blue semiconductor laser [1]. Due to the shorter wavelength of the blue light, it can afford higher data storage densities than previously. Going from infrared to blue light quadruplicates the amount of data that can be stored in a given area [2]. We developed a Transverse Magneto-Optical Kerr Effect magnetometer for the characterization of thin films at short wavelength. As an alternative to the blue diode lasers that are not commercially available yet, we used ultra-bright light emitting diodes (LED) as light sources.

Depending on the geometry of the configuration, the magneto-optical Kerr effect is named polar, longitudinal or transverse. In the case of light incident in a linearly-polarized state, the polar and longitudinal Kerr effects induce an elliptically polarized reflected state with the main axis of polarization rotated away from the plane of the incident polarization [3]. In the present work, we focus our interest on the transverse Kerr effect that is a completely independent mode. This effect corresponds to change in the reflectivity of the p-polarized light. In the pure configuration no rotation of the polarization occurs. The relative change of reflectivity $\Delta R/R$ is proportional to the in-plane magnetization component perpendicular to the plane of incidence. The Kerr effects present a spectral dependence and they are sensitive to the magnetization in the region of the skin depth, that is typically 10-20 nm in most metals [4].

The magnetic media for magneto-optical recording are based on rare earth (RE)-transition metal (TM) amorphous thin films. The commercially available magneto-optical devices work in the polar Kerr effect configuration. In the visible spectrum, the polar signal decreases while decreasing wavelength for alloys containing heavy rare earths and yttrium [5, 6, 7]. The major part of the magneto-optical effect of RE-TM films comes from TM at long wavelengths but at short ones the RE contribution is relatively important [8].

Sample Preparation

The samples were deposited on glass and silicon substrates, respectively, for the magneto-optical study and for analyzing their composition by Rutherford backscattering analysis. The magneto-optical active layer was dc magnetron co-sputtered from elemental targets. Two silicon nitride protecting layers were deposited by reactive rf magnetron sputtering. The amorphous state of the samples was obtained by cooling the rotating substrate holder with liquid nitrogen. The deposition rates were around 0.2 nm/s for the metallic layers and 0.01 nm/s for silicon nitride layers. The base pressure was ~ 10^{-7} Torr and the sputtering pressure was $5x10^{-3}$ Torr of Ar. The typical thicknesses were around 33 nm, 66 nm and 120 nm, respectively, for silicon nitride, ferri-ferromagnetic (Gd-Co, Y-Co, Co, Fe) and sperimagnetic layers (Dy-Co, Ho-Co). The measured compositions of RE-Co layers were: $Y_{43}Co_{57}$, $Gd_{38}Co_{62}$, $Gd_{40}Co_{60}$, $Dy_{37.7}Co_{62.3}$, $Dy_{38}Co_{62}$, $Dy_{41}Co_{59}$ and $Ho_{33}Co_{67}$. In order to induce a well-defined in-plane easy axis, the samples were annealed under a 400 kA/m in-plane magnetic field, in a 10^{-5} mbar vacuum furnace, during 2 hours at 300°C.

LED based TMOKE magnetometer

Callegaro and Puppin [9] demonstrated that high intensity light-emitting diodes are an excellent alternative to fixed-wavelength magneto-optical measurements. The advantages offered by LEDs are:

- the use of incoherent light reduces the noise due to the speckles or other interference fringes which are extremely sensitive to the mechanical stability of the setup,
- low power consumption reduces power-on delays due to thermal stabilization,
- low price and highly stable intensity of this kind of light source,
- the relatively narrow bandwidth (~20 nm).
- amplitude modulation is easily achieved.

We used LEDs whose wavelengths (and intensities) are: blue 450 nm (0.6 cd), green 563nm (0.2 cd), red 660 nm (1.2 cd) and, sometimes, we also used yellow 585 nm (0.16 cd).

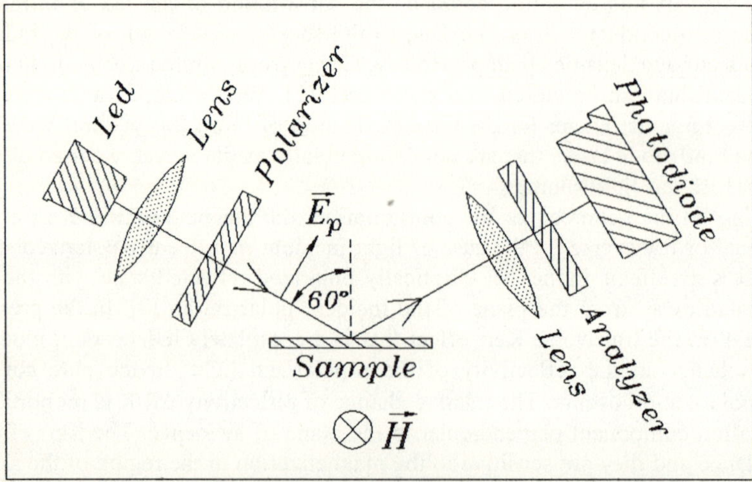

Figure 1. Optical diagram of the TMOKE-LED. The LED is modulated at 1 kHz and the photodiode is connected to a lock-in amplifier synchronized with the LED. A pair of Helmholtz coils generates the in-plane magnetic field (H).

The LED based TMOKE magnetometer (TMOKE-LED) apparatus consists of a LED modulated at 1 kHz by a function generator, the rotating dichroïc polarizer and analyzer, two focusing lenses, a Si photodiode connected to a lock-in amplifier. A pair of Helmholtz coils applies the magnetic field in the plane of the film up to 12 kA/m. The apparatus is shown schematically in Fig. 1. The angle of incidence of the p-polarized light beam is 60° relative to the normal of the sample. After reflection by the sample it crosses the analyzer which is also p-polarized and the resulting intensity is measured with the photodiode. This version of the experimental setup works at room temperature. From a practical point of view, this equipment corresponds to an adaptation of another TMOKE setup, working with a red laser diode [10].

Measurements and Discussion

Fig. 2 represents hysteresis loops obtained with the TMOKE-LED from the sample $Gd_{40}Co_{60}$. It can be seen that the noise corresponding to the green LED measurement is higher than to the blue and red ones. This is due to the low level of the signal measured with the green LED. In fact, the signals were around 10^{-2} to 10^{-3} V for the blue and red LEDs, 10^{-4} V for green and 10^{-5} V for yellow LEDs.

It can be seen that the loops for the $Gd_{40}Co_{60}$ sample are inverted if compared to the pure Co loop. This behavior was observed also for all Dy-Co and Gd-Co samples and it is probably due to the higher RE sub-lattice magnetization relative to the Co one [11]. For these samples, the concentration of Co is lower than in the case of the compensation composition at room temperature. By the other side, usual hysteresis loops were measured for the $Ho_{33}Co_{67}$ sample whose compensation temperature is below the room temperature. These results agree with that reported in Ref. 5, 8, and 12.

The hysteresis loops represented in Fig. 2 exhibit the same shape for the three LEDs, but the amplitude ($\Delta R/R$) at the magnetic saturation depends on the wavelength. It appears that the sample demonstrates a spectral dependence. Let us define $(\Delta R/R)_{sat}$ as the relative variation of the reflectivity for the saturation magnetic field. In Fig. 3, the module $(\Delta R/R)_{sat}$ is plotted for all samples as a function of the wavelength. In fact, the hysteresis loops for Gd-Co and Dy-Co are

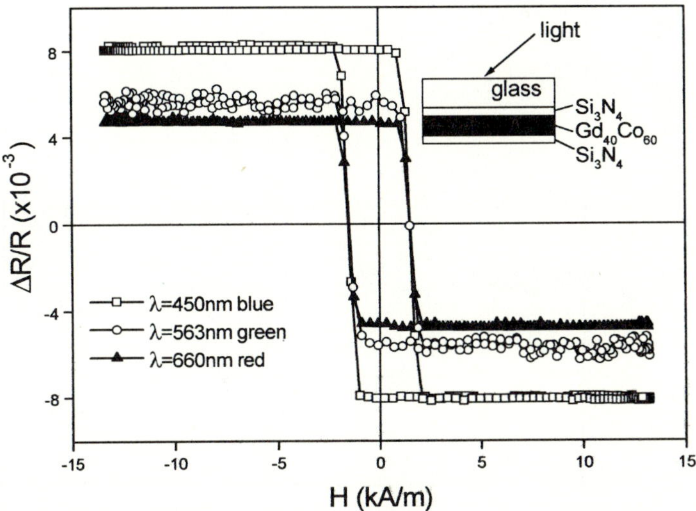

Figure 2. The hysteresis loops measured for a Gd-Co sample, for three wavelengths (λ). The inversion of the loops means the RE sub-lattice magnetization is higher than the Co one. All hysteresis loops were performed from the substrate side.

effectively inverted, giving a negative value. The results in this figure show that the TMOKE signal for the amorphous films increases while the wavelength decreases. This result is, surprisingly, in opposition to the spectral dependence of the polar Kerr effect for the same kind of films [7, 8, 12]. From the application point of view, this would be an interesting result, even if it is not well understood yet. The red light energy is 1.85 eV, high enough to excite the magnetic electrons of the Co, but not enough to extract a magnetization signal from the deeper 4f magnetic electrons of the RE atoms. For this wavelength the reduction by 73% of the TMOKE signal relative to the Co, can be considered as corresponding approximately to a kind of a volume averaged contribution of the Co atoms to the RE-Co amorphous thin films signal. The saturation TMOKE signal in function of the wavelength is almost constant for Co sample. While for Fe, it increases while the wavelength increases. The signal from the Fe film is higher than for all others Co based samples. For the composition around 60% of Co, it can be seen in Fig. 3 that TMOKE signal decreases from Dy-Co to Gd-Co and getting the minimum value for Y-Co samples. A similar result was obtained for polar effect by Hansen and collaborators [7]. These results also show that the difference between the TMOKE signal for each amorphous samples in relation to the Co one, increases when wavelength is shortened. The same result was reported by Ganshina [13], for amorphous $GdCo_2$ films. It indicates that the contribution of the RE is more important in the blue region of the spectra [8].

A rigorous calculation of the effect could not be performed as the optical constants of the

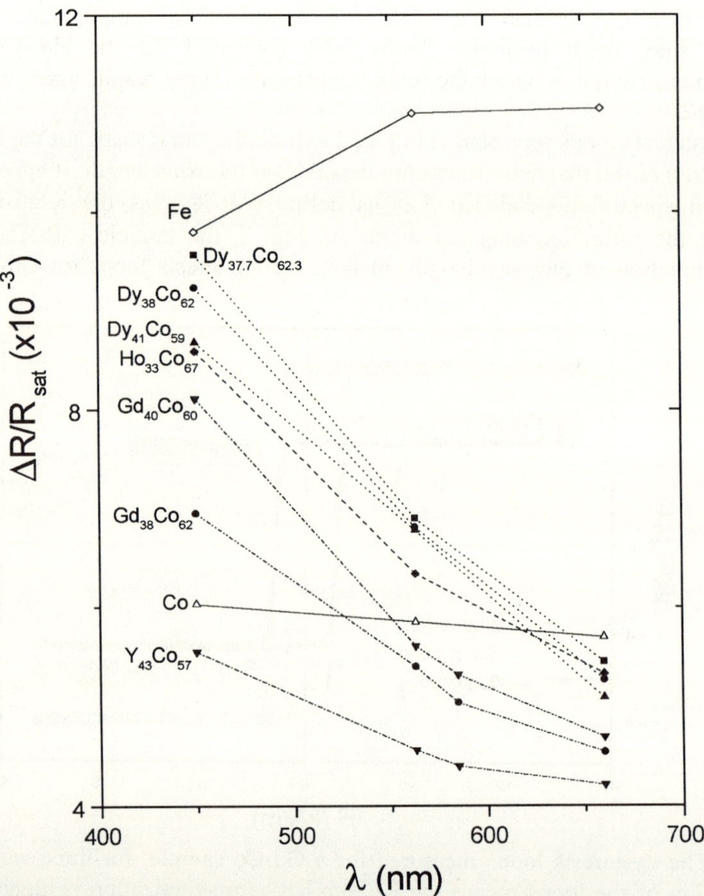

Figure 3. Saturation TMOKE signal (ΔR/R) in function of wavelength (λ) of the incident light for RE-Co amorphous and Co and Fe thin films.

RE-TM compounds were unknown yet. The changes in $\Delta R/R$ can be due to an increase of the magnetic contribution ΔR but it can also be attributed to a decrease of R. This is a classical artificial mode of enhancement of the effect by the means of an antireflecting coating [14]. The reflectivity coefficients were measured for all samples. Whatever the thicknesses of the protecting layers it appears a general behaviour: reflectivity was the lower for the red light and the higher value for the green with an intermediate value for blue light. This shows that the magnetic contribution ΔR is increased while wavelength decreases. This result was compared to previous measurements of polar Kerr effect published by Honda et al [15] on Dy-Co and Gd-Co and Choe et al [8] on Gd-Co compounds. Measurements of Dy_xCo_{1-x} alloys for lower x were reported by Honda. They showed θ_K increasing with λ but also an evolution towards an inversion of the slope of θ_K (= $f(\lambda)$) while increasing the Dy ratio x. In the case of Gd-Co compounds, θ_K exhibited a constant slope as a function of λ for x = 0.4. In all cases no correction of the Faraday effect of the substrate was applied. Choe et al [8] found a constant slope for $\theta_K = f(\lambda)$ for $Gd_{35}Co_{65}$ but the measurements indicate that this slope decreases while the Gd concentration increases.

Conclusion

The measurements that we performed show that the ultra-bright LEDs are a good light source for MO measurements. Our experimental data for Co based amorphous films give evidence that the transverse Kerr effect increases while decreasing wavelength, in the visible spectra, for yttrium and heavy rare earths like Dy, Ho and Gd. For Co, the transverse signal is almost constant and for Fe it increases when the light goes from blue to red. For a composition around 60 at. % Co, the Dy contribution is higher than Gd's one that is, itself, higher than Y contribution for MO signal. The comparison of the results of RE-Co samples with results of Co sample, lead to conclude that the rare earth contribution raises at low wavelengths. Moreover, it appears that the characteristic quantity of the TMOKE might be complemented by a specific figure of merit like $\theta_K\sqrt{R}$ in the case of the polar Kerr effect.

This work was supported by the Brazilian agencies FAPESP and CNPq.

References

[1] S. Nakamura, Physics World, February 1998, p. 31.
[2] P. Ball, New Scientist, 29 March 1997, p. 28.
[3] J.M. Florczak and E.D. Dahlberg, J. Appl. Phys., 67-12 (1990), p. 7520.
[4] P.Q.J. Nederpel and W.D. Martens, Rev. Sci. Instrum. 56-5 (1985), p. 687.
[5] P. Hansen, D. Raasch and D. Mergel, J. Appl. Phys., 75-10 (1995), p. 5267.
[6] S. Uchiyama, X. Y. Yu and S. Tsunashima, J. Phys. Chem. Solids, 56-11 (1995), p. 1557.
[7] P. Hansen, S. Klahn, C. Clausen, G. Much and K. Witter, J. Appl. Phys., 67-5 (1990), p. 3194.
[8] Y.J. Choe, S. Tsunashima, T. Katayama and S. Uchiyama, J. Magn. Soc. Jpn 11-S1 (1987), p. 273.
[9] L. Callegaro and E. Puppin, Rev. Sci. Instrum. 66-11 (1995), p. 5375.
[10] T. Yonamine, Y. Souche and A.D. Santos, J. Appl. Phys., 79-8 (1996), p. 5266.
[11] A.P.B. Tufaile, M.Sc. Thesis, Universidade de São Paulo, Brazil, 1996.
[12] K. Lee and N. Heiman, AIP Conf. Proc., 18 (1974), p. 108.
[13] E. A. Ganshina, private communication, Moscow State University.
[14] A. V. Sokolov, Optical Properties of Metals, Blackie and Son, London (1967).
[15] S. Honda and M. Yoshiyama, Jpn. J. Appl. Phys., 27 (1988), p. 1687.

Materials Science Forum Vols. 302-303 (1999) pp. 120-124

Low Temperature Magnetic Characterization of Sperimagnetic Films by TMOKE

A.P.B. Tufaile and A.D. Santos

Instituto de Física, Universidade de São Paulo, CP 66318, CEP 05389-970, São Paulo, Brazil

Keywords: Magneto-Optics, Thin Films, Sperimagnetism, Amorphous Materials

Abstract

Using the Transverse Magneto-optical Kerr effect, we have studied the sperimagnetic Dy-Co amorphous thin films. We developed a low temperature TMOKE magnetometer to measure the in-plane hysteresis loops for thin films, in the temperature range from 10 to 300K. These hysteresis loops show quite interesting curves and can give new insights on the understanding of the sperimagnetism in RE-TM amorphous thin films.

Introduction

The magneto-optical magnetometry is recognized as the easiest way to quickly characterize the hysteresis loops of magnetic thin films. In some special cases, as for rare earth (RE)-transition metal (TM) alloys, using red light, we get the hysteresis loops only of the TM element in the alloy [1-3]. In the past, using this characteristics of the TMOKE, we have studied the macroscopic ferri-ferromagnetic transition [2] occurring on the amorphous $YCo_2/GdCo_2$ bilayers. In the Ref. [2], we presented the room temperature experimental data. In order to analyze the temperature dependence [3] of this phenomenon, we developed a low temperature TMOKE (TMOKE-lt) magnetometer. Here we will present our TMOKE-lt magnetometer and after the study of the sperimagnetic Dy-Co amorphous thin films. The main application of the RE-TM amorphous thin films is as magneto-optical recording media [4]. Usually they present sperimagnetism [5,6] and its compensation temperature (T_{comp}) is the best condition to invert its magnetization during the recording process [4]. In this paper, we show a new experimental insight on the Co sub-lattice magnetic contribution to the sperimagnetism of RE-TM amorphous thin films, in the temperature range near T_{comp}.

Sample Preparation

Co-deposition using electron-beam sources (or for some samples, magnetron sputtering) produced the samples from elemental targets. The glass and Si substrates were cooled to liquid N_2, giving amorphous films. The thickness and composition were determined by Rutherford backscattering spectroscopy. The samples were field annealed in a 10^{-5} mbar vacuum furnace, by 3 hours at 300°C, under a 0.5 T in-plane magnetic field. To the magnetic characterization we used a vibrating sample magnetometer (VSM) in a 9 T superconducting magnetic coil and the TMOKE-lt magnetometer. All the data presented in this paper was obtained from the amorphous film of composition $DyCo_{2.3}$ and thickness 110 nm.

TMOKE-lt magnetometer

The instrumentation based on the magneto-optical Kerr effects is quite useful to the magnetic characterization of thin films. There are three different Kerr effects (polar, longitudinal and transversal) in reflection [7,8], where the transversal is a completely independent mode, permitting to obtain the in-plane component of the magnetization. It corresponds to the relative variation of the reflectivity (ΔR/R) and in our geometry is proportional to the magnetization component parallel to the applied magnetic field.

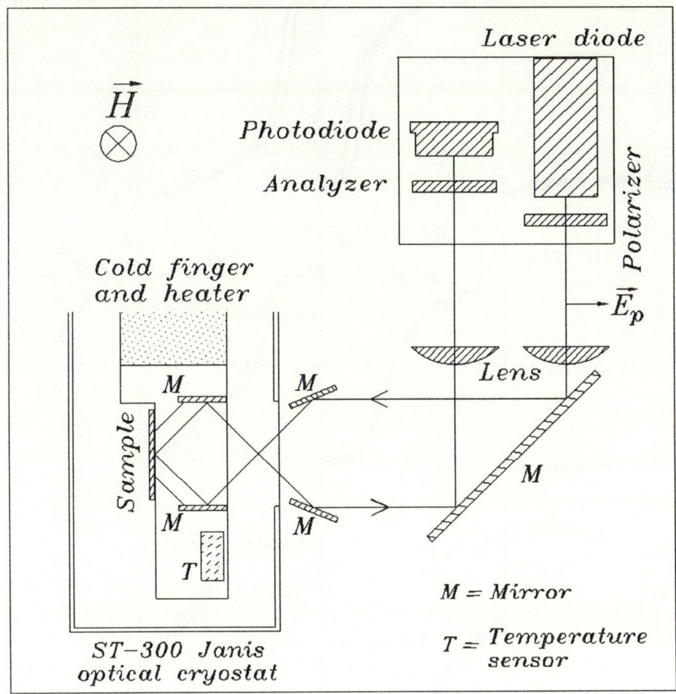

Figure 1: Schematic representation of the optical setup of the low temperature TMOKE magnetometer.

The TMOKE-lt magnetometer we developed is based in a laser diode (λ= 670 Å) modulated at 5 kHz. We utilized a Janis Optical Cryostat, model ST300, in the pole gap of a 2 T electromagnet. As is described in the Fig. 1, the p-polarized incident light beam makes 45° to the sample normal. All the optical components were assembled inside a 500 mm long, 25x50 mm^2 rectangular aluminum tube. Inside the tube we have the laser diode, the rotating polymer polarizer and analyzer, the photodiode, the mirrors and two lenses addressed to conjugate the focus between the laser-diode, the sample and the photodiode. As the sample is 20 mm far from the optical window of the cryostat, we included two mirrors in the sample holder, to keep the incident angle at 45°. The magnetic field is in-plane and the photodiode signal is analyzed by a lock-in amplifier connected to a PC microcomputer. The TMOKE-lt works from 6 to 400K without any optical or mechanical realignment, it can operate in the transverse and longitudinal MOKE modes, and can measure hysteresis loops and magneto-optical transversal susceptibility (χ_t) [9].

Results and Discussion

In the Fig. 2, we present for the temperature range from 10 to 250K, the TMOKE hysteresis loops obtained for the Dy-Co amorphous film. As TMOKE is an indirect method of magnetic characterization, it always needs interpretations. In our case, the hysteresis loops corresponds only to the magnetic contribution of the Co sub-lattice. Even the loops being quite complicated, we can extract some interesting information. Below 200K the hysteresis loops are inverted, indicating that the Co sub-lattice magnetic moment is lower than the Dy one. As is already known for RE-TM amorphous alloys the coercivity consistently raises just to T_{comp} [4], and then abruptly reduces to a quite low value. Above 200K the hysteresis loops for low magnetic field presents normal sense. In spite of that, it shows a new slow inversion of the magnetization direction of the Co sub-lattice at a magnetic field higher than the coercive field, as can be seen in the Fig. 2, from d) to f). In the Fig. 2e), superimposed to the TMOKE curve at 238K, we show the magnetization curve obtained by VSM at the temperature of 230K. There, we can see no visible effects of this inversion of the Co sub-lattice on the magnetization curve. Meaning the total magnetic moment of the Dy sub-lattice is also inverting its sense.

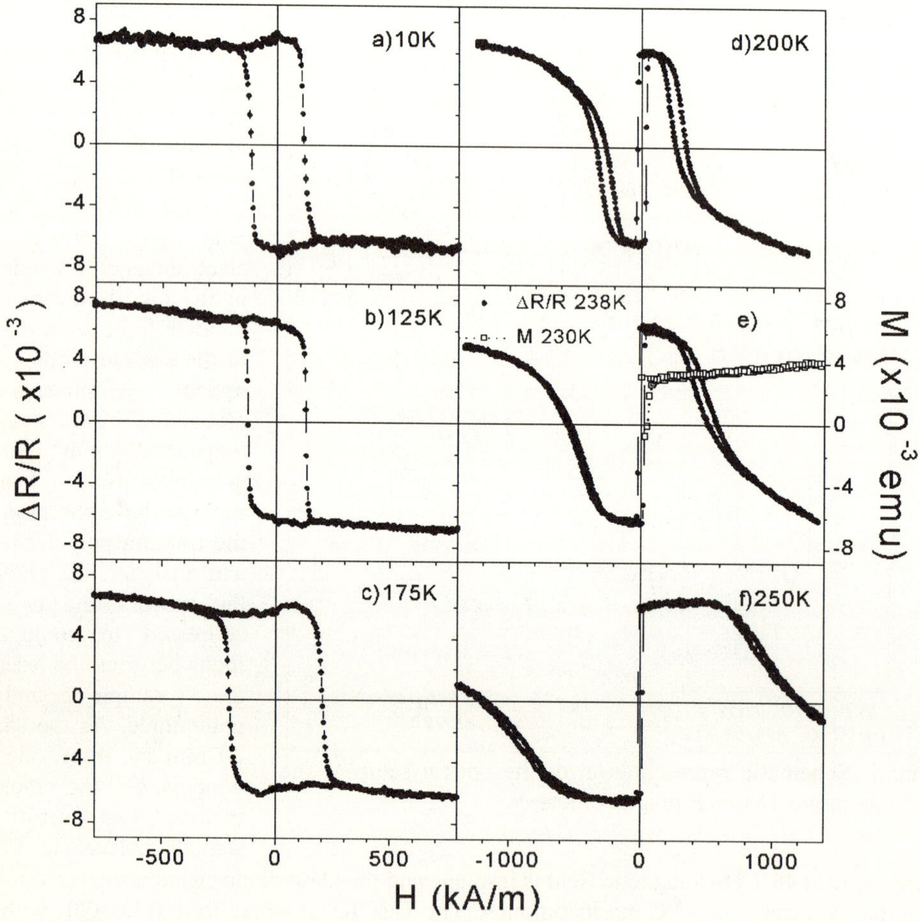

Figure 2: TMOKE hysteresis loops of the Co sub-lattice for the $DyCo_{2.3}$ amorphous film. In e) is superimposed the magnetization curve (□) to 230K.

We can observe also, what we call the H_c for T< 200K, becoming a kind of transition field, H_t, at temperatures higher than T_{comp}. H_t is about .4 T near T_{comp} e raises to about 1.5 T at room temperature. This transition field appears in samples produced by e-beam and magnetron sputtering, and was observed in Dy-Co, Ho-Co and even in a ferrimagnetic Gd-Co amorphous thin film. It is not well understood yet, but seems to be an intrinsic phenomenon and not related to the possible heterogeneity in the samples. In the Fig. 3, we present a new series of measurements done on the same sample, but not in the same place. We have the TMOKE hysteresis loops for temperatures in the range from 205 to 220K. In this series we obtained at 211K what we can considerate the TMOKE equivalent to the T_{comp}. In this temperature we have the inversion of the Co magnetization in the remanence. Considering the light beam is about tenths of 1 mm^2 and, the strong on

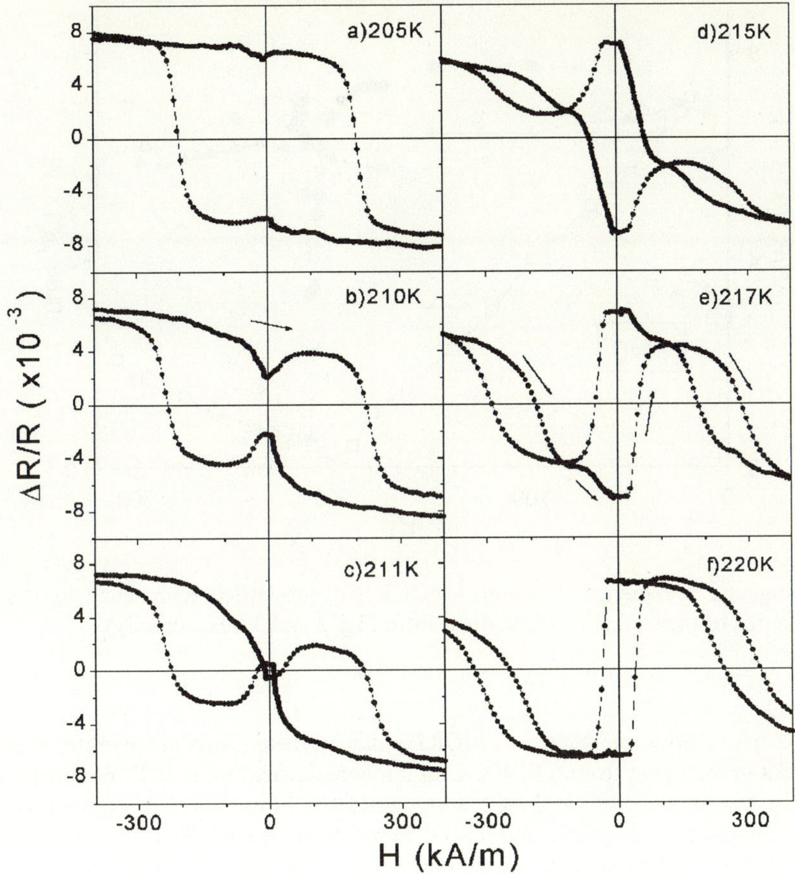

Figure 3: TMOKE hysteresis loops of the Co sub-lattice for the $DyCo_{2.3}$ amorphous film.

composition T_{comp} dependence, this higher value of T_{comp} indicates some heterogeneity in the local stress, anisotropy or composition of the sample.

In the Fig. 4 we present the magnetic remanence (□) in function of the temperature. These measurements were done by saturating the sample at 4.2K, removing the magnetic field and then raising the temperature just to the room temperature. Together with the magnetic remanence in the Fig. 4, we put the TMOKE Co sub-lattice remanence, obtained from Fig. 2 and 3. For the first experimental series (▲) we have a roughly monotonic decreasing of the remanence, but to the second series (●), we can see clearly the remanent magnetic moment of the Co sub-lattice going to zero at 211K. This figure also show the strong difference on the temperature dependence of both, the TMOKE and magnetic remanence. Using mean field theory [10], we have fitted [11] this experimental results and shown that the slowly decreasing TMOKE data corresponds only to the magnetic contribution of the Co sub-lattice, that is almost constant below room temperature. Finally, returning to the Fig. 3, the hysteresis loops near T_{comp} present several details that were not analyzed, and need an extensive study in new samples, in order to differentiate what is intrinsic to the sperimagnetism or a specific characteristic of each sample.

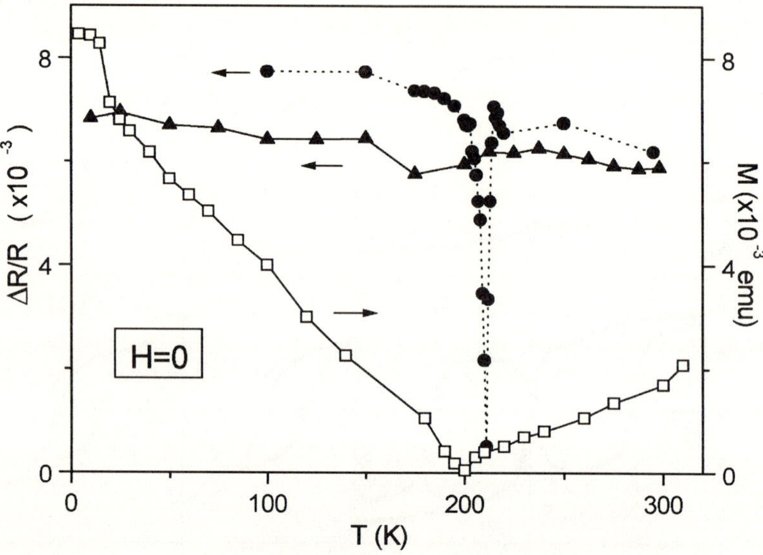

Figure 4: Magnetic remanence (□) and TMOKE Co sub-lattice remanence to first (▲) and second (●) series of measurements (obtained from Fig. 2 and 3, respectively).

Conclusions

We have developed a low temperature TMOKE magnetometer, easy to operate, stable and very precise. It works in the range from 6 to 400K, in a magnetic field up to 2 T. We analyzed TMOKE Co sublattice hysteresis loops of the Dy-Co amorphous thin films. There is a quite good correlation between T_{comp} obtained by magnetic and TMOKE methods. Above T_{comp}, we show an interesting transition, not present in the magnetic measurements, that can help us to understand in more details the sperimagnetism of the RE-TM amorphous thin films.

This work was supported by the Brazilian agencies FAPESP and CNPq.

References

[1] K. H. J. Buschow, in "Ferromagnetic Materials" (E.P. Wohlfarth and K. H. J. Buschow, ed. Elsevier Science Publishers B. V., 1988), vol. 4, p. 493.

[2] T. Yonamine, Y. Souche and A. D. Santos, J. Appl. Phys., 79(8) (1996), p. 5266.

[3] T. Yonamine, A.P.B. Tufaile, Y. Souche and A.D. Santos, J. Appl. Phys., 81(8) (1997), p. 5722.

[4] A.B. Marchant, "Optical Recording: A technical overview" (ed. Addison-Wesley Publishing Co, NY, 1990).

[5] J.M.D. Coey, J. Chappert, J.P. Rebouillat and T.S. Wang, Phys. Rev. Lett., 36-17 (1976), p. 1061.

[6] T. Kaneyoshi, "Introduction to Amorphous Magnets" (ed. World Scientific, Singapore, 1992).

[7] Z.J. Yang and M.R. Scheinfein, J. Appl. Phys., 74 (1993), p.6810.

[8] J.M. Florczak and E.D. Dahlberg, J. Appl. Phys.,67-12 (1990), p. 7520.

[9] A.D. Santos, L.G.C. Melo, C.S. Martins, F.P. Missell, Y. Souche, F.L.A. Machado and S.M. Rezende, J. Appl. Phys., 79(8) (1996), p. 6546.

[10] R. Hasegawa, B.E. Argyle and L.-J. Tao, AIP Conf. Proc., 18 (1974), p. 110.

[11] A.P.B. Tufaile, M.Sc. Thesis, Universidade de São Paulo, Brazil, 1996.

Materials Science Forum Vols. 302-303 (1999) pp. 125-133

Colossal Magnetoresistance in Manganites: Role of the Electronic Phase Segregation

M.R. Ibarra and J.M. De Teresa

Dpto. de Física de la Materia Condesada&ICMA, Universidad de Zaragoza-CSIC, ES-50009-Zaragoza, Spain

Keywords: Colossal Magnetoresistance, Manganites, Magnetotransport, Magnetovolume, Metal-Insulator Transition

ABSTRACT

The huge intrinsic magnetoresistance found in mixed-valence manganites has as origin the existence of a metal-insulator transition. Usually, this transition is accompanied by a para-ferromagnetic transition. In the paramagnetic phase the formation of dynamic phase segregation in the form of magnetic polarons provides the mechanism for carrier localisation. The insulator state can also be achieved in the antiferromagnetic state or when the long-range ferromagnetic interaction is not extended across the whole sample, giving rise to the existence of static phase segregation in which metallic ferromagnetic clusters coexist with insulating paramagnetic, antiferromagnetic or spin-glass regions. In such insulating regions, charge ordering can also be present. The application of a magnetic field can increase the size of the ferromagnetic clusters up to a critical field at which they start to percolate and, consequently, the carrier mobility is strongly increased producing giant magnetoresistance.

In this communication we report on experiments of magnetovolume effects and neutron diffraction including SANS (small angle neutron scattering) giving evidence for such effects. They can be explained by the interplay of magnetic long-range interaction and the large electron-phonon interaction observed in these compounds. The relevance of this interaction for the electronic phase segregation has been proved by means of the oxygen isotopic effect.

1 INTRODUCTION

The recent upsurge of the investigation of manganese oxides is framed within the investigations on new materials with giant magnetoresistance (GMR). A magnetoresistive effect occurs in all metals as a consequence of the Lorentz force acting on moving electrons. This effect is very small (<1%) and positive, i.e. the electronic scattering increases with the applied magnetic field. However, in the case of magnetic systems this scattering can be drastically modified due to the field effect on the magnetic moments. The existence of an intrinsic spin moment provides the mechanism for the giant magnetoresistance (50%) found in magnetic multilayers [1]. This effect is also observed in bulk magnetic single-layered alloys like FeRh [2]. The GMR has also been found in granular systems in which ferromagnetic nanoparticles are embedded in a non-magnetic metal film [3]. The phenomenon of GMR has been broadly explained using a simple two-current model [4]. In this model two independent conduction channels are considered. These two channels are defined by the spin-up and spin-down conduction electrons, the total current being the addition of the current in both channels.

Recently, GMR has been found in a different kind of compounds: thin films of ferromagnetic manganese-based oxides [5]. The figures for the magnetoresistance (MR) ratios were as high as 10^4% in bulk polycrystals of these compounds [6]. The effect was called colossal magnetoresistance (CMR). This huge value for the MR can be increased even up to seven orders of magnitude as the nature of the CMR is related to the metal-insulator transition which takes place in these compounds. In mixed valence manganites of formula $R_{1-x}A_xMnO_3$ the substitution of a R^{+3} ion by an alkaline earth A^{+2} gives rise to the appearence of 3d holes (Mn^{+4}, $3d^3$) in a narrow band with e_g symmetry. This is due to the octahedral O^{-2} environment of the Mn, which splits the five-fold orbital degeneracy of the ground state in a triplet ground state t_{2g} and a e_g doublet. It is well established that a strong intra-atomic Hund coupling giving a local magnetic moment with the maximum multiplicity (also called high-spin state) exists. The itinerant character of the e_g electron (or hole) is provided by the covalent nature of the Mn-O bond, which produces a net charge transfer between Mn ions. This mechanism gives rise to an indirect ferromagnetic interaction (Zener-type) called double exchange [7] and explains the existence of long-range ferromagnetic order in these mixed-valence manganese oxides. This interaction coexists with the antiferromagnetic superexchange, present in all 3d oxides. Other interactions should also be considered in order to explain not only the magnetic behaviour but also the magnetotransport properties, i.e. the electron-phonon interaction. This interaction is particularly strong and relevant in the series $La_{1-x}Ca_xMnO_3$ as has been experimentally demonstrated by thermal expansion and magnetostriction measurements [8], the isotopic effect [9] and infrared spectroscopy [10]. At high temperatures, in the paramagnetic insulating phase the strong electron-phonon coupling provides the mechanism for the carrier localization in the form of small lattice polarons as proposed in Holstein's model [11-14]. A narrow polaronic band is formed at temperatures higher than $\Theta_D/2$ (Θ_D is the Debye temperature). At low temperatures the broadening of this band enhances the itinerancy of the carrier . The electron-phonon interaction also influences the ferromagnetic interaction [9]. Goodenough proposed a phenomenological approach in which a vibrational contribution should be considered as a result of the dynamical Jahn-Teller effect [15]. This last mechanism does not imply charge transfer between Mn^{+3} ions but a resonant state in which the local moments tend to be aligned.

The lifetime of the lattice polaron at specific lattice sites is longer by some orders of magnitude than the characteristic electronic hopping time of the eg electron between Mn ions. As a consequence, the charge (eg electron) trapped at the polaron time scale, in the meantime, visits the neighbouring Mn^{+4} ions. This charge transfer (Mn^{+4}-O^{-2}-Mn^{+3}) gives rise to a short-range ferromagnetic interaction. The entity formed by the small lattice polaron+ferromagnetic cluster is referred to as magnetic polaron. These magnetic polarons form in the paramagnetic phase and constitute a sort of *dynamic electronic phase segregation,* which is responsible for the intrinsic magnetoresistance in CMR perovskites [16, 17].

When the La ion is substituted by another rare earth ion, electronic and structural changes can take place. The accommodation of a smaller A ion in a cubic perovskite produces a tilting and buckling of the oxygen octahedra which give rise to orthorhombic or rhomboedral structures in which the Mn-O-Mn bond angle is smaller than 180°. This distortion of the structure has a strong influence on the ferromagnetic interaction, this interaction being weaker as such an angle becomes smaller. This effect has been broadly studied by Jirak [18] during the last two decades. It has been broadly observed that the weakening of this interaction brings about a vaste phenomenology, which we have recently reviewed [19].

Other mechanism which is effective in order to weaken the ferromagnetic interaction is the existence of Mn vacancies. This is achieved when $LaMnO_3$ perovskites are prepared in an oxygen atmosphere. The excess of oxygen can not be accommodated in the perovskite structure and equal amount of Mn and La vacancies are produced [20,21]. The effect of these vacancies is first to increase the amount of Mn^{+4}, which gives rise to an increase of the ferromagnetic interaction. This is true up to a value of 14% of Mn^{+4}, which corresponds to the existence of a 2.5% of Mn vacancies. However, for a larger amount of vacancies the coherence of the long-range order is broken and a cluster glass state is achieved at 5% of vacancies when the Mn^{+3}/Mn^{+4} ratio attains the value of 30% [19,22].

In both cases, i.e large lattice distortions or Mn vacancies, the magnetotransport, magnetic and magnetovolume effects can be explained considering the existence of *static phase segregation* at low temperature. In these cases the phase segregation consists, as will be described herein, of ferromagnetic clusters embedded in a paramagnetic matrix. There exist other examples in which the ferromagnetic clusters coexist with charge-ordered regions, for instance in the $La_{1-x}Ca_xMnO_3$ [14, 19, 23] series and in $(La_{0.5}Nd_{0.5})_{2/3}Ca_{1/3}MnO_3$ [24]. The effect of the field is to increase the cluster size up to a critical value in which they percolate and the ferromagnetic state is reached. We have also studied the role played by the electro-phonon interaction in the existence of the phase segregation described above. With such an aim we have studied the oxygen isotope effect in some relevant compounds [24, 25].

In this contribution we report several experimental results showing the existence of electronic phase segregation in manganese perovskites and we try to argue that this hypothesis can explain all the observed phenomenology.

2 EXPERIMENTAL PROCEDURES

The samples were prepared either by standard sol-gel procedure or by solid-state reaction (more details can be found in [19] and references therein). The quality of the samples was checked by x-ray diffraction and scanning electron microscopy. The magnetization was measured using a vibrating sample magnetometer and a SQUID using steady magnetic fields up to 12 and 5 Tesla respectively. a.c. initial susceptibilty measurements were performed using a mutual inductance bridge. The thermal expansion was measured within the temperature range 4 K-1000 K. Below room temperature the strain gauge technique was used. For high temperatures the expansion was measured using a cryofurnace and the "push rod" method. The standard four-contact technique was used to measure the electrical resisitivity. The parallel and perpendicular to the field magnetostriction were measured using the strain-gauge technique using steady magnetic fields up to 12 Tesla and pulsed fields up to 14 Tesla. The neutron diffraction experiments were performed at the high-flux reactor of the Institute Laue Langevin in Grenoble. The D1B powder diffractometer with a wavelength of 2.52 Å was used for thermal scans in the temperature range 2.5 K-600 K. Information about the structural changes and magnetic transitions and structures was obtained by refinement of the neutron spectra using the FULLPROF program. The SANS experiments were performed at the D16 using a wavelength of 4.5 Å and with a resolution in the transfer momentum vector **Q** of 5×10^{-3} Å^{-1} within the range 0.013 Å^{-1}<Q<0.65 Å^{-1}. The versatility of this instrument allows applying magnetic fields up to 5 Tesla and rotating the multidetector an angle of $2\Theta=100°$, which permits also obtaining the thermal and field dependence of the intensity of Bragg peaks.

3 MAGNETOVOLUME

First, we would like to emphasize why a section on magnetovolume effects should be considered in a general description and explanation of the relevant magnetic and magnetotransport properties of manganites. Little atention has been paid in the past to this kind of investigations. The main source of magnetovolume effects in magnetic solids with stable local magnetic moment are the volume dependence of the crystal electric field interaction and the spatial dependence of the exchange interaction. Other important source for this effect is the change of local magnetic moment as a consequence of longitudinal magnetic fluctuations, giving rise to the invar and anti-invar effect. However, in our pionnering work on magnetovolume effects in mixed-valence CMR manganites we discovered a large effect in the paramagnetic phase that could not be explained considering the abovementioned mechanisms [14]. We found a large contribution in the paramagnetic phase over the usual lattice thermal expansion, which is given by the Grüneisen law (see Fig. 1a). This contribution vanishes at the para-ferromagnetic transition, which marks the change from a polaronic to a metallic behaviour of the resistivity as shown in Fig. 1b. This extra and novel anharmonic contribution was associated with the charge localization, which gives rise to a state of higher volumen. This spontaneous effect is suppressed by the application of a magnetic field as can be observed in Fig. 1a. In a simple generalization of these results one can consider the upper thermal expansion curve (high volume state HV) as the state in which the charges are localized and the lower curve (low volume state LV) as the metallic state. Consequently, the spontaneous or field-induced metal→insulator transition is associated with a LV→HV transition. Moreover, in this compound there exists a very similar field dependence of the volume magnetostriction and the resistivity that makes reliable the former assumption (see Fig. 2).

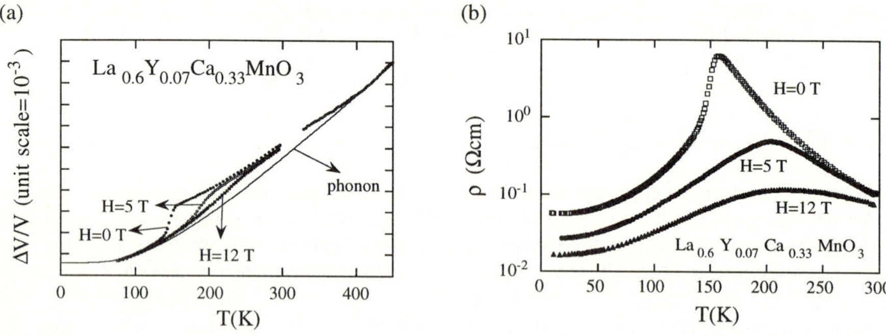

***Fig. 1 (a)** Volume thermal expansion under several applied magnetic fields (0, 5 and 12 Tesla) and calculated phonon Grüneissen-type contribution for the compound* $La_{0.6}Y_{0.07}Ca_{0.33}MnO_3$. ***(b)** Thermal dependence of the resistivity at applied magnetic fields of 0, 5, and 12 Tesla in the same compound.*

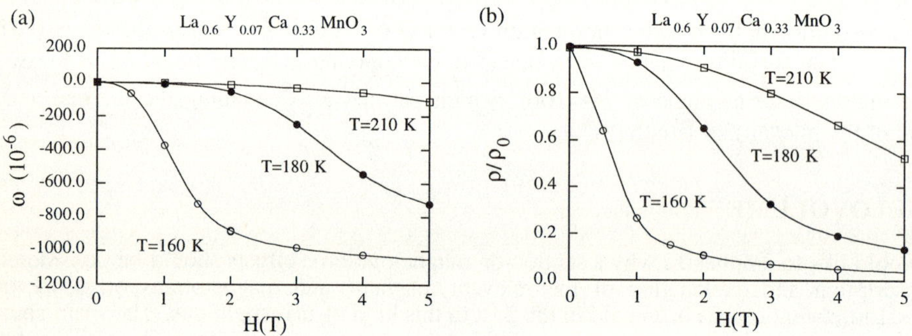

***Fig. 2 (a)** Volume magnetostriction at fixed temperatures in the paramagnetic regime (160, 180 and 210 K) for the compound* $La_{0.6}Y_{0.07}Ca_{0.33}MnO_3$. ***(b)** For the same compound, resistivity as a function of the magnetic field at the same temperatures.*

After this first finding, we have produced a large amount of experimental work corroborating these ideas [19]. We will show in the next section that the localised charges within the paramagnetic phase form magnetic polarons [16]. In Fig. 3 we show the thermal expansion curves obtained for the series $(La_{1-x}Tb_x)_{2/3}Ca_{1/3}MnO_3$. Two samples of the same nominal concentration but slightly different carrier concentration (#1~33% Mn^{+4} and #2~28% Mn^{+4}) show the same behaviour observed in the compound described previously. The thermal expansion curves collapse into one HV curve in the polaronic regime and in a LV curve when the metallic state is achieved. At high Tb doping rate (x>0.33) the thermal expansion results can be interpreted as the polaronic state remains down to the lowest temperature and the metallic state is not achieved. In fact, that is observed in these compounds as found in resistivity measurements (see Fig. 3b). The sample x=0.25 constitutes an interesting case. For this compound the thermal expansion at low temperature shows values which are between the HV and LV curves. One can interpret that the insulator-metal transition is not complete in the whole sample coexisting at low temperature insulator with ferromagnetic (metallic) regions. This is in agreement with the resistivity value at 4 K (see Fig. 3b), which is an intermediate value between the insulating compound, x=0.33 ($\rho \sim 10^{12}$ Ωcm), and the conducting x=0 ($\rho \sim 10^{-3}$ Ωcm). These results support the model of the existence of a two-volume state, one HV-insulating and the other, LV-conductive.

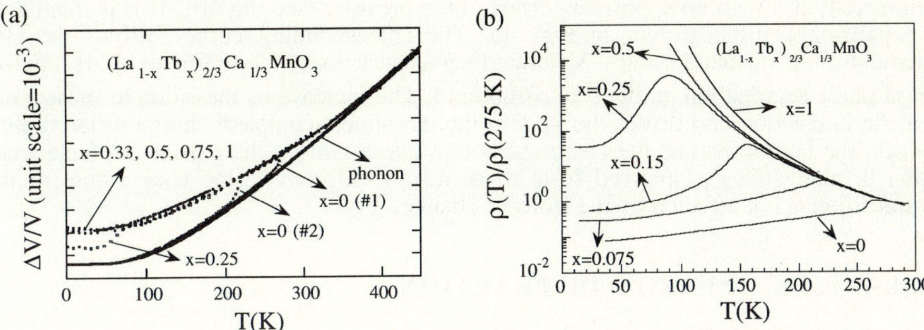

*Fig. 3 (a) Volume thermal expansion curves for selected compounds of the series $(La_{1-x}Tb_x)_{2/3}Ca_{1/3}MnO_3$ (symbols). The phonon contribution is represented by a continous line. **(b)** Thermal dependence of the normalized resistivity for selected compounds of the same series.*

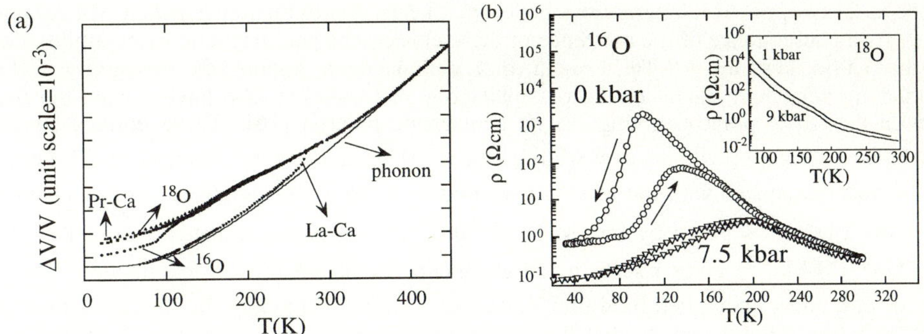

*Fig. 4 (a) Volume thermal expansion curves for selected compounds with fixed carrier concentration (Mn^{+4}/Mn^{+3}=0.5). The compound $Pr_{2/3}Ca_{1/3}MnO_3$ (Pr-Ca) is a prototype for the insulating state; $La_{2/3}Ca_{1/3}MnO_3$ (La-Ca) is a prototype for a compound with metallic behaviour at low temperatures. ^{18}O and ^{16}O stands for the corresponding isotopic compounds of $(La_{0.5}Nd_{0.5})_{2/3}Ca_{1/3}MnO_3$. The phonon contribution is represented by a continous line. **(b)** Thermal dependence of the resistivity at different pressures of the $(La_{0.5}Nd_{0.5})_{2/3}Ca_{1/3}MnO_3$ isotopic compounds.*

We have also studied the relevance of the electron-phonon interaction on the phase segregation [24, 25]. In that sense, we have studied the isotopic effect on the thermal expansion and volume magnetostriction. We will focus on the results found in the compound $(La_{0.5}Nd_{0.5})_{2/3}Ca_{1/3}MnO_3$. This compound, when prepared with natural oxygen (mainly ^{16}O), has a transition from a localised (below 400 K) to a charge-ordering state at T_{CO}=200 K. At T_C=100 K this compound becomes ferromagnetic [26]. The thermal expansion results are shown in Fig. 4a, where they can be compared with the HV curve corresponding to the insulator state at low temperatures ($Pr_{0.7}Ca_{0.3}MnO_3$) and LV metallic $La_{2/3}Ca_{1/3}MnO_3$. The behavior is very smilar to the previously described

$(La_{0.75}Tb_{0.25})_{2/3}Ca_{1/3}MnO_3$. We can make the same assumption, i.e. the low temperature is governed by the coexistence of ferromagnetic-metallic and charge ordered-insulating regions. Indeed, this behaviour would explain the thermal dependence of the resistivity shown in Fig. 4b where we can see that the application of a pressure of 4.5 kbar completely drives the system into the metallic state. After the ^{18}O oxygen isotope exchange, the compound becomes insulator even at the lowest temperatures [24, 26], with no significant effect of the pressure (see Fig 4b). This is manifested in the thermal expansion results, shown in Fig. 4a. The corresponding curve follows the HV curve, characteristic of the insulator state. Volume magnetostriction measurements [24] confirmed the existence of phase segregation in the ^{16}O compound. The increase of the isotopic mass weakens the ferromagnetic interaction and drives the system into an almost complete charge-ordered phase. As a consequence, the large effect of the electron-phonon interaction on the existence of electronic phase segregation in manganites is inferred from these results. Moreover, the temperature of the charge ordering transition is not affected by the isotopic change.

4 SMALL-ANGLE NEUTRON DIFFRACTION

We have used the SANS facilities at the high-flux reactor of the ILL. The used instrument (D16) allows determining magnetic correlation lengths, ξ, within the interval 10 Å<ξ<100 Å. Using this technique we can obtain ξ by fitting the q-dependence of the neutron intensity to a lorentzian function: $I=I_0/(q^2+(1/\xi)^2)$. In Fig. 5a we show the thermal dependence of ξ for the compound $La_{0.60}Y_{0.07}Ca_{0.33}MnO_3$, without field and under applied field of 5 Tesla. In both cases we can see how ξ diverges at T_C=150 K. This is an expected behaviour, when the ferromagnetism sets in. The ferromagnetic regime is associated with a metallic state in which the resistivity drops in four orders of magnitude as ξ diverges. At temperatures above T_C, ξ keeps approximately constant ≈14 Å up to well above T_C. As a consequence of the existence of these clusters the paramagnetic susceptibility does not follow the Curie law up to $T\approx 2T_C$. In this paramagnetic regime the resistivity follows a semiconducting behaviour characteristic of small polaron formation. We have related the magnetic cluster with the lattice polaron giving rise to a magnetic polaron [16]. These entities constitute a dynamic phase segregation with an average life time $\tau>10^{-5}$ s at a specific lattice site [27]. During this time the e_g hole (associated with the Mn^{+4}) is shared with the Mn^{+3} neighbour with a characteristic hopping time $\tau_h<10^{-9}$ s given by the 3d bandwidth. This electronic transfer within the magnetic polaron (Mn^{+3}-O-Mn^{+4}) gives rise to a ferromagnetic interaction (double exchange). The thermal dependence of ξ under an applied field of 5 Tesla shows similar behaviour. In this case ξ diverges at a higher temperature. This suggests that the long range ferromagnetic interaction is stable at higher temperatures. The field dependence of ξ is shown in Fig. 5b. In this figure we can observe how at temperatures above T_C the size of the magnetic polarons increases with field and at a determined value of the magnetic field ξ diverges and, simultaneously, the resistivity and volume decrease drastically. We interpret these effects as a result of the percolation of the magnetic clusters above a critical size which brings about the insulator-metal transition responsible for the colossal magnetoresistance in mixed-valence manganites.

We have reported in the previous section several compounds of the series $(La_{1-x}Tb_x)_{2/3}Ca_{1/3}MnO_3$ in which the increase of the Tb concentration produces a weakening of the ferromagnetic interaction. This is due to the dependence of the charge transfer with the Mn-O-Mn bond angle. Within the concentration range 0<x<0.25 the compounds experience an insulator-metal transition (as shown in Fig. 3b). However, for larger doping concentrations the samples are insulators at the lowest temperature. The compound x=0.33 was characterised as an insulator cluster glass at low temperature with a freezing temperature of 50 K [27]. The thermal dependence of ξ (shown in Fig. 5c) indicates the existence of magnetic clusters down to 2 K. Below 50 K the ferromagnetic clusters, of average size 17 Å, are frozen and the resistance increases up to more than 10^{13} Ω at 4 K. This is a new ground state in which this form of static electronic phase segregation gives rise to an insulator state. The application of a large enough magnetic field (H>4 Teslas) is able to induce the metallic state. We assume that at these high values of the magnetic field the clusters are aligned and the electronic hopping between clusters is possible and, consequently, a new cuasi-ferromagnetic percolated state is formed. From our SANS experiment we can see how ξ diverges at 4 Teslas but we can not warranty an infinite value of ξ as should occur in a true ferromagnet. The same uncertainty is present in our measurements of ξ in the compound ξ=0.25. It clearly seen in Fig. 5c an spontaneous divergence of ξ at T_C≈60 K. From the high value of the resistivity found in this compound at low temperature and the intermediate value of the volume between the LV (metallic) and HV (insulator) (see Fig. 3b), we consider the low temperature ground state of this compound as a cuasi-ferromagnet with an average magnetic correlation length larger than 100 Å. The application of a magnetic field saturates the sample and the metallic ferromagnetic state is achieved when the large clusters percolate and extend across the whole sample.

The behaviour of the mixed valence compounds $LaMnO_{3+\delta}$ can be accounted for in a similar way. In Fig. 5d we display the thermal dependence of ξ for the compound δ=0.07 and δ=0.15. The compound δ=0.07 shows a divergence of ξ at T_C=150 K. The thermal dependence of the SANS intensity (no reported here, see reference [22]) shows a significant contribution well below T_C down to 4 K. This behaviour and the magnetotransport measurements indicate the existence of a cuasi-ferromagnet insulator ground state at low temperature, which can saturate at 1 Tesla. The compound δ=0.15 is characterised by a large amount of Mn vacancies (5%) which is enough to destroy the long-range magnetic interaction. In Fig. 5b, it is clear how at low temperature ξ remains with a finite value of ≈19 Å. This ground state at low temperature has been characterised as an insulator cluster-glass with a freezing temperature of 50 K. A huge magnetoresistance is found in this compound which can be explained as in the aforementioned Tb-doped compound (x=0.33).

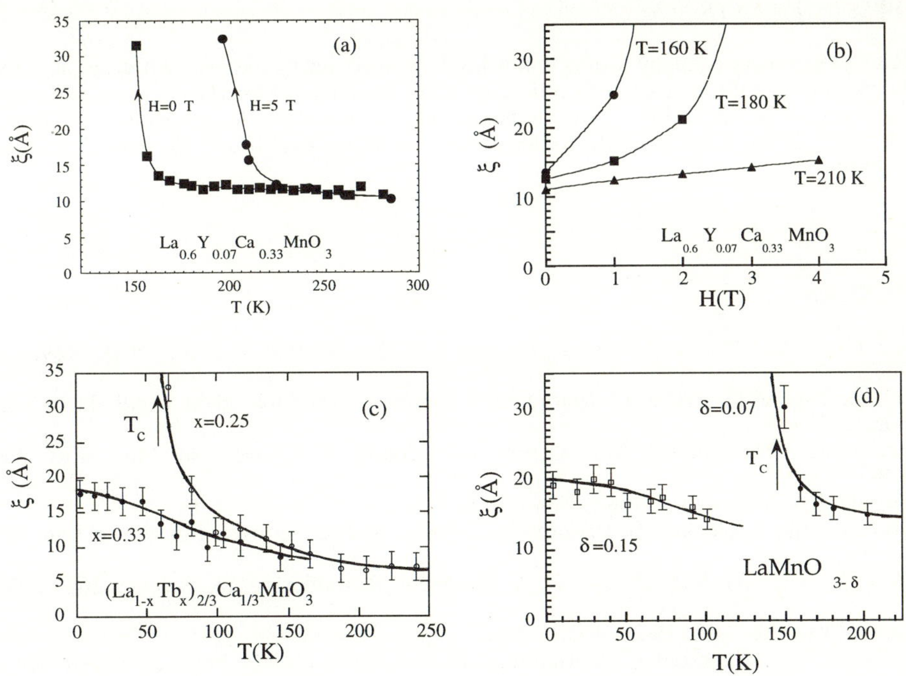

Fig. 5 *Magnetic correlation length, ξ, for selected compounds as a function of the temperature or the magnetic field. The onset of ferromagnetism brings about the divergence of ξ.*

5. CONCLUSIONS

We have established the existence of a ground state at low temperature in which ferromagnetic clusters of average ferromagnetic correlation length of ≈15-20 Å are embedded in a paramagnetic matrix. This is the source of giant magnetoresistance at low temperatures in some mixed-valence manganites. Another source of giant magnetoresistance occurs at high temperatures, i.e. in the paramagnetic phase. In this case we have established that the existence of dynamic phase segregation in the form of magnetic polarons is the responsible for the intrinsic colossal magnetoresistance found near room temperature in these compounds. There exists still another source of magnetoresistance based on the existence of the new ground state found in $(La_{0.5}Nd_{0.5})_{2/3}Ca_{1/3}MnO_3$ [24]. This compound has a transition from a paramagnetic charge-localised to paramagnetic charge ordered state at T_{CO}≈210 K. Below T_C≈200 K metallic ferromagnetic and charge-ordered paramagnetic phases coexist. Below T_N≈170K the charge ordered regions undergo a para-antiferromagnetic transition. This model explains the existence of an intermediate volume at the lowest temperatures between the LV and HV states. In order to check the relevance of the electron phonon interaction on the existence of this ground state, we performed the same characterisation on a sample in which the ^{16}O isotope oxygen was changed by ^{18}O isotope. We consider that the effect of the change on the oxygen mass is to modify the lattice dynamics and, consequently, the electron-phonon interaction. In the ^{18}O compound the charge ordered antiferromagnetic state is favoured at low temperature, becoming an insulator , i.e. the ferromagnetic interaction is weaker in this compound.

To sum up, experiments of thermal expansion and magnetostriction combined with microscopic techniques as neutron diffraction have allowed us to ascribe the existence of colossal magnetoresistance found in several mixed-valence manganites to the existence of electronic phase segregation, either dynamic at high temperatures (magnetic polarons) or static at low temperatures (cuasi-ferromagnet or cluster-glass).

ACKNOWLEDGEMENTS

This article is a summary of many results in this field obtained during the last two years as a result of a friendly and fruitful collaboration with colleagues at the Institute of Materials Science in Zaragoza: J. Blasco, J. García, P.A. Algarabel, C. Marquina, B. García-Landa and A. del Moral; C. Ritter at the ILL (Grenoble); G.-M. Zhao and H. Keller at Universitat of Zurich and Z. Arnold at Institut of Physics (Praha).

REFERENCES

[1] M.N. Baibich, J.M. Broto, A. Fert, F. Nguyen Van Dan, F. Petroff et al., Phys. Rev. Lett. **61** (1988) 2472.
[2] P.A. Algarabel, M.R. Ibarra, C. Marquina, J. Galibert, M. Iqbal et al. Appl. Phys. Lett. **66** (1995) 3062.
[3] A. Berkowitz, J.R. Mitchell, M.J. Carey, A.P. Young, S. Zhang et al. Phys. Rev. Lett. **68** (1992) 3745.
[4] A. Fert and I.A. Campbell J. Phys. F. **6** (1976) 849.
[5] R. von Helmolt, J. Wecker, B. Holzapfel, L. Shultz, and K. Samwer, Phys. Rev. Lett. **71** (1993) 2331.
[6] S. Jin, T.H. Tiefel, M. McCormack, R. A. Fasrnacht, R. Ramesh et al., Science **264** (1994) 413.
[7] C. Zener, Phys. Rev. **82** (1951) 403.
[8] M.R. Ibarra, P.A. Algarabel, C. Marquina, J. Blasco, and J. García, Phys. Rev. Lett. **75** (1995) 3541.
[9] G. Zhao, K. Conder, H. Keller, and K.A. Müller, Nature **381** (1996) 676.
[10] K.H. Kim, J.Y. Gu, H.S. Choi, G.W. Park, and T.W. Noh, Phys. Rev. Lett. **77** (1996) 1877.

[11] T. Holstein , Annals of Physics **8** (1959) 343.
[12] G.F. Dionne, Technical report 1029 Licoln Laboratory, Massachusetts Institute of Technology (1996).
[13] M. Jaime, M.B. Salamon, M. Rubinstein, R.E. Treece, J.S. Horwitz et al., Phys. Rev. B **54** 11914 (1996).
[14] M.R. Ibarra, J.M. De Teresa, J. Blasco, P.A. Algarabel, J. García et al., Phys. Rev. B **56** (1997) 8252.
[15] J.B. Goodenough, A. Wold, J. Arnot and N. Menyuk, Phys. Rev. B **124** (1961) 373.
[16] J.M. De Teresa, M.R. Ibarra, P.A. Algarabel, C. Ritter, C. Marquina et al. Nature **386** (1997) 256.
[17] J. B. Goodenough and J.S. Zhou, Nature **386** (1997) 229.
[18] Z. Jirak, S. Krupicka, Z. Simsa, M. Dlouha and S. Vratislav, J. Magn. Magn. Mater. **53** (1985) 153.
[19] M.R. Ibarra and J.M. De Teresa in "Transition metal oxides" Edt. C.N.R. Rao and B. Raveau, World Scientific (1998).
[20] M. Verelst, N. Rangavittal, C.N.R. Rao and A. Rousset, J. Solid State Chem. **104** (1993) 74.
[21] J. A. M. Roosmalen and E.H.P. Cordjunke, Solid. State Chem. **110** (1994) 109.
[22] C. Ritter, M.R. Ibarra, J.M. De Teresa, P.A. Algarabel, C. Marquina et al. Pys. Rev. B **56** (1997) 8902.
[23] E. L. Nagaev, Phys. Stat. Sol. (b) **186** (19994) 9 and Uspekhi Fizicheeskikh Nauk **39** (1996) 791 englished translated.
[24] M.R. Ibarra, G-M. Zhao, J.M. De Teresa, B. García-Landa, Z. Arnold et al. Phys. Rev. B **57**, 7446 (1998).
[25] B. García-Landa, M.R. Ibarra, J.M. De Teresa, G-M. Zhao, K. Conder et al. Sol. Stat. Commun. **105** (1998) 567
[26] G-M. Zhao, H. Keller, K. Conder, A. Shegenlaya and K.A. Müller, Solid State Commun. **104** (1997) 57
[27] Z. Kapusta, P. Riedy, M.R. Ibarra, J.M. De Teresa, M. Viret and J.M.D. Coey to be published.

Materials Science Forum Vols. 302-303 (1999) pp. 134-138

Raman Scattering in CMR Manganites

E. Granado[1], N.O. Moreno[1], A. García[1], J.A. Sanjurjo[1], C. Rettori[1], I. Torriani[1], S.B. Oseroff[2], J.J. Neumeier[3], K.J. McClellan[3], S.-W. Cheong[4] and Y. Tokura[5]

[1] Instituto de Física "Gleb Wataghin", UNICAMP, 13083-970, Campinas-SP, Brazil

[2] San Diego State University, San Diego, CA 92182, USA

[3] Los Alamos National Laboratory, Los Alamos, New Mexico 87545, USA

[4] AT&T Bell Laboratories, Murray Hill, New Jersey 07974, USA

[5] Department of Physics, University of Tokyo 113, Japan

Keywords: Colossal Magnetoresistance, Frequency Shifts, Raman Effect

Abstract. Polarized Raman spectra of single and polycrystalline $La_{1-x}Sr_xMnO_3$ ceramic samples were studied as a function of temperature. For the rhombohedral $LaMnO_{3.1}$ and $La_{0.7}Sr_{0.3}MnO_3$, the observed Raman peaks were associated to modes arising from the folding of the Brillouin zone under lattice deformation. For the orthorhombic $LaMnO_{3.0}$, the Raman spectra are consistent with the Pnma structure and show an anomalous *softening* of the 494 and 604 cm^{-1} modes below the antiferromagnetic ordering temperature, $T_N \simeq 140$ K.

1 Introduction

The concomitant metal-insulator (MI) and ferromagnetic-paramagnetic (FM-PM) transitions in the $R_{1-x}A_xMnO_{3+\delta}$ (R = Rare Earths; A = Alkaline Earths) perovskites have recently attracted much attention. Particularly, for $0.1 < x < 0.5$ giant negative magnetoresistance near room T has been reported,[1] making these materials very attractive for future magnetic sensing applications. Lattice polaron effects, enhanced by a strong Jahn-Teller coupling in Mn^{3+}, were invoked to explain the reduction in electron kinetic energy near and above the FM ordering temperature, T_c. The study of the phonon spectra in these compounds, by means of Raman experiments, may provide valuable information on the lattice dynamics of these materials.

In this work we studied: *i)* low T polarized Raman spectra in single crystals of $La_{0.7}Sr_{0.3}MnO_3$, *ii)* low T unpolarized Raman spectra in polycrystalline $LaMnO_3$; and *iii)* the T dependence of the Raman spectra in a polycrystalline sample of $LaMnO_{3.0}$.

Samples of $LaMnO_{3+\delta}$ treated in oxygen present rhombohedral crystal structure, space group D_{3d}^6 ($R_{\bar{3}C}$), with two formulas per unit cell. The 30 vibrational degrees of freedom per unit cell result in the following modes at the zone center:

$$\Gamma(D_{3d}^6) = 2A_{1u} + 3A_{2g} + A_{1g} + 4A_{2u} + 4E_g + 6E_u \quad (1)$$

where the A_{1g} and the four E_g modes are Raman active. The Raman active modes are associated to R(1/2,1/2,1/2)-point Brillouin zone boundary modes in the cubic phase, which become zone center modes because the Brillouin zone is halved in the rhombohedral phase.

$LaMnO_{3.0}$ shows O′ orthorhombic ($\mathbf{b}/\sqrt{2} < \mathbf{c} < \mathbf{a}$) structure, space group D_{2h}^{16} (Pnma). The primitive unit cell has four formula units giving 60 Γ-point vibrational modes. Factor group analysis yields the following symmetry distribution of these modes at the zone center:

$$\Gamma(D_{2h}^{16}) = 8A_u + 10B_{1u} + 8B_{2u} + 10B_{3u} + 7A_g + 5B_{1g} + 7B_{2g} + 5B_{3g} \quad (2)$$

where the $7A_g + 5B_{1g} + 7B_{2g} + 5B_{3g}$ modes are Raman active.

The perovskites $R_{1-x}A_xMnO_{3+\delta}$ (R=Rare Earths; A=Alkaline Earths) show decreasing deviations from the ideal cubic structure upon doping. The (LaSr)MnO_3 system has D_{2h}^{16} symmetry for $x < 0.17$ and rhombohedral symmetry, D_{3d}^{6}, for $x > 0.17$.

2 Experimental Details

Ceramic samples were prepared by standard methods. The oxygen content of the $LaMnO_{3+\delta}$ samples was controlled by annealing them in oxygen and argon atmospheres. The single crystal of $La_{0.7}Sr_{0.3}MnO_3$ was grown by the floating zone method. The structure and phase purity were checked by X-ray powder diffraction and the crystals orientation determined by the conventional Laue method. Lattice parameters were obtained using a X-ray Rietveld refinement program.[2] The refined profiles were taken in a Rigaku R200 diffractometer and a rotating anode generator with Cu $K\alpha$ radiation. Magnetization (*dc/ac*) measurements have been taken in a Quantum Design PPMS magnetometer. Electron spin resonance (ESR) experiments were carried out in a Varian X-band spectrometer using a room T rectangular TE_{102} cavity and a He gas flux T controller. The Raman measurements were performed in a backscattering geometry using a Jobin Yvon T64000 triple spectrometer equipped with a cryogenic charge-coupled device camera, using the 514.5 nm line of an argon ion laser. The samples were mounted on the cold finger end of a closed-cycle He refrigerator. The laser power incident on the samples was $\lesssim$ 15 mW focused in a diameter of $\sim$ 50 μm.

3 Results and Discussion

Table I gives the crystal structure and room temperature lattice parameters for all the studied samples.

Figs. 1(a) and 1(b) show, respectively, the Raman spectra at T = 10 K on fresh surfaces of an oxygen annealed polycrystalline sample of $LaMnO_{3.1}$ and of a single crystal of $La_{0.7}Sr_{0.3}MnO_3$ at different polarizations. Both samples are rhombohedral, D_{3d}^{6} space group. Table II gives, for our samples and some isomorphous rare earth aluminates,[3, 4] the correlation between the observed D_{3d}^{6} zone-center Raman active phonons and the R(1/2,1/2,1/2)-point phonons of the O_h^1 perovskite structure. Spectra taken on surfaces exposed to the air during few weeks present a broad peak at $\sim$ 600-650 cm^{-1} in almost all our samples, which we attribute to disorder induced scattering.

Fig. 2 shows the Raman spectra at 15 K of an argon annealed polycrystalline ceramic sample of $LaMnO_{3.0}$. This sample is orthorhombic, D_{2h}^{16} space group (Pnma), with a large deviation from the cubic perovskite structure. Figs. 3(a) and 3(b) show, respectively, for the three most intense Raman peaks of Fig. 2 , the T dependence of their frequency and linewidth. Notice that the 494 and 604 cm^{-1} peaks present an anomalous *softening* of the frequency for T $\lesssim$ 150 K. No anomaly was observed in the T dependence of the linewidth for these peaks. Recent lattice dynamic calculations for the orthorhombic $LaMnO_3$ [5] indicate that the 604 cm^{-1} peak corresponds to an in-phase *stretching* mode of the oxygen cage. The same calculations indicate that the 494 cm^{-1} peak is associated to an out-of-phase *bending* mode of the oxygen cage. Finally, the 290 cm^{-1} peak is attributed to an out-of-phase x-rotation of the oxygen cage, and therefore does not involve internal distortions of the MnO_6 octahedra.[5]

Figs. 4(a) and 4(b) show, respectively, the ESR and *dc/ac*-susceptibility data for the $LaMnO_{3.0}$ sample of Fig. 2. In agreement with earlier neutron scattering[6] experiments, the data in Fig. 4 show that our $LaMnO_{3.0}$ sample experience an antiferromagnetic (AFM) and weak ferromagnetic (WFM) ordering at T $\simeq$ 140 K.

It is known that in some AFM and FM transition metal oxides compounds, optical modes present frequency anomalies below the ordering temperature, associated to magnetostrictive effects.[8, 9] The anomalous *softening* shown in Fig. 3(a) for the 494 and 604 cm^{-1} modes in $LaMnO_{3.0}$ may also be caused by a change in the lattice parameters at $T_N \lesssim 150$ K. Recent neutron diffraction experiments in stoichiometric $LaMnO_3$ measured the T dependence of the lattice parameters.[7] The small anomalies observed in the **a** and **c** (Pnma) lattice parameters at T ≈ 140 K may be the origin for the anomalous *softening* of the 494 and 604 cm^{-1} modes.

4 Conclusions

In summary, we have studied the Raman spectra of the $La_{1-x}Sr_xMnO_{3+\delta}$ manganites with different degrees of deviations from the ideal cubic perovskite structure. For rhombohedral samples, the main modes were assigned to R-point modes of the cubic structure, which become Raman active due to the folding of the Brillouin zone. For strongly distorted orthorhombic samples a wide variety of modes appear, but less than the 24 modes allowed for this symmetry. For the argon annealed $LaMnO_{3.0}$ sample the 494 and 604 cm^{-1} modes present an anomalous *softening* for $T \lesssim 150$ K, of ~ 1 and ~ 8 cm^{-1}, respectively. These *softenings* were associated to the AFM ordering experienced by this sample and observed by ESR and *dc/ac*- susceptibility measurements.

5 Acknowledgments

This work was supported by FAPESP Grant No 95/4721-4, São Paulo-SP-Brazil and NSF-DMR No 9705155 and NSF-INT No 9602928.

References

[1] R. von Helmolt, J. Wecker, B. Holzapfel, L. Schultz and K. Samwer, Phys. Rev. Lett. **71**, 2331 (1993).

[2] D.B. Wiles and R.A. Young, J. Appl. Crystallogr. **14**, 149 (1981).

[3] J.F. Scott, Phys. Rev. **183**, 823 (1969).

[4] R.T. Harley, W. Hayes, A.M. Perry and S.R.P. Smith, J. Phys. C: Solid State Phys. **6**, 2382 (1973).

[5] M.N. Iliev, M.V. Abrashev, H.-G. Lee, V.N. Popov, Y.Y. Sun, C. Thomsen, R.L. Meng, and C.W. Chu, Phys. Rev. **B57**, 2872 (1998).

[6] E.O. Wollan and W.C. Koehler, Phys. Rev. **100**, 545 (1955).

[7] C. Ritter, M.R. Ibarra, J.M. De Teresa, P.A. Algarabel, C. Marquina, J. Blasco, J. García, S. Oseroff, S-W. Cheong, Phys. Rev. **B56**, 8902 (1997).

[8] M. Udagawa, K. Kohn, N. Kashizuka and T. Tsushima, Solid State Commun. **16**, 779 (1975).

[9] K.H. Kim, J.Y. Gu, H.S. Choi, G.W. Park and T.W. Noh, Phys. Rev. Lett. **77**, 1877 (1996).

Table 1. Structures and lattice parameters of the studied compounds at room temperature.

Sample	Space Group	a (nm)	$b/\sqrt{2}$ (nm)	c (nm)	α (°)
$La_{0.7}Sr_{0.3}MnO_3$	D^6_{3d}	0.5434(1)			60.437
$LaMnO_{3.1}$	D^6_{3d}	0.5475(1)			60.583
$LaMnO_{3.0}$	D^{16}_{2h}	0.5724(1)	0.5442(1)	0.5534(1)	

Table 2. Correlation for the Raman active phonons at the point Γ in the D^6_{3d} Brillouin zone and at point R in the O^1_h Brilloiun zone. A tentative assignment of the Raman phonon frequencies at 10 K in $La_{0.7}Sr_{0.3}MnO_3$, together with published data in other rhombohedral ABO_3 perovskites are also shown.

Approximate eigenvector	O^1_h mode symmetry	D^6_{3d} mode symmetry	Phonon frequencies (cm^{-1}) $La_{0.7}Sr_{0.3}MnO_3$ 10 K	$LaMnO_{3.1}$ 10 K	$PrAlO_3$ 212 K	$NdAlO_3$ 300 K
Rotation of oxygen cage	$F_{2u}(R)$ (Γ_{25})	$E_g(\Gamma)$	42		45	50
		$A_{1g}(\Gamma)$	199	217	222	241
Motion of A ion	$F_{1u}(R)$ (Γ_{15})	$E_g(\Gamma)$		115	164	
Internal vibration of oxygen cage	$E_u(R)$ (Γ'_{12})	$E_g(\Gamma)$				
	$F_{1u}(R)$ (Γ_{15})	$E_g(\Gamma)$	426	431	506	509

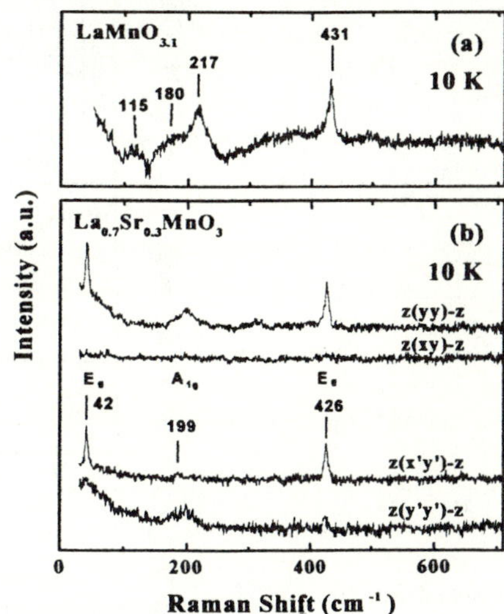

Fig. 1: Raman spectra, at T = 10 K, of the rhombohedral samples: $LaMnO_{3.1}$ polycrystalline (a) and $La_{0.7}Sr_{0.3}MnO_3$ single crystal for different polarizations. The x, y and z directions correspond to cubic [100] axes (Mn-O-Mn), and x'and y' indicate the cubic [110] directions (La-O-La).

This article was processed using the LaTeX macro package with TTP style

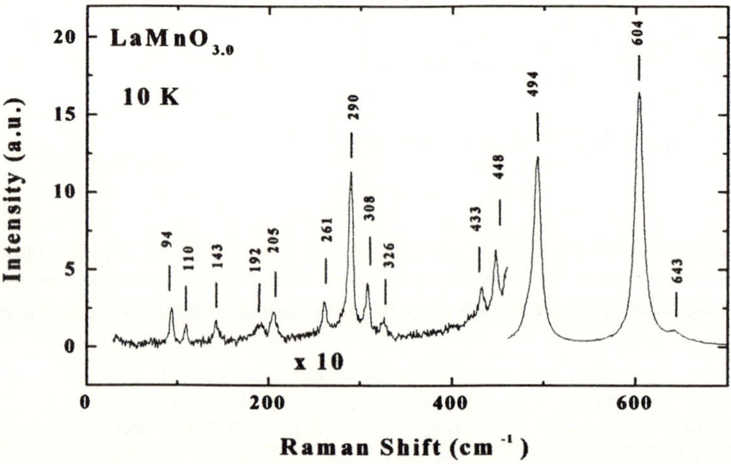

Fig.2: Raman spectrum, at T = 10 K, of the orthorhombic $LaMnO_{3.0}$ argon annealed polycrystalline ceramic sample.

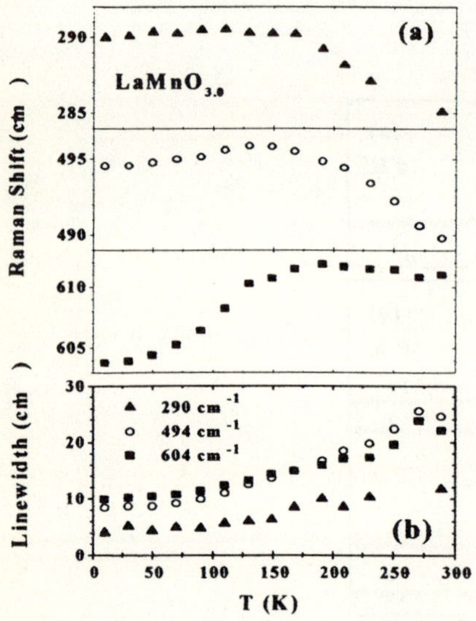

Fig. 3: T dependence of the: Raman shift (a) and linewidth (b) for the 290, 494, and 604 cm^{-1} peaks of Fig. 2.

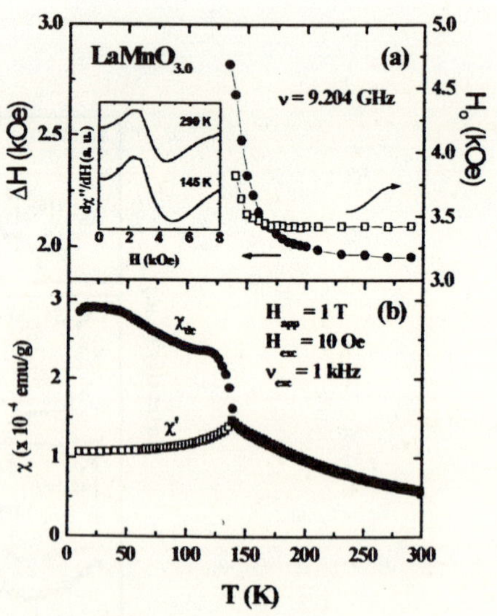

Fig. 4: T dependence of the: (a) ESR linewidth and resonance field and; (b) *dc* (χ_{dc}) and *ac* (χ')-susceptibilities for the sample of Fig. 2.

Materials Science Forum Vols. 302-303 (1999) pp. 139-143

Simultaneous Measurements of Resistance and Elastic Modulus in $La_{2/3}Ca_{1/3}MnO_3$

A.A. Ghilarducci[1], H.R. Salva[1], R.D. Sánchez[2] and C. Vázquez[3]

[1] Comisión Nacional de Energía Atómica, Centro Atómico Bariloche and Instituto Balseiro, 8400 S.C. de Bariloche, Argentina

[2] Departamento de Física-F.C.E. y N., Universidad de Buenos Aires, Argentina

[3] Departamento de Física Aplicada, Universidad de Santiago de Compostela, Spain

Keywords: Colossal Magnetoresistance, Anelasticity, Internal Friction

Abstract- We have measured the anelastic constants of the magnetoresistant perovskite $La_{2/3}Ca_{1/3}MnO_3$ around the metal-insulator transition in an inverted forced torsion pendulum (5 Hz) and by the vibrating reed technique (2 Khz). At torsion pendulum, the resistance of the sample was simultaneously measured during thermal cycles with and without applying a small magnetic field. The torsion and bending ("shear" and "Young"modulus) measurements were qualitatively similar. The modulus becomes harder at the transition while the internal friction value is greater in the vibrating reed experiment.There was no detected hysteresis cycling the temperature at 0.05 K/min around the transition temperature (260 K). But, if we applied a magnetic field of 320 Gauss an hysteresis of 0.7 K is observed which we suppose is related with the magnetic domains when the sample orders. There is a correlation between the resistance and torsion measurements which do not depend in the applied magnetic field.

1 Introduction

In the last years great interest was focused to the manganese perovskites[1][2] because of their important magnetoresistance and possible applications. Lot of work has been done over the magnetoelectric properties, structure, oxygen content, etc., but not much attention was paid for the elastic properties of this materials. In this report we describe our measurements in $La_{2/3}Ca_{1/3}MnO_3$, one of the most studied composition because of the high temperature metal-insulator transition and magnetoresistive properties.

2 Experimental set-up

The ceramic sample was prepared by the conventional method of successive annealings at 1300 C during 60, 50 and 100 hours, grinding each cycle and pressed at 5 Tn. Finally it was placed during 24 hours in an oxygen atmosphere at 1200 C. Then it was cut in a suitable shape for each experiment. The torsion measurements were made in an inverted forced pendulum capable of measuring in the 4-300 K and 0.001-10 Hz ranges. The internal friction is obtained as the phase angle between the generation and the detection signals in a signal analyser. These signals are sinusoids describing the oscillating movement of the sample. The shear modulus is the relationship between the absolute values of the generation-detection signals. The magnetic field was provided by a coil which could produced up to 600 Gauss in short term and 350 Gauss in long term.The sample (1.5x1.5x8 mm^3) was mechanically clamped at both ends and the electric contacts were made with silver paint at the clamps. Resistance measurements were done on two wires.

The vibrating reed experiments were performed in a conventional system where the Young modulus is proportional to the square frequency and the internal friction to the inverse of the amplitude of resonance. The sample (0.1xlx6 mm^3)was glued at one end with silver paint.

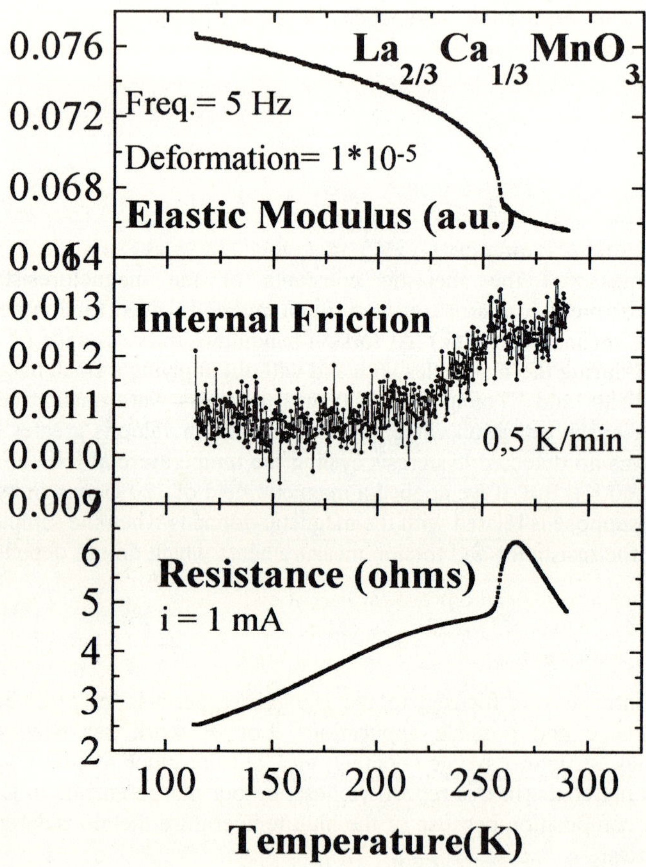

Fig. 1. Simultaneous measurement of elastic variables and resistance

3 Results

Our first temperature scan at 0.5 K/min is shown in figure 1. There, just around the rnetal-insulator transition, a modulus step is seen at nearly the same temperature where the resistance drops. The intensity of the internal friction peak at the transition is very poor, probably due to the small deformation used. Higher amplitudes were not available because of sample dimensions and exciting current. Measuring frequency scans were also performed at different temperatures (255, 265 and 290 K) and no anomalous behavior was detected.

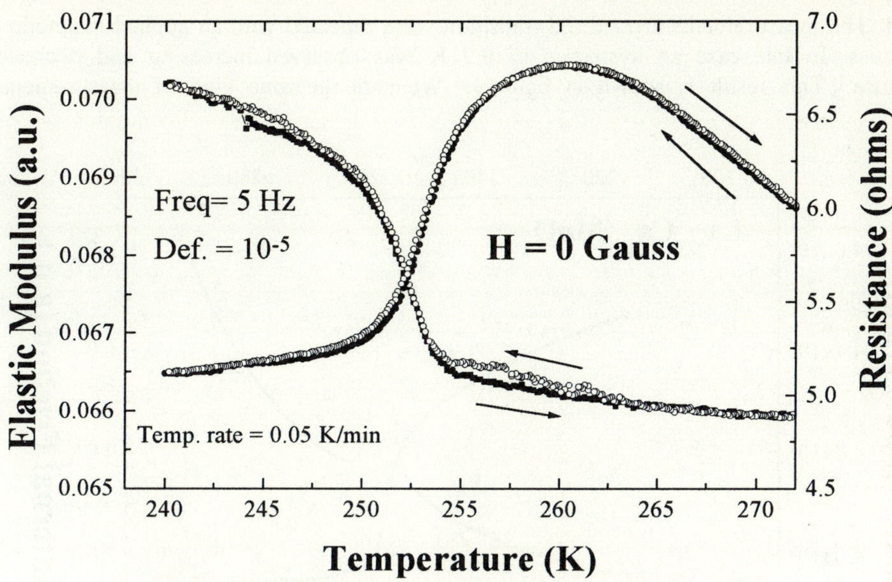

Fig 2. Elastic modulus and resistance at cero magnetic field

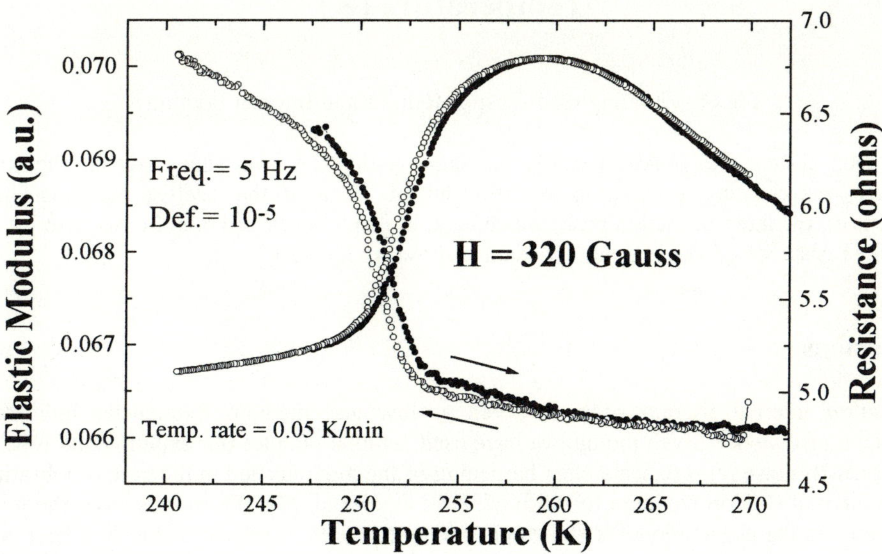

Fig. 3. Elastic modulus and resistance with applied magnetic field

Detailed measurements were made around the transition temperature at very slow temperature rate (0.05 K/min) looking for any hysteresis effect (figure 2), but, in this case, no hysteresis was observed. The measurements around the transition were repeated with an applied magnetic field of 320 Gauss. In this case an hysteresis of 0.7 K was observed increasing and decreasing the temperature. This result is shown in figure 3. We made the same kind of measurements in the

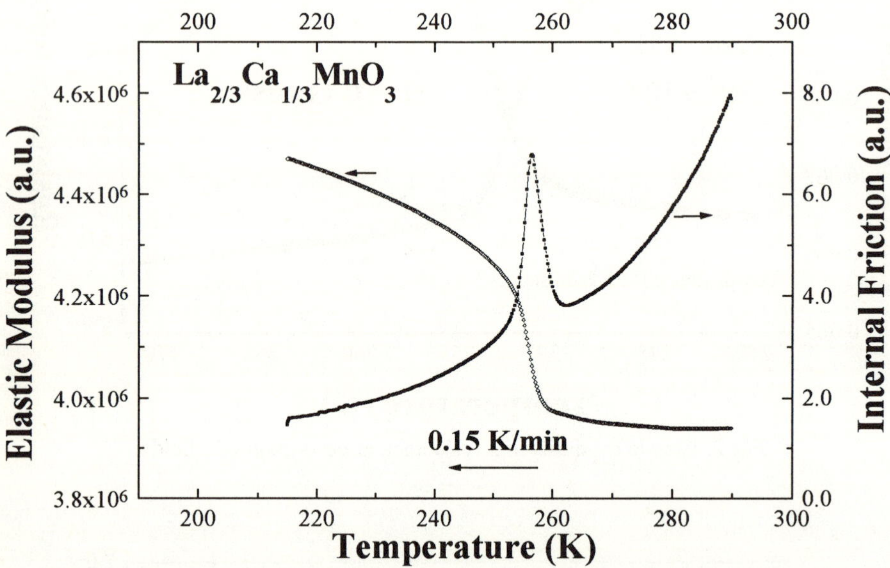

Fig. 4. Vibrating reed Elastic modulus and Internal Friction

vibrating reed system and obtained nearly the same results, i.e. a modulus step at the transition temperature, representing a variation in the order of 10% of the relative modulus.The great difference is on the internal friction peak, which here, is a well developed one. In this case we did not measure the resistance of the sample. This data is shown in figure 4.

4 Conclusions

The transition internal friction peak observed at low and medium frequencies indicate a no dependence on frequency. Even though we have used ceramic samples our experimental results have enough signal-to-noise ratio to see a clear hardening of the modulus and in the case of vibrating reed a peak on internal friction compare to those of Z.B. Guo et al. [3]. We suppose that the hysteresis cycle obtained in the elastic modulus with applied magnetic field, is related to the spin ferromagnetic ordering in the sample. We have not simultaneous measurements of magnetization and resistance, but in this case an interesting feature is the fact that the rate of change in elastic modulus is proportional to the rate of change in the resistance. This is directly related in the hysteresis loop in both quantities (elastic modulus and resistance). The relation between the rates of change in magnetization and resistivity is a consequence of the double exchange interaction responsible of the ferromagnetic order. So, when the resistivity changes, the magnetisation does so and here we show that the rnagnetostrictive effect is very irnportant.

On another hand K. Li et al.[4] have shown that the elastic modulus obeys the relation

$$\Delta E = \tanh(\Delta E^{*}T/T_c)$$

Where ΔE is the change in the elastic modulus and Tc the transition temperature.This is the relation obtained for the magnetisation of a ferromagnet below the transition temperature in a Langevin model. M.R. Ibarra et al.[5] have also shown through thermal expantion measurements that below Tc the sample has an important contraction, which could be traduced as an important hardening of the elastic modulus. This facts let us suppose the hardening of the elastic modulus is due to the important magnetoelastic interaction in this cornpound.
Detailed measurements are necessary to take advantage of this kind of measurements.

References

[1] Schiffer P., Ramirez h. P., Bao W. and Cheong S.W., Phys.Rev.Lett.**75** (18) (1995) 3336-3339.

[2] Coey J.M.D., Viret M. and Ranno L., Phys.Rev.Lett.**75** (21) (1995) 3910-3913.

[3] Z.B. Guo, Y.W. Du, J.S. Zhu, H. Huang, W.P. Ding and D. Feng, Phys.Rev.Lett **78**(6), (1997) 1142.

[4] K. Li, X. Li, C. Liu, Z. Zhu, J. Du, D. Hou, X. Nie, J. Zhu and Y. Zhang, Phys.Rev.B **56**(21) (1997) 13662-13665.

[5] M.R. Ibarra, P.A. Algaralbel, C. Marquina, J. Blasco and J. Garcia, Phys.Rev.Lett. **75**(19) (1995) 3541-3544.

Materials Science Forum Vols. 302-303 (1999) pp. 144-148

Transport and Magnetic Properties in $La_{0.67}Ca_{0.33}MnO_3$ Thick Films

C. Hart[1], O. Arés[1] and R. Escudero[2]

[1] Superconductivity Lab. IMRE, Havana University, Vedado 10400, Havana, Cuba

[2] Low Temperature Lab., Materials Research Institute, UNAM, Mexico DF, Mexico

Keywords: Manganites, Colossal Magnetoresistance, Thick Films

ABSTRACT: We present transport and magnetic measurements in thick films of the perovskite $La_{0.67}Ca_{0.33}MnO_3$ on alumina substrates, made by a conventional paint-on method. In our case the temperature dependence of resistance is characterized by the existence of a broad peak (at 180 K) far below the Curie temperature (240K). The samples show a strong ferromagnetism, while the magnetoresistance coefficient exhibits a general diminishing monotonic behavior with the temperature. We discuss these results, considering the magnetic disorder related with possible local variation of the Mn^{4+}content and its relation with the transport properties.

INTRODUCTION

The perovskite-like $LaMnO_3$ exhibits both strong ferromagnetism (FM) and metallic conductivity, when La^{3+} ions are partially replaced by A^{2+} valence ions (A=Ca, Ba, Sr, Pb and Cd.) [1]. It is reported that this substitution creates a Mn^{3+}- Mn^{4+} mixed valence state, creating mobile charge carriers which couple with the Mn spins [2]. On the other hand in these materials the resistivity (ρ) changes by orders of magnitude when a magnetic field (H) is applied i.e. the magnetoresistance (MR) is colossal (CMR) [3]. The large MR of these compounds is also connected to the strong dependence of ρ on the spin structure and the spin order. The double exchange model (DE) gives an illustrative picture of the correlation of the magnetic order with the electric transport [4]. As is known, the antiferromagnetic phase (AFM) [5] is governed by superexchange (SE) via the oxygen p-shell and possesses a semiconductor behavior with relatively high values of the resistivity and without MR phenomena. That depends on the concentration (x) of the divalent metal (A); in general, for a given range of concentration x (0.21-0.3), ρ exhibits a sharp peak at a temperature T_p near to the Curie temperature (T_c) [6]. The MR has a peak near T_c [3,6]. Although this interpretation is comprehensive for the magnetic transition associated with the insulator to metal transition, there remain many discrepancies between the theoretical predictions and the experimental results for the MR phenomena [7]. In our case, in particular, $La_{0.66}Ca_{0.33}MnO_3$, as it is reported in other works which have used the same composition [8], both "critical" temperatures (T_c and T_p) are somewhat separated; the resistivity shows a broad transition far below T_c and in general the MR decays monotonically, with a little peak near T_c. On the other hand, our samples exhibits a strong ferromagnetism (M=200 emu/g at the saturation field of 5 kOe), which suggests a well defined ferromagnetic phase in apparent contradiction with the transport measurements. In this work, we propose some hypotheses to explain this apparent contradictory behavior.

We have prepared thick films on alumina substrate by a simple paint-on method. The thick film technology could represent the basis for a broad range of applications, where lower cost should be advantageous [9].

EXPERIMENTAL PROCEDURE

Ceramic $La_{2/3}Ca_{1/3}MnO_3$ powder was prepared by standard solid state reaction techniques, starting from the appropriate mixture of oxides. The mixed powder was heated in air at 1100°C for 24 h with intermediate grinding. The paste prepared by mixing the powder and an organic binder, was painted carefully on the alumina substrate. The thick film obtained (40 μm thick) was fired at 1200°C for 24 h in an oxygen atmosphere. The average thickness of the film after the final treatment was ≈ 16 μm, as estimated from electron microscopy measurements (See Fig.1). The calculated resistivity at 300K was about 1Ω-cm. XRD analysis of the film revealed a single crystallographic phase. Magnetotransport have been carried out using a standard 4-probe dc method .The magnetic field (up to 10 kOe) is applied in the plane of the substrate .The dc magnetic susceptibility was measured using a SQUID magnetometer.

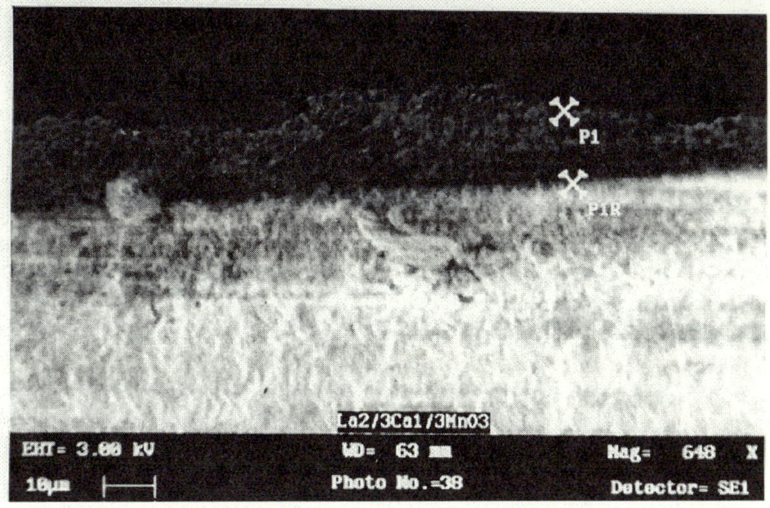

Fig.1 SEM picture of the $La_{2/3}$ $Ca_{1/3}MnO_3$ thick film. The crossings show the film thickness (~6μm).

RESULTS AND DISCUSSIONS

In Fig. 2 are shown the temperature dependence of the resistance at zero magnetic field and magnetic susceptibility at 0.1 kOe. The R vs T curve exhibits a broad peak at 180K as has been reported previously for these materials. The films show a semiconductor behavior (negative dR/dT) above T_p and metallic behavior (positive dR/dT) below $T_c = 246$ K [3,6]. This evident disagreement between T_p and T_c has been reported and related, between others factors, to FM clusters in an AFM matrix for large x values (x = 0.46) [6], and due to inhomogenities [10]. In particular, reference [6] proposes an effective two medium theory in which the resistivity of the sample is described in the two phase regions; one FM (β phase) and the another one AFM (α or γ depending on the resistivity values), fitting ρ in the AFM as $\rho_1 = \exp(T_0/T)^{1/4}$ and in the FM region as $\rho_2 = a + bT$.

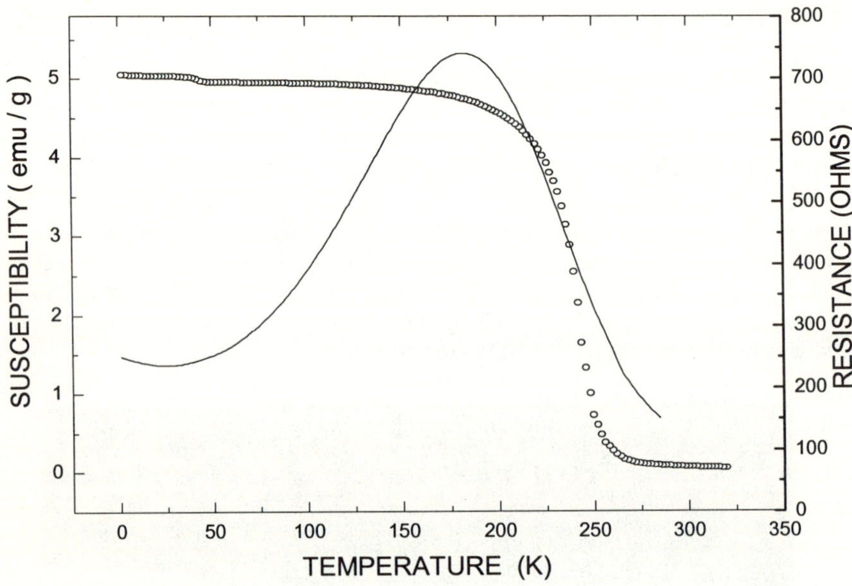

Fig.2 Magnetic susceptibility (χ) and resistance (R) vs temperature for a La-Ca-Mn-O thick film.

According to the proposed equation, the opposite temperature dependence of ρ_1 and ρ_2 would rise to the broad peak far below Tc. Using this model we can adjust our experimental results considering a small FM volume fraction p of ≈ 35%. Using the magnetization measured at low temperatures and an applied field of 10 kOe (200 emu/g) (see Fig.4), it is possible to calculate that our samples have practically the theoretical magnetization per Mn atom, about 4 Bohr magnetons for the x= 1/ 3 composition [11]. This experimental result, combined with the fact that our composition is 0.33 and not 0.46, as in Ref [6], made it difficult to suppose that we have a mix of AFM and FM phases. Although T_p is 66 K below of T_c, the slope dR/dT begins to decrease at approximately T = T_c (See Fig. 2). Therefore, the disagreement between T_p and T_c could be attributed to a percolation phenomenon, i.e. even though it is certain that each Mn^{3+}/Mn^{4+}pair contributes one free carrier to metallic conduction, this conduction does not appear below a threshold value of x or T. Microscopic disorder or inhomogeneities [10] lead to the formation of FM islands, which are not connected sufficiently to give a conduction path for a temperature T ($T_p<T<T_c$). For $T=T_p$ the conduction of the free carriers is established by percolation through FM regions. In samples with smaller disorder the percolation is logically achieved at a T near to T_c. In Fig. 3 the MR behaviour is shown for samples for a maximum applied magnetic field of 10 kOe. The MR coefficient has a general decreasing monotonic characteristic, except for a little peak near Tc. For single crystals and sintered bulk samples [6], where the MR coefficient has only a defined peak at Tp and it is negligible at low temperatures. Similar decreasing MR dependence such as that obtained in this work has been reported for polycrystalline $La_{2//3}Ba_{1/3}$ MnO_3 [12]. In this case it is suggested an additional source of MR related to the domains in the grain boundaries, i.e., an "extrinsic source" of MR. Although at $T \rightarrow 0$ the MR of the FM phase is very small, the magnetic inhomogenities of our samples, related to the mentioned local variations of x, can lead to the formation of the zones (for example canted spin structures) which could be easily oriented at low temperatures. This argument seems to be in correspondence with the high magnetization show in our samples.

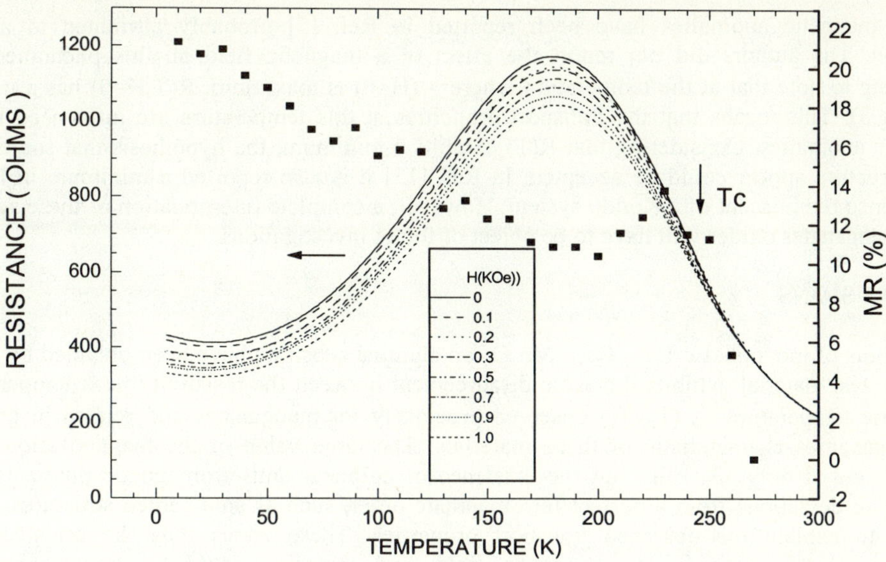

Fig. 3 Resistance Vs Temperature at different applied magnetic field and the MR coefficient vs Temperature at 10 kOe.

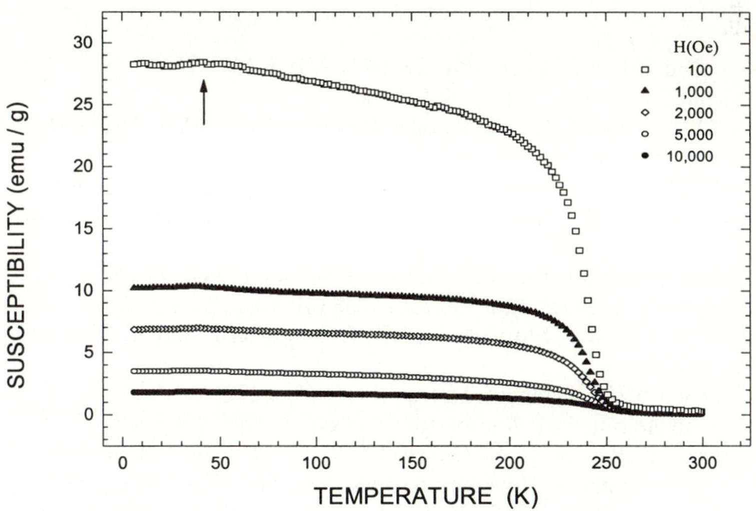

Fig.4. Magnetic susceptibility (10^{-2}) vs temperature at different applied magnetic field. The arrow shows the magnetic anomaly present at low temperature.

An interesting feature appears at low temperature (See Fig.4). This structure disappears with increasing applied magnetic field. These phenomena could be related to some sort of magnetic transition (for example, the development of a canted spin structure).

Similar magnetic anomalies have been reported in Ref. [5] probably attributed to an AFM transition. The authors did not report the effect of a magnetic field on this phenomena. It is interesting to note that at the temperature where χ (H=0) is maximum, R(T,H=0) has a minimum (See Fig.3). This means that the transport properties at this temperature are influenced by these magnetic anomalies. Considering that R(T) exhibits a minimum, the hypothesis that some canted spins structure appear could be accepted. In Ref. [13] it is also reported a minimum in R=R (T) dependence reminiscent of a Kondo system. However, a complete interpretation of these anomalies in the manganates oxides will have to be object of future investigations.

CONCLUSIONS

Thick films of monophasic $La_{2/3}$ $Ca_{1/3}$ Mn O_3 on alumina substrate have been obtained by paint on method. The material exhibits the same disagreement between the resistivity peak temperature T_p and Curie temperature T_c ($T_p<T_c$) observed previously for manganates and related in general to inhomogeneities characteristic of these materials. The large value of the magnetization per Mn atoms found at only 1T, rules out the existence of collinear antiferromagnetic phase. However, zones with deviations from a perfect ferromagnetic order, such as spin canted structures, must be present to explain the observed transport properties. These zones must be oriented at low temperature and relatively low magnetic field, and therefore could be responsible for the magnetoresistance observed at low temperature. The magnetic signature found at low temperatures which disappear at 1T, could be another evidence of the softness of non perfect ferromagnetic order present in our samples.

REFERENCES

[1] G.H.Jonker and J.H van Santen, *Physica* **16** (1950), p.337.
[2] P.G. de Gennes, *Phys.Rev.* **118** (1960), p.141.
[3] S. Jin, M. Mc Cormac, T. H. Tiefel and R.Ramesh, *J.Appl.Phys.* **76** (1994), p.6929.
[4] C. Zener, *Phys.Rev.* **81** (1951) p.440.
[5] A.Nossovf, J.Pierre, V.Vassliev and Ustimov, *J. Condens.Matter* **8** (1996), p.66.
[6] G.H. Rao, J. Rsum, Y.Z. Sum, J.L. Zhang and J.K. Liang, *J.Condens.Matter* **8** (1996), p.66 p.5393.
[7] Yasuo Endoh, Cozuma Hirota, *J.Phys.Soc. of Japan* **8** (1997), p. 66.
[8] R.D. Sánchez and J. Rivas. *Appl.Phys. Lett.* **68** (1996), p.1.
[9] Ll. Balcells, R. Enrich, J. Mora, A. Calleja. J. Fontcuberta, and X. Obradors, *Appl. Phys. Lett.* **68** (1996), p.1486.
[10] C.N.R. Rao and A.K. Cheelham, *Science* **4** (1996), p.272.
[11] J.E. Nuñez-Regueiroand M. Kadin, *Appl. Phys. Lett.* **68** (1996), p.2747.
[12] R. Shereekala, M. Rajeswari, K. Ghosh, J.Y. Gu, C. Kwon. Z. Trajanovic, T. Boettcher, R.L. Greene, R. Ramesh and T. Venkatesan, *Appl.Phys. Lett.* **71** (1997), p.282.
[13] Y.X. Jia, Li lu, K. Khanzeni, D. Yen, C.S. Lee and A. Zeul, *Solid StateCommun.* **94** (1995), p.197.

Materials Science Forum Vols. 302-303 (1999) pp. 149-158

Unconventional Properties of $Ln_{2-x}Ce_xCuO_{4-y}$ (Ln=Pr, Nd, Sm, Eu) Compounds

R.F. Jardim

Instituto de Física, Universidade de São Paulo, C.P. 66318, 05315-970, São Paulo, Brazil

Keywords: Electron-Doped Superconductors, Electrical-Resistance, Surface Resistance, Granularity

Abstract

We report on the synthesis, structural, and general physical properties of electron-doped superconductors with general formula $Ln_{2-x}Ce_xCuO_{4-y}$; Ln = Pr, Nd, Sm, Eu. This family of oxides provide a remarkable opportunity to study the relationship between structural and physical properties since, by moving along the 4f rare earth series, the evolution of several transport and magnetic properties can be nicely correlated to the steric effects associated with the lanthanide contraction. For example, there is a continuos decrease of the superconducting critical temperature T_{ci} with decreasing ionic radius of the lanthanide, i. e., $T_{ci} \sim 25$ K for Pr, ~ 20 K for Sm, and ~ 13 K for Eu. In addition, superconductivity has not been found in the $Gd_{2-x}Ce_xCuO_{4-y}$ system. Compounds with these stoichiometries also exhibit a great variety of macroscopic physical phenomena including metal-insulator transition, granular superconductivity, and superconductor-insulator transition. These cooper oxide materials are yet another class of magnetic superconductors in which antiferromagnetic order of the Ln ions; Ln = Nd, Sm; coexists with superconductivity. In this work we focus on two unconventional properties of these cuprates: (1) an increased of the electrical resistance below the superconducting transition which has been observed in granular $Sm_{2-x}Ce_xCuO_{4-y}$; and (2) the appearance of granular superconductivity at temperatures for which there is no evidence for a bulk Meissner effect. The first feature is associated with an abrupt increase of the electrical resistance below the superconducting transition T_{ci}. We have found that the resistance excess, $\Delta R(T,B)$, is proportional to $(T/T_{ci})^{-4}$ close to T_{ci} and shows a crossover to a $(T/T_{ci})^{-1/4}$ dependence at lower temperatures. We argued that the increase of ΔR close to T_{ci} is due to the depletion of the concentration of normal carriers as predicted by the semi-phenomenological two-fluid theory of superconductivity. The second feature has been observed by using field-dependent; $0 \leq B \leq 7$ T; rf surface resistance measurements. The data are described with an extended form of the Coffey-Clem theory and discussed within a granular superconducting scenario. These measurements also indicate that near T_{ci} $-dB^c_{c2}/dT \sim 0.4$ T/K, for B || c axis.

1. Introduction

The occurrence of high T_c superconductivity in layered copper oxides [1] has generated an enormous amount of interest in these remarkable materials. A broad and concentrated interdisciplinary research effort on both basic and applied aspects of high temperature superconductivity is currently underway on a worldwide scale. Although similar in many respects to conventional superconductors, these high-T_c materials exhibit many unusual features which both challenge theory and imbue these materials with potential for wide-ranging technological applications. Noteworthy examples include extremely high supercoducting critical temperatures T_{ci}, small coherence lengths, enormous upper critical magnetic fields, and large anisotropy of various properties due to the layered structure. Prior to 1989, the prevailing view was that the charge carriers responsible for superconductivity in all high T_c oxides with $T_{ci} > 30$ K were mobile holes confined to the conducting CuO_2 planes, the basic building blocks of these layered materials [2]. However, this view has been challenged by the discovery of new superconducting materials $Ln_{2-x}M_xCuO_{4-y}$; Ln = Pr, Nd, Sm, Eu; M = Ce, Th; $0.0 \leq x \leq 0.2$; [3-6] in which the charge carriers are believed to be electrons which reside within the conducting CuO_2 planes [7].

Superconducting properties in these cuprates are obtained by doping the parent compound

$Ln_{2-x}M_xCuO_{4-y}$ by a variety of means, some of them which are categorized in Table I along with the highest onset superconducting critical temperatures T_{ci} for each compound (x = 0.15) and lattice parameters *a* and *c* of their tetragonal T'-phase structure.

Table I

Ln	M	T_{ci} Onset (K)	*a* (Å)	*c* (Å)	Reference
Pr	Ce	25	3.9611	12.2166	[3]
Nd	Ce	22	3.9469	12.0768	[3]
Sm	Ce	19	3.9216	11.9093	[3]
Eu	Ce	13	3.9067	11.8447	[8]
Gd	Ce	0	3.9011	11.8401	[9]
Pr	Th	23	3.9712	12.2256	[8]
Nd	Th	19	3.9539	12.1181	[4]
Sm	Th	7	3.9251	11.9507	[5]

Table I - The T'-phase $Ln_{1.85}M_{0.15}CuO_{4-y}$-based electron-doped superconductors and their onset superconducting critical temperatures and lattice parameters *a* and *c* .

The largest group of these materials has the chemical formula $Ln_{2-x}M_xCuO_{4-y}$, where M^{+4} ions are substituted by Ln^{+3} ions, thereby donating electrons from outside the CuO_2 planes, which results in T_{ci}'s as high as ~ 25 K for x ≈ 0.15 and y ≈ 0.02. The electron-doped $Ln_{2-x}M_xCuO_{4-y}$ compounds crystallize in a tetragonal "T'-phase" structure consisting of CuO_2 planes in which the copper ions are surrounded by a square planar arrangement of oxygen ions. The combined results shown in Table I reveal that decreasing the ionic radius of the lanthanide results in an appreciable decrease of the superconducting critical temperature T_{ci} of the system. For example, T_{ci} is ~ 25 K for $Pr_{1.85}Ce_{0.15}CuO_{4-y}$ and it decreases monotonically with decreasing ionic radius of the Ln, being essentially zero for $Gd_{1.85}Ce_{0.15}CuO_{4-y}$. Such a decrease of T_{ci} is accompanied by a reduction of the volume of the unit cell of the T'-phase structure. Thus, it seems that there is a definite correlation between superconductiviy and steric effects in this series, as discussed elsewhere [10]. Besides these interesting features, some of these electron-doped superconductors as $Ln_{2-x}M_xCuO_{4-y}$; Ln = Sm, Nd; [11, 12] show coexistence between superconductivity and magnetism at low temperatures.

In this work we focus on the effect of superconductivity on the macroscopic conduction behavior of polycrystalline samples of the electron-doped superconductor $Ln_{2-x}Ce_xCuO_{4-y}$ [3]. In general, these materials are best described by being comprised of small superconducting islands embedded in an insulating matrix [13, 14] - a structure similar to a granular superconductor. We have performed electrical resistance R(T), magnetoresistance R(T, B), magnetization M(T), and field-dependent rf surface-resistance $R_s(T, B)$ measurements in these compounds. From these experimental results, we have found an abrupt increase of R(T) below the superconducting transition T_{ci}. This increase in the magnitude of R(T) is rapidly suppressed by the application of large magnetic fields [15]. By considering the excess of resistance ΔR below T_{ci}, we have found that, for $0.6 < (T/T_{ci}) < 1$, ΔR(T, B) is proportional to $(T/T_{ci})^{-4}$ in accordance with the semi-phenomenological two-fluid theory of superconductivity. Also, from the results of field-dependent; $0 \leq B \leq 7$ T; noncontact rf surface-resistance measurements, we confirm the onset of granular superconductivity in the high-temperature superconductor $Sm_{2-x}Ce_xCuO_{4-y}$; $0.15 \leq x \leq 0.18$; at temperatures for which there is no evidence for a bulk Meissner effect. All these unconventional properties are discussed within the framework of a granular superconducting model.

2. Experimental procedure

Polycrystalline samples of $Ln_{2-x}Ce_xCuO_{4-y}$ were prepared using the sol-gel route [16]. All the pellets were sintered in air and at 980 °C for 48 h, cut in parallelepiped-shape 2x2x8 mm^3 bars, reduced in Ar at 950 °C for 18 hours and quenched to room temperature in 2 h.

Powder x-ray diffraction, using a Phillips PW1710 diffractometer employing CuK_α radiation, revealed single phase materials with the T'-structure. Conventional four-wire electrical resistance and magnetoresistance measurements were performed in applied magnetic fields B as high as 8.8 T and temperatures as low as 4.2 K using a general purpose commercial variable-temperature and magnetic field cryostat (Physical Properties Measurements System - Quantum Design) employing excitation current densities ranging between $6x10^{-6} \leq J \leq 1x10^{-2}$ A/cm^2. Zero-field-cooled (ZFC) and field-cooled (FC) magnetic susceptibility curves were obtained in a custom-made vibrating sample magnetometer from 4.2 to 30 K and typical magnetic field of 2 Oe. Meissner fractions were estimated from the theoretical density of the $Ln_{1.85}Ce_{0.15}CuO_{4-y}$ unit cell with no demagnetizing corrections.

For the microwave experiments, the samples were mounted on the bottom center of a TE_{101} microwave cavity, in the region of maximum rf magnetic field, and minimum rf electric field. The resonant frequency of the cavity was either 12.9 or 9.86 GHz, and the microwave source was "frequency locked" to the cavity frequency. The static magnetic field was applied in the sample plane, and could be rotated relative to the fixed direction of the current density, **J**. Two configurations are of special interest, those with the applied field parallel and perpendicular to **J**. In these configurations, the Lorentz force on fluxons is minimum (nominally zero), and maximum, respectively. Changes in the resistive dissipation arising from the temperature and field dependence of the surface resistance were detected by measuring the changes $\Delta R_s(\mathbf{B}, T) = R_s(\mathbf{B}, T) - R_s(B = 0, T)$ or $\Delta R_s(B = 0, T) = R_s(B = 0, T) - R_s(B = 0, T = T_0)$ where T_0 is a low starting temperature. At a sufficiently low temperature, the surface resistance is expected to be very small, but not zero. Changes in the surface resistance are determined by measuring changes in the power reflected from the cavity, using a point contact diode. The temperature-dependent empty cavity "background" was measured and found to be small and temperature independent for 3 - 30 K.

Since the electrical resistivity (ρ) of the samples was measured as a function of temperature, these data provided a calibration for the microwave system, operating at an angular frequency ω. The expression for surface resistance in the normal is well known; $R_s = (\mu_0\omega\rho/2)^{1/2}$. A second calibration point was obtained from the low temperature response at B = 0; the surface resistance was assumed to have a small residual value ~ 0. The resulting calibration is believed accurate to ± 5 %.

3. Increased Resistance below the Superconducting Transition T_{ci}

Figure 1 displays the temperature dependence of the electrical resistivity, $\rho(T)$, and magnetization M(T) for a sample of $Sm_{1.83}Ce_{0.17}CuO_{4-y}$ subjected to an excitation current of 1 mA. A number of important features can be identified in this curve. First, the samples displays a semiconductor-like behavior for temperatures above ~ 20 K. Secondly, $\rho(T)$ exhibits a local minimum in its magnitude below $T_{ci} \approx 17.5$ K, without any evidence of a zero resistance state at low temperatures. These features provide evidence that with decreasing temperature superconductivity develops at T_{ci} within superconducting islands embedded in an insulating matrix. However, the $\rho(T)$ data suggest that this sample has a very small superconducting volume fraction since the zero resistance state is not observed even at low temperatures. On the contrary, with decreasing temperature the development of a rapid increase of $\rho(T)$ is observed well below T_{ci}. It is also important to note that the superconducting transition temperature T_{ci} of these Sm-based compounds is always found to be close to 20 K [12-14]. In this and other samples, T_{ci} is slightly lower probably due to the highly disordered nature of the materials.

We have also identified a partial drop in the R(T, B = 0) curve at $T_{ci} \approx 17.5$ K as the onset of the superconductivity in these compounds, as shown in Fig.1 (a). This figure also shows the computed resistance, $\Delta R(T, B)$, exceeding the normal state resistance value expected from a simple polynomial extrapolation of R(T, B = 0) down to low temperatures. Such an extrapolation, obtained by a second order polynomial fit of the R(T, B = 0) data from 65 K to 30 K, as well as the definition of ΔR are displayed in Fig. 1 (a).

Further evidence for superconductivity below $T_{ci} \sim 17.5$ K is provided by the temperature dependence of the magnetization M(T) shown in Fig. 1 (b). This compound exhibits appreciable diamagnetism below $T_{ci} \sim 17.5$ K with comparable screening (ZFC) and Meissner (FC) signals. This feature suggests that superconducting islands or clusters are essentially isolated in the insulating matrix. Also, we have inferred the superconducting fraction of this sample by computing the value of M(T = 5 K) in applied magnetic field of 2 Oe and found that superconducting volume fraction is ~ 6%. This value of superconducting volume is rather small, probably below the percolation threshold [17], and further suggests that this sample is comprised of isolated superconducting islands embedded in a nonsuperconducting host.

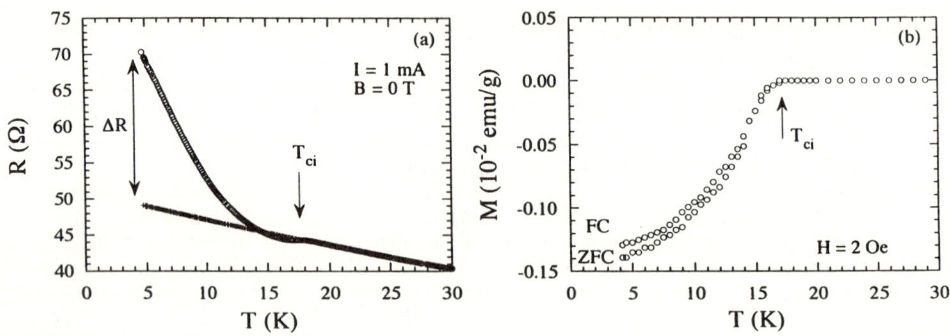

Figure 1. Temperature dependence of the electrical resistance R(T) together with the second order polynomial fit of the data (a), and magnetization M(T) (b). The M(T) data show appreciable diamagnetism below $T_{ci} \approx 17.5$ K.

For a better understanding of the underlying mechanism responsible for the increase in R(T) below T_{ci}, we have performed measurements of R(T) under applied magnetic fields H < 8.8 T, as shown in Fig. 2. For temperatures above $T_{ci} \approx 17.5$ K, we have found that R(T) is almost insensitive to the applied magnetic field. However, below T_{ci}, appreciable negative magnetoresistance is observed for increasing applied magnetic field. A careful inspection of these results also shows that such decrease in the magnitude of R(T) is particularly pronounced at low temperatures. For example, at T = 5 K, increasing magnetic field from B = 0 to B = 8.8 T causes a decrease in R(T) from ~ 70 to ~ 50 Ω.

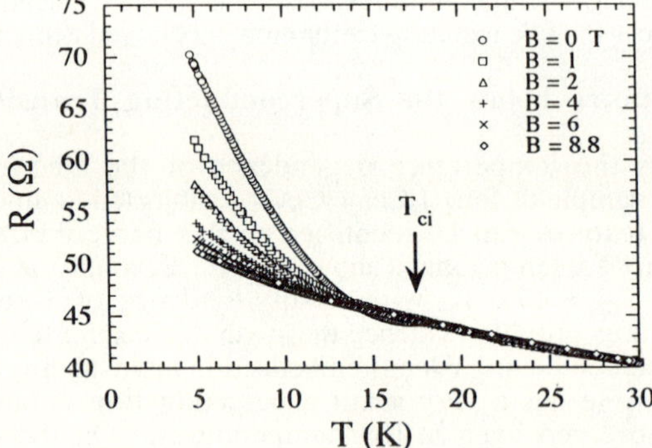

Figure 2. Temperature dependence of the electrical resistance R(T) in applied magnetic field as high as 8.8 T for a polycrystalline sample of $Sm_{1.87}Ce_{0.13}CuO_{4-y}$. T_{ci} denotes the superconducting critical temperature.

All the transport properties described above can be combined with the expected morphology of these samples resulting in a simple picture involving single particle tunneling and Josephson coupling. These granular samples can be described by a collection of isolated metallic islands which undergo a superconducting transition at T_{ci}. As the superconducting volume fraction is bellow the percolation threshold, the zero resistance state is not achieved even at low temperatures. Instead, an abrupt increase in R(T) is observed below T_{ci}. However, the development of superconductivity within isolated islands depletes the concentration of normal charge carriers, as predicted by the semi-phenomenological two-fluid theory [2, 10]. According with this theory, electrons start to condense out into a superfluid at the superconducting transition temperature T_{ci}. The proportion of superfluid increases as the temperature is lowered and no normal fluid would remain at very low temperatures. It is also well known that the proportion of electrons which remains in the normal fluid below T_{ci} must vary as $(T/T_{ci})^4$ [18]. Thus, the excess of electrical resistance of the system below T_{ci} would be increased by the factor of $(T/T_{ci})^{-4}$. Hence, one would expect ΔR(T, B) to increase gradually by a factor of $(T/T_{ci})^{-4}$ just below the superconducting transition.

To test and gain insight into these important mechanisms, we have analyzed the $\Delta R(T, B)$ vs $(T/T_{ci})^{-4}$ curves for several applied magnetic fields and in the temperature range $0.6 < (T/T_{ci}) < 1$, as shown in Fig. 3. At the first glance, it seems that the behavior of $\Delta R(T, B)$ versus $(T/T_{ci})^{-4}$ is rather complicated for several reasons. One of them concerns the negative value of ΔR for zero applied magnetic field in the temperature range $0.8 \leq (T/T_{ci}) \leq 1$. However, this is the expected behavior of these granular samples since a fractional drop in the magnitude of R(T) at T_{ci} is always observed (see Fig. 1). Indeed, the deviation of ΔR from the $(T/T_{ci})^{-4}$ behavior, $0.8 \leq (T/T_{ci}) \leq 1$, is observed for all curves obtained in applied magnetic fields below 1 T, as shown in the inset of Fig. 3. Such a deviation of ΔR from the $(T/T_{ci})^{-4}$ behavior occurs because coupling between superconducting islands would form superconducting clusters just below T_{ci}. This kind of short-range phase coherence of the superconducting order parameter would be affected by the application of magnetic fields leading a different temperature dependence of $\Delta R(T, B)$. Such a feature is evident from the data displayed in the inset of Fig. 3 in the range $0.8 \leq (T/T_{ci}) \leq 1$.

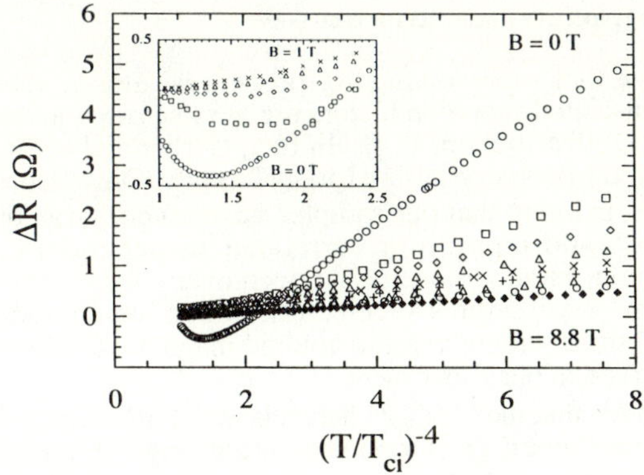

Figure 3. The resistance excess ΔR as a function of $(T/T_{ci})^{-4}$ in the temperature range $0.6 < T < 1$. The figure shows data for the sample subjected to several applied magnetic fields up to 8.8 T. An expanded view of the data close to $\Delta R \sim 0$ is shown in the inset.

We have also observed that $\Delta R(T, B) \propto (T/T_{ci})^{-4}$ in a larger temperature range $0.6 \leq (T/T_{ci}) \leq 1$, as shown in Fig. 4. In fact, from a careful analysis of these curves we have found that this temperature range increases monotonically with increasing applied magnetic field. For higher applied magnetic fields, $3 \leq H \leq 8.8$ T, $\Delta R(T, B)$ can be written as $\Delta R(T, B) = A + B(T/T_{ci})^{-4}$ in the temperature range $0.6 \leq (T/T_{ci}) \leq 1$, as expected by the two - fluid theory. The discussion here involves some additional aspects since superconductivity close to T_{ci} is suppressed by a large magnetic field. By taking this into account, $\Delta R(T, B)$ would strongly decrease, approaching zero for fields exceeding the upper critical field B_{c2}. This would occur essentially due to Cooper pair breaking effects which increase the density of states and lower the sample resistance, resulting in a gradual increase of the number of normal charge carriers available for the transport properties. Such a behavior has been also observed in our experiments. However, it is important to stress here that, even for B = 8.8 T, we have found that ΔR is not strictly zero. An explanation for this behavior is two fold:

(1) It is well known that the upper critical magnetic fields of electron-doped superconductors are very anisotropic [19]. For example, it has been found that $Nd_{1.85}Ce_{0.15}CuO_{4-y}$ has $dB_{c2}/dT \approx$ -8.85 T/K for $B \perp c$ and $dB_{c2}/dT \approx$ - 0.4 T/K for $B \parallel c$ [19(a)]. Similar values were found in $Sm_{1.85}Ce_{0.15}CuO_{4-y}$ [19(b)]. The value of dB_{c2}/dT parallel to the *a* - *b* plane is rather high and combined with the misorientation of the crystallites within the granular material would cause a large distribution of B_{c2} in polycrystalline samples. Thus, an application of a magnetic field of B = 8.8 T would not be enough to completely suppress the superconductivity close to T_{ci} in all crystallites of our samples.

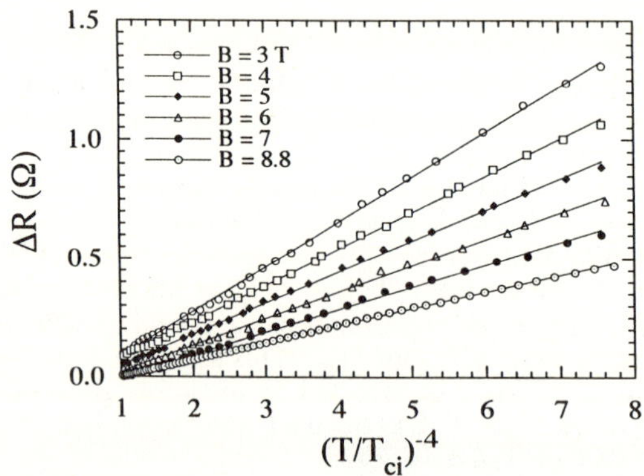

Figure 4. The resistance excess ΔR vs. $(T/T_{ci})^{-4}$; $0.6 < T < 1$; for several applied magnetic fields up to 8.8 T. The figure also shows the respective linear fits for each curve.

(2) It is also known that superconducting properties, mostly the superconducting critical temperature T_{ci}, of electron-doped superconductors are very sensitive to small fluctuations of either cationic (Ce) and anionic (O) distributions [14]. In fact, Gerber and co-authors [14] have estimated the width of such distribution in polycrystalline $Ln_{1.85}Ce_{0.15}CuO_{4-y}$ (Ln = Pr, Nd, Sm, Eu) to be of order of ~ 10 K. Indeed, it is likely that our samples have a very large distribution of T_{ci}. Such a large distribution in T_{ci} would imply in a correlated dispersion in B_{c2} values, generating superconducting regions in the neighborhood of T_{ci} even in applied magnetic fields as high as 8.8 T. Thus, by considering these two features of electron-doped superconductors, one is leading to conclude that ΔR would decrease with increasing applied magnetic field without approaching zero in the high field limit, as observed in our experiments.

It is important to notice that the $(T/T_{ci})^{-4}$ dependence of ΔR, even in applied magnetic fields as high as 8.8 T, is limited to the $0.6 < (T/T_{ci}) < 1$ temperature range. For lower temperatures, $(T/T_{ci}) < 0.6$, fits of the data revealed that $\Delta R(T, B) = \Delta R_0(B)(T/T_{ci})^{-1/4}$. Thus, for values of $(T/T_{ci}) < 0.6$, other mechanisms are likely to contribute to changes in the magnitude of ΔR. In fact, it seems that there is a competition between these mechanisms, each one of them becoming more or less evident depending on the magnitude of the applied magnetic field, as discussed in detail elsewhere [20].

4. Granular Superconductivity and Bulk Meissner Effect

Previously, we have reported the observation of two resistive superconducting transitions [13, 14] in polycrystalline samples of $Sm_{2-x}Ce_xCuO_{4-y}$; $0.15 \leq x \leq 0.18$; the transition at the highest temperature denoted by T_{ci} was identified by a fractional drop in the resistivity, followed by a plateau at lower temperatures. This partial resistive transition was observed as much as several degrees above the bulk transition which occurs at a temperature denoted by T_{cj}. We have attributed the transition at T_{ci} to the nucleation of superconductivity in small isolated regions with estimated dimensions varying from 6 - 300 Å [14]. At lower temperatures denoted by T_{cj}, which vary systematically with the Ce content x, magnetization data showing a Meissner effect, and dc resistivity measurements, showing vanishing resistivity, are correlated. The transition at T_{cj} was attributed to Josephson coupling of the superconducting granules. This interpretation of the resistivity data has been discussed elsewhere [15] providing the motivation to employ field and temperature surface resistance measurements to examine this system with a technique that depends neither on electrical continuity nor on bulk screening currents. Two features, a relatively small B_{c2} and intrinsic granularity, combine to make these materials attractive candidates for microwave surface resistance experiments; the application of a large (7 T) magnetic field will induce large changes in the surface resistance through flux-flow and phase-slip resistivities. In the following, we present experimental results obtained using a non-contact surface resistance method which conclusively demonstrate that the superconductive onset temperature is T_{ci}.

It is well known that the magnetoresistance of high temperature superconductors in the normal state is very small, while the application of a field in the superconducting state of these type-II materials invariably leads to substantial power dissipation at microwave frequencies, even if the sample exhibits excellent dc flux pinning. This results because the fluxons may be forced to vibrate with very small amplitudes ~ Å, even while in a pinning center. In the simplest view, vibration of the normal core provides an obvious dissipation mechanism, as is described by Bardeen-Stephen [21] flux-flow. Further, if the material contains defects which lead to the isolation of small superconducting regions, these barriers may act as Josephson junctions through which the flux may pass. Such normal barriers are natural flux-pinning sites. These sites give rise to phase-slip resistivity as described by the Tinkham-Lobb [22] model as extended by Blackstead [23] and applied to a variety of materials. It is not unusual for high temperature superconductors to display a complete lack of flux-flow resistivity, while exhibiting phase-slip resistivity in field-dependent dc resistivity measurements. In the case of anisotropic crystals the phase-slip resistivity has axial symmetry, relative to the applied field, while the planar flux-flow contribution varies with $\sin^2\phi$ ($\phi \angle$ **J**, **B**), and vanishes with **J** parallel to B($\phi = 0$). If the applied field is applied in a nonplanar direction, the flux-flow also depends on the angle of the applied field relative relative to the c-axis, but is always maximum for $\phi = \pi/2$.

In the case of polycrystalline samples, the average resistivity, and surface resistance, is dominated by the largest contribution to the resistivity tensor, which arises from those crystallites oriented with their *c* axes parallel to the applied magnetic field. The resistivity is typically seen to arise principally from two dissipation mechanisms: phase-slip and flux-flow. Each of these contributions is anisotropic, and for crystallites aligned with the applied field in their *a* - *b* planes, each is much smaller than for field alignment parallel to the *c* axis. The flux-flow and phase-slip resistivities exhibit similar anisotropy, with both contributions minimized with B in the crystallite plane. However, each of these contributions to the electrical resistivity exhibits different dependence on the relative orientations of the applied magnetic field, and the current direction. This feature was addressed through the use of Monte Carlo-like approach, in which the average response of a collection of randomly oriented crystallites is computed, using the average resistivity for the system [24].

For twinned orthorhombic or tetragonal samples, the comparatively small flux-flow contribution varies as $\sin^2(\phi)$, as the field is rotated relative to the current density, while phase-slip resistivity is isotropic in the *a* -*b* plane. In a microwave cavity experiment, the direction of the current density is determined by the cavity geometry and mode excited; the current density is limited to a small penetration depth and cannot "percolate" through the sample, as is possible in dc measurements. If the sample consists primarily of small superconducting regions, with linear dimensions ~ ξ (the coherence length) that are well separated, no field dependence for temperatures below but near to T_{ci} is expected. This results because such small superconducting regions are too small to be penetrated by a normal core and remain superconducting. If flux fails to penetrate the superconducting granules, no flux flow may result, and there will be no phase slip either. As the temperature is reduced and the superoconducting regions grow and link-up, both flux-flow and phase-slip will observed *before* T_{cj} is reached. If the superconducting granules do not achieve percolation, bulk screening currents will be inhibited, and no bulk Meissner effect is anticipated. The field for which fluxon penetration will occur will be nearer B_{c2} than B_{c1}, for small superconducting granules, and will approach B_{c1} only as the granule grows to a size commensurate with the penetration depth λ. The minimum temperature for which field-dependent losses are detected will vary with the size distribution of the superconducting granules, and so will be a function of the preparation conditions and composition dependence which lead to the granularity. In the following, the model for surface resistance, as given by Coffey and Clem [25], and extended to include a contribution to the resistivity from a granular material with junctions, will be employed. Details of this model are given elsewhere [24].

In Fig. 5, the electrical resistivity $\rho(T)$ and magnetization M(T) for four samples of varying Ce content is provided, as a function of temperature and current level. The data indicate that as the Ce content decreases, the resistivity increases systematically. In addition, the initial transition temperature T_{ci} increases, and the width of the plateau region between T_{ci} and T_{cj} decreases. At the higher current level, the width of the lower (bulk) transition, which has an onset at T_{cj}, increases. The temperatures T_{cj} are correlated with the magnetization data while the temperature T_{ci} determined from the resistivity data do not correlate. Careful inspection of the data reveals that diamagnetism only at temperatures below T_{cj}, the bulk transition temperature of the system. The lack of significant diamagnetism in the temperature interval $T_{cj} \leq T \leq T_{ci}$ has been attributed to the presence of superconductivity with an onset temperature T_{ci} in small "islands" with typical dimensions of 6 - 300 Å [26].

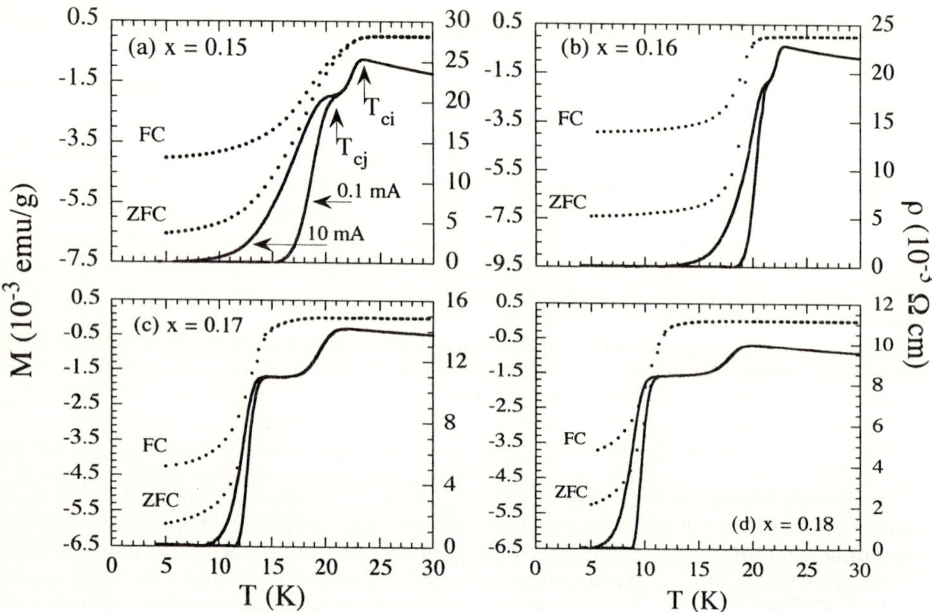

Figure 5. Temperature dependence of magnetization M (left axis) and electrical resistivity ρ (right axis) in polycrystalline samples of $Sm_{2-x}Ce_xCuO_{4-y}$; $0.15 \leq x \leq 0.18$. Magnetization measurements were performed in an applied magnetic field of 1 Oe. The critical temperatures T_{ci} and T_{cj}, as well as, both excitation currents are marked in (a).

In Fig. 6, the results both for $R_s(B = 0)$ and $R_s(B = 1\ T)$ for the flux-flow and phase-slip configurations are given, along with the fits to data as given by the model [24], for the composition $Sm_{1.85}Ce_{0.15}CuO_{4-y}$. The data clearly show that the onset of field-induced dissipation occurs for temperatures slightly lower than T_{ci}, but higher than T_{cj}. The transition, as seen in the surface resistance is quite broad, but is described quite accurately by the model in which the superconducting onset temperature is close to 28.5 K. The model for $\Delta R_s(B = 1\ T)$ is less successful in describing the data, but does have the correct trends and approximate size. The figure includes an indication of the resistive onset as measured with low density dc excitation current.

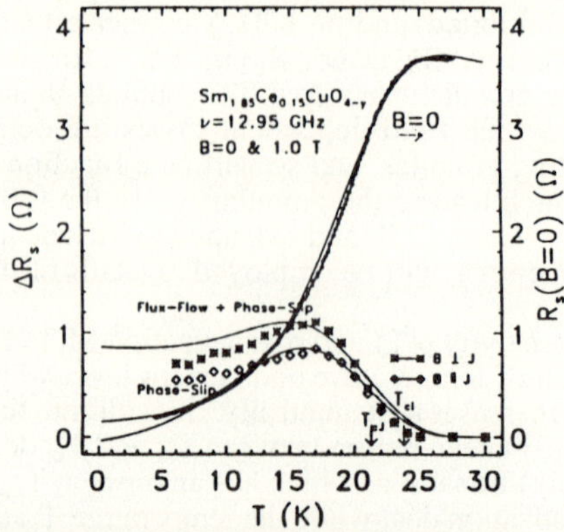

Figure 6. Surface resistance R_s at B = 0 (right axis) and field induced changes in the surface resistance induced by application of B = 1 T, $\Delta R_s(B, T) = R_s(B, T) - R_s(B = 0, T)$ (left axis) for two different field-current configurations, for the composition $Sm_{1.85}Ce_{0.15}CuO_{4-y}$. The data for B = 0 are accurately described by the model in Ref. 24 (solid lines) while the data for the field induced changes are less accurately described, particularly at low temperatures. The peak in $\Delta R_s(B, T)$ reflects the feature that the differences between $R_s(B = 1\ T)$ and $R_s(B = 0)$ are largest near (but below) the transition temperature. If the axially symmetric phase-slip contribution to the dissipation were smaller, the peak would be much more pronounced, and nearer to the transition temperature. These results indicate that the dissipation is dominated by effects due to granularity.

Data for the largest Ce concentration studied, x = 0.18, as a function of temperature with B = 0, and at an applied field of B = 1 T parallel and perpendicular to the current density are given in Fig. 7. In this case, the data suggest that the superconducting granules manifest a size distribution which has a smaller average size than the samples with smaller Ce concentrations, a result which is consistent with the relative dopant levels. Of some interest is the behavior of the surface resistance as compared to the dc electrical resistivity. The resistivity data, for temperatures greater than T_{ci}, exhibit a negative slope, while the surface resistance data in the same temperature range have a positive slope. We conclude that the traces of superconductivity observed in the microwave measurements in this temperature range dominate the tendency toward semiconductor behavior.

We have carried out measurements of the surface resistance to higher fields, for the sample with x = 0.15 [24]. We observed that the application of a 7 T magnetic field is inadequate to drive the sample completely normal. This even though the temperature (T = 23.5 K) is several K below the onset temperature determined from the fitting parameter, but is ~ 1 K higher than the onset temperature determined from dc measurements. We concluded from the fitting procedure that $B_{c20} \approx 10.56$ T. Using the fit value for B^c_{c20} gives $-dB^c_{c2}/dT \sim 0.4$ T/K, which is very near the value of 0.43 T/K as determined by Almasan et al. [27] in similar compounds.

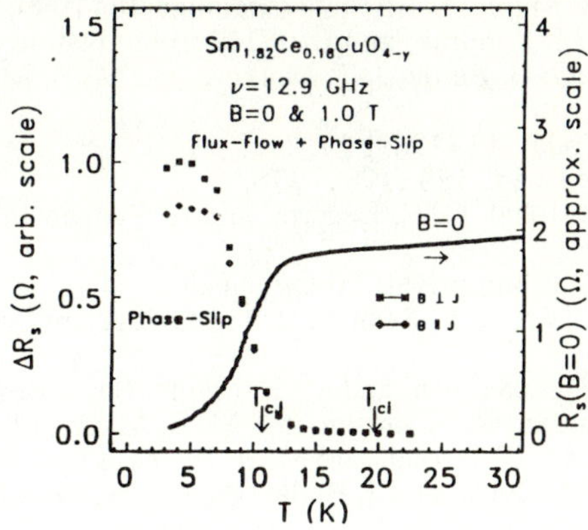

Figure 7. Surface resistance for B = 0 and 1 T with the field parallel and perpendicular to the rf current density, for the composition $Sm_{1.82}Ce_{0.18}CuO_{4-y}$. The data also indicate that the onset of superconductivity occurs at a temperature well above T_{cj}. The results with B = 1 T as a function of temperature also indicate a relatively small renormalized flux-flow contribution.

5. Conclusions

In conclusion, we have observed an abrupt increase in the magnitude of the electrical resistance below the superconducting transition $T_{ci} \approx 17.5$ K in granular samples of $Sm_{1.83}Ce_{0.17}CuO_{4-y}$. By computing the resistance excess ΔR(T, B), $0 \leq B \leq 8.8$ T, below the superconducting transition temperature, we have found that ΔR is proportional to $(T/T_{ci})^{-4}$ close to T_{ci}. We argued that the increase of ΔR close to T_{ci} is mostly due to the depletion of the concentration of normal charge carriers as predicted by the semi-phenomenological two-fluid theory of superconductivity. We also found a crossover of ΔR to an $(T/T_{ci})^{-1/4}$ dependence at lower temperatures and have discussed this behavior in the framework of a superconducting granular scenario.

The application of a magnetic field, in conjunction with the rf magnetic field, enabled unambiguous confirmation of the superconductive onset temperature. The superconducting onset temperature observed with this technique was slightly higher than could be determined from the conventional dc measurement. This is in part due to the temperature dependence of the normal state resistivity, which has a small negative slope. Previously, it has been shown that multiphase samples of Hg-Ba-Ca-Cu-O and Bi-Sr-Ca-Cu-O [28] exhibit distinct, easily identified, temperature dependent features, both for B = 0, and B > 0. No such behavior was observed for these samples. It is concluded that these samples consist of only one superconducting phase. The microwave surface resistance experiments presented here unambiguously demonstrate that for these $Sm_{2-x}Ce_xCuO_{4-y}$

samples, the superconducting onset is coincident with the onset temperature determined by dc electrical resistivity measurements, even though there is no evidence for a macroscopic Meissner effect at these temperatures.

Acknowledgments

This paper would not have possible without the results of and stimulating discussions with H. Blackstead, M. J. R. Sandim, P. A. Suzuki, C. H. Westphal, E. A. Early, D. Stroud, M. B. Maple, S. Spagna, S. Tripp, and R. Sager. The support of the Brazilian agency FAPESP under contract No. 96/8416-4 is gratefully acknowledged. The author is Brazilian Research Council CNPq fellow under Grant No. 304647/90-0.

References

1. J. G. Bednorz and K. A. Müller, Z. Phys. B **64** (1986) 189.
2. A. W. Sleight, Science **242** (1998) 1519.
3. Y. Tokura, H. Takagi, and S. Uchida, Nature **337** (1989) 345.
4. J. T. Markert and M. B. Maple, Solid State Commun. **70** (1989) 145.
5. A. C. W. P. James, S. M. Zahurak, and D. W. Murphy, Nature **338** (1989) 240.
6. E. A. Early, N. Y. Ayoub, J. Beille, J. T. Markert, and M. B. Maple, Physica C **160** (1989) 320.
7. A. Khurana, Physics Today **42** (1989) 17.
8. J. T. Markert et al., Physica C **158** (1989) 178.
9. R. F. Jardim, C. H. Westphal, C. C. Becerra, and A. Paduan-Filho, Phys. Rev. B **45** (1992) 10485.
10. P. A. Suzuki and R. F. Jardim, private communication.
11. Y. Dalichaouch, B. W. Lee, C. L. Seaman, J. T. Markert, and M. B. Maple, Phys. Rev. Lett. **64** (1990) 599.
12. M. C. de Andrade, Y. Dalichaouch, and M. B. Maple, Phys. Rev. B **48** (1993) 16737.
13. E. A. Early, C. C. Almasan, R. F. Jardim, and M. B. Maple, Phys. Rev. B **47** (1993) 433.
14. R. F. Jardim, L. Ben-Dor, D. Stroud, and M. B. Maple, Phys. Rev. B **50** (1994) 10080.
15. A. Gerber, T. Grenet, M. Cyrot, and J. Beille, Phys. Rev. Lett. **65** (1990) 3201; R. F. Jardim, M. C. de Andrade, E. A. Early, M. B. Maple, and D. Stroud, Physica C **232**, 145 (1994).
16. R. F. Jardim, L. Ben-Dor, and M. B. Maple, J. of Alloys and Compounds **199** (1993) 105; P. A. Suzuki and R. F. Jardim, Physica C **267** (1996) 153
17. D. J. Bergman and D. Stroud, Solid State Physics: Advances in Research and Applications, H. Ehrenreich and D. Turnbull, eds (Academic 1992), Vol. 46, pp. 148.
18. B. S. Chandrasekhar, Superconductivity vol. 1, edited by R. D. Parks, (Dekker, New York, 1969); C. J. Adkins, J. M. D. Thomas, and M. W. Young, J. Phys. C: Solid State Phys. **13** (1980) 3427.
19. (a) Y. Hidaka, and M. Suzuki, Nature **338** (1989) 635; (b) C. C. Almasan, S. H. Han, E. A. Early, B. W. Lee, C. L. Seaman, and M. B. Maple, Phys. Rev. B **45** (1992) 1056.
20. M. J. R. Sandim, P. A. Suzuki, S. Spagna, S. Tripp, R. E. Sager, and R. F. Jardim, Physica C **289** (1997) 265.
21. J. Bardeen and M. J. Stephen, Phys. Rev. **140**, A1197 (1965).
22. M. Tinkham and C. J. Lobb, in *Solid State Physics 42*, H. Ehrenreich and D. Turnbull, eds. (Academic Press 1989) pp. 91.
23. H. A. Blackstead, Phys. Rev. B **47**, 11411 (1993).
24. H. A. Blackstead, R. F. Jardim, P. Beeli, D. B. Pulling, and A. K. Heilman, Phys. Rev. B **57** (1998) 3683.
25. M. W. Coffey and J. W. Clem, Phys. Rev. B **45** (1972) 10527.
26. E. A. Early, R. F. Jardim, C. C. Almasan, and M. B. Maple, Phys. Rev. B **51** (1995) 8650.
27. C. C. Almasan, S. H. Han, E. A. Early, B. W. Lee, C. L. Seaman, and M. B. Maple, Phys. Rev. B **45** (1992) 1056.
28. H. A. Blackstead, D. B. Pulling, M. Paranthaman, and J. Brynestad, Phys. Rev. B.**51** (1995) 3783; H. A. Blackstead, D. B. Pulling, and H. Sato, Physica Status Solidi (a) **140** (1993) 509.

Materials Science Forum Vols. 302-303 (1999) pp. 159-163

Effects of Oxygen on the Phase Diagram of $La_{1.85}Sr_{0.15}Cu_{4-\delta}$

P.N. Lisboa Filho[1*], S.M. Zanetti[2], E.R. Leite[2] and W.A. Ortiz[1]

[1] Grupo de Supercondutividade e Magnetismo, Departamento de Física, Universidade Federal de São Carlos, Brazil

[2] Laboratório Interdisciplinar de Electroquímica e Cerâmica, Departamento de Química, Universidade Federal de São Carlos, Brazil

Keywords: Superconductivity, Magnetism, Chemical Environment

Abstract *In this work we study the effects of oxygen on the magnetic response of* $La_{1.85}Sr_{0.15}CuO_{4-\delta}$ *ceramic samples. Oxygen-overdoped samples, with* δ *ranging beyond the superconducting phase limits, were prepared and measured, leading to the conclusion that overdoped samples present all characteristic features of a spin-glass. The irreversible behavior of the system and the time-relaxation of the remnant magnetization below the glass limit are also discussed.*

Since the discovery of high temperature superconductivity (HTS) by Bednorz and Müller in 1986 [1], a variety of mechanisms have been invoked to explain the high temperature superconducting state. After a decade of studies, a vast scenario is now envisaged, although the yoke of the problem remains still an open issue.

HTS compounds have similar structures: while $La_{2-x}Sr_xCuO_{4-\delta}$ (LSCO), $Bi_2Sr_2Ca_{n-1}Cu_nO_X$ (BSCCO) and $TlBa_2Ca_{n-1}Cu_nO_{2n+3}$ (TBCCO) have planes of Cu and O atoms, $YBa_2Cu_3O_7$ (YBCO) have planes and chains of Cu and O. It is well known that Cu-O planes play a major role in generating superconductivity, while the chains act as electron reservoirs which can be filled or emptied either by changing the oxygen stoichiometry or through other types of doping.

The quaternary compounds $La_{2-x}Sr_xCuO_{4-\delta}$, which probably represent the simplest form of copper oxide superconductors, exhibit a rich phase diagram, including antiferromagnetism (AF), spin-glass ordering (SG), and superconductivity (SC). All of these phases have been extensively studied [2-5]. In the SC range, $0.06 < x < 0.21$, the effects of oxygen concentration were also studied [6,7] to determinate the optimum value of δ and its influence on superconductivity. For samples with $x = 0.15$, known to be those with the highest T_c, the optimum δ is about 0.07 [6], with a very narrow range for the occurrence of superconductivity. It has also been shown that Sr- and O-induced holes are equivalent ways of disrupting the antiferromagnetic long-range order [7]. This encourages one to verify whether x and δ are also corresponding variables in other regions of the phase diagram.

In fact, Radaelli and coworkers [8] have studied the influence of oxygen content on the the formation of spurious phases and on the properties of samples of superconducting LSCO prepared via the solid state reaction route. As they have pointed out, it is very difficult to vary δ, and their study covers only a small region around the optimum oxygen concentration. A similar study, by Buffeteau et al. [9] with samples obtained by a sol-gel route, has indicated a greater oxygen permeability. They have also detected a non-superconducting phase in oxygen undoped samples.

This work is part of an effort to understand the interplay between x and δ and, moreover, which role is individually played by each of them in the occurrence of superconductivity in the LSCO system. We thus restrict the present study to values of x in the vicinity of 0.15. In order to achieve control over δ, one has to bear in mind that chemical diffusion of oxygen in LSCO is very difficult, even at high temperatures [10]. For this reason, we decided to follow a modified sol-gel technique, described below, which allows for a better control of oxygen content and reveals the influence of chemical environment on the magnetic response of the sample. We explore here the region of the magnetic phase diagram corresponding to samples overdoped with oxygen, including in the study the existing correlations among δ and the chemical environment.

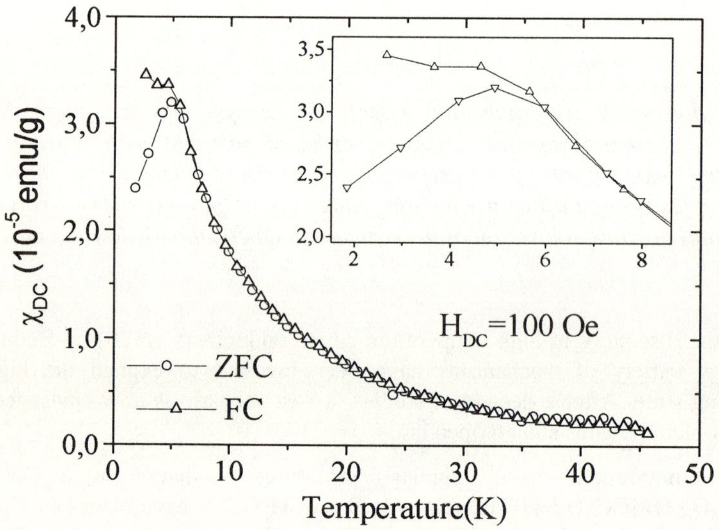

Figure 1 - ZFC and FC runs at H = 100 Oe, for an overdoped sample (4-δ = 4.2). The inset enhances the hysteretic region below the irreversibility temperature T_i = 6 K.

Preparation of HTS bulk materials has been extensively explored [11]. The most common approach, known as the solid-state reaction route, requires several cycles of thermal treatments and grindings. Alternatively, routes using polymeric precursors may lead to high quality ceramics with the advantage of homogeneous mixing, excellent stoichiometric control and the production of active particles of very small size [12-14].

The samples studied in this work were prepared by a modified polymeric precursors technique, known as Pechini Method [15]. Stoichiometric La_2O_3, $SrCO_2$ and CuO were dissolved together in distilled water, powdered citric acid being then added in the ratio of 1/3 for metals.

Finally, 1,0 ml of ethylene glycol per gram of sample is added. The resulting blue solution was neutralized to ph~7 with dietylamina, in order to avoid bicarbonate formation. This solution is turned into a gel, and subsequently decomposed to a solid by heating at 400°C. The resulting powder was calcinated in vacuum at 900°C for 24h, with two intermediary grindings, in order to prevent formation of a $La_2SrCu_2O_6$ impurity phase [8]. This powder was then pressed into pellets and sintered in air at 1125°C for 12h.

Structural characterization obtained by X-ray diffraction (XRD) using a Siemens diffractometer model D5000, showed that all samples are single phase. Three samples of powder heat-treated at 800°C for one hour and dissolved in nitric acid were analyzed with an inductively coupled plasma spectrometer, by Thermo Jarrel Ash, model AtomScan 25, and corfimed by atomic absorption spectrophotometry using a Cerius/Intralab instrument model AA 12/1475. The mean values obtained by these chemical analyses of metal contents lead to a chemical formula $La_{1.850}Sr_{0.148}Cu_{0.986}O_{4-\delta}$.

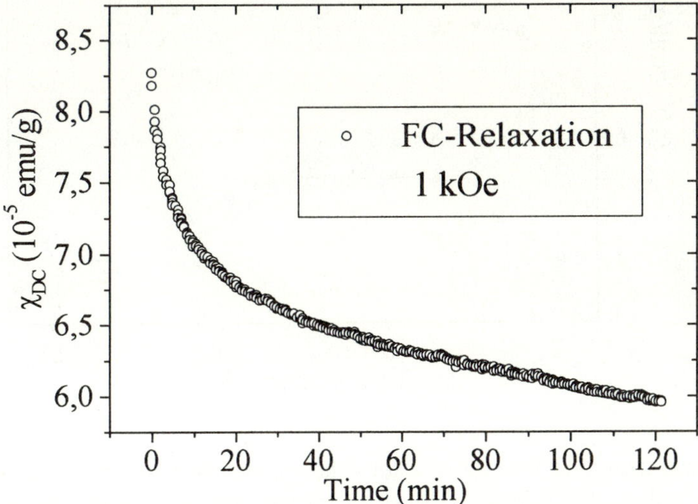

Figure 2 - Field-cooled (FC) relaxation of χ_{DC}, measured at T = 1.8 K after field-cooling the sample in 1 kOe. The field was turned off at time t = 0

Measurements of the magnetization as a function of temperature, M(T), were taken in a fixed applied field H, with a SQUID magnetometer, MPMS-5S, supplied by Quantum Design. Each point of $\chi_{DC}(T) = M(T)/H$ resulted from two scans over a length of 4 cm, with 24 point readings per scan. Magnetic measurements were taken in two different manners: (i): ZFC - warming after zero-field cooling the sample; (ii) FC - measuring while cooling in a field.

Figure 1 shows ZFC and FC runs, taken at 100 Oe, of $\chi_{DC}(T)$ for an overdoped sample with $4-\delta = 4.2$. The magnetic response is positive and exhibits a hysteretic behavior below the irreversibility temperature $T_i = 6$ K, as can be clearly seen in the inset. Above T_i, the curve is well-fitted by a Curie-Weiss law, up to room temperature. There is also a maximum in the ZFC curve, usually associated with the freezing temperature (T_f) in spin-glass systems.

In order to further examine other characteristic features of spin-glasses, we have measured the time relaxation of χ_{DC} for T below and above T_i. As expected, there is no relaxation above T_i, whereas below the irreversibility limit, the system presents a typical spin-glass time-dependent response, well described by the commonly observed stretched exponential [16]. Figure 2 presents a field-cooled (FC) relaxation of χ_{DC}, measured at T = 1.8 K after field-cooling the sample in 1 kOe, which was turned off at time t = 0.

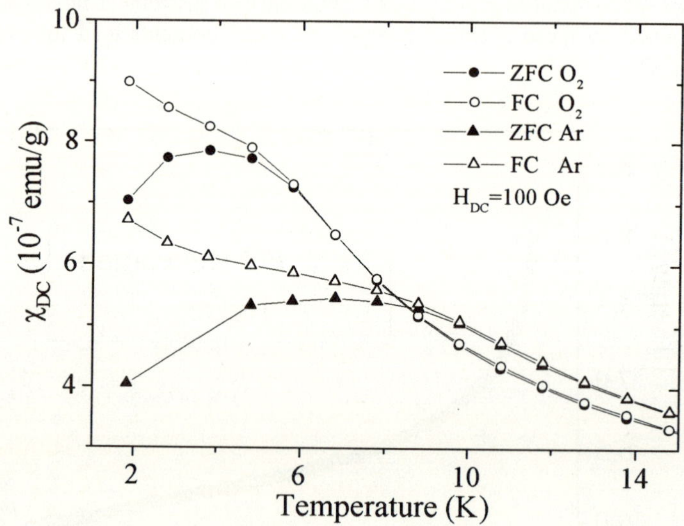

Figure 3 - Sample treated in O_2 at 700°C for 4 hours, and then in Ar, at the same temperature, for 8 hours. The effect on the extension of the spin-glass phase is appreciable.

We have also heat-treated samples of $La_{1.850}Sr_{0.148}Cu_{0.986}O_{4-\delta}$ under several atmospheric conditions, using different gases at various temperatures. Figure 3 is an example of these studies, where a twin sample of that shown in Fig.1 was treated in O_2 at 700°C for 4 hours, and then in Ar, at the same temperature, for 8 hours. It shows how easily one can effectively control the extension of the spin-glass phase, despite the extreme difficulty to bring the sample to exhibit superconductivity.

The importance of chemical environment changes and their relation with the charge transfer mechanisms leading to superconductivity, has been studied to determine the effect of Sr-doping in the atomic arrangement [17,18]. It is also well-known that oxygen-doping could lead to a change in the valence of copper [19]. As we progress in establishing correspondences between δ

and x, preliminary electron paramagnetic resonance studies are being conducted to determine whether changes in the chemical environment in Cu-O planes, i.e., changes of copper valence, can be detected and related to different routes of sample processing.

This work has financial support of brazilian agencies FAPESP and CNPq. Two of the authors are personally supported by FAPESP. The authors wish to thank Mr. Jovino Araujo de Souza for the AAS measurements and Roberto Weider de Assis Franco for preliminary EPR studies.

References

[1] J.G. Bednorz and K. A. Müller, Z. Phys. B 64 (1986), p.189

[2] A. Aharony, R. J. Birgeneau, A. Coniglio, M. A. Kastner, H. E. Stanley, Phys. Rev. Lett. 60 (1988), p.1330

[3] G. Shirane, R. J. Birgeneau, Y. Endoh, M. A. Kastner, Physica B 197 (1994), p.158

[4] F. C. Chou, N. R. Belk, M. A. Kastner, R. J. Birgeneau , A. Aharony, Phys. Rev. Lett. 75 (1995), p.2204

[5] R. J. Gooding, N. M. Salem, R. J. Birgeneau, F. C. Chou, Phys. Rev. B 55 (1997), p.6360

[6] E. Takayama-Muromachi and David Rice, Physica C 177 (1991), p.195

[7] C. Y. Chen, E. C. Branlung, C. C. Bae, K. Yang, M. A. Kastner, A. Cassanho, R. J. Birgeneau, Phys. Rev. B 51 (1995), p. 3671

[8] P. G. Radaelli, D. G. Hinks, A. W. Mitchell, B. A. Hunter, J. L. Wagner, B. Dabrowski, K. G. Vandervoot, H. K. Viswanathan, J. D. Jorgensen, Phys. Rev. B 49 (1994), p.4163

[9] B. B. Buffeteau, T. Hargreves, B. Grevin, C. Marin, Physica C 294 (1998), p.55

[10] Y. Idemoto, M. Ohara, K. Fueki, Physica C 229 (1994), p. 361

[11] M. Kakihana, Journal of Sol-Gel Technology 6 (1996), p.7

[12] H. H. Wang, Inorg. Chem. 26 (1987), p.1476

[13] C. T. Chu and Bruce Dunn, J. Am. Ceram. Soc. 70 (1987),p. 375

[14] P. A. Lessing, Ceramic Bulletin 68 (1989), p.1002

[15] M. P. Pechini, U.S. Patent 3 330 697 (1967)

[16] J. A. Mydosh, in *Spin glasses: an experimental introduction*; ed. by Taylor and Francis, (1993), p. 96.

[17] D. Haskel, E. A. Stern, D. G. Hinks, A. W. Mitchell, J. D. Jorgensen, Phys. Rev. B 55 (1997), p.6360

[18] C. T. Chen, Phys. Rev. Lett. 68 (1992), p.2543

[19] H. Shimizu, K. Fujihara, K. Hatada, Physica C 288 (1997), p.190

*Corresponding author: ppnl@iris.ufscar.br – fax: +55.16.261.4835

Materials Science Forum Vols. 302-303 (1999) pp. 164-168

Raman-Scattering and Weak-Ferromagnetism Studies in $Eu_{2-x}Pr_xCuO_4$

P.G. Pagliuso[1], E. Granado[1], J.A. Sanjurjo[1], C. Rettori[1], I. Torriani[1], D. Rao[2], S.B. Oseroff[2], J. Sarrao[3] and Z. Fisk[3]

[1] Instituto de Física "Gleb Wataghin", Universidade Estadual de Campinas, Campinas, SP, Brazil

[2] San Diego State University, San Diego, CA 92182, USA

[3] National High Magnetic Field Laboratory, Florida State University, Tallahassee, FL 32306, USA

Keywords: Weak-Ferromagnetism, Forbidden Raman Modes, Lattice Distortions

Abstract. We have performed magnetization and Raman experiments in single crystals of $Eu_{2-x}Pr_xCuO_4$ ($0 \lesssim x \lesssim 1$) and the results were interpreted in terms of local distortions within the CuO_2 planes. We find that samples grown in Pt crucibles contain Pt impurities, that are partially responsible for these distortions. We observed: *i*) the appearance of two *forbidden* Raman modes (fRM) and *ii*) a weak-ferromagnetic (WF) component when the crystals are field cooled in the *ab*-plane. The temperature where the WF component is first observed, T_{WF}, the coercive field, H_C, and the intensity of the fRM decrease as the Pt content decreases. For crystals with a similar Pt content, the above quantities are larger for samples with smaller lattice unit cell volume, V. We concluded that the Pt impurities and/or the reduction of V beyond a *critical* value, V_c, are the origin of these orthorhombic distortions which in turn are responsible for the fRM and WF observed in these compounds.

1 INTRODUCTION

The Eu_2CuO_4 compound is particularly interesting among the antiferromagnetic (AF) rare-earth (RE) cuprates, RE_2CuO_4 (RE = Pr, Nd, Sm, Eu and Gd) of T′-(I4/mmm) tetragonal structure, because it is at the boundary between superconductivity (SC) and weak-ferromagnetism (WF).[1,2] The light RE compounds do not show WF, and become superconductors when appropriate electron doping ions are incorporated into the lattice ($RE_{2-x}M_xCuO_4$ with M = Ce^{4+}, Th^{4+} and $x \approx 0.15$) and treated in reducing atmospheres.[3] However, for Gd and heavier RE compounds, SC can not be induced.[1, 4, 5] Instead, WF is observed in them.[6] Eu_2CuO_4 grown in alumina crucibles and CuO flux shows no WF, and SC can be induced on it. But, Eu_2CuO_4 grown in Pt crucibles and PbO flux shows WF when field cooled (FC) in the *ab*-plane, and SC is not observed. [2] In all the RE_2CuO_4 compounds, the Cu moments order AF in the *ab*-plane at $T_N \approx$ 240 - 290 K.[5] The presence of WF has been attributed to a canting of the Cu moments in the *ab*-plane, away from their perfect AF alignment. In the T′-(I4/mmm) structure WF order is forbidden, and lattice distortions in the CuO_2 planes were invoked to account for it.[2] These distortions result in a canting of the Cu moments and allow for an antisymmetric Dzyaloshinsky-Moriya exchange interaction between adjacent Cu moments, giving rise to the observed WF.[6]

In this report we present data taken in samples grown in different crucibles (Pt and Al_2O_3) and fluxes (PbO and CuO). The analysis of these results lead us to conclude that Pt impurities and/or the reduction of the lattice unit cell volume beyond a *critical* value, V_c, are the origin of the distortions responsible for the fRM and WF observed in these compounds.

2 EXPERIMENTAL DETAILS

Crystals of $Eu_{2-x}Pr_xCuO_4$ were grown from nominal stoichiometric mixtures of the respective oxides in the following crucibles/fluxes: Pt/PbO, Pt/CuO and Al_2O_3/CuO. The single crystals were of plate-like shape, with the *c*-axis perpendicular to the large face. Typical crystal sizes were 4x3x0.3 mm^3. Electron-microprobe analysis shows Pt content of 2 - 5 at.% of Cu, and no traces of Pb for crystals grown in Pt/PbO. For crystals grown in Pt/CuO the Pt content was found to be less than 1 at.% of Cu. The Raman measurements were carried out in backscattering geometry with the samples mounted on the cold finger of a close-cycle He refrigerator. The x-ray powder diffraction experiments were performed in a Rigaku R200 diffractometer at room T. Magnetization measurements were made in a Quantum Design dc SQUID magnetometer.

3 RESULTS AND ANALYSIS

Figure 1a shows the hysteresis loops of the *ab*-plane dc-magnetization obtained after FC in 50 kOe, $M_{FC}(H_a,T)$, at $T = 100$ K for the applied field interval $H_a = \pm 100$ Oe. WF is observed in all these samples. The WF disappears for $x > 0.5$ for crystals grown in Pt/PbO, and is not present for any value of x when they are grown in Al_2O_3/CuO. Magnetization data for FC and zero field cooling (ZFC) up to $H_a = \pm 50$ kOe were reported previously in Ref. 2. Figure 1 shows that the remnant magnetization, $M^r_{FC}(T)$ and $H_C(T)$ depends on the crystal preparation conditions. Larger values of $M^r_{FC}(T)$ and $H_C(T)$ are obtained for crystals with: *i*) higher Pt content ($x = 0$) and *ii*) smaller V ($x \to 0$, and similar Pt content).

$M_{FC}(H_a, 100\text{ K})$ shows a nearly linear increase with the applied field, H_a, for $H_a \gtrsim 30$ Oe (see Fig. 1 and Ref. 2). That results from the REs magnetic susceptibility, $dM_{FC}/dH_a \approx \chi_{RE}(100\text{ K})$.[8] Hence, for $H_a \gtrsim 30$ Oe, $M_{FC}(H_a,T)$ can be approximated to:

$$M_{FC}(H_a,T) = M^r_{FC}(T) + \chi_{RE}(T)H_a, \tag{1}$$

If we assume that $M^r_{FC}(T)$, is only the result of the WF component due to the canting of the Cu moments in the *ab*-plane, $M^{Cu}_{FC}(T)$, Eq. 1 can be rewritten as:

$$M_{FC}(H_a,T) = M^{Cu}_{FC}(T) + \chi_{RE}(T)H_a, \tag{2}$$

For ZFC and neglecting the contribution from $\chi_{RE}(T)H_i(T)$, the *ab*-plane total dc-magnetization can be written as:

$$M_{ZFC}(H_a,T) = M^{Cu}_{ZFC}(H_a,T) + \chi_{RE}(T)H_a. \tag{3}$$

Figure 1b gives, for samples showing WF, the *ab*-plane ZFC dc-magnetization of the Cu ions, $M^{Cu}_{ZFC}(H_a,T) = M_{ZFC}(H_a,T) - \chi_{RE}(T)H_a$, at $T = 100$ K between $H_a = \pm 50$ kOe. For each sample, $\chi_{RE}(T)H_a$ was obtained from the high field of the $M_{FC}(H_a,T)$ data in Figure 1a. The data show that crystals, with: *i*) higher Pt content ($x = 0$), and *ii*) smaller V ($x \to 0$, and similar Pt content) are easy to magnetize, i.e., show larger $M^{Cu}_{ZFC}(H_a,T)$.

Figure 1c shows the T dependence of $M_{FC-ZFC}(H_a,T) = M_{FC}(H_a,T) - M_{ZFC}(H_a,T)$ for $H_a \leq 1$ kOe. For these H_a values, $M^{Cu}_{ZFC}(H_a,T)$ is negligible[2] (see Fig. 1b) and $M_{FC-ZFC}(H_a,T)$, according to Eqs. 2 and 3, give an upper limit for the WF component of the Cu moments, $M^{Cu}_{FC}(T)$:

$$M_{FC-ZFC}(H_a,T) = M^{Cu}_{FC}(T) - M^{Cu}_{ZFC}(H_a,T) \approx M^{Cu}_{FC}(T), \tag{4}$$

In all cases it was possible to fit the experimental data to:

$$M_{FC-ZFC}(H_a,T) \approx M_{FC}^{Cu}(T) = M_{FC}^{OCu}[1 - T/T_{WF}]^{\beta}, \tag{5}$$

where M_{FC}^{OCu} is the saturation of the WF component at $T = 0$, T_{WF} is the T where 3D long range FM correlations first appears, and β the exponent which was found to be $\approx 0.54(2)$. This suggest that mean-field theory accounts for the 3D long range WF ordering of the Cu moments. From M_{FC}^{OCu} we estimate the *average* number of Bohr magnetons per Cu ions to be 5.5(3) x10^{-3} μ_B/Cu and 1.3(2) x10^{-3} μ_B/Cu for the crystals with $x = 0$ grown in Pt/PbO and Pt/CuO, respectively. It is known that T_N is $\approx$ 265 K and $\approx$ 284 K for Eu_2CuO_4 and Pr_2CuO_4, respectively. [10] Although lower values for T_N were reported for crystals grown in Pt/PbO [9], our results suggest that $T_{WF} < T_N$ for our compounds. However, this should be taken with care since we have not measured T_N in our crystals. Figure 1c shows that M_{FC}^{OCu} is smaller and T_{WF} lower for crystals with: *i*) lower Pt content ($x = 0$), which is consistent with the absence of WF in samples grown in Al_2O_3/CuO ($M_{FC}^{OCu} \rightarrow 0$; $T_{WF} \rightarrow 0$), and *ii*) larger V ($x = 0.25$, 0.50, and similar Pt content). Notice that the crystal with the largest V ($x = 1$), shows no WF ($M_{FC}^{OCu} \rightarrow 0$; $T_{WF} \rightarrow 0$).[2]

Figure 2a shows the XX-polarized Raman spectra at T = 12 K for the samples studied. The allowed Raman mode, B_{1g}, for the T′-structure [2] is observed at $\sim$ 324 cm^{-1} in all our crystals. The relative intensity of the *f*RM at $\sim$ 413 cm^{-1}, B_{1g}^*, to that of the allowed B_{1g} peak, $I(B_{1g}^*)/I(B_{1g})$, strongly depends on the compound and the crystal preparation conditions.[2] Larger $I(B_{1g}^*)/I(B_{1g})$ ratios are found for crystals with: *i*) higher Pt content ($x = 0$) and *ii*) smaller V ($x \rightarrow 0$, and similar Pt content (see Figs. 1 - 2). For Gd_2CuO_4 grown in Al_2O_3/CuO and Pt/PbO (179.2 A$^3 \lesssim$ V $\lesssim$ 179.9 A^3), where highly correlated ordered orthorhombic distortions in the CuO_2 planes have been found,[12] even larger $I(B_{1g}^*)/I(B_{1g})$ ratios[11] were observed. However, for Eu_2CuO_4 crystals grown in Al_2O_3/CuO (V $\approx$ 181.1 A^3) a very small $I(B_{1g}^*)/I(B_{1g})$ ratio is found (see Fig. 2 and Ref. 2).

Figure 2b presents the T dependence of the $I(B_{1g}^*)/I(B_{1g})$ ratio for the crystals grown in Pt/PbO ($T_{WF} \approx$ 237 K) and Pt/CuO ($T_{WF} \approx$ 151 K) with $x = 0$, and grown in Pt/PbO ($T_{WF} \approx$ 107 K) with $x = 0.5$. The data show that the *number* of distortions, responsible for the B_{1g}^* peak, decreases as T increases and only a small number of distortions remain above T_{WF}. Notice that Figure 1 and Figure 2 show that the 3D FM ordering appear at $I(B_{1g}^*)/I(B_{1g})$ $\approx$ 1. This ratio may correspond to that *critical number*.

4 CONCLUSIONS

The T′-structure compounds with a unit cell volume smaller than a *critical value* $V_c \approx$ 181.1 A^3 (close to the V value of Eu_2CuO_4 grown in Al_2O_3/CuO) present local distortions within the CuO_2 planes, allowing for WF and *f*RM.[1, 6, 11] However, if Cu ions are substituted by Pt impurities from the crucible, favoring the formation of local distortions in the CuO_2 planes, a slightly higher value for V_c is then found.

The results of this work and that of our previous report,[2] as well as those from others in $Gd_{2-x}Sm_xCuO_4$, Gd_2CuO_4 and Eu_2CuO_4 compounds,[6] lead us to conclude that there is a delicate instability of the T′-structure at the middle of the RE series. Our data on Eu_2CuO_4 crystals grown by different techniques show that this instability appears at a *critical* lattice unit cell volume close to that of Eu_2CuO_4 grown in Al_2O_3/CuO ($V_c \approx$ 181.1 A^3). Based on the x-ray[1] and neutron diffraction[12] results, we associate this instability with orthorhombic distortions involving a displacement of the O(1) oxygen atoms within the CuO_2 planes. Therefore, we claim that these distortions are responsible for the observed WF and *f*RM.

In summary, our Raman and magnetization measurements have shown that the distortions in the T′-structure are caused by: *i*) the reduction of the lattice unit cell volume beyond a *critical* value ,V_c, and *ii*) the substitution of Cu ions by Pt impurities incorporated from the Pt crucible during the crystal growth.

5 ACKNOWLEDGMENTS

This work was supported by FAPESP Grant No 97/03065-1 and No 95/4721-4, São Paulo-SP-Brazil and NSF-DMR No 9117212, No 9016241, and No 9501529. C. Rettori acknowledged financial support from FAPESP-Brazil.

References

[1] T. Schultz, R. Smith, A. Fondado, C. Maley, T. Beacom, P. Tinklenberg, J. Gross, C. Saylor, S. Oseroff, Z. Fisk, S. W. Cheong, and T. E. Jones, J. Apply. Phys. **75**, 6723 (1994);

[2] A.D. Alvarenga, D. Rao, J.A. Sanjurjo, E. Granado, I. Torriani, C. Rettori, S. Oseroff, J. Sarrao, and Z. Fisk, Phys. Rev. B**53**, 837 (1996).

[3] Y. Tokura, H. Takagi, and S. Uchida, Nature (London) **337**, 345 (1989).

[4] H. Okada, M. Takano, and Y. Takeda, Physica C**166**, 111 (1990).

[5] P. Bordet, J. J. Capponi, C. Chaillout, D. Chateigner, J. Chenavas, Th. Fournier, J. L. Hodeau, M. Marezio, M. Perroux, G. Thomas, and A. Varela, Physica C**185-189**, 539 (1991).

[6] S.B. Oseroff, D. Rao, F. Wright, D.C. Vier, S. Schultz, J.D. Thompson, Z. Fisk, S. W. Cheong, M.F. Hundley, and M. Tovar, Phys. Rev. B**41**, 1934 (1990).

[7] J.T. Markert, E.A. Early, T. Bjϕrnholm, S. Ghamaty, B.W. Lee, J.J. Neumeier, R.D. Price, C.L. Seaman, and M.B. Maple, Physica C**158**, 178 (1989).

[8] R.D. Zysler, M. Tovar, C. Rettori, D. Rao, H. Shore, S.B. Oseroff, D.C. Vier, S. Schultz, Z. Fisk, and S.-W. Cheong, Phys. Rev. B**44**, 9467 (1991).

[9] G.B. Martins, D. Rao, J.A. Valdivia, M.A. Pires, G.E. Barberis, C. Rettori, P.A. Venegas, S. Oseroff, and Z. Fisk, Phys. Rev. B**51**, 11909 (1995)

[10] T. Chattopadhyay, J.W. Lynn, N. Rosov, T.E. Grigereit, S.N. Barilo, and D.I. Zhigunov, Phys. Rev. B**49**, 9944 (1994); I.W. Sumarlin, J.W. Lynn, T. Chattopadhyay, S.N. Barilo, D.I. Zhigunov, and J.L. Peng, Phys. Rev. B**51**, 5824 (1995).

[11] M.A. Laguna, M.L. Sanjuan, A. Butera, M. Tovar, Z. Fisk, and P. Canfield, Phys. Rev. B**48**, 7565 (1993).

[12] M. Udagawa, Y. Nagaoka, N. Ogita, M. Masada, J. Akimitsu, and K. Ohbayashi, Phys. Rev. B**49**, 585 (1994).

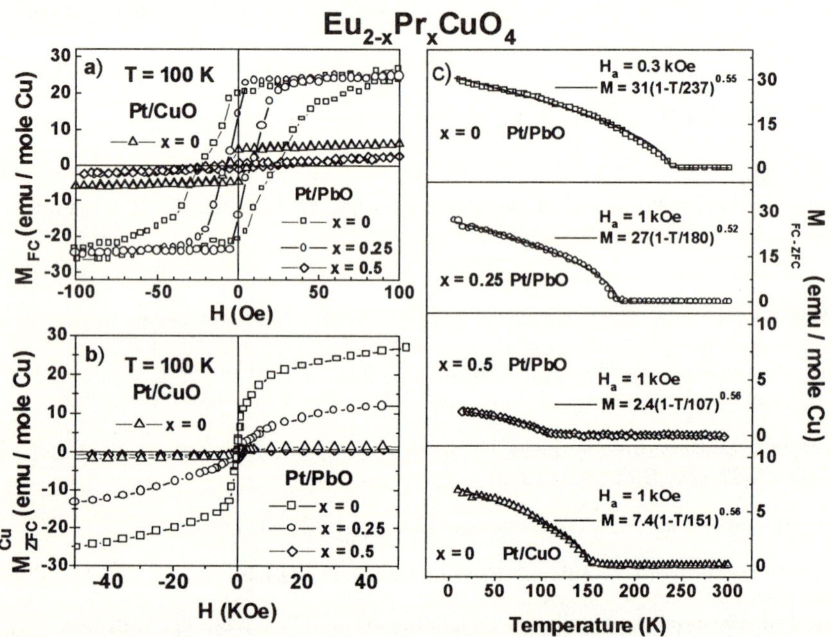

Fig. 1. a)Hysteresis loops of the *ab*-plane FC (50 kOe) total dc-magnetization, M_{FC} (H_a, T), at T = 100 K between H_a = ± 100 Oe, for the samples showing WF. b) Cu ions *ab*-plane ZFC dc-magnetization at 100 K between H_a = ± 50 kOe, , for the samples showing WF. c) T dependence of Cu ions *ab*-plane ZFC dc-magnetization, M_{FC-ZFC} (H_a, T), for $H_a \leq 1$ kOe, for the samples showing WF. The solid lines are the best fittting to Eq.5 (se text).

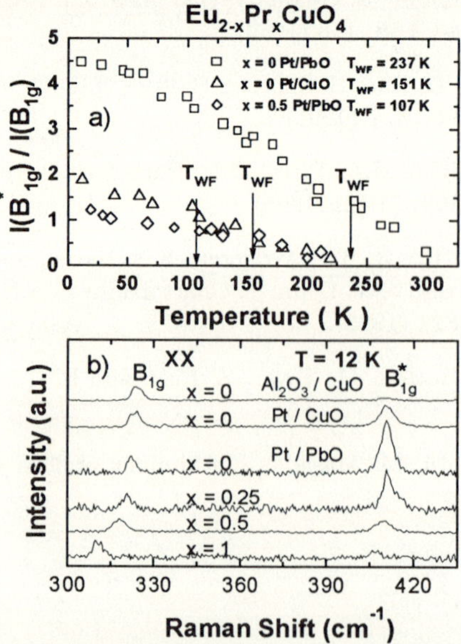

Fig. 2. A) T dependence of the I (B_{1g}^*)/I(B_{1g}) ratio for the $Eu_{2-x}Pr_xCuO_4$ crystals grown in Pt/PbO ($T_{WF} \approx 237$ K) abd Pt/CuO, ($T_{WF} \approx 151$ K) with x = 0, and Pt/PbO ($T_{WF} \approx 107$ K) with x = 0.5. b) XX-polarized Raman spectra at T = 12 K for the $Eu_{2-x}Pr_xCuO_4$ crystals studied in this work.

Materials Science Forum Vols. 302-303 (1999) pp. 169-178

Investigating Giant Magnetoresistance in Granular Alloys by Experiments and Simulations

M. Knobel, E.F. Ferrari and F.C.S. Silva

Instituto de Física "Gleb Wataghin", Universidade Estadual de Campinas-UNICAMP, C.P. 6165, Campinas 13083-970 S.P., Brazil

Keywords: Giant Magnetoresistance, Granular Alloys, Monte Carlo Simulations

Abstract The relationship between giant magnetoresistance and magnetization in granular materials is reviewed. Experimental data on melt-spun Cu-Co ribbons are explained in terms of a number of structural and magnetic parameters. Although these parameters can be theoretically manipulated, the macroscopic results can be obtained only by means of extensive computer calculations. The simulations are performed using a Monte Carlo-Metropolis algorithm, and they are extremely helpful to understand the role of the different terms in the magnetic and transport properties of these challenging systems.

I. Introduction

Granular alloys are formed by single-domain ferromagnetic grains embedded in an immiscible metallic matrix. Although the phenomenon of giant magnetoresistance (GMR) in these granular systems has been studied for many years [1,2], there are still several properties that remain unexplained. For example, it is yet not clear which is the origin of the small hysteresis present in the granular samples, if it is due to small blocked grains present in the material or if it is related to magnetic correlations among grains. Also, the real relationship between the resistivity of the sample and its magnetization is still unknown. Generally speaking, the connection between GMR and magnetization is extremely complex, depending on the distribution of grain sizes (and consequently on their average distance), the interfacial structure and the magnetic interactions among grains. Besides some technological importance, the relationship between GMR and magnetization can give important hints about several fundamental aspects of magnetism and electronic transport at the nanometer scale.

Gittleman *et al.* [3] were the first to show that the change in resistivity with the applied magnetic field is proportional to the correlation of the magnetic moments of neighboring grains averaged over all configurations. This fact was promptly detected in the new granular materials by Xiao *et al.*, who noted that the apparent dependence of GMR on the magnetic field is a direct consequence of the field dependence of the magnetization and that the GMR data plotted against the magnetization ressembles an inverted parabola. It is usually observed that magnetoresistance data displaying rather large hysteresis when represented as a function of the magnetic field seem to lie in a single curve when plotted against the magnetization. However, the relationship between GMR and the magnetization is not as simple as in a single superparamagnetic case [3-5].

We have developed a model to calculate the equilibrium magnetic and electrical properties of granular alloys as functions of the concentration, temperature and applied magnetic field [6]. The simulations are performed using a Monte Carlo-Metropolis algorithm, using as input a number of parameters obtained from different experiments. In the model we consider a log-normal distribution of particle sizes and the different contribution to the electronic scattering from the boundary and from the bulk of the magnetic grains [7]. The anisotropy constants of the blocked particles and the average coupling among neighboring grains are also considered in the simulations. In this way, it is possible to reproduce rather accurately the experimental data obtained in real systems (Cu-Co melt-spun alloys), getting, as an output, relevant information about magnetic (coupling strength, dominant type of interaction), electric (average electronic mean free path, ratio of boundary to volume scattering cross sections) and structural parameters (size distribution and density of magnetic grains).

II. Relationship between GMR and Magnetization

The starting point to describe the relationship between GMR and magnetization is a model developed by Gittleman *et al.* who, in 1972 [3], used a superparamagnetic framework to describe the effect of an external field on the resistance in granular Ni-SiO_2 films. The model considered few initial hypothesis, aimed to give a simple picture of the MR. They supposed that the grains had an unique size, and that the field-dependent term of conductivity was proportional to the degree of correlation among neighboring grains, averaged over all possible configurations. Ignoring the scattering inside each grain, the conductivity σ can be written as:

$$\sigma(H,T) = \sigma_0(T) + \sigma_1(T)\langle\vec{\mu}_1 \cdot \vec{\mu}_2\rangle / \mu^2, \qquad (1)$$

where σ_0 is the conductivity without the application of a magnetic field and σ_1 is a proportionality factor. Therefore, the magnetoresistance can be written as:

$$\frac{\Delta\rho}{\rho_0} = -\frac{\sigma_1}{\sigma_0}\frac{\langle\vec{\mu}_1 \cdot \vec{\mu}_2\rangle}{\mu^2}. \qquad (2)$$

Considering that the correlation among the grains' magnetic moments is just owing to the applied magnetic field, one easily obtain that $\langle\vec{\mu}_1 \cdot \vec{\mu}_2\rangle = \mu^2 [L(\mu H/kT)]^2$, being $L(\mu H/kT)=M/M_s$ the Langevin function and $M/M_s=m$ the reduced magnetization. Hence, one promptly verifies that the magnetoresistance depends on magnetization following a quadratic law:

$$\frac{\Delta\rho}{\rho_0} = -\frac{\sigma_1}{\sigma_0}\left(\frac{M}{M_s}\right)^2. \qquad (3)$$

Although the preceeding arguments imply that in a granular system the magnetoresistance is proportional to the square of magnetization, the experiments show that the *MR* vs. M/M_s curves usually do not follow a parabolic law, and strong deviations are clearly visible. In fact, deviations from the model are expected if some of the initial hypothesis do not correspond to reality. Let us discuss some of them:

(a) *The system is composed of individual magnetic grains with the same moment dispersed in a non-magnetic metalic matrix*. The system cannot be considered as formed by individual grains if the fraction of the magnetic element is high enough to permit percolation. Even for systems with small fraction of magnetic material there will always be a rather broad distribution of grain sizes, and therefore the grains will not present the same magnetic moment. The simple existence of a distribution of grain sizes is enough to cause a considerable deviation from the parabolic behavior [7].

(b) The grains are basically single domain particles. If the grains are larger than a critical size of the order of few nanometers, there is a energetically favourable tendency to form several magnetic domains. However, in the GMR granular systems the great majority of the grains are well bellow this critical size.

(c) *The anisotropy energy is negligible when compared with the thermal energy kT. Therefore, there is no hysteresis and all particles are in the so-called superparamagnetic state.* If there were few grains with dimensions greater than 10 nm (Fe, Co), their anisotropy energy would be comparable to the thermal energy at room temperature, and these grains would remain "blocked". Consequently, the magnetization curve would display some hysteresis (even if the coercivity and remanence are small). At lower temperatures this situation is more important, leading to a gradual disappearance of superparamagnetism. Therefore, it is necessary to take into account a distribution of anisotropy axes and energies (of crystalline or shape origin), which will certainly alter the single superparamagnetic parabolic law.

(d) The conductivity depends on the applied field just through correlation among magnetic moments of neighboring grains, averaged over all possible configurations. The average correlation among magnetic grains depends on the mean free path (l) of the conduction electrons. If l was much lower than the average distance between grains, the MR would be drastically reduced. On the other hand, if l was much greater than their distance, it would be necessary to consider correlation among moments which are relatively far from each other. In this case, it would be also important to consider the so-called spin diffusion length, which gives an idea of the time interval in which the spin carries the memory of the its last scattering [8].

(e) The electronic scattering does not depend on the grain geometry. The GMR in granular systems is rooted in spin-dependent scattering of the type electron-impurity which occurs inside the grains and/or at the interfaces between the magnetic grains and the non-magnetic metallic matrix. In this case, the geometry of the interface, its structure and its extension are fundamental parameters to consider in an appropriate model [9].

(f) The grains do not interact among themselves. The interaction among magnetic grains, of dipolar or indirect RKKY origin, gives rise to short-range correlations between neighboring magnetic grains. The effect of these correlations can be noticeable when the electronic mean free path l is of the order of the distance between different grains

Therefore, as pointed out above, the reason for the deviation from a quadratic law, $GMR \propto (M/M_S)^2$, can have several origins. For example, one knows that in real systems there is a broad distribution of magnetic particles sizes [7,9]. Also, an important hypothesis to be questioned is the absence of correlation among particles. There is strong evidence that the interactions among grains can play a relevant role in the magnetic and transport properties of granular materials [4, 5, 10-12]. One important drawback to be considered is that the structural characterization of these granular systems is rather difficult, hindering an appropriate analysis of experimental data. From the theoretical point of view, the complexity of the system does not admit an analytic solution, and intensive computer simulations are necessary to clarify the mechanisms responsible for the magnetic and magnetotransport responses of these challenging materials.

The effect of several parameters on the magnetization and magnetoresistance has been partially investigated by several authors. El-Hilo *et al.* [13] used the Monte Carlo method for determining the magnetoresistance dependence on the mean intergranular distance (or particle concentration) with the simple expression obtained by Gittleman *et al.* [3]. As mentioned before, the influence of the log-normal distribution of magnetic moments and the ratio of the boundary to the volume scattering cross sections on the magnetization and magnetoresistance was examined recently by ourselves [7]. Hickey *et al.* [14] and Wiser [15] have developed fenomenological models which explains the almost linear variation of the magnetoresistance with the magnetization as a consequence of existing correlations between blocked and superparamagnetic particles. The blocking effect on the magnetization curves was simulated by a Monte Carlo method in a work by Dimitrov and Wysin. [16]. Allia *et al.* [5, 17, 18] have proposed a fenomenological model that takes into account explicitly the correlation arising from the dipolar and exchange interactions. Recently, Vargas *et al.* [12, 19] have studied, also by the Monte Carlo method, the role played by dipolar interactions and the ratio of the electron mean-free-path to the intergranular distance on the transport properties of granular systems.

In order to investigate the effect of all these factors on GMR in granular alloys, we have used a Monte Carlo algorithm with Metropolis dynamics in a canonical ensemble. The model is provided with a number of parameters that must be determined from empirical data, namely, the mean intergranular distance, the mean magnetic moment and the standard deviation characterizing a log-normal distribution, the ratio of the boundary to the volume scattering cross sections, the anisotropy constants, the average exchange coupling between neighboring grains, and finally the electron mean-free-path in the metallic matrix. After these parameters are known from experimental data, the equilibrium magnetic properties of the granular alloy can be calculated as a function of the concentration, temperature, and applied magnetic field.

III. Experimental Results

Continuous ribbons of $Cu_{88}Co_{12}$ were prepared by melt-spinning on a Cu-Zr wheel (He atmosphere) at the Istituto Elettrotecnico Nazionale Galileo Ferraris, Torino, Italy. The as-quenched samples were submitted to dc Joule heating in vacuum to induce the precipitation of Co grains [5]. Magnetization and magnetoresistance curves were obtained at room temperature. The magnetization was measured using a home-made alternating gradient magnetometer with the applied field varying between ±15 kOe. The magnetoresistance was measured using a conventional four-contact technique in the parallel configuration, i.e., with the magnetic field applied in the plane of the ribbon, parallel to the bias current, in the range ±10 kOe.

Table I. Structural and magnetic parameters obtained from the magnetization curves (see details about the procedure in [7]). mean magnetic moment $\langle\mu\rangle$, standard deviation σ, saturation magnetization M_S, grain density N, mean intergranular distance $\langle D\rangle$ and mean radius $\langle r\rangle$.

Sample	Joule heating current (A)	$\langle\mu\rangle$ (μ_B)	σ (μ_B)	M_S (emu/g)	N (10^{18}grains/cm^3)	$\langle D\rangle$ (nm)	$\langle r\rangle$ (nm)
a	4.0	2453	2661	14.0	5.5	5.7	1.6
b	5.0	3662	2750	17.4	4.6	6.0	1.8
c	6.0	6740	6009	17.3	2.5	7.4	2.2

Figure 2 shows the hysteresis loops measured on three samples (*a*, *b*, and *c*) after annealing under currents of 4.0, 5.0, and 6.0 amperes, respectively. Both the high-field magnetization and the remanence are observed to increase with the annealing current, indicating an increased fraction of precipitated Co and a larger mean size of Co grains. Nevertheless, the magnetization curves maintain their superparamagnetic characteristics (slow tendency towards saturation and small inner loop area). Solid lines are the fitting of the magnetization function of a system of superparamagnetic grains with a log-normal distribution of moments (see details in [7]). From the extrapolated values of the specific saturation magnetization, we calculated the grain densities and the mean intergranular distances. The mean radii are determined under the assumption that complete precipitation of Co has occurred in samples *b* and *c* (due to the equal values of their saturation magnetization). Table 1 summarizes the structural data obtained through the fittings. The results are in good agreement with experimental data found in the literature and give further support to the current interpretation of the annealing effects on the structural properties of Cu-Co granular alloys [11, 20-25]. The magnetoresistance curves are shown in Fig. 2. The highest values for the GMR ratio at high fields are observed for the sample *b* whose granular structure is intermediary between those of samples *a* and *c*. It is worth noting that we failed in our attempt to fit the experimental curves using the model of uncorrelated moments exposed in Refs. [7, 26]. We obtained the best fitting on curve *c* with the parameter $\alpha = 28\mu_B^{1/3}$ (solid line), although it has displayed a fast tendency to saturation. The much worse fittings on curves *a* and *b* (not shown) had indicated to us that correlations can be more intense in these samples.

IV. Simulations

One thousand and one hard spheres were randomly positioned within a box. Each sphere represent a magnetic particle, whose magnetic moment was generated following a log-normal distribution function. Also, a random distribution of magnetic easy axes was considered, to account for the uniaxial anisotropy of the Co clusters. We used Marsaglia and Zaman's algorithm [27] to produce uniformly distributed pseudo-random 32 bit values.

According to this arrangement, the total energy of the *i*th grain is given by:

$$E_i = \varepsilon_1 \mu_i \sin^2 \alpha_i + \varepsilon_2 \mu_i \sin^4 \alpha_i - \vec{\mu}_i \cdot \vec{H}_i, \qquad (4)$$

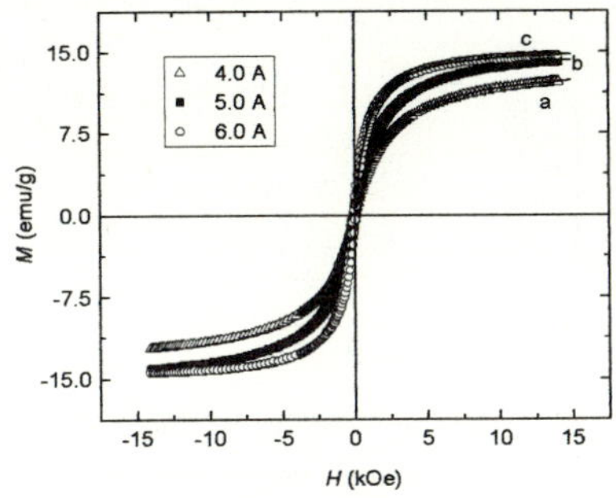

Figure 1. Magnetization curves measured in samples *a-c*. The solid lines are the fits considering just a distribution of magnetic moments. The parameters obtained from these fits are displayed in Table 1.

Figure 2. Magnetoresistance curves measured in the same samples of Fig. 1. The full line is the fit of the data for sample *c* considering the model developed in ref. [7].

where $\varepsilon_{1,2}=K_{1,2}/I_s$ are the anisotropy constants divided by the saturation magnetization of the bulk material and $\vec{H}_i$ is the total field acting on grain *i*. $\vec{H}_i$ is the vector sum of the external field $\vec{H}\,(=H\,\hat{z})$ applied along the *z* axis and the dipolar and indirect exchange fields arising from the interaction with other magnetic grains:

$$\vec{H}_i = H\,\hat{z} + \sum_{j \neq i} \left[\frac{3(\vec{\mu}_j \cdot \vec{r}_{ij})\vec{r}_{ij}}{r_{ij}^5} - \frac{\vec{\mu}_j}{r_{ij}^3} \right] + \sum_{r_{ij} < \Lambda} J_{ij}(r_{ij})\,\vec{\mu}_j \qquad (5)$$

The coefficient $J_{ij}(r_{ij})\vec{\mu}_j$ is the indirect exchange coupling between the grains i and j whose distance r_{ij} is smaller than a distance Λ comparable to the electron mean-free-path.

The indirect exchange is an effective interaction arising from the Ruderman-Kittel-Kasuya-Yosida (RKKY) interaction between all the pairs of atomic magnetic moments, localized one in each grain. The sum of RKKY couplings is carried out over all the $\mu_{i,j}/\mu_{atom}$ atoms in the grains [19]. However, since it is unreasonably difficult to calculate all the indirect exchange couplings between 500500 different pairs of grains, we determined their distribution function through curve fitting on a set of thousands of couplings calculated as if the grains were spherical clusters of randomly placed atoms. In this way, we obtained a single-parameter distribution function:

$$f(J) = \frac{3\sqrt{3}}{8\pi C}\frac{1}{1+\left|J/C\right|^{3/2}}, \qquad (6)$$

such that $f(J)\,dJ$ is the fraction of coefficients between J and $J + dJ$ and C is a parameter determined through curve fitting. Once the distribution is known, we generate indirect exchange couplings between all the pairs of grains according to this distribution function. Figure 3 shows a set of calculated couplings and the distribution curve fitting for a sample of 170 grains.

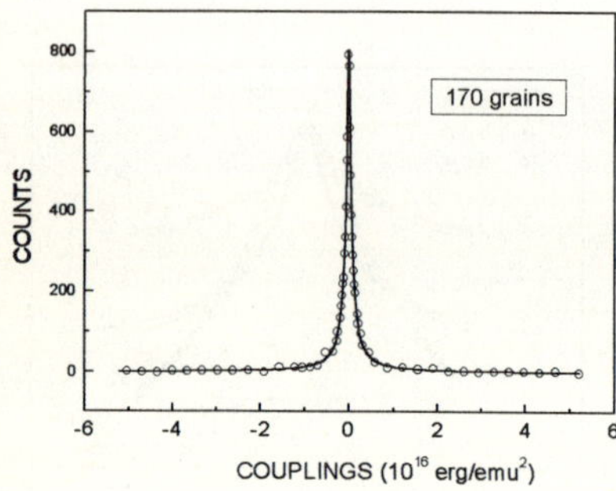

Figure 3. Distribution function of RKKY couplings among 170 spherical Co grains formed of randomly placed atoms. The solid line is a fit using equation (6).

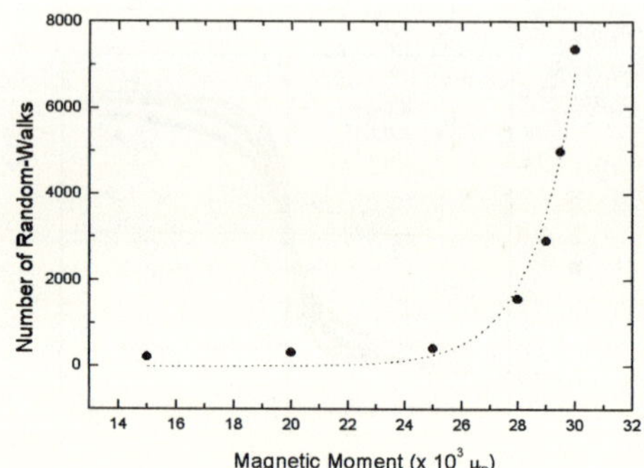

Figure 4. Average number of random-walks necessary to invert magnetic moments of different magnitudes. The points were obtained after 1.5×10^5 random-walks per moment, at 290 K, considering uniaxial anisotropy.

The magnetization and magnetoresistance calculations were based on the Metropolis *et al.* algorithm where the thermodynamic averages are taken after thousands of turns per momentum. The magnetization is simply the mean:

$$M = \left\langle \frac{1}{L_x L_y L_z}\sum_i \mu_{iz} \right\rangle, \qquad (7)$$

where L_x, L_y and L_z are the box edges. On the other hand, the magnetoresistance is calculated as the following average:

$$MR = \left\langle A \sum_i \sum_{r_{ij}<\Lambda} \left(1+\alpha\,\mu_i^{-1/3}\right)\left(1+\alpha\,\mu_j^{-1/3}\right)\vec{\mu}_i \cdot \vec{\mu}_j \right\rangle, \tag{8}$$

where the constant A is just a proportionality factor, the parameter α reflects the ratio of the scattering cross section at the boundaries of the grains to the scattering cross section within the grains (see [7]) and the parameter Λ is the correlation range between neighboring magnetic moments, which is proportional to the electron mean-free-path in the metallic matrix. Equation (8) was derived from a model proposed by Zhang and Levy [26] in which an expression for the magnetoresistance in granular alloys is obtained using the formalism for layered structures with currents perpendicular to the plane of the layers and taking into consideration the spin-dependent scattering within the magnetic grains and at their boundaries. The parameter Λ was introduced following the conclusion of Vedyayev *et al.* [28] that the conductivity of granular alloys is not self-averaging in the local limit when the grain sizes and/or distances are comparable to the mean-free-path. In this case, the spin-dependent size-effects determine the magnetoresistance amplitude in analogy with the current in plane geometry in layered structures. This conclusion agrees with other phenomenological approaches [29, 30]. Since the indirect exchange interaction and the magnetic moment correlation are both mediated by the conduction electrons, the same parameter Λ appears in Eqs. 05 and 08. It is worth noticing that, as a consequence of the condition $r_{ij}<\Lambda$, the grains with radius larger than Λ do not contribute to the magnetoresistance, although they contribute to the magnetization. Therefore, the direct relationship between these magnetic quantities breaks down in the local limit.

Simulations are in excellent agreement with the experimental magnetization curves. Figure 5 shows a large scale view of the hysteresis loops calculated under various assumptions in order to make a comparison with the hysteresis loop measured on sample *a*. The complete simulation includes the moment distribution, the anisotropies, and the interactions. The dipolar and indirect exchange interactions have caused a small reduction in the magnetization, which indicates that the interactions have a net antiferromagnetic effect on neighboring grains. As can be seen in Fig. 5, there are large discrepancies when solely the moment distribution is considered (dashes) or the moments are supposed to be all of the same value (dots). Similar results have been obtained for samples b and c. More details can be found in Ref. [6].

The average value of the components of the dipolar field which act in each one of the magnetic moments is null at equilibrium owing to the thermal fluctuations. This fact doesn't mean, however, that the dipolar field is not taken into account in the expression for the free energy in the exponent of the canonical distribution. On the contrary, the total energy of the system appears in the exponent and the averages are performed to all possible configurations. The Monte Carlo-Metropolis method reproduces the equilibrium of the canonical ensemble, and therefore it takes into account the thermal fluctuations. To do this it is just necessary that the number of Monte Carlo steps be large. Each Monte Carlo step is the process corresponding to the consecutive introduction of random modifications in all the degrees of freedom of the system [31]. If the attempt frequency is known, the number of steps can be properly converted into time units [31]. In our simulations we have rotated each moment randomly 1500 times, enough to give a null average of the dipolar field.

In thermodynamic equilibrium there is no hysteresis. If the MC-metropolis method simulates the equilibrium, where does the hysteresis come from? If the new configuration of the system is generated independently of the previous orientation, the system will rapidly escape from any metastable position, and there will be no hysteresis. Therefore, it is necessary to generate new positions in such a way as to obtain hysteresis. Obviously, there is not just one single procedure which could be determined empirically. It is unknown, for real cases, how a macroscopic system goes to equilibrium. Therefore, it is just a question of choosing a constraint in the Metropolis algorithm. In our case we have limited the new orientation of the magnetic moments as the end-point of a constrained random-walk rotation, which consists of a series of small rotations around a fixed number of randomly chosen axes, restricted by the condition that the difference between the energy at the new positions and the energy at the initial position must be smaller than the thermal energy of 25 *kT*, for a normal measurement which lasts of the order of 100 s. This method was chosen just for velocity in the calculations, which is a crucial factor in our simulations. In fact, the Metropolis algorithm is extensively used because it is rapid, once it gives probability one to any event that reduces the energy of the system. In other words, the Metropolis algorithm doesn't care if there is an energy barrier to go from one state to other, it operates as if there would be no barriers between two energetic states. The random-walk method introduced seems to be adequate to reproduce the blocking conditions. Each random-walk consists in a certain number *r* of small rotations in turn of *r* axes randomly chosen, which generate in sequence *r* new positions, each one holding the condition that the difference in the total energy of the system between the new configuration and the old one must not be greater than the characteristic thermal energy *kT*. The random-walk was chosen because it explores all the possible ways, swapping the limits of the blocked region. The simulations have shown that the random-walk applied in small grains makes them to rotate almost freely (few number of MC steps), when there is no applied field. In the case of large grains, the simulations requires a large number of steps to perform a moment reversal, of course, if we introduce the uniaxial anisotropy. Therefore, the superparamagnetic entities change rapidly the magnetic moment orientation, whereas the large particles are blocked at the considered temperature. For example, at 290 K, a moment of 30000 μ_B inverts with 7360 random walks, but a moment of 33000 μ_B does not flip even with 88600 random-walks. Even more, the number of random walks seems to increase exponentially with the absolute value of the magnetic moment, which confirms that the number of random-walks give us an estimate of the relaxation time (see Fig. 4). We believe that this is an original contribution to the MC-Metropolis model, in order to describe off-equilibrium systems.

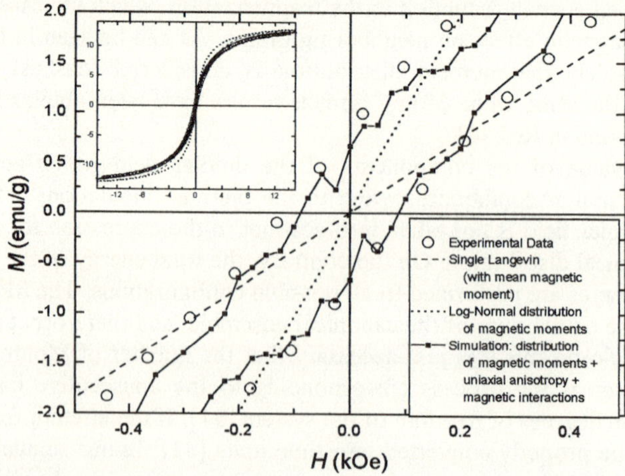

Figure 5. Comparison between the experimental data (open circles), and the different simulations, considering a single Langevin function (dashed line), a distribution of Langevin functions (dotted line), and the complete calculation (black points), which takes into account not only the anisotropy, but also the interactions among magnetic particles.

Despite our initial expectations and our efforts to obtain good fittings, the discrepancies between the simulated magnetoresistance curves and the experimental data for samples *a* and *b* still persist. However, in the present simulations the GMR behavior of both samples is different. For sample *a*, the calculated GMR results appear to be much larger than the measured one, while for sample *b*, it is much smaller. As before, the agreement is much better in the case of sample *c*. The differences between the GMR curves calculated using the present model and the experimental data cannot be explained either by the superparamagnetic behavior of small grains or by the correlations among neighboring particles, because all these factors were included in our calculations. Another interpretation that is gaining acceptance in the literature attributes the maximum of the GMR amplitude as a function of the average radius of the grains to the breakdown of the self-averaging hypothesis considered in the Zhang and Levy model [26] (also known as spin dependent size effect [28]) when the grain sizes and/or intergranular distances are comparable to the electron mean-free-path. In this case, the current lines deviate from the regions with higher resistivity and the magnetotransport in granular alloys is intermediate between those for current perpendicular to plane (CPP) and current in plane (CIP) geometries in multilayers [28, 30]. In our case, although the experimental GMR curves were not successfully reproduced, the simulations give strong evidence that the crossover of the magnetotransport from CPP- to CIP-type must be taken into account in future developments of the model.

Acknowledgments

The authors wish to express their gratitude to the group of Prof. J.E. Schmidt (Porto Alegre), where some magnetic measurements were performed. We acknowledge several discussions with P. Allia (Torino), J. D'Albuquerque e Castro (Rio de Janeiro) and D. Altbir (Santiago de Chile). We thank A. Stantero, P. Tiberto and F. Vinai from the IENGF, Torino, Italy, for making the as-quenched samples available to us. This research was performed using the equipment of CENAPAD-SP - Centro Nacional de Alto Desempenho em São Paulo/UNICAMP/FINEP-MCT and was supported by CAPES, CNPq and FAPESP.

References

[1] A. E. Berkowitz, J. R. Mitchell, M. J. Carey, A. P. Young, S. Zhang, F. E. Spada, F. T. Parker, A. Hutten, and G. Thomas, Phys. Rev. Lett. 68 (1992) p.3745.

[2] J. Q. Xiao, J. S. Jiang, and C. L. Chien, Phys. Rev. Lett. 68 (1992) p. 3749.

[3] J. I. Gittleman, Y. Goldstein, and S. Bozowski, Phys. Rev. B 5 (1972) p. 3609.

[4] J. F. Gregg, S.M. Thompson, S.J. Dawson, K. Ounadjela, C.R. Staddon, J. Hamman, C. Fermon, G. Saux and K. O'Grady., *Phys. Rev. B* **49** (1994) p.1064.

[5] P. Allia, M. Knobel, P. Tiberto, and F. Vinai, Phys. Rev. B 52 (1995) p.15398.

[6] E.F. Ferrari, F.C.S. Silva, M. Knobel, A.D.C. Viegas, A.R. Morrone and J.E. Schmidt. (to be published).

[7] E.F. Ferrari, F.C.S. Silva and M. Knobel, Phys. Rev. B 56 (1997) p. 6086.

[8] F. Badia, X. Batlle, A. Labarta, M.L. Watson, A.B. Johnston and J.N. Chapman, J. Appl. Phys. 82 (1997) p.677.

[9] W. Wang, F. Zhu, J. Weng, J. Xiao and W. Lai, Appl. Phys. Lett. 72 (1998) p.1118.

[10] D.A. Eastham, Y. Qiang, T.H. Maddock, J. Kraft, J.-P. Schille, G.S. Thompson., J.Phys. Cond. Matter 9 (1997) L497.

[11] A.D.C. Viegas, J. Geshev, L.S. Dorneles, J.E. Schmidt and M. Knobel, J. Appl. Phys. 82 (1997) p.3047.

[12] P. Vargas, D. Altbir, J. D'Albuqerque e Castro and U. Raff, J Phys.-Condens. Matter 9 (1997) p.9931.
[13] M. El-Hilo, K. O'Grady, and R. W. Chantrell, J. Appl. Phys. 76 (1994) p. 6811.
[14] B.J. Hickey, M.A. Howson, S.O. Musa, and N.Wiser., Phys. Rev. B 51 (1995) p.667.
[15] N. Wiser, J. Magn. Magn. Mater. 159 (1996) p.119.
[16] D. A. Dimitrov and G. M. Wysin, Phys. Rev. B 54 (1996) p. 9237.
[17] P. Allia, P. Tiberto and F. Vinai, Philos. Mag. B 76 (1997) p.447.
[18] P. Allia, P. Tiberto and F. Vinai, J. Appl. Phys. 81 (1997) p.4599.
[19] P. Vargas and D. Altbir, J. Magn. Magn. Mater 167 (1997) p.161.
[20] R. von Helmolt, J. Wecker, and K. Samwer, Phys. Stat. Sol. 182 (1994) K25.
[21] R. H. Yu, X. X. Zhang, J. Tejada, J. Zhu, M. Knobel, P. Tiberto, P. Allia, and F. Vinai., J. Appl. Phys. 78 (1995) p.5062.
[22] A. E. Berkowitz, J. R. Mitchell, M. J. Carey, A. P. Young, D. Rao, A. Starr, S. Zhang, F. E. Spada, F. T. Parker, A. Hutten, and G. Thomas, J. Appl. Phys. 73 (1993) p.5320.
[23] R. Busch, F. Gärtner, C. Borchers, P. Haasen, and R. Bormann, Acta Mater. 44 (1996) p.2567.
[24] H. J. Blythe, G. A. Jones, and V. M. Fedosyuk., J. Mater. Science 31 (1996) p.6431.
[25] M. G. M. Miranda, G. J. Bracho Rodriguez, A. B. Antunes, M. N. Baibich, E. F. Ferrari, F. C. S. da Silva, and M. Knobel, J. Magn. Magn. Mater. (in press).
[26] S. Zhang and P. M. Levy, J. Appl. Phys. 73 (1993) p.5315.
[27] G. Marsaglia and A. Zaman, Annals of Appl. Probability, Vol. 1, No.3 (1991).
[28] A. Vedyayev, B. Mevel, N. Ryzhanova, M. Tshiev, B. Dieny, A. Chamberod, and F. Brouers, J. Magn. Magn. Mater. 164 (1996) p.91.
[29] M. Rubinstein, Phys. Rev. B 50 (1994) p.3830.
[30] R. Y. Gu, L. Sheng, D. Y. Xing, Z. D. Wang, and J. M. Dong, Phys. Rev. B 53 (1996) p.11685.
[31] J.M. González, R. Ramirez, R. Smirnov Rueda, J. González, Phys. Rev. B 52 (1995) p.16034.

Materials Science Forum Vols. 302-303 (1999) pp. 179-183

Magnetic Characterization of Granular CuFeNi Alloys

C.S. Martins, H.R. Rechenberg and F.P. Missell

Instituto de Física, Universidade de São Paulo, São Paulo, S.P., Brazil

Keywords: Giant Magneto Resistance, Granular CuFeNi

ABSTRACT

Giant magnetoresistance was studied in alloys of composition $Cu_{80}Fe_{10}Ni_{10}$, $Cu_{80}Fe_{15}Ni_5$, $Cu_{80}Fe_5Ni_{15}$, and $Cu_{80}Fe_{20}$, produced by melt-spinning and annealed in the temperature range 300-600°C. Samples were characterized by Mössbauer spectroscopy at several temperatures. At both 78 K and 300 K, spectra for all CuFeNi compositions showed a singlet associated with superparamagnetic Fe nanoparticles, but at 4.2K the sextet characteristic of ferromagnetic Fe was visible. From the hyperfine field reduction we estimated the radius of the Fe particles. Magnetization data for $Cu_{80}Fe_{10}Ni_{10}$ at 300 K were compared with a model which features a log-normal distribution of non-interacting superparamagnetic particles. Thus the mean magnetic moment and the particle density were determined. The magnetoresistance ratio MR was compared to a model which considers bulk and interface scattering.

INTRODUCTION

After the discovery of giant magnetoresistance (GMR) in granular solids[1,2], it has become increasingly clear that the size distribution of the magnetic nanoparticles plays an important role in the phenomenon. For a random distribution of noninteracting moments of the same magnitude, the magnetoresistance ratio MR = $(\rho(H) - \rho(H{=}0)) / \rho(H{=}0)$ should be a quadratic function[3] of the relative magnetization M/M_s. However, many granular systems do not follow the expected quadratic behavior and various explanations have been proposed for this. Some consider the role of magnetic interactions, while others refer to the distribution of magnetic particles. In any case, a careful consideration of this problem should furnish important insights into magnetotransport in granular solids[4].

In the present paper we determined GMR in granular $Cu_{80}(Fe,Ni)_{20}$ alloys[5-7]. The magnetoresistance ratio MR and the relative magnetization M/M_s were studied in terms of a model which employs a log-normal distribution of non-interacting superparamagnetic particles. Mössbauer spectroscopy furnished additional information about the size of the magnetic nanoparticles.

II. EXPERIMENT

Alloys of composition $Cu_{80}Fe_{10}Ni_{10}$, $Cu_{80}Fe_{15}Ni_5$, $Cu_{80}Fe_5Ni_{15}$, and $Cu_{80}Fe_{20}$ were melt-spun on a copper wheel in a He atmosphere at wheel speeds=40 m/s. Compositions reported here are in atomic percent, in contrast with Refs. 5-7 where weight percent was used. The resulting ribbons, about 1 mm wide and 30 μm thick, were then annealed in an Ar atmosphere at various temperatures in the range 300-600°C to maximize the magnetoresistance.

Magnetization measurements were made with a vibrating sample magnetometer (VSM) mounted on a 90 kOe superconducting solenoid. Magnetoresistance was measured with an ac bridge with the magnetic field applied in the ribbon plane. Some measurements of the anisotropic magnetoresistance (AMR) were made in fields up to 20 kOe. For $Cu_{80}Fe_{10}Ni_{10}$, the

AMR was found to be on the order of 0.5% at 300K. A standard, constant acceleration spectrometer with a $^{57}CoRh$ source was used for the Mössbauer spectroscopy.

III. GMR AND MÖSSBAUER RESULTS

Of the four alloys studied, $Cu_{80}Fe_{10}Ni_{10}$ showed the largest values of the magnetoresistance ratio MR $=(\rho\ (H) - \rho\ (H=0))\ /\ \rho\ (H=0)$. In figure 1, we show MR as a function of the applied magnetic field at 4.2 K for $Cu_{80}Fe_{10}Ni_{10}$ ribbons annealed for two hours at the indicated temperatures. Our largest value of MR = -17% was obtained for an annealing temperature of 400°C, in agreement with Ref. 7. However, our value of MR is smaller than that reported in Ref. 7, in part, because those authors chose a different normalization for this quantity. For the higher annealing temperatures, the field dependence of MR appears to be nearly linear at high fields.

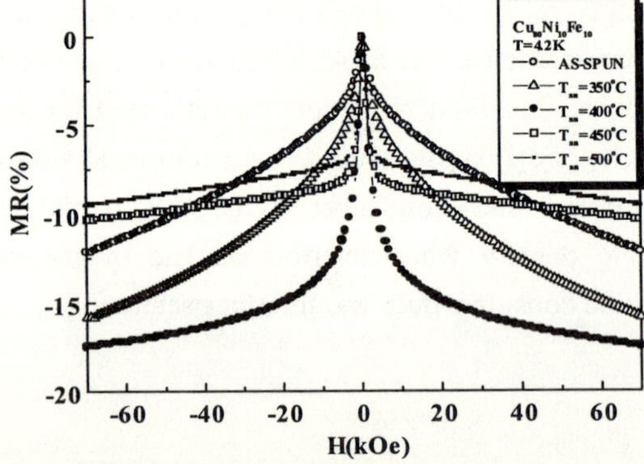

Figure 1 - Magnetoresistance ratio MR vs. magnetic field H for $Cu_{80}Fe_{10}Ni_{10}$ at 4.2 K for different annealing temperatures.

To characterize the magnetic state of the nanoparticles in these ribbons, all samples were characterized by Mössbauer spectroscopy. At both 78K and 300K, Mössbauer spectra for all CuFeNi ribbons, both as-cast and annealed, showed a singlet associated with superparamagnetic Fe. The 300K spectrum for $Cu_{80}Fe_{10}Ni_{10}$ is shown in figure 2a. At 4.2 K, however, the temperature is below the blocking temperature for the Fe nanoparticles and the characteristic Fe sextet appears as shown in figure 2b. The solid curve superimposed on the data of figure 2b corresponds to a fit made using a Gaussian distribution of hyperfine fields. The Gaussian was centered at 314 kOe and had a width of 82 kOe. For comparison, we note that α-Fe has a hyperfine field of about 340 kOe at 4.2 K. As we discuss below, magnetization measurements at 300 K on $Cu_{80}Fe_{10}Ni_{10}$ allowed us to estimate 2.1 μ_B/Fe for this material, characteristic of α-Fe. Thus, both magnetization and Mössbauer measurements indicate the presence of nanoparticles with composition close to that of α-Fe.

The observed hyperfine field reduction can be explained by the collective oscillation model for small particles[8]. From the expression $H_{part}/H_{bulk} = 1 - kT\ /\ 2KV$, we obtained an estimate for the anisotropy energy barrier KV/k = 26 K. This value yields a Mössbauer blocking temperature of order 15 K, which is consistent with our experiments. Assuming $K = 4.8 \times 10^4$ J/m^3, we obtain a mean radius 1.2 nm for the Fe nanoparticles. This is at least an order of magnitude smaller than the value reported[9] for Fe nanoparticles in splat-cooled and annealed $Cu_{75}Fe_{25}$. To understand this discrepancy, we prepared a sample of $Cu_{80}Fe_{20}$ by melt spinning. The Mössbauer spectrum of the as-cast sample at 300 K showed a singlet superimposed upon a well-defined sextet, indicating the presence of many large Fe particles with blocking temperatures above 300 K. Apparently the presence of Ni in the alloy facilitates the formation of a solid solution and results in a greater dispersion of the Fe in the Cu matrix.

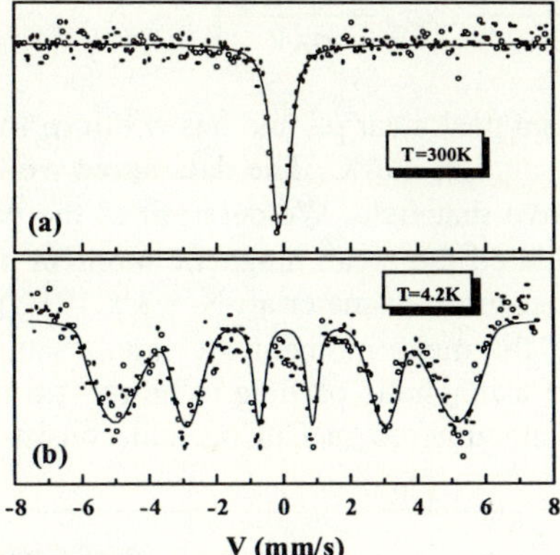

Figure 2 - Mössbauer spectrum for $Cu_{80}Fe_{10}Ni_{10}$ ribbon annealed at 400°C for two hours. Measurements at (a) 300 K and (b) 4.2 K.

IV. GMR VS. MAGNETIZATION

As mentioned in the introduction, a study of the dependence of MR on the reduced magnetization can furnish a useful characterization of the material as well as important insights into the mechanisms responsible for this effect. This paper neglects interactions between particles, but considers a log-normal distribution of magnetic moments associated with the superparamagnetic particles. When the magnetic moment distribution function is used to fit the magnetization curve, one obtains two parameters associated with the distribution curve, which can then be used to calculate the mean magnetic moment and the particle density. Using the same distribution function in a model[10] for the MR provides information on electron scattering within the magnetic grains and at magnetic-nonmagnetic interfaces.

Thus our first task is to determine the normalized curves of magnetization and MR. For this study, we considered $Cu_{80}Fe_{10}Ni_{10}$ because the MR was largest for this sample. However, we considered T = 300 K, since the ferromagnetic hysteretic contribution to M was very small at this temperature and it was straightforward to determine the anhysteretic magnetization curve. Since the magnetization had not saturated at 70 kOe, data were fitted to the expression $M = M_S (1 - a / H)$ in order to obtain the saturation magnetization M_S. We obtained M_S = 19 emu/g (2.1 μ_B/Fe), as mentioned above. The MR data were also fitted to a similar expression to determine the saturation magnetoresistance MR_S: $MR = MR_S (1 - b / H)$. If $MR \sim (M/M_S)^2$, then the constant b should equal 2a. In fact, we obtained MR_S = - 5.5% and b = 2a. The normalized magnetoresistance $MR_n \equiv MR/MR_S$ was plotted as a function of the normalized magnetization M/M_S and the result is nearly a parabola. Deviations from the parabolic form are subtle and must be characterized with care[4].

The normalized magnetization data up to 70 kOe were then compared with an expression[4] wherein the magnetization of a distribution of superparamagnetic grains is given in terms of the Langevin function L(x) = coth x - 1/x:

$$M(H,T) = \int_0^\infty \mu\, L(\frac{\mu H}{kT}) f(\mu)\, d\mu \qquad (1)$$

The magnetic moment distribution function f(μ) is a log-normal function divided by μ:

$$f(\mu) = \frac{N}{\sqrt{2\pi}\,\sigma}\frac{1}{\mu}\exp\left[-\frac{\ln^2(\mu/\mu_o)}{2\sigma^2}\right] \qquad (2)$$

This distribution function has a peak near μ_o and has width σ. In figure 3 we show experimental data for $Cu_{80}Fe_{10}Ni_{10}$ annealed at 400°C. The data agree well with the model below 20 kOe and deviate only slightly above that field. We determined the parameters of f(μ), $\mu_o = 450\ \mu_B$ and σ = 0.93, and then calculated the mean magnetic moment per grain $\langle\mu\rangle = 693\ \mu_B$ and the number of magnetic grains per gram of material: $N = 3 \times 10^{18}$/g for our $Cu_{80}Fe_{10}Ni_{10}$ at 300 K. Using the above value of the mean moment ⟨μ⟩ and a value of 2.1 μ_B/atom for α-Fe, we estimated the mean radius of a magnetic particle to be r = 1.6 nm, in good agreement with the radii of magnetic particles encountered in CuCo[4] and consistent with our Mössbauer results presented above.

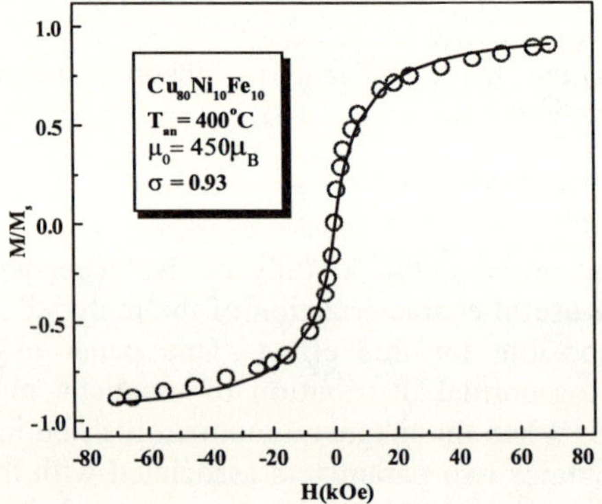

Figure 3 - Magnetization data for $Cu_{80}Fe_{10}Ni_{10}$ ribbon annealed at 400°C (circles) compared with model which assumes lognormal distribution of non-interacting superparamagnetic particles.

In Ref. 4, the distribution of magnetic moments was used in a model proposed by Zhang and Levy[10] for granular films based upon the formalism for the magnetoresistance in layered structures. Both bulk and interface scattering contribute to the magnetoresistance by means of the factor $\mu + \alpha\,\mu^{2/3}$, where the parameter α is proportional to the ratio between the interface and bulk scattering cross sections and has been given in Eq. (9) of Ref. 4. We tried fitting our MR data to this model, using the values of μ_o and σ determined from the magnetization, and treating α as an ajustable parameter. We used

$$MR(H,T) = -\frac{A}{N^2}\left[\int_0^\infty (\mu + \alpha\,\mu^{2/3})\,L\!\left(\frac{\mu H}{kT}\right) f(\mu)\,d\mu\right]^2 \qquad (3)$$

where both A and α are coefficients which do not depend upon H and N is the number of grains per unit volume. The result is shown in figure 4. There is good agreement for magnetic fields below about 15 kOe, after which the agreement is somewhat worse. Larger values of α do not bring the calculated (dashed) curve into better agreement with the data. Therefore, it seems that a further discurssion of MR in these materials must take account interactions between particles

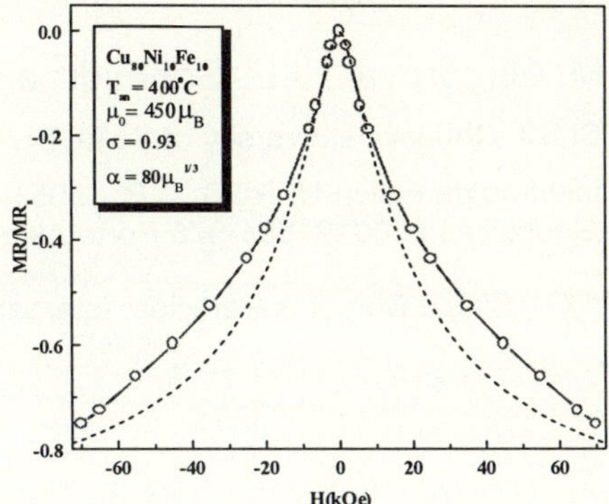

Figure 4 - Normalized magnetoresistance for $Cu_{80}Fe_{10}Ni_{10}$ ribbon annealed at 400 °C (circles) compared with curve calculated using Eq.(3) (dashed curve).

V. CONCLUSIONS

Both magnetization and Mössbauer measurements indicate the presence of nanoparticles in $Cu_{80}Fe_{10}Ni_{10}$ with composition close to that of α-Fe and effective radius slightly larger than 1 nm. This size is considerably smaller than that found for Fe particles in splat-cooled[9] $Cu_{75}Fe_{25}$, but is comparable to values encountered in $Cu_{90}Co_{10}$. Thus $Cu_{80}Fe_{10}Ni_{10}$ and $Cu_{90}Co_{10}$ seem to be physically comparable and it is interesting to compare MR results in these materials. The magnetization data for $Cu_{80}Fe_{10}Ni_{10}$ at 300 K were found to be in good agreement with curves calculated assuming a distribution of non-interacting superparamagnetic Fe particles. However, the MR data agreed with the proposed model only for low fields.

Acknowledgments - FAPESP, CNPq, CAPES, and FINEP provided financial support. The authors wish to thank S.A. Romero for the melt-spun ribbons.

References

1. A. E. Berkowitz, J. R. Mitchell, M. J. Carey, A. P. Young, S. Zhang, F. E. Spada, F. T. Parker, A. Hutten, and G. Thomas, *Phys. Rev. Lett.* **68** (1992) 3745.
2. J. Q. Xiao, J. S. Jiang, and C. L. Chien, *Phys. Rev. Lett.* **68** (1992) 3749.
3. J. I. Gittleman, Y. Goldstein, and S. Bozowski, *Phys. Rev. B* **5** (1972) 3609.
4. E. F. Ferrari, F. C. S. da Silva, and M. Knobel, *Phys. Rev. B* **56** (1997) 6086.
5. L. H. Chen, S. Jin, T. H. Tiefel, S. H. Chang, M. Eibschutz, and R. Ramesh, *Phys. Rev. B* **49** (1994) 9194-9197.
6. L. H. Chen, Y. K. Yang, and Y. H. Chang, *IEEE Trans. Magn.* **31** (1995) 4012-4014.
7. L. H. Chen, S. Jin, T. H. Tiefel, S. Y. Liao, and Y. D. Yao, *J. Appl. Phys.* **79** (1996) 5599-5601.
8. S. Mørup, H. Topsøe, and J. Lipka, *J. Physique* **37** (1976) C6-287.
9. M. S. Rogalski, M. M. Pereira de Azevedo, and J. B. Sousa, *J. Magn. Magn. Mater.* **163** (1996) L257-L263.
10. S. Zhang and P. M. Levy, *J. Appl. Phys.* **73** (1993) 5315-5319.

Materials Science Forum Vols. 302-303 (1999) pp. 184-188

Magnetization Processes of Small γ-Fe_2O_3 Particles in Nonmagnetic Matrix

V. Masheva[1], M. Grigorova[1], J.E. Schmidt[2] and M. Mikhov[2]

[1] Faculty of Physics, "St. Kl. Ohridski" University of Sofia, BU-1126 Sofia, Bulgaria

[2] Instituto de Física-UFRGS, C.P. 15051,
Av. Bento Gonçalves 9500, 91501-970 Porto Alegre, Brazil

Keywords: Fine Particles, γ-Fe_2O_3, Single Domains Particles, Interparticle Interactions

Abstract Magnetization processes of small (≈ 350Å) and nearly spherical particles of γ-Fe_2O_3 mixed with SiO_2 were studied by initial, hysteresis and remanence magnetization curves. The non-monotonous dependence of the interparticle interactions on the magnetic phase concentration observed was related to changes in the domain status of some of the particles from single-domain to multi-domain ones.

1 Introduction

Magnetic properties of fine γ-Fe_2O_3 particles are an object of permanent interest and a number of papers on these properties were published (for review see Ref. [1], and the list of references, given in Ref. [2]). Hysteresis loops and remanence curves of small γ-Fe_2O_3 particles dispersed in Al_2O_3 with different packing densities were recently studied and an unusual behavior of increasing the effective interparticle interaction with increasing the degree of dilution was observed [3]. The aim of the present work is to study more accurately and in more details the magnetization processes of a random system of γ-Fe_2O_3 particles whose sizes are nearly single-domain, and which are mixed with SiO_2 particles with the same size.

2 Experimental

The magnetic powder, used in the present study, was characterized by X-ray diffraction, TEM, and Mössbauer spectroscopy in Ref. [3]. It consists of particles with cubic spinel-type (Fd3m) structure which are slightly elongated with an average aspect ratio of $\langle c/a \rangle \approx 1.2$. The particle size distribution has a broad maximum at about 350 Å. The particles are in pure ferrimagnetic state without any detectable superparamagnetic contributions at room temperature (RT).

Powders of γ-Fe_2O_3 and SiO_2 with weight concentrations, X of γ-Fe_2O_3, varying between 1.0 and 0.006, were mixed with assistance of oleic acid as a surfactant. Magnetization curves were measured by alternating gradient field magnetometer in fields up to 6 kOe at RT. The saturation magnetizations were determined by VSM.

3 Results and discussion

The RT saturation magnetization, measured for the pure γ-Fe_2O_3 powder is $\sigma_s = 74.2$ emu/g, in coincidence with the reported data of Ref. [4]. The coercivity of $H_c = 256 \pm 3$ Oe was measured for $X = 1$. It increases up to 280 ± 3 Oe for X down to 0.05, and then decreases slightly for further decreasing X, as shown in Fig. 1. The linear increase of the coercivity with decreasing X can be attributed to the existence of uniaxial single-domain particles in the samples [5]. It has been found [6] that the coercivity can obey the relation: $H_c(p) = H_c(0)(1 - k_0 p)$, for $p < 0.33$, where: p is the volume fraction of the magnetic phase in the sample, $k_0 = [(2N_c + N_a)/(N_c \times N_a)] \times 4\pi/3$, and N_c and N_a are the demagnetizing factors along the polar (c) and equatorial (a) axes of the particles, considered as prolate spheroids. For the particles studied with $\langle c/a \rangle \approx 1.2$, the value of $k_o^{calc} = 3.03$ is calculated. Taking into account the shape, the volume and the weight concentration X of the samples, one can find that $p < 0.3$ for all the samples, so that the above formula can be used. The slope of the experimentally obtained $H_c(p)$ dependence is characterized by $k_o^{exp} = 2.2$. The difference between the calculated and measured values of k_o is a reasonable one, because, as already stated, not all of the particles can be considered as uniaxial single-domain ones, and the existence of multi-domain particles leads to a decrease of k_o. The contribution of particles with cubic anisotropy seems to be negligible, because, as recently shown [7], even an aspect ratio of $\langle c/a \rangle \geq 1.04$ is enough to introduce pure uniaxial anisotropy in γ-Fe_2O_3 particles. The deviation from this linear law is an indication that some other mechanism, apart from the single-domain behavior, start to play some more substantial role at low concentrations.

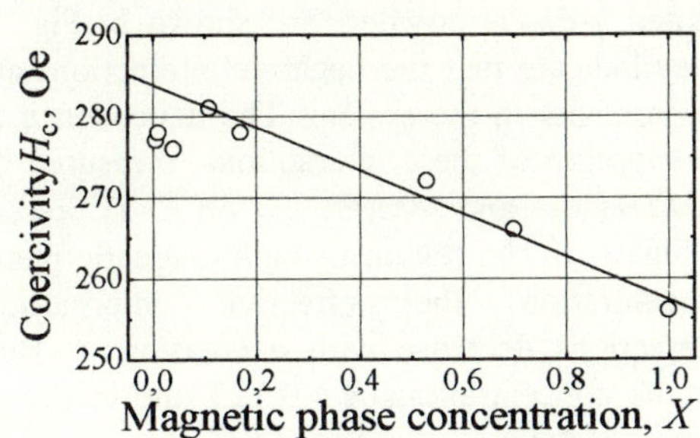

Fig. 1 The concentration dependence of H_c. The line represents the linear fit with $H_c = 284 - 27 \times X$.

The squareness, $S = M_r/M_s$ of 0.36 (M_s is the saturation moment and M_r is the remanence after saturation) was obtained for $X = 1$. It is practically independent on the magnetic phase concentration down to $X = 0.01$, but tends to decrease very slightly below this concentration. S is lower than the value of 0.5 for the ideal Stoner - Wohlfarth (S-W) system. As far as the RT Mössbauer spectrum does not indicate any superparamagnetic contribution, this low value of S could be attributed to the existence of multi-domain particles as well as to agglomeration of single-domain ones.

The other parameters characterizing the remanence state such as: remanence coercivities H_r, defined by $M_d(H_r) = 0$, and H'_r, defined by $M_r(H'_r) = 0.5\ M_r(\infty)$; coercivity factor CF, defined as $CF = (H_r - H_c)/H_c$; and interaction field factor IFF, defined as $IFF = (H'_r - H_r)/H_c$ [8, 9], have also substantially higher values than for the ideal S-W system. Both CF and IFF also exhibit some peculiarity in the range of concentration lower than $X < 0.01$.

The effective interparticle interactions for the samples with different magnetic phase concentrations were additionally studied by two methods. The first one is the widely used δM(H) plots, based on the Wohlfarth relation: $\delta M(H) = M_d(H) - [M_r(\infty) - 2 \times M_r(H)]$, where $M_r(H)$ and $M_d(H)$ are the values of isothermal remanence magnetization and direct current demagnetization remanence in a given field normalized to saturation magnetization remanences $M_r(\infty)$ [10]. The second method, proposed in Refs. [11, 12], uses $\delta M^a(H)$ plots, based on the set of coefficients of Fourier description of initial magnetization curve and entire hysteresis loop. As already shown for a number of systems,

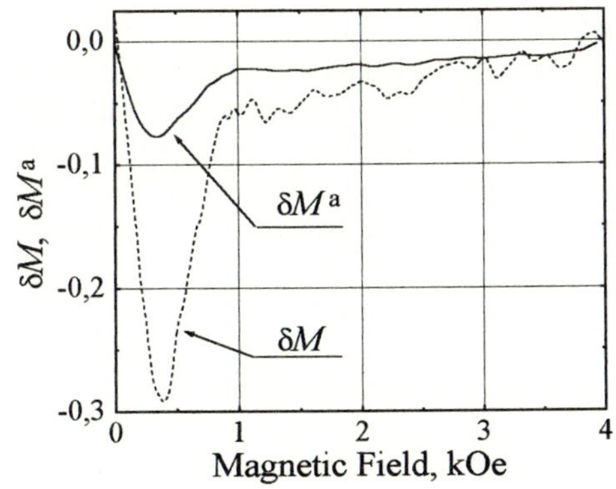

Fig. 2 $\delta M^a(H)$ and $\delta M^a(H)$ plots for non-diluted powder of γ-Fe_2O_3.

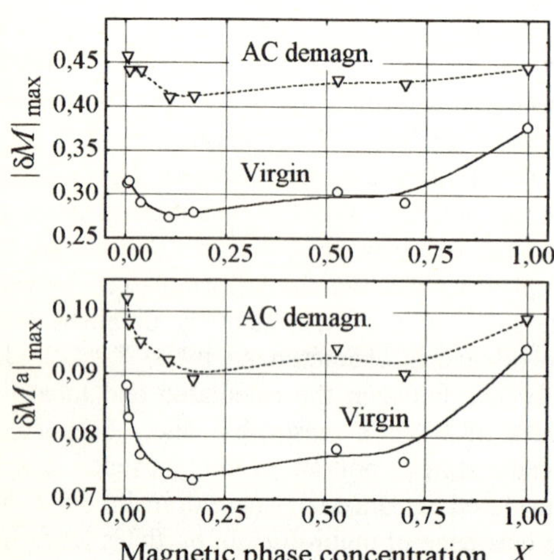

Fig. 3 The dependence of $|\delta M^a(H)|_{max}$ and $|\delta M(H)|_{max}$ on X, measured for virgin and AC demagnetized samples.

$|\delta M^a(H)|_{max}$ plots follow very well the $\delta M(H)$ ones. Both, $\delta M(H)$ and $\delta M^a(H)$ plots for non-diluted γ-Fe_2O_3 powder, are shown in Fig. 2. They indicate that the negative interactions are predominant in the system. The dependence of the strength of these interactions, measured by $|\delta M^a(H)|_{max}$ and $|\delta M(H)|_{max}$, on X is present in Fig. 3. In the region of high magnetic phase concentration the effective interparticle interactions decrease with decreasing X. But, for the concentrations $X < 0.125$, the values of both $|\delta M^a(H)|_{max}$ and $|\delta M(H)|_{max}$ start to increase as the concentration decreases. It should be noted that a similar behavior in the region of low X was also expressed by the interaction field factors *IFF*, and that in the same region H_c deviates from the linear law.

All these facts can be explained as follows. The critical single-domain size of γ-Fe_2O_3 particles has been studied and reported in many papers [1, 13, 14, 15]. It has been shown experimentally in Ref. [4] that the critical size for coherent magnetization rotation for γ-Fe_2O_3 particles is $\approx$ 400 Å. In our case, the maximum in the particle size distribution function is located $\approx$ 350 Å, but particles with sizes up to 600 Å (and maybe more) also exist. Thus, the samples studied consist of single-domain, multi-domain, and pseudo-single-domain particles. With decreasing X, the distances between the particles increase, the interparticle interactions are reduced, which leads to a decrease of both $|\delta M_a(H)|_{max}$ and $|\delta M(H)|_{max}$. This effect is predominant for the higher concentrations, $X > 0.02$. At lower concentrations the distances between particles become so large that some of them, which contribute as single-domain ones in the closer packed samples, change their domain state from a single-domain to a multi-domain one. The common interpretation of δM and δM_a plots is that its sign and value characterize the interparticle interactions. The positive δM indicate the positive interparticle interaction and vice versa. But, on a whole, a system with predominant positive interactions is easier to magnetize than to demagnetize, while the system with predominant negative interactions is easier to demagnetize than to magnetize. Let consider the simplest case of a particle consisting of two equivalent domains with antiparalel magnetic moments in the initial demagnetized state. Any other state of the same particle, characterized with some remanence magnetization, will be in a state of higher energy than the initial one. Thus, from the point of view of the minimum energy, such a particle is, in general, easier to be demagnetized than to be magnetized, and it would contribute to the negative δM. There are some particles in the ensemble, that behave as single-domain particles in the samples with high magnetic phase concentration, but which behave as multi-domain ones in the samples with low enough concentrations. These changes produce a small but well established increase of the effective interparticle interactions, measured by $|\delta M^a(H)|_{max}$ and $|\delta M(H)|_{max}$, as well as some peculiarities in the concentration behavior of *IFF* and *Hc*.

Such a change of the magnetic single-/multi-domain state of fine particles as a function of the interparticle distances has been predicted theoretically in Ref. [16] and observed experimentally in Refs.[6, 15] on the base of the coercivity and anhysteretic remanence measurements. Very recently the time-variation of the remanence in a system of fine spherical γ-Fe_2O_3 particles was attributed to the same phenomena [17].

4 Conclusions

The magnetization behavior of fine (≈ 350 Å in diameter) and nearly spherical γ-Fe_2O_3 particles, dispersed in SiO_2 with different magnetic phase concentrations was studied at RT. It was found, by analyzing the concentration dependencies of $\delta M(H)$ and $\delta M^a(H)$ plots, a tendency for the systems with low magnetic phase concentrations to be easier to demagnetize than to magnetize in comparison with the systems with high magnetic phase concentrations. This effect was explained by the change of magnetic state of some of the particles from a single-domain one at high concentrations to a multi-domain one at low concentrations.

It was also demonstrated that the Fourier analysis of initial magnetization and hysteresis curves is a very sensitive and useful tool for investigation of the magnetization processes. Its main advantage is that it is based on the easier and faster magnetization measurements, while the $\delta M(H)$ plots need more complicated and time consuming remanence measurements. Furthermore, $\delta M^a(H)$ plots carry information about the entire hysteresis loop and the initial magnetization curve while $\delta M(H)$ plots and *IFF* are obtained from remanence states only.

Acknowledgments

The work is supported by the Bulgarian Foundation for Scientific Investigation under grant Ф - 453/1994 and by FAPERGS, Brazil.

The authors are grateful to Dr. J. Geshev for the helpful discussions and to Mr. V.Kostadinov for the technical assistance.

References

[1] Bate G. (1986), "Recording Materials" in *Ferromagnetic Materials*, ed. by E.P.Wohlfarth, North-Holland Publ. Co., vol. 2, (1968) pp. 381-507.

[2] E.Schmidbauer, and R.Keller, (1996), "Magnetic properties and rotational hysteresis of Fe_3O_4 and γ-Fe_2O_3 particles ≈ 250 nm in diameter", *J. Magn. Magn. Mater.*, **152** (1969), pp. 99-108.

[3] M.Grigorova, V.Masheva, D.Nihtianova, V.Rusanov, L.Bozukov, V.Gerganov, and M.Mikhov, "Hysteresis loopds and remanence curves of small γ-Fe_2O_3 particles with different packing densities", in *Magnetic Hysteresis in Novel Magnetic Materials*, ed. by G.C.Hadjipanayis, NATO ASI Series E: Applied Sciences, Kluwer Academic Publishers, **338** (1997), pp. 357-361.

[4] A.E.Berkowitz, W.J.Schuele, and P.J.Flanders, "Influence of crystallite size on the magnetic properties of acicular γ-Fe_2O_3 particles", *J. Appl. Phys*, **39** (1968), pp. 1261-1263.

[5] B.D.Cullity, "Introduction to Magnetic Materials", Addison-Wesley Publ. Co., Inc, (1972), p. 389.

[6] A.H.Morish, and L.A.K.Watt, "Effect of the interaction between magnetic particles on the critical single-domain size", *Phys. Rev.*, **105** (1957), pp. 1476-1478.

[7] J. Geshev, A.D.C. Viegas, and J.E. Schmidt, "Negative Remanent Magnetization of Fine Particles," *Journal of Applied Physics*, in press.

[8] E.P.Wohlfarth, "Relations between different modes of acquisition of the remanent magnetization of ferromagnetic particles", *J. Appl. Phys.*, **29** (1958), pp. 595-596.

[9] A.R.Corradi, and E.P.Wohlfarth, "Influence of densification on the remanence, the coercivity and the interaction field of elongated γ-Fe_2O_3 powders", *IEEE Trans. Magn.*, **MAG-14** (1987), pp. 861-863.

[10] P.E.Kelly, K. O'Grady, P.I.Mayo, and R.W.Chantrell, "Switching mechanism in cobalt - phosphorus thin films", *IEEE Trans. Magn.*, **MAG-25** (1989), pp. 3881-3883.

[11] V.Masheva, J.Geshev, and M.Mikhov, "Fourier analysis of hysteresis loops and initial magnetization curves: application to the singular-point-detection method", *J. Magn. Magn. Mater.*, **137** (1994), pp. 350-357.

[12] J.Geshev, V.Masheva, and M.Mikhov, "Interaction effects detected by Fourier analysis of hysteresis loops and initial magnetization curves", *IEEE Trans. Magn.*, **MAG-30** (1994), pp. 863-865.

[13] A.H.Morish, and S.P.Yu, "Dependence of the coercive force on the density of some iron oxide powders", *J. Appl. Phys.*, **26** (1955), pp. 1049-1055.

[14] A.Aharoni, "Perfect and imperfect particles", *IEEE Trans. Magn.*, **MAG-22** (1986), pp. 478-483.

[15] D.J.Dunlop, "Grain-size dependence of anhysteresis in iron oxide micropowders", *IEEE Trans. Magn.*, **MAG-8** (1972), pp. 211-213.

[16] E.Kondorskii, "The nature of high coercivity in fine-dispersed ferromagnets and theory of single-domain structure" (in russian), *Izvest. Akad. Nauk S.S.S.R. (ser. Fiz.)*, **16** (1952), pp.398-411.

[17] M.P.Morales, M.El-Hilo, and K.O'Grady, "Time dependence effects and reversal mechanisms in γ-Fe_2O_3 particles", *J. Magn. Magn. Mater.*, **140-144** (1995), pp. 2211-2212.

Materials Science Forum Vols. 302-303 (1999) pp. 189-193

Anomalous Remanence of Granular CoCu

J. Geshev, A.D.C. Viegas and J.E. Schmidt

Instituto de Física, UFRGS, 91501-970 Porto Alegre, RS, Brazil

Keywords: Remanent Magnetization, Competing Anisotropies, Fine Particles

Abstract. Magnetization and remanence data for Co(6Å)/Cu(14Å)×15 multilayers, evaporated at room temperature on a Si (111) substrate, and subsequently annealed under argon atmosphere at 250°C for 1 hour, were obtained. The magnetization measurements show an easy direction parallel to the film. The experimentally obtained reduced in-plane dc demagnetization remanence curve for the annealed sample and for field applied nearly perpendicular to the easy axis, has a very unusual behavior: a clear minimum in the remanence versus the field dependence is seen with amplitude much higher than 1. In the framework of the model presented here, this effect is explained as caused by a portion of the sample, having *negative remanent magnetization.*

1 Introduction

The Co-Cu system has been widely investigated, mainly after the discovery of the giant magnetoresistance effect in magnetic multilayers and subsequently in several other kinds of systems including granular systems [1,2] and multilayers composed of immiscible elements submitted to proper annealing treatment. Thermally treated multilayers, in some cases, represent an intermediate state between multilayer and granular structures, as the as-made samples are progressively modified by the annealing.

Very recently, the correlation between giant magnetoresistance and magnetic interactions has been investigated in some Co-Ag and Co-Cu multilayer and granular systems [3-5], as the interaction effects were estimated by using remanent magnetization measurements. This technique, widely used in fine-particle systems during the last decade, is based on a comparison of the principal remanence curves, the isothermal remanent magnetization curve, $M_r(H)$ and the dc demagnetization remanence curve, $M_d(H)$. The $M_r(H)$ curve is obtained on an initially demagnetized sample by applying a positive magnetic field H which is then removed and the remanence M_r is measured. A larger field is then applied and the process is repeated until saturation is reached. The $M_d(H)$ curve is measured by first saturating the sample in a positive field and then measuring the remanence M_d after application of progressively larger negative fields. Kelly *et al.* [6] studied the interactions in granular thin films, using the so-called δM plot,

$$\delta M = M_d(H) - M_r(\infty) + 2M_r(H). \qquad (1)$$

δM=0 for all H for a non-interacting system.

For uniaxial anisotropy particles, negative δM values are often taken to indicate "demagnetizing" interactions in the sense that the interactions have the effect to stabilize the demagnetized state, for example by formation of flux closure structures. This is normally the case

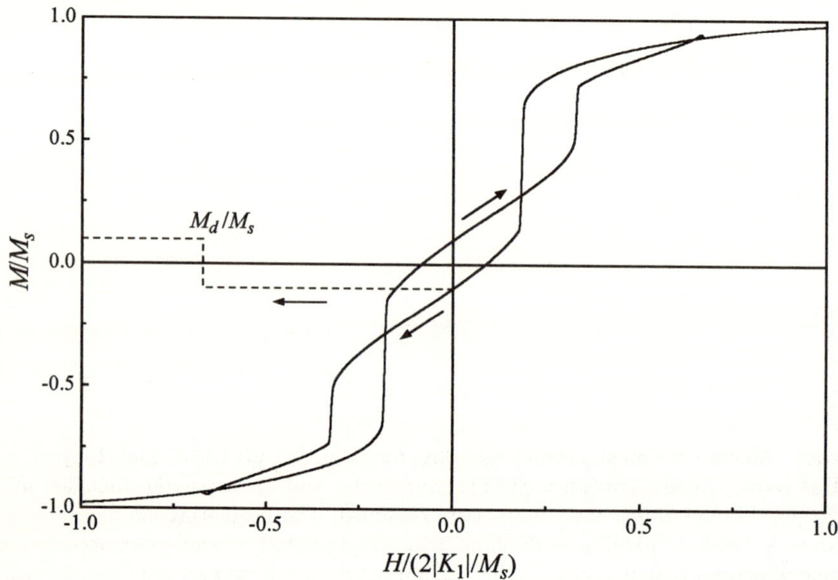

Fig. 1. Hysteresis loop and M_d/M_s curve for a particle with $K_1<0$, $K_u/|K_1|=0.5$, uniaxial anisotropy direction given by $[\bar{1}\bar{1}2]$, and **H** direction given by $\theta = 95°$ and $\psi = 147.5°$.

for particulate recording media. Positive values of δM have been attributed to interactions promoting the magnetized state.

$M_d(H)$ vs. $M_r(H)$ plots, calculated for the case of randomly orientated non-interacting spherical particles with cubic anisotropy (3 and 4 easy magnetization axes) are nonlinear in a "positive" sense with the curves concave downwards, leading to positive δM plots [7]. Ultra thin films provide conditions where uniaxial and cubic anisotropies coexist. Recent studies on such particulate systems [8,9] revealed some interesting features, e.g. the negative remanent magnetization of some particles. Fig. 1 shows a representative hysteresis loop and $M_d(H)/M_s$ curve for a particle with cubic magnetocrystalline anisotropy (with negative first-order anisotropy constant, K_1), and where co-exists an uniaxial anisotropy (with anisotropy constant K_u), whose direction is given by $[\bar{1}\bar{1}2]$; Here M_s is the particle's saturation magnetization, $K_u/|K_1| = 0.5$ and **H** direction given by $\theta = 95°$ and $\psi = 147.5°$. It is seen that the hysteresis loop is inverted, i.e. in the low-field region the two branches of the loop are crossed, resulting in negative remanence and coercivity.

In this paper, we present experimental magnetization and remanence data for thermally treated Co(6Å)/Cu(14Å)×15 multilayers. The unusual experimental remanence curves were fitted by using model calculations, assuming that uniaxial and cubic anisotropies coexist in these samples.

2 Experimental

Co(6Å)/Cu(14Å)×15 multilayers were evaporated at room temperature on a Si (111) substrate, and subsequently annealed under argon atmosphere at 250°C for 1 hour. The X-ray diffraction data for these thermally treated samples showed that the Co layers are fcc with the [111] direction out of plane and, due to the very small Co layer thickness, they are discontinuous, composed by very fine Co grains. The conventional X-ray diffraction shows a pure fcc Cu

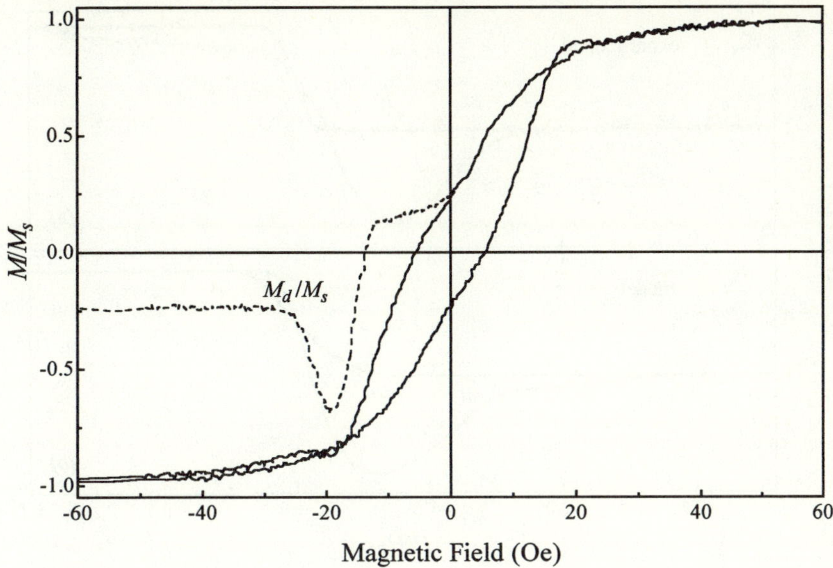

Fig. 2. Experimental magnetization and remanence data for an initially dc demagnetized sample; **H** direction is approximately perpendicular to the texture axis.

structure. Although bulk hcp Co is stable below the allotropic transformation temperature of about 420°C, there is no evidence of hcp Co phase in our samples; the fcc phase of fine Co particles is usually stable at ambient temperature (see, e.g. Sato *et al.* [10] and the references therein). Due to the small size of the Co particles and the high lattice coherency between the fcc Co phase and the Cu matrix, this Co phase could not be distinguished.

3 Results and Discussion

The magnetization measurements were performed using an alternating gradient magnetometer. The samples show an easy direction parallel to the film, which can be attributed to strain, induced during the deposition. The experimental in-plane hysteresis loop and the corresponding dc demagnetization remanence curve for an initially dc demagnetized annealed sample, and for magnetic field applied nearly perpendicular to the easy axis are plotted in Fig.2. The dc demagnetized state is obtained by first saturating the sample in a positive direction, applying the remanent coercive field in the negative direction and then returning the field to zero. It can be seen that the $M_d(H)$ curve shows a very unusual behavior: a clear minimum is seen with amplitude much higher than $M_r(\infty)$. By decreasing the angle between the in-plane easy axis and the field direction, a gradual decrease of the minimum depth is observed, and the remanence curve for field applied parallel to the easy axis does not show minimum.

In the framework of the model used here, this effect can be explained as caused by the contributions of some particles having negative remanent magnetization. Consider a system consisting of non-interacting single-domain particles whose anisotropy is made up of a cubic magnetocrystalline and an uniaxial components. Let the direction cosines of the magnetization vector $\mathbf{M}_s$ of such a particle be α_1, α_2 and α_3 and those of the uniaxial anisotropy l, m and n, referred to the cube axes. Neglecting the thermal activation effects and considering coherent

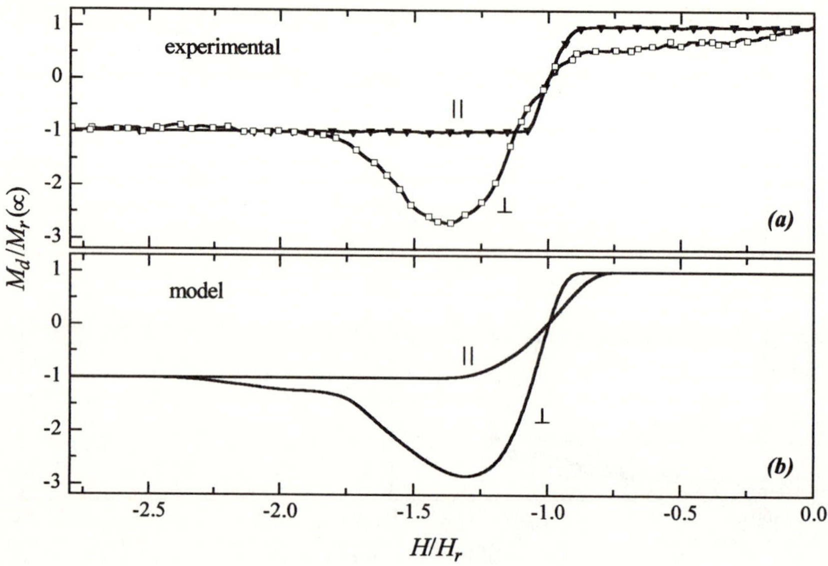

Fig. 3. (a) Experimental $M_d(H/H_r)$ curves measured parallel (||) and perpendicular ($\perp$) to the texture axis. (b) The corresponding fitting curves.

magnetization rotations only, the total reduced free energy of the particle can be written as

$$\eta(\gamma,\vartheta) = \frac{E}{2|K_1|V} = \frac{1}{2}\mathrm{sign}(K_1)(\alpha_1^2\alpha_2^2+\alpha_2^2\alpha_3^2+\alpha_3^2\alpha_1^2) - \frac{K_u}{2|K_1|}(\alpha_1 l+\alpha_2 m+\alpha_3 n)^2 - h\cos\phi. \quad (2)$$

E is the total energy, V the particle volume, h the reduced magnetic field ($= HM_s/2|K_1|$), and $\cos\phi$ is given by

$$\cos\phi = \cos\gamma\cos\theta + \sin\gamma\sin\theta\cos(\vartheta-\psi). \quad (3)$$

Here γ and ϑ are the spherical coordinates of $\mathbf{M}_s$, and θ and ψ are those of $\mathbf{H}$; ϕ is the angle between $\mathbf{M}_s$ and $\mathbf{H}$.

When $K_1 > 0$, there are six easy directions along the principal cube axes, and for $K_1 < 0$ there are eight minima of the type $\langle 111\rangle$. As $K_u/|K_1|$ increases, these minima move towards the direction $[l,m,n]$ and, in general, they reach this direction only when $K_u/|K_1| \to \infty$.

In the presence of an applied field, the energy function can be minimized only numerically. The two-variable minimization procedure used in the present work is described elsewhere [7,9].

The experimental $M_d(H)$ curves measured parallel (||) and perpendicular ($\perp$) to the texture axis, as well as the corresponding model calculated curves are shown in Fig. 3 [here H_r is the remanence coercivity, i.e. $M_d(H_r) = 0$]. In the calculations it is assumed that $[\bar{1}\bar{1}2]$ is the easy axis in the (111) plane (which is the assumed to be the sample's plane), and the ratio between the uniaxial and the cubic magneto-crystalline anisotropy constants is equal to 0.44. This value is sufficiently high so the number of the minimum directions is reduced to two [11]. A slight disorder of the particle orientations is allowed as well by taking into account the contribution the particles with $\theta \in (90°, 110°)$ and $\psi \in (135°, 155°)$, i.e. $\mathbf{H}$ is nearly perpendicular to $[\bar{1}\bar{1}2]$ in the (111) plane. It can be seen, that the model curves fit very well the experimental ones. The minimum in the experimental $M_d(H)$ ($\perp$) curve in Fig. 3(a) is thus explained as a result of the remanent magnetization contributions of particles that have magnetic response to the applied field as that in Fig. 1: each one of these particles has negative remanence for fields up to a certain value (the anisotropy field for the particle). For higher (negative) magnetic fields, the remanent magnetization is positive, resulting in an increase of the total $M_d(H)$.

4 Conclusions

Some magnetic properties of thermally treated Co(6Å)/Cu(14Å)×15 multilayers, evaporated at room temperature on a Si (111) substrate, were investigated. A clear anomalous minimum in the experimental in-plane dc demagnetization remanence curve for fields applied nearly perpendicular to the easy axis was observed. It was demonstrated that this phenomenon can be explained as caused by the contributions of some particles, having negative remanent magnetization. Such a remanence can appear when uniaxial and cubic anisotropies coexist.

Acknowledgments

This work has been supported by Fundação de Amparo à Pesquisa do Estado do Rio Grande do Sul (FAPERGS, Brazil), Conselho Nacional de Desenvolvimento Científico e Tecnológico (CNPq, Brazil), and Financiadora de Estudos e Projetos (FINEP, Brazil).

The authors thank to Mr. Angelo Morrone for his fundamental help in developing the alternating gradient magnetometer.

References

[1] A.E. Berkowitz, J.R. Mitchell, M.J. Carey, A.P. Young, S. Zhang, F.E. Spada, F.T. Parker, A. Hutten and G. Thomas, Phys. Rev. Lett. 68(25) (1992), p. 3745.

[2] Q. Xiao, J.S. Jiang and C.L. Chien, Phys. Rev. Lett. 68(25) (1992), p. 3749.

[3] A.D.C. Viegas, J. Geshev, L.F. Schelp and J.E. Schmidt, J. Appl. Phys. 82(5) (1997), p. 2466.

[4] A.D.C. Viegas, J. Geshev, L.S. Dorneles, J.E. Schmidt and M. Knobel, J. Appl. Phys. 82(6) (1997), p. 3047.

[5] A.D.C. Viegas, J.Geshev, J.E. Schmidt and E.F. Ferrari, J. Appl. Phys. 83(11), 1 June (1998).

[6] P. E. Kelly, K. O'Grady, P. I. Mayo and R. W. Chantrell, IEEE Trans. Magn. 2 (1989), p. 3881.

[7] J. Geshev and M. Mikhov, J. Magn. Magn. Mater. 104–107 (1992), p. 1569.

[8] A. S. Arrott, *Nanomagnetism*, edited by A. Hernando (Kluwer, Dordrecht, 1993), p. 73.

[9] J. Geshev, A. D. C. Viegas and J. E. Schmidt, J. Appl. Phys. 84(3), 1 August (1998).

[10] H. Sato, O. Kitakami, T. Sakurai, Y. Shimada, Y. Otani and K. Fukamichi, J. Appl. Phys. 81 (1997), p. 1858.

[11] D. G. Tonge and E. P. Wohlfarth, Phil. Mag. 3 (1958), p. 536.

This article was processed using the LaTeX macro package with TTP style

Materials Science Forum Vols. 302-303 (1999) pp. 194-201

Modelling of Giant Magnetoimpedance from 1 Hz to 10 GHz

A. Yelon, D. Ménard, M. Britel and P. Ciureanu

Groupe de recherche en physique et technologie des couches minces (GCM), Département de génie physique, Ecole Polytechnique de Montréal, CP 6079, Succursale centre-ville, Montréal, Québec H3C 3A7, Canada

Keywords: Magnetoimpedance, Wires, Fibers, Metals, Ferromagnetic Resonance

Abstract

We have developped a technique which makes it possible, in principle, to take account of all of the physical phenomena which may play a role in a giant magnetoimpedance (GMI) experiment. This technique is a version of the classical approach to the modelling of physical effects : establishment and solution of the dispersion relations of the material, followed by satisfaction of the boundary conditions. We have also found simple approximations which yield a good description of the behavior, and a clear picture of the phenomena which are principally responsible for it. Thus far, we have applied the technique to the solution for a relatively simple situation : a short fiber or wire in an axial dc field larger than the anisotropy or coercive fields. Agreement with experiment in this case is excellent, from the lowest frequencies up to 6 GHz. We believe that GMI is an extremely powerful technique for characterizing ferromagnetic metals, providing significantly more information than ferromagnetic resonance experiments.

I. Introduction

It has been recognized [1-4] since 1994 that the large changes in a.c. impedance of magnetic fibers, wires, and ribbons which are observed when an applied d.c. field is varied, called the giant magnetoimpedance (GMI) effect, is due to the fact that the impedance, Z, is inversely proportional to the skin depth, δ, which is dependent upon the magnetic field. This dependence is the result of the dependence of the transverse magnetic permeability upon d.c. field :

$$\delta = \left(c^2/2\pi\omega\sigma\mu\right)^{1/2}. \quad (1)$$

In Eq. 1, c is the velocity of light, ω is the angular frequency, and σ is the conductivity. Since the permeability may vary greatly as a function of the static field or the frequency, the impedance will also vary considerably as a function of these parameters. Thus, the problem of explaining the GMI response of a sample is the same as the problem of finding its permeability, μ.

Much of the work which has been done on GMI has been in the low frequency range. At low enough frequencies, μ will have its static value, so that the behavior can be predicted using a static analysis, such as critical curves, as has been done in a number of cases [5,6]. However, μ is , in general, a function of both field and frequency, and will depend upon all of the magnetic properties of the material, upon the conductivity, and upon the geometry. Various phenomenological models [2, 7-9] have been presented to take account of some, or all of these effects, particularly at low fields, when the effect of domain structure can make any detailed analysis difficult. The difficulties in these case arise from two sources : the problem of determining the static domain structure of the sample, and the challenge of performing a complete analysis of behavior when the properties or magnetic structure of the sample are inhomogeneous.

We have pointed out [10,11] that a rigorous calculation of GMI response consists of simultaneously solving Maxwell's (M) equations and the Landau-Lifschitz (LL) equation of motion for the magnetization. There are several ways in which such calculations can be performed [12]. We believe that here are great advantages in following the straightforward approach, of obtaining the dispersion relations from these equations, and then substituting the solutions obtained from those relations into the boundary conditions. In what follows, we explain the method in some detail, and present our recent calculation of GMI in a cylinder with a d.c. field parallel to its axis, large enough to produce a uniform d.c. magnetization. We then describe the approximate techniques currently under development which permit us to obtain the essential features of the predictions in a simple fashion. This is followed by a brief description of the experimental technique which permits measurement from very low frquencies up to 7 GHz. We then present a comparison between the model and experimental results. We discuss the material parameters which can be obtained **through** comparison with the exact calculation, and the possible origins of discrepancies between the model and the experiment. We discuss the limitations of the model, and compare our approach with other possible modelling methods. Finally, we briefly consider the general interest of GMI as a characterisation technique.

II. Modelling Philosophy

As we have pointed out above, we believe that there are considerable advantages to analyzing the behavior of a ferromagnetic sample by using the most straighforward approach available. This consists of the solution of the equations of motion to obtain the dispersion relations, the solution of the dispersion relations to obtain the wave vectors, and the substitution of the wave vectors into the boundary conditions, in order to extract the impedance, or any other physical information of interest. This procedure will be illustrated by an example in the following section. For a metallic, ferromagnetic material, without high order interactions, such as magnetoelasticity, there will, in general, be four pairs of wave vectors, corresponding to two polarizations of the « electromagnetic » (EM) waves, and two « spin » wave polarizations, although there will be an admixing of spin components in the electromagnetic wave in the most general case, and visa versa.

Among the advantages of this approach are the fact that the procedures to be followed are clear. It is relatively easy to find solutions, either exact or approximate, in which some of the waves are neglected, and to evaluate the physical significance and impact of this neglect. Having found solutions for relatively simple problems, it is fairly straightforward to introduce more complicated geometries, interactions and static magnetic arrangements. It is possible to build models for non-linear behavior upon solutions for the linearized problem. Finally, there exists a substantial literature of such problems which have already been solved.

We have shown [10] that models of GMI and of ferromagnetic resonance (FMR) behavior are formally equivalent. That is, both involve the same calculations. In the case of FMR, when the alternating magnetic field is applied from the exterior, the quantity of interest is the surface impedance, Z_s. The two impedances, Z and Z_s, are related by a simple geometric factor [10,13]. Thus, if we know the complete solution of the one, we know the solution of the other. For purposes of understanding FMR experiments, many calculations have been performed, usually in the linearized regime for saturated magnetization. Among the problems for which solutions are to be found in the literature are the isotropic ribbon with d.c. field normal to the plane of the ribbon [14], field in the plane of the ribbon [15-18], and field at an arbitrary angle to the plane [19], as well as the single-crystal ribbon with field in an arbitrary direction [20]. The solution for ribbons with magnetoelastic interactions is also known [21,22]. In this case, the equations for the mechanical diplacements must be solved along with the M and LL equations, and there may be up to seven pairs of roots of the dispersion relations. The published calculations take account of a number of possible boundary conditions, including multilayer samples [23,24]. In addition, there exist a few calculations for samples with domain structures [25], and for high a.c. fields [26]. We shall discuss such problems in Sec. V.

III. GMI in a wire

One problem which had never been investigated, using the approach which we have just described, was that of GMI in a wire with a d.c. field large enough to saturate the magnetization parallel to its axis. As this problem is relatively simple to solve, we have chosen as our first application of the method described above [13]. Assuming that the length, l, of the wire is much greater than its radius, a, so that demagnetizing factor along the length is small, the solution we have obtained applies from the lowest frequencies to frequencies at which the sample length approaches one quarter of the EM wavelength. Under this assumption, the bulk and surface impedances are related by [13] :

$$Z = \frac{\ell}{2\pi a} Z_s . \tag{2}$$

We have found that the analysis is very similar to the analysis of the ribbon with d.c. field parallel to the current [15]. The essential difference is that the a.c. field and magnetization components are described by Bessel functions, rather than by exponentials. Since these components must be zero at the center of the wire, $r = 0$, only the functions $J_0(kr)$ and $J_1(kr)$ can play a role in the solution. Allowing for the fact that the k in the Bessel functions plays a role which is different from that played by the k in the exponentials, the dispersion relations are exactly those found by Ament and Rado [15]. As in the case of the ribbon, there are only three roots which are of significance for GMI. The fourth corresponds to an EM wave polarized parallel to the current direction. That is, the values of k which are of importance for GMI may be obtained from a bicubic equation :

$$K^6 + c_1 K^4 + c_2 K^2 + c_3 = 0. \tag{3}$$

In Eq. 3, $K = k\varepsilon\delta_0$, where δ_0 is the skin depth without magnetic effects and

$$\varepsilon^2 = A / \left(2\pi M_s^2 \delta_o^2\right), \tag{4}$$

with A the exchange constant and M_s the saturation magnetization. The coefficients in Eq. 3 are given in refs. [13] and [15].

The determinantal equations which account for the satisfaction of the boundary conditions are not identical in the case of wide ribbons and cylindrical wires, but are very similar. If a is allowed to go to infinity, they become identical. Thus, knowing the solution of one problem helps greatly in the solution of the other.

This is true despite the fact that the physical situations are, in fact, quite different. The oscillating spin components, which are identical for all positions equally distant from the ribbon surface, are out of phase for different positions with the same value of r. This is due to the fact that in the ribbon case, we are dealing with an essentially uniform precession mode, whereas the mode in the wire is a spin wave of wavelength $2\pi r$. This is illustrated in Fig. 1.

In Figure 2 we show the result of a typical calculation of Z as a function of frequency for one applied field. At low frequencies, we see the expected $\omega^{1/2}$ behavior discussed above. At high frequencies we see that the impedance increases far more rapidly, reaching a peak at the FMR frequency for that applied field. As we shall see, this result is in excellent agreement with experiment. A plot such as Fig. 2, with frequency as the principal variable is the clearest way of showing what physical phenomena are involved. In particular, we see that both the real and imaginary part of the impedance are proportional to $\omega^{1/2}$ at low frequency, as expected from our discussion above. We recognize that for the field of 320 Oe, the peak at 4491 MHz (for which the normalized frequency is 0.21) corresponds to FMR. If we wish to examine applications, it is the magnetoimpedance at fixed frequency which should be plotted as a function of field, as in Fig. 4. We shall compare the prediction with experiment, below.

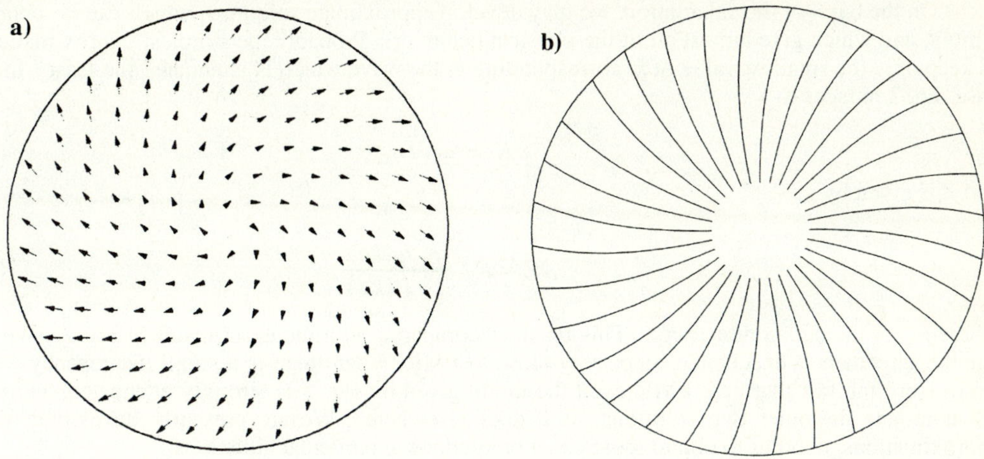

Figure 1. Radial dependent solution, for the alternating part of the magnetization, of the form $J_1(kr)\exp(i\omega t)$: a) projection of the magnetization in the plane normal to the applied field at an arbitrary time; b) equiphase lines or spin wave fronts, not shown in the center, since the r.f. magnetization goes to zero.

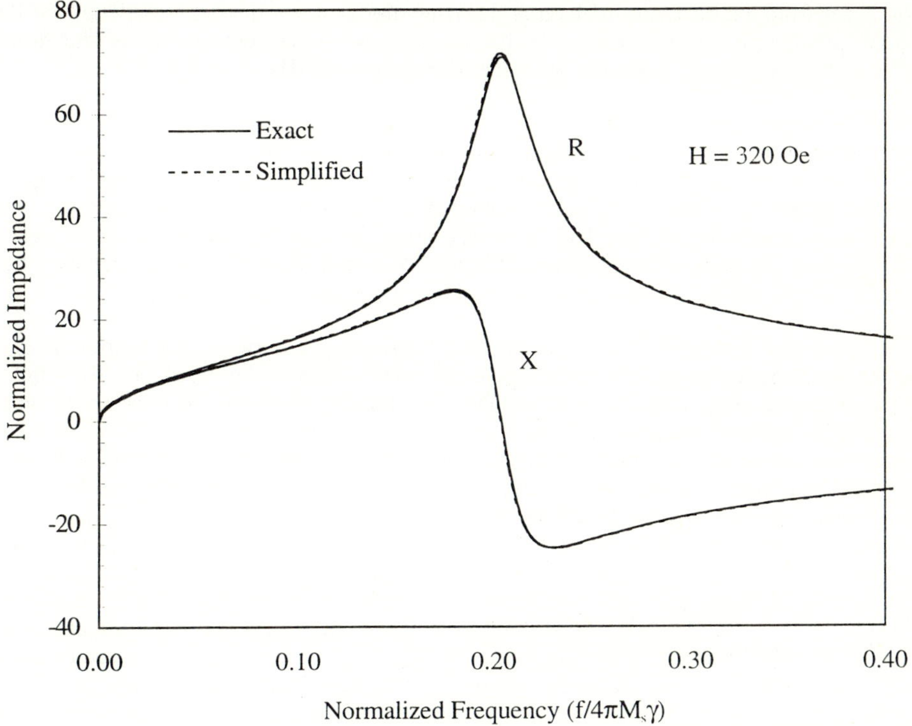

Figure 2. Comparison between the exact and simplified solution for the impedance spectra at 320 Oe.

On the basis of the calculation, we may develop approximate solutions, which can be obtained simply, and which give almost all of the physical behavior. Probably the simplest approximation is to keep only the smallest value of k, corresponding to the wave which is attenuated the least. In this case, Eq. 3 reduces to :

$$K_2 = -c_2 / c_3, \tag{5}$$

and Z is given by :

$$Z = R_{dc} \frac{ka}{2} \frac{J_0(ka)}{J_1(ka)}, \tag{6}$$

where R_{dc} is the d.c. conductivity. This result is compared with the exact result in Fig 2. We see that the agreement is excellent, except very close to FMR, when there is a small discrepancy. This arises from the fact that near FMR even the mode given by Eq. 5 is strongly attenuated, and does not dominate the other two solutions, as it does elsewhere. We are presently developing other approximations, in order to obtain more exact predictions in particular situations.

IV. Experimental Techniques And Results

In order to measure the magnetoimpedance over a wide frequency range, we have employed a coaxial line, in which the magnetic wire is the central conductor, in a short circuited configuration, connected to a network analyzer. The construction of the coaxial section so that it can be matched to a 50 ohm line has been described elsewhere [27-29]. The frequency range of measurement is not limited by the electronics, since modern network analyzers can yield reliable data which would permit determination of the real and imaginary parts of the impedance, up to at least 50 GHz. The limitation arises from the necessity of having the wire shorter that a quarter wavelength of the EM wave, and of avoiding cavity resonances of the coaxial stub. At present, we are confident that our data can be compared, at high precision, with the model, up to 6 GHz.

We have found [13] that it is relatively easy to obtain a parameter set which gives excellent agreement between the model and experiment for all frequencies when the field is large (Fig. 3). However, there can be significant disagreement between the model and experiment at the lowest fields, and the amplitude of the experimental impedance value is sometimes lower than the predicted value, when the combination of field and frequency bring the sample close to the FMR peak. The origins of the two discrepancies are quite different. We shall consider the discrepancy between amplitudes at high field here, and the low field problem in the following section.

When the field is smaller, there is a significant discrepancy between the amplitudes and the positions, of the experiment and the model, near resonance. This is shown in Fig. 4, which more closely resembles the usual way of showing GMI as a function of field, the most useful approach for most practical applications, than do Figs. 2 and 3. For a given set of parameters, the resonance peak height is predicted to be virtually independent of field, whereas it is observed to be lower at lower fields than at higher fields. We believe that this is due to small inhomogeneities in sample properties : anisotropy fields and directions and perhaps also of M_s. At high field, regions of slightly varying properties will be brought into resonance within a small range of frequencies, whereas they will not at lower fields. This question is presently being investigated, and we may eventually be able to characterize the uniformity of sample properties using GMI amplitude as a measurement criterion. In contrast, the discrepancies between model and experiment at the very lowest fields are not systematic, and arise from the failure of our assumption of saturation of the uniform magnetization (since we assumed isotropy). This constitutes one of many challenges to a complete model.

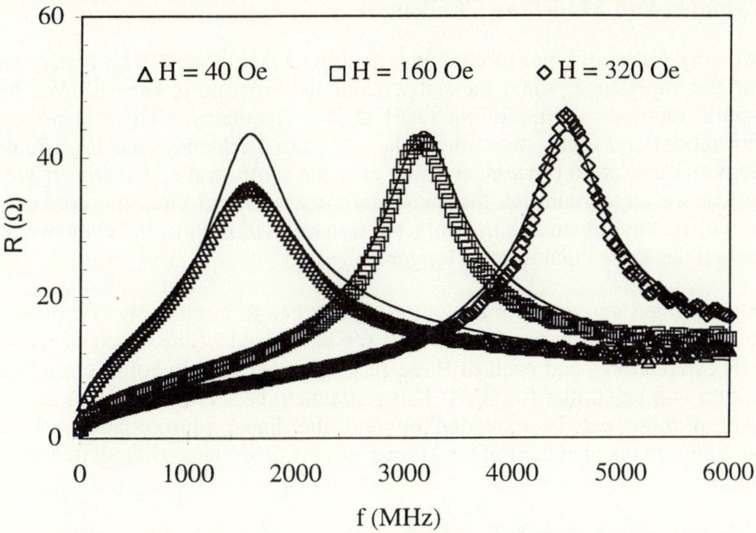

Figure 3. High frequency impedance spectra of a CoFeSiB wire for three applied fields. The dots are the experimental data and the solid lines are the theoretical fits with the parameters given in ref. 13.

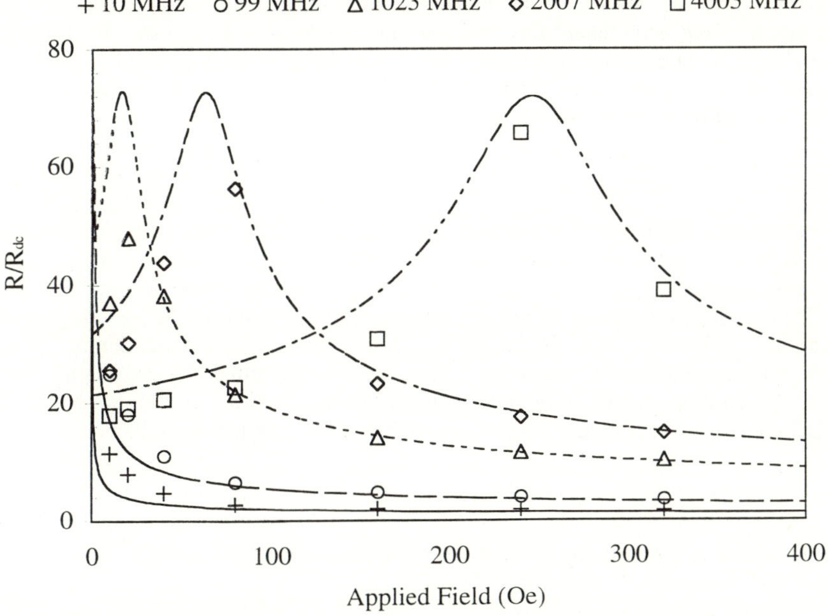

Figure 4. Calculated GMI response for several frequencies for the same wire as in Fig. 3. The open symbols are the experimental data.

V. Competing Models And Modelling Challenges

There are two very different types of challenge facing a GMI model. The first is whether it is the best way to find the impedance, when the static magnetic structure is known. We have already mentioned quasi-static methods of modelling GMI at low frequency. There is no contradiction between the current model and quasi-static methods. At zero frequency, the LL equation simply reduces to the energy minimization process, and will give the same result. Clearly, if we are at low enough frequency that we are certain that the quasi-static method will yield the correct result, it is unlikely that there will be any advantage in using the dynamic equations, but they work well, and continue to work when the static equations no longer do.

As we have mentioned above, there are other approaches to a relatively complete solution of the M equations and the LL equation than that which we have used. Since each of these has been widely used in FMR calculations, and each of these has its advantages in some circumstances, it is likely that the situation will be similar for GMI. This remains to be seen, as more model calculations are published. All of these can be extended beyond the linear aproximation, at the cost of mathematical complexity, to take account of non-linear effects, encountered at switching thresholds, or at high a.c. currents.

The most difficult problem is modelling GMI at fields sufficiently low that the magnetization is not saturated, but is divided into domains. The problem is twofold. It is difficult enough calculating the permeability when the magnetization structure is known, although there have been calculations of FMR for relatively simple domain structures [25]. It is even more difficult to ascertain the correct magnetization distribution in a sample whose domain structure has not been observed directly, and for which the properties are not known exactly. This can only be done by bootstrapping models and experiments until they converge. Under these circumstances, any approach which yields information which can clarify the possiblities should be considered. Recent approaches have included the modelling of plausible domain structures, which effects such as magnetoelatic coupling have been taken into account, and in which wall motion and rotation have been considered [7,9,30]. Micromagnetic models have also recently been applied to GMI [31,32]. Finally, equivalent circuit models have been used [8, 33] to aid the analysis of magnetization processes in GMI. All of these may contribute to our understanding of GMI.

VI. Conclusion

We have presented a GMI model which provides excellent agreement with experiment under appropriate circumstances. While much work remains in the development of this, and of alternative models, this success permits us to make use of GMI as a characterization technique. We have shown [13, 34, 35] that GMI can easily be used to measure Ms of a sample, if we assume that the gyromagnetic ratio, γ, is known, or to determine γ if Ms is measured independently. The same approach permits the determination of anisotropy fields. Complete fitting [13] permits the extraction of other parameters, including conductivity and magnetic damping. Thus, in addition to its practical applications, GMI promises to become a major scientific tool for magnetism studies.

References

[1] L. V. Panina, K. Mohri, K. Bushida and M. Noda, J. Appl. Phys. **76** (1994), p. 6198.
[2] R.S. Beach and A.E. Berkowitz, J. Appl. Phys. **76** (1994), p. 6209.
[3] K.V. Rao, F.B. Humphrey, and J.L. Costa-Krämer, J. Appl. Phys. **76** (1994), p. 6204.
[4] J. Velazquez, M. Vasquez, D-X. Chen and A. Hernando, Phys. Rev. B **50** (1994), p. 16737.
[5] F.L.A. Machado and S.M. Rezende, J. Appl. Phys. **79**, 6558 (1996).
[6] R. S. Beach, N. Smith, C.L. Platt, F. Jeffers et A. E. Berkowitz, Appl. Phys. Lett. **68** (1996), p. 2753.
[7] L.V. Panina, K. Mohri, T. Uchiyama, M. Noda, and K. Bushida, IEEE Trans. Magn. **31** (1995), p. 1249.
[8] R. Valenzuela, M. Knobel, M. Vasquez and A. Hernando, J. Phys. D: Appl. Phys. **28** (1995), p. 2404.
[9] D. Atkinson and P.T. Squire, IEEE Trans. Magn. **33** (1997), p. 3364.

[10] A. Yelon, D. Ménard, M. Britel and P. Ciureanu, Appl. Phys. Lett. **69** (1996), p. 3084.
[11] A. Yelon, M. Britel, D. Ménard and P. Ciureanu, Physica A **241** (1997), p. 439.
[12] L. D. Landau et E. M. Lifshitz, Electrodynamics of Continuous Media, Pergamon, Oxford, (1975).
[13] D. Ménard, M. Britel, P. Ciureanu and A. Yelon, Submitted. to J. Appl. Phys.
[14] A.I. Akhiezer, V.G. Bar'yakhtar, and M.I. Kaganov, Usp. Fiz. Nauk. 72 (1960), p. 3. Sov. Phys. Usp. 3 (1961), p. 661.
[15] W.S. Ament and G.T. Rado, Phys. Rev. **97** (1955), p. 1558.
[16] J.R. MacDonald, Phys. Rev. **103** (1956), p. 280.
[17] D.S. Rodbell, Physics **1** (1965), p. 279.
[18] A. Yelon, G. Spronken, T. Buithieu, R.C. Barker, Y.-J. Liu, T. Kobayashi, Phys. Rev. B **10** (1974), p. 1070.
[19] C. Vittoria, R.C. Barker, and A. Yelon, J. Appl. Phys. **40** (1969), p. 1561.
[20] C. Vittoria, G.C. Bailey, R.C. Barker, and A. Yelon, Phys. Rev. B 7(1973), p. 2112.
[21] T. Kobayashi, R.C. Barker, J.L. Bleustein, and A.Yelon, J. Appl. Phys. 40 (1973), p. 1561.
[22] C. Vittoria, J.N. Craig, and G.C. Bailey, Phys. Rev. B 10 (1974), p. 3945.
[23] G. Spronken, A. Friedmann, and A. Yelon, Phys. Rev. B 15 (1977), p. 5141.
[24] C. Vittoria, J. Appl. Phys. 57, 3712 (1985); Phys. Rev. B 32 (1985), p. 1679.
[25] J.O. Artman, Phys. Rev. **105** (1957), p. 62.
[26] R.I. Joseph and E. Schlömann, J. Appl. Phys. **38** (1967), p. 1915.
[27] P. Ciureanu, C. Akyel, M. Britel, J. Gauthier, A. Yelon, P. Rudowski, G. Rudkowska, and J.O. Ström-Olsen, IEEE Trans. Magn. **32** (1996), p. 4678.
[28] D. Ménard, M. Britel, V.P. Paramonov, A.S. Antonov, P. Ciureanu, P. Rudkowski, J.O. Ström-Olsen and A. Yelon, J. Appl. Phys. **81** (1997), p. 4032.
[29] A. Antonov, A. Granovsky, A. Lagar'kov, N. Perov, N. Usov and T. Furmanova, Physica A **241** (1997), p. 420.
[30] D. Atkinson and P.T. Squire, J. Appl. Phys., in press.
[31] A. Antonov, A. Dykhne, A. Lagar'kov and N. Usov, Physica A **241** (1997), p. 425.
[32] N. Usov, A. Antonov, A. Dykhne and A. Lagar'kov, J. Phys.: Condens. Matter. **10** (1998), p. 2453-2463.
[33] K.L. Garcia and R. Valenzuela, J. Appl. Phys., in press.
[34] P. Ciureanu, M. Britel, D. Ménard, A. Yelon, C. Akyel, M. Rouabhi and R.W. Cochrane, J. Appl. Phys., in press.
[35] D. Ménard, D. Frankland, P. Ciureanu, A. Yelon, M. Rouabhi, R.W. Cochrane, T.A. Ovari, and H. Chiriac, J. Appl. Phys., in press.

Materials Science Forum Vols. 302-303 (1999) pp. 202-208

Highly Asymmetric Giant Magnetoimpedance

F.L.A. Machado*, A.R. Rodrigues, A.A. Puça and A.E.P. de Araujo

NUMAG, Departamento de Física, UFPE, 50670-901 Recife, Pernambuco, Brazil

Keywords: Giant Magnetoimpedance, Amorphous, Soft-Ferromagnetism, Magneto-Transport

Abstract Enhanced and highly asymmetric giant magnetoimpedance (GMI) spectra, measured in pieces of ribbons of an amorphous alloy under an applied dc bias-electrical current, are reported. The asymmetry depends on the magnitude of the dc bias-current and on its relative direction to the dc applied magnetic field. This novel effect allows one to determine both the magnitude and the direction of a magnetic field from a magnetoimpedance data. A theoretical model, used to simulate the enhancement and the asymmetry, is also discussed.

Introduction The magnetoimpedance measured in ribbons of soft-ferromagnetic materials was found to be largely enhanced and dependent on the frequency (*f*) and magnitude of the ac electrical current (I_{ac}) [1-3]. It was also found that it varies strongly with the applied magnetic field (*H*) in the small field regime (less than 10 *Oe*). Because of the large values obtained and in association to the giant magnetoresistance (GMR) effect observed in multi-layers [4] and granular materials [5,6], the effect was called giant magnetoimpedance (GMI). The dependence of the GMI spectra with I_{ac} and *f* is being extensively investigated in magnetic materials with different compositions, structures and shapes. There have been reports on longitudinal and transversal GMI measured in ribbons, wires, and films of amorphous, nano- and poli-crystalline high permeability materials [7-12]. There are also great potentials for applications of the GMI in the development of new magnetic sensor technologies [13,14].

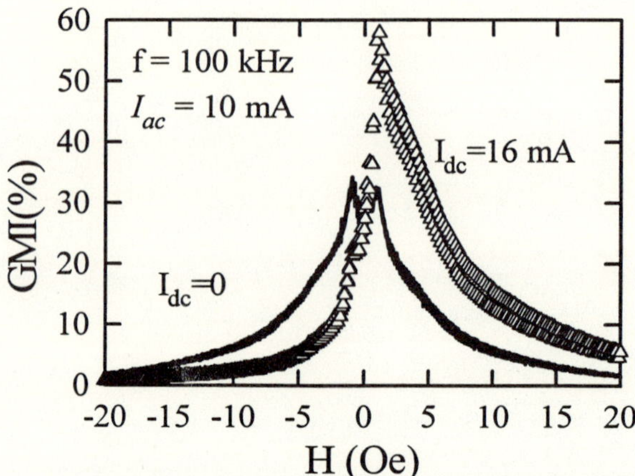

Fig. 1. GMI spectra for zero bias dc electrical current (solid line) and for I_{dc}=16 *mA* (opened triangles). The width of the solid line is due to the hysteresis.

In this work we report, for the first time, GMI spectra measured under the influence of an applied dc bias electrical current. We found that the GMI become highly asymmetric, it is enhanced and the asymmetry depend on the relative direction of the bias dc electrical current and *H*. A theoretical model used for simulation of the data will be discussed shortly after the experimental results are presented.

Techniques and experimental results The measurements were made in pieces of as-quenched, melt-spun ribbons, of the $Co_{70.4}Fe_{4.6}Si_{15}B_{10}$ alloy. The samples used throughout the experiments were 1 *cm* long (*L*), 1.4 *mm* wide (*l*), and 40 *μm* thick (*t*). The measurements were carried out at room temperature for an ac electrical current of 10 *mA* of magnitude and frequency varied up to 30 *MHz*. A four-probe lock-in detection technique in the *Z*-θ mode was used to measure the GMI. The applied dc magnetic field and the bias dc current were varied in the interval $-40 \leq H \leq 40$ *Oe* and $-150 \leq I_{dc} \leq 150$ *mA*, respectively. Both the ac and the dc electrical current where applied along the direction of the *H*, i.e., along the length and in the plane of the sample.

Fig. 1 shows the GMI ($=10^2$ [*Z*(*H*)-*Z*(40 Oe)]/*Z*(40 Oe)) spectra for zero dc bias current (solid line) and for I_{dc}=16 *mA* (opened triangles). The biased spectrum shows an unexpected strong asymmetry characterized by an enhancement of one of the peaks while the other one is diminished.

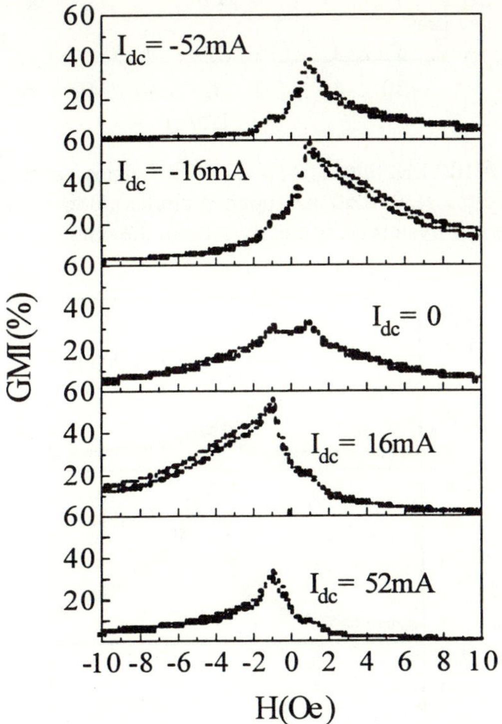

Fig. 2. GMI spectra obtained for several values of I_{dc} showing that the strong asymmetry depends on the direction the bias dc electrical current is applied and on its magnitude.

The enhancement factor and the asymmetry depend on the magnitude and direction of the bias dc current as shown in Fig. 2. Also, higher the bias current higher are the values of *H* one needs to apply in order to saturate the GMI. The maximum values of the GMI spectra are plotted as a function of I_{dc} in Fig. 3 for *f*=100 *kHz* and I_{ac}=10 *mA*. The zero bias GMI (≈30%) initially increases with I_{dc} reaching a maximum value (≈70%) for I_{dc}≈25 *mA* and diminishes for higher I_{dc}. A hysteresis is also observed in the I_{dc}-dependence of the GMI (Fig. 3): the closed circles are data taken while I_{dc} is initially varied from 0 to -150 *mA*; next, the opened circles were taken varying I_{dc} from 0 to 60 *mA*; finally, I_{dc} was swept from 0 to -74 *mA* (opened triangles) and from 0 to 76 *mA* (closed triangles). The GMI value is still large (70%) for the largest value of I_{dc} (=150 *mA*) used in our measurements (only the portion of the data below 75 *mA* is shown in the figure).

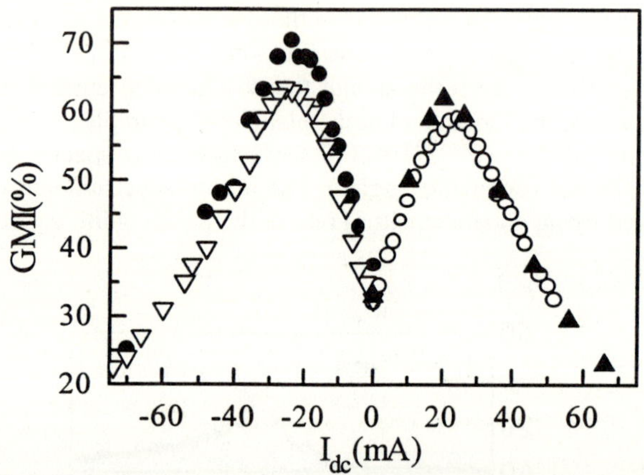

Fig. 3. GMI versus I_{dc} for f=100 kHz and I_{ac}=10 *mA*. The data were taken while I_{dc} was swept from: 0 to -150 *mA* (closed circles), 0 to 60 *mA* (opened circles), 0 to -74 *mA* (opened triangles), and 0 to 76 *mA* (closed triangles). A hysteresis is clearly seen in the data.

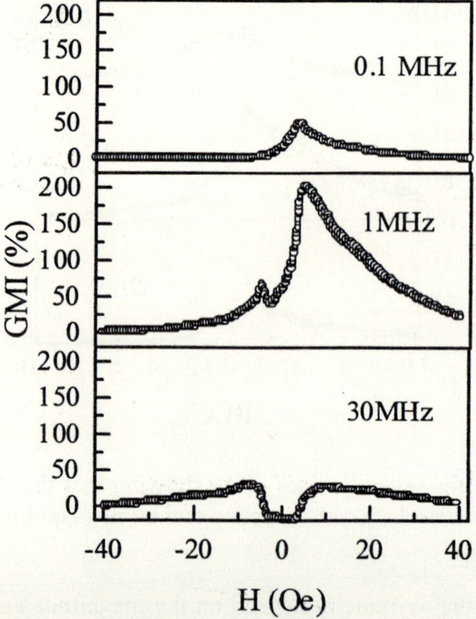

Fig. 4 GMI spectrum for three values of f and I_{dc}=25 *mA*

For comparison, the dc magnetoresistance measured in soft-magnetic materials are usually less than 0.1%. The asymmetry persists in the high frequency regime but it diminishes as f is increased. The two peaks become nearly equal for f close to 20 MHz and the asymmetry is inverted above this value of frequency (Fig. 4) The maximum value of the GMI was also measured as a function of f for three values of the bias dc current: 0, 25 and 50 *mA*. These results are shown in Fig. 5. The curves

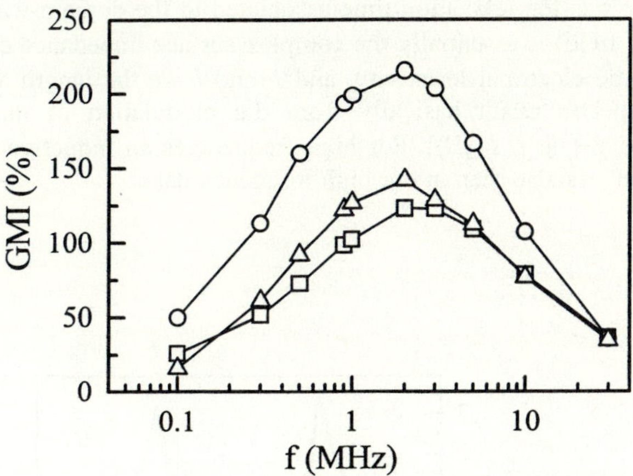

Fig. 5. GMI versus f for three values of I_{dc}: 0 (squares), 25 (circles), and 50 mA (triangles). The solid lines are guides to the eyes.

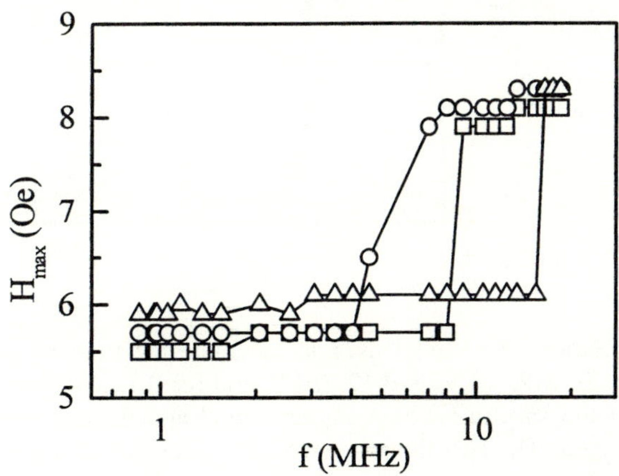

Fig.6. H_{max} plotted as a function of f for: I_{dc}=0 (triangles), 5 (squares) and 10 *mA* (circles). Note that H_{max} increase its value for frequencies above some characteristic value of f which depend on the value of I_{dc}. The solid lines are guides to the eyes.

have an bell shape with a maximum around f=2 *MHz*. At this frequency the highest value of the GMI (210% for H=6 *Oe*) is also obtained for I_{dc}=25 *mA* (see Fig. 3). These results indeed show that the GMI is enhanced when one applies the bias dc current. The value of H which correspond to the position of the peak in the GMI spectrum (H_{max}) is nearly independent of I_{dc} for low frequencies but it shows a somewhat strong variation at high frequencies. The peaks shift about 2 *Oe* for f above 4 *MHz* for I_{dc}=10 *mA*. Fig. 6 shows three curves of H_{max} versus f for zero bias and for I_{dc}=5 and 10 *mA*.

A theoretical model for the GMI The skin-depth ($\delta=[2\rho/(\mu_T\omega)]^{1/2}$) in high magnetic permeability ($\mu_T \approx 10^5\mu_0$) material can be extremely small (20-2000 *nm*) even for low values of $f(=\omega/2\pi)$. This causes the electrical current to be driven closer to the surface of the sample. The low frequency GMI ($\omega\tau<<1$, where τ is the relaxation time associated to the domain-wall motion induced by the alternating magnetic field) is essentially the complex surface impedance effect[15]: $Z_s=(1-i)\rho L/2l\delta$. ρ (=160 $\mu\Omega cm$) is the electrical resistivity, and L and l are the length and width of the ribbons, respectively. The spectra result basically from the modulation of the skin-depth through its dependence on f and $\mu_T(=\mu_T(f, I_{ac}, H))$. For high frequencies an inductive factor become important [9]. This contribution is also seen in our high frequency data.

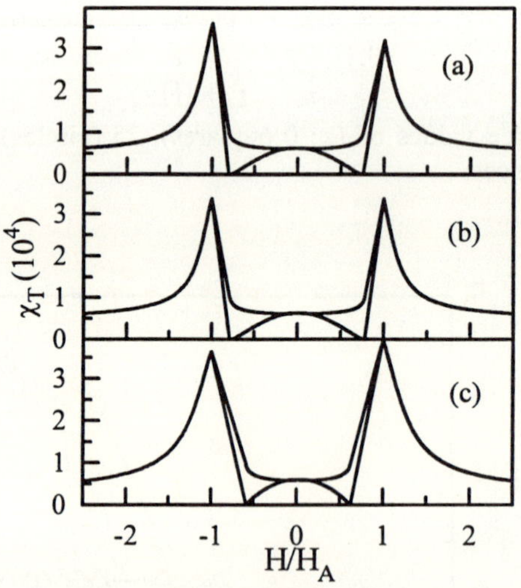

Fig. 7. Computer simulation of a model used to calculate the transversal magnetic susceptibility. The simulation was made with (*a*) h_{dc}=-0.12, (*b*) 0, and (*c*) 0.12 *Oe*, and H_A = 6 *Oe*, h_{ac}=0.24 *Oe*, and $H_O=h_{dc}$. Note that the spectra become asymmetrical and the asymmetry is inverted when the direction of h_{dc} is changed. The two lines in each curve are due to a hysteresis resulting from the domain configuration while sweeping *H* before and after the field induced transition.

We have calculated in ref. 15 the transversal magnetic susceptibility ($\mu_T=1+4\pi\chi_T$) of ribbons of soft-ferromagnets. Our theoretical results successfully explained the *H*- and *f*-dependence of the GMI. In our model we considered a transversal in plane induced anisotropy (H_A) which are produced, for instance, by cooling the ribbons in a magnetic field of the order of 2 *kOe*. H_A break the domains in a regular pattern of domains transversal to the length of the ribbon [16]. Besides the anisotropy term, we considered the Zeeman energies associated to *H* and to h_{ac}. A parabolic energy term associated to the displacement of the domain walls is also considered. χ_T is obtained by minimizing the overall energy that gives the configuration of the domains for each value of H and h_{ac}. The results can be summarized as following. The domains initially pointing in the direction of h_{ac} can be continuously rotated until they become completely aligned with H. However, the domains initially aligned against h_{ac} can only be rotated up to a critical value of H. For higher values of H these domains suffers a field induced transition and the magnetization become uniform pointing along the direction of the domains initially aligned with h_{ac}. This field induced transition, which can also be observed by measuring the magnetization [17], produce an enhancement on the transversal susceptibility. These results correctly explain the two symmetrical peaks on the magnetic field dependence of the GMI. However, in order to account for the asymmetry in the GMI peaks due to the dc electrical current I_{dc} we found necessary to make two changes in our model. First we replace H by an effective dc magnetic field H_{eff} ($H_{eff} = \sqrt{H^2 + h_{dc}^2}$) which makes an angle β (=$artan(H/h_{dc})$) with the anisotropy field. Secondly, we introduce a new Zeeman energy associated to a dc magnetic field H_O pointing along the direction I_{dc} is flowing. H_O could be associated to the magnetic field that yields a torque in the domain wall as a reaction to the Lorentz force caused by the passage of an electron through the wall. The existence of such a magnetic field was proposed by L. Berger [18,19] and it is given by:

$$H_O = \left(\frac{a}{t}\right)\left(\frac{M_S R_S}{\rho}\right) h_{dc} \tag{1}$$

where a is the wall spacing, M_S (= 8.6 kOe) is the saturation magnetization and R_S (=0.43 $\mu\Omega cm kOe^{-1}$) is the anomalous Hall coefficient [7]. Using the values of the parameters for our samples we find that the proportionality factor between H_O and h_{dc} ($H_O = N\, h_{dc}$) as being N=0.05 for a=100 μm [16]. However, for as quenched samples the wall spacing can be of the order of the width of the sample ($a \approx 1$ mm) which gives $N \approx 1$. After the inclusion of these new terms in our model and using the same procedure followed in Ref. 15 one may calculate the transversal magnetization, the corresponding susceptibility, and the GMI. Fig. 7 shows three curves obtained by a computer simulation of the transversal magnetic susceptibility obtained by our model with the modifications proposed above. The curves were obtained with h_{ac}=0.24 Oe, H_A = 6 Oe, N=1, and h_{dc} = -0.12 (a), 0 (b) and 0.12 Oe (c). Note that the peaks are symmetric when there is no dc electrical current and became asymmetrical for $I_{dc} \neq 0$ as observed in our experimental data (Fig. 2). The asymmetry does also switch place when the direction of I_{dc} is changed. We also modified the model used to calculated the GMI in wires and found the same kind of asymmetry shown in Fig. 7.

Discussions and conclusions The samples used in the present paper are not annealed in a magnetic field. Thus, a single anisotropy as supposed in our simple model is not expected. Even so, the asymmetry and some enhancement of the GMI are observed if we use a value of N close to what is expected for such samples. We believe that our model needs improvements and the parameters are varied in order to completely account for the features resulting from the bias current. At the present there is a lack of a theoretical model that calculates the GMI for a general configuration of domains. Finally, the dependence of the GMI with the magnitude of I_{dc} can be understood as following. Depending on the sign and on the magnitude of I_{dc} it will be easier to rotate some of the domains while the others will be more difficult to be rotated. Thus, the dc electrical current influences the transversal magnetization and, consequently, the GMI. For high values of I_{dc} a complete alignment of the domains is expected resulting in a uniform rotation of the magnetization where no field induced transition is observed. The inversion in the asymmetry and the strong variation in the position of the maximum in the GMI indicate that a new mechanism might become important in high frequencies. Indeed, we should not expect the domain walls following the very high frequency ac magnetic field. For these frequencies the existence of a bias dc current seems to become not important as shown in Fig. 5. We also investigated both theoretically and experimentally the effect of applying H making small angles with the direction of the ac current but we did not find the asymmetry as presented above.

In summary, we found that the magnitude of the two peaks of the GMI spectra become highly asymmetric when measured with a dc bias electrical current. The asymmetry depends on the relative direction of the dc bias current to the direction of the applied dc magnetic field. The shape of the GMI versus I_{dc} is similar to that of the GMI versus H but it is highly enhanced from an initial value of 32% at zero bias current to 70% for f=100 *kHz* and reaches 210% at 2 *MHz* for I_{dc}=25 *mA* The asymmetry generated by the dc bias current on the GMI allows one to measure both the magnitude and direction of an applied magnetic field. A theoretical model, which shows a similar anomalous behavior as seen in GMI under the bias dc current, was presented.

Acknowledgments We thank Professor F.P. Missell from University of São Paulo for providing the samples used in this work and K.C. Mendes for helping us in some measurements. This work was partially supported by CNPq, CAPES, FACEPE and FINEP (Brazilian Agencies).

References

[1] F.L.A. Machado, B.L. da Silva, and E. Montarroyos, J. Appl. Phys. 73(1993), p 6387.
[2] F.L.A. Machado, B.L. da Silva, S.M. Rezende and C.S. Martins, J. Appl. Phys. 75(1994), p. 6563.
[3] F.L.A. Machado, C.S. Martins and S.M. Rezende, Phys. Rev. B15(1995), p. 3926.
[4] M.N. Baibich, J.M. Broto, A. Fert, F.N. Van Dau, F. Petroff, Phys. Rev. Lett. 61(1988), p. 2472.
[5] A.E. Berkowitz, J.R. Mitchell, M.J. Carey, A.P. Young, S. Zhang, F.E. Spada, F.T. Parker, A. Hutten, and G. Thomas, Phys. Rev. Lett. 68(1992), p. 3745.
[6] J.Q. Xiao, J.S. Jiang, and C.L. Chien, Phys. Rev. Lett. 68(1992), p.3749.
[7] K.C. Mendes, F.L.A. Machado, L.G. Pereira, S.M. Rezende, F.C. Montenegro, M.V.P. Altoé, and F.P. Missell, J. Appl. Phys. 79(1996), p. 6555.
[8] K.C. Mendes and F.L.A. Machado, J. Mag. Mag. Matls. 177-181(1998), p. 111.
[9] L.V. Panina and K. Mohri, Appl. Phys. Lett. 65(1994), p. 1189.
[10] M. Knobel, M.L. Sanches, C. Gomes-Polo, P. Marin, M. Vasques, and A. Hernando, J. Appl. Phys. 79(1996), p. 1646.
[11] R.S. Beach and A.E. Berkowitz, J. Appl. Phys. 76(1994), p. 6209.
[12] R. L. Sommer and C.L. Chien, Appl. Phys. Lett. 67(1995), p. 3346.
[13] K. Inada, K. Mohri, and K. Inuzuka, IEEE Trans. on Magn. 30(1994), p. 4623.
[14] R. Valenzuela, M. Vazquez, and A. Hernando, J. Appl. Phys. 79(1996), p. 6549.
[15] F.L.A. Machado and S.M. Rezende, J. Appl. Phys. 79(1996), p. 6558.
[16] A.D. Santos, L.G.C. Melo, C.S. Martins, F.P. Missell, Y. Souche, F.L.A. Machado, and S.M. Rezende, J. Appl. Phys. 79(1996), p. 6546.
[17] E. Montarroyos, J.R.L. de Almeida, F.L.A. Machado, and S.M. Rezende, J. Mag. Mag. Matls 104-107(1992), p. 149.
[18] L. Berger, J. Phys. Chem. Solids 35(1974), p. 947.
[19] L. Berger, J. Appl. Phys. 49(1978), p. 2156.

* Corresponding author: Tel. +55 81 2718450; fax: +55 81 2710359; e-mail: flam@df.ufpe.br.

Materials Science Forum Vols. 302-303 (1999) pp. 209-218

Hysteretic Behavior and Anisotropy Fields in the Magneto-Impedance Effect

M. Vázquez, J.P. Sinnecker and G.V. Kurlyandskaya

Instituto de Magnetismo Aplicado, UCM/RENFE, and Instituto de Ciencia de Materiales, CSIC, P.O. Box 155, ES-28230 Las Rozas (Madrid), Spain

Keywords: Magnetization Process, Magneto-Impedance, Amorphous and Nanocrystalline Materials

Abstract

An overview of the Magnetoimpedance effect (MI) in a series of magnetic materials with different geometries (wires, ribbons, tubes and thin films) and structures (crystalline, nanocrystalline and amorphous) is given. Two types of MI, namely single peak, SP, and two peak,TP, are discussed. The observation of both types of behaviours is general and depends on the amplitude and the frequency of a.c. current required to measure the impedance. A correlation between MI hysteresis and hysteresis loops is presented. The influence of the a.c. current frequency, magnetic anisotropies and irreversible magnetization processes upon the hysteretic MI effect is discussed

Introduction.-

The first and main consequence of the application of a d.c. magnetic field to a ferromagnetic material is the modification of its magnetic state via the magnetization process occurring by domain wall displacements and magnetization rotation within domains. In fact, the magnetic state at a given applied d.c. field can be characterized by the magnetic response to an additional a.c. magnetic excitation, which is represented by a differential permeability. This permeability is determined by a set of conditioning circumstances associated to the magnetic characteristics of the material itself.

In addition, the application of a magnetic field can first modify the transport properties via the magnetoresistance effect (scattering of electrons) which will not be treated here, but also the total impedance itself through an electromagnetic classical effect. Usually, the impedance is determined by the so-called four points technique which requires the sample to be submmitted to an electrical current, resulting in a given measurable voltage through it. The impedance is thus evaluated from the relationship between that voltage and the applied current.

Nevertheless, depending on the frequency of the exciting current one can obtain various results. In the limit of very low frequencies, the current just generates a circumferential magnetic field, which eventually acts superimposed to the above-mentioned d.c. applied field. As a consequence of the temporal variation of this circular field, changes of the circular component of magnetization are induced so resulting in an extra electromotive force at the ends of the sample according to the Faraday-Lenz law. Such changes of the circumferential component of magnetization are responsible for the so-called magneto-inductive effect (M_ϕ & H_ϕ magnetic hysteresis loops) and for the Matteucci effect (M_ϕ & H_z loops)[1,2].

For higher frequencies, typically above 100 kHz, there is an effective reduction of the cross section through which the current is flowing owing to the generated eddy currents[3]. As a consequence of the skin effect, the current flows through a sheath close to the surface of the sample. In this case, only a radial fraction of the sample is involved in the high frequency magnetization process, and there is a correlation between the penetration depth and the magnetic permeability via the expression:

$$\delta = \sqrt{\frac{1}{\pi \sigma \nu \mu}} \qquad (1)$$

where σ is the electrical conductivity, ν the frequency of the current and μ the circular or transverse permeability.

In turn, the total impedance,Z, is related to that penetration length according to:

$$Z = R_{DC}\left(\frac{ka}{2}\right)\frac{J_0(ka)}{J_1(ka)} \qquad (2)$$

where a is the radius of wire, and J the Bessel functions. Consequently, upon application of a d.c. magnetic field, H, the accompanying change of permeability finally results in an effective modification of the impedance.

The experimental techniques to perform magnetic and magnetoimpedive measurements are slighthly different for the various analysed samples so that, in the following we refer to indicated references for detailed description of experimental procedures. Just to avoid uncertainties, we will consider the present definition for the magneto-impedance ratio (GMI):

$$\frac{\Delta Z}{Z} = \frac{\left[Z(H) - Z(H_{max})\right]}{Z(H_{max})} \times 100\% \qquad (3)$$

where $Z(H)$ is the impedance measured at the dc field H and $Z(H_{max})$ is the impedance measured at the maximum dc field.

Finally, at the limit of very high frequencies, above 100 MHz, the skin effect penetration depth can become comparable to atomic distance and resonance effects are observed, which is also out of the scope of the present work[4].

The giant magnetoimpedance, GMI, effect discovered recently in magnetic wires and other extremely soft magnetic materials is attracting much attention because of its technological applications[5-12]. Most of such applications deal with magnetic field sensor devices (i.e., current and position sensors) and magnetoelastic sensors (i.e., stress sensor and magnetostriction evaluation)[13]. Looking for these applications, it is required to determine which are the optimum conditions for which GMI is most effective. In particular, which is the optimum range of frequency to obtain large ratio GMI to noise, or the maximum field sensitivity for the specific intervale of field to be sensed. In this regard, it is quite important to know the differential transverse permeability response to applied d.c. fields, which is basically determined by general magnetic characteristics as domain structure and spatial distribution of magnetic anisotropy. In connection with such magnetic characteristics there are some effects as irreversibilities in the magnetization process or disaccommodation or magnetic after-effects which in principle represent an obstacle for a proper use of MI effect in technological applications[14,15].

In this work, we are collecting experimental results concerning the different behaviours observed for the MI as well as its irreversibility in various types of materials and analyse its correlation with the anisotropy fields and the hysteretic behavior of the corresponding loops for various kinds of materials as: water-quenched amorphous and nanocrystalline wires, policrystalline wires, glass-coated amorphous microwires, stress and field annealed nanocrystalline ribbons, electroplated tubes and processed sputtered thin films. This large variety of materials is considered to point out the generality of the presented results. In section 2, we proceed with a general discussion about the various domain structures and magnetic anisotropies in the studied materials, while in section 3 we describe the two types of MI behaviors as associated to the transverse anisotropy field. In section 4, we will introduce the hysteretic behavior and correlate it with the transverse permeability in the above mentioned materials and introduce the irreversibility field. Finally, in section 5 a final discussion is presented on the correlation between the hysteresis in the magneto-impedance and the magnetization process. The potential influence of that hysteretic behavior in technological devices is shortly commented.

Domain structure and a.c. circular magnetization process under bias d.c. magnetic field.-

According to Eq. 2 the circular permeability and its change with dc axial field is determining the actual amplitude of magnetoimpedance. This permeability will be determined by the particular domain structure of the material. Although we will consider also the cases of ribbons, tubes and films we restrict ourselves in this section to domain structrures of wires. A further analysis can be accordingly extended to the other geometries.

Owing to its cylindrical geometry we consider 3 major or ideal configurations for a wire with axial, circumferential and radial easy axis.A general information on domains in wires can be found elsewhere[16,17]. The first case is that of having an axial anisotropy which results ideally in a single-domain structure and square axial hysteresis loop with a given switching field at which the magnetization reverses between two stable configurations of domains. The presence of an applied d.c. reverse axial field nearly as large as the switching field has the effect of quasi-compensate the axial magnetic anisotropy. Magnetization feels quasi-free for an outstanding response under the action of incremental fields. If the easy axis is not fully axial but with some circular component (helical direction) such incremental field, will result in a noticeable circular permeability. As indicated in the scheme of Fig. 1a, the circumferential permeability peaks for a d.c. axial field close to the switching field. For quite small or very large d.c. fields, the result is nevertheless very modest. In short, there is an intrinsic irreversible and assymetric character for the circumferential permeability with the applied d.c. axial field. In this framework, we can introduce an "irreversibility field", H_{irrv}, corresponding to that switching field. Actually, this circular permeability appears from the existance of some helical component of the magnetization and is essentially irreversible. Considering the magneto-impedance effect, a peak when increasing the axial bias field is expected when the switching field is reached.

The second case is when the easy axis of intrinsecal anisotropy is circumferential and the domain structure is a bamboo-like structure as indicated in Fig. 1b. Now the dc axial field plays a role to balance the circumferential anisotropy so that, a large peak for the circumferential permeability is obtained just when the applied field reaches the anisotropy field, $H = H_k$. This permeability behaves nearly reversibly with increasing and decreasing d.c. field but with some small hysteretic component inherent to the magnetization rotation. Accordingly, the MI ratio should exhibit a maximum value for a bias field equaling the circular anisotropy field.

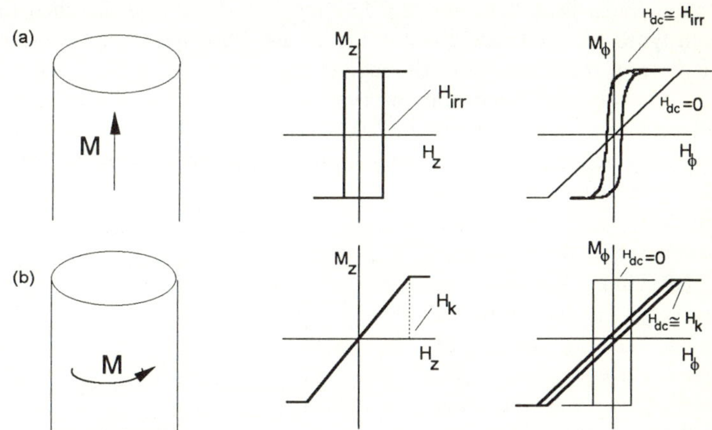

Figure 1: Schematic view of the effect of a dc axial applied field on the axial and circular magnetization curves for wires with (a) axial and (b) circular easy magnetization directions.

Finally, in the case of having strong enough radial anisotropy, a radial network of domains is expected. Now, the bias field balances the radial anisotropy field, and a large circumferential permeability is expectable when, as in the first case, there is some circular component now for the radial anisotropy. Both kinds of behaviors in the MI are now expected: peaks for the irreversible field and for the anisotropy field.

The two types of magnetoimpedance behavior.-

Provided that the amplitude and frequency for the current flowing along the sample are fixed, it is currently spoken of two types of magnetoimpedance behaviors upon increasing the d.c. axial magnetic field[10]. In the following we will label them as: i) Single-peak, SP, behavior where a continous decrease of the magneto-impedance ratio is observed, and ii) Two-peaks, TP, behavior where the maximum magneto-impedance ratio is shifted from zero up to a given applied field followed by the monotonic decrease. Moreover, the field at which this maximum is observed has been currently ascribed or correlated to the transverse magnetic anisotropy field. The occurrence of these two behaviors, apparently different, has been discussed recently[18].

We will explore here the consequences of the following reasoning: as the maximum MI ratio is observed for a given d.c. applied field when there is a maximum change of circumferential permeability (i.e., the a.c. permeability reaches a maximum), the first mentioned behavior would correspond to the case where the easy magnetization direction for the magnetic anisotropy is nearly parallel to wire`s axis. But even in this case some finite longitudinal field is required to roughly attain the magnetization saturation. Such a finite field is ascribed to the inherent (although maybe quite small) irreversible component of the axial magnetization process which has been described in the preceeding section. Consequently, in this case the maximum MI ratio would correspond to that finite field at which the irreversible magnetization process is completed. This argumentation predicts that SP behavior is a particular case of the TP if we zoom in detail enough into the very small field region. So, in the following we are considering two axial fields determining the maximum MI ratios namely, the axial field equaling the transverse anisotropy field, H_k, and the so-labelled irreversibility field, H_{irr}, corresponding to the field ascribable to the completion of the irreversibilites of the axial magnetization process.

In order to check that, we first consider the probably most traditional magnetoimpedive material as is a negative-vanishing magnetostriction Co base amorphous wire. As reported in early papers on GMI, this outstandingly soft material can be considered in its as-cast state as an example for the single-peak behavior. So, looking into detail, we searched for the two-peaks behavior[19]. Both types of results are given in Fig.2a and 2b respectively where it is unveiled the importance of the plotting scale to observe the second one.

On the other hand, the observation of these SP or TP behaviors depends upon the frequency and amplitude of the a.c. current[20,5]. This is clearly shown in the figures 3a and 3b. In Fig. 3a, the frequency dependence of the magnetoimpedance ratio is shown for a policrystalline permalloy wire after being submmitted to a particular treatment. While for relatively low frequency, a SP like behavior is observed, TP characteristics appear at the range of higher frequencies. As observed, not only the value of the maximum MI ratio changes with frequency but also the applied field at which the maximum is observed shifts towards higher field with increasing frequency. This shifting has been recently studied in detail in glass coated amorphous microwires[21]. The shifting of the peaks with increasing frequency towards higher bias fields in

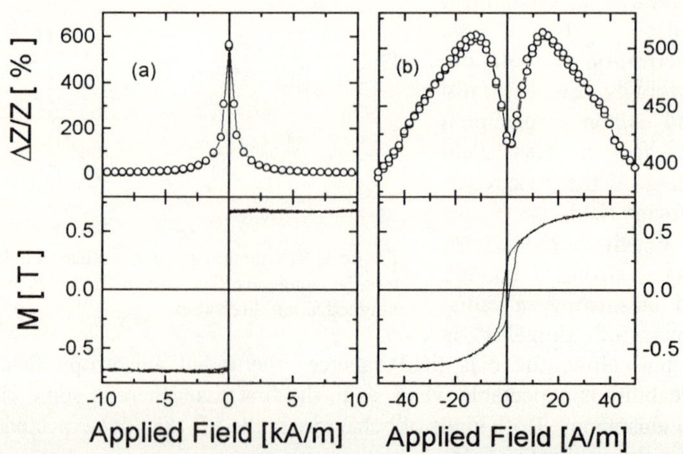

Figure 2 : Magnetoimpedance and axial hysteresis loops of as cast $Co_{68.15}Fe_{4.35}Si_{12.55}B_{15}$ amorphous wires showing a Single Peak (a) and a Two Peak (b) behavior.

the TP behavior is ascribed to the increasing average magnetoelasic anisotropy in the region involved in the skin-effect penetration depth[22]. In Fig. 3b, a large modification of the MI aspects with the amplitude of the a.c. current is shown for a cold-drawn and subsequently stress-annealed Co rich amorphous wire[23]. In fact, the picture evolves from SP to TP when decreasing the amplitude. The TP behavior is actually a consequence of not reaching the circular coercivity. In the case of the sample of the Fig. 3b, the domain structure is purely circumferential. These modifications are discussed in a recent work as arising from the particular characteristics of the transverse permeability at the working conditions. In short, and as far as the appearance of SP and TP behaviors with frequency and and amplitude, the drift of the field for the maximum MI with frequency would indicate the increase of the anisotropy field ascribeable to the region containing the penetration depth. In turn, the amplitude of the current determines a critical value for which the coercivity is reached in the circular magnetization process under the biasing effect of the axial d.c. field[1].

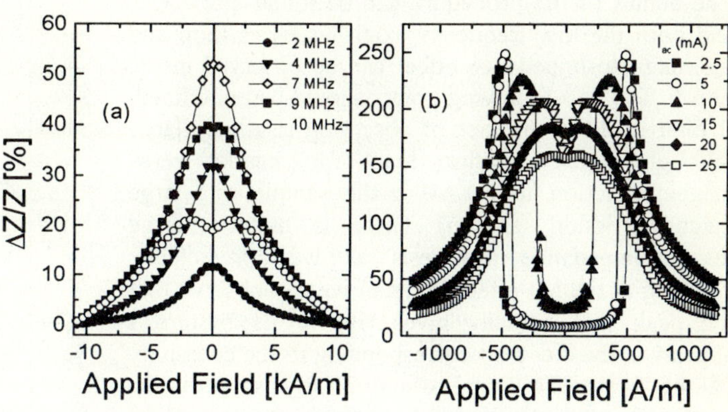

Figure 3 : (a) Frequency dependence of the magnetoimpedance measured in Carpenter HyMu80® alloy ($Ni_{80}Mo_{4.2}Fe_{bal}$ permalloy). The sample is obtained after a cold-drawn process is carried out in the as-cast wire. (b) Magnetoimpedance dependence with the amplitude of the a.c. current for a cold-drawn and subsequently stress-annealed Co rich amorphous wire.

Hysteretic Magneto-Impedance.-

The appearance of some hysteretic behavior in the magnetoimpedance effect has been reported in some previous works[24-28,5]. But no general or systematic study has been performed. Nevertheless, this is an undesirable effect for most pourposes especially in technological applications as for instance in sensor devices. Consequently, the study to determine under which conditions it appears seems very justified.

As mentioned above, there has been a few reports mentioning the GMI hysteresis. But which is its origin?. We know that the origin of the MI phenomenon is related to skin effect penetration depth and its modification with an applied dc field through the change of the circumferential permeability. But, as we have discussed in the preceeding section, that change in the permeability must follow the normal hysteretic behavior accompanying the magnetization processes. It means that an apparent hysteresis for the GMI should appear in a case when there is large hysteretic behavior of the magnetization process. This point is actually difficult to observe since precisely, the appearance of GMI requires very soft magnetic behavior and quite large values of permeability which in principle is in contradiction with a large hysteretic behavior.

An experiment was performed to observe large hysteretic GMI[29]. It was done in an Fe base amorphous wire with large and positive magnetostriction. In its as-cast state it exhibits two main regions with axial (core) and radial (sheath) easy magnetization directions so that, practically no MI effect is observed. A thermal treatment was performed in order to relaxe the radial anisotropy and increase the circumferential permeability[30]. Moreover, partial crystallization was even achieved. The details of the procedure can be found elsewhere, and both the low frequency axial hysteresis loop and the magneto-impedance effect can be observed in the Fig. 4. The two following points should be underlined: i) First of all is the fact of observing relatively large (altough much smaller than for nearly zero-magnetostriction alloys) MI in this sample with large magnetostriction, and ii) There is quite a large magnetoimpedance hysteresis as was predictable. Looking at the MI effect in detail we can observe that the peak that actually gives rise to its hysteresis appears at the d.c. field corresponding to the coercivity of the hysteresis loop, which clearly support the previous assumption about the correlation of an irreversibility field, H_{irr}.

That case is really a very peculiar one, but we wanted also to rediscover the hysteretic MI in more common magnetoimpedive materials but with differentiated magnetization processes. The first case we have analysed is that one of a sample with a well defined transverse anisotropy field but also exhibiting irreversibilities. In figure 5, the magnetoimpedance effect and the axial hysteresis loop are shown for a material with improved technological capabilities as a nearly non-magnetostrictive Co base glass-coated amorphous microwire[31,13]. As observed, there is a clear transverse anisotropy of magnetoelastic origin determined by the internal stresses frozen in during the fabrication process. Besides this characteristic, there is also an hysteretic component in the magnetization process. In this case, the MI shows the TP behavior but if analysed in detail, a small, but detectable and repetitive, hysteresis appears between the two peaks. This hysteresis can be certainly adscribed to the hysteresis of the loop, and here, the irreversibility field, H_{irrv}, roughly equals the transverse anisotropy field, H_k.

The next step has been to find a material with well defined transverse anisotropy exhibiting also hysteresis but with H_k well separated from H_{irr}. This is the case of the Fe base amorphous alloy (now with ribbon shape) with particular composition so that a fine nanocrystalline microstructure is stable. The particular treatment was to devitrify the amorphous alloy under an applied tensile stress. More details about how this hysteresis loop and the particular domain structure are determined are given elsewhere. The results are plotted in the Fig.6 where both the axial hysteresis loop and the hysteretic magnetoimpedance behavior are given

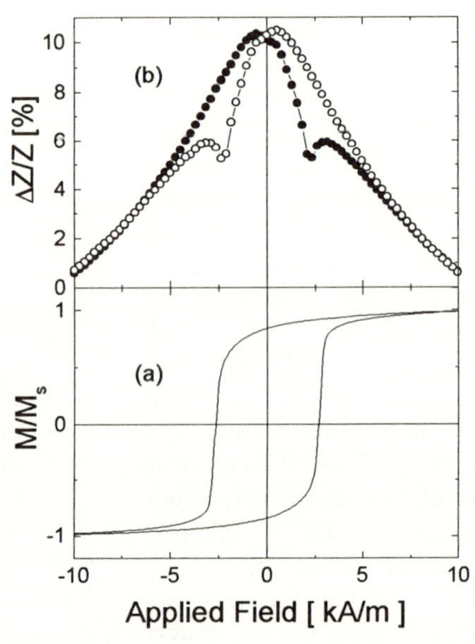

Figure 4: Magnetoimpedance and hysteresis loops for a $Fe_{77.5}Si_{7.5}B_{15}$ current annealed wire.

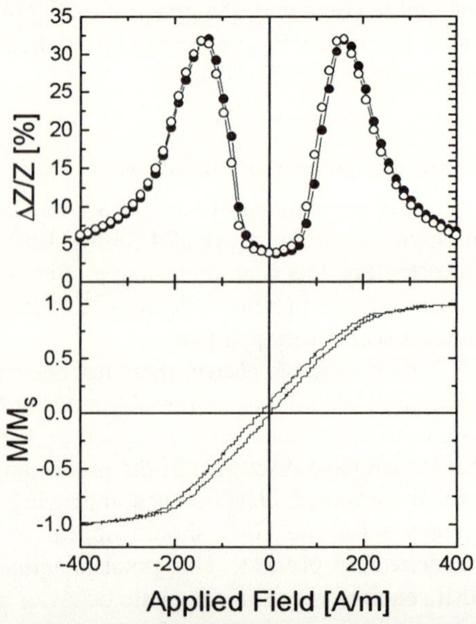

Figure 5: Magnetoimpedance and hysteresis loop measured in glass covered $Co_{68.5}Mn_{6.5}Si_{10}B_{15}$ microwire. A slightly hysteretical behavior in the MI curve can be observed.

and compared[32]. A correlation between irreversibility and anisotropy fields clearly followed.

A novel typed of materials, electroplated magnetic tubes, has been recently reported to show MI effect[33]. Additonal experimental work on these materials is being extensively followed by us in a number of novel materials as: Co base (Fig. 7)[34] and NiFe (Fig.8)[35] soft tubes electrodeposited on cupper wire. Fig. 7 corresponds to an as obtained sample, and from the combined evaluation of hysteresis loop and observed domain pattern, a radial magnetic anisotropy is deduced. Looking at the MI, a hysteresis is observed with roughly $H_{irrv} = H_k$. This behavior somehow remembers us to that one shown in Fig. 4 for the treated Fe base wire. The presence in both cases of a radial component of anisotropy could be associated to the large observed MI hysteresis. In the case of Fig. 8, the sample was submmitted to an axial-field annealing

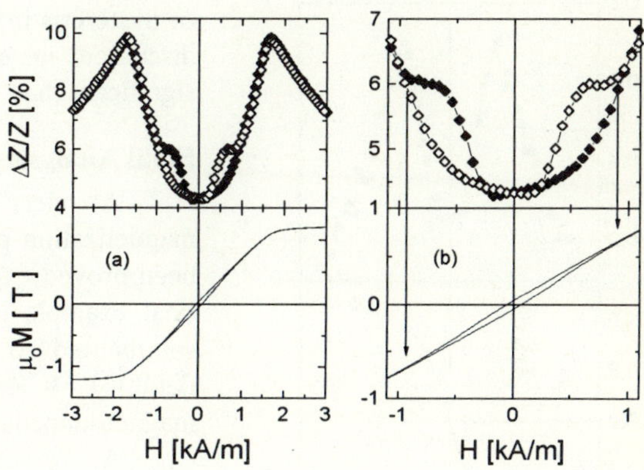

Figure 6 : Magnetoimpedance and hysteresis loops in stress annealed $Fe_{73.5}Cu_1Nb_3Si_{13.5}B_9$ ribbons.

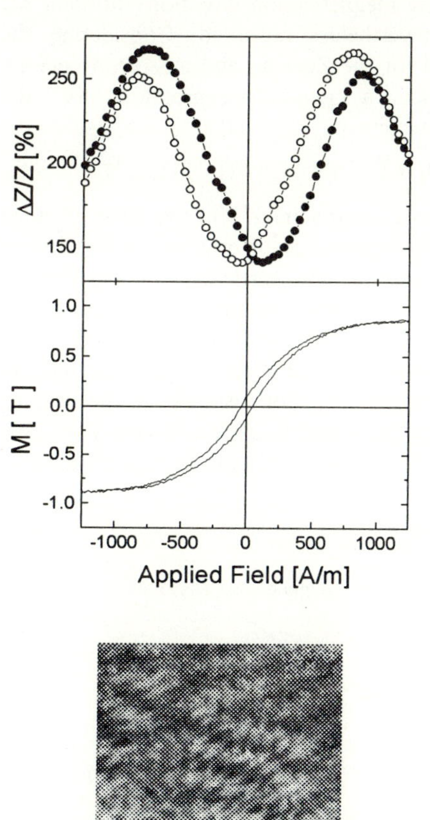

Figure 7: Hysteresis loops, magnetoimpedance and MFM microscopy surface domain structure image obtained for a Cu wire with a 9μm CoP.

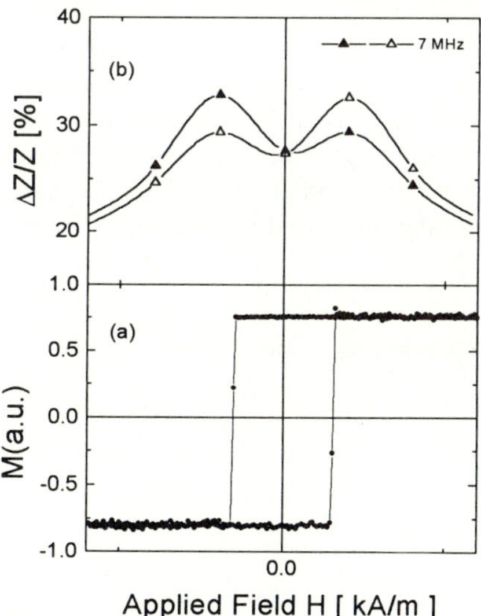

Figure 8: (a)Hysteresis Loops and (b) magnetoimpedance in Cu wires with an eletrodeposited layer of NiFe.

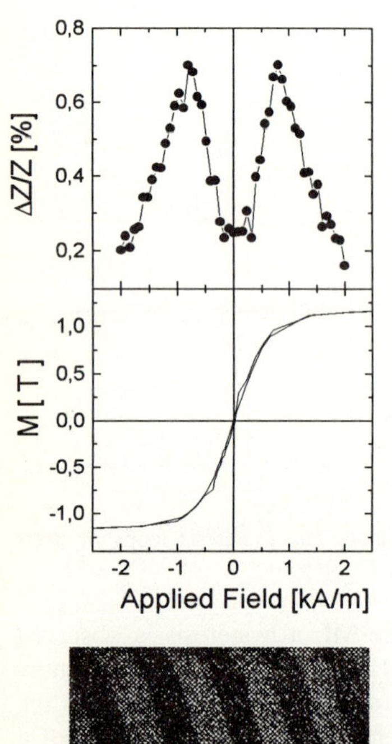

Figure 9 : Hysteresis loops, magnetoimpedance and Kerr effect surface domain structre obtained for a CoFeB sputtered thin film.

so inducing an axial magnetic anisotropy as can be deduced from the axial hysteresis loop. It is remarkable the appearance of hysteresis in the MI also for this type if anisotropy as was discussed in section 2. This hysteresis is anyway less significant than that shown in Fig. 7.

Final Analysis and Discussion

A clear correlation between domain structure, magnetization process and magneto-impedance behavior has been proved experimentally along the preceeding sections. A final example is given in Fig. 9 for a sputtered thin film submmitted to a thermal treatment under a.c. transverse field[36]. It exhibits a well defined anisotropy and TP magnetoimpedance behavior.

A comparison between domains structure, magnetization process and MI has been reported previously by several authors[18,25,37,38]. Depending on the orientation of the easy magnetization direction, different MI characteristics are nevertheless observed. Considering the equilibrium condition for the Zeeman and anisotropy energy terms, we introduce a general expression for the circumferential permeability as:

$$\mu = \alpha M_s \left((H - \alpha H_k)^{-1} + (H - \beta H_{irrv})^{-1} \right) \quad (4)$$

where α is a parameter approaching 1 in the case of small amplitudes for the transverse anisotropy and a.c. current, while β takes into account the particular pinning centers structure. According to this general expression local maximum values are expectable when the applied field reaches either H_k or H_{irrv}. Nevertheless an additional complexity must be taken into account arising from the non-homogeneous distribution of magnetic anisotropies[13]. For instance, in the case of amorphous wires, a radial distribution for the transverse permeability is expected (see Fig. 10a) coming from the intrinsecal distribution of internal stresses. Such distribution can be smoothed (Fig.

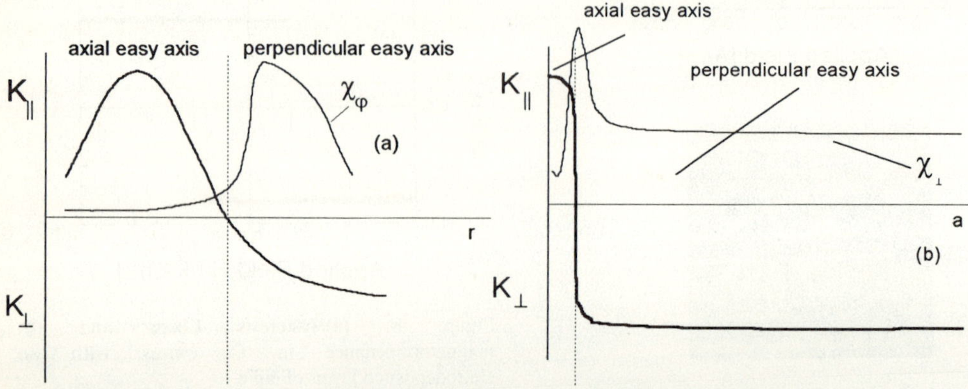

Figure 10: Schematic view of the magnetic anisotropy constant K profile and trasnversal susceptibility in (a) amorphous as cast wires and (b) nanocrystalline annealed ribbons with transversal anisotropy.

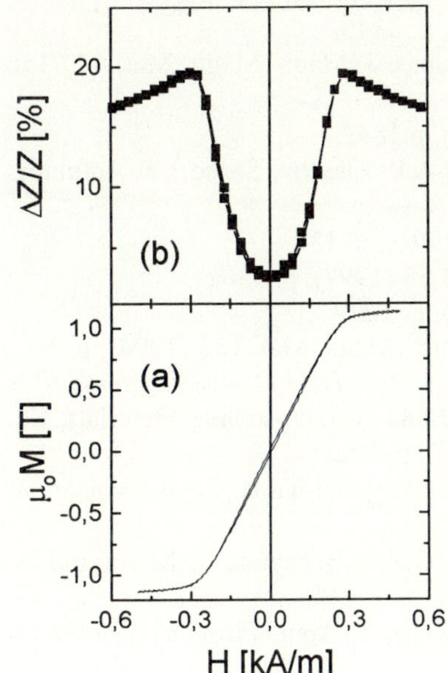

Figure 11 : (a) Hysteresis and (b) magnetoimpedance measured in a field annealed nanocrystalline ribbon.

10b) in the cases when the samples are submmitted to adequate treatments and in particular to magnetic field annealing. This is shown in Fig. 11 for a ribbon which has been submmitted to a transverse field annealing while devitrifying from amorphous structure into a nanocrystalline structure embedded in a residual amorphous matrix[39].

Conclusions.

As a general conclusion, we present a systematic study in different samples showing the correlation between magnetization process, domain structure and magnetoimpedance ratio.

We have also established and characterised the existence of SP and TP behaviors for any sample which can be observed depending on the frequency and amplitude of the a.c. current flowing along the samples. The observation of hysteresis in the magnetoimpedance is on the other hand clearly correlated to the irreversible magnetization processes.

As far as the use of this MI phenomenon in samples for sensing elements in sensor devices is concerned, a SP behavior with vanishing hysteresis is in general desirable. In order to have this particular behavior, we can finally conclude that the easy magnetization direction should be essentially parallel to the transverse field created by the a.c. current, and that the strength of such a transverse anisotropy must be as small and homogeneous as possible.

Acknowledgments

The authors are deeply endebted to many researchers at the Instituto de Magnetismo Aplicado for graciously supplying experimental material presented in this repport. In particular thanks are given to highly qualified Ph.D. students D. Garcia, J.L. Muñoz, J.M. García-Beneytez, J.J. Freijo and J.M. Garcia, to Dr. A.Asenjo and Dr. A. Mendez (ABB, Raleigh, U.S.A), to Dr. A.Garcia Arribas(Universidad de Oviedo) and to Dr. D.-X. Chen and Prof. A. Hernando. Dr. J.P. Sinnecker wishes to thank the Brasilian agency FAPESP and Dr. G. Kurlyandskaya to the Spanish Ministery of Education for finantial support during their stays at the IMA. The work has been supported by the Spanish CICYT under project MAT/95-0273.

References

[1] J. Velázquez, M. Vázquez, D.-X. Chen and A. Hernando, Phys. Rev. B 50 (1994), p. 16737.

[2] K. Mohri, T. Kohzawa, H. Yoshida and L.V. Panina, IEEE Trans. Magn. 28 (1992), p. 3150.

[3] L.D. Landau, E.M. E.M. Lifshift and L.P.Pitaevskii, *Electrodynamics of Continuous Media* (Butterworth-Heinemann, Washington DC, 1995), p. 210.

[4] P. Ciureanu, M. Britel,, D. Menard, C. Akyel, A. Yelon, M. Rouabhi, R.W. Cochrane, P. Rudkowski and J.O. Strom-Olsen, J. Appl. Phys. (in press).

[5] L.V. Panina, K. Mohri, T. Uchiyama and M. Noda, IEEE Trans. Magn. 31 (1995), p. 1249.

[6] R.S. Beach and A.E. Berkowitz, Appl. Phys. Letters, 64 (1994), p. 3652.

[7] K.V. Rao, F.B. Humphrey and J.L. Costa-Kramër, J. Appl. Phys. 76 (1994), p. 6204.

[8] M. Knobel, M.L. Sánchez, J. Velázquez and M. Vázquez, J. Phys.: Condens. Matter. 7 (1995), p. L-115.

[9] F.L.A. Machado, C.S. Martins and S.M. Rezende, Phys. Rev. B51 (1995), p. 3926.
[10] R.L. Sommer and C.L. Chien, Appl. Phys. Letter 67 (1995), p. 3346.
[11] M. Tejedor, B. hernando, M.L. Sánchez and A. García-Arribas. J. Magn. Magn. Mat. 157/158 (1996), p. 141.
[12] W. Ku, F. Ge and J. Zhu, J. Phys. D : Appl. Phys. 30 (1997), p.1842.
[13] M. Vázquez, M. Knobel, M.L. Sánchez, R. Valenzuela and A.P. Zhukov, Sensors & Actuators A 59 (1997), p. 20.
[14] L.V. Panina and K. Mohri, J. Magn. Magn. Mat. 157/158 (1996), p. 137.
[15] M. Knobel, M.L. Sartorelli and J.P. Sinnecker, Phys. Rev. B 55 (1997), p. 3362.
[16] M. Vázquez and D.-X. Chen, IEEE Trans. Magn. 31 (1995), p. 1229.
[17] P.T. Squire, D. Atkinson, M.R.J. Gibbs and S. Atalay, J. Magn. Magn. Mat. 132 (1994), p. 10.
[18] D.-X. Chen, J.L. Muñoz, M. Vázquez and A. Hernando, Phys. Rev. B 57 (1998) (in press).
[19] M.L. Sánchez, M. Knobel and D.-X. Chen, Nanostructured and Non-crystalline Materials, Ed. M. Vázquez and A. Hernando (World Scientific, Singapore, 1995), p. 592.
[20] V.E. Makhotin, B.P. Shurukhin, V.A. Lopatin, P.Yu. Marchukov and Yu.K. Levin. Sensors & Actuators A, 25-27 (1991), p. 759.
[21] M. Vázquez, A.P. Zhukov, P. Aragoneses, J. Arcas, J.M. Garcia-Beneytez, P. Marin and A. Hernando. IEEE Trans. Magn. 34 (1998) (in press).
[22] M. Vázquez, J.M. García-Beneytez, J.P. Sinnecker and L. Li, J. Appl. Phys. 83 (1998) (in press).
[23] J.J. Freijo, V.R. Ramanan, M. Vázquez and A. Mendez, (to be published).
[24] R.L. Sommer and C.L. Chien, J. Appl. Phys. 79 (1996), p. 5139.
[25] D. Menard, D. Frankland, P. Ciureanu, A. Yelon, M. Rouabhi, R.W. Cochrane, H. Chiriac and T.A. Ovari, J. Appl. Phys. (in press)
[26] R.L. Sommer and C.L. Chien, Appl. Phys. Letter 67 (1995), p. 857.
[27] K.C. Mendes, F.L.A. Machado, L.G. Pereira,S.M. Rezende, F.C. Montenegro, M.V. Altoé and F.P. Missell, J. Appl. Phys. 79 (1996), p. 6555.
[28] F.L.A. machado, B. Lopes da Silva and E. Monterroyos, J. Appl. Phys. 73 (1993), p. 6387.
[29] J.P. Sinecker, P. Tiberto, G.V. Kurdlyandskaya, E.H.C. Sinnecker, M. Vázquez and A. Hernando. J. Appl. Phys. (to be published).
[30] Y. Takemura, H. Tokuda, K. Komatsu, S. Masuda, T. Yamada, K. Kakuno and K. Saito. IEEE Trans. Magn. 32 (1996), p. 4947.
[31] M. Vázquez and A.P. Zhukov, J. Magn. Magn. Mat. 160 (1996), p. 223.
[32] M. Vázquez, G.V. Kurlyandskaya, J.L. Muñoz, A. Hernando, N.V. Dimitrieva, V.A. Lukshina and A.P. Potapov. J. Magn. Magn. Mat. (in press).
[33] R.S. Beach, N. Smith, C.L. Platt, F. Jeffers and A.E. Berkowitz. Appl. Phys. Letter 68 (1996), p. 2753.
[34] J.P. Sinnecker, A. García-Arribas, J.M. García, A. Asenjo and M. Vázquez (to be published).
[35] D. Garcia, G.Kurlyandskaya, M. Vázquez, F.I. Toth and L.K. Varga, (to be published).
[36] D. Garcia, J.L. Muñoz, G. Kurlyandskaya, M. Vázquez, M. Ali and M.R.J. Gibbs, IEEE Trans. Magn. (in press).
[37] A.D. Santos,L.G.C. Melo, C.S. Martins, F.P. Missell, Y. Souche, F.L. Machado and S.M. Rezende, J. Magn. Magn. Mat.
[38] D. Atkinson and P.T. Squire, J. Appl. Phys. (in press).
[39] G.V. Kurlyandskaya, J.M. García-Beneytez, M. Vázquez, J.P. Sinnecker, V.A. Lukshina and A.P. Potapov, J. Appl. Phys. (in press).

Materials Science Forum Vols. 302-303 (1999) pp. 219-223

Domain Wall Oscillations in GMI Configuration

L.G.C. Melo and A.D. Santos

Instituto de Física, Universidade de São Paulo, CP 66318, 05389-970 São Paulo, SP, Brazil

Keywords: Domain Wall, Amorphous Ribbons, Magneto-Impedance, Magneto-Optics

Abstract

In this work we characterize the domain wall movements in $Co_{70.4}Fe_{4.6}Si_{15}B_{10}$ amorphous ribbons. Transversely field annealed samples were studied by MOKE methods. We used a TMOKE magnetometer to get hysteresis loops and MOKE microscopy to obtain the domain structure. In both cases, an alternative electric current was applied through the long axis of the samples, in the frequency range from 10 Hz to 1 MHz. A strong amplitude variation in the domain wall displacement was observed in function of the current frequency. Also the hysteresis loops shows a reduction on the coercive field to frequencies going from DC to 10 kHz, and an increasing value to higher frequencies. These impressive experimental results are analyzed in terms of the light and electric current penetration depth (skin effect), and of the damping on domain wall movements.

Introduction

The giant magneto-impedance (GMI) effect in high permeability, low magnetostriction amorphous ribbons [1] and wires [2] of composition $Co_{70.4}Fe_{4.6}Si_{15}B_{10}$ has attracted some attention due to its both scientific and technological character. The magneto-impedance is more typically obtained when a DC magnetic field, Hdc, is applied parallel to an AC electric current, Iac, and both are perpendicular to the in-plane easy magnetization axis. For very soft magnetic materials, the measured variation of impedance may be understood in terms of its proportionality to the transverse permeability, μ_t, through the penetration depth, $\delta=c(2\pi\omega\sigma\mu_t)^{-1/2}$, where $\omega=2\pi f$ is the frequency of the current and σ the conductivity of the material. At low frequencies, $f< 10kHz$, μ_t can be described in terms of two contributions: the domain wall (χ_{dw}) and the magnetization rotation (χ_{rot}) susceptibilities [3]. The rotational mechanism responsible for the GMI is considerate as dominant at frequencies of the driving current above 1 MHz. The phenomenological model of Atkinson and Squire[4] accounts quite good for this behavior of μ_t for intermediate values of frequency, in terms of a domain wall stiffness parameter, β. For the very high frequency cases, it must be take account of the dynamical effects of magnetization related to the atomic magnetic moment of the material.

In order to get fundamental information to understand the GMI effect, in this work we characterize the domain wall (DW) movements. We studied transversely field annealed amorphous ribbons of $Co_{70.4}Fe_{4.6}Si_{15}B_{10}$. The samples used here, present high permeability, low coercivity and a very high DW mobility. This last characteristic of our samples is decisive to be possible the domain observations in high frequency. Normally the DW study is done statically or in low frequency [5,6] range, what means our results represent a quite unusual utilization of the MOKE domain observation. During the observation, the sample was submitted to an electric current in the frequency range from 10 Hz to 1 MHz. In another special utilization of the MOKE, we measure the TMOKE hysteresis loops to the sample submitted to the same AC electric current. All these experimental results will be analyzed in terms of the light and electric current penetration depth (skin effect), and of the damping on DW movements.

Experimental Details

In this experiment we used $Co_{70.4}Fe_{4.6}Si_{15}B_{10}$ amorphous ribbons of about 1 mm wide and 0.03 mm thick, cut into pieces of 20 mm in length. The samples were field annealed in a vacuum furnace (10^{-5} mbar), at 300°C, during 20 min. in an in-plane magnetic field of 0.5 T, perpendicular to the ribbon axis. The induced transverse anisotropy produced a regular transverse stripe domain structurc, as was already reported by us [5] and also previously by others authors [6]. For magneto-optical (MO) measurements, the ribbons were mechanically polished using 1/4 μm diamond paste.

The domain images were achieved by MOKE, using a Carl Zeiss Jenapol optical microscope coupled to a 10^{-3} lux Hamamatsu vidicon video camera. The video signal is processed by a DT2867 Data Translation image processing board working in a PC microcomputer. For all DW images presented here we used the same sample position. It was submitted to a 10mA square wave current from an HP3325b function generator. This electric current generates a magnetic field transversal to the ribbon axis. The measurements were done to the current in the range from 10 Hz to 1 MHz. We obtained the images by averaging 64 frames (~2.1 s) of the video signal, after the subtraction in real time of the magnetically saturated background. To an static condition (without the AC current), we have two main gray levels, corresponding to both sense of the magnetization in the domain structure. But, in the case of an applied AC current, there are the oscillations of the DW. However, due to the persistence of the tube camera we can not detect this movement of the DW, but only to get a time averaging of the domain structure. Then, we have a third intermediary averaged gray level, corresponding to the region of DW displacement. Beside of this persistence effect we have also the ~2s of frame averaging, leaving to the impossibility of detecting any instantaneous picture of the moving domain structure, to all the frequencies used in this work. As a rule to the pictures in this work, the ribbon axis and the current propagation direction are always the horizontal one.

We measured the hysteresis loops in a transverse magneto-optical Kerr effect (TMOKE) magnetometer. In this system a *p*-polarized incident light wave from a modulated (f= 1 kHz) diode laser (λ=670nm) is reflected by the sample, detected by a photodiode and analyzed by a lock-in amplifier. This is not a classical hysteresis loop measurement, because similarly to the DW observation, the sample is excited by a sinusoidal electric current passing in its longitudinal direction. The current effect on the hysteresis loops will be shown and discussed in the next section.

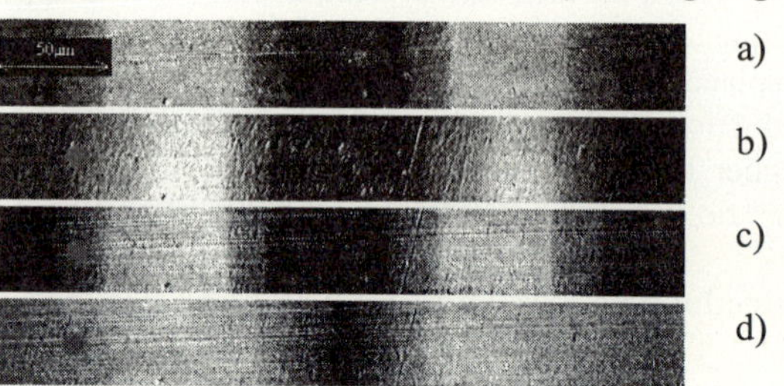

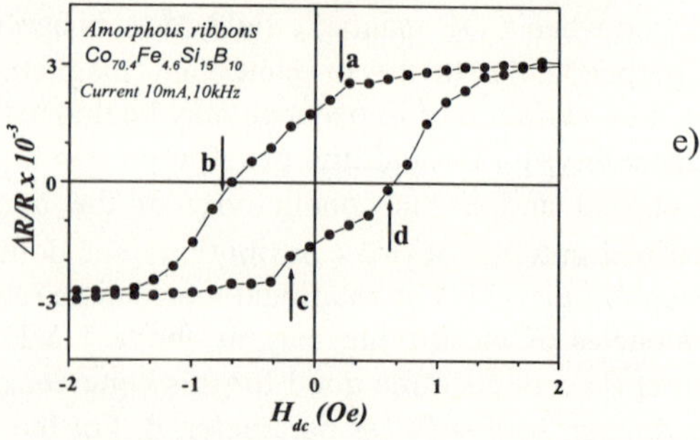

Fig.1. Domain observations and hysteresis loop to the sample submitted to an AC current of 10mA and 10kHz. The dc magnetic field and the direction of current are horizontal in the images. The figures a), b), c), and d) relates to the points marked in the figure e).

Results and Discussion

Using the typical experimental arrangement of the GMI measurements, we analyze the electric current effects on the DW structure. In the Fig. 1, we present some images of the domain structure when the ribbon is submitted to both the AC current and the longitudinal magnetic field, Hdc. We applied a (10mA, 10kHz) square wave current. Starting from the positive saturation magnetic field, the domain structure arrives at Hdc ~ 0.20 Oe. This domain structure can be seen in the Fig. 1a. There, the intermediate gray level represents the region of the displacement of the DW, as explained before. When Hdc increases negatively to –0.55 Oe, Fig. 1b, the limits of the displacement of the DW becomes less definite, and with a little more negative value of Hdc the structure vanishes. After, we repeated these measurements departing from the negatively saturated condition. At Hdc ~ –0.20 Oe the transverse domain structure was obtained again, Fig. 1c. For a Hdc = 0.60, quite near to the Hc, almost all of the domains disappeared, Fig. 1d. This indicates the beginning of the range of exclusively magnetization rotation process. So, the magnetic field range to obtain a domain structure is quite small. In the Fig. 1e, we present the corresponding MOKE hysteresis loops, where the arrows indicate the magnetic field limits to the DW observation.

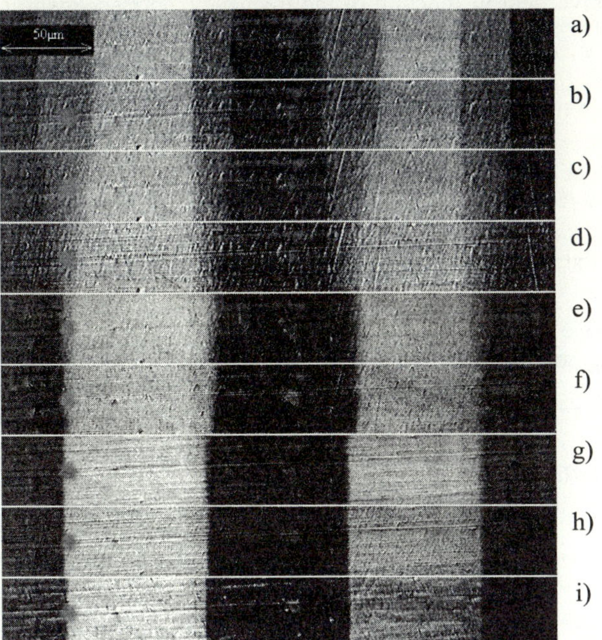

Fig.2. Domain structure to sample submitted to a longitudinal square wave current of amplitude 10mA. The intermediate gray level corresponds to the oscilation region of the domain walls. a) 10 Hz, b) 1 kHz, c) 10 kHz, d) 25 kHz, e) 50 kHz, f) 100 kHz, g) 200 kHz, h) 500 kHz and i) 1 MHz.

Now, we will analyze the current frequency depedence on the DW displacements. We made a series of measurements, presented in the Fig. 2. There, we have fixed the current to 10 mA and variated the frequency from 10 Hz to 1 MHz. The magnetic field was put to zero (Hdc= 0), to keep it in the condition to have domain walls. At low frequencies we can clearly distinguish the region corresponding to the displacements of domain walls. It is characterized by the intermediate gray level, Fig. 2a-d. In the Fig. 2a and b, respectively to 10 Hz and 1 kHz, we see the same amplitude of the DW displacement. But, to higher frequencies, this displacement reduces monotonically. This is seen in the picture at 10 kHz, Fig. 2c, where the length of the DW displacement reduced to ~80% of that in Fig. 2b. As can be seen in the Fig. 2c-i, this DW displacement goes to about zero at the frequency of 1 MHz. At this frequency the resulting domain structure became the same as in the not applied current remanent state. At the same time that the displacement shrinks, it becomes also blurrier. We have three hypothesis to explain these high frequency effects. The first is the so called skin effect, that corresponds to a reduction of the current (and also the magnetic field) penetration depth, δ , in function of the frequency. This is a consequence of the very high permeability and relatively low resistivity of our sample. The second is the damping of the DW movement, due to the induced eddy-currents in its region. Finally, we have the possibility of some experimental limitation in the video and sampling system. The video system works at 30 frames/s, while the current frequency arrives to 1 MHz.

Trying to resolve this question, we decided to investigate the influence of the oscillating current upon the magnetization process through its hysteresis loops, obtained by TMOKE magnetometry. The magnetic field Hdc was applied parallel to the longitudinal axis of the sample and perpendicular to the domain structure. The Hdc sweep frequency was 10^{-2}Hz. We employed for the current the same setup as that utilized for the images, but using a sinusoidal current. The Fig. 3a shows the hysteresis loops without any current. The coercive field, Hc_o, is 1.25 Oe. The loop is marked by a range of irreversible magnetization process, characterized by its coercivity and related to the displacement of the domain walls; followed by a range of magnetization rotation closed to its saturation. In Fig. 3b the sample was submitted to a current of 10mA, 10Hz. In this case, the Hc reduced to ~50% of Hc_o, and continues to decrease, just to 40%, at the frequency of 1kHz (Fig. 3d). According to the Fig. 2a and b, the wall movements are undamped at this frequencies. That means, the domain walls follow quite good the current-generated alternative magnetic field, Hac. By the other side, the coercivity depends mainly on the energy profile of the pinning regions, and the domain walls are passing through the pinning regions of the material more easily. Then, we can considerate that the reduction of the coercive field in function of the frequency, corresponds to an

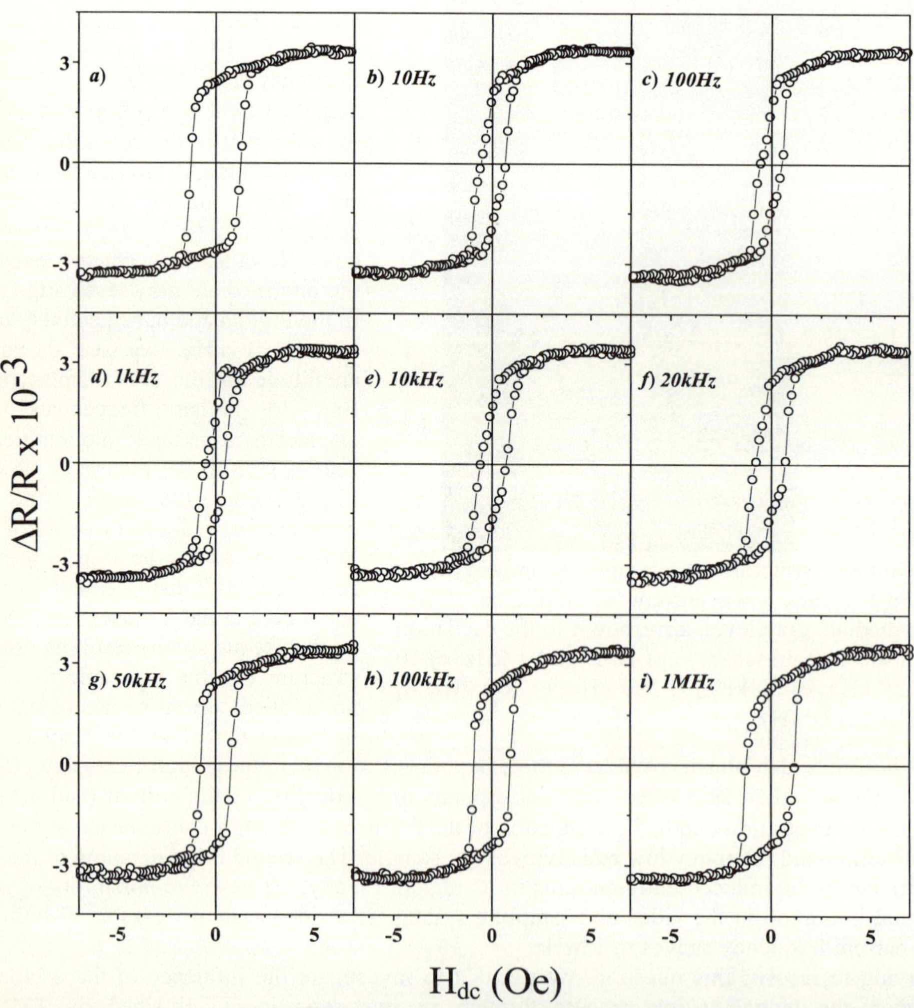

Fig. 3: TMOKE Hysteresis loops obtained passing a longitudinal sinusoidal electric current of 10 mA, through the long axis of the ribbon.

effective reduction of the potencial barrier of the pinning regions. Meaning that, as high the frequency is, as far of the quasi-equilibrium condition the DW is and less effective the pinning regions are. However, from 10kHz to 1 MHz, Fig. 3e–i, we can see the Hc growing gradualy with frequency. In Fig. 3i, at 1MHz, the original hysteresis loops of Fig. 3a is practically recuperated. This consistency with the results of Fig. 2 show us that this is an intrinsic effect and eliminates the possibility of any equipment limitation. This high frequency deleterious effect in Fig. 2 and 3 are consistent with both hypothesis, of the skin effect and the damping of the DW movement. Using in the expression to δ, the frequency and the permeability to the light beam and the applied electric current, we can show that the penetration depth to the light beam is at least 30 times lower than the current penetration depth. Meaning that by MOKE measurements, we are always getting information from the superficial region (hundreds of Å) of the sample, where certainly there is magnetic field enough to oscillate the domain walls. Concluding, the stable domain structure in the Fig. 2i and the hysteresis loop in the Fig. 3i, showing the coercivity of the same value of Hc_0, is a clear consequence of the damping of the DW movements, and does not correspond to the skin effect.

Conclusions

At relatively small applied magnetic field, from Kerr-effect images presented in Fig. 2, in conjunction with hysteresis loops magnetometry in longitudinal GMI configuration in Fig. 3, we completely characterized the range of frequencies, that the damping of DW movement becomes important. This range starts at ~10 kHz, and for 1 MHz there is practically no oscillations of the domain walls. At low frequencies, the DW oscillations due to an AC driving current acts upon the coercivity of the material, reducing it by an effective reduction of the pinning potential barrier. These results can be useful to better understanding the GMI phenomenon at frequencies below 1MHz, for samples presenting high DW mobility.

This work was supported by the Brazilian agencies FAPESP and CNPq.

References

[1] F.L.A. Machado, B.L.da Silva, S.M. Rezende and C.S. Martins, J.Appl.Phys., 75(1994), p.6563.
[2] R.S. Beach and A.E. Berkowitz, Appl.Phys.Lett., 64(1994), p.3652.
[3] L.V. Panina, K. Mohri, T. Uchiyama, M. Noda and K. Bushida, IEEE Trans.Magn., MAG-31(1995), p.1249.
[4] D. Atkinson and P.T. Squire, 7th Joint MMM-Intermag Conf., San Francisco, January 6-9, 1998, paper CQ-03.
[5] A.D. Santos, L.G.C. Melo, C.S. Martins, F.P. Missell, Y. Souche, F.L.A. Machado and S.M. Rezende, J. Appl. Phys., 79(1996), p.6546.
[6] J.D. Livingston, W.G. Morris and T. Jagielinski, J. Appl. Phys. 55(1984), p.1790.

Materials Science Forum Vols. 302-303 (1999) pp. 224-228

Creep-Induced Magnetic Anisotropy of Amorphous Ribbons Studied by Magneto-Impedance and Magnetic Resonance

L. Kraus[1,3], A.N. Medina[1], F.G. Gandra[1], M. Knobel[1] and V. Haslar[2]

[1] Instituto de Física "Gleb Wataghin", Universidade Estadual de Campinas-UNICAMP, C.P. 6165, 13083-970 Campinas S.P., Brazil

[2] Institute of Physics, Academy of Sciences of the Czech Republic, Na Slovance 2, CZ-18040 Praha 8, Czech Republic

[3] on leave from [2]

Keywords: Giant Magneto-Impedance, Amorphous Ribbons, Magnetic Properties

Abstract Longitudinal (LMI) and transversal (TMI) magneto-impedance effects and the ferromagnetic resonance (FMR) were investigated in amorphous $Co_{67}Fe_4Cr_7Si_8B_{14}$ ribbons with hard-ribbon-axis anisotropy induced by stress-annealing. The magnitude of effective anisotropy field H_K was controlled by the stress applied during annealing (0 - 800 MPa). The LMI exhibits peaks centered close to $\pm H_k$, whereas the TMI shows a broad plateau around $H = 0$. The in-plane FMR measured on small disks spark-cut from the ribbons exhibit the usual resonance peak at high fields and a more complicated absorption spectrum at low fields, where the samples are not saturated. Generally the low field spectrum is composed of two peaks, the intensity of which depends on the orientation of magnetic field with respect to the ribbon axis. The experimental results are compared with numerical calculations based on the electrodynamic theory of GMI. The behavior of magnetoimpedance and FMR can be explained by fluctuations of local easy axis directions.

I. Introduction

The effect of giant magneto impedance (GMI) is being attracting much attention because of its prospective practical applications. The GMI is based on skin-effect, the phenomenon well known from the classical electrodynamics. It has been shown, both experimentally [1-3] and theoretically [2,3], that the highest magnitudes of GMI can be obtained in materials with weak magnetic anisotropy perpendicular to the direction of driving current. In Co-rich amorphous alloys, with small magnetostriction constant, such anisotropy can be easily induced by stress-annealing [4]. The aim of this work was to investigate the influence of creep-induced magnetic anisotropy in a Co-rich amorphous ribbon on GMI effect and ferromagnetic resonance (FMR). The experimental results are compared with the predictions of a theoretical model based on a simultaneous solution of Maxwell equations and the Landau-Lifshitz equation of motion.

II. Experimental procedure and results

Amorphous $Co_{67}Fe_4Cr_7Si_8B_{14}$ ribbon, 10 mm wide, was prepared by the planar flow casting. Five pieces of the ribbon were stress-annealed for 1 hour at 380°C with different stress σ_a applied along the ribbon length. Samples 7 cm long, denoted here as A, B, C, D, E (corresponding to the stress σ_a = 0, 200, 400, 600, 800 MPa), were cut from the central parts of the annealed ribbon pieces. Magneto-impedance was measured for different frequencies (10kHz - 120kHz) with the

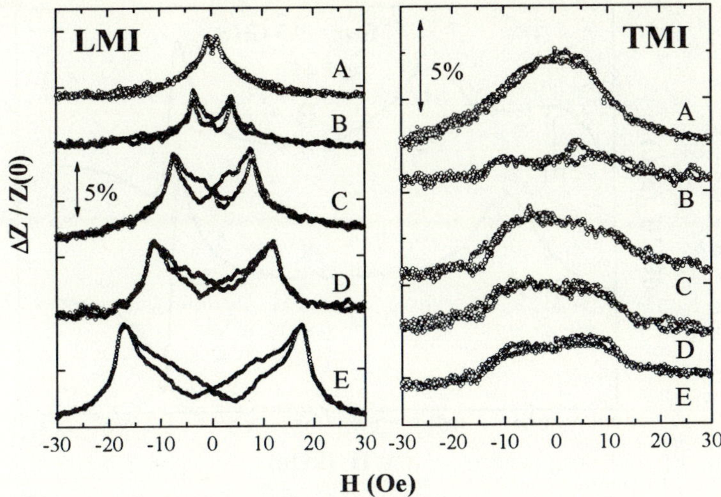

Fig. 1. Longitudinal (LMI) and transversal (TMI) magneto-impedance effect for the samples stress-annealed with different applied stress. (frequency: 100 kHz, current: 10mA).

in-plane magnetic field applied either longitudinally (LMI) or transversally (TMI) to the direction of driving current (identical to the ribbon axis). The ferromagnetic resonance experiments were carried out on 3 mm disk samples spark-cut from the same ribbons. For the measurements the Varian EPR spectrometer working at the frequency of 9.3 GHz was used. The disk was located at the center of rectangular TE_{102} cavity with both the microwave and the DC magnetic fields in the disk plane.

The results of the longitudinal (LMI) and transversal (TMI) magneto-impedance measurements at 100 kHz on the samples annealed with different σ_a are shown in Fig. 1. The LMI curves show two distinct peaks while the TMI exhibit a broad plateau around $H = 0$. As the stress σ_a increases the positions of LMI peaks shift and follow the anisotropy fields $\pm H_K$. The amplitude of LMI slightly increases with σ_a reaching the maximum value about 10%. The frequency dependence of the amplitude is approximately linear, in the investigated range 10 kHz - 120 kHz. The amplitude of TMI is less than 5% in all the samples. The central plateau on TMI curves broadens with increasing H_K but its height does not seem to be correlated with σ_a.

The typical resonance spectra for two different disk samples are shown in Fig. 2, where the derivative of absorption measured with the field H making 45^0 with the ribbon axis is plotted. Besides the normal FMR mode, observed at high magnetic fields, the spectra also show typical low-field structure consisting generally of two peaks denoted as P1 and P2. From the position of the FMR peak the saturation magnetization J_s and the anisotropy field H_K can be determined. Using the spectroscopic splitting factor $g = 2.1$ one obtains $J_s = 0.4996$ T, which is very close to the value 0.5 T determined by VSM. The angular dependence of resonance field is shown in Fig. 3. Because the FMR linewidth is relatively large the resolution of H_{res} is not very good. Only for the sample E, annealed with the highest stress, the $\cos 2\phi$ dependence (superimposed on a random function) can be distinguished. From this dependence the value $H_K \approx 14$ Oe was estimated, which is in good agreement with 17.9 Oe determined from the magnetization and LMI measurements.

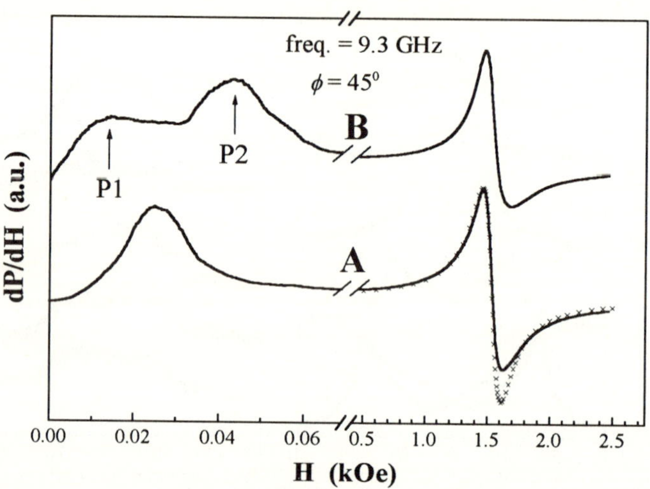

Fig. 2. Resonance spectra for samples A and B measured with in-plane field H making the angle ϕ= 45° with the ribbon axis. The crosses show the theoretical fit of resonance curve (J_s = 0.4996 T,.α = 0.03, A = 5 10^{-12} J/m).

The resonance linewidth can be used for determination of the damping parameters. Assuming the Gilbert type of damping and the exchange stiffness constant A = 5 10^{-12} J/m [5] one gets for the linewidth of 160 Oe, observed in the sample A, the Gilbert damping parameter α = 0.03. The theoretically calculated resonance curve with the above parameters A and α is plotted by the crosses in Fig. 2.

The low-field part of absorption spectra depends on the anisotropy field H_K and the orientation of magnetic field in the ribbon plane. In the sample A, annealed without stress, one peak at about H = 30 Oe is observed. Its position only slightly changes with the angle ϕ. In the stress-annealed samples the two peaks P1 and P2 are generally observed. The angular dependence of low field spectra for the sample E, with the highest H_K, is shown in Fig. 4. For H parallel to the hard magnetic axis only the first peak (P1) is observed below 40 Oe. With increasing angle ϕ the amplitude of P1 decreases. The second peak (P2) occurs above 40 Oe for H close to the easy axis direction. It should be noted that the position of the maximum of P2 well corresponds to the knee on the magnetization curve measured in the disk with magnetic field along the easy axis direction.

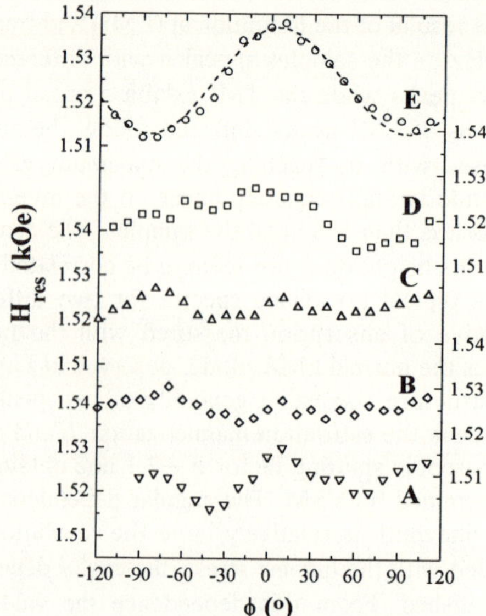

Fig. 3. The angular dependence of FMR resonance field. (ϕ is the angle between H and ribbon axis)

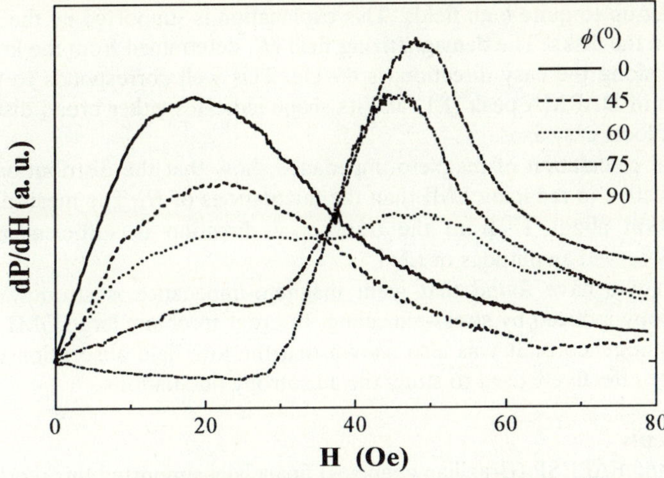

Fig. 4. The the low field absorption spectra of the sample E for different orientations of magnetic field in the ribbon plane.

IV. Discussion

The experimental results of both the magneto-impedance and FMR measurements can be discussed within the frame of electrodymanic theory of GMI for a planar film conductor [6]. If the easy axis of uniaxial magnetic anisotropy is perpendicular to the direction of driving current and the DC magnetic field the theory gives extremely high amplitudes of GMI. For example, if the exchange-conductivity effect is neglected the GMI magnitude is determined only by the ratio δ_0/δ_m, where δ_0 the nonmagnetic skin depth and δ_m the "magnetic relaxation depth" are given by the equations

$$\delta_0 = \sqrt{\frac{2\rho}{\mu_0 \omega}} \qquad \delta_m = \sqrt{\frac{\alpha \rho}{\gamma J_s}}, \tag{1}$$

where ρ is the resistivity, α the Gilbert damping parameter, ω the circular frequency and γ the gyromagnetic ratio. Using $\rho = 1.5$ μΩm, $J_s = 0.5$ T and $\alpha = 0.03$ the amplitude of GMI effect calculated for the frequency of 100 kHz is 1666%. If the exchange interaction, which reduces the GMI, is taken into account the theoretical value is 406 % but still nearly two orders higher than the experimental one. To explain the discrepancy between the theory and experiment the fluctuations of anisotropy field H_K and/or easy axis directions has been proposed [2,3].

The low field microwave absorption at frequencies higher than the natural resonance frequency (so called the DARMA peak) can be a powerful tool for investigation of anisotropy distribution [7]. Without fluctuations of anisotropy and with field applied along the hard axis, a sharp step on the derivative of absorption should be observed at $H = H_K$. The dispersion of local H_K or easy axis directions causes widening of the transition and shift of the DARMA peak to lower fields. The low field spectra observed in our disk samples, however, are more complicated than those reported for thin films [7]. This is probably because of relatively large demagnetizing field of disks. We attribute the first peak (P1) to the unresolved FMR mode in nonsaturated state, i.e. to DARMA. The appearance of the second peak (P2) for the field direction close to the easy axis can be related to some closure domains at the edges of disk, which may persist up to quite high fields. This explanation is supported by the hysteresis loop measurements on the disks. The demagnetizing field H_d, determined from the knee of the M-H curve measured along the easy direction, is 49 Oe. This well corresponds to the position of P2. The position of DARMA peak (P1) and its shape indicates either broad distribution of H_K and/or spread of local easy axes.

The numerical calculation of magneto-impedance show that the distribution of local easy axes is more effective in reducing LMI than the fluctuations of H_K The mean deviation of the easy axis direction about 1° from the transversal direction is sufficient to explain the experimentally observed amplitudes of LMI.

In conclusion, we have found that giant magneto-impedance is strongly dependent on magnetic anisotropy induced by stress-annealing. In order to obtain large GMI the anisotropy must be very homogeneous. It was also shown that the low field absorption of microwaves (DARMA) can be effectively used to study the anisotropy fluctuations.

Acknowledgements
CAPES, CNPq and FAPESP (Brazilian agencies) financially supported this work.

References

[1] K.V. Rao, F.B. Humphrey and J.L. Costa-Krämer, J. Appl. Phys. 76 (1994), p. 6204.
[2] L.V. Panina, K. Mohri, T. Uchiyama, M. Noda and K. Bushida, IEEE Trans. Magn. 31 (1995), p. 1249.
[3] D. Atkinson and P.T. Squire, IEEE Trans. Magn. 33 (1997), p. 3364.
[4] O.V. Nielsen, IEEE Trans. Magn. 21 (1985), p. 2008.
[5] Z. Frait and D. Fraitová and L. Pust, J. Magn. Magn. Mater. 54-57 (1986), p. 1307.
[6] L. Kraus, submitted to J. Magn. Magn. Mater.
[7] M.Rivoire and G.Suran, J. Appl. Phys. 78 (1995), p. 1899.

Materials Science Forum Vols. 302-303 (1999) pp. 229-233

Influence of Induced Anisotropy and Magnetostriction in the Magnetoimpedance and Its Aftereffect in CoFeSiB Amorphous Ribbons

K.R. Pirota[1], J. Schoenmaker[1], M.L. Sartorelli[1], M. Knobel[1], J. Gutierrez[2] and J.M. Barandiarán[2]

[1] Instituto de Física "Gleb Wataghin", Universidade Estadual de Campinas-UNICAMP, C.P. 6165, 13083-970 Campinas S.P., Brazil

[2] UPV/EHU, Depto. de Electricidad y Electronica, Apdo. 644, ES-48080 Bilbao, Spain

Keywords: Giant Magnetoimpedance, Amorphous Ribbons, Magnetic Properties

Abstract Large giant magnetoimpedance (GMI) ratios and negligible GMI relaxation are desirable behaviours to a good sample for magnetic sensor applications. Both effects were measured in as-quenched and stress-annealed amorphous ribbons with different magnetostriction values. The specific thermal treatment employed drastically reduces the GMI aftereffect of the sample, while it keeps large GMI response, therefore improving its perspectives in applications.

I. Introduction

Since 1994, when it was discovered, giant magnetoimpedance (GMI) has attracted great interest owing the perspectives for applications in technological devices. Namely, GMI is a huge variation of the impedance of a soft magnetic sample when an external magnetic field is applied on it. More recently, in 1997, a non-negligible aftereffect of the impedance was observed after a sudden change in the applied magnetic field, which can be an important drawback in possible applications of the GMI phenomenon [1,2]. The study of the effect of induced anisotropies and magnetostriction values on this relaxation effect can furnish some hints about its origin, and can indicate some ways to produce samples with vanishingly small aftereffect, but maintaining large GMI ratios.

The results shown in this paper indicate a strong effect of induced magnetic anisotropies in the GMI, because the GMI profile is strongly altered after a specific stress-annealing. Also, the total GMI amplitudes and sensitivities are modified by the thermal treatments. The GMI aftereffect is drastically reduced after the annealing, and therefore it is possible to produce samples with large GMI ratios and very low aftereffect, suitable for technological applications in magnetic sensors.

II. Experimental details

In this work we used the amorphous ribbons listed in Table 1, which were produced by melt-spinning. In the table, each as-quenched sample has a corresponding saturation magnetostriction value λ_s, which was measured by the SAMR method.

We studied the as-quenched ribbons as well as annealed samples, which were submitted to a pre-annealing of 1 hour at 360°C followed by one hour stress-annealing at 340°C (applied tensile stress of 400 MPa) [3]. This thermal treatment produces a shift on the λ_s values, as well as a magnetic anisotropy perpendicular to the ribbon axis, as can be seen in Ref. [3].

The 8 cm long strips were positioned perpendicular to the earth magnetic field in a Helmholtz coil system (magnetic field up to 100 Oe). An ac current (I<12 mA, f<1MHz) was generated by an EG&G 5302 lock-in amplifier which also reads the in-phase and quadrature components of the voltage induced in the sample. The flowing current was constantly monitored in a standard resistor connected in series with the sample, and kept to a fixed value by a GPIB controlled system [4]. The GMI was measured in all samples, with driving currents of 1, 5 and 8 mA, and frequencies of 100, 500 and 900 kHz

Table 1. Composition and saturation magnetostriction constant λ_s of the as-quenched samples studied in this work.

Sample	Nominal composition	λ_s ($\times 10^{-6}$)
a	$(Fe_{0.04}Co_{0.96})_{70}Si_{12}B_{18}$	-0.48
b	$(Fe_{0.045}Co_{0.955})_{70}Si_{12}B_{18}$	-0.35
c	$(Fe_{0.047}Co_{0.953})_{70}Si_{12}B_{18}$	-0.3
d	$(Fe_{0.048}Co_{0.952})_{70}Si_{12}B_{18}$	-0.28
e	$(Fe_{0.049}Co_{0.951})_{70}Si_{12}B_{18}$	-0.2
f	$(Fe_{0.05}Co_{0.95})_{70}Si_{12}B_{18}$	-0.16
g	$(Fe_{0.053}Co_{0.947})_{70}Si_{12}B_{18}$	-0.1
h	$(Fe_{0.057}Co_{0.943})_{70}Si_{12}B_{18}$	≈ 0

In GMI measurements, the driving current on the sample was kept constant while the in-phase and quadrature components of the impedance were monitored as a function of the applied field. A typical curve of GMI (defined here as GMI≡1-$Z(H)/Z(H=0)$), is shown in Figure 1 (sample *a*), for a drive current of 5 mA with frequency f=900 kHz. Note that the total variation goes beyond 30% after the saturation is almost reached.

In the GMI aftereffect measurements, a fixed magnetic field of 100 Oe is applied for 3 s, then it is switched off. The decay of the field occurs continuously, achieving zero in less then 5 ms. This time is defined as t=0, and therefore one can define the absolute impedance decay $\delta Z = |Z(t=0.25s)| - |Z(t=16s)|$. Also, the quantity $dvz = \delta z \cdot i$, which represents the absolute variation of the voltage, is studied as function of the frequency (100 and 900 kHz) and amplitude of the driving current I.

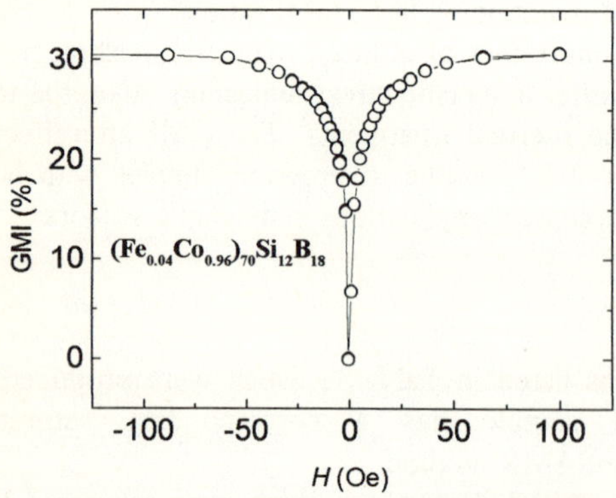

Figure 1. Typical Giant Magnetoimpedance curve.

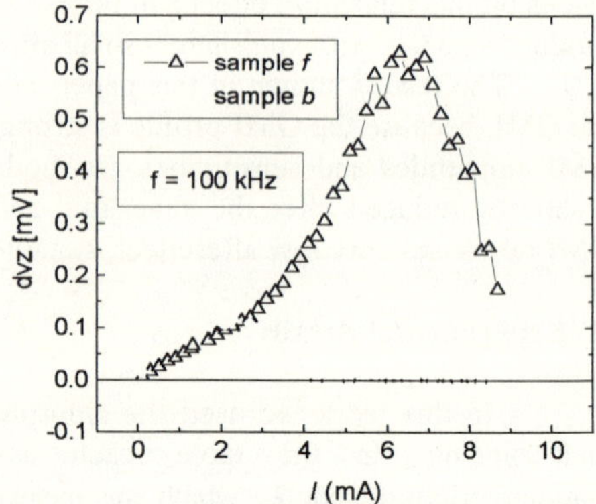

Figure 2. Current dependence of the absolute voltage decay *dvz*, between 0.25 s and 16 s after the aplication of a saturating magnetic pulse.

For example, Figure 2 shows a typical behaviour of $dvz = \delta z \cdot i$ as a function of the driving current I for the as-quenched samples (*b*) and (*f*). Here the frequency of the current I was kept constant at 100 kHz. All the *dvz vs I* curves in this frequency display a similar maximum, where the 180° domain-wall displacement amplitude is thought to correspond to approximately half of the domain-wall width [5,6].

III. Results and Discussions

Figure 3 shows two curves that represent the behaviour of GMI (working with I=5 mA and f=900 kHz) in the sample *f*, as-quenched and after the annealing described in section II. Both curves were normalised according to our previous definition, and the variation of the as-quenched curve reach 50% while the annealed one just achieves values up to 30%. Note in the inset that the impedance displays an asymmetric behaviour probably related to the hysteretic properties of the sample.

One can observe a peak structure in the GMI behaviour of all samples annealed which is not visible in the as quenched ones. This fact can be related to the well-known induced transverse anisotropy by stress annealing, which results in a domain structure basically consisting of transverse stripe domains.

We observe a strong suppression of relaxation (aftereffect) in the ribbons that were submitted to the annealing. Figure 4 shows the behaviour of *dvz* vs. drive current I for the same ribbon reported in Figure 3. The frequency of the drive current was kept constant at f=900 kHz and we have measured *dvz* until nearly 8 mA. In both cases we don't observe a maximum at this working frequency, but in the case of the as-quenched sample, there is a tendency to saturation. In the annealed sample, there is no measurable aftereffect. Note that the values obtained to *dvz* in the as-quenched sample are up to 5.5 mV, which is much larger than in the case shown in figure 2, for f=100 kHz.

In figures 5(a) and 5(b), we compare the maximum GMI as a function of the λ_s value of each sample, between annealed and as-quenched ribbons. Maximum values of GMI (up to 65% for f=900kHz) are reached around nearly zero magnetostriction ribbons in both cases. As one can see, the thermal treatment does not affect the general dependence of GMI on the drive current frequency and magnetostriction negative values (only some annealed ribbons have positive magnetostriction values). Both figures show the increase of f and the decrease of λ_s improving the GMI effect.

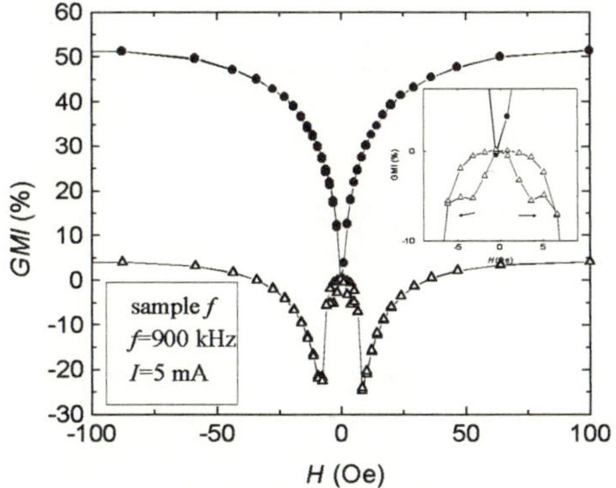

Figure 3. GMI curves of sample *f* for the as-quenched ribbon (•) and the annealed one (Δ).

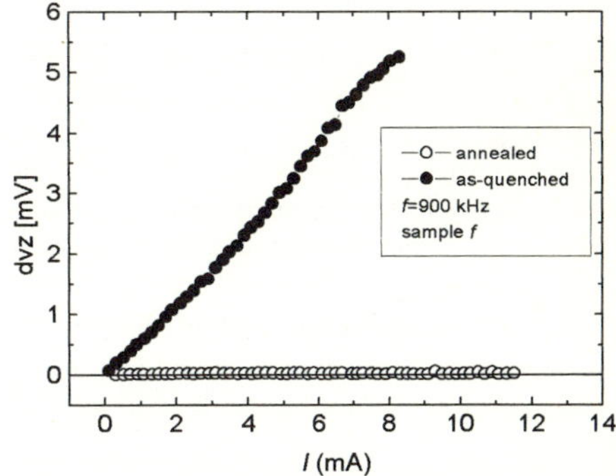

Figure 4. current dependence of the absolute voltage decay dvz, for the sample *f*, before and after annealing.

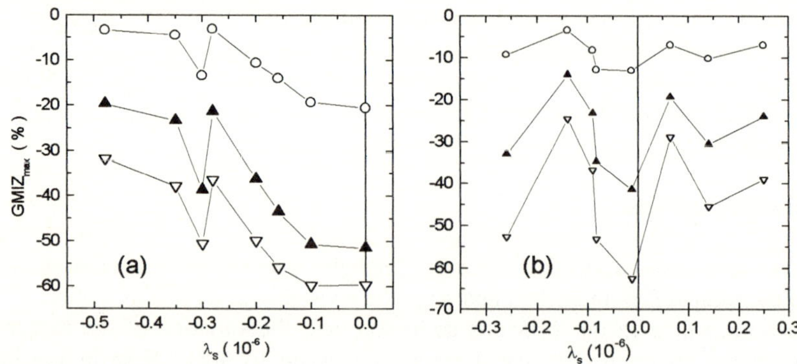

Figure 5: Maximum GMI values as a function of magnetostriction for (a) as-quenched samples and (b) annealed samples. In both we kept the driving current amplitude at 8 mA, and the driving current frequency of 100 kHz (o), 500 kHz (Δ) and 900 kHz (∇).

The dependence of the maximum aftereffect with the λ_s values can be seen in figures 6(a) (as-quenched ribbons) and 6(b) (annealed ribbons) with 100 kHz driving current frequency. We measured *dvz* as a function of the driving current amplitude for 900 kHz but the curves did not display a maximum, and the physical origin of this fact is still not clear. One can realise that the GMI aftereffect almost vanishes in the annealed ribbons, except for the sample *h*, which had already low aftereffect in as-quenched state.

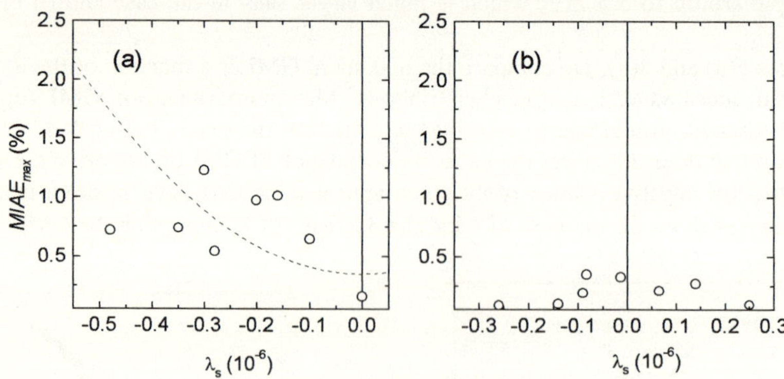

Figure 6: Maximum aftereffect as a function of magnetostriction for (a) as-quenched samples and (b) annealed samples. Here, the driving current frequency was kept constant at 100 kHz.

Although rather scattered, the experimental data on as-quenched samples indicate that there is a relationship between the maximum aftereffect and the magnetostriction value. In Fig. 6(a) we plotted a square law (dashed line), just as a guide to the eyes, which is expected from the few theoretical models that explain the magnetic permeability aftereffect in amorphous ferromagnets [5,6].

The samples must have good response to the magnetic field associated with negligible aftereffect if one wants to apply them to magnetic sensors. The induced anisotropy, obtained in the thermal treatments, improved the GMI properties as well as reduced its relaxation characteristics. However, the GMI profile was significantly altered by the thermal treatment, which can hinder some specific applications. Further investigations are being performed in order to understand the role of induced anisotropies in the magnetization processes and in the giant magnetoimpedance response.

Acknowledgements

CAPES, CNPq and FAPESP (Brazilian agencies) financially supported this work.

References:

[1] M. L. Sartorelli, M. Knobel, J. Schoenmaker, J. Gutierrez e J. M. Barandiarán, Appl. Phys. Lett. 71 (1997) p. 2208.
[2] M. Knobel, M. L. Sartorelli and J. P. Sinnecker, Phys. Rev. B 55 (1997) p. 1829.
[3] O. V. Nielsen, J. R. Petersen, B. Hernando, J. Gutierrez and F. Primdhal, Anales de Física (B) 86 (1990) p. 271.
[4] J.P. Sinneker, M. Knobel, M.L. Sartorelli and J. Schoenmaker, J. Physique (in press).
[5] P. Allia and F. Vinai, Philos. Mag. B 61 (1990) p. 763.
[6] P. Allia, C. Beatrice and F. Vinai, in: Magnetoelastic Effects and Applications, ed.: L. Lanotte (Elsevier, Amsterdam, 1993), p. 47.

Materials Science Forum Vols. 302-303 (1999) pp. 234-238

Role of Soft Magnetic Properties on the Sensitivity of the Giant Magneto-Impedance Effect

H. Chiriac[1], T.A. Óvári[1], C.S. Marinescu[1] and M. Knobel[2]

[1] National Institute of R&D for Technical Physics, 47 Mangeron Blvd., RO-6600 Iaşi 3, Romania

[2] Instituto de Física "Gleb Wataghin", UNICAMP, 13083-970 Campinas, São Paulo, Brazil

Keywords: Giant Magneto-Impedance Effect, Amorphous and Nanocrystalline Soft Magnetic Materials

Abstract We present results on the giant magneto-impedance effect in amorphous and nanocrystalline wires, glass-covered wires, and ribbons. The analysis of the experimental data shows that improvements in the soft magnetic properties of such materials compensate the absence of suitable domain structures, usually required for a sensitive magneto-impedance response. The magneto-impedance effect in nanocrystalline materials is in some cases larger than in amorphous ones with favorable magnetic structures.

1. Introduction

The giant magneto-impedance (GMI) effect consists of a large variation of the high frequency impedance of a magnetic conductor when it is subjected to a small dc magnetic field. This effect was found appropriate for sensor applications [1]. The most sensitive GMI response was reported in soft magnetic materials with very small negative magnetostriction, such as cold-drawn amorphous $Co_{68.15}Fe_{4.35}Si_{12.5}B_{15}$ wires [2].

According to the existing results, a material that displays a large and sensitive GMI effect, has to comply as much as possible with the following two conditions: (i) - it has to have good soft magnetic properties, i.e. high permeability and low coercive force, and (ii) - it has to have an appropriate domain structure, usually a transverse one for ribbons (in-plane, transverse to the ribbon axis and to the applied dc field as well), and a circumferential one for wires. In the real case, it is difficult to obtain materials in which both conditions are fully accomplished. Thus, $Co_{68.15}Fe_{4.35}Si_{12.5}B_{15}$ amorphous wires, ribbons, and glass-covered wires, are very close to the ideal case due to their very small negative magnetostriction ($\lambda_S \approx -1 \times 10^{-7}$), which simultaneously ensures good soft magnetic properties and the formation of a favorable domain structure, due to the magnetomechanical coupling between internal stresses induced during preparation and magnetostriction [2]. Consequently, CoFeSiB amorphous materials display a sensitive GMI response.

The aim of the paper is to analyze the influence of soft magnetic properties on the sensitivity and magnitude of the GMI response. Thus, we investigated the GMI response of as-cast CoFeSiB amorphous ribbons, and we monitored the changes induced by annealing in the magnitude of their GMI ratio, defined as $\Delta Z/Z = [Z(H_{dc} = 0) - Z(H_{dc} = 1 \text{ kA/m})]/Z(H_{dc} = 0)$, knowing that annealing relieves internal stresses, and thus, the initial favorable domain structure is altered, but the soft magnetic properties are simultaneously improved. These results are compared with those measured on Fe-based nanocrystalline ribbons. We have also performed a study of the GMI effect in CoFeSiB amorphous cold-drawn and glass-covered wires, the results being compared with those obtained for Fe-based nanocrystalline glass-covered wires.

2. Experiment

$Co_{68.15}Fe_{4.35}Si_{12.5}B_{15}$ amorphous glass-covered wires and ribbons have been prepared by rapid quenching from the melt. Fe-based ribbons ($Fe_{73.5}Cu_1Nb_3Si_{13.5}B_9$, $Fe_{90}Hf_7B_3$, and $Fe_{90}Zr_7B_3$) and glass-covered wires ($Fe_{73.5}Cu_1Nb_3Si_{13.5}B_9$) have been obtained in amorphous state by rapid solidification, and then properly annealed (1h at 550°C, in vacuum) in order to obtain the nanocrystalline state. Annealing of CoFeSiB amorphous ribbons was performed for 10 min. by conventional furnace heating in vacuum, at 300, 350, and 400°C, using a heating rate of 10°C/min. For the experimental investigations, we used ribbons with 18 μm in thickness and 320 μm in width, and glass-covered wires with the metallic core diameter of 15 μm and the glass cover thickness of 5 μm). All the investigated samples were 40 mm in length. The CoFeSiB cold-drawn amorphous wire with 30 μm in diameter, obtained from a conventional in-rotating-water quenched amorphous wire, has been kindly supplied by Prof. K. Mohri, from the Nagoya University, Japan.

GMI measurements were performed using a driving ac current with the maximum value of 15 mA and frequencies up to 10 MHz, with a Hameg HM 205-3 digital oscilloscope coupled to a computer which allowed automatic frequency control, data acquisition and processing. The maximum value of the axial dc magnetic field H_{dc} for these measurements was 1 kA/m, value that allows the saturation of the samples, and over which Z remains constant at a given frequency of the ac current. The changes in permeability and coercive force have been continuously monitored during annealing.

3. Results and discussion

Fig. 1 shows the frequency dependence of $\Delta Z/Z$ for an as-cast CoFeSiB amorphous ribbon, and for ribbons annealed at 300, 350, and 400°C, respectively.

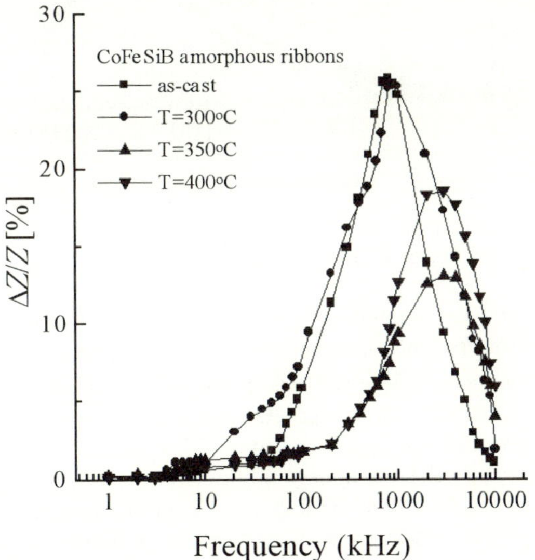

Fig. 1. Frequency dependence of the GMI ratio, $\Delta Z/Z = [Z(H_{dc} = 0) - Z(H_{dc} = 1 \text{ kA/m})]/Z(H_{dc} = 0)$ for as-cast and annealed CoFeSiB amorphous ribbons.

One observes that the maximum value of $\Delta Z/Z$ slightly decreases after annealing at 300°C, while for the sample annealed at 350°C it decreases to 50%, reaching a minimum value. The maximum GMI ratio is improved after annealing at 400°C, as compared to the 350°C annealing.

The GMI ratio is the largest for the as-cast ribbon (over 25%, at ~600 kHz), due to the existence of a transverse domain structure [3]. Annealing at 300 and 350°C alters this domain structure together with the improvement of soft magnetic properties by means of stress relief. However, for both these heat treatments, $(\Delta Z/Z)_{max}$ decreases as the annealing temperature increases, being shifted to higher frequencies. This behavior of the maximum GMI ratio indicates that the alteration of the transverse domain structure influences in a stronger way the GMI response of amorphous ribbons, as compared to the improvement of their soft magnetic properties. Nevertheless, annealing at 400°C is less effective from the point of view of decreasing the $(\Delta Z/Z)_{max}$ value, showing that the improvement of the soft magnetic properties becomes important as concerns the GMI response over a given value of the annealing temperature, even in the absence of a favorable domain structure.

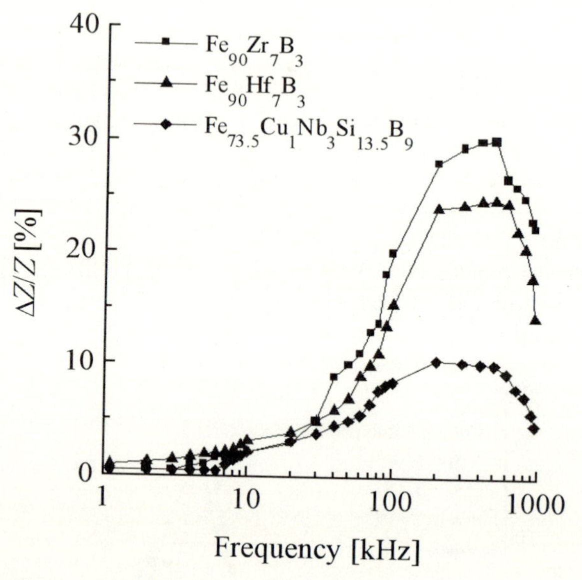

Fig. 2. Frequency dependence of the GMI ratio, $\Delta Z/Z = [Z(H_{dc} = 0) - Z(H_{dc} = 1\ \text{kA/m})]/Z(H_{dc} = 0)$, for the $Fe_{73.5}Cu_1Nb_3Si_{13.5}B_9$, $Fe_{90}Hf_7B_3$, and $Fe_{90}Zr_7B_3$ nanocrystalline ribbons.

Fig. 2 shows the frequency dependence of $\Delta Z/Z$ for $Fe_{73.5}Cu_1Nb_3Si_{13.5}B_9$, $Fe_{90}Hf_7B_3$, and $Fe_{90}Zr_7B_3$ nanocrystalline ribbons. One observes that the sensitivity of the GMI effect is almost similar in the three types of nanocrystalline ribbons in the frequency range 1 – 30 kHz, where $\Delta Z/Z$ is less than 5%. At higher frequencies, the $Fe_{90}Zr_7B_3$ nanocrystalline ribbon has the most sensitive GMI effect (over 30%), followed by the $Fe_{90}Hf_7B_3$ (over 25%) and $Fe_{73.5}Cu_1Nb_3Si_{13.5}B_9$ (over 10%) ones. Thus, the GMI ratio of FeZrB nanocrystalline ribbons exceeds the value obtained for CoFeSiB amorphous ones (25%), while the GMI ratio of FeHfB nanocrystalline ribbons is equal to that obtained in the latter. These results are in agreement with the larger values of transverse permeability found in these two nanocrystalline ribbons as compared to the CoFeSiB amorphous ones. The differences in the values of $\Delta Z/Z$ for nanocrystalline ribbons with different compositions can be mainly attributed to the differences in the electrical resistivities of these alloys [4]. It is worthwhile to mention that the maximum GMI ratios in these ribbons correspond to frequencies below 1 MHz.

Fig. 3 illustrates the frequency dependence of the GMI ratio for CoFeSiB amorphous glass-covered wires, wires after glass removal, and cold-drawn amorphous wires. One observes that cold-drawn amorphous wires display the most sensitive GMI effect, followed by amorphous wires after glass removal. It was expected to have a more sensitive effect in cold-drawn amorphous wires at the beginning of the frequency range, since the skin effect appears at lower frequency in such wires due to their larger diameters. Moreover, cold-drawn CoFeSiB amorphous wires display an appropriate magnetic structure, i.e. circumferential, in their entire volume, and large circumferential permeability [2]. Amorphous wires after glass removal display a larger GMI ratio than glass-covered ones due to their superior soft magnetic properties, i.e. larger permeability and lower coercivity, originating in the stress relief determined by glass removal [5].

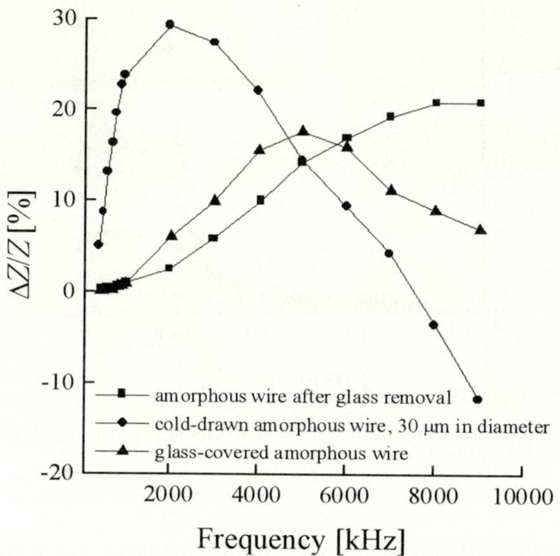

Fig. 3. Frequency dependence of the GMI ratio for CoFeSiB amorphous glass-covered wires, wires after glass removal, and cold-drawn amorphous wires.

Fig. 4. shows the frequency dependence of the GMI ratio, $\Delta Z/Z$, for FeCuNbSiB wires after glass removal, with the annealing temperature as a parameter. We investigated only wires after glass removal due to their improved soft magnetic properties originating in the stress relief produced by glass removal, as compared to glass-covered ones [6]. After annealing at 500°C, the maximum MI ratio increases to about 14% due to further stress relief produced by annealing. Once the nanocrystalline phase appears (annealing for 1 h at 550°C), the maximum MI ratio reaches about 28%. Annealing over 550°C results in a deterioration of the wire's soft magnetic properties due to the increase in size of the crystalline grains.

Nanocrystalline wires without glass have a much more sensitive GMI response as compared to glass-covered ones, since they are magnetically softer (the coercive force is about 100 A/m for the first ones, and almost 400 A/m for the latter). The GMI response of nanocrystalline wires without glass is even more sensitive than that of CoFeSiB amorphous wires after glass removal, their maximum GMI ratio being about 20%.

The sensitivity of the GMI response of nanocrystalline wires after glass removal is very close to that of cold-drawn CoFeSiB amorphous ones. The tendency of these thin wires (15 μm in diameter) to reach a similar magnitude of the GMI effect (~28 – 30%) like the cold-drawn ones (30 μm in diameter), emphasizes once again the close relationship between good soft magnetic properties and a sensitive GMI effect, since at close values of the resistivities, it is obvious that the differences in the circumferential permeability play an important role in the magnitude of the skin effect. In addition, the frequency that corresponds to $(\Delta Z/Z)_{max}$ in nanocrystalline wires (~ 4MHz) is closer to that of cold-drawn amorphous ones (~ 2 MHz), than to that of amorphous wires after glass removal (~ 9 MHz).

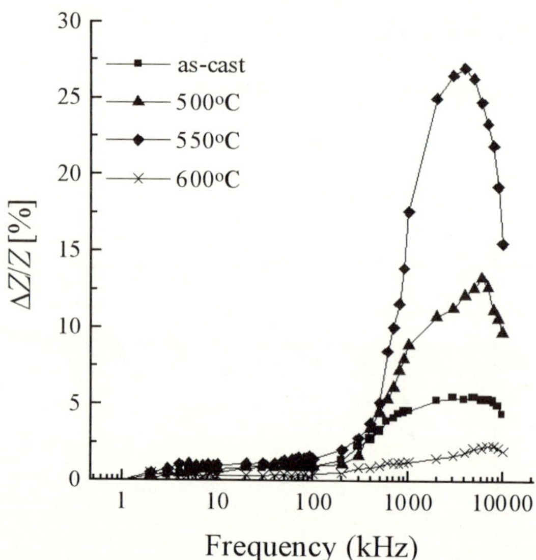

Fig. 4. Frequency dependence of the GMI ratio for FeCuNbSiB wires after glass removal, with the annealing temperature as a parameter.

4. Conclusion

The magnitude and sensitivity of the GMI effect in any magnetic conductor is determined by the existence of a favorable domain structure together with good soft magnetic properties.

The existence of an appropriate domain structure, i.e. transverse for ribbon shaped materials and circumferential for wire shaped ones, results in a sensitive GMI effect, even in the absence of outstanding soft magnetic properties. On the other hand, the existence of such magnetic structures does not represent an absolute criterion to be followed in the selection of an appropriate material for GMI-based applications.

The absence of suitable domain structures is well compensated by very good soft magnetic properties, usually induced by proper annealing. Thus, Fe-based nanocrystalline soft magnetic ribbons and wires display a large and sensitive GMI effect, their response being comparable or even larger than that of as-cast amorphous ones.

References

[1] L.V. Panina, K. Mohri, T. Uchiyama, K. Bushida, and M. Noda, in Nanostructured and Non-Crystalline Materials, edited by M. Vázquez and A. Hernando (World Scientific, Singapore, 1995), p. 461.

[2] L.V. Panina, K. Mohri, K. Bushida, and M. Noda, J. Appl. Phys. 76 (1994), p. 6198.

[3] H. Chiriac, F. Vinai, T.-A. Óvári, C.S. Marinescu, F. Barariu, and P. Tiberto, Mat. Sci. Eng. A 226-228 (1997), p. 646.

[4] M. Knobel, H. Chiriac, J.P. Sinnecker, S. Marinescu, T.-A. Óvári, and A. Inoue, Sensors and Actuators A: Physical 59 (1997), p. 256.

[5] H. Chiriac, Gh. Pop, T.-A. Óvári, and F. Barariu, IEEE Trans. Magn. 32 (1996), p. 4872.

[6] H. Chiriac, T.-A. Óvári, Gh. Pop, and F. Barariu, J. Appl. Phys. 81 (1997), p. 5817.

Materials Science Forum Vols. 302-303 (1999) pp. 239-243

Temperature Distribution in a Joule Effect Annealed Amorphous Glass-Covered Wire

H. Chiriac[1], M. Knobel[2] and T.A. Óvári[1]

[1] National Institute of R&D for Technical Physics, 47 Mangeron Blvd., RO-6600 Iaşi 3, Romania

[2] Instituto de Física "Gleb Wataghin", UNICAMP, 13083-970 Campinas, São Paulo, Brazil

Keywords: Amorphous Glass-Covered Magnetic Wires, Joule Effect Annealing

Abstract We present results on the Joule effect heating of amorphous glass-covered magnetic wires. A simple theoretical model that allows us to determine the radial distribution of temperature in an amorphous glass-covered wire subjected to Joule effect annealing is presented. The experimental verification of the model was performed on $Fe_{77.5}Si_{7.5}B_{15}$ amorphous glass-covered wires by measuring the temperature dependence of the electrical resistance for a furnace annealed sample and the dependence of resistance on the annealing current for a Joule heated sample.

1. Introduction

Amorphous glass-covered wires (AGCW), with diameters of 2 to 40 μm and the glass cover thickness of 1 to 20 μm, display remarkable magnetic properties, which make them suitable for sensor applications. Their magnetic properties and behavior depend on composition and dimensions – metallic wire diameter, glass cover thickness, and their ratio – through the magnitude and distribution of internal stresses induced during preparation [1]. This fact offers large tailoring possibilities for their basic magnetic quantities, on one hand directly during preparation – by adjusting the technological parameters that determine the wire dimensions, and on the other, through different types of annealing, e.g. heat treatment, stress-annealing, stress-field annealing, etc.

Annealing primarily determines stress relief, acting in this way on the stress distribution, and changing the magnetoelastic anisotropy, which is important in these wires, but structural changes may also occur, e.g. the nanocrystalline phase formation [2], resulting in the appearance of other types of anisotropy or in changes of the magnetostriction constant.

There are two ways of performing annealing: (1) classical furnace annealing, which exhibits some disadvantages, such as the nonuniform heating of long samples and the difficulties in directly monitoring the changes in the electrical and magnetic properties of the annealed samples, and (2) Joule effect annealing, where the samples are annealed by passing a dc current through them.

The aim of this paper is to determine the temperature distribution in an AGCW annealed by means of the Joule effect, in order to easily calculate the temperature that corresponds to a given annealing current, and to investigate the uniformity of the temperature within the cross section of such a wire. With this aim, we used the Fourier heat conduction equation with the appropriate boundary conditions required by the existence of the metal-glass interface and by the surface heat losses.

The experimental verification of the model was performed on $Fe_{77.5}Si_{7.5}B_{15}$ AGCW, by measuring the temperature dependence of the electrical resistance for classical furnace annealing, and the same dependence for Joule heating.

2. Theoretical analysis

Let us consider an amorphous glass-covered wire (AGCW) with the radius of the metallic core a, and the total wire radius (metal + glass) b. Fig. 1 shows a schematic diagram of such a wire. The use of cylindrical coordinates is required by the problem's symmetry. We assume that the relevant heat transfer occurs only on the radial direction.

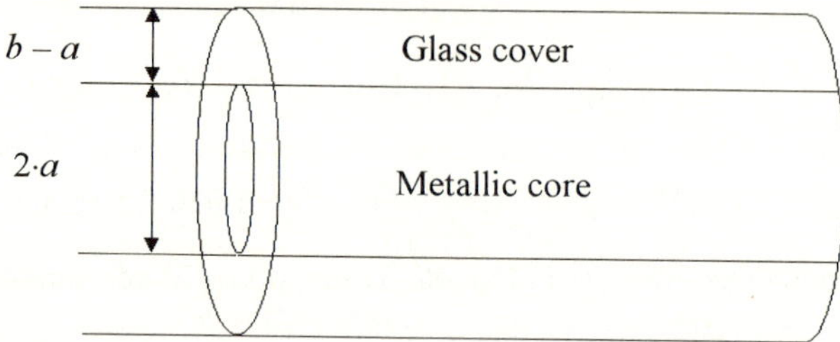

Fig. 1. Schematic diagram of an amorphous glass-covered wire: a – radius of the metallic core; b – wire radius.

The Fourier heat conduction equation for both regions of the wire (metal and glass) are:

– for the metal ($0 \le r \le a$):
$$\frac{dT_m(r)}{dr} = -\frac{1}{2} \cdot \frac{j^2 \cdot \rho}{k} \cdot r \tag{1}$$

– for the glass ($a \le r \le b$):
$$\frac{dT_g(r)}{dr} = \frac{B}{r} \tag{2}$$

where m refers to the metal and g refers to the glass, j being the current density in the metallic wire, ρ its resistivity, and k its thermal conductivity. B is a coefficient that depends on the heat generated in the metal by Joule effect.

The general solutions for both regions are:

– metal:
$$T_m(r) = C - \frac{1}{4} \cdot \frac{j^2 \cdot \rho}{k} \cdot r^2 \tag{3}$$

– glass:
$$T_g(r) = A + B \ln r \tag{4}$$

in which A and C are coefficients resulting from the boundary conditions.

In order to determine the exact solutions, we impose the following boundary conditions:

(i) The heat flux from the metal – glass interface must be continuous:
$$k \cdot \frac{dT_m}{dr} = k_1 \cdot \frac{dT_g}{dr}$$
k_1 being the thermal conductivity of the glass.

(ii) At the metal – glass interface, the temperatures from both regions must be equal:
$$T_m(a) = T_g(a)$$

(iii) At the outer surface of the wire, only radiative heat loss occurs:
$$\left.\frac{dT_g}{dr}\right|_{r=b} = -\frac{\sigma \cdot \varepsilon}{k_1} \cdot \left(T^4(b) - T_0^4\right)$$

where: σ is the Stefan-Boltzmann constant, ε – the emissivity, and T_0 – the room temperature.
We obtain from *(i)*:
$$B = -\frac{1}{2} \cdot \frac{j^2 \rho}{k_1} \cdot a^2 \tag{5}$$

From *(iii)* we have:
$$T(b) = \left(T_0^4 + \frac{1}{2} \cdot \frac{j^2 \rho}{\sigma\varepsilon} \cdot \frac{a^2}{b}\right)^{1/4} \tag{6}$$

Taking into account Eq. (4), that for $r = b$ becomes $T(b) = A + B\ln b$, we obtain the coefficient A:

$$A = \frac{1}{2}\frac{j^2\rho}{k_1}a^2 \ln b + \left(T_0^4 + \frac{1}{2}\frac{j^2\rho}{\sigma\varepsilon}\frac{a^2}{b}\right)^{1/4} \quad (7)$$

From *(iii)* and (7) we obtain coefficient C:

$$C = \frac{1}{4}\frac{j^2\rho}{k}a^2 + \frac{1}{2}\frac{j^2\rho}{k_1}a^2 \ln\left(\frac{b}{a}\right) + \left(T_0^4 + \frac{1}{2}\frac{j^2\rho}{\sigma\varepsilon}\frac{a^2}{b}\right)^{1/4} \quad (8)$$

Thus, the solutions (3) and (4) become:

$$T_m(r) = \frac{1}{4}\frac{I^2\rho}{\pi^2 a^2 k}\left[1 - \left(\frac{r}{a}\right)^2 + 2\frac{k}{k_1}\ln\left(\frac{b}{a}\right)\right] + \left(T_0^4 + \frac{1}{2\pi^2 a^2 b}\frac{I^2\rho}{\sigma\varepsilon}\right)^{1/4} \quad (9)$$

$$T_g(r) = \frac{1}{2}\frac{I^2\rho}{\pi^2 a^2 k_1}\ln\left(\frac{b}{r}\right) + \left(T_0^4 + \frac{1}{2\pi^2 a^2 b}\frac{I^2\rho}{\sigma\varepsilon}\right)^{1/4} \quad (10)$$

in which $I = j\cdot\pi\cdot a^2$.

Using Eqs. (9) and (10), we found the radial distribution of temperature in an AGCW for a given value of the dc current. Fig. 2 illustrates this distribution for I = 17.8 mA, in the case of an $Fe_{77.5}Si_{7.5}B_{15}$ AGCW with a = 9.0 μm and b = 18.0 μm. The parameters used in our calculations are: k = 30 $Wm^{-1}K^{-1}$, k_1 = 1.177 $Wm^{-1}K^{-1}$, ρ = 1.3 × 10^{-6} Ωm, T_0 = 0°C. Instead of the product $\sigma\cdot\varepsilon$, we used the so-called wire loss parameter P, its value for the FeSiB alloy being 2.5 × 10^{-7} $Wm^{-2}K^{-4}$, also used in a previous work in which the temperature distribution during Joule heating of in-rotating-water quenched FeSiB amorphous wires is calculated in a similar approach [3].

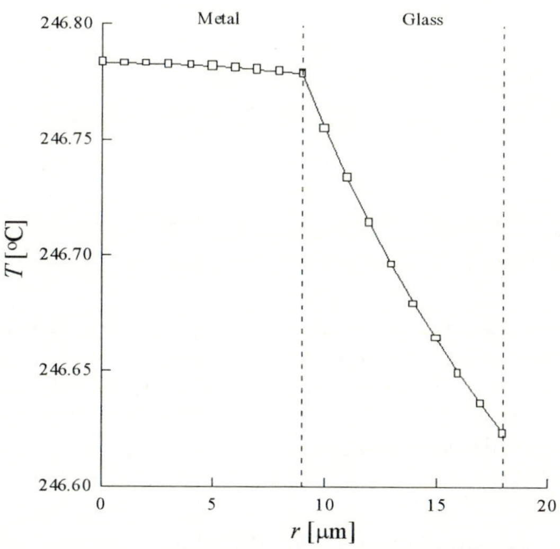

Fig. 2. Calculated radial distribution of temperature in an $Fe_{77.5}Si_{7.5}B_{15}$ amorphous glass-covered wire with a = 9 μm and b = 18 μm; the value of the annealing dc current is I = 17.8 mA.

P does not contain only the radiative heat losses, since in this case its value would be unrealistic from the physical point of view (it corresponds to an emissivity $\varepsilon > 1$), but it also contains the conductive and convective heat losses that take place in the wire, and which are not considered in our simple model.

One observes that the temperature difference between the center of the metallic wire and its surface is only 0.0047°C, and between its center and its outer surface (glass) is 0.16°C. The temperature difference between the surface of the metal and the wire surface (glass) is 0.1553°C. Thus, it can be considered that the temperature is constant in the wire's cross section, the Joule effect annealing ensuring in this way a uniform heating of a sample.

3. Experimental results. Discussion

We checked our calculated results by determining experimentally the temperature in the wire for a given value of the annealing current, based on the temperature dependence of the electrical resistance. Thus, we measured the temperature dependence of the resistance for a classical furnace annealed sample, and separately the dependence of resistance on the annealing current for a Joule heated sample. The maximum annealing temperature for the in-vacuum furnace annealing was 530°C, below the crystallization point (T_X = 535°C), and the maximum value of the dc current for the Joule effect annealing was 27 mA. The heating rate was about 10°C/min in both cases. From these two curves, we obtained by interpolation an experimental curve that gives the temperature in the wire as function of the annealing current.

Based on Eq. (9), for r = 1 μm, we obtained the theoretical dependence of temperature on the annealing current.

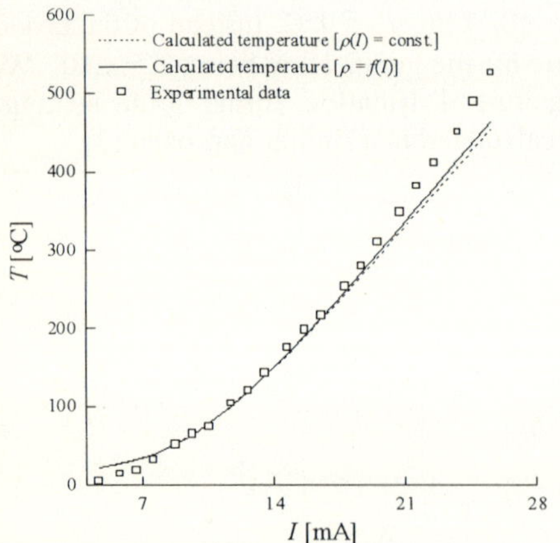

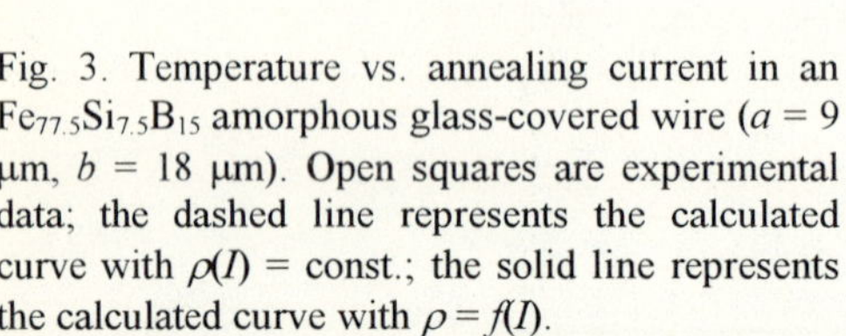

Fig. 3. Temperature vs. annealing current in an $Fe_{77.5}Si_{7.5}B_{15}$ amorphous glass-covered wire (a = 9 μm, b = 18 μm). Open squares are experimental data; the dashed line represents the calculated curve with $\rho(I)$ = const.; the solid line represents the calculated curve with $\rho = f(I)$.

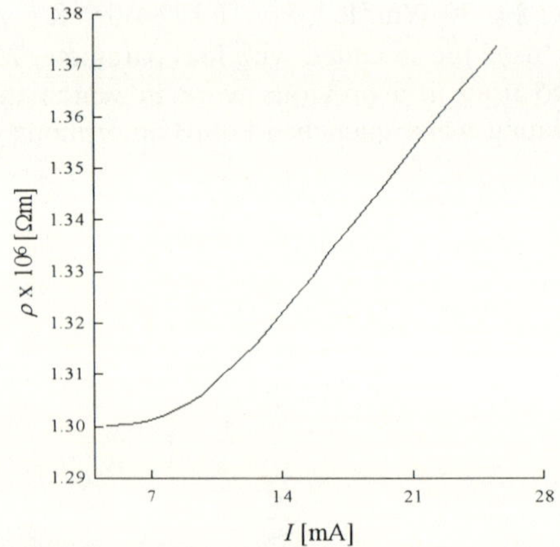

Fig. 4. Calculated dependence of the electrical resistivity on the value of the annealing current (after [4]) for the FeSiB amorphous glass-covered wire with a = 9 μm and b = 18 μm.

Fig. 3 illustrates the experimental and theoretical relationships between the annealing current and temperature in the metallic region of an AGCW. There are two calculated curves: one calculated according to Eq. (9) without considering the dependence of resistivity on the annealing current, and a second one, calculated by considering this dependence according to [4] (Fig. 4). Both calculated curves have similar shapes, and there is no significant quantitative difference between the obtained temperature values (Fig. 3). This fact confirms the validity of our model.

The experimental data is in agreement with the model only if we consider the changes in the material parameters that occur during structural changes – metastable phase relaxations – which inherently take place in any annealed amorphous alloy. Accordingly, Eq. (9) becomes:

$$T_m(r,I)=\frac{1}{4}\frac{I^2\rho}{\pi^2a^2k}\left[1-\left(\frac{r}{a}\right)^2+2\frac{k}{k_1}\ln\left(\frac{b}{a}\right)\right]+\left(T_0^4+\frac{1}{2\pi^2a^2b}\frac{I^2\rho}{P}R(I)\right)^{1/4} \qquad (11)$$

in which P represents the heat losses ($P = 2.5 \times 10^{-7}$ Wm^{-2}K^{-4}), and $R(I)$ the influence of metastable phase relaxations. The best fit was obtained for $R(I) = (I^2/2\sqrt{2}) \times 10^4$. Fig. 5 shows the current dependence of the relaxation factor R.

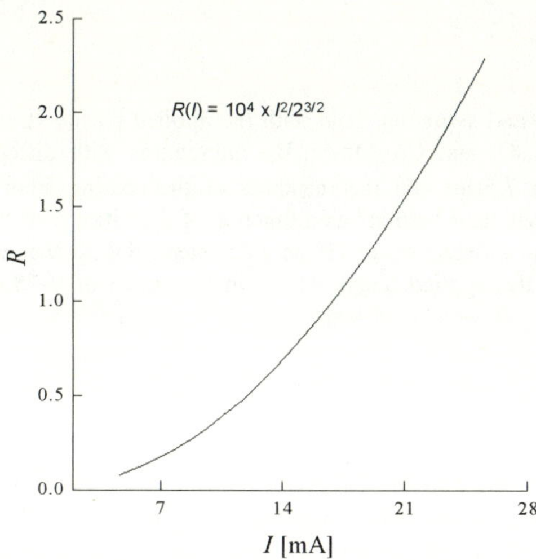

Fig. 5. Estimated dependence of the wire's relaxation factor R on the value of the annealing current I.

Thus, the influence of structural changes increases as the value of the annealing current increases, and it has to be considered.

4. Conclusion

A simple model that allows the calculation of temperature in a Joule effect annealed amorphous glass-covered wire is presented. The temperature in the cross section of a Joule annealed sample is constant. The correlation between experimental and calculated data indicates the requirement of considering the metastable phase relaxations that occur in the amorphous wire during annealing.

References

[1] H. Chiriac and T.-A. Óvári, Prog. Mater. Sci. 40 (1996), p. 333.
[2] H. Chiriac, T.-A. Óvári, Gh. Pop, and F. Barariu, J. Appl. Phys. 81 (1997), p. 5817.
[3] M. Knobel, P. Allia, C. Gómez-Polo, H. Chiriac, and M. Vázquez, J. Phys. D: Appl. Phys. 28 (1995), p. 2398.
[4] H. Chiriac and I. Aştefănoaei, Phys. Stat. Sol. (a) 153 (1996), p. 183.

Materials Science Forum Vols. 302-303 (1999) pp. 244-248

Stress Dependence of Switching Field in Ultra-Thin Amorphous Wires

A.F. Cobeño[1], P. Aragoneses[2], J.M. Blanco[2], L. Dominguez[2], A. Zhukov[3,4], V. Larin[4] and J. González[3]

[1] Departamento de Electrónica y Telecomunicaciones, EUITI, ES-20600 Eibar (Guipúzcoa), Spain

[2] Departamento de Física Aplicada I, Escuela Universitaria de Ingeniería Técnica Industrial, ES-20011, San Sebastián, Spain

[3] Departamento de Física de Materiales, Facultad de Ciencias Químicas, P.O. Box 1072, ES-20080 San Sebastián, Spain

[4] "Amo Tec" LTD, B-d Dacia 15/78, Kishinev, Moldova

Keywords: Amorphous Microwires, Magnetic Bistability, Switching Field

ABSTRACT

The variations of remanence and switching field with the applied tensile stress have been measured in glass coated amorphous $Fe_{70}B_{15}Si_{10}C_5$ and $Co_{68}Mn_7Si_{10}B_{15}$ microwires with different radius of metallic nucleus ranging from 1.9 μm up to 7.5μm and the thickness of the coating from 3.8 μm up to 10.8 μm. The experimental results of the switching field H* as a function of the ratio, r, of the thickness of glass coat to the radius of the microwire show a reduction of H* as r decreases due to the reduction in the average internal stresses. H* increases with the applied tensile stress in the range of 0-750MPa, the increase being more significant as r decreases.

INTRODUCTION

In the las few years, the Taylor-Ulitovsky method has been employed for preparation of tiny glass coated amorphous microwires with magnetic bistability [1-3]. The range of the diameter of the metallic nucleus is from l μm to 20 μm and the glass coating thickness is between 1-10 μm. It has been found that these amorphous microfibers also exhibit magnetic bistability (MB) even for samples with a length of a few millimetres and for nearly zero-magnetostrictive alloys [1]. On the other hand, microwires of negative magnetostriction compositions do not exhibit magnetic bistability (MB)[1,4], as was observed in "commercial" amorphous wires [5-7]. The observed compositional dependence of magnetic bistability and enhanced values of the remanence were explained as mainly due to additional stresses arising from the outer glass cover due to differences in thermal expansion coefficients with the metallic nucleus [1,8-10]. It was experimentally confirmed in [4] by appearance of MB from unhysteretic hysteresis loop after glass removal in Co-rich microwires. Depending on the thickness of the glass coating the switching field is, at least, one order of magnitude higher than for melt-spun wires [1,11,8]. Enhanced soft magnetic properties and magneto-impedance were found recently in nearly-zero $Co_{68}Mn_7Si_{10}B_{15}$ microwires after adequate heat treatments [10,12].

A few non-systematic measurements (only for 3 very different geometries of Fe-rich microwires) of the effect of external tensile stresses on switching field have been performed [13] for Fe-rich microwires. In the case of the conventional Fe and Co-rich amorphous wires, H* is very sensitive to the thermal treatments and external stress [7]. It was concluded, that the experimental data on H* for Fe-rich wires fit a $\sigma^{1/2}$ law (σ: applied tensile stress) for applied stresses larger than the internal ones. Co-rich wires did not show such a dependence. Consequently, a nucleation mechanism was invoked to explain the MB process in Fe-rich wires and domain wall propagation in Co-rich wires [5-7].

The aim of this work is to present and analyse new data concerning magnetic properties of Fe and Co-rich microwires, and to study the effect of external tensile stresses on their magnetic properties.

EXPERIMENTAL

Amorphous microwires of nominal compositions $Fe_{70}B_{15}Si_{10}C_5$ and $Co_{68}Mn_7Si_{10}B_{15}$ were obtained by the Taylor-Ulitovsky technique [1] based on the Taylor method [14]. This method essentially consists in drawing a Pyrex-like glass containing the molten metal to a wire. In this way, tiny wires are obtained with a metallic nucleus covered by insulating glass coating. Four samples with different radius of the metallic nucleus and different thickness of the insulating coating have been investigated for each composition (see details in Table 1).

Table1. Compositions and geometry of investigated samples.

Alloy	Label	Metallic nucleus radius, r_m, μm	Glass coating thickness, t, μm	Ratio $r = t/(r_m+t)$
$Fe_{70}B_{15}Si_{10}C_5$	A(Fe)	7.5	4.4	0.37
$Fe_{70}B_{15}Si_{10}C_5$	B(Fe)	6.5	4.0	0.38
$Fe_{70}B_{15}Si_{10}C_5$	C(Fe)	5.4	6.0	0.53
$Fe_{70}B_{15}Si_{10}C_5$	D(Fe)	3.0	8.5	0.74
$Co_{68}Mn_7Si_{10}B_{15}$	A(Co)	6.4	4.0	0.38
$Co_{68}Mn_7Si_{10}B_{15}$	B(Co)	5.4	6.0	0.53
$Co_{68}Mn_7Si_{10}B_{15}$	C(Co)	3.1	8.5	0.73
$Co_{68}Mn_7Si_{10}B_{15}$	D(Co)	1.9	9.0	0.825

Details concerning experimental set-up can be found elsewhere [7]. Chemical etching of the external glass coating was done in dilute hydroflouric acid.

EXPERIMENTAL RESULTS

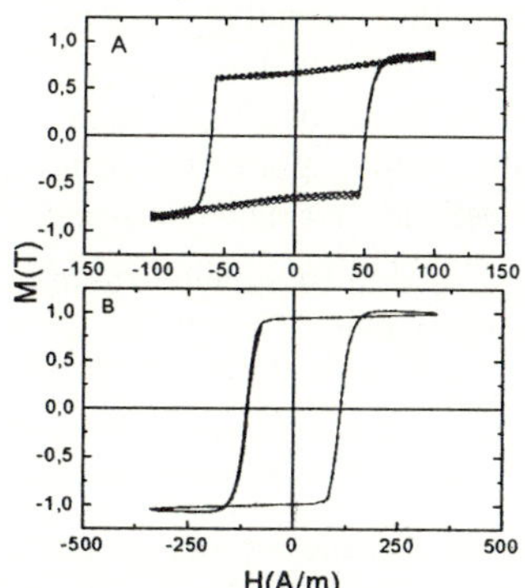

Fig. 1. Hysteresis loops of B(Co)(A) and A(Fe)(B) samples.

Fig 1 (A, B) shows examples of the axial hysteresis loops for the Co-rich and Fe-rich microwires respectively. For both compositions rectangular hysteresis loops are observed. Dependencies of H* on the ratio r of the thickness of glass coating to the radius of the microwire are shown in figure 2. These values of H* increase with r and they are clearly larger than that reported one in [5-7] for a Fe-rich amorphous wire with a diameter of 125 μm (about 8 A/m).

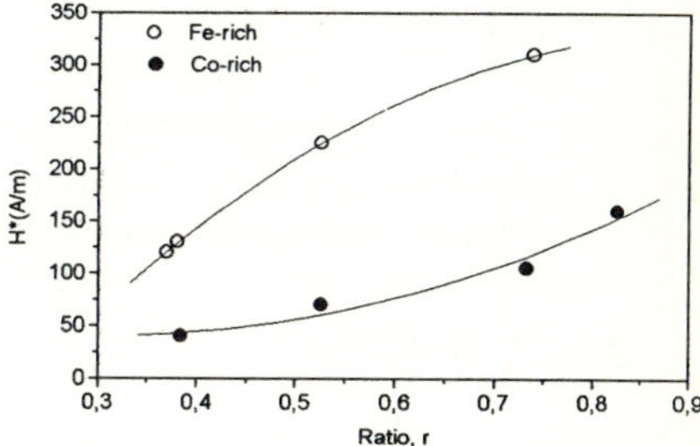

Fig. 2. Effect of the ratio r on the critical field H* of Fe-rich and Co-rich microwires.

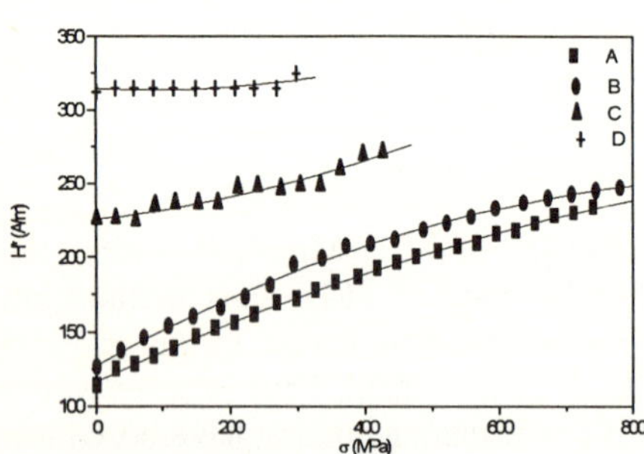

Fig. 3. Effect of external stresses, σ, on H* of Fe-rich microwires (labels from Table I).

Figures 3 and 4 show the tensile stress dependence of the switching field for the samples studied. The switching field H* increases monotonically with the applied stress for the Fe-rich microwires. Generally H*(σ) dependence for Co-rich samples is rather different: there is a range of σ, where the switching field is roughly independent of σ and then H* starts to increase. This critical value H* increases with decrease of the metallic nucleus radius (see Fig.4). This evolution of H* with the applied tensile stress, σ_a, in microwires is quite different from the behaviour of H* in Fe-rich amorphous wires obtained by the rotating-water technique, especially in the range of small applied stress where a minimum of H* was found in conventional Fe-rich wires for a certain σ_a [5-7].

For Fe, an increase in H* is obtained with σ_a in the range of 0-750MPa, such an increase being more significant as r decreases. Fig. 5 represents changes of the switching field resulting from chemical removal of the glass coating in Fe-rich and Co-rich samples. As was expected, some reduction of the switching field was observed in both cases. However, as in the case of the tensile stress dependence of the switching field, some saturation was observed in the case of Co-rich samples after initial decay of H*. Besides, Co-rich microwires with similar geometry have lower values of H*, reflecting smaller magnetoelastic energy of that alloy due to smaller values of the magnetostriction constant.

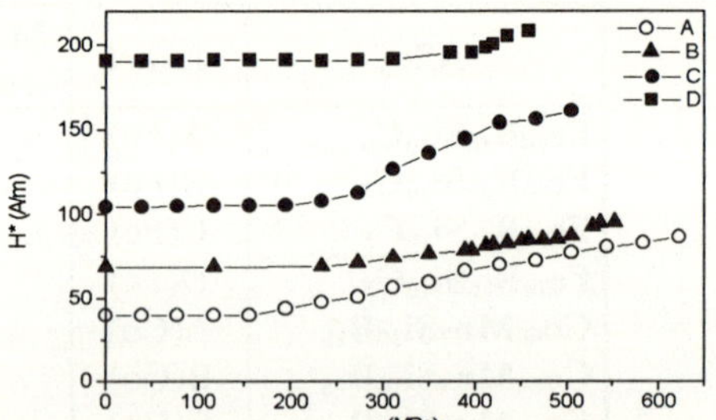

Fig. 4. Effect of external stress, σ, on H* of Co-rich microwires (labels from Table I).

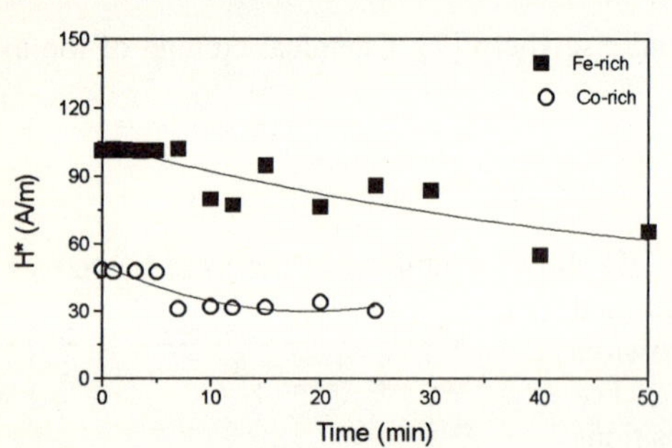

Fig. 5. Effect of chemical etching on H* of Fe-rich A(Fe) and Co-rich A(Co) samples.

DISCUSSION

Two important points should be brought out. First is the contribution of the internal stresses. The critical field should be proportional to the energy required to form the domain wall γ involved in the remagnetization process. The wall energy is related to the magnetoelastic anisotropy and then to the applied tensile stress by [6]:

$$H^* \approx H_N \propto \gamma \propto \frac{\left[A\frac{3}{2}\lambda_S(\sigma_a+\sigma_r)\right]^{1/2}}{\cos\alpha} \qquad (1)$$

where α is the angle between magnetization and axial direction. A is the exchange energy constant, λ_s is the saturation magnetostriction constant and σ_r is the residual tensile stress. Consequently, H* must be proportional to $\sigma_a^{1/2}$ for σ_a larger than σ_r and cos $\alpha \approx 1$. Such proportionality is only observed for Fe-rich samples having small values of the r parameter (see Fig.3).
Another mechanism which can be involved in H* (r) dependence is the size effect: with a decrease of the radius of metallic nucleus the demagnetization factor decreases too and therefore the size (the penetration depth) of the closure domains also decreases. This was shown experimentally in [3]. Classical approach of magnetic bistability requires critical size of reversed domains which arises from the counterbalance between surface (domain wall) and volume (magnetostatic) energies. Therefore, when the radius of the sample is very small, higher external magnetic fields are needed to reach the critical size of the demagnetisation nucleus whose volume is limited by the sample dimensions.

Let now discuss stress dependence of the switching field. As was mentioned above, there is no minimum observed in H* (σ_a) dependences for the compositions investigated. This difference can be explained according to Eq.(1) from the existence of elevated quenched-in internal stresses mainly due to the glass coating. Besides, the stress dependence of the switching field for Fe-rich compositions in case of smallest radius of the metallic nucleus is rather weak (see Fig. 3), comparable to H*(σ) for higher r ratios. If proportionality of H*(σ) can be considered, where σ is the total (applied and internal) stress, then a contribution of the internal stresses makes the H*(σ) dependence weaker. It should be mentioned here, that the observed H*(σ) dependence for different r and its explanation is not in agreement with experimental data of [13], where in similar Fe-rich compositions the increase of r resulted in stronger H*(σ) dependence.

Another point is that the existence of such high quenched-in internal stresses seems to be able to change significantly the value of magnetostriction constant [15-16]. It has been experimentally found that

$$\lambda(\sigma)= \lambda(0)-B\sigma \qquad (2)$$

where $\lambda(0)$ is the saturation magnetostriction constant without applied stresses and B is a positive coefficient of order 10^{-10} MPa. So, if internal stresses are of order of 10^3 MPa, as were estimated in [1,8,9], and the magnetostriction constant is around 10^{-7} (as expected for the Co-rich composition investigated here), the second term of (2) is almost of the same order as the first one. Perhaps this point is important in the case of Co-rich compositions where the magnetostriction constant is positive and close to zero. One can thus expect a change of sign of the magnetostriction constant under the effect of internal quenched-in stresses.

CONCLUSIONS

The following conclusions can be presented:

- The effect of external tensile stresses on switching field behaviour was investigated in two different microwires of Fe-rich and Co-rich compositions with magnetic bistability.
- The switching field of Co-rich compositions depends on the sample geometry (radius of metallic nucleus and thickness of glass coating), as was previously found in Fe-rich compositions.
- Glass coating removal results in drastic change of the switching field in both compositions indicating existence the strong internal frozen-in stresses.
- A strong effect of tensile stress on switching field value was found for both compositions. The well-known square root dependence of H* on σ was found in Fe-rich compositions for the thicker metallic nucleus.

- Observed experimental behavior was explained in terms of dependence of the nucleation field on both external and internal stresses and taking into account a dependence of the magnetostriction constant on the strength of the total stresses.

ACKNOWLEDGEMENTS

One of the authors, A. Zhukov, wishes to thank the Gobierno Vasco for financial support. The work has been supported by the Gobierno Vasco under project PI 95/78.

REFERENCES

[1] Vázquez M., Zhukov A., J. Magn. Magn. Mat. 160 (1996) 223.

[2] Hagiwara M., Inoue A., "Rapidly Solidified Alloys" (Ed H.H.Liebermann, New York, Marcel Dekker 1993) 141

[3] González J., Murillo N., Larin V., Barandiarán J.M., Vázquez M, Hernando A., IEEE Trans. Magn, Mag-33 (1997) 2362.

[4] C.F. Catalan, V.M. Prida, J. Alonso, M. Vázquez, A. Zhukov, B. Hernando and J. Velázquez, Rapidly Quenched and Metastable Materials, Supplement to the Mat. Sci. Eng.A (1997) 438.

[5] Severino A.M., Gómez-Polo C., Marín P., Vázquez M., J. Magn. Magn. Mat. 103 (1992) 117.

[6] Vázquez M., Hernando A., J.Phys.D:Appl. Phys. 29 (1996) 939.

[7] González J., Blanco J.M., Vázquez M., Barandiarám J.M., Rivero G., Hernando A., J.Appl.Phys.70 (1991) 6522

[8] Chiriac H., Ovari T.A. and Pop Gh., Phys. Rev. 52 (1995) 10105

[9] Velázquez J., Vázquez M. and Zhukov A. J. Mat. Res. 11 (1996) 2499

[10] Baranov S.A., Larin V.S., Torcunov A.V., Zhukov A.P., Vázquez M. in : M. Vázquez and A. Hernando (Eds.), Nanocrystalline and Non-Crystalline Materials, World Scientific, Singapore, 1995, p. 567.

[11] A. P. Zhukov, M. Vázquez, J. Velázquez, H. Chiriac and V. Larin, J. Magn and Magn, Mater 151 (1995) 132.

[12] Marin P., Arcas J., Zhukov A., Vázquez M and Hernando A., in : G.C. Hadjipanayis (ed.), Magnetic Hysteresis in Novel Magnetic Materials, 1997, Kluwer Academic Publishers, Netherlands, p.743.

[13] Chiriac H., Ovari T.A. and Gh. Pop, J. Magn. Magn. Mat . 157/158 (1996) 227

[14] Taylor G.F., Phys.Rev. 24 (1924) 6555

[15] J.M. Barandiaran, A. Hernando, V. Madurga, O.V. Nielsen, M. Vázquez and M. Vázquez - López, Phys. Rev. 35 (1987) 5066

[16] J. M. Blanco, P.G. Barbon, J. González, C. Gomez-Polo and M. Vázquez, J. Magn. Magn. Mater. 104-107 (1992) 137

Materials Science Forum Vols. 302-303 (1999) pp. 249-253

Relationship between Structural and Magnetic Properties of Soft Nanocrystalline Materials

N. Murillo[1], J. González[1] and J.M. González[2]

[1] Dpt. de Física de Materiales, Facultad de Química, ES-20009 San Sebastián, Spain

[2] Instituto de Ciencia de Materiales, CSIC, Cantoblanco, ES-28049 Madrid, Spain

Keywords: Nanocrystalline Materials, Coercivity and Magnetic Hysteresis

Abstract

The structural and magnetic properties of $Fe_{73.5}Si_{13.5}B_9Cu_1(Ta/Nb)_3$ alloy ribbons after heat treated carried out (by using a conventional furnace and the Joule heating method) at temperatures and current densities when a partial devitrification to a nanocrystalline structure occurs have been studied. Room temperature magnetization work (W_m) and coercivity (H_c) were evaluated for several nanocrystalline samples having different grain size (D). Variations of W_m with the thermal treatment parameters are interpreted in the framework random anisotropy model in terms of the relaxation processes as well as in those pinning centres according to the annealing conditions and the nature of the nanocrystallization process. From our experimental results of $H_c(D)$ it was found a D^{3-4} law which could be explained by extending the original random anisotropy model.

Introduction.

In the development of soft magnetic materials over the last few years, an excellent nanocrystalline soft magnetic alloy called FINEMET has been discovered. The typical production process is to anneal the precursor amorphous alloy ribbon at about 540 °C leading to crystallization and forming Fe(Si) bcc grains of 10-15 nm. Such fine grains give rise to excellent soft magnetic properties. According to the random anisotropy model proposed by Herzer [1], these excellent properties result from the formation of a nanocrystalline structure where the intergrain distance is smaller than the exchange length, leading to an extremely low average anisotropy. The measurable macroscopic properties (e.g., the magnetic permeability, μ, and the coercive field, H_c) should strongly depend on the grain size D (μ, $H_c \propto D^6$ if one neglects the induced anisotropy) [2]. Besides the grain size, other factors contribute to the soft magnetic behaviour: such as the absence of FeB type compounds

[3], and the magnetostriction value in both the FeSi crystallites and in the remaining amorphous phase [3-5] which gives rise to the very low total magnetostriction in the nanocrystalline materials. In this work we present some results of the magnetization work (W_m) and coercive field (H_c) of nanocrystalline $Fe_{73.5}Cu_1X_3Si_{13.5}B_9$ (X= Nb and Ta) with different degree of the nanocrystalline state.

Experimental Procedure

Amorphous alloys with the composition given above were prepared by the melt-spinning technique. Dimensions of the ribbons were typically of about 2 mm x 20 μm of cross-section and they were cut with a length of 10 cm for magnetic measurements.
Thermal treatments carried out by conventional furnace and current-annealing technique were used in order to obtain the nanocrystalline state. The magnetization work and the coercive field were measured in the as-cast and annealed samples. The coercive field was evaluated from measurement of the quasi-static hysteresis loop. The magnetization work was determined from the longitudinal magnetization curve. The average grain size, D, was determined by X-Ray Diffraction and (TEM) Transmission Electrons Microscopy.

Results and Discussion.

Fig. 1 shows the evolution of the magnetization work, W_m, with the density of the current annealing (J_{ann}) of the Nb-containing sample for annealing of 1 and 10 minutes. The decrease of W_m with J_{ann} should be ascribed to the relaxation process of internal stresses because such relaxation giving rise to a significant decrease of the effective magnetic anisotropy. The grouth of W_m at J_{ann} = 45 A/mm^2 can be explained in terms of an increase of the local magnetic anisotropy (inside the grains) leading to some decoupling of the magnetic moments through the interface. This process should be ascribed to the nucleation of the first crystals. The consequent growth of these crystals after their nucleation result in a decrease of W_m. Observed minimum of W_m can be ascribed to the optimum nanostructure. Such decrease could be understood as a reduction of the effective magnetic anisotropy.

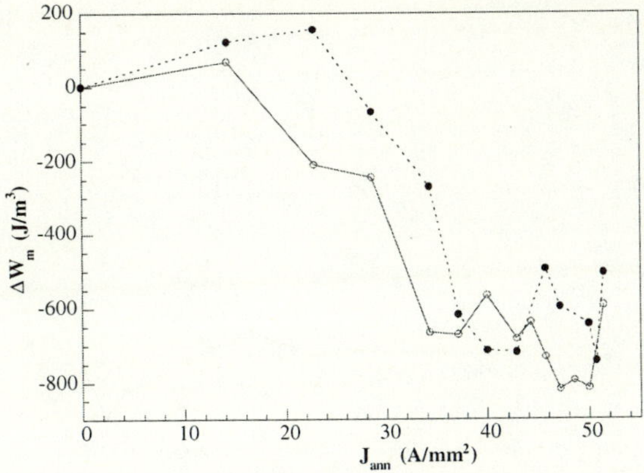

Figure 1. Magnetization Work, W_m, in sample $Fe_{73.5}Si_{13.5}B_9Cu_1Nb_3$ annealed at different current densities during 10 min (O) and 1 min (●).

Fig. 2 and fig. 3 show the variations of H_c submitted to thermal treatments (furnace and current-annealing) for the two alloys investigated. The initial decrease of H_c should be ascribed to the relaxation of internal stress leading to a decrease of the magnetic anisotropy and consequently the domain wall energy. The subsequent increase of H_c could be related with the precipitation of the first nucleus with grain size less than 10 nm (and small relative volume of this crystalline phase) acting as destabilization agents of the exchange coupling of the magnetic moments in the amorphous phase [6]. The growth of bcc Fe(Si) nanocrystals (grain size below 20 nm) and relative volume of crystalline phase is linked to the second decrease of the coercive field. The final increase of H_c is connected to the growth of grains (with size over 35 nm) and the precipitation, moreover, of boride compounds (mainly Fe_2B) with large magnetocrystalline anisotropy (430 $kJ.m^{-3}$) [7].

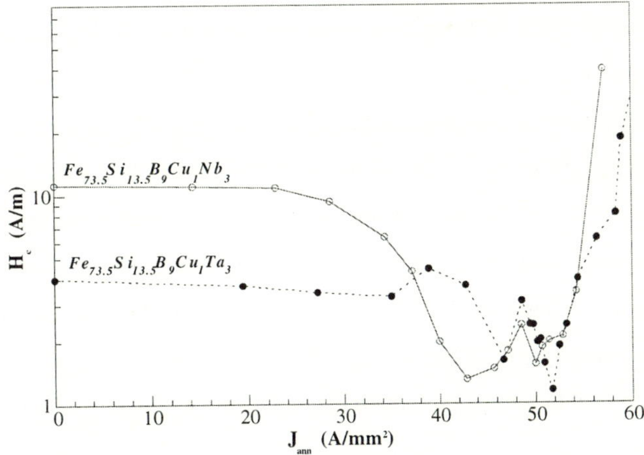

Figure 2. Coercive force in sample $Fe_{73.5}Si_{13.5}B_9Cu_1X_3$ annealed at different current densities during 1 min: X = Nb (O) and X = Ta (●).

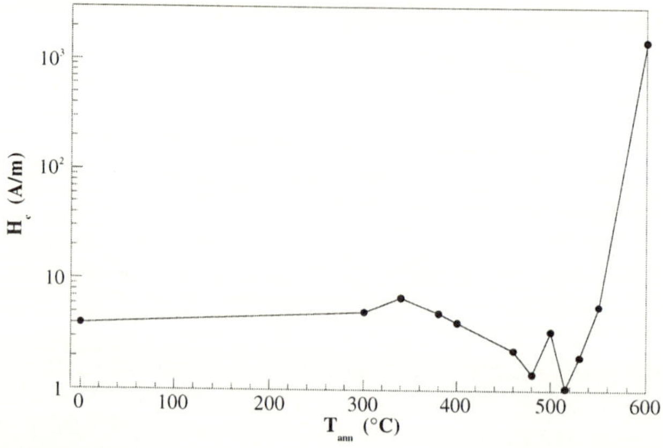

Figure 3. Coercive force in sample $Fe_{73.5}Si_{13.5}B_9Cu_1Ta_3$ annealed at different current densities during 1 min (●).

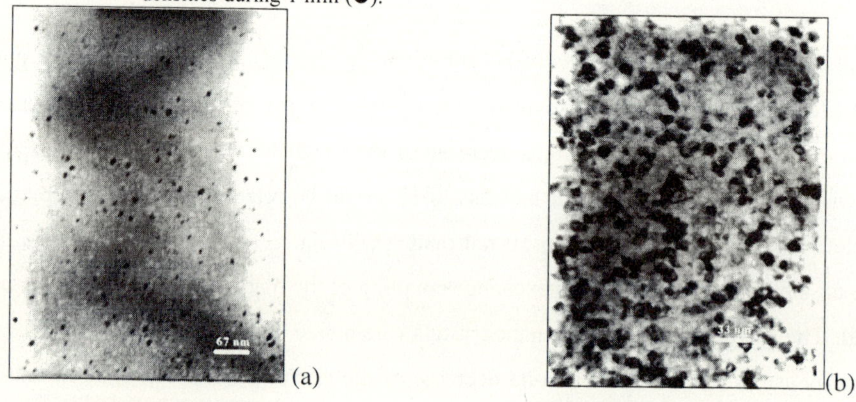

Figure 4. Transmision Electron Microscopies of $Fe_{73.5}Si_{13.5}B_9Cu_1Nb_3$ alloy annealed from room temperature to 540 °C.

In order to investigate the effect of the grain size on the coercivity, fitting our data on coercive field to a power law, we have obtained the dependence of H_c (D) in the two compositions according our experimental data. Experimental results of this dependence are shown in figure 4, leading to $H_c \approx D^{3-4}$, which results to be quite different to that of the well-known D^6. Recently, Suzuki et al. [8] have reported dependence of H_c with the third power of D which was explained by extending the original random anisotropy model. Such deviation of D^6 law could be associated with the presence of long-range uniaxial anisotropy which could modify the length exchange correlation value (with some anisotropic character becoming from the crystal-amorphous interface) and also to the presence of other interactions (different to that of exchange) [9] which can give some degree of decoupling of the crystalline grains.

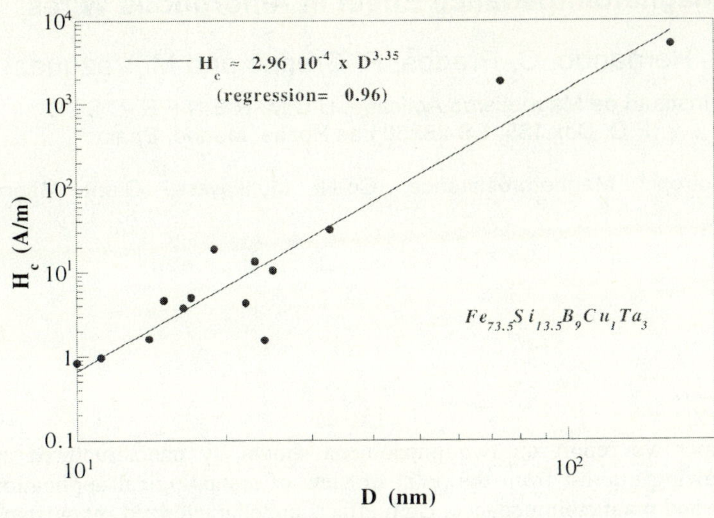

Figure 5. Coercive force in with grain size in samples $Fe_{73.5}Si_{13.5}B_9Cu_1X_3$ (X= Ta and Nb).

Acknowledgements

We gratefully thank to Dr. J. Echeverria from CEIT (San Sebastián, Spain) for make the Transmision Electronic Microscopies.

References

[1] Y. Yoshizawa, S. Oguma and K. Yamauchi, J. Appl. Phys. 64 (1988), p. 6044.
[2] G. Herzer, IEEE Trans. Magn. Mag-26 (1990), p. 1397.
[3] G. Herzer, Mater. Sci. Engin. A 133 (1991), p. 1.
[4] M. Knobel, R. Sato Turtelli and H.R. Rechenberg, J. Appl. Phys. 71 (1992), p. 6008.
[5] J. González, N. Murillo, J.M. Blanco, J.M. González and T. Kulik, J. Appl. Phys. 76 (1994), p. 1131
[6] M. Vázquez, P. Marín, H. A. Davies and A. O. Olanfijana, Appl. Phys. Lett. 64 (1994), p. 3184.
[7] A. Iga, Y. Tamara and A. Yamase, J. Phys. Soc. Japan 21 (1966), p. 404.
[8] K. Suzuki, G. Herzer and J. M. Cadogan, J. Magn. Magn .Mat. (in press).
[9] J. M. González, N. Murillo, J. González, J. M. Blanco and J. Echeverria, J. Mat. Res. 11 (1996), p. 512.

Materials Science Forum Vols. 302-303 (1999) pp. 254-262

Anisotropic Magnetoresistance in Co-Ni Multilayers and Giant Magnetoimpedance Effect in Amorphous Wires

A. Hernando, C. Prados, P. Crespo and M. Vázquez

Instituto de Magnetismo Aplicado, U.C.M.-R.E.N.F.E.-C.S.I.C.,
P.O. Box 155, ES-28230 Las Rozas, Madrid, Spain

Keywords: Anisotropic Magnetoresistance, Co-Ni Multilayers, Giant Magnetoimpedance, Amorphous Wires

Abstract.

In this paper we report on two phenomena shown by nanostructured and amorphous materials with growing interest from the point of view of technological applications: anisotropic magnetoresistence and magnetoimpedance. Both effects are characterized by outstanding variations of resistance and impedance, respectively, for low applied fields which enable the construction of magnetic field sensors. In this paper we present an overview of these phenomena and their applications.

Introduction.

In recent years, research on magnetism has been focused on heterogeneous materials which exhibit a wide range of magnetic properties from the very soft amorphous alloys to the Rare earth-Transition metal-Metalloid type permanent magnets.

It has been shown that the important aspect to take into account in heterogeneous systems is the strength of exchange coupling between their different components [1]. A critical parameter of these systems is the structural compositional fluctuation length (l), that can be varied from a few angstrom to some microns. The relationship between this parameter and the magnetic exchange correlation length (L), or the electron mean free path (λ) will determine the resulting magnetic and electric behaviour respectively. If $L>l$, the magnetization process is collective giving rise to a soft magnetic material whereas if $L<l$, each phase magnetizes separately, resulting in a harder material. On the other hand, if $l<\lambda$, the resulting conductivity will no longer be the conductivity of the individual components generating such new kind of magnetoelectric effect as the so-called giant magnetoresistance (GMR). Since its discovery few years ago (Baibich [2]), this effect has attracted a large interest because of its potential technological applications. In general, the magnetoresistance, MR, can be described as the relative changes in the resistivity of the material under the application of dc magnetic fields. Among the different types of MR observed in magnetic materials three should be noted: anisotropic (AMR), giant (GMR) and colossal magnetoresistance (CMR). AMR [3], which appears in traditional 3d magnetic materials is anisotropic with respect of the field direction. MR is positive (i.e. resistance increases with the applied field) when the magnetic field is applied parallel to the current, and negative when perpendicular. On the other hand, GMR and CMR, which have been discovered more recently in multiphase systems and perovskite oxides, respectively [2, 4], are characterized by large MR factors and by their isotropy with respect to the field direction (both are always negative). The very different macroscopic transport properties of AMR, GMR and CMR materials are a consequence of the distinct underlying microscopic physics. The intrinsic origin of AMR is the spin-orbit coupling, this changes the shape of the electron cloud and creates a local anisotropy in each domain; GMR is due to spin-dependent scattering of electrons [5-8] and CMR comes from the superexchange between Mn^{3+} and Mn^{4+} through oxygen atoms.

As concerns with amorphous materials, some works have also reported huge values of MR using ac currents for samples in the shape of wire or ribbon. In this case, the effect was also called ac magnetoresistance since a small ac current (frequency in the range from 10-1000 Hz) flows along the sample so that the changes in the resistance are measured as a function of an external dc magnetic field. These first results were tentatively explained with similar arguments as for the GMR observed in heterogeneous systems, although it seems now that this so called ac-GMR can be rather explained considering the circular magnetization process of wire or ribbon shaped amorphous alloys when working at relatively low frequencies of the flowing current. In the case of higher frequencies (above 0.1 MHz) the effect is related with the well-known classical skin effect, which in these ferromagnetic materials is closely connected with the circular magnetization process. In this range of frequencies, changes of both real and imaginary parts of the impedance are detected when an dc magnetic field is applied, resulting in a dramatic change in impedance. This effect has been called Giant magnetoimpedance (GMI) [9].

In this work firstly we will report on the appearance of extraordinary anisotropic magnetoresistance factors in Co/Ni sputtered multilayers. Magnetic softness, large anisotropic magnetoresistance and a special contact configuration give rise to magnetoresistance factors as large as 130% at room temperature and under applied fields of 40 Oe. Secondly, the GMI effect observed in Co-rich amorphous wires will be described. These wires, obtained by the in-rotating-water technique meet the conditions (geometry, domain structure, domain wall mobility, low and negative magnetostriction,...) for an efficient GMI phenomenon to occur.

Experimental

Different sets of multilayered samples have been grown by the planar magnetron sputtering technique on Si (111) water cooled substrates. Samples will be designated as follows: $(Ni_xCo_y)_n$, where x and y mean the thickness in Å of Ni and Co respectively and n is the number of bilayers deposited. Typical dimensions and number of bilayers are x, y=25-50 Å and n=40, respectively. Detailed description of growing conditions and structural characterization can be found in reference [10].

Hysteresis loops have been measured also at room temperature using a vibrating sample magnetometer. MR measurements have been carried out with the four lead technique using a Fluke 8842a multimeter and pressure contacts in magnetic fields up to 5000 Oe at room temperature.

Amorphous ferromagnetic wires of composition $(Co_{0.94}Fe_{0.06})_{72.5}B_{15}Si_{12.5}$, were prepared by the in-rotating water technology and kindly provided by UNITIKA Ltd. Japan. GMI measurements have been carried out by measuring the change of the ac voltage, in phase and out of phase with the constant ac driving current, as a function of a dc applied magnetic field.

Anisotropic magnetoresistance in Ni/Co multilayers

Ni/Co multilayers exhibit soft magnetic properties and they behave as an exchange coupled material. Typical hysteresis loop is shown in fig 1. Coercive field is below 5 Oe and there is no presence of any step in the magnetization curve.

Magnetoresistance measurements have been carried out in three sets of ferromagnetic/ferromagnetic multilayers $(Co_xNi_y)_{40}$ with x=25, y=25; x=25, y=50; x=50, y=50, using the four lead technique.

The four lead setup consists of four Cu strips forming a square of 3 mm. Different contacts geometries have been used as it is shown in fig. 2. Samples were cut in circular shape with the electric current and the applied field always in the film plane: (a) -longitudinal- $(V/I)_l$, driving the current along one of the sides of the square parallel to the magnetic field, I_{mp}, and measuring the voltage V_{no}. (b) – transversal - $(V/I)_t$, driving the current along one of the sides transverse to the magnetic field I_{mn}, and getting the voltage drop in the other V_{op}. (c) – cross - $(V/I)_c$, driving the current along one of the diagonals, I_{np}, and measuring the voltage drop along the other V_{mo}. In all cases the current I is kept constant at 1 mA [10].

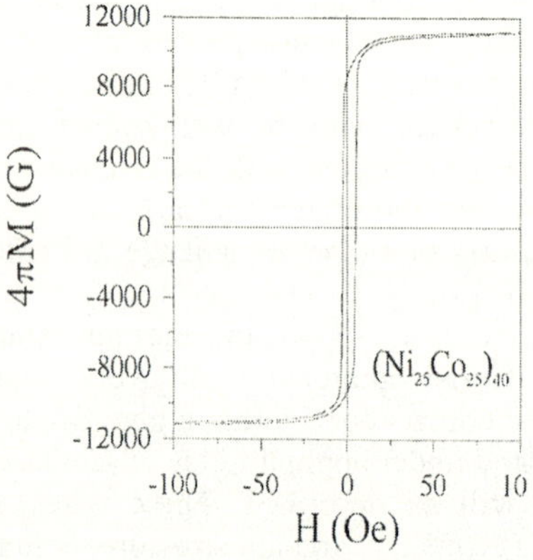

Fig.1. A typical hysteresis loop of a Ni/Co multilayer. Coercive field is around 5 Oe.

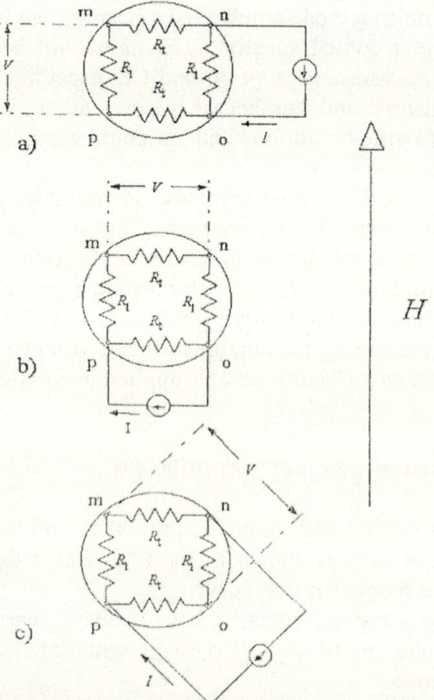

Fig.2. Resistor arrangement for the different measuring configurations: a) longitudinal, b) transversal and c) cross configuration.

Magnetoresistance loops at room temperature in the three above mentioned configurations have been recorded. The results for $(Ni_{25}Co_{50})_{40}$ are shown in fig. 3 (not shown). It is worth remarking the high value of the ordinary AMR ratio exhibited by these samples (about 6%) with very low applied field. The values obtained for other compositions are displayed in table 1 along with other magnetic properties obtained by means of hysteresis loops.

Moreover, this high value has been enhanced by using the cross configuration up to values of 130% giving rise, together with the soft magnetic properties exhibited for example in fig 1, to sensitivities of 6% Oe^{-1} (see Fig. 4). Three features come together in these samples to account for this extraordinary MR: the magnetic softness (compared to traditional GMR multilayers and granular solids, or the other magnetic/magnetic superlattices grown with other techniques), the large ordinary AMR and the cross configuration of current-voltage drop which enhances the AMR effect.

The electric behaviour of the four-lead setup can be easily modeled by a four resistor arrangement as shown in fig. 2. By elemental circuital laws, it is found that the magnitudes $(V/I)_l$, $(V/I)_t$ and $(V/I)_c$ are related with R_l and R_t through:

$$(V/I)_l = \frac{1}{2}\frac{R_l^2}{R_l + R_t}$$

$$(V/I)_t = \frac{1}{2}\frac{R_t^2}{R_l + R_t} \qquad (1)$$

$$(V/I)_c = \frac{1}{2}\left|(R_l - R_t)\right|$$

In the case of the cross configuration the sample can be assumed to be a Wheatstone bridge. Supposing a perfectly homogeneous sample and perfectly symmetric contacts and no applied field, R_t would be equal to R_l and according to eq. 1 there would be no voltage drop between m and o. After applying a magnetic field, and due to the anisotropic character of the ordinary magnetoresistance, R_t decreases and R_l increases giving rise to a voltage drop in m-o.

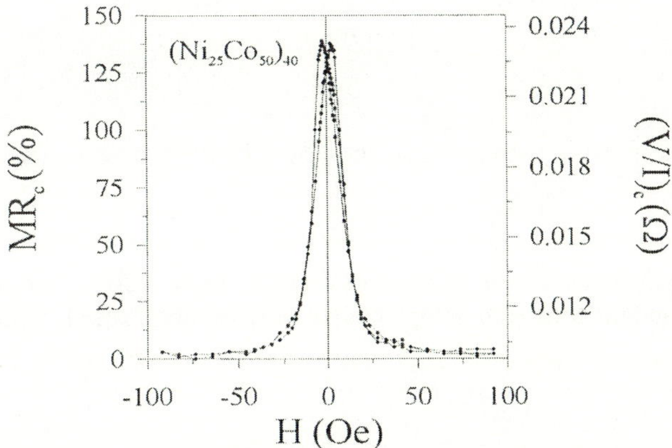

Fig.4. Magnetoresistance loop for a $(Ni_{25}Co_{50})_{40}$ sample in the cross configuration.

Further MR measurements were performed in different sets of samples with different interface areas and layer thicknesses in order to study the effect of the interfaces on the MR. The samples grown for such study were a) $(Ni_{25}Co_{50})_{40}$ already characterised; b) $(Ni_{1000}Co_{2000})_1$; c) $(Ni_{25}Co_1)_{40}$ and d) $(Ni_{1000}Co_0)_1$. The 1 Å of Co does not mean a uniform Co layer of 1 Å but a deposition time proportional to that thickness [11]. This time is insufficient to fill a complete layer, hence giving rise to an impurity layer.

The magnetoresistance measurements have been carried out as described above using the longitudinal and transverse configurations. MR loops are plotted in fig. 5, 6 and 7. At sight of these figures, interesting comparisons can be made. Similar behaviour appears between samples c) and d): when Co impurities are inserted in the sample the MR factor is enlarged by a factor of two. This fact points out that since MR increases by 100% with respect to the MR in the Ni sample, the scattering in the Ni bulk must be of the same order of magnitude as that of the Co impurities, the influence of the interfaces being corroborated this way. To reinforce this idea it is also worth noting that although sample c) does not have whole Co layers, the value of the MR is close to that of the multilayer a).

A consequence of the Co exchange penetration in the Ni layer is the magnetic softness of the Ni/Co multilayers presented here. The magnetic softness along with the large AMR factor, greatly due to the scattering at the interfaces make the Ni/Co multilayered system a candidate for technological applications.

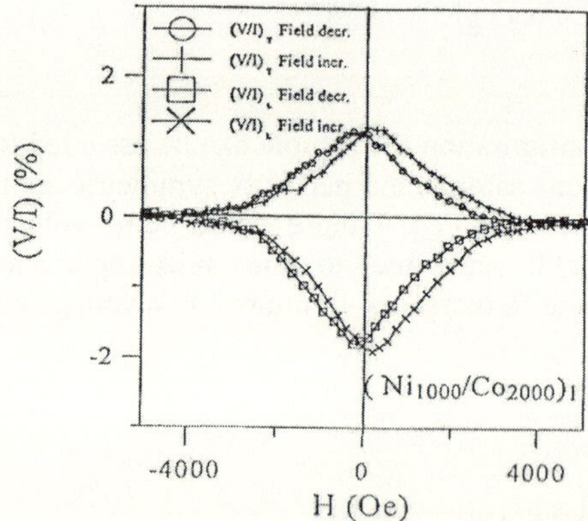

Fig.5. Magnetoresistance loop for a $(Ni_{1000}Co_{2000})_{40}$ multilayer in the transverse (up) and longitudinal (low) configurations.

In order to greatly enhance the magnetoresistance -up to 130%-, we also propose a cross configuration in the standard four-lead setup, with an eye on applications such as field or current sensors.

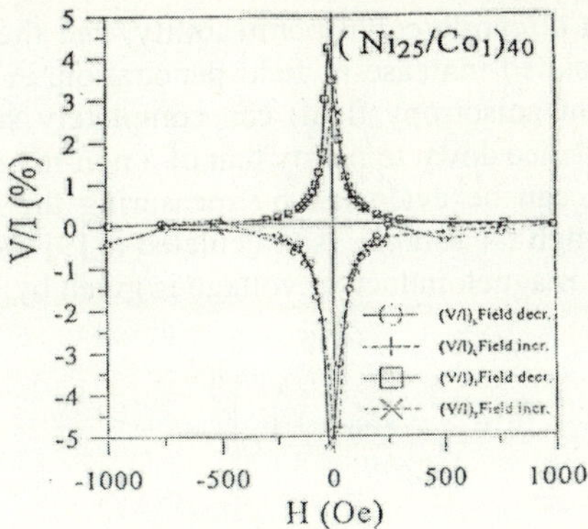

Fig.6. Magnetoresistance loop for $(Ni_{25}Co_1)_{40}$.

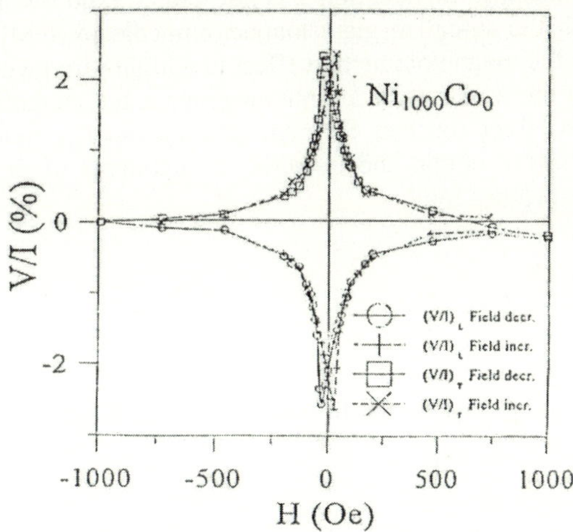

Fig.7. Magnetoresistance loop for $(Ni_{1000}Co_0)$.

Giant Magnetoimpedance in amorphous wires.

The giant magnetoimpedance effect (GMI) has attracted great interest recently by its technological applications in magnetic sensors and devices. GMI can be described as a large relative change of the ac impedance response (up to around 300%) of a ferromagnetic material when it is submitted to a dc magnetic field. GMI has been observed in ferromagnetic wires and ribbons, whose magnetic structure includes a considerable fraction of magnetic domains with circumferential orientation and hence large circumferencial permeability. Both high frequency and high permeability lead to the skin effect, by reducing the field penetration to a small cross section near the surface of the material. The application of a dc magnetic field in the axial direction of the wire or ribbon results in a decrease in circumferential permeability, and therefore, in a decrease of its impedance response, by the associated increase in field penetration. A magnetic field sufficiently strong (greater than the equivalent anisotropy field) can completely supress the circumfenrential magnetization, reducing the impedance down to nearly that of a non-magnetic material.

Generally, the impedance can be evaluated by measuring the voltage at the ends of the sample when a current passes through the sample. As calculated in [9] the maximum relative change in impedance originating from the magnetoinductive voltage is given by:

$$\frac{Z_{max}}{Z_{min}} = \left(1 + \left(\frac{\pi \mu_o R^2 \chi_{\phi max}\, \nu}{2\rho}\right)^2\right)^{1/2} \qquad (2)$$

Therefore, the variation in impedance increases with the circular susceptibility χ_ϕ, and the frequency, ν, of the current flowing through the wire and decreases with increasing resistivity, ρ. Some of the technological applications based in this magnetoinductive effect can be found in reference [12].

However at high enough frequencies the equation 2 is not longer valid owing to the skin effect. In this case we are dealing with the so-called giant magnetoimpedance (GMI) effect, which can be considered as an extension of the magnetoinductive effect to a higher frequency range.

An a.c. current flowing along the wire creates an inhomogeneous a.c. magnetic field. In the case of very high frequency, the skin effect confines the penetration of such a field to a reduced surface shell giving rise to changes in the ohmic and inductive components of the impedance. The skin penetration depth, δ, can be expressed by:

$$\delta = \left(\frac{\rho}{\chi_\phi\, \nu}\right)^{1/2} \qquad (3)$$

The total impedance, Z, is related to that penetration depth through the expression:

$$Z = \frac{1}{2} R_{dc} (kR) \frac{J_o(kR)}{J_1(kR)} \qquad (4)$$

being R_{dc} the resistance for ν=0 and J_i the Bessel functions with k=(1+I)(R/δ).

According to eq. 4, changes in impedance can be produced by modification of the resistivity and the permeability. The circular magnetic permeability is strongly reduced by the presence of an external biasing magnetic field. Consequently the penetration depth increases with the external field and the total impedance decreases in a parallel way. Alternatively, changes in permeability can be performed by applying mechanical stresses or by modifying the resistivity, for example via nanocrystallization. Figure 8 shows the change in impedance in a nearly non magnetostrictive Co based wire as a function of d.c. axial field when performing the experiment at a constant a.c. current through the wire (10 mA r.m.s.) and at different working frequencies. Fig 9 shows the change in total voltage at the end of the wire when initially a given a.c. current is set and then the d.c. field is applied. Note that in that case the current remains no longer constant after the application of the field.

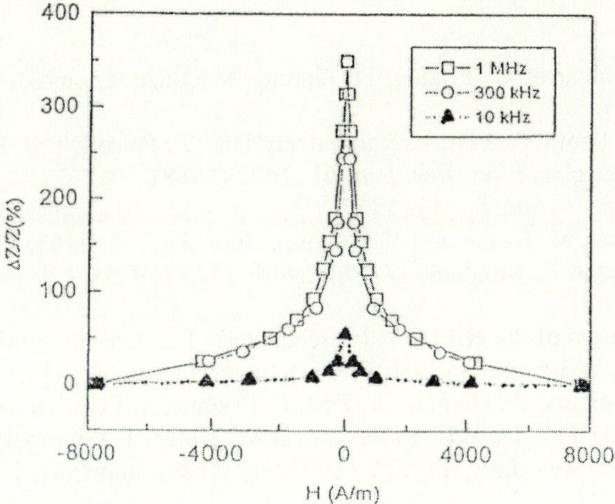

Fig.8. d.c. axial field dependence of GMI effect at various working frequencies and constant a.c. current (10 mA r.m.s.) flowing through the wire.

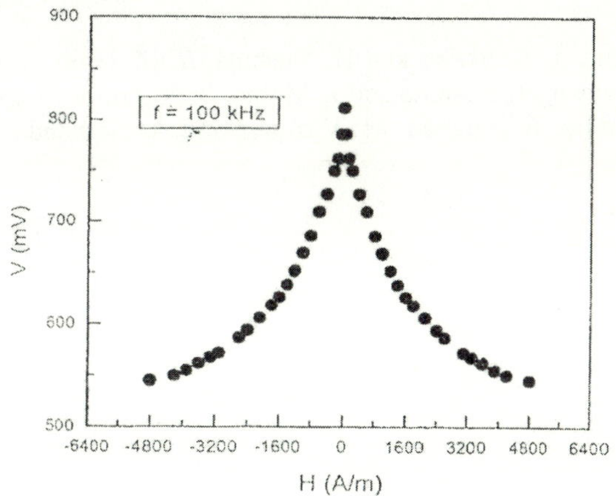

Fig.9. GMI voltage at the ends of the wire as a function of the dc applied field in the case where the a.c. current (10 mA r.m.s.) is initially set but not kept constant during the application of the dc field.

This effect can be practically used in magnetic sensors for detecting d.c. magnetic fields. It should be noted that large GMI is obtained by applying fields of the order of units of Oe or less, which is several orders of magnitude smaller than the field required to observe GMR effect. Also, the change in GMI at the lowest magnetic fields is outstandingly sensitive. Several attempts are presently in progress to develop this kind of magnetic field sensor [13]. Alternatively GMI effect can be used for measuring electrical current just by detecting the magnetic field created by such a current.

REFERENCES

[1] A. Hernando, I. Navarro, C. Prados, D. García, M. Vázquez and J. Alonso. *Phys. Rev. B* **53**(13), 8223 (1996).

[2] M.N. Baibih, J.M. Broto, A. Fert, F. Nguyen van Dau, F. Petroff, P.E. Etienne, G. Creuzet, A. Friedich, and J. Chazelas, *Phys. Rev. Lett.* **61**, 2472 (1988).

[3] W. Thompson, *Prod. R. Soc.* **8**, 546 (1857).

[4] K. Chanara, T. Ohno, M. Kasai, and Y. Kozono, *Appl. Phys. Lett.* **63**, 1990 (1993).

[5] J.W.F. Dorlejin and A.R. Miedema, *J. Phys. F* **5**, 1543 (1975); J.W.F. Dorlejin, Philips Res. Resp. **31**, 287 (1976).

[6] J. Mapps, Proceedings of the NATO Advanced Study Institute on Applied Magnetism, edited by R. Gerber, C.D. Wright and G. Asti Erice (Kluwer, Dordrecht, 1992) pp 377-403.

[7] F. Petroff A. Barthélémy, A. Hamzic, A. Fert, P. Etienne, S. Lequien and G. Creuzet, *J.Magn. Magn.Mater.* **93**, 95 (1991); E.E. Fullerton, D.M. Kelly, J. Guimpel, I.K. Schuller and Y. Bruynseraede, *Phys. Rev. Lett.* **68**, 859 (1992); S. Zhang and P.M. Levy, Phys. Rev. B **43**, 11048 (1991).

[8] M. Suzuki and Y. Taga, *Phys. Rev. B* **52**, 361 (1995).

[9] J. Velázquez, M. Vázquez, D. -X. Chen and A. Hernando. *Phys. Rev. B*, 50 16737 (1994).

[10] C. Prados, D. García, F. Lesmes, J. J. Freijo, A. Hernando. *Appl.Phys. Lett.* 67 718 (1995).

[11] F. Lesmes, A. Salcedo, J.J. Freijo, D. García, A. Hernando and C. Prados, *Appl. Phys. Lett.* **69**, 1 (1996).

[12] K. Mohri, K.Kawashima, T. Kohzawa and H. Yoshida, *IEEE Trans. Magn.* **29**, 1245 (1993); K.Kawashima, T. Kohzawa, H, Yoshida and K.Mohri, *IEEE Trans. Magn.* **29**, 3168 (1993).

[13] R. Valenzuela, J. J. Freijo, A. Salcedo, M. Vázquez and A. Hernando. *J. Appl. Phys.* 81(8) 4301 (1997).

Materials Science Forum Vols. 302-303 (1999) pp. 263-272

Physics of Nanocrystalline Magnetic Materials

M. Dahlgren and R. Grössinger

Institut für Experimentalphysik, TU-Vienna, Wiedner Hauptstrasse 8-10, AU-1040 Vienna, Austria

Keywords: Exchange Coupling, Anisotropy Field, Remanence, Nanocrystalline Materials

Abstract

The mechanism of exchange coupling will be discussed in single phase hard magnetic materials as well as in nanocomposites. Temperature dependent anisotropy field H_A measurements will be presented on Pr-Fe-B samples. Additionally the temperature dependence (4.2K < T < 300K) of the saturation magnetisation, of the remanence and of the coercivity will be presented. From the ratio J_r/J_S the degree of enhancement can be deduced. At lower temperatures a decoupling of the grains was observed which becomes also visible in the $H_A(T)$ behaviour. The influence of the exchange coupling on the Curie temperatures as well as on the spin reorientation temperature for Nd-Fe-B and Pr-Fe-B based samples is shown.

1. Introduction

Nanocrystalline materials exhibit excellent soft and hard magnetic properties. In both cases the exchange coupling acts over the grain boundaries; the main difference results from the ratio exchange length to grain size. For soft magnetic materials the magnetic exchange length is larger then the grain size D, therefore an averaging of the magnetocrystalline anisotropy occurs which leads to an extremely low coercivity and a very high permeability. Well known examples are the Finemet type material which is based on Fe-Cu-Nb-B-Si as well as Fe-Zr-Cu-B [1,2]. In both cases according to the random anisotropy model of Alben-Becker-Chi [3] the anisotropy of the nanocrystalline Fe-Si respectively the α-Fe is averaged and is reduced by a factor D^6. Additionally in the nanocrystalline state these materials exhibit a low magnetostriction which is a result of the superposition of the negative Fe-magnetostriction and the positive magnetostriction of the amorphous matrix [4]. In hard magnetic nanocrystalline materials the magnetic exchange length is smaller than the grain size but still comparable. This leads to a remanence enhancement which makes these materials interesting for applications. The remanence enhancement was first observed in Nd-Fe-B-Si [5] and later also in many other compounds [6,7]. Exchange coupled hard magnetic materials can be divided into two main groups:

- single phased nanocrystalline
- nanocomposites, also known as spring magnets (contains a hard + soft magnetic phase)

Those materials consist of isotropic hard and soft magnetic grains with diameters ranging from 20 to 50 nm. Previous investigations showed that when the grain sizes decrease under ~50 nm they exhibit magnetic properties which are different from those of microcrystalline magnetic materials [8].

For applications the most interesting feature is an enhanced remanence for isotropic grains with uniaxial easy axis. Here the ratio M_r/M_s is higher than the expected 0.5 for noninteracting unaixial grains according to [9] (Eq. 1).

$$J_r = J_s \frac{\int_0^{2\pi}\int_0^{\pi/2} \cos\vartheta \sin\vartheta d\vartheta d\varphi}{\int_0^{2\pi}\int_0^{\pi/2} \sin\vartheta d\vartheta d\varphi} = \frac{1}{2} J_s \qquad \text{(Eq. 1)}$$

Additionally they exhibit a single phase type magnetic hysteresis loop instead of the expected two phase type hysteresis loop that is measured for microcrystalline magnets. Also the shape of the recoil curves is characteristic for spring type magnets [10].

It was shown already in previous works that due to the exchange coupling the anisotropy field, the on-set temperature for FOMP [11], the Curie temperature [12] and the spin reorientation temperature [13] are affected. Up to now these effects were demonstrated on Nd-Fe-B based material, however there the spin reorientation complicates the situation. In order to overcome these problems we concentrated in the present work on Pr-Fe-B based samples where first the exchange coupling between hard magnetic grains and secondly the exchange coupling between hard and soft magnetic grains will be discussed.

2. Exchange Coupling

2.1 Exchange Coupling Between Hard Magnetic Grains

If two grains with different direction of their easy axis are exchange coupled then the magnetic moment will be tilted out of the easy axis direction. This phenomenological picture of the exchange interaction between two grains is sketched in Fig. 1. By applying this simple picture to an agglomeration of grains we would expect that the grains are interacting with their neighbours resulting in an increase of the remanence.

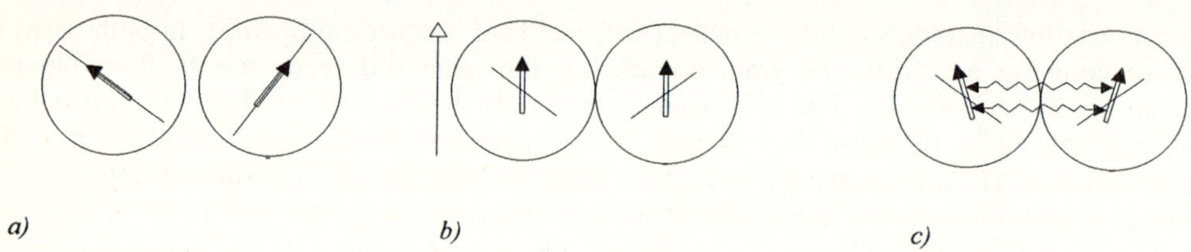

a) *b)* *c)*

Figure 1 A schematic picture of the influence of exchange coupling between two hard magnetic grains with different easy axis directions. a) Two isolated grains. b) Two grains under an applied field high enough to fully align the magnetic moment. c) Two grains whose grains are tilted out of the easy axis due to exchange interaction.

The coupling between two hard magnetic grains is governed by a competition between the exchange energy and the magneto-crystalline anisotropy energy, since $\frac{K}{J_s^2/4\mu_0}$ is much larger than 1. A representative parameter that would describe the exchange coupling would be the exchange length ℓ_K which is proportional to $(A/K)^{1/2}$. For $Nd_2Fe_{14}B$ the exchange length is 3 - 4 nm and the influence of the exchange length decays exponentially (Eq. 2).

$$\beta(x) = \beta_0 \exp\left(-\frac{x}{\ell}\right) \qquad \text{(Eq. 2)}$$

where ℓ is the extension at which distance the disturbance $\beta(x)$ is 1/e of the original disturbance

By assuming an exponential decay of the exchange coupling the simple picture in Figure 1 has to be revised. A more realistic picture is shown in Figure 2 where the exponential decay is taken into account. The nanocrystalline materials that have been reported with an enhanced remanence have grain sizes ranging between 20 to 30 nm. Comparing the exchange length of 3-4 nm with grain sizes of 20-30 nm it follows that only parts of the grain are coupled.

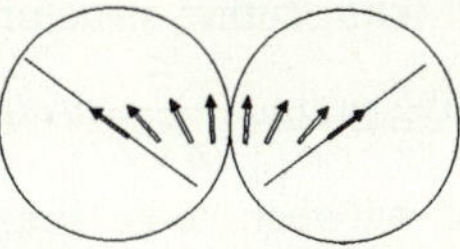

Figure 2 A sketch of the distribution of coupling between two hard magnetic grains

For ℓ_K between 3-4 nm and D between 20-30 nm a measure of the exchange coupling gives R_{1D} ranging from 20% to 40% of the grain. A real magnet consist of 3D grains, therefore the coupling has to be considered on a 3D level. Then the extension of the exchange coupling can be described as the ratio of the coupled volume of the grain to the total grain volume.

$$R_{3D} = \frac{V_{coupl}}{V_{tot.}} = \frac{(D/2)^3 - (D/2 - \ell_K)^3}{(D/2)^3} = 1 - \left(1 - 2\frac{\ell_K}{D}\right)^3 \qquad \text{(Eq. 3)}$$

For ℓ_K between 3-4 nm and D between 20-30 nm (Eq. 3) gives a measure of the exchange coupling R_{3D} ranging from 50 % to 80 % of the grains.

The quantum mechanical mechanism of the coupling between the grains is not considered in the micromagnetic model because the exchange energy is not treated on an atomic scale.

The exchange interaction for the rare-earth atoms is of the RKKY type whereas for the Fe an itinerant exchange is responsible. For $Nd_2Fe_{14}B$ the exchange between the Fe (3d-3d interaction) electrons, the Nd (4f-4f interaction) electrons and the Fe-Nd (3d-4f interactions) electrons have to be considered, but the itinerant 3d-electrons may be more important.

2.2 Interaction Between Soft and Hard Magnetic Grains

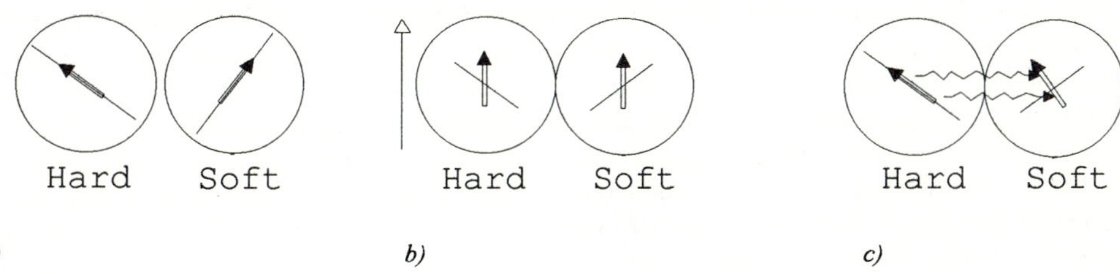

Figure 3 A schematic picture of the influence of interaction between a hard and a soft magnetic grain with different direction of the easy axis a) Two isolated grains. b) Two grains under an applied field high enough to fully align the magnetic moment c) Two grains where the soft magnetic grain is polarised by the hard magnetic grain.

In the case of nanocomposites the magnetisation of the soft magnetic phase shall be higher than that of the hard magnetic phase in order to be attractive for technical applications. In this case the polarisation of the soft magnetic grains enlarges the remanence and causes an enhancement even larger than that due to the exchange coupling among the hard magnetic grains. The remanence enhancement due to soft magnetic grains is phenomologically shown in Fig. 3.

For an agglomerate of hard and soft magnetic grains the magnetic moment of the soft magnetic grains will be polarised in the average direction of the surrounding magnetic moment of the hard magnetic grains. This will lead to an enhancement of the remanence.

It should be noted that for α-Fe with cubic structure the ratio remanence to saturation polarisation is for isotropic non-interacting grains $\frac{6}{\pi\sqrt{2}} arctg \frac{1}{\sqrt{2}} \approx 0.839$ [14]

Taking into account the stray field between α-Fe grains the remanence to saturation ratio becomes smaller and can approach 0,5. Therefore to examine if an enhanced remanence is due to interaction between the hard magnetic grains with uniaxial symmetry (hexagonal or tetragonal) and soft magnetic α-Fe grains with cubic structure the ratio remanence to saturation has to exceed as an upper limit the theoretical one based on a simple addition of the two contributions of noninteracting grains as shown in Eq 4.

$$\frac{J_r^{tot}}{J_s^{tot}} = \frac{1}{J_s^{tot}}\left(\nu_H \frac{1}{2} J_s^H + (1-\nu_H)\frac{6}{\pi\sqrt{2}} arctg\left(\frac{1}{\sqrt{2}}\right) J_s^S\right) \qquad \text{(Eq. 4)}$$

where ν_H is the volume fraction of the hard magnetic grains, J_s^H is saturation polarisation of the hard magnetic grains and J_s^S is saturation polarisation of the soft magnetic grains.

To examine from the micromagnetic point of view how the polarisation of the soft magnetic grains occurs the different energy contributions have to be studied. If the soft magnetic grains would be of several 100 nm, then the magnetic state would be determined by a competition between the stray field energy and the exchange energy ($\frac{K}{J_s^2/4\mu_0}$ is much smaller than 1). But when the grains become smaller (under 50 nm) the stray field energy seems to be negligible in this case even exchange coupling between the nanocrystalline soft magnetic grains has to be assumed. One example for that are nanocrystalline soft magnetic materials. Their excellent soft magnetic properties are understood from an averaging of their anisotropy constant due to an exchange coupling of several grains. Here the stray field energy has no influence and exchange coupling is given by ℓ_K. ℓ_K is the exchange length that is determined from a competition between the exchange energy A and the magneto-crystalline anisotropy energy. This can also be understood by the fact that stray field energy is a long range interaction with a classical dipolar character whereas the exchange coupling is a short range interaction. In the case for nanocrystallites the short range interaction seems to be more important which acts on a quantum mechanical scale

The main difference compared to exchange coupling for only hard magnetic grains is that the coupling inside the soft magnetic grain is mediated by the 3d-electrons. This is also a hint that the coupling mechanism between the grains should be mediated by the 3d-electrons as well for the soft-hard grains coupling as for the hard-hard grains.

The exchange hardening of the soft magnetic phase by the hard magnetic phase can be understood if the exchange coupling between the soft and hard magnetic grains acts as an exchange stiffness of the soft magnetic spins. Kneller and Hawig [15] developed a 1D model to describe how the exchange hardening occurs. The soft phase is effectively hardened by the magneto-crystalline anisotropy of the hard phase, because the large magneto-crystalline anisotropy results in a large domain wall energy density for the hard phase. Essentially the same process is described in a numerical model by Schrefl *et al.* [16] for soft magnetic grains embedded in a hard magnetic matrix.

3. Anisotropy field

It is not an easy task to determine the anisotropy field in these intrinsic isotropic samples. In addition, the strong magnetic exchange interaction among nanocrystallites makes the measurements even more difficult. In this respect, the singular point detection (SPD) technique [17] provides a unique method in determining directly the anisotropy field of isotropic samples. In Fig. 4 the temperature dependencies of the anisotropy field measured for $Pr_9Fe_{85}B_6$, $Pr_{11.76}Fe_{82.36}B_{5.88}$ and $Pr_{18}Fe_{76}B_6$ are shown.

It is worthwhile to note that the anisotropy field measured by using the SPD technique corresponds only to that of the hard magnetic phase. In the case of $Pr_9Fe_{85}B_6$ which comprises of $Pr_2Fe_{14}B$ and α-Fe, that is the $Pr_2Fe_{14}B$ phase. The reduction of the anisotropy field in the nanocrystalline $Pr_{12}Fe_{14}B$ grains is a strong indication of the presence of magnetic exchange interactions (see also Fig. 1 and 2). In case of the microcrystalline sintered Pr-Fe-B magnets the magnetic exchange interaction among the micro-sized grains is insignificant to influence the easy magnetisation direction (EMD) of the grains. Thus the anisotropy field measured in the microcrystalline magnets corresponds to the anisotropy field of isolated $Pr_2Fe_{14}B$ grain. However, when the average grain size is reduced to nanometer size (see $D_V < 50$ nm), the exchange interaction among the hard magnetic grains may become so strong that the EMD´s of nanocrystallites can be slightly tilted without the presence of external fields. The actual direction of the final EMD could be the energetically compromise of the surrounding nanocrystallites. This slightly tilted EMD gives a naturally reduction of the anisotropy field. This reduction of H_A shows also in the SPD peak of the d^2M/dt^2 versus H measurement where the peak indicates the position of H_A (see Fig. 5). From Fig. 4 the systematic decrease of $H_A(T)$ with increasing coupling between the grains going from too high Pr content (18% Pr) to a stoichiometric sample (12% Pr) to a low Pr content material (9% Pr) becomes clearly visible for higher temperatures. For lower temperatures a decoupling of the grains occur causing equal values of H_A for all three samples. This is also a proof that the differences in H_A as measured at room temperature are not a result of a disturbed surface.

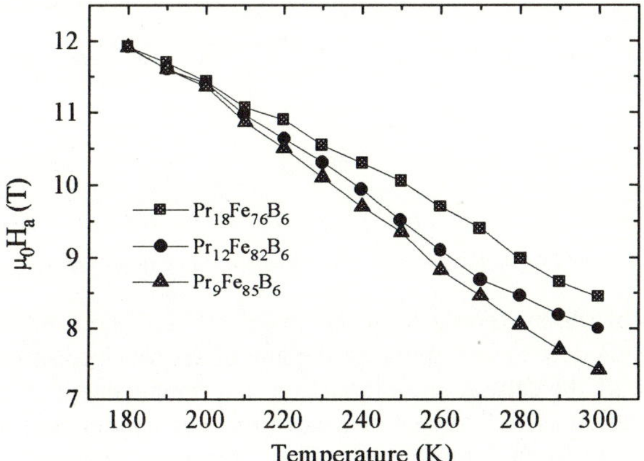

Figure 4 The temperature dependence of the anisotropy field for hot pressed mechanically alloyed Pr-Fe-B.

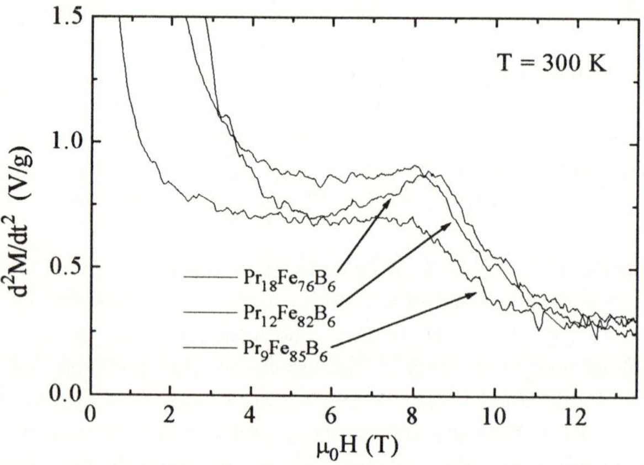

Figure 5 SPD peak measured at 300 K for mechanical alloyed hot pressed Pr-Fe-B samples with nano-sized grains.

4. Magnetisation and Remanence

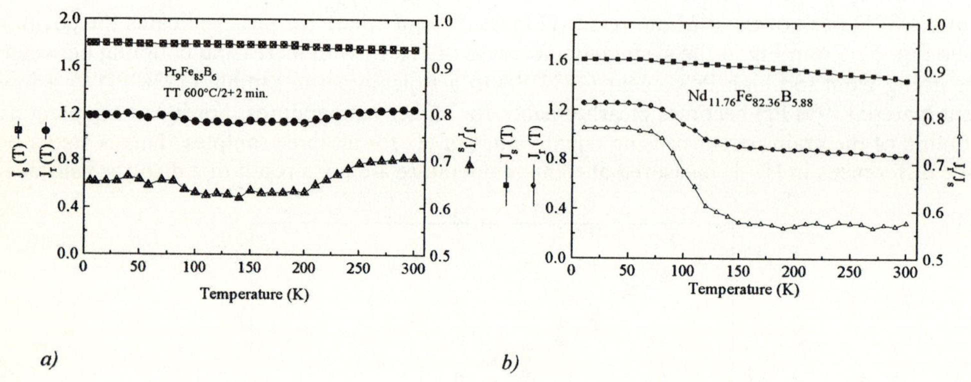

Figure 6 Temperature dependence of the saturation polarisation J_s, the remanence J_r and the ratio J_r/J_s for nanocrystalline a) mechanically alloyed $Pr_9Fe_{85}B_6$. and b) rapidly quenched stochiometric $Nd_{11.76}Fe_{82.36}B_{5.88}$.

From the measurements of the hysteresis loop the coercivity and the remanence J_r is obtained. The saturation polarisation was determined applying the law of approach to saturation. Fig. 6 shows the temperature dependence of the saturation polarisation, the remanence and the ratio J_r/J_S of the sample with low Pr concentration as well as that of a stochiometric nanocrystalline Nd-Fe-B sample as a function of the temperature. As a comparison in Fig. 7 the same is shown for the sample with high Pr content (18% Pr). In a previous work these properties were already studied on the Nd-Fe-B system [11]. Because of the spin reorientation which happens in $Nd_2Fe_{14}B$ at 135K an easy cone is formed at low temperatures which is caused by a reduction and change of sign of the first order anisotropy constant. Therefore below 135 K J_r/J_S increases due to the easy cone situation (see Fig. 6b). The aim of these investigations was to search for a decoupling of the nanocrystalline grains due to the generally increasing anisotropy at low temperatures. For this purpose the Pr-Fe-B system is more suitable. For the $Pr_9Fe_{85}B_6$ sample a decrease of J_r/J_S with decreasing temperature was found which indicates a decoupling due to the increase of the anisotropy at low temperatures. A similar behaviour was found for the stochiometric sample. It is worth to note that for the $Pr_{18}Fe_{76}B_6$ sample where the grains should be decoupled due to the existence of a nonmagnetic Pr rich phase, a J_r/J_s ratio of about 0.5 was found for the whole temperature range.

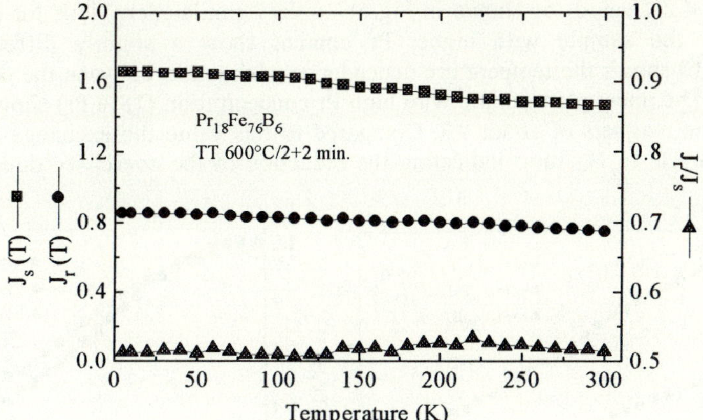

Figure 7 Temperature dependence of the saturation polarisation J_s, the remanence J_r and the ratio J_r/J_s for mechanically alloyed nanocrystalline $Pr_{18}Fe_{76}B_6$.

From the temperature dependence of the magnetisation and of the anisotropy the temperature dependence of the exchange length can be obtained (see Fig. 8). The exchange length generally decline with decreasing temperature which supports the decoupling which was found for low temperatures.

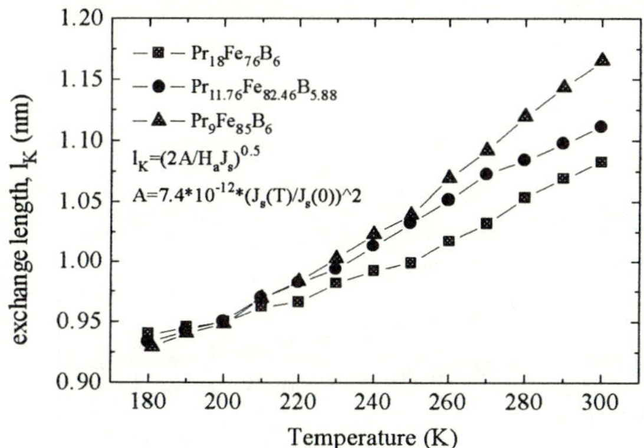

Figure 8 The temperature dependence of the exchange length, ℓ_K, obtained for an exchange energy constants $A=7.4*10^{-12}$ J/m.

5. Coercivity

Generally the character of the temperature dependence of the coercivity in exchange coupled material is not significant. However plotting the temperature dependence of the coercivity normalised to the 4.2K value– as shown in Fig. 9 - a very similar behaviour for all three samples was found. Only the sample with higher Pr content show a slightly different temperature dependence. Fig. 10 shows the temperature dependence of the ratio between the coercivity and the anisotropy field. The uncoupled sample with high Pr concentration (18% Pr) shows the usual ratio for microcrystalline magnets of about 0.3. Compared to this value the exchange coupled samples exhibit a very low ${}_iH_C$ to H_A ratio indicating the reduction of the coercivity due to the exchange coupling.

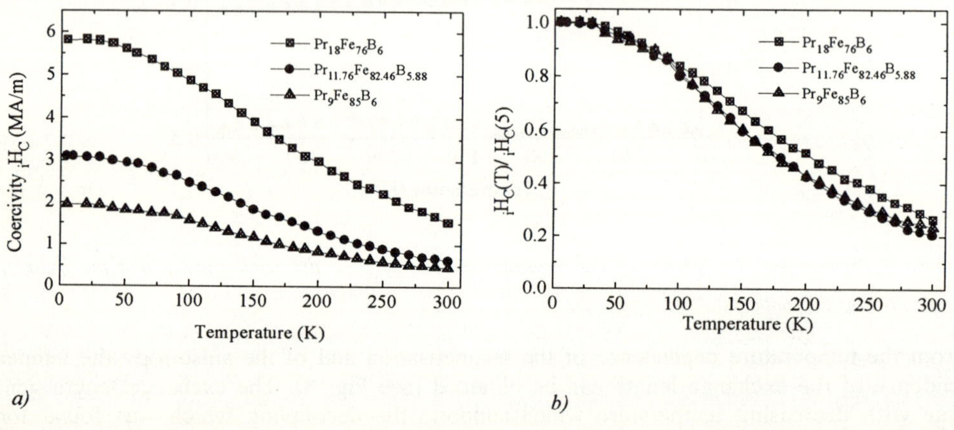

Figure 9 Temperature dependence of a) the coercivity and b) the reduced coercivity ${}_iH_C(T)/ {}_iH_C(5)$ measured on mechanically alloyed Pr-Fe-B samples.

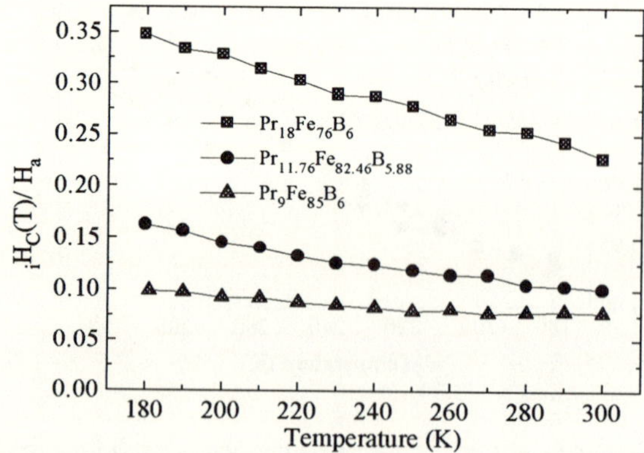

Figure 10 Temperature dependence of the ratio coericivity to anisotropy field for mechanically alloyed nanocrystalline Pr-Fe-B samples.

6. Curie-Temperature and spin reorientation

Table 1 Spin reorientation temperatures of $Nd_2Fe_{14}B$ with different degree of grain sizes. T_{SR} is the spin reorientation temperature, H_C the coercivity at 300 K, D_V the average grain size evaluated by analysing the X-ray diffraction line broadening.

Materials	T_{SR}(K)	H_C(MA/m)	D_V(nm)
Single crystalline $Nd_2Fe_{14}B$	135	0	>>10^4
Polycrystalline $Nd_2Fe_{14}B$	135	0	~10^4
Sintered $Nd_{15}Fe_{75}B_8$	135	~0.80	~5000
HDDR treated $Nd_{16}Fe_{76}B_8$	129	~0.80	200~300
Melt-spun $Nd_2Fe_{14}B$	117	~0.74	~19

From a systematic study of the spin behaviour temperature as a function of the grain size for Nd-Fe-B magnets a clear tendency for a decreasing spin reorientation temperature with decreasing grain size of $Nd_2Fe_{14}B$ was found (see Table 1). According to calculations based on a phenomenological model [18] the anisotropy constant is changed due to the presence of the strong inter-grain magnetic coupling. In the case of $Nd_2Fe_{14}B$, the spin reorientation temperature corresponds to that temperature where K_1=0. The phenomenological calculations [18] show that, in nanocrystalline $Nd_2Fe_{14}B$, the temperature where K_1 becomes zero is decreased.

Table 2 T_C of the $Pr_2Fe_{14}B$ phase measured by ac-susceptibility for different compositions after different heat treatments.

	$Pr_9Fe_{85}B_6$	$Pr_{11.76}Fe_{82.46}B_{5.88}$	$Pr_{18}Fe_{76}B_6$
Nanocrystalline	306°C	309°C	297°C, 293°C
1st heat treatment T = 700°C, t =30 min	302°C, 289°C	296°C	294°C
2nd heat treatment T = 1000°C, t =15 h	295°C, 287°C	294°C, 291°C	295°C

Since the exchange interaction between the grains influences the anisotropy and, through the anisotropy constant K_1, the spin reorientation temperature, we would expect that the exchange energy A would also be influenced. The relation between the mean exchange energy A between atoms and T_C can be estimated from the equation:

$$2ZAJ(J+1)=k_B T_C \qquad \text{(Eq. 5)}$$

where Z is the mean coordination number of the atoms, J the total angular momentum quantum number and k_B the Boltzman-constant. If we assume that the exchange interaction between the grains has the same origin as the exchange interaction between the atoms, information on the exchange energy should be obtainable by measuring T_C as a function of composition and grain size.

For $Pr_2Fe_{14}B$ a value for T_C = 292°C and for $Nd_2Fe_{14}B$ T_C = 312 - 315°C [19] was reported. For ac-susceptibility measurement - which was calibrated measuring pure Ni - T_C was determined from the peak of the first derivative. Here, as in Nd-Fe-B samples [12], for some of the samples two or even three clearly detectable peaks close to T_C of the hard magnetic phase were observed. From all ac-susceptibility measurements a decrease of T_C was found in the nanocrystalline state comparing to the microcrystalline state for the low rare earth content and the stochiometric samples.

7. Summary

A survey of the coupling mechanisms between nanocrystalline grains for single phased material as well as for nanocomposites was given. It was shown that for nanocomposites the so-called remanence enhancement has to be checked taking into account also the contribution of the cubic soft magnetic phase before a real exchange coupling can be assumed. For different Pr-Fe-B based samples at low temperatures a decoupling of the nanocrystalline grains was found. This was demonstrated measuring the temperature dependence of the anisotropy field as well as of the J_r/J_S ratio. In the case of Nd-Fe-B based material the spin reorientation causes at low temperatures an upturn of J_r/J_S which can be understood by the easy cone situation. The exchange length of the Pr-Fe-B based samples also decreases strongly with decreasing temperature supporting thus the idea of decoupled grains. In Nd-Fe-B based material a changed spin reorientation temperature was found in nanocrystalline material. Also the Curie temperature changes slightly due to the exchange coupling.

Acknowledgement

This work was supported by the Austrian Science Foundation under the projects P10945-PHY and P11524-PHY.

References

[1] Y. Yoshizawa, S.Oguma, K.Yamauchi, J. Appl.Phys. **64** (1988) 6044

[2] K.Suzuki, A. Makino, A.Makino, A.Inoue, T.Masumoto, J. Appl. Phys. **70** (1991) 6232

[3] R.Alben, J.J.Becker, M.C.Chi, *J. Appl. Phys.*, **49** (1978) 1653.

[4] G.Herzer, J. of Magn. Magn. Mat. **112** (1992) 258

[5] J.E.Keem, G.B. Clement, A.M.Kadin and P.G.McCallum, *Proc. Conf. ASM Materials Week 87*, ed. Salsgiver J.A., American Society for Metals (1987) p. 87.

[6] R. Coehoorn, D.B.de Mooij, J.P.W.B Duchateau and K.H.J Buschow, *J. de Physique*, **C8** (1988) 669.

[7] A. Jha, H.A.Davies, R.A. Buckley, J. Magn. Magn. Mater. 80 (1989) 109

[8] H.A.Davies, J.F.Liu, G.Mendoza; *Proceedings of the 9th International Symposium on Coercivity and Anisotropy*, ed. F.P. Missell *et al.* (1996) vol 2 p.251.

[9] E.C.Stoner and W.P.Wohlfarth, *Phil. Trans. Roy. Soc. (London)*, **A240** (1948) 5299.

[10] P.G. McCormick, J. Ding, E.H.Feutrill and R. Street *J. Magn. Magn. Mater.*, **157-158** (1996) 7.

[11] M. Dahlgren, X. C. Kou, R. Grössinger, J. Wecker, *Proceedings of the 9th International Symposium on Coercivity and Anisotropy*, ed. F.P. Missell *et al.* (1996) vol 2 p.307.

[12] M. Dahlgren, R.Grössinger, E. de Morais, S. Gama, G. Mendoza, J.F.Liu, H.A.Davies, IEEE Trans. Magn. (1997) 3895.

[13] X. C. Kou, M. Dahlgren, R. Grössinger and G. Wiesinger, *J. Appl Phys.* **81** (1997) 4428

[14] E.F.Kneller, Ferromagnetismus, Springer-Verlag, Berlin (1962)

[15] E.F.Kneller and R.Hawig, *IEEE Trans. Magn.* **27** (1991) 3588

[16] T.Schrefl, R.Fischer, J.Fidler and H.Kronmüller, *J. Appl. Phys.* **76** (1994) 7053.

[17] G.Asti and S.Rinaldi, *J.Appl. Phys.*, **45** (1974) 3601.

[18] Z. D.Zhang, X. C.Kou, and R.Grössinger, J. Magn. Magn. Mater. (submitted).

[19] M.Sagawa, S. Fujimura, N.Togawa, H.Yamamoto and Y.Matsuura, *J. Appl. Phys.*, **55** (1984) 2083.

Materials Science Forum Vols. 302-303 (1999) pp. 273-277

Investigation of Interaction Effects in a Nanocrystalline $Nd_2Fe_{14}B$+αFe Composite System

O.V. Billoni[1], S.E. Urreta[1] and L.M. Fabietti[1,2]

[1] Facultad de Matemática, Astronomía y Física, Universidad Nacional de Córdoba, Ciudad Universitaria, 5000 Córdoba, Argentina

[2] CONICET

Keywords: Nanocrystalline Magnetic Composites, Remanence Enhancement, Exchange Interaction

Abstract

The interactions responsible for the observed properties in a two-phase Nd_2Fe_{14} B + αFe composite are investigated. These interactions are demonstrated by irreversible magnetization curves strongly dependent on the magnetic configuration of the system. Their effects are analyzed by a) considering the changes in the effective distribution of critical fields for reversion of the magnetization in different configurations and b) assuming that they are described by a uniform mean interaction field superimposed to the applied field. Both descriptions lead to confirm the results of micromagnetic calculations in the sense that the magnetic behavior of these fine grained composites is governed by exchange interactions. As the grain size decreases, two competitive effects are found to determine the grain size dependence of coercivity: the exchange hardening of the soft phase and the exchange weakening of the hard phase.

Introduction

Isotropic composite magnets consisting of magnetically hard and soft (but high spontaneous polarization) particles may show remarkably high energy products if they are optimally structured. In these composite magnets the soft magnetic grains cause a high magnetization and the hard magnetic grains induce a large coercivity, provided that the particles are regular and small (a few nanometers) [1].

Micromagnetic calculations [2-4]indicate that the high remanence of nanocrystalline composite magnets may be attributed to a) the increased spontaneous magnetization of the soft magnetic phase and b) remanence enhancement owing to the interparticle exchange interactions. They also predict that two competitive effects determine the grain size dependence of coercivity: 1) magnetic inhomogeneities owing to short range exchange interactions between hard magnetic grains favor the nucleation of reversed domains. The nucleation field of hard magnetic grains coupled by short-range exchange interactions decreases with decreasing the grain size. 2) Exchange interactions between different phases suppress the nucleation of reversed domains within the soft magnetic particles. The nucleation field of soft grains embedded within a hard magnetic phase increases as the grain size decreases. The nucleation of reversed domains within the soft phase particles may initiate a cascade type demagnetization process so the coercive field becomes determined by the nucleation field of the soft magnetic particles and thus it will increase with decreasing grain size.

The aim of this work is to investigate the interactions responsible for the observed properties in a two phase Nd_2Fe_{14} B + αFe composite for different grain size distributions and volume fraction of soft phase.

Experimental Procedures

Two-phase composites were produced by melt spinning a master $Nd_{1.6}Fe_{14}B$ alloy at different roll velocities [5]. The composition, determined from X-ray data and the saturation polarization values, was f=0.18 volume fraction of αFe for all velocities. The mean grain sizes of the phases were determined applying the Scherrer formula to the X-ray peaks, after decomposition of the spectrum in the range of 2θ between 41° and 45°. Magnetic measurements were performed in a VSM magnetometer with a maximum applied field of 1.5 T, at room temperature. The samples were saturated by a pulsed field of 5T [6]. Individual flakes and massive samples (ribbons bonded in a multilayer configuration) were measured with their lengths parallel to the field. .

Results

The demagnetization curves - shown in Figure 1- illustrate how the squareness of the hysteresis diagram becomes largely improved by reducing the mean grain size of the phases. In these composites the mean grain size of the soft phase is close to one half the size of the hard phase

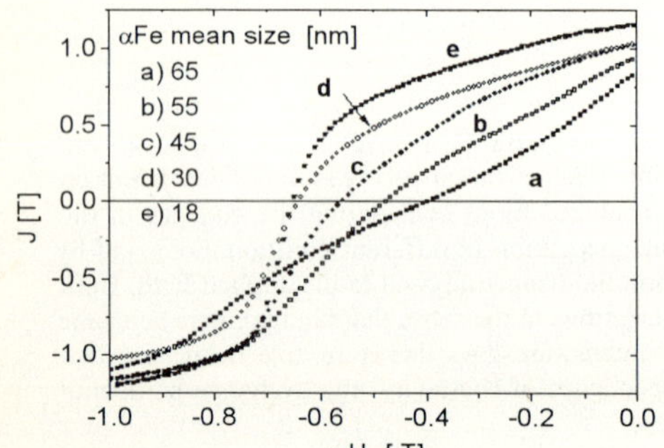

Figure 1: Major demagnetization curves for the composites investigated.

Figure 2 shows the dependence of the remanent polarization J_R, the coercive field ${}_JH_C$ and the remanence enhancement factor E_H ($=\frac{2J_R}{J_S}-1$) on the mean grain size of the soft phase in the composite with 0.2 volume fraction of αFe. It may be observed that both coercivity and remanence largely increase as the mean grain size of the soft magnetic phase reduces below 30nm. In this range of sizes, E_H reaches values of 0.30, indicating the presence of strong exchange coupling between the grains.

The interactions in these fine magnetic particle systems are often studied by comparing the remanence after recoil curves obtained for the magnet in different magnetic configurations. Two types of remanence curves may be considered: the demagnetization remanence $J_r^d(H_a)$ and the initial (thermally demagnetized) remanence curve $J_r^{th}(H_a)$. $J_r^d(H_a)$ is obtained by demagnetization via applying and removing an increasing field H_a to a previously saturated sample with a saturation remanence $J_R = J_r^d(0)$. The $J_r^{th}(H_a)$ is obtained from recoil curves during the initial magnetizing curve.

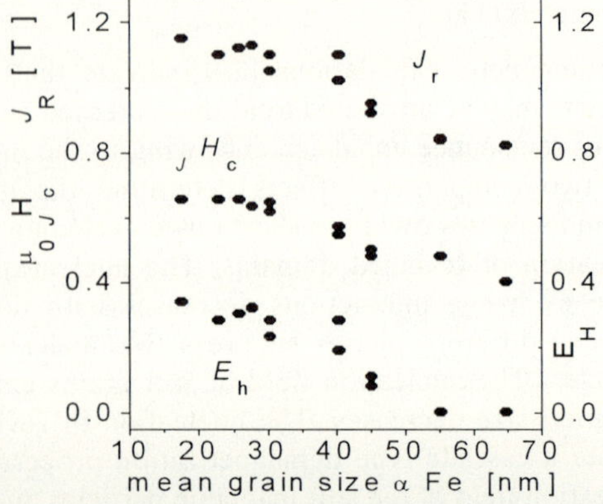

Figure 2: Grain size dependence of the magnetic properties of the composite.

These remanence curves $J_r(H_a)$ are closely related to the irreversible magnetization produced by the application of an external field H_a to the system in two states with different magnetic configurations. In the thermally demagnetized state, the magnetic moments of the particles are randomly distributed, regardless of their switching fields. In the configuration which leads to $J_r^d(H_a)$ the magnetic moments of the particles are preferentially oriented in the saturation direction (highly anisotropic orientation distribution). Then, the difference in the irreversible magnetization obtained in these two situations should be related to interactions. Figure 3 schematizes the irreversible magnetization obtained for the composite in these magnetic configurations.

In a system of non-interacting particles, these magnetic configurations are equivalent and the following relation holds [7]:

$$2I^{th} = I^d \qquad (1)$$

where $I^{th} = J_r^{th} / J_R$ and $I^d = 1-(J_r^d / J_R)$. In presence of interactions, relation *(1)* is not satisfied and the magnitude δI, defined as

$$\delta I = 2I^{th} - I^d \qquad (2)$$

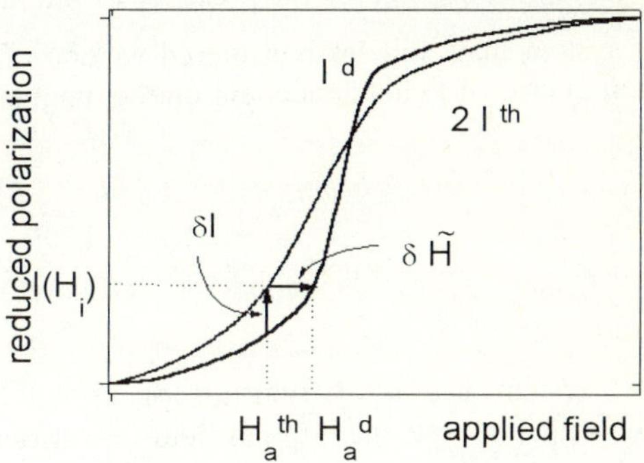

Figure 3: Irreversible magnetization curves strongly dependent on the magnetic configuration of the system.

is often taken as a measure of the deviations from the behavior of a system of non interacting particles. Figure 4 shows δI as a function of the applied field H_a for three composites (20% αFe) with different grain sizes and for a fourth one with 30% αFe and a mean grain size of the soft phase of 22nm. It may be observed that in all the cases δI changes its sign for applied fields close to the knee in the demagnetization curve associated with the reversion of hard grains (see Fig.1). The positive values of δI at low fields indicate that in the anisotropic configuration, intergranular interactions work to maintain a high value of J until a given critical field. The positive peak is then associated with the fraction of irreversible processes occurring at a particular field in the magnetizing mode which are inhibited by interactions in the demagnetizing mode. The negative peak, on the other hand, corresponds to the fraction of these processes in excess - induced by interactions- relative to the initial configuration. It describes how interactions promote - in the anisotropic configuration - the reversion of magnetization by an avalanche type process in a larger extent than in the thermal configuration. In this picture, the results shown in Figure 4 indicate that as the mean grain size decreases, both, the critical field for the onset of reversion and the tendency to demagnetization by a cascade type process, increase. For the composite with a larger volume fraction of soft phase the critical field is lower but the behavior of the hardest grains is comparable to that of the composite with a lower volume fraction of αFe and similar grain size.

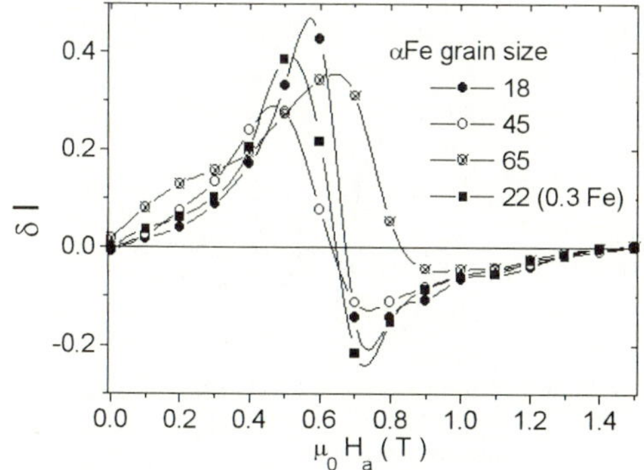

Figure 4: Deviation δI from the behavior expected in a system of single domain non interacting grains

Discussion

In a system consisting of single domain particles with uniaxial anisotropy, each particle may be characterized by its critical (nucleation) field at which the magnetization switches irreversibly.

Then, for a given change in the internal field dH_i the change in the irreversible magnetization of the ensemble may be expressed as:

$$dJ_{irr} \cong 2 < j > f(H_i)dH_i \tag{3}$$

where $<j>$ is a mean change in the polarization produced by the reversion of each particle and $f(H)$ is the distribution function of critical fields in the system. Without interactions between the particles, and assuming a demagnetizing factor $D=0$, the internal field $H_i \equiv H_a$ and the change in the irreversible magnetization for both configurations $2I^{th}$ and I^d are identical; i.e. $\delta I=0$.

In terms of the equation *(3)* the presence of interactions may be described in two ways (Figure 3):

a) the system may still be considered as one of non interacting particles but with a distribution function of critical fields dependent on the magnetic configuration:

$$\delta H_i = \delta H_a = 0 \quad ; \quad \delta I = \int_0^{H_a} [\, f_{th}^* - f_d^* \,] dH \neq 0 \tag{4}$$

where f_{th}^* and f_d^* are the apparent (effective) distributions associated with each magnetic configuration.

b) The system has a distribution of critical fields which is independent of the magnetic configuration but now, the internal fields are strongly dependent on the actual configuration;

$$\delta I = 0 \quad ; \quad \delta H_i = \delta H_i(H_a, conf) \tag{5}$$

In description a), for a given applied field H_a, the magnetization obtained for different configurations differs in δI. This difference is attributed to two different distributions of critical fields, which are given approximately by the irreversible susceptibilities measured in each magnetic configuration. Figure 5 shows the irreversible susceptibility χ_a^{th} and χ_a^d for two composites with different grain sizes. It may be observed that the effective distributions associated with non isotropic magnetic configurations are narrower and the peak values are shifted to higher fields and this effect increases as the grain size decreases. The effective hardening of the softest grains is consistent with a strong exchange coupling between different phases.

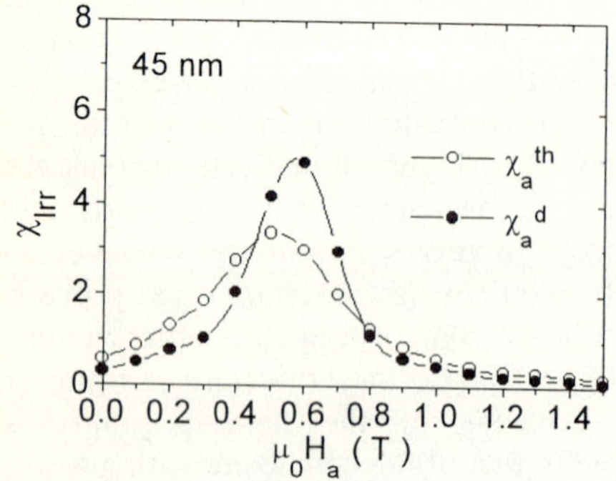

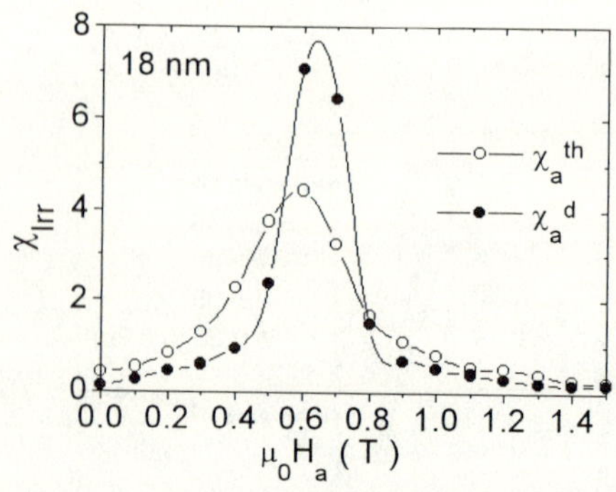

Figure 5: Irreversible susceptibilities related to the effective distributions of critical fields shown by the composite with different grain sizes for the two magnetic configurations investigated. The softest grains in the distribution become hardened in the anisotropic configuration as a result of the exchange coupling. This effect increases as the grain size decreases.

In the approach b) the same distribution of critical fields is assumed for both magnetic configurations. Then, the condition $\delta I = 0$ is obtained only when the same internal fields are acting on both configurations. A simple description of the internal field for each magnetic configuration as a function of the applied field is obtained assuming that:

$$H_i^{conf} = H_a + \widetilde{H}^{conf} \tag{6}$$

where $\widetilde{H}$ is a uniform mean interaction field dependent on the configuration. Then, for a given value of polarization I the same internal field is expected for both configurations

$$H_a^d + \widetilde{H}^d = H_a^{th} + \widetilde{H}^{th} \tag{7}$$

and then $\Delta\widetilde{H}^{th\rightarrow d} = \widetilde{H}^d - \widetilde{H}^{th} = H_a^{th} - H_a^d$ is the difference in the mean interaction field at a given applied field between the two different magnetic configurations. Figure 6 shows the values $\Delta\widetilde{H}^{th\rightarrow d}$ resulting for two composites with different grain sizes. The negative values of $\Delta\widetilde{H}^{th\rightarrow d}$ are associated with the exchange hardening of the soft phase, which becomes improved by reducing the grain size. The positive values, on the other hand, describe how efficient is the exchange interaction between hard grains to promote a cascade process. The sharp transition observed for the optimally quenched composite (18 nm) takes place at the coercive field, which is the result of a competition between the effect of exchange hardening of the soft phase and a simultaneous exchange weakening of the hard phase. This may be explained considering that, in these as quenched microstructures, the mean grain sizes of the phases are not independent. Grain sizes of αFe leading to optimal hardening are always associated to hard grain sizes promoting strong exchange coupling.

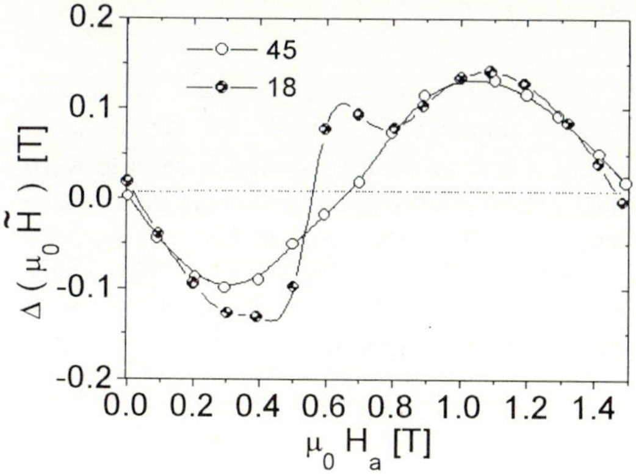

Figure 6: Effective mean interaction field (from eqn7) describing the exchange hardening of the softest grains (negative) and the exchange weakening of hardest grains (positive).

Acknowledgements

This work has been partly supported by CONICET, SECYT-UNC and CONICOR, Argentina.

Bibliography

1. J. Bauer, M. Seeger, A. Zern and H. Kronmüller , J. Appl. Phys. **80** (3), (1996) 1667.
2. H. Kronmüller, T. Schrefl, R Fischer, and J Fidler , Proc. 8th Int. Symposium on Magnetic Anisotropy and Coercivity , RE-TM Alloys, Birmingham, (1994) 1.
3. T. Schrefl, J. Fidler and H. Kronmüller , Phys. Rev. B.-**49** n 9 (1994) 6100.
4. R. Fischer, T. Screfl, H. Kronmüller and J. Fidler , J. Magn. Magn. Mat. **150** (1995) 329.
5. O.V. Billoni, S. E. Urreta, L. M. Fabietti and H. R. Bertorello. To be published in J. Magn Mag Mat. 1998.
6. O. Evequoz; Internal Report, FaMAF, Universidad Nacional de Córdoba, Argentine. (1995).
7. E. P. Wohlfarth , J. App. Phys. **29** (1958) 595.

Materials Science Forum Vols. 302-303 (1999) pp. 278-283

Rietveld Analysis and Structural Characterization of Well-Coupled and Partially-Coupled Flash Annealed Nanocrystalline PrFeB

R.K. Murakami, A.M.P. Passaro, C.S.M. Partiti and V. Villas-Boas

Instituto de Física, Universidade de São Paulo, São Paulo, SP, Brazil

Keywords: Permanent Magnets, Nanocrystalline Magnets, Flash-Annealing

Abstract

Enhanced-remanence nanocrystalline $Pr_{4.5}Fe_{77}B_{18.5}$ alloys were produced by flash annealing. Good magnetic coupling is obtained for samples annealed above 560°C, while for temperatures below 540°C the alloys showed phase separation. A coupled and an partially-coupled sample were carefully examined. A Rietveld analysis of the X-ray diffraction results showed that the main phase of both samples were $Pr_2Fe_{14}B$. This result is consistent with a Rietveld analysis of neutron diffraction data and with Curie temperature measurements results for both samples. Mössbauer spectroscopy and atomic force microscopy were also performed in order to understand the phase distribution and the differences between structural properties in these materials.

I. Introduction

In the late 1980s one of the most exciting additions to permanent magnet materials resulted from the discovery of exchange spring magnets [1]. These magnets are nanocrystalline materials consisting of a hard magnetic rare earth-based intermetallic, exchange-coupled to a soft magnetic phase. They have been the subject of much recent interest due to the possibility of their presenting enhanced remanence and large energy products in magnets containing large volume fractions of a soft magnetic phase. Enhanced remanence behavior ($M_r/M_s = 0.75$) was reported by Coehoorn *et al.* [2, 3] in melt-spun and furnace-annealed alloys with compositions around $Nd_4Fe_{78}B_{18}$. This remanence value is higher than would be expected ($M_r/M_s = 0.50$) for a random collection of non-interacting particles with uniaxial anisotropy and it has been discussed [4] in terms of two suitably dispersed and exchange-coupled ferromagnetic phases. In fact the nanocrystalline magnetic material reported by Coehoorn *et al.* [2, 3] is a more complex system composed of three phases: the main phase Fe_3B which is soft magnetic, $Nd_2Fe_{14}B$ the hard magnetic phase and α-Fe.

In spite of their interesting properties, NdFeB nanocrystalline magnets tend to have low H_c values. In previous papers [5-7], we showed that materials with improved coercivity could be produced by rapid, high current crystallization of amorphous ribbons, the so called flash annealing process. The flash annealing process, also known as dc-Joule heating process, is a non-conventional annealing technique, characterized by a fast release of thermal energy to the material resulting from passing a current through the sample [8]. It induces the formation of non-equilibrium phases possibly displaying better physical properties with respect to furnace annealling. In the case of nanocrystalline $R_4Fe_{78}B_{18}$, we found that flash-annealed ribbons presented higher coercivities and higher remanence ratios than furnace-annealed ribbons, showing that flash annealing improves applications-related properties of these nanocrystalline magnetic materials [5 - 7].

The aim of the present work is to examine the differences between the structural properties in two different samples: (i) a sample annealed at 600°C that showed the highest coercive field H_c and remanence ratio M_r/M_s and which we will refer to as the well-coupled sample (WC), and

(ii) a sample annealed at 520°C that showed phase separation in the second quadrant demagnetization curve and which will be referred to as the partially-coupled sample (PC). By comparing the coupled and the partially-coupled materials, we hope to understand the conditions necessary to achieve good coupling between the phases present in these materials. The samples were studied by X-ray diffraction, atomic force microscopy (AFM), Mössbauer spectroscopy and magnetic measurements.

2. Experiment

Partially amorphous ribbons of nominal composition $Pr_{4.5}Fe_{77}B_{18.5}$ were produced by melt-spinning in He on a mild steel wheel. The flash annealings were carried out in a vacuum of 10^{-5} torr for 30 s and for temperatures 500-640°C. Details about the treatment and the determination of the sample temperature can be found in previous papers [6, 9].

Magnetization measurements were made at room temperature using a vibrating sample magnetometer mounted in a 20 kOe electromagnet.

X-ray diffraction data from samples WC and PC were collected at room temperature using a Philips PW1710 diffractometer ($\lambda = 1.5405$Å) of the Laboratório de Caracterização Tecnológica of the EPUSP. The scanning was performed in the interval $20° < 2\theta < 80°$, using a rate of 0.05°, measuring the counts for 40 seconds. The data were refined using the DBWS-9411 program developed by Young et al [10] for Rietveld analysis of the X-ray data.

Mössbauer spectra were measured at room temperature with a conventional constant-acceleration spectrometer and a Rh (^{57}Co) source.

The atomic force microscopy (AFM) measurements for samples WC and PC were performed at room temperature with a Digital Instruments - Nanoscope III, having J-type scan heads in the contact mode, utilizing a NanoProbe with 100 μm triangular Si_3N_4 microcantilevers with thick legs. The effective spring constant as specified by the supplier was 0.58 N/m.s.

3. Results and discussion

In a previous work, Curie temperature measurements and a Rietveld analysis of neutron diffraction data of these materials showed that, the phases present in sample WC are Fe_3B, $Pr_2Fe_{14}B$ and α-Fe, while in sample PC we also found the $Pr_2Fe_{23}B_3$ phase [11, 12].

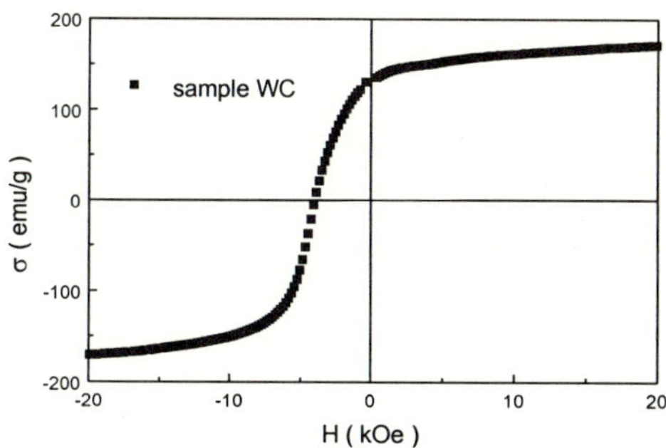

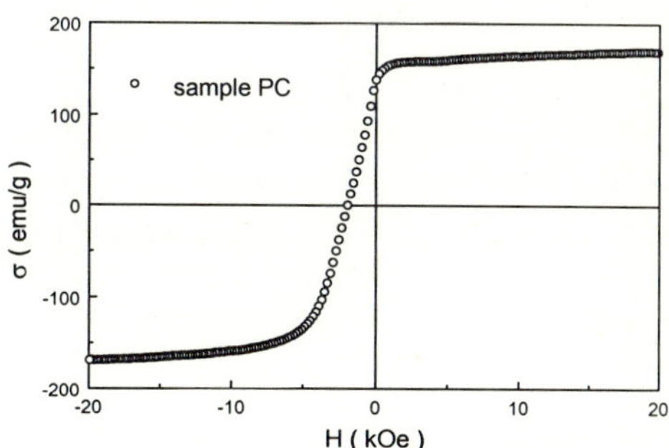

Fig. 1 : Specific magnetization σ vs. applied field H for sample WC and sample PC.

Excellent magnetic coupling was observed for all the samples annealed at temperatures 560-640°C. In figure 1 we show the specific magnetization σ as a function of the applied magnetic field H for sample WC and for sample PC. For sample WC, we obtained the best magnetic properties, which are H_c = 4.3 kOe and M_r/M_s = 0.72. For the samples which were flash-annealed at temperatures lower than 560°C, the magnetic coupling between phases deteriorated [9, 11].

A Rietveld analysis of X-ray diffraction data showed that the main phases of both samples were the same: $Pr_2Fe_{14}B$ and Fe_3B. The Rietveld analysis of X-ray diffraction data indicates for sample WC the presence of approximately (60±3) wt% $Pr_2Fe_{14}B$, (39±3) wt% Fe_3B and approximately 1 wt% α-Fe while for sample PC (55±5) wt% $Pr_2Fe_{14}B$, (31±4) wt% Fe_3B, (13±2) wt% $Pr_2Fe_{23}B_3$ and approximately 1 wt% α-Fe. This result for the mass percentage of each phase is in complete agreement with a Rietveld analysis of neutron diffraction data performed in these samples [11, 12].

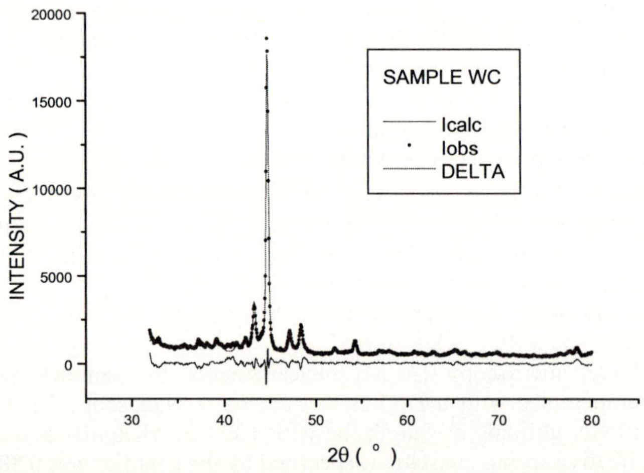

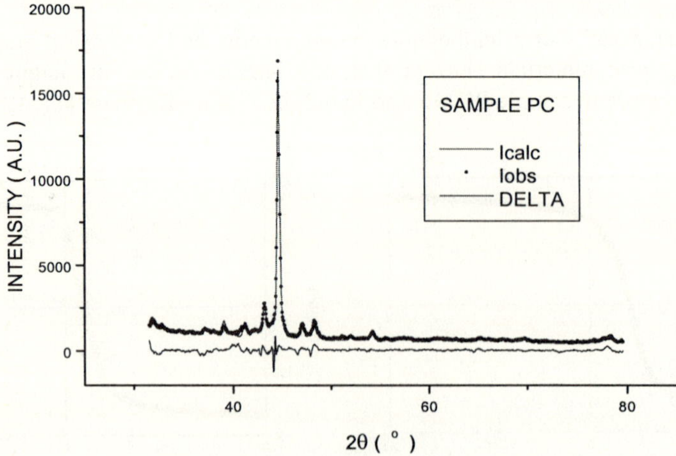

Fig. 2 : Best fit to Rietveld analysis of the X-ray data from samples WC and PC

The Rietveld analysis of both X-ray diffraction and neutron diffraction data of our samples are quite different from that usually reported for $Nd_4Fe_{78}B_{18}$. The percentages reported for conventionally annealed $Nd_4Fe_{78}B_{18}$ are 60 - 70 wt% Fe_3B, 20 - 36 wt% $Nd_2Fe_{14}B$ and 4 - 10 wt% α-Fe [2, 13]. On the other hand, recent results obtained for the Rietveld analysis of the neutron diffraction data of a conventionally annealed $Pr_{4.5}Fe_{77}B_{18.5}$ show that the phase composition of this sample is nearly the same as that of the WC sample [14]. This result contradicts the suggestion that the difference between the WC sample and $Nd_4Fe_{78}B_{18}$ should be attributed to the very different types of annealings carried out [11].

Room temperature Mössbauer spectra for samples WC and PC are shown in figure 3. The continuous line shown is a guide to the eye only.

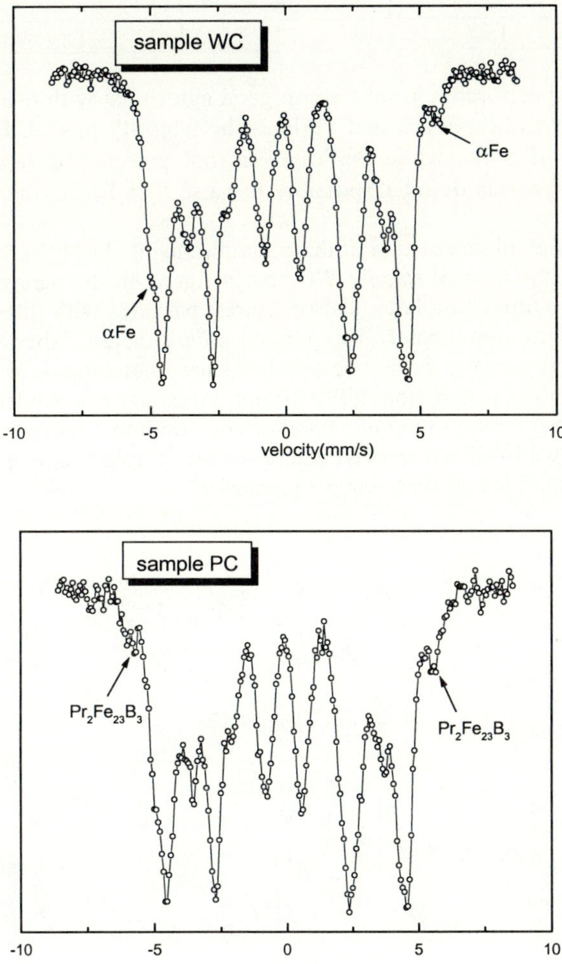

Fig. 3 : Mössbauer spectra for samples WC and PC

These spectra are difficult to fit due to the great number of Fe sites in the various phases present. Therefore we chose the alternative method of summing calculated spectra, corresponding to various phase in variable proportions and comparing the resulting simulation to the experimental spectra. The Mössbauer parameters used in the simulations were those published for the Fe_3B [15] and $Pr_2Fe_{14}B$ [16]. The $Pr_2Fe_{23}B_3$ parameters were assumed to be similar to those reported by Buschow and co-workers [17] for the similar compound $Nd_2Fe_{23}B_3$. The most prominent spectral lines were used for a first identification, and a first estimate of the best relative areas was made. The best simulated spectrum was then obtained by successive approximations. The relative amounts of Fe atoms present in each phase were then calculated from the relative spectral areas of simulated spectra and the results are in Table 1.

Table 1 - Relative amounts of Fe atoms present in each phase in %.

Sample	$Pr_2Fe_{14}B$	Fe_3B	α-Fe	$Pr_2Fe_{23}B_3$
WC	56	43	1	
PC	52	36		12
$Nd_4Fe_{78}B_{18}$ [2,3]	15	73	12	

The Mössbauer spectroscopy results are in good agreement with neutron diffraction and X-ray data, both in identifying $Pr_2Fe_{14}B$ and Fe_3B as the majority phases. In sample WC we also found a small amount of α-Fe, while $Pr_2Fe_{23}B_3$ is not present. In sample PC we identified $Pr_2Fe_{23}B_3$, while no α-Fe was detected, possibly because it is below the detection limit of this method.

We also studied the microstructure of these samples using AFM. In figure 4a, we present the AFM image of the nanostructure of sample WC and in figure 4b the nanostructure of sample PC. For sample WC we have three morphological structures: particles with diameters of about 20 nm; rounded agglomerates with aproximately 100 nm and a dark material that occupies the interstices between the grains. In this sample, the smallest particles are frequently in or over the agglomerates. In sample PC the grain sizes varied from 20 to 40 nm, randomly surrounded by the dark material. Some areas suggest the beginning of a preferential orientation into circular structures. Atomic force microscopy indicated a well-ordered microstructure for the coupled sample, but a highly irregular microstructure in the case of the partially-coupled material.

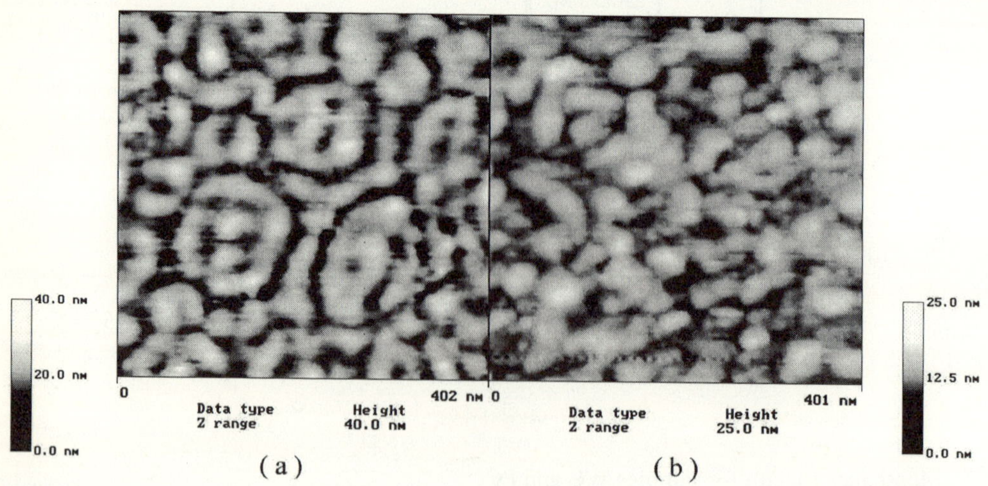

Fig. 4: Atomic force microscopy images for (a) sample WC and (b) sample PC.

4. Conclusions

We have produced samples of well-coupled and partially-coupled $Pr_{4.5}Fe_{77}B_{18.5}$ by flash annealing. Atomic force microscopy indicated a well-ordered microstructure for the coupled sample, but a highly irregular microstructure in the case of the partially-coupled material. A Rietveld analysis of the X-ray diffraction data and the Mössbauer spectroscopy results are consistent with a Rietveld analysis of neutron diffraction data of these materials. They showed that the main phases of both samples were the same ($Pr_2Fe_{14}B$ and Fe_3B) and also that the mass percentage of each phase is the same obtained by the three analyses.

Acknowledgements

R. K. Murakami wishes to thank J. A. Coaquira, C. Paiva-Santos and F. M. S. Carvalho for helping her in the Rietveld analysis. The authors also acknowledge useful conversations with H. R. Rechenberg, W. B. Yelon and F. P. Missell, and the financial support of FAPESP, CNPq, and FINEP.

References

[1] R. W. McCallum, A. M. Kadin, G. B. Clemente, and J. E. Keem, *J. Appl. Phys.* **61** (1987) 3577.
[2] R. Coehoorn, D. B. de Mooij, J. P. W. B. Duchateau, and K. H. J. Buschow, *J. de Phys:* **49** (1988) C8-669.
[3] R. Coehoorn, D. B. de Mooij, and C. de Waard, *J. Magn. Magn. Mater.* **80** (1989) 101.
[4] E. F. Kneller and R. Hawig, *IEEE Trans. Magn:* **27** (1991) 3588.
[5] M. V. P. Altoé, M. S. Lancarotte, H. R. Rechenberg, F. P. Missell, and J. M. González, *IEEE Trans. Magn.* **31** (1995) 3614.
[6] V. Villas-Boas, S. A. Romero, and F. P. Missell, in *Magnetic Anisotropy and Coercivity in Rare-Earth Transition Metal Alloys* , edited by F. P. Missell, V. Villas-Boas, H. R. Rechenberg and F. J. G. Landgraf (World Scientific, Singapore, 1996) pp. 31.
[7] V. Villas-Boas, S. A. Romero, and F. P. Missell, *J. Appl. Phys.* **81** (1997) 4434.
[8] P. Allia, P. Tiberto, M. Baricco, and F. Vinai, *Rev. Sci. Instrum.* **64** (1993) 1053.
[9] "Nanocrystalline magnetic materials obtained by flash annealing", R. K. Murakami and V. Villas-Boas, to be published in *Materials Research*..
[10] R- A. Young, A. Sakthivel, T. S. Moss and C. Paiva Santos, *J. Appl. Cryst.* **28** (1995) 366.
[11] "Magnetic and structural characterization of flash annealed nanocrystalline PrFeB", R. K. Murakami, A. M. P. Passaro, W. B. Yelon, F. P. Missell and V. Villas-Boas, to be published at the *Proceedings of the Fifteenth International Workshop on Rare-Earth Magnets and Their Applications* (Dresden, Germany, september 1998).
[12] "Neutron diffraction studies of magnetic materials", W. B. Yelon, to be published in *Advanced Materials Research.*
[13] S. Hirosawa and H. Kanekiyo, *Proc. 13th Int. Workshop on RE Magnets & their Applications* (Birmingham, UK,1994) pp. 87.
[14] "Magnetic and microstructural properties of nanocrystalline PrFeB", R. K. Murakami, M. Chen, W. B. Yelon and V. Villas-Boas, in preparation.
[15] J. Jiang, I. Dézsi, U. Gonser and J. Weissmüller, J. of Non-Crystalline Solids **116** (1990) 247.
[16] F. E. Pinkerton and W. R. Dunham, *J. Appl. Phys.* **57** (1985) 4121.
[17] K. H. J. Buschow, D. B. de Mooij and H. M. Van Noort, J. of Less-Common Met. **125** (1986) 135.

Materials Science Forum Vols. 302-303 (1999) pp. 284-288

Hybrid Magnets

M. Emura, A.C. Neiva and F.P. Missell

Instituto de Física, Universidade de São Paulo, C.P. 66318, 05315-970 São Paulo, S.P., Brazil

Keywords: Bonded Magnet, Ba-Ferrite, $Nd_2Fe_{14}B$ + α-Fe, Magnetic Interactions

Abstract

Injection-molded, oriented hybrid magnets consisting of mixtures of $BaO.6Fe_2O_3$ and MQP-Q (exchange-coupled $Nd_2Fe_{14}B$ + α-Fe) are compared to bonded magnets made only from ferrite or MQP-Q. The magnetic fractions of the hybrid magnets consist of 80, 60, or 40 wt% ferrite blended with 20, 40, or 60 wt% MQP-Q. The microstructure was investigated by X-ray diffraction and scanning electron microscopy. Magnetic interactions were studied via IRM and DCD remanence curves. In contrast to what is observed in exchange spring magnets, the IRM and DCD susceptibilities of all magnets present peaks of nearly the same width, centered at roughly the same value of the magnetic field, consistent with weak dipolar interactions between magnetic particles. A separation of the magnetization into reversible and irreversible components was performed using results of the moving Preisach model. This analysis is in agreement with conclusions obtained from IRM and DCD remanence curves.

I. Introduction

Bonded magnets constitute the fastest growing segment of the permanent magnet market [1,2]. These materials are produced by encapsulating a magnetic powder in a resin or polymer and then compacting or molding the material to the final part shape. The commercially important bonded magnets are based primarily upon ferrite or NdFeB. However, low-neodymium, exchange-coupled alloys are also interesting for bonded magnets because of their good corrosion resistance and low saturation fields. The Magnequench alloy MQP-Q ($Nd_2Fe_{14}B$ + α-Fe) has recently been used with barium ferrite to produce hybrid magnets with interesting coercive properties [2]. For a bonded magnet whose magnetic fraction consists of 20 wt% MQP-Q and 80 wt% ferrite, the reversible temperature coefficient of coercivity is approximately zero around and above room temperature. Furthermore, varying the composition of the magnetic fraction allows one to substantially increase the saturation moment of the hybrid magnet. We recently presented results on the structural and magnetic properties of these materials [3]. Since bonded magnets present an ideal opportunity to study systems in which dipolar fields should be the predominant form of interaction between particles, we consider again the question of interactions in these materials. We performed a careful separation of the magnetization into reversible and irreversible components which confirmed conclusions obtained from remanence curves.

II. Experiment

Five commercial bonded magnet compositions, kindly provided by J. Ormerod and S. Constantinides of The Arnold Engineering Co., were considered in this work. Three anisotropic hybrid magnets, produced by injection molding, had magnetic fractions consisting of 80, 60, or 40 wt% Ba ferrite blended with 20, 40, or 60 wt% of MQP-Q melt-spun ribbons and a binder of Nylon 12. These are designated as samples 2401, 2402, and 2403, respectively. An anisotropic magnet consisting only of Ba ferrite is designated as 1060, while an isotropic magnet whose magnetic fraction is MQP-Q is identified as 2203. For the latter two magnets, polyamide was used as a binder. The saturating field requirements as well as the temperature dependence of the coercivity for these magnets have been published previously [2]. For all magnets we were able to reproduce the catalog density by assuming the volume fraction of the magnetic material to be 60%.

The alignment of these magnets was characterized with a Philips MDP 1880 powder diffractometer using Cu K_α radiation. The microstructure was examined with an optical microscope as well as with a Leica Cambridge F440 scanning electron microscope (SEM). Magnetization M measurements between 4.2 and 300 K were made with a vibrating sample magnetometer (VSM) mounted in a 9T superconducting coil, while measurements above 300 K were made in a furnace mounted on a 2T electromagnet. In all cases M vs. H was corrected using an appropriate demagnetization factor. Typical magnetization curves for several of the magnets are shown in Fig. 1.

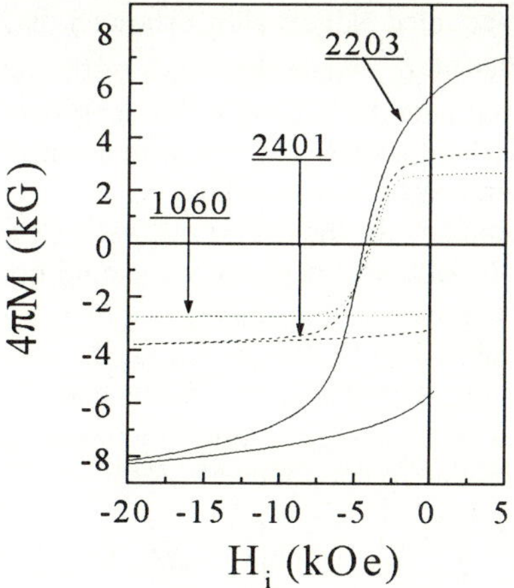

Fig.1 Magnetization M vs. internal magnetic field H_i for three bonded magnets.

III. Results and discussion

In the hybrid magnets considered here, the Ba ferrite grains present a high degree of alignment as can be verified by x-ray diffractograms in which the (0 0 k) peaks of this phase are prominent. Optical metallography and SEM observations show the ribbons of MQP-Q to be randomly dispersed in the ferrite-Nylon 12 matrix [3]. A useful way to quantify the degree of alignment of these samples is in terms of the ratio $r = M_r/M_S$, where M_r is the measured remanence and M_S is the saturation magnetization, obtained by extrapolating high field M values to infinite fields using $M = M_S(1 - a/H)$. This ratio decreases as the ferrite fraction of the magnet decreases. A value of $r = 0.91$ was encountered in magnet 1060, while $r = 0.54$ was found in sample 2203, indicative of exchange coupling between the magnetic phases of this sample.

A magnetic evaluation of magnetostatic interactions was made using remanence curves. The IRM remanence curve $M_r(H)$ started from an ac-demagnetized state, with the remanent magnetization being measured as a function of increasing positive applied field until positive saturation was obtained. The DCD curve $M_d(H)$ was measured from a positive saturation condition, with the remanence measured as a function of increasing negative field, until negative saturation was reached. For any system of single domain non-interacting particles, Wohlfarth showed that $M_d(H) = M_r(\infty) - 2\,M_r(H)$. One can then compare the IRM and DCD susceptibilities, $\chi_r = dM_r/dH$ and $\chi_d = dM_d/dH$, for the magnets studied. For noninteracting systems, the two curves should be of the same width, should have maxima at the same values of the magnetic field and the amplitudes should differ by a factor of 2, as can be seen by differentiating the Wohlfarth relation. It is most remarkable that these predictions are roughly obeyed for the five magnets studied, even though four of them contain varying amounts of exchange-coupled MQP-Q [3]. A recent study [4] of several exchange-coupled NdFeB alloys showed that the amplitude of χ_d was up to 50 times greater than that of χ_r, while the width of the former was about 10 times less than the latter. On the other hand, CoNiCr thin films approached the ideal behavior for large values of the Cr underlayer thickness [5], consistent with a reduction in exchange-coupling between grains with increasing Cr underlayer.

Although the results obtained for the hybrid magnets are more or less in agreement with our expectations for dipolar-coupled magnets, we should note that $M_r(H)$ and $M_d(H)$ are only approximations to the irreversible magnetization. A more complete analysis, based upon the moving Preisach model [6,7], shows that the reversible magnetization can be obtained from experimentally determined reversible and irreversible susceptibilities. Using experimental values of χ^i_{rev} and χ_{irr} along the initial magnetization curve and demagnetization curve, it is possible to determine $M_{rev}(H_i)$

$$M_{rev}(H_i) = \int_0^{H_i} \chi^i_{rev}(H_i') \left[1 + k_m \chi_{irr}(H_i')\right] dH_i' \quad (1)$$

where k_m is the moving parameter from the moving Preisach model [8]. The irreversible magnetization can be evaluated from

$$M_{irr}(H_i) = M(H_i) - M_{rev}(H_i) \quad (2)$$

where $M(H_i)$ are the experimental magnetization data. The magnetization was separated into M_{rev} and M_{irr} in this fashion and the results are shown in Fig. 2. For the calculations, k_m values in the range 15-20 were used for all magnets. We see that the reversible component appears to be larger in the magnet containing more MQP-Q. This may be the case, but may also be due to difficulties in determining the reversible component of the oriented (ferrite) component of the magnet.

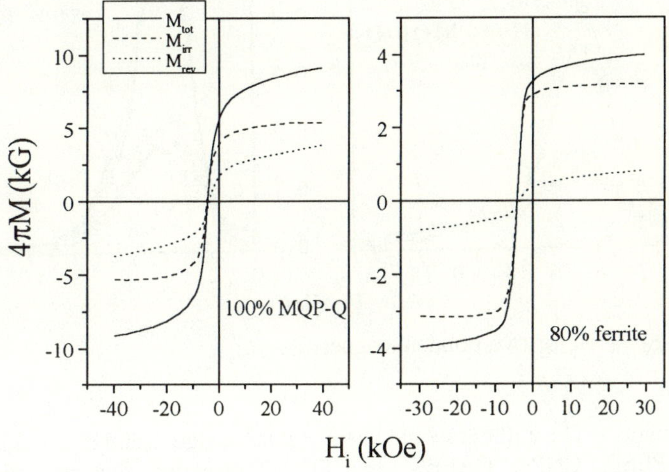

Fig. 2 Separation of the magnetization into reversible M_{rev} and irreversible M_{irr} components for two bonded magnets.

Having determined the irreversible magnetization by the procedure outlined above, χ_r and χ_d were obtained by differentiating the appropriate parts of the magnetization curve. In Fig. 3, the results are compared to the susceptibilities obtained from the remanence curves. As can be seen in Fig. 3, there is general agreement on all features of the susceptibility curves, although the susceptibility values obtained by differentiating M_{irr} are consistently lower than those obtained from the remanence curves.

IV. Conclusions

We have studied hybrid magnets consisting of oriented Ba-ferrite and non-oriented MQP-Q ribbons. In these bonded magnets, the magnetic interactions seem to be predominantly weak dipolar interactions as seen from the curves of χ_r and χ_d vs. H_i which are in general agreement with what one expects for non-interacting particles. Separation of the magnetization into M_{rev} and M_{irr}, using results of the moving Preisach model, indicates, for the five magnets studied here, a trend whereby the reversible component of magnetization increases with increasing fraction of MQP-Q. There is good agreement between results obtained from remanence curves and those obtained from the separation of the magnetization into M_{rev} and M_{irr}.

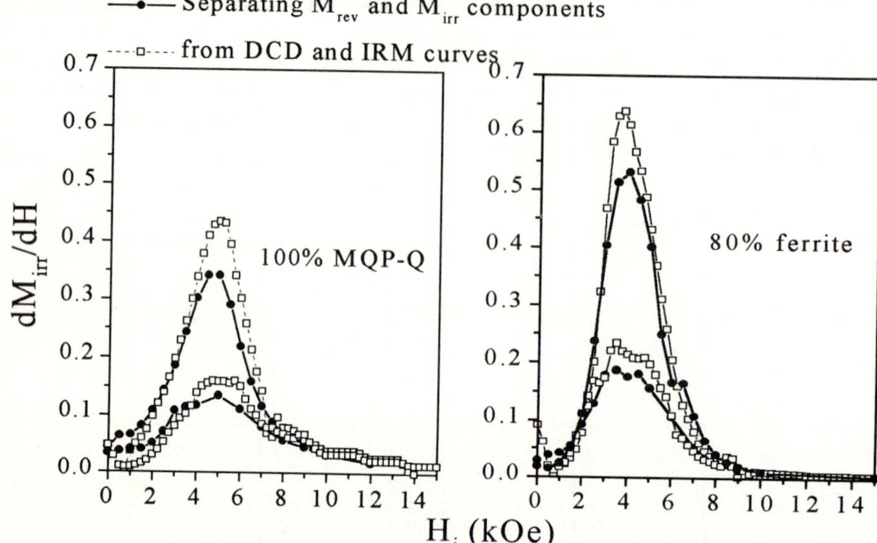

Fig. 3 χ_r and χ_d vs. H_i for two bonded magnets.

Acknowledgments – The authors would like to acknowledge useful discussions with D. R. Cornejo. FAPESP, CNPq, CAPES, and FINEP provided financial support. X-ray diffraction and SEM work was performed at the Laboratório de Caracterização Tecnológica - EPUSP.

References

1. J. J. Croat, *J. Appl. Phys.* **81** (1997) p. 4804.
2. J. Ormerod and S. Constantinides, *J. Appl. Phys.* **81** (1997) p. 4816.
3. M. Emura, A. C. Neiva, F. P. Missell, K. L. Babcock, J. Ormerod, and S. Constantinides, *J. Appl. Phys.* **83** (1998) p. 7127.
4. L. H. Lewis, D. O. Welch, and V. Panchanathan, *J. Appl. Phys.* **81** (1997) p. 4422.
5. P. I. Mayo, K. O'Grady, P. E. Kelly, J. Cambridge, I. L. Sanders, T. Yogi, and R. W. Chantrell, *J. Appl. Phys.* **69** (1991) p. 4733.
6. D. R. Cornejo, Ph.D. Thesis, Universidade de Sao Paulo, 1998.
7. D. R. Cornejo and F. P. Missell, this conference.
8. V. Basso and G. Bertotti, *IEEE Trans. Magn.* **30** (1994) p. 64.

Materials Science Forum Vols. 302-303 (1999) pp. 289-293

Mössbauer Spectroscopy Study of ^{57}Fe in $R_3Fe_{29-x}T_x$ (R=Y, Ce, Nd, Sm, Gd, Tb and Dy; T=V and Cr)

L.C.C.M. Nagamine[1], I. Souza Azevedo[1], E. Baggio-Saitovitch[1], H.R. Rechenberg[2], X.F. Han[1,3] and L.Y. Lin[3]

[1] Centro Brasileiro de Pesquisas Físicas, Rua Dr. Xavier Sigaud 150, 22290-180, Rio de Janeiro, Brazil

[2] Instituto de Física, Universidade de São Paulo, P.O. Box 66318, 05316-970 São Paulo, Brazil

[3] Institute of Semiconductors, Chinese Academy of Sciences, Beijing 100083, P.R. China

Keywords: ^{57}Fe Mössbauer Spectra, Magnetic Properties, Hyperfine Field, $R_3Fe_{29-x}T_x$, Fe Magnetic Moment

Abstract. ^{57}Fe Mössbauer spectra for the Fe atoms in the $R_3Fe_{29-x}T_x$ (R=Y, Ce, Nd, Sm, Gd, Tb, Dy; T=V, Cr) compounds were collected at 4.2 K. The analysis of Mössbauer spectra was based on the results of magnetization and neutron powder diffraction measurements. The average Fe magnetic moments at 4.2 K, deduced from our data, are in accord with magnetization measurements. The average hyperfine field of $Tb_3Fe_{29-x}Cr_x$ (x=1.0, 1.5, 2.0, and 3.0) decreases with increasing Cr concentration, which is also in accordance with the variation of the average Fe magnetic moment in the $Tb_3Fe_{29-x}Cr_x$ compounds.

1 Introduction

A new class of rare earth-iron (R-Fe) compounds stabilized by the presence of a third element (transition metal T), located at the iron rich corner in R-Fe-T phase diagrams, has been discovered[1]. This class was recognized as having the $Nd_3(Fe,Ti)_{29}$-type stoichiometry and structure with the monoclinic symmetry[2] and belonging to the A2/m space group[3] by x-ray powder diffraction (XRD), and then confirmed by neutron powder diffraction (NPD)[4]. Subsequent works on the same series with interstitial nitrides indicated that the $Sm_3(Fe,Ti)_{29}N_y$ [5] compounds can be regarded as new candidates for hard magnet application, which further aroused a great upsurge in intensive studies on this novel compounds.

In order to understand the magnetic structure and the intrinsic magnetic properties of $R_3Fe_{29-x}T_x$, it is important to do a systematic study on the ^{57}Fe hyperfine interactions of these compounds. In this work, the ^{57}Fe Mössbauer effect spectroscopy on $R_3Fe_{29-x}T_x$ (R=Y, Ce, Nd, Sm, Gd, Tb, and Dy; T=V and Cr) samples was measured at 4.2 K, and the theoretical fit of these spectra was performed considering the several crystallographic sites for the Fe atoms.

2 Experimental methods

The polycrystalline samples of $R_3Fe_{29-x}T_x$ (R=Y, Ce, Nd, Sm, Gd, Tb, and Dy; T=V and Cr) were prepared and characterized as it was described in our previous work[6, 7]. The ^{57}Fe Mössbauer spectra were collected with sample and γ-ray source (25 mCi of $^{57}Co/Rh$) immersed in liquid He. The spectra were analyzed with four and five independent subspectra based on the results of XRD[3], neutron powder diffraction[4], and magnetization measurements[6, 7].

3 Results and discussion

To simplify fitting of the spectra, the magnetic nonequivalent Fe sites were neglected and some constraints were used. So, the 11 Fe sites, present in the $R_3Fe_{29-x}Cr_x$ compounds, were considered together into five groups for R=Y, Ce, Nd, Sm, Tb, and Dy, except in the case of R=Gd (four groups), relative to the Fe- and R-atom numbers of nearest neighbor (NN) and the average bond length.

Figure 1 and 2 show ^{57}Fe Mössbauer spectra with the theoretical curves and the set of subspectra, for $R_3Fe_{29-x}Cr_x$ (R=Y, Ce, Nd, Sm, Gd, Tb, and Dy) and $Tb_3Fe_{29-x}Cr_x$ (x=1.0, 1.5, 2.0, and 3.0) compounds, obtained at 4.2 K.

Table 1 gives, for comparison, those five and four groups of the 11 Fe real sites, the Fe occupancy number (FON) of five or four groups of Fe sites, together with an initial area ratio as used in the fitting procedure for $Nd_3Fe_{29-x}Cr_x$ (x=5.48[4] and 4.5) and $Gd_3Fe_{288.0}Cr_{1.0}$ compounds. Table 2 gives the fitted values of the isomer shift, the quadrupole splitting and the hyperfine field for the five types of Fe sites for R=Y, Ce, Nd, Sm, Tb and Dy, and for the four types of Fe sites for the R=Gd, and the average hyperfine fields of $R_3Fe_{29-x}Cr_x$ compounds at 4.2 K. The relative area ratio of the subspectra are also listed in this table after taking into account the Cr occupancy of the Fe sites mentioned above.

The isomer shifts and quadrupole splittings found for these "groups" of Fe sites did not show any regular variation with R atomic number, due to the varying Cr concentration (x) in our samples. On the other hand, the average hyperfine field increased for increasing values of the rare earth effective spin (g_J-1)J, indicating a significant spin polarization induced by the neighboring rare earth.

The isomer shift values are similar for the set of $R_3Fe_{29-x}Cr_x$ (R=Ce, Nd and Sm) compounds and between the compounds $Tb_3Fe_{28.0}Cr_{1.0}$ and $Dy_3Fe_{27.4}Cr_{1.6}$ probably due to the near Cr concentration and the near electron configuration of Fe($3d^5 4s^1$) and R($4f^n 6s^2$). The isomer shift values for $R_3Fe_{29-x}Cr_x$ (R=Ce, Nd, Sm, Gd, Tb and Dy) are smaller than those for $Y_3Fe_{27.2}Cr_{1.8}$ relative to these groups of Fe sites. This can come from the main contribution of $6s^2$ electrons for R ions which increase the density $|\phi(0)|_a^2$ of s-electrons at Fe nuclear by molecular-orbital hybridization.

From the results obtained for the ^{57}Fe hyperfine field of $R_3Fe_{29-x}Cr_x$, we roughly have:

$$B_{hf}(1) > B_{hf}(2) > B_{hf}(3) > B_{hf}(4) > B_{hf}(5)$$

which is in agreement with the Fe- and R-atom NN numbers in the order at these Fe "type" sites, except for $B_{hf}(2)$. $B_{hf}(1)$ corresponds to the Fe dumbbell sites with the largest Fe-atom NN numbers (13) and the largest Fe magnetic moment. $B_{hf}(2)$ corresponds to the Fe_5 and Fe_{10} sites with the smallest Fe-atom NN numbers (9) but with the largest R-atom NN numbers (3), which is in accordance with the transferred spin polarization mentioned above.

The weighted average hyperfine fields B_{hf} of $R_3Fe_{29-x}Cr_x$ are listed in table 2. Assuming a ratio of 14.5 T/μ_B between B_{hf} and μ_{Fe} obtained from ^{57}Fe Mössbauer spectroscopic studies in the R_xFe_y alloys[8], the average Fe magnetic moment, $\mu^{MS}(1)$, of $R_3Fe_{29-x}Cr_x$ can be estimated as shown in column 8 in table 2. If a ratio of 13.8 T/μ_B deduced from our results for $Y_3Fe_{27.2}Cr_{1.8}$ is used instead, the average Fe magnetic moment, $\mu^{MS}(2)$, is obtained as shown in column 9 in table 2. It can be seen that the average Fe magnetic moment deduced from Mössbauer data corresponds to the results of magnetization measurement at 4.2 K, μ^{exp}, placed in column 10 for each compound[7].

Similarly, the average hyperfine fields for $R_3Fe_{29-x}V_x$ (R=Y and Gd) obtained by fitting their Mössbauer spectra at 4.2 K are 26.3 and 29.6 T, which correspond to 1.81 and 2.04 μ_B/Fe respectively, assuming a ratio of 14.5 T/μ_B between B_{hf} and μ_{Fe} in the R_xFe_y alloys[8]. It is in

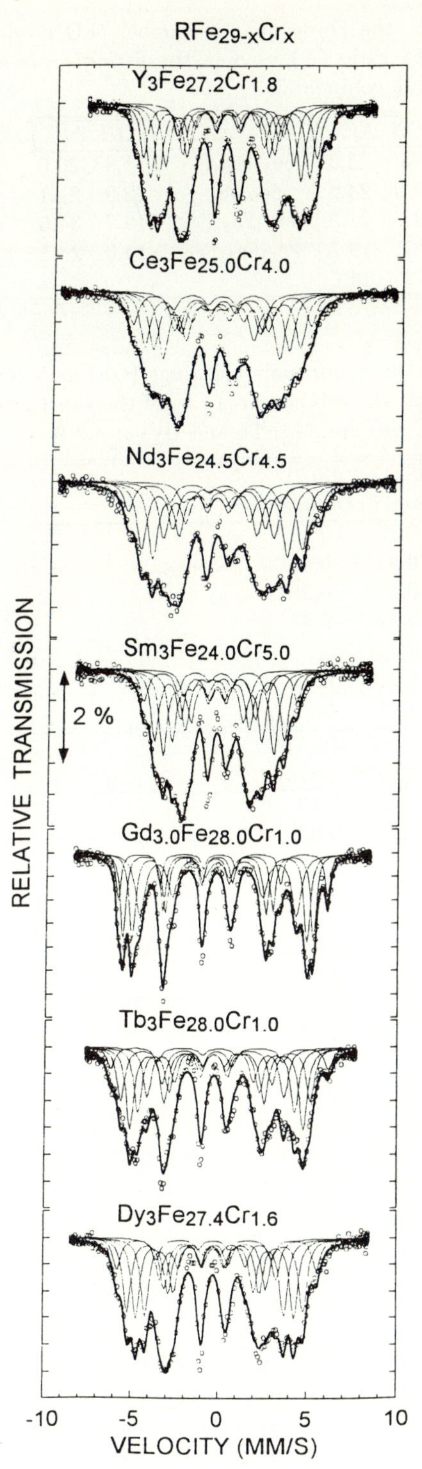

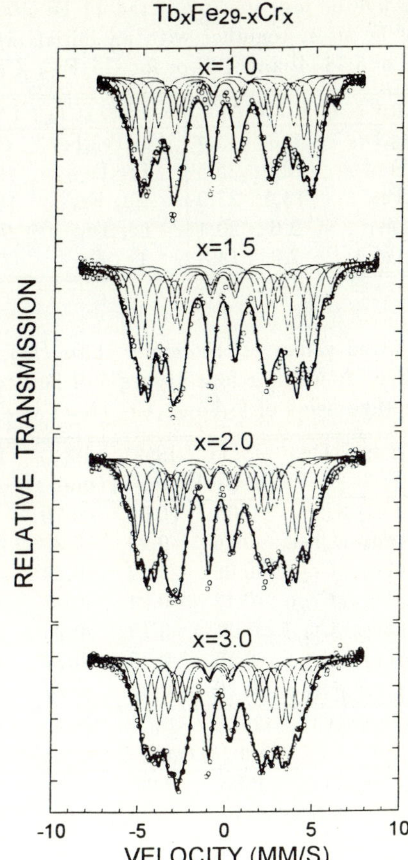

Fig.2 The observed ^{57}Fe Mössbauer spectra, the fitting curves and the sets of subspectra of $Tb_3Fe_{29-x}Cr_x$ (x=1.0, 1.5, 2.0, and 3.0) at 4.2 K.

Fig.1 The observed ^{57}Fe Mössbauer spectra, the fitting curves and the sets of subspectra of $R_3Fe_{29-x}Cr_x$ (R= Y, Ce, Nd, Sm, Gd, Tb, and Dy) at 4.2 K.

Table 1. The five or four groups of the 11 Fe sites, the Fe occupancy number (FON) of five or four "types" of Fe sites, together with an initial area ratio (A) used in the fitting procedure for $Nd_3Fe_{29-x}Cr_x$ (x=5.48 [4] and 4.5) or for $Gd_3Fe_{28.0}Cr_{1.0}$ compounds.

$Nd_3Fe_{23.52}Cr_{5.48}$	FON	A(%)	$Nd_3Fe_{24.5}Cr_{4.5}$	FON	A(%)	$Gd_3Fe_{28.0}Cr_{1.0}$	FON	A(%)
Fe_3,Fe_6,Fe_7	5.3	11.4	Fe_3,Fe_6,Fe_7	6.5	13.3	Fe_3,Fe_6,Fe_7	11.4	20.0
Fe_5,Fe_{10}	12.0	25.5	Fe_5,Fe_{10}	12.0	24.5	Fe_5,Fe_{10},	12.0	21.1
Fe_8,Fe_9	13.1	27.9	Fe_8,Fe_9	13.7	27.8	Fe_4,Fe_8,Fe_9	19.7	34.5
Fe_1,Fe_{11}	9.6	20.4	Fe_1,Fe_{11}	9.7	19.7	Fe_1,Fe_2,Fe_{11}	14.0	24.5
Fe_2,Fe_4	7.0	14.8	Fe_2,Fe_4	7.2	14.6			

Table 2. The fitted values of the isomer shifts (IS), the quadrupole splittings (QS) and hyperfine fields (B_{hf}) of five "types" (or four "types") of Fe sites, the relative area (A) of the subspectra and the average hyperfine fields of $R_3Fe_{29-x}Cr_x$ (R=Y, Ce, Nd, Sm, Gd, Tb and Dy) at 4.2 K.

$R_3Fe_{29-x}Cr_x$	IS(1) (mm/s)	IS(2) (mm/s)	IS(3) (mm/s)	IS(4) (mm/s)	IS(5) (mm/s)
$Y_3Fe_{27.2}Cr_{1.8}$	0.05	-0.08	-0.10	-0.12	-0.14
$Ce_3Fe_{25.0}Cr_{4.0}$	-0.26	-0.22	-0.30	-0.32	-0.43
$Nd_3Fe_{24.5}Cr_{4.5}$	-0.06	-0.22	-0.28	-0.32	-0.28
$Sm_3Fe_{24.0}Cr_{5.0}$	-0.12	-0.24	-0.28	-0.32	-0.47
$Gd_3Fe_{28.0}Cr_{1.0}$	0.00	-0.19	-0.20	-0.21	
$Tb_3Fe_{28.0}Cr_{1.0}$	-0.16	-0.20	-0.24	-0.25	-0.27
$Dy_3Fe_{27.4}Cr_{1.6}$	-0.20	-0.20	-0.26	-0.24	-0.20

$R_3Fe_{29-x}Cr_x$	QS(1) (mm/s)	QS(2) (mm/s)	QS(3) (mm/s)	QS(4) (mm/s)	QS(5) (mm/s)
$Y_3Fe_{27.2}Cr_{1.8}$	0.03	0.06	0.04	0.10	0.13
$Ce_3Fe_{25.0}Cr_{4.0}$	-0.04	-0.06	0.01	0.05	0.03
$Nd_3Fe_{24.5}Cr_{4.5}$	-0.08	0.11	0.08	0.05	0.32
$Sm_3Fe_{24.0}Cr_{5.0}$	-0.10	0.07	0.06	-0.01	0.24
$Gd_3Fe_{28.0}Cr_{1.0}$	0.12	-0.01	0.02	0.20	
$Tb_3Fe_{28.0}Cr_{1.0}$	-0.07	-0.12	0.12	0.10	0.08
$Dy_3Fe_{27.4}Cr_{1.6}$	0.12	-0.02	0.03	0.04	0.33

$R_3Fe_{29-x}Cr_x$	A(1)	A(2)	A(3)	A(4)	A(5)
$Y_3Fe_{27.2}Cr_{1.8}$	0.09	0.26	0.26	0.26	0.13
$Ce_3Fe_{25.0}Cr_{4.0}$	0.09	0.24	0.27	0.27	0.13
$Nd_3Fe_{24.5}Cr_{4.5}$	0.13	0.27	0.30	0.22	0.08
$Sm_3Fe_{24.0}Cr_{5.0}$	0.12	0.25	0.30	0.25	0.08
$Gd_3Fe_{28.0}Cr_{1.0}$	0.13	0.26	0.32	0.29	
$Tb_3Fe_{28.0}Cr_{1.0}$	0.08	0.20	0.30	0.24	0.18
$Dy_3Fe_{27.4}Cr_{1.6}$	0.11	0.25	0.25	0.26	0.13

$R_3Fe_{29-x}Cr_x$	$B_{hf}(1)$ (T)	$B_{hf}(2)$ (T)	$B_{hf}(3)$ (T)	$B_{hf}(4)$ (T)	$B_{hf}(5)$ (T)	B_{hf} (T)	μ_1^{MS} μ_B/Fe	μ_2^{MS} μ_B/Fe	μ^{exp} μ_B/Fe
$Y_3Fe_{27.2}Cr_{1.8}$	34.1	29.9	26.8	23.8	19.6	26.5	1.83	1.92	1.92
$Ce_3Fe_{25.0}Cr_{4.0}$	25.4	22.3	19.5	16.8	12.7	19.1	1.32	1.50	1.57
$Nd_3Fe_{24.5}Cr_{4.5}$	27.3	22.9	19.3	15.8	11.5	19.9	1.37	1.60	1.74
$Sm_3Fe_{24.0}Cr_{5.0}$	27.4	23.2	19.8	16.1	12.1	20.0	1.38	1.64	1.48
$Gd_3Fe_{28.0}Cr_{1.0}$	37.2	33.8	27.8	31.1		31.5	2.17	2.22	2.03
$Tb_3Fe_{28.0}Cr_{1.0}$	38.5	33.6	30.7	27.8	24.3	30.1	2.08	2.12	1.90
$Dy_3Fe_{27.4}Cr_{1.6}$	35.3	31.1	24.4	28.0	20.6	27.7	1.91	1.99	1.92

accordance with the results of magnetization measurements at 4.2 K, 1.81 and 1.92 μ_B/Fe[7], for the $Y_3Fe_{27.4}V_{1.6}$ and $Gd_3Fe_{28.4}V_{0.6}$ compounds.

The average hyperfine fields at 4.2 K for $Gd_3Fe_{29-x}T_x$ (T=V and Cr) is larger than that of the other members of the series, which is in agreement with the largest average Fe magnetic moment for Gd compounds, as shown by the magnetization measurements[7].

The average hyperfine fields obtained at 4.2 K for $Tb_3Fe_{29-x}Cr_x$, 30.7, 29.9, 26.8 and 26.5 T corresponding to x=1.0, 1.5, 2.0 and 3.0, linearly decrease with increasing the Cr concentration. Thus their average Fe magnetic moments are 2.12, 2.06, 1.85 and 1.83 μ_B/Fe, respectively. It is in accordance with the results of magnetization measurements at 4.2 K: 1.90, 1.80, 1.79 and 1.64 μ_B/Fe, and supports the suggestion that the 3:29 structure is an ordered combination of the 2:17 and 1:12 structures in a ratio of 1:1, and the average Fe moment at 4.2 K in the $Tb_3Fe_{29-x}Cr_x$ compounds in the range $1.0 \leq x \leq 3.0$ can be expressed as

$$M_{Fe}(3:29) = M_{Fe}(2:17) - x[M_{Fe}(2:17) - M_{Fe}(1:12)]/X_{max}$$

where M_{Fe}(2:17) and M_{Fe}(1:12) are the average Fe moments in 2:17 and 1:12 compounds, and X_{max} is the highest Cr concentration included[6].

This empirical formula seems interesting, and it indicates that there exists a correlation on intrinsic magnetic properties between the R_2Fe_{17}, $RFe_{12-x}T_x$, and $R_3Fe_{29-x}T_x$ structures, and also implies that a higher average Fe atomic magnetic moment, i.e., a higher saturation magnetization, can be achieved for the R_2Fe_{17} compounds unlike $RFe_{12-x}Cr_x$. This can be understood since the increasing of Cr concentration in the 2:17, 3:29, and 1:12 compounds leads to a decrease of the average Fe magnetic moments. This may be due to the variations of spin up and spin down Fe-3d density of states at the Fermi level, E_F, associated with the mixture of the Cr valence electrons.

In conclution we may say that in the 2:17, 3:29, and 1:12 compounds the average Fe magnetic moments decrease with increasing of Cr concentration.

Acknowledgments: Project supported by the exchange program between CNPq - Brazil and State Committee of Science and Technology - People's Republic of China.

References

[1] S. J. Collocott, R. K. Day, J. B. Dunlop, and R. L. Davis, *Proc. 7th Int. Symp. on Magnetic Anisotropy and Coercivity in Rare-Earth-Transition Metal Alloys*(Canberra, 1992), p.437

[2] H. S. Li, J. M. Cadogan, R. L. Davis, A. Margarian, and J. B. Dunlop, *Solid State Commun.*, **90** (1994) 487

[3] O. Kalogirou, V. Psycharis, L. Saettas, and D. N. Niarchos, *J. Magn. Magn. Mater.*, **146** (1995) 335

[4] W. B. Yelon and Z. Hu, *J. Appl. Phys.*, **79** (1996) 1330

[5] F. M. Yang, B. Nasunjilegal, J. L. Wang, J. J. Zhu, W. D. Qin, N. Tang, R. W. Zhao, B. P. Hu, Y. Z. Wang, and H. S. Li, *J. Phys.: Conden. Matter*, **7** (1995) 1679

[6] X. F. Han, H. G. Pan, H. L. Liu, F. M. Yang, and Y. W. Zheng, *Phys. Rev. B*, **56** (1997) 8867

[7] X. F. Han, F. M. Yang, H. G. Pan, Y. G. Wang, J. L. Wang, H. L. Liu, N. Tang, R. W. Zhao, and H. S. Li, *J. Appl. Phys.*, **81** (1997) 7450

[8] P. C. M. Gubbens, J. H. F. van Apeldoorn, A. M. Vander Kraan, and K. H. J. Buschow, *J. Phys. F*, **4** (1974) 921

Materials Science Forum Vols. 302-303 (1999) pp. 294-298

Crystallographic and Magnetic Properties of Hydride $R_3Fe_{29-x}T_xH_y$ (R=Y, Ce, Nd, Sm, Gd, Tb and Dy; T=V and Cr)

X.F. Han[1,2], L.Y. Lin[1], E. Baggio-Saitovitch[2], R.G. Xu[3], X.H. Wang[3] and H.G. Pan[3]

[1] Material Science Center, Institute of Semiconductors, Chinese Academy of Sciences, P.O. Box 912, Beijing 100083, P.R. China
E-Mail: xfhan@novell.cat.cbpf.br and xfhan@red.semi.ac.cn

[2] Centro Brasileiro de Pesquisas Físicas, Rua Dr. Xavier Sigaud, no. 150-Urac-RJ, Brasil
E-Mail: elisa@novell.cat.cbpf.br

[3] Material Department of Zhejiang University, Hangzhou 310027, P.R. China

Keywords: Magnetic Properties, Hydride, Hydrogenation, Saturation Magnetization, Curie Temperature, First-Order Magnetization Process, $R_3Fe_{29-x}T_xH_y$

Abstract. The crystallographic and intrinsic magnetic properties of hydride $R_3Fe_{29-x}T_xH_y$ (R=Y, Ce, Nd, Sm, Gd, Tb, and Dy; T=V and Cr) have been investigated. The lattice constants and the unit cell volume of $R_3Fe_{29-x}T_xH_y$ decrease with increasing R atomic number from Nd to Dy, except for Ce, reflecting the lanthanide contraction. Regular anisotropic expansions, mainly along the a- and b-axis rather than along the c-axis, are observed for all the compounds upon hydrogenation. Hydrogenation leads to an increase in Curie temperature. First-order magnetization processes (FOMP) occur in magnetic fields of around 1.5 T and 4.0 T at 4.2 K for $Nd_3Fe_{24.5}Cr_{4.5}H_{5.0}$ and $Tb_3Fe_{27.0}Cr_{2.0}H_{2.8}$, and around 1.4 T at room temperature for $Gd_3Fe_{28.0}Cr_{1.0}H_{4.2}$. Abnormal crystallographic and magnetic properties of $Ce_3Fe_{29-x}T_xH_y$ suggest that the Ce ion is non-triply ionized.

1 Introduction

Due to the great interest in high performance permanent magnetic materials, intensive investigations on the R-Fe-T phase diagrams (R=rare earth, T=transition metal as a stabilizing element) located at the iron rich corner, led to the discovery of the novel $R_2(Fe,T)_{19}$ phase[1, 2]. Its real stoichiometric composition and crystallographic structure were determined as a $Nd_3(Fe,Ti)_{29}$-type structure with the monoclinic symmetry[3] and the A2/m space group[4] by x-ray powder diffraction (XRD) and then confirmed by neutron powder diffraction[5, 6]. Subsequent works on its interstitial nitrides and carbides indicated that the $Sm_3(Fe,T)_{29}N_y$ (T=Ti[7, 8, 9], V[10, 11], Cr[10, 12, 13, 14], and Mo[15]) compounds and the $Sm_3(Fe,T)_{29}C_y$(T= Ti[9, 16] and Cr[17]) compounds can be regarded as new candidates for hard magnet application. The crystallographic and magnetic properties of the $R_3(Fe,T)_{29}$ compounds and their nitrides have been studied extensively by some researchers in other papers.

Up to now, little attention was focused on the investigation of hydride $R_3Fe_{29-x}T_xH_y$. Only the studies of $Nd_3(Fe,Ti)_{29}H_y$[18] and $Sm_3(Fe,T)_{29}H_y$ (T=Ti, V, and Cr)[10] were reported. In this work, a systematic study on the phase formation and intrinsic magnetic properties of the hydrides $R_3Fe_{29-x}T_xH_y$ (R=Y, Ce, Nd, Sm, Gd, Tb, and Dy; T=V and Cr) has been performed because the metal hydrides are interesting candidates for applications in the areas of heat pumps, cooling systems, and thermal storage medium.

2 Sample preparation and experimental methods

In order to obtain the saturation magnetization and magnetic anisotropy field at 4.2 K and room temperature (RT), the procedure for sample preparation and measurement have been the same as already referred in our previous works[20]. Previously to the preparations of the hydride phase, the single-phased 3:29 alloys were pulverized into fine powders with a size of 100-1000 μm. Hydrogenation was performed by heating the fine powders, at 5 atm of hydrogen pressure, at 573 K for R=Ce, Nd, Sm, Gd, and Tb or at 673 K for R=Y and Dy for a period of 2-4 hours in a high pressure container.

The x-ray diffraction (XRD) patterns and thermomagnetic analysis (TMA) show that each hydride of $R_3Fe_{29-x}T_x$ was single phase with the same structure as the parent compound, except for a little amount of α-Fe. The hydrogen concentration in each hydride $R_3Fe_{29-x}T_xH_y$ was deduced from the atmosphere pressure difference, before and after hydrogenation, according the gas state equation $PV = nRT$ at a constant temperature.

3 Results and discussion

Figure 1 shows XRD patterns with Cu $K\alpha$ radiation for some of the $R_3Fe_{29-x}V_xH_y$ (R=Y, Ce, Tb, and Dy) compounds. The crystallographic parameters of the hydrides were obtained by fitting the XRD patterns with monoclinic symmetry and $A2/m$ space group. Figure 2 displays thermomagnetic curves for $R_3Fe_{29-x}V_xH_y$ in a low field of 0.04 T. Table 1 gives the crystallographic parameters, the unit cell volume and density ρ, the Curie temperature T_C, the saturation magnetization M_S and anisotropy field H_A at 4.2 K and at room temperature (RT) for the hydrides and their parent compounds.

It can been seen that the lattice constants and the unit cell volume of $R_3Fe_{29-x}T_x$ and $R_3Fe_{29-x}T_xH_y$ decrease generally with increasing R atomic number from Nd to Dy, except for Ce, reflecting the lanthanide contraction. This leads to an increase in density. Abnormal crystallographic and magnetic properties may originate from the Ce ions being in a non-trivalent state. It is noticeable that $Ce_3Fe_{29-x}T_x$ and $Gd_3Fe_{28.4}V_{0.6}$ show the maximum storage of hydrogen in the series of compounds. So they can be considered as interesting candidates for further investigation in order to develop new hydrogen storage medium.

The regular anisotropic expansions of $R_3Fe_{29-x}T_x$ are observed upon hydrogenation along the a-, b-, and c-axis for each hydride, such as $\frac{\delta a}{a} \simeq \frac{\delta b}{b} > \frac{\delta c}{c}$, which indicate that it occurs mainly in the basal c-plane rather than along the c-axis for the all hydrides. These results are in agreement with the hypothesis that hydrogen atoms mainly occupy the $4e_1$ and $4e_2$ sites which are located in the basal plane referred to the 1:5 cell[10].

As an example, figure 3 displays the magnetic isotherms at 4.2 K for the magnetically-aligned samples of $Nd_3Fe_{24.5}Cr_{4.5}H_{5.0}$ measured with a VSM. First-order magnetization process (FOMP) occurs in magnetic field of around 1.5 T at 4.2 K for $Nd_3Fe_{24.5}Cr_{4.5}H_{5.0}$.

It can be seen that hydrogenation leads to improvements in the saturation magnetization at RT compared with the parent compounds owing to the increase in Curie temperature as shown in table 1. The saturation magnetization of $Y_3Fe_{27.4}V_{1.6}H_{2.2}$ and $Y_3Fe_{27.2}Cr_{1.8}H_{3.8}$ at 4.2 K is 57.3 $\mu_B/f.u.$ and 43.0 $\mu_B/f.u.$ corresponding to an average Fe magnetic moment of 2.09 μ_B/Fe and 1.58 μ_B/Fe, respectively. Hydrogenation leads to an increase of 0.28 μ_B/Fe but a decrease of 0.34 μ_B/Fe in the average Fe magnetic moment compared with that in the $Y_3Fe_{27.4}V_{1.6}$ and $Y_3Fe_{27.2}Cr_{1.8}$ at 4.2 K, respectively.

The decreases in the anisotropy field of both hydrides $Y_3Fe_{27.4}V_{1.6}H_{2.2}$ and $Gd_3Fe_{28.4}V_{0.6}H_{6.9}$ at 4.2 K and at RT indicate that hydrogenation decreases the anisotropy field of the Fe-sublattice in the $R_3Fe_{29-x}V_x$ (R=Y, Ce, Nd, Sm, Gd, Tb, and Dy) compounds.

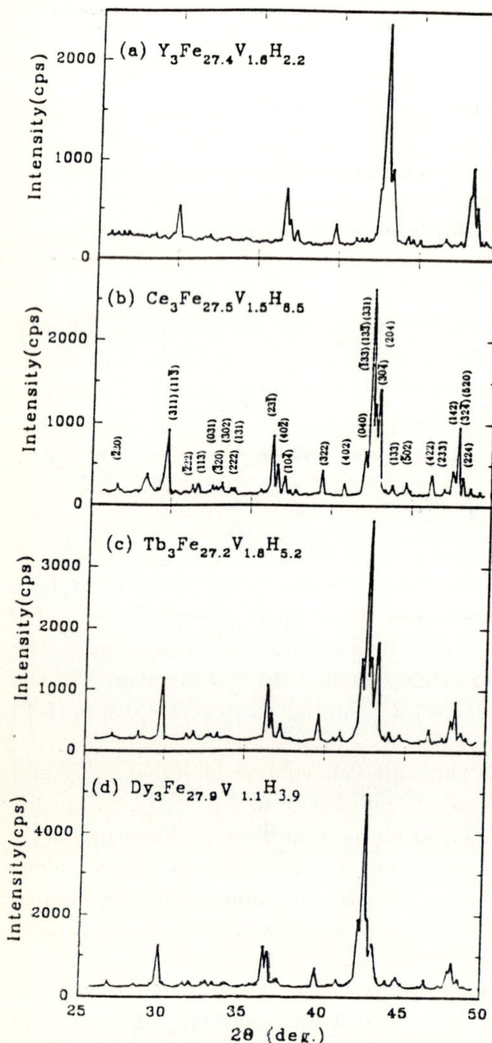

Fig.1 XRD patterns with Cu $K\alpha$ radiation for $R_3Fe_{29-x}V_xH_y$ (R= Y, Ce, Tb, and Dy).

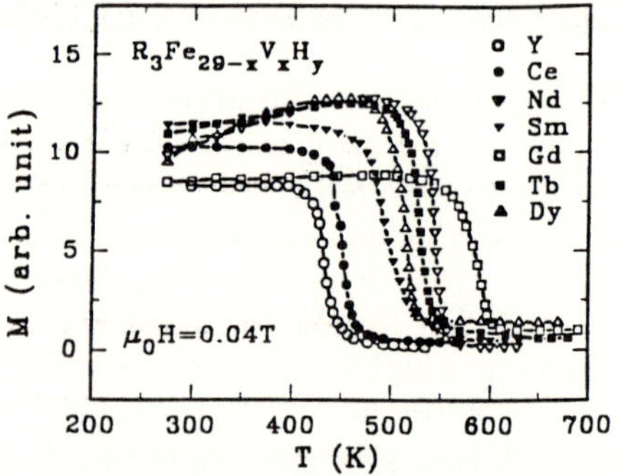

Fig.2 Thermomagnetic curves for $R_3Fe_{29-x}V_xH_y$ (R=Y, Ce, Nd, Sm, Gd, Tb and Dy) in a low field of 0.04 T.

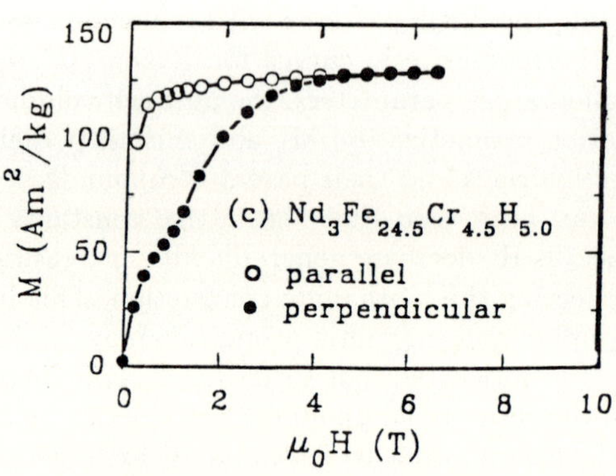

Fig.3 Magnetic isotherms at 4.2 K for $Nd_3Fe_{24.5}Cr_{4.5}H_{5.0}$ measured with a VSM in an external field applied either parallel or perpendicular to the alignment direction of the powder samples.

Table 1. The lattice parameters and the unit cell volume V=abcsinβ derived from x-ray analysis data with the $A2/m$ space group; the x-ray density ρ derived from the lattice constants, the Curie temperature T_C, the saturation magnetization M_S and the anisotropy field H_A at 4.2 K and at RT for $R_3Fe_{29-x}T_x$ and $R_3Fe_{29-x}T_xH_y$ (R=Y, Ce, Nd, Sm, Gd, Tb, and Dy; T=V and Cr).

$R_3Fe_{29-x}T_x(H_y)$	a (nm)	b (nm)	c (nm)	β (deg.)	V (nm^3)	ρ (g/cm^3)	T_C (K)	$M_S^{4.2K}$ (μ_B/f.u.)	M_S^{RT} (μ_B/f.u.)	$\mu_0H_A^{4.2K}$ (T)	$\mu_0H_A^{RT}$ (T)
$Y_3Fe_{27.4}V_{1.6}$	1.0560	0.8482	0.9656	97.00	0.8584	7.267	435	49.6	38.6	8.3	1.7
$Ce_3Fe_{27.5}V_{1.5}$	1.0553	0.8495	0.9675	96.72	0.8614	7.837	350	50.5	32.8	5.5	1.2
$Nd_3Fe_{27.0}V_{2.0}$	1.0647	0.8574	0.9738	96.85	0.8826	7.685	455	54.7	46.8	12.8	2.4
$Sm_3Fe_{26.7}V_{2.3}$	1.0605	0.8546	0.9708	96.86	0.8735	7.831	502	42.7	33.8	8.4	4.3
$Gd_3Fe_{28.4}V_{0.6}$	1.0601	0.8514	0.9690	97.01	0.8680	7.991	518	33.6	34.7	11.2	3.6
$Tb_3Fe_{27.2}V_{1.8}$	1.0556	0.8487	0.9670	96.88	0.8600	8.061	479	23.1	23.3	17.2	5.2
$Dy_3Fe_{27.9}V_{1.1}$	1.0562	0.8485	0.9664	96.86	0.8598	8.127	437	23.8	—	—	—
$Dy_3Fe_{27.7}V_{1.3}$	1.0561	0.8478	0.9661	96.86	0.8588	8.123	449	22.7	24.7	12.8	9.4
$Y_3Fe_{27.4}V_{1.6}H_{2.2}$	1.0602	0.8506	0.9682	96.99	0.8661	7.211	445	57.3	46.8	5.0	1.2
$Ce_3Fe_{27.5}V_{1.5}H_{6.5}$	1.0675	0.8592	0.9752	96.99	0.8877	7.633	461	52.0	47.1	4.0	1.0
$Nd_3Fe_{27.0}V_{2.0}H_{4.0}$	1.0727	0.8632	0.9783	97.10	0.8988	7.559	554	57.1	52.5	5.4	0.8
$Sm_3Fe_{26.7}V_{2.3}H_{4.8}$	1.0709	0.8604	0.9757	97.27	0.8918	7.679	522	47.3	48.6	10.0	5.0
$Gd_3Fe_{28.4}V_{0.6}H_{6.9}$	1.0718	0.8588	0.9768	97.00	0.8924	7.798	598	39.4	43.4	7.6	3.2
$Tb_3Fe_{27.2}V_{1.8}H_{5.2}$	1.0646	0.8563	0.9707	97.17	0.8780	7.916	542	23.0	32.3	6.6	2.7
$Dy_3Fe_{27.9}V_{1.1}H_{3.9}$	1.0628	0.8544	0.9691	97.14	0.8732	8.006	527	26.2	39.1	8.0	2.1
$Y_3Fe_{27.2}Cr_{1.8}$	1.0570	0.8474	0.9653	96.98	0.8582	7.273	427	52.2	38.6	3.4	1.4
$Ce_3Fe_{25.0}Cr_{4.0}$	1.0527	0.8484	0.9668	96.67	0.8576	7.771	317	39.5	19.7	2.5	0.6
$Nd_3Fe_{24.5}Cr_{4.5}$	1.0615	0.8556	0.9714	96.90	0.8759	7.685	413	49.9	32.2	9.1	1.2
$Sm_3Fe_{24.0}Cr_{5.0}$	1.0585	0.8521	0.9684	96.90	0.8671	7.857	424	36.0	29.4	9.6	2.7
$Gd_3Fe_{28.0}Cr_{1.0}$	1.0604	0.8515	0.9686	96.95	0.8682	7.985	507	35.9	32.8	7.7	3.6
$Tb_3Fe_{28.0}Cr_{1.0}$	1.0585	0.8492	0.9675	96.92	0.8634	8.049	465	26.3	30.2	9.2	3.2
$Tb_3Fe_{27.0}Cr_{2.0}$	1.0559	0.8494	0.9666	96.89	0.8607	8.059	470	21.3	22.0	9.8	3.7
$Dy_3Fe_{27.4}Cr_{1.6}$	1.0556	0.8475	0.9655	96.91	0.8574	8.138	442	22.6	26.3	14.1	4.6
$Y_3Fe_{27.2}Cr_{1.8}H_{3.8}$	1.0625	0.8528	0.9686	97.18	0.8708	7.183	488	43.0	40.3	3.6	1.5
$Ce_3Fe_{25.0}Cr_{4.0}H_{5.4}$	1.0640	0.8563	0.9731	96.87	0.8800	7.593	398	38.6	34.0	1.2	0.7
$Nd_3Fe_{24.5}Cr_{4.5}H_{5.0}$	1.0714	0.8620	0.9770	97.07	0.8955	7.559	488	37.9	40.8	4.0	1.6
$Sm_3Fe_{24.0}Cr_{5.0}H_{4.2}$	1.0658	0.8582	0.9721	97.11	0.8823	7.730	474	37.6	32.7	10.3	2.9
$Gd_3Fe_{28.0}Cr_{1.0}H_{4.2}$	1.0664	0.8580	0.9724	97.17	0.8828	7.869	571	37.6	42.1	8.4	2.9
$Tb_3Fe_{28.0}Cr_{1.0}H_{4.2}$	1.0647	0.8558	0.9712	97.20	0.8780	7.931	554	25.0	34.9	9.6	1.5
$Tb_3Fe_{27.0}Cr_{2.0}H_{2.8}$	1.0614	0.8537	0.9676	97.15	0.8700	7.984	530	22.6	33.0	10.6	4.3
$Dy_3Fe_{27.4}Cr_{1.6}H_{2.1}$	1.0593	0.8508	0.9666	97.07	0.8646	8.078	508	21.9	32.5	11.6	2.7

No obvious changes in the anisotropy field are observed for $Y_3Fe_{27.2}Cr_{1.8}H_{3.8}$ at 4.2 K and at RT. The increases of the anisotropy field for $Sm_3Fe_{29-x}T_xH_y$ at 4.2 K and at RT mainly originate from the larger increase of the Sm-sublattice's anisotropy field (or crystalline electric field coefficient A_2^0) after hydrogenation.

Acknowledgments: Project supported by the exchange program between CNPq-Brazil and State Committee of Science and Technology-People's Republic of China.

References

[1] S. J. Collocott, R. K. Day, J. B. Dunlop, and R. L. Davis, *Proc. 7th Int. Symp. on Magnetic anisotropy and Coercivity in Rare-Earth-Transition Metal Alloys* (Canberra, 1992), p.437

[2] Ye V. Shcherbakova, G. V. Ivanova, A. S. Yermolenko, Ye V. Belozerov, and V. S. Gaviko, *J. Alloys Compounds*, **182** (1992) 199

[3] H. S. Li, J. M. Cadogan, R. L. Davis, A. Margarian, and J. B. Dunlop, *Solid State Commun.*, **90** (1994) 487

[4] O. Kalogirou, V. Psycharis, L. Saettas, and D. N. Niarchos, *J. Magn. Magn. Mater.*, **146** (1995) 335

[5] W. B. Yelon and Z. Hu, *J. Appl. Phys.*, **79** (1996) 1330

[6] Z. Hu, W. B. Yelon, O. Kalogirou, and V. Psycharis , *J. Appl. Phys.*, **80** (1996) 2955

[7] F. M. Yang, B. Nasunjilegal, J. L. Wang, J. J. Zhu, W. D. Qin, N. Tang, R. W. Zhao, B. P. Hu, Y. Z. Wang, and H. S. Li, *J. Phys.: Conden. Matter*, **7** (1995) 1679

[8] B. P. Hu, G. C. Liu, Y. Z. Wang, B. Nasunjilegal, R. W. Zhao, F. M. Yang, H. S. Li, and J. M. Cadogan, *J. Phys.: Conden. Matter*, **6** (1994) L197

[9] K. H. Müller, J. B. Dunlop, A. Handstein, B. Gebel, and P. A. P. Wendhausen, *J. Magn. Magn. Mater.*, **157/158** (1996) 117

[10] K. Koyama, H. Fujii, and S. Suzuki, *J. Magn. Magn. Mater.*, **161** (1996) 118

[11] X. F. Han, F. M. Yang, Q. S. Li, M. C. Zhang, S. Z. Zhou, *J. Phys.: Conden. Matter*, **10** (1998) 151

[12] S. Suzuki, S. Suzuki and M. K. Awasaki, *IEEE Trans. on Magn.*, **31** (1995) 3695

[13] X. F. Han, F. M. Yang, H. G. Pan, Y. G. Wang, and Y. Z. Wang, *J. Appl. Phys.*, **83** (1998) 4366

[14] Y. Z. Wang, B. P. Hu, G. C. Liu, H. S. Li, X. F. Han, and C. P. Yang, *J. Phys.: Conden. Matter*, **9** (1997) 2787

[15] H. G. Pan, F. M. Yang, Y. Chen, X. F. Han, N. Tang, C. P. Chen, and Q. D. Wang, *J. Phys.: Conden. Matter*, **9** (1997) 2499

[16] B. P. Hu, G. C. Liu, Y. Z. Wang, B. Nasunjilegal, N. Tang, F. M. Yang, H. S. Li, and J. M. Cadogan, *J. Phys.: Conden. Matter*, **6** (1994) L595

[17] Y. Z. Wang, B. P. Hu, G. C. Liu, H. S. Li, X. F. Han, and C. P. Yang, *J. Phys.: Conden. Matter*, **9** (1997) 2793

[18] D. H. Ryan, J. M. Cadogan, A. Margarian, and J. B. Dunlop, *J. Appl. Phys.*, **76** (1994) 6150

[19] F. M. Yang, X. F. Han, F. R. de Boer, and H. S. Li, *J. Phys.: Conden. Matter*, **9** (1997) 1339

[20] X. F. Han, F. M. Yang, H. G. Pan, Y. G. Wang, J. L. Wang, H. L. Liu, N. Tang, R. W. Zhao, and H. S. Li, *J. Appl. Phys.*, **81** (1997) 7450

Materials Science Forum Vols. 302-303 (1999) pp. 299-303

Magnetic Analysis of $RE_{80}Fe_{15}C_5$ (RE=Nd, Pr) and $Nd_{80}Fe_{15}(B_{1-x}C_x)_5$ ($0.85 \leq x \leq 1.00$) Alloys

G. López[1], P.H. Domingues[1], V. Barthem[1] and J.L. Sánchez Ll.[2]

[1] Instituto de Física, Universidade Federal do Rio de Janeiro, Centro de Tecnologia, Bloco A, P.O. Box 68528, 21945-970, Rio de Janeiro, Brazil

[2] Laboratorio de Magnetismo, Facultad de Física-IMRE, Universidad de La Habana, La Habana 10400, Cuba

Keywords: Nd-Fe-B and Pr-Fe-B Alloys, Ac-Susceptibility, Curie Temperature

Abstract

The temperature dependence of saturation magnetization, σ_s, intrinsic coercivity, iHc, and ac-magnetic susceptibility, χ_{ac}, from 4.2 to 300 K were studied for $RE_{80}Fe_{15}C_5$ samples. The magnetic analysis of $Nd_{80}Fe_{15}(B_{1-x}C_x)_5$ alloys with very low B-content ($0.85 \leq x \leq 1.00$) was also carried out. The latter confirms that the phase D_1 only forms for x = 1.00 after annealing at 600°C. A comparison of the $\chi_{ac}(T)$ curves for $RE_{80}Fe_{15}X_5$, X = B, C, suggests that phase D_1 is different from the ternary carbide $RE_2Fe_{14}C$. This contrasts with the results observed for $RE_{80}Fe_{15}B_5$ (RE = Nd, Pr) samples where the annealing at 600°C transforms A_1 into $RE_2Fe_{14}B$. With the decrease of the temperature the coercive fields of $RE_{80}Fe_{15}C_5$ alloys strongly increases (31.2 kOe for RE = Nd and 39.3 kOe for RE = Pr at 10K) confirming the high magnetocrystalline anisotropy of the D_1 phase.

1. Introduction

In previous works we performed thermomagnetic and microstructural studies in $RE_{80}Fe_{15}(B_{1-x}C_x)_5$ alloys with $0.00 \leq x \leq 1.00$ and RE = Nd, Pr [1,2]. It was observed that A_1 is the main magnetic phase formed in as-cast samples. In some alloys a stable phase, referred in the literature as phase A_3, also exists (A_3 appears up to x = 0.5 for RE = Nd (T_c = 292°C) and up to x = 0.75 for RE= Pr (T_c = 264°C)). Annealing of $RE_{80}Fe_{15}(B_{1-x}C_x)_5$ alloys at 600°C leads to the formation of the following stable magnetic phases: a) $RE_2Fe_{14}B$ (together with A_3 as minor phase in the above mentioned composition intervals); and, b) an unknown hard magnetic phase, labeled by us as D_1, for x = 1.00 (iHc = 2.2 kOe at room temperature). The latter shows T_c = 246°C for RE = Nd and T_c = 230°C for RE = Pr. A SEM analysis was also carried out in order to characterize the $RE_{80}Fe_{15}C_5$ alloys in as-cast state and after 2 hours of annealing. D_1 phase is believed to be a needle-like phase with Fe/RE ratio larger than one.

In the present paper we report the temperature dependence of saturation magnetization, intrinsic coercivity and AC susceptibility from 4.2 to 300 K of annealed $RE_{80}Fe_{15}C_5$ (RE= Nd, Pr) alloys. In addition, we investigated $Nd_{80}Fe_{15}(B_{1-x}C_x)_5$ samples with very low B-content ($0.85 \leq x \leq 1.00$) in both, the as-cast state and after 2 hours of annealing at 600°C.

2. Experimental

Alloys were prepared by arc melting the pure elements under an Ar high purity atmosphere. The raw materials were at least 99.9% pure (Fe and C from Alpha Products Co. and Nd, Pr and B from Aldrich Chemical Co.). Samples were wrapped in tantalum foils, annealed at 600°C for two hours in closed quartz ampoules and then quenched in water. Magnetic measurements were done on nearly spherical samples by using an EG&G-PAR model 4500 VSM (H_{max} = 90 kOe). The d.c. thermomagnetic analysis (TMA) curves were recorded above room temperature at H_{ext} = 200 Oe; the Curie points were inferred from the minimum in the $d\chi_o/dT$ vs. T curves. The $\chi_{ac}(T)$ curves were measured on bar-shaped samples (1 x 1 x 5 mm^3 aproximately) by using a Lakeshore model 7225 susceptometer applying an AC excitation magnetic field of 0.50 Oe of peak value and frequency of 500 Hz.

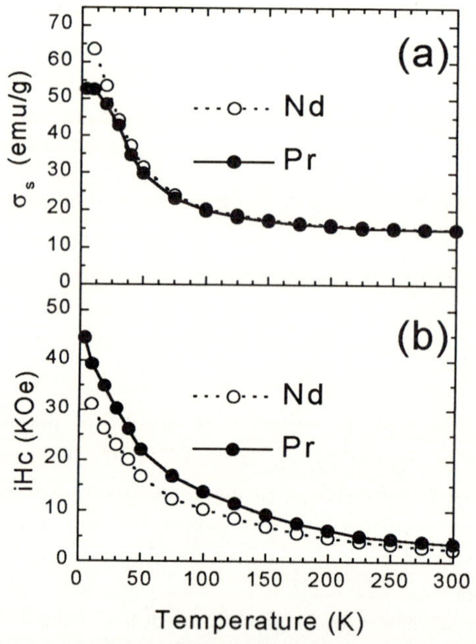

Fig. 1 Saturation magnetization (a) and intrinsic coercive field (b) as a function of the temperature for $RE_{80}Fe_{15}C_5$ (RE = Nd, Pr) alloys. The measurements were done for H_{max} = 90 kOe.

3. Results and discussion

3.1 $RE_{80}Fe_{15}C_5$ (RE= Nd, Pr)

Fig. 1 shows the temperature dependence of the saturation magnetization and coercive field for annealed $RE_{80}Fe_{15}C_5$, RE = Nd, Pr, alloys. As can be observed, the behavior of these physical properties for both systems is quite similar.

In anisotropic materials when the first order anisotropy constant, K_1, becomes greater than K_2, the anisotropy field can be written as $H_a=2K_1/M_S$. The dependence showed by σ_s and iHc with the decreasing of temperature reflects that the magnetocrystalline anisotropy and the anisotropy field of the D_1 phase strongly increase. Grieb et al [3] reported in the ternary phase diagram of Nd-Fe-C system the formation of a stable hard magnetic phase, that they called λ, in the temperature range 600-700°C.

In Fig. 2 the $\chi_{ac}(T)$ dependence for annealed $Nd_{80}Fe_{15}X_5$ (X = B or C) alloys is reported. For $Nd_{80}Fe_{15}B_5$ a transition appears at 135K, this matches well with the spin-reorientation temperature reported for $Nd_2Fe_{14}B$ phase [4]. In $Nd_2Fe_{14}C$ this phenomenon occurs at 120K. It is worthy of mention that in $Nd_{80}Fe_{15}C_5$ no event was detected around this temperature. The latter favors the suggestion that D_1 is different from 2-14-1. Taking into account that in the RE-rich alloys studied the magnetic phases precipitate in a RE-matrix we have also measured $\chi_{ac}(T)$ curves for pure RE metals. For pure Nd transitions were observed at about 10K and 30K. They are also present in the $\chi_{ac}(T)$ curves for $Nd_{80}Fe_{15}X_5$ alloys with X = B or C. The other events observed in the curves occur at 69K and 60K (for X = B and C, respectively); however, they do not appear in the pure Nd sample. Fig. 3 shows the $\chi_{ac}(T)$ plots obtained for $Pr_{80}Fe_{15}X_5$ alloys, X = B or C, and Pr metal. The behavior observed below 40K can be associated with the Pr matrix while the transition observed at 54K for $Pr_{80}Fe_{15}B_5$ should have the same origin as those showed by $Nd_{80}Fe_{15}X_5$ alloys with X = B or C.

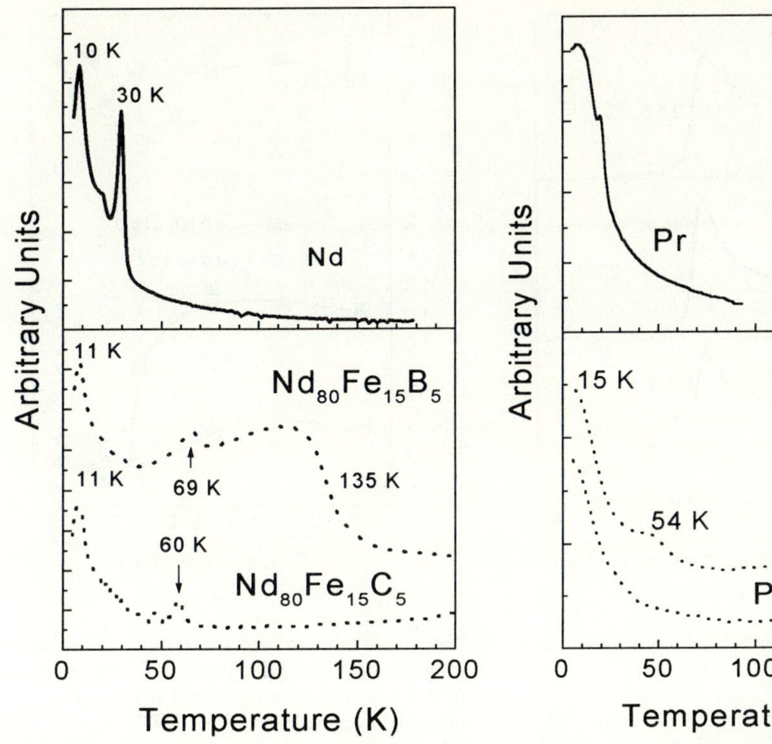

Fig. 2 ac magnetic susceptibility as a function of the temperature for: (a) Nd metal; (b) $Nd_{80}Fe_{15}X_5$ (RE = B, C) after annealing at 600°C.

Fig. 3 ac susceptibility as a function of the temperature for: (a) Pr metal; (b) $Pr_{80}Fe_{15}X_5$ alloys with X = B, C, after annealing at 600°C.

3.2 $Nd_{80}Fe_{15}(B_{1-x}C_x)_5$ (0.85 ≤ x ≤ 1.00)

In order to get more information on the formation of the D_1 phase, we have investigated $Nd_{80}Fe_{15}(B_{1-x}C_x)_5$ samples with very low B-content: 0.85 ≤ x ≤ 1.00.

3.2.1 As-cast samples

Fig. 4 shows the typical TMA curves for the as-cast $Nd_{80}Fe_{15}(B_{1-x}C_x)_5$ samples with 0.85 ≤ x ≤ 1.00. Well-defined drops in the initial magnetic susceptibility, χ_o, were observed at 245°C. This transition can be attributed to the presence of metastable A_1 phase. In all cases the Hopkinson effect, characterized by the existence of a well defined maximum in the $\chi_o(T)$ curve near T_c, is observed. This is a feature commonly exhibited by A_1 [5]. For x = 0.90 another transition was obtained at 290°C which is attributed to the stable phase A_3. Similar curves were measured for $RE_{80}Fe_{15}(B_{1-x}C_x)_5$, RE= Nd [1] and Pr [2], samples in the composition range 0.00 ≤ x ≤ 0.75. In these papers it was also emphasized that the Curie point of A_1 does not change with the replacement of B for C, suggesting that C is not significantly soluble in this phase to produce a variation in this magnetic property.

The dependence of σ_s and iHc with C-content for as-cast samples is reported in Fig.5. The values of both physical quantities are in agreement with the previously reported results [1, 2].

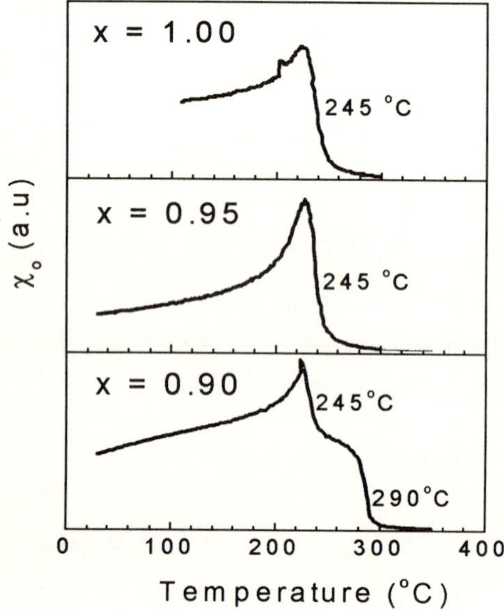

Fig. 4 TMA curves for $Nd_{80}Fe_{15}(B_{1-x}C_x)_5$ alloys in the as-cast state($0.90 \leq x \leq 1.00$)

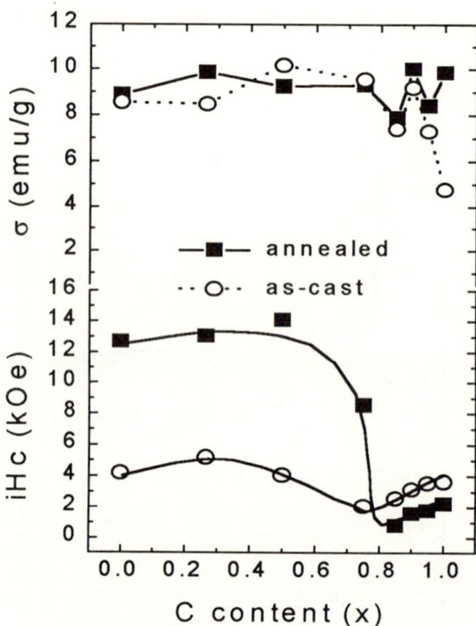

Fig. 5 Saturation magnetization and intrinsic coercive field for $Nd_{80}Fe_{15}(B_{1-x}C_x)_5$ alloys with $0.75 \leq x \leq 1.00$: a) in the as-cast state; b) after 2h of annealing at 600°C

3.2.2 Annealed samples

Fig. 6 shows the typical TMA curves for annealed $Nd_{80}Fe_{15}(B_{1-x}C_x)_5$ alloys with $0.85 \leq x \leq 1.00$. For x = 1.00 two magnetic phases were detected with T_c = 50°C and T_c = 240°C, respectively. According to these T_c values phases can be identified as Nd_2Fe_{17} [6] and D_1 [1]. For $0.85 \leq x \leq 0.90$ an unidentified magnetic phase with T_c = 227 °C is observed together with A_3. A slight knee is only detected at 245°C for x = 0.95. Therefore, small concentrations of B inhibit the formation of D_1

The dependence of σ_s and iHc with the C content for annealed alloys provides additional information about the nature of the phases formed, as well as of the phase transformation that had occurred in the specimens (Fig. 5). For $0.85 \leq x \leq 0.95$ the σ_s values do not change appreciably upon C addition. The drop of iHc for x above 0.75, as well as the relatively low values showed by samples with $0.85 \leq x \leq 0.95$, reflects that in this composition range a phase different from $Nd_2Fe_{14}B$ is formed.

4. Conclusions

$RE_{80}Fe_{15}X_5$ (RE= Nd, Pr, and X = B, C) alloys show complex $\chi_{ac}(T)$ curves in agreement with their multiphase constitution. The differences between $\chi_{ac}(T)$ plots for $Nd_{80}Fe_{15}X_5$ (X = B, C) alloys suggest that D_1 is a phase different from $RE_2Fe_{14}C$. The high uniaxial magnetic anisotropy of D_1 was confirmed.

The metastable phase A_1 is the main magnetic phase formed in the as-cast $Nd_{80}Fe_{15}(B_{1-x}C_x)_5$

alloys with $0.85 \leq x \leq 1.00$. This phase coexists with the stable phase A_3 (T_c=290°C) up to x = 0.90. It seems that A_3 has a minor influence on the coercivity. Our results suggest that C is not appreciably soluble in A_1, in agreement with data reported by us in Refs. [1, 2].

The annealing of $Nd_{80}Fe_{15}(B_{1-x}C_x)_5$ alloys leads to the decomposition of A_1 and the formation of an unknown stable phase having T_c = 227°C and A_3 phase It was observed that for a concentration of C around x = 0.75-0.80 the coercive field falls in both, annealed and as-cast ingots. Above this composition iHc increases regularly with the increase of C concentration.

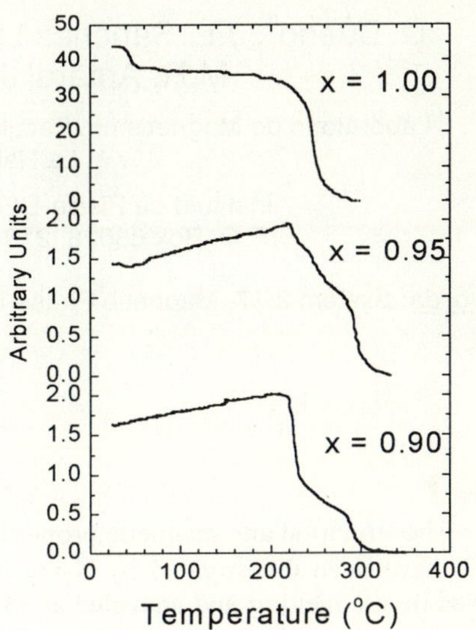

Fig. 6 TMA analysis for $Nd_{80}Fe_{15}(B_{1-x}C_x)$ alloys after 2h of annealing at 600°C ($0.90 \leq x \leq 1.00$).

Acknowledgements

This work was partially supported by the Brazilian agencies FINEP, FUJB, CNPq and FAPERJ. G. López is grateful to CNPq-CLAF for the doctoral scholarship. J.L. Sánchez thanks to FAPERJ and TWAS for supporting his visit to the IF-UFRJ.

* On leave of absence from Laboratorio de Magnetismo, Facultad de Física-IMRE, Universidad de La Habana, La Habana 10400, Cuba.

References

[1] J.L. Sánchez Ll., G. López, E. Estevez and J. Fidler. J. Alloys and Compounds, 243 (1996) 120

[2] J.L. Sánchez Ll., G. López, E. Estevez and J. Fidler. J. of Alloys and Compounds 264 (1997) 252.

[3] B. Grieb, E. Th. Henig, G. Martinek, H.H. Stadelmaier and G. Petzow, IEEE Trans. Mag., MAG-26 (1990) 1367.

[4] X.C. Kou and R. Grössinger. J. Mag. Mag. Mat. 95 (1991) 184

[5] J.L. Sánchez, F. Leccabue, F. Bolzoni, R. Panizzieri and Xue Rong Hua J. Mag. Mag. Mat. 84 (1990), 79.

[6] K.H.J. Buschow. Rep. Prog. Phys., 40 (1977) 1179.

Materials Science Forum Vols. 302-303 (1999) pp. 304-308

Phase Analysis and Magnetic Properties of $RE_2(Fe_{1-x}Mo_x)_{17}$ (RE=Nd, Gd) Alloys

D. Bueno[1], J.L. Sánchez Ll.[1], A. Pentón[1], H. Salim de Amorim[2], M.R. Amaral Jr[2] and P.H. Domingues[2]

[1] Laboratorio de Magnetismo, Facultad de Física-IMRE, Universidad de La Habana, La Habana 10400, Cuba

[2] Instituto de Física-UFRJ, Centro de Tecnologia, Bloco A, P.O. Box 68528, 21945-970, RJ, Rio de Janeiro, Brasil

Keywords: System 2:17, Magnetic Phase-Diagram

Abstract

The structural and magnetic properties of $RE_2(Fe_{1-x}Mo_x)_{17}$ alloys (RE= Gd and Nd; $0.00 \leq x \leq 0.10$) have been investigated by X-ray diffraction and magnetic techniques. The samples were prepared by arc melting and annealed at 1373 K. The study of thermomagnetic analysis curves and XRD patterns reveals that the phase composition is quite similar for both alloy systems. The Rietveld method was employed to determine the phase composition and lattice parameters. In the $0.00 \leq x \leq 0.03$ composition range, the samples are almost single phase with the rhombohedral Th_2Zn_{17}-type crystal structure; with the substitution of Fe for Mo a slight lattice expansion is induced. For higher Mo concentrations, this phase co-exists with a certain amount of spurious phases: a) 1:12-type compound for RE= Nd, and b) 1:12- and 3:29-type compounds for RE= Gd. This suggests that the solubility of Mo in 2:17 is low. The Curie temperatures do not show remarkable variations when Fe is replaced by Mo. In $Gd_2(Fe,Mo)_{17}$, the increase in the Mo content leads to a decrease in the saturation magnetization values while in $Nd_2(Fe,Mo)_{17}$ a smooth increase is observed. A possible explanation for this behavior, in terms of a different site occupancy of Mo atoms in the structure, is discussed.

1. Introduction

Recently, much attention has been devoted to the monoclinic intermetallic phase $RE_3(Fe,M)_{29}$ with M=Ti, V, Cr, Mn, Mo [1, 2]. The structure of the new phase is intermediate between the rhombohedral Th_2Zn_{17} and the tetragonal $ThMn_{12}$, and it can be understood as an alternate stacking of 2:17 and 1:12 blocks. More recent papers [3] suggested that these compounds are better described by the A2/m space group, which is a minimal non isomorphic supergroup of the $P2_1/c$ one. The formation of this structure strongly depends on the content of the stabilizing element, annealing temperature and cooling rate. For this reason it is difficult to produce single-phase samples. Generally, the samples obtained are formed by a mixture of phases, where the $RE_3(Fe,M)_{29}$ is the main matrix phase and $RE_2(Fe,M)_{17}$ and/or $RE(Fe,M)_{12}$ are impurity phases.

In a previous work, Sánchez et al [4] analyzed the formation of $Nd_3Fe_{27.5}Mo_{1.5}$ compounds. Here the rhombohedral 2:17 phase was one of major phases obtained for certain annealing and cooling rate conditions. In order to obtain a better understanding of the relationship between these phases, we study the formation of the 2:17 phase in $RE_2(Fe_{1-x}Mo_x)_{17}$ RE= Nd, Gd, type alloys. The preparation conditions as well as structure and magnetic properties are reported.

2. Experimental details.

Samples with nominal compositions $RE_2(Fe_{1-x}Mo_x)_{17}$ for RE=Nd, Gd and x=0.01, 0.03, 0.05, 1.0 were prepared by arc melting under a static argon atmosphere. Purity of raw materials was at least 99.9 %. The as-cast samples were wrapped in tantalium foils and annealed in argon in quartz tubes at 1373 K for 48 h. Annealing was followed by a water quenching.

X-ray diffraction studies were performed on powered samples in step-scanning mode with: a) an automated Siemens type-F diffractometer with a Philips PW1830/25 high-stability high voltage generator. A graphite monochromator was used to select Co Kα-doublet. b) a Karl Zeyss HZG4 diffractometer with a Co source. The structure refinement was performed with the DBWS-9411 program developed by Young et al. [5]. For the iron atoms corrections were used for the anomalous scattering effect. The values used for Δf ' and Δf " were obtained from the procedure described by Cromer [6].

Magnetic measurements were performed in a vibrating sample magnetometer, Oxford model 3001 with a maximum applied field of H_{max}= 16kOe. Thermomagnetic analysis (TMA) curves were recorded in a thermomagnetic analyzer in a constant external magnetic field of H_{ex}=200 Oe and with a heating rate of 3°/min. Curie temperatures were inferred from the maximum in the $d\chi_o/dT$ versus T curves.

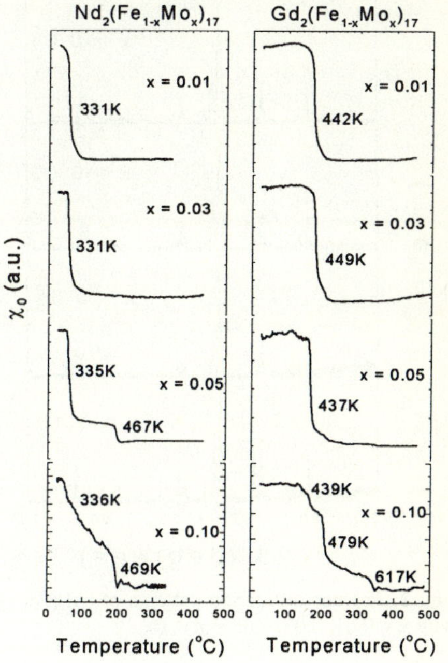

Fig. 1: Thermomagnetic curves for $RE_2(Fe_{1-x}Mo_x)_{17}$ alloys (RE = Nd, Gd) with x = 0.01, 0.03, 0.05 and 0.10

3. Results and discussion

Fig. 1 shows the TMA curves recorded for both series of samples. As can be seen, a quite similar phase constitution of magnetic phases is obtained for both alloy systems. The samples are almost single phase up to x = 0.05. Well defined drops of the initial susceptibility at 442 K, 449 K and 437 K for Gd, and 331 K and 335 K for Nd may be associated to 2:17 phase for they are similar to those reported for this phase [7, 8], and considering the probably entrance of the Mo atoms in the structure. In both cases there was a remanence of α-(Fe,Mo) for the compositions x = 0.01 and x = 0.03. In samples with x = 0.05 the transition found at 467 K for Nd may be associated to the 1:12 phase instead of α-(Fe,Mo), whereas for Gd no other important transitions were detected. A more complex phase constitution is obtained for the samples with x = 0.10. According to their T_c values, phases can be identified as: a) 2:17 with T_c = 439 K for Gd and T_c = 366 K for Nd, respectively, b) 1:12, T_c = 617 K for Gd and T_c = 569 K for Nd, respectively, and c) 3:29 T_c = 481 K for Gd [9].

Since the TMA is not a quantitative technique, accurate x-ray diffraction patterns were collected for Rietveld processing in order to have more precise information on the phase composition. Figs. 2 and 3 show the corresponding patterns for both series of samples. Again, a quite similar behavior is obtained for both alloy systems. Table 1 summarizes the results of the quantitative XRD phase analysis, together with the T_c values of the phases detected by TMA. The refinements were made considering that, in the rhombohedral 2:17 phase, Mo atoms only occupy the 6c sites of the structure.

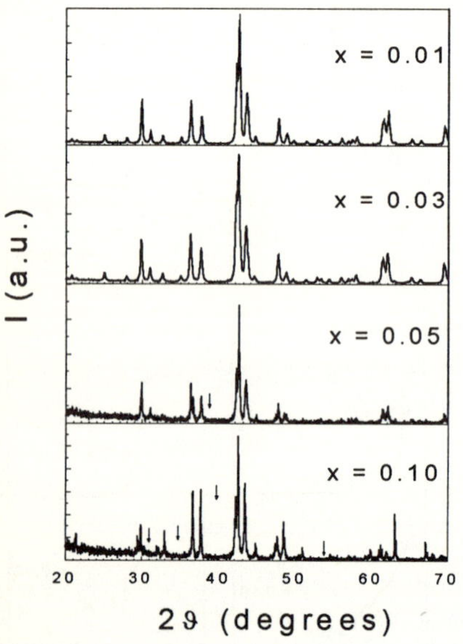

Fig. 2 Powder diffraction patterns of $Nd_2(Fe_{1-x}Mo_x)_{17}$ with x = 0.01, 0.03, 0.05 and 0.10

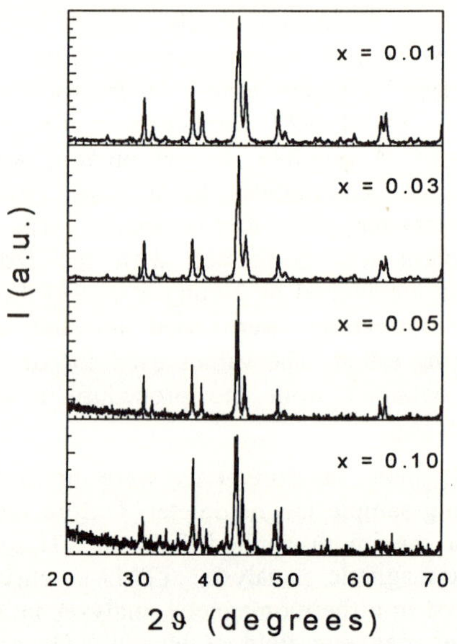

Fig 3 Powder diffraction patterns of $Gd_2(Fe_{1-x}Mo_x)_{17}$ with x = 0.01, 0.03, 0.05 and 0.10

Table 1 - Curie temperatures and phase composition for $RE_2(Fe_{1-x}Mo_x)_{17}$ samples (RE = Nd, Gd) with x = 0.01, 0.03, 0.05 and 0.10

Compound	x	Phases	%wt	T_c(K)	R_{wp}
$Gd_2(Fe_{1-x}Mo_x)_{17}$	0.01	2:17	99.6	442	13.2
		α-Fe	0.4	>600	
	0.03	2:17	99.8	449	14.2
		α-Fe	0.2	>600	
	0.05	2:17	87.1	437	12.03
	0.10	2:17	23.91	439	15.9
		1:12	58.51	617	
		α-Fe	17.54	479	
$Nd_2(Fe_{1-x}Mo_x)_{17}$	0.01	2:17	97.8	331	11.87
		α-Fe	2.0	>600	
	0.03	2:17	98.4	331	10.83
		α-Fe	1.5	>600	
	0.05	2:17	98.7	335	13.1
		1:12	1.25	4676	
	0.10	2:17	18.76	336	16.01
		1:12	76.66	469	

All the samples for both series are almost 2:17 single phase with the rhombohedral Th_2Zn_{17}-type crystal structure up to x= 0.03. For higher Mo content, some important amount of secondary phases is obtained. Thus, the solubility limit for Mo in the rhombohedral 2:17 structure

Table 2: Refined lattice parameters and cell volumes for the $RE_2(Fe_{1-x}Mo_x)_{17}$ samples (RE = Gd, Nd) with x = 0.01, 0.03. The values for x = 0.00 were taken from [7,8].

Compound	x	a(Å)	c(Å)	V(Å^3)
$Gd_2(Fe1\text{-}xMo_x)_{17}$	0.00	8.497	12.420	776.5
	0.01	8.529(2)	12.450(4	784.3
	0.03	8.530(3)	12.462(4	785.1
$Nd_2(Fe_{1-x}Mo_x)_{17}$	0.00	8.576	12.458	793.5
	0.01	8.573(2)	12.471(4	793.8
	0.03	8.581(2)	12.497(4	796.9

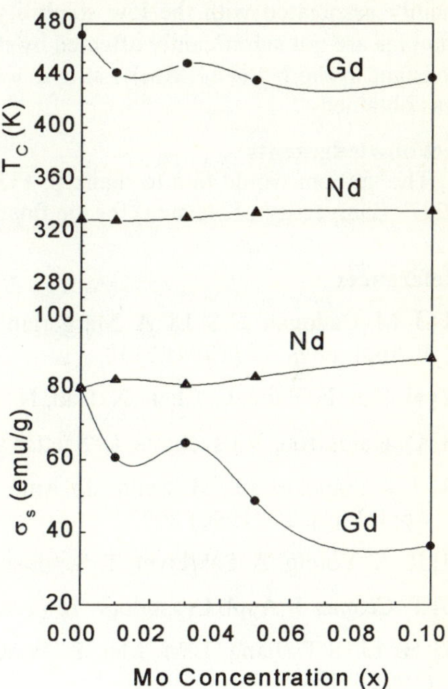

Fig.4: Dependence of σ_s and T_c with the molybdenum content

can be estimated as x = 0.05. This result is consistent with those obtained for $Sm_2(Fe_{1-x}Ti_x)_{17}$ and $Sm_2(Fe_{1-x}V_x)_{17}$ type alloys [10]. In these cases, the solubility limits reported are x = 0.03 for Ti and x = 0.06 for V. These elements show similar values for metallic radii, so similar solubility limits should be expected.

The trends in the occupancies may be understood, at least in part, on the basis of the nearest neighbor environment of each site. The unit cell of the rhombohedral 2:17 structure has 4 sites for Fe. This leads to a total of 68 Fe atoms. If we suppose a homogeneous distribution of the Mo atoms for a composition like x = 0.03 we obtain 2.04 atoms in each unit cell. Taken into account this result and the fact that Mo has a higher metallic radius than Fe, a preferential entrance of Mo in the dumbbell 6c sites of the structure can be expected since those have the largest Wigner-Seitz volume. Similar results were observed for Ti in $Nd_2(FeTi)_{17}$ compounds [11]. The low concentration of Mo makes possible the substitution in non-adjacent sites to 6c, avoiding the short 6c-6c bond distance, as was obtained for Al and Ga atoms in [12]

Table 2 shows the refined cell parameters and the corresponding volumes for the samples up to x = 0.03. For comparison, the parameters for pure 2:17 compounds, taken from [8, 13], are also included. The lattice parameters increase slightly with x, particularly in the $Nd_2(Fe_{1-x}Mo_x)_{17}$ samples. This result may be mainly associated with the low solubility of Mo in the structure.

Fig. 4 shows the dependence of σ_s and T_c with the Mo content. The values for x = 0.00, were taken from [8, 14]. For the $Nd_2(Fe_{1-x}Mo_x)_{17}$ alloys, both properties remain almost constant. These results are consistent with the slight increment obtained for the cell parameters and the low incorporation of Mo in the structure. In the case of $Gd_2(Fe_{1-x}Mo_x)_{17}$ a similar behavior is observed for T_c values while a remarkable decrease in σ_s is obtained. The decrease in σ_s may be associated with a different site occupation and thus to a variation in the interactions of the transition-metal sublattice.

4. Conclusions

The solubility found for Mo atoms in $RE_2(Fe_{1-x}Mo_x)_{17}$ alloys RE= Nd and Gd with the rhombohedral 2:17 structure is very low. Single phases samples are obtained up to x= 0.03. For higher Mo content, some substantial amount of secondary phases is obtained. The lattice parameters slightly increase with x, particularly in the $Nd_2(Fe_{1-x}Mo_x)_{17}$ samples. This result may be mainly associated with the low solubility of Mo in the structure. The Curie temperature of the samples are not significantly affected by the substitution. The values obtained for σ_s remains almost constant in the $Nd_2(Fe_{1-x}Mo_x)_{17}$ alloys, while in the case of $Gd_2(Fe_{1-x}Mo_x)_{17}$ a remarkable decrease was obtained.

Acknowledgments

The authors would like to thank to FUJB, FAPERJ, FINEP, CNPq and "Concurso Alma Mater 1997" (University of Havana) for the financial support given.

References

[1] J. M. Cadogan, H.S. Li, A. Margarian, J.B. Dunlop, D. H. Ryan, S. J. Collocot and R. L. Davis. J. Appl. phys. 76 (1994) 6138

[2] H. Pan, F. Yang, C. Chen, X. Han, N. Tang and Q. Wang. J. Mag. Mag. Mat. 161 (1996) 177

[3] O. Kalogirou, V.Psycharis, L. Saettas and D. Niarchos. J. Mag. Mag. Mat. 146 (1996) 335

[4] J. L. Sánchez Ll., H. Salim de Amorim, M. R. Amaral Jr. and P.H. Domingues. J. Phys. D: Appl. Phys. 29 (1996) 2677

[5] R. A. Young, A. Sakthivel, T. S. Moss, and C. Paiva-Santos J. Appl. Crystalogr. 28 (1995) 366

[6] T. Cromer J. Appl. Crystalogr. 16 (1983) 437

[7] H. Li, S.T. Jiang, L.M. Mei, P. W. Gao, J.H. Dong, and X.H. Ding J. Mag. Mag. Mat. 163 (1996) 125

[8] N.Q. Huang, B.M. Ma, W.E. Wallace and S.G. Sankar (unpublished)

[9] H. Pan, F. Yang, C. Chen, X. Han, N. Tang, J. Hu, J. Wang, R. Zhao, K. Zhou, and Q. Wang J. Mag. Mag. Mat. 159 (1996) 352

[10] B.P. Hu, X.L. Rao, J.M. Xu, G.C. Liu, F.Cao, X.L. Dong, H. Li, L. Yin, and Z.R. Zhao J. Mag. Mag. Mat. 114 (1992) 138

[11] Z. Hu and W.B. Yelon J. Appl. Phys. 76 (1994) 6147

[12] J.G. Long, K.G. Marasinghe, S. Mishra, A.O. Pringle, Z. Hu, W.B. Yelon, D.P. Middleton, K.H.J. Buschow, and F. Grandjean. J. Appl. Phys. 76 (1994) 5383

[13] M. Katter, J. Wecker, C. Kuhrt, L. Shultz, X.C. Kou, and R. Grossinger J. Mag. Mag. Mat. 111 (1992) 293

[14] W. Xlang-Zhong, K. Donnelly, J.M.D. Coey, B. Chevalier, J. Etourneau and T. Berlureau. J. Mat. Science 23 (1988) 329

Materials Science Forum Vols. 302-303 (1999) pp. 309-313

Magnetic Properties of Nd-Dy-Fe-B-Nb Sintered Magnets Produced Using the Hydrogen Decrepitation Process

H. Takiishi, E.P. Soares, L.F.C.P. Lima and R.N. Faria

Instituto de Pesquisas Energéticas e Nucleares IPEN-CNEN, CEP 05422-970, São Paulo, SP, Brazil

Keywords: NdFeB Magnets, Hard Magnets, Hydrogen-Decrepitation Process, Dy, Nb Magnets

Abstract

Sintered permanent magnets with a nominal composition $Nd_{14.5}Dy_{1.5}Fe_{76}B_7Nb$ have been prepared using hydrogen decrepitation process. A study on the relationship between processing and magnetic properties has been carried out. For a particular processing condition, the intrinsic coercivity reached the maximum value of 15.6 kOe. The energy product of the best Nb-containing sintered magnet achieved approximately 36 MGOe.

Introduction

Niobium addition to Nd-Fe-B magnets results in an improvement in coercivity, squareness factor and thermal stability[1-4]. The substitution of Dy for Nd in Nd-Fe-B sintered magnets also results in an increase in the anisotropic field and coercivity[5,6]. Previous investigation of these Nb containing Nd-Dy-Fe-B magnets used conventional powder metallurgy process[7]. In the present work, sintered Nb-containing Nd-Dy-Fe-B magnets have been prepared via the hydrogen decrepitation (HD) process[8]. In order to establish the optimum processing conditions[9], a study of the relationship between the magnetic properties of the sintered magnets and the milling time has been carried out for the Nd-Dy-Fe-B-Nb magnets.

Experimental

Alloy ingots were prepared by induction melting the pure constituents under argon atmosphere. The alloy was poured into rectangular water cooled copper mould. The as-cast microstructure of the starting alloy with the familiar columnar grain structure is shown in Fig. 1.

In order to produce the magnets via the HD process small pieces of the bulk ingot were placed in a hydrogenation vessel which was evacuated to backing-pump pressure and hydrogen was then introduced to a pressure of 2 bar. The HD material was then transferred to a roller ball-mill and milled using cyclohexane as the milling medium. The resultant fine powder decrepitated was then dried and transferred to a small cylindrical rubber tube, pulsed in a magnetic field of 60 kOe and isostatically pressed at 200 MPa. The green compacts were vacuum sintered at 1050°C for 1 hour and furnace cooled (~3.5°C/min.). The sintered magnets were then magnetized and their demagnetization curves determined.

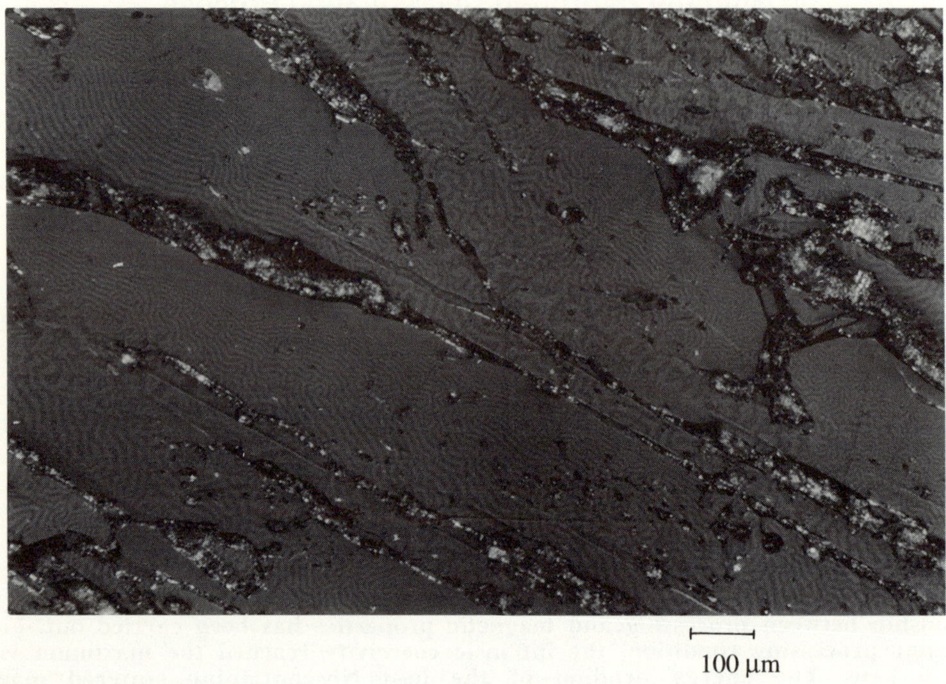

Fig.1. Microstructure of the $Nd_{14.5}Dy_{1.5}Fe_{76}B_7Nb$ ingot.

Results and Discussion

Figs. 2 and 3 show how the coercivity and remanence are affected by the milling time. An initial increase in both properties is observed, with material milled for 12 hours having a iHc greater than 15 kOe. The intrinsic coercivity of these Nb-containing magnets passes through a maximum and then decreases dramatically. The remanence, unlike iHc, goes through a flat maximum and then decreases drastically. Total deterioration of the magnetic properties is observed when the milling time reaches 60 hours.

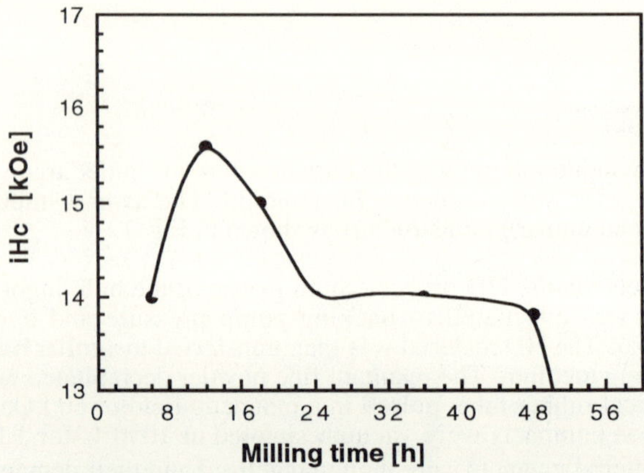

Fig.2. Variation of iHc with the milling time.

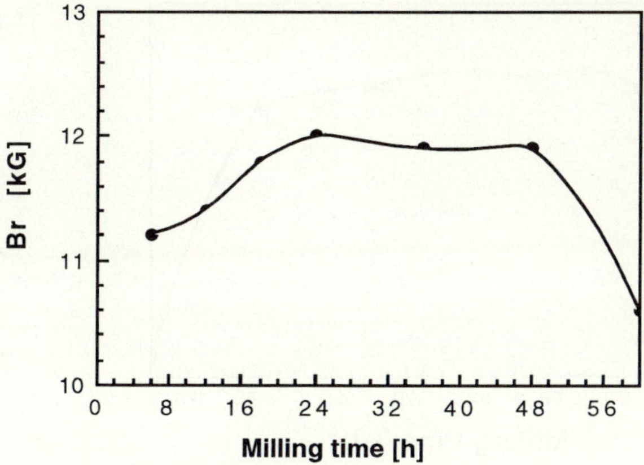

Fig.3. Variation of Br with the milling time .

The effects of milling time on the energy product and squareness factor are shown in Figs.4 and 5, respectively. In the same way, there is an initial enhancement of the energy product, reaching a maximum value (35.8 MGOe) for the material milled for 24 hours, followed by a decrease.

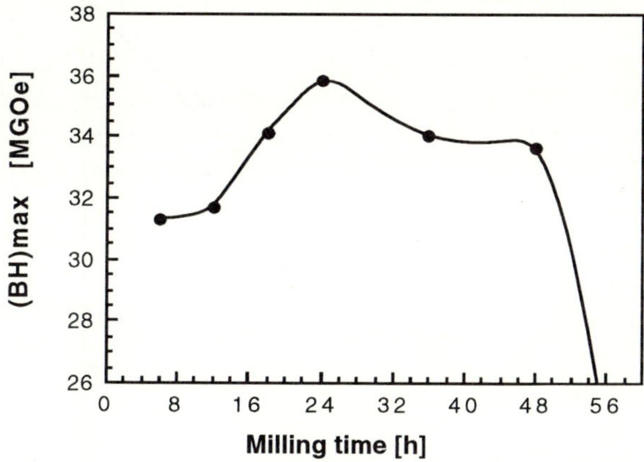

Fig.4. Variation of (BH)max with the milling time.

The preparation of sintered magnets from RE-Fe-B powders becomes difficult when powders with fine particle size are to be sintered. Christodoulou et al[10] reported a dramatic decrease in remanence and intrinsic coercivity of the sintered magnet when the particle size passes below a critical value. This behaviour was attributed to the oxidation of the fine powder. It has also been reported[11] that overmilling reduces the intrinsic coercivity by causing damage to the particle surface.
The second quadrant demagnetization curve for a $Nd_{14.5}Dy_{1.5}Fe_{76}B_7Nb$ HD sintered magnet is shown in Fig. 6 (best iHc). A very nice loop shape is obtained in the magnet prepared with the powder milled for 12 hours (SF=0.91). A summary of magnetic properties is given in Table 2. Microstructural studies are underway in an attempt to explain the coercivity behaviour observed in this work.

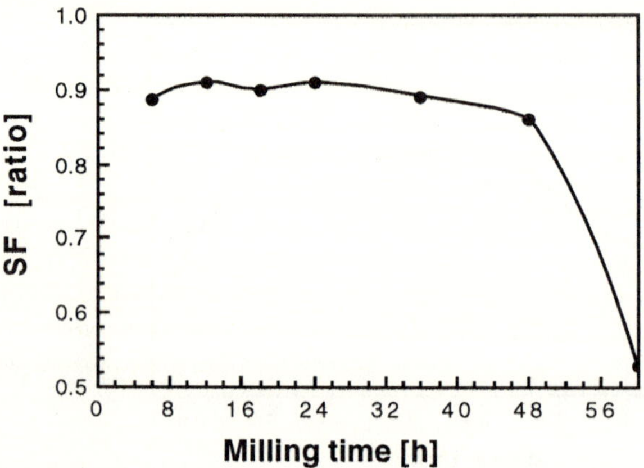

Fig.5. Variation of squareness factor with milling time.

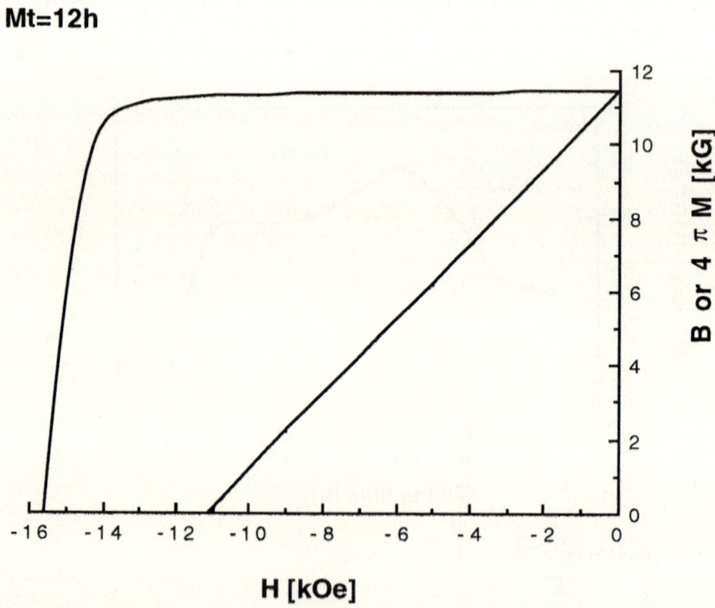

Fig.6. Demagnetization curve for a $Nd_{14.5}Dy_{1.5}Fe_{76}B_7Nb$ magnet prepared using the HD powder milled for 12 hours.

Table I. Magnetic properties of the Nd-Dy-Fe-B-Nb HD magnets.

Milling Time [Hours]	Br [kG]	iHc [kOe]	(BH)max [MGOe]	bHc [kOe]	SF ratio
6	11.2	14.0	31.3	11.1	0.89
12	11.4	15.6	31.7	11.1	0.91
18	11.8	15.0	34.1	11.5	0.90
24	12.0	14.0	35.8	11.8	0.91
36	11.9	14.0	34.0	11.6	0.89
48	11.9	13.8	33.6	11.4	0.86
60	10.6	5.6	14.0	5.4	0.53

Average error: Br±0.5 , iHc±0.5 , (BH)max:±0.9

Conclusions

$Nd_{14.5}Dy_{1.5}Fe_{76}B_7Nb$ HD magnets prepared from the ingot material exhibit a maximum in iHc when the HD powder is milled for 12 hours. Good overall magnetic properties have been achieved for the sintered magnet prepared from the hydrided powder milled for 24 hours (Br = 12 kG, iHc = 14 kOe, bHc = 11.8 kOe, (BH)max = 35.8 MGOe and SF = 0.91). There was a total degradation in all magnetic properties when the hydride powder is milled for 60 hours.

Acknowledgments

Many thanks are due to the Comissão Nacional de Energia Nuclear and Fundação de Amparo à Pesquisa do Estado de São Paulo. Thanks are due to GfE for the provision of the alloys.

References

[1] W. Rodewald and B. Wall, J. Magn. Magn. Mater. 8 (1989), p. 57
[2] Y. Xiao, K. J. Strnat, H. F. Mildrum and A. E. Ray, 9th International Workshop on Rare-Earth Magnets and Their Appl.,Bad Soden, FRG (1987), p. 467
[3] M. Tokunaga, H. Harada and S. R. Trout, IEEE Trans. Magn. 23, 5 (1987), p. 2284
[4] M. Tokunaga, H. Kogure, M. Endoh and H. Harada, IEEE Trans. Magn. 23, 5.(1987), p. 2287
[5] D. Plusa, J. J. Wyslocki, B. Wyslocki and R. Pfranger, J. Less-Common Metals, 133 (1987), p. 231
[6] W. Rodewald, J. Less-Common Metals, 111 (1985), p. 771
[7] W. Gong, Q. Li and L. Yin, J. Appl. Phys. 69 (8) (1991), p. 5512
[8] I. R. Harris, J. Less-Common Metals, 131 (1987), p. 245
[9]. R. N. Faria, J. S. Abell, and I. R. Harris, J. Alloys and Compounds, 177 (1991), p. 311
[10] C. N. Christodoulou, J. Schlup and G. C. Hadjipanayis, J. Appl. Phys., 61 (8) (1987), p. 3760
[11] J. Ormerod, Ed. I. V. Mitchell, Brussels, Belgium, CEE (1984), p.69

Materials Science Forum Vols. 302-303 (1999) pp. 314-318

Kinetics and Thermodynamics of the Solid-HDDR Process in the $Nd_{13}Fe_{80.5}B_{6.5}$ Alloy

C.E. Rodríguez Torres[1], F.H. Sánchez[1], M.B. Fernández van Raap[1], L. Mendoza Zélis[1], F. Actis[2] and H. Nagel[3]

[1] Departamento de Física, Universidad Nacional de La Plata, c.c. 67 (1900) La Plata, Argentina

[2] Centro de Investigación de Materiales y Metrología, Instituto Nacional de Tecnología Industrial, c.c. central 884, 500 Cordoba, Argentina

[3] Nagel & Krönert Eng. GmbH, Südfeld 2, DE-59174 Kamen, Germany

Keywords: Hard Magnets, HDDR, Mössbauer

Abstract

Mössbauer spectroscopy was applied to the quantitative determination of the reaction progress during some steps in the hydrogen assisted solid preparation of Nd-Fe-B based hard magnetic materials, where the signals coming from the disproportionated and recombined phases can be properly resolved. New results on the kinetics and activation energy of the recombination process, as well as on the temperature dependence of the disproportionation reaction on $Nd_{13}Fe_{80.5}B_{6.5}$ and $Nd_{14.07}Ga_{0.98}Zr_{0.07}B_7Fe_{77.88}$ alloys are presented. These results are compared with recent studies performed on an alloy of nominal composition $Nd_{13.67}Co_{15.74}Al_{0.77}Ga_{0.27}Zr_{0.03}Fe_{62.20}B_{7.33}$. It is found that the main observed difference lies in the disproportionation rate which is slower for the additionated alloys.

Introduction

The hydrogenation, disproportionation, desorption and recombination (HDDR) process is a promising technique for producing Nd-Fe-B based hard magnets[1,2]. Isotropic as well as anisotropic[3] magnets can be prepared at a laboratory scale with relative ease. However, in order to scale the process to a continuous industrial one, some problems must be solved first. One step in order to better face situations involving hydrogen and heat transport, is to study the thermodynamics and kinetics at the laboratory scale as previously reported for $Nd_{13.67}Co_{15.74}Al_{0.77}Ga_{0.27}Zr_{0.03}Fe_{62.20}B_{7.33}$ [4]. In this paper we extend this study to $Nd_{13}Fe_{80.5}B_{6.5}$ and compare kinetic and thermodynamic data -such as time dependence and activation energy of the DR reaction- for the two alloys. We also search for temperature intervals where this part of the process may be manipulated so as to control the disproportionation degree. In principle, incomplete disproportionation will leave matrix grains with the original crystalline orientation, preserving anisotropy information. We expect that the appropriate design of this process may provide a way to prepare magnets with the desired degree of anisotropy by the HDDR route. In order to differentiate the influence of additives, we performed HD experiments at a fixed temperature (860 °C) with the previously mentioned alloys and with $Nd_{14.07}Ga_{0.98}Zr_{0.07}B_7Fe_{77.88}$.

All these studies were carried out using Mössbauer effect (ME) spectroscopy on ^{57}Fe together with x-ray diffraction (XRD). Through the determination of short range order around Fe atoms, the former technique allows the identification of the various phases present during the HDDR process and a quantitative determination of the reaction progress in cases where the signals coming from the disproportionated and recombined phases can be properly resolved.

Experimental

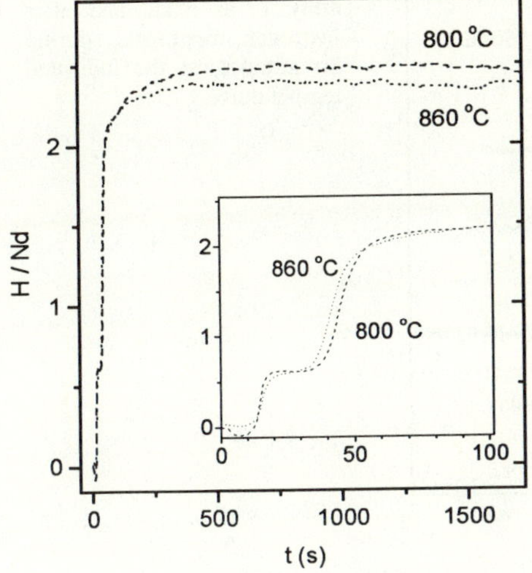

Fig. 1- Ratio between the number of H atoms absorbed during treatments at 800 and 860 °C and the total number of Nd atoms, as a function of time, for alloy 1.

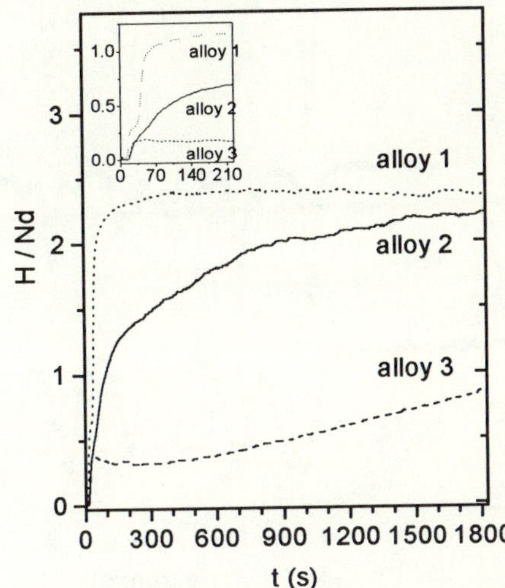

Fig. 2- Ratio between the number of H atoms absorbed during the treatment at 860 °C and the total number of Nd atoms, as a function of time, for alloys 1, 2 and 3.

An ingot with nominal composition $Nd_{13}Fe_{80.5}B_{6.5}$ (alloy 1) was cut into pieces. For each batch, 174 mg of the ingot were evacuated to a base pressure of less than 6 Pa into a quartz tube and then treated during 30 minutes, under H_2 at a pressure of $1.19x10^5$ Pa, at 800 or 860 °C. The pressure P and temperature T during these treatments were digitally recorded every second. Alloys 2 ($Nd_{14.07}Ga_{0.98}Zr_{0.07}B_7Fe_{77.}$) and 3 ($Nd_{13.67}Co_{15.74}Al_{0.77}Ga_{0.27}Zr_{0.03}Fe_{62.20}B_{7.33}$) were treated under the same conditions at 860 °C. From the T-P information the number of absorbed H_2 moles during HD was calculated (Fig. 1 and 2).

To determine the recombination kinetics for alloy 1, the fully disproportionated system (30 min at 800 °C, $P_{H2} \approx 120$ kPa) was evacuated and kept at the same temperature during different additional times (5, 10, 15, 20, 30, 45, 50, 60, 75 and 90 s).

The ^{57}Fe Mössbauer effect (ME) measurements were performed at room temperature under transmission geometry with a standard constant acceleration spectrometer having a ^{57}Co*Rh* radioactive source. ME spectra were fitted with Lorentzian curves with respect to the off-resonance background.

Results and discussion

Fig. 3 shows the ME spectrum of the ingot alloy 1 along with those obtained after treatments with hydrogen at 800 and 860 °C. The spectra show that complete disproportionation of the ϕ-phase has occurred [5]. These results agree with the absorption curves shown in Fig. 1. There, for 800 and 860 °C treatments, two absorption processes are evident: the first one associated with the hydridization of the Nd –rich, grain-border phase and the hydrogenation of the ϕ-phase (nearly completed in about 35 s), and the second one corresponding to the disproportionation reaction of the latter phase.

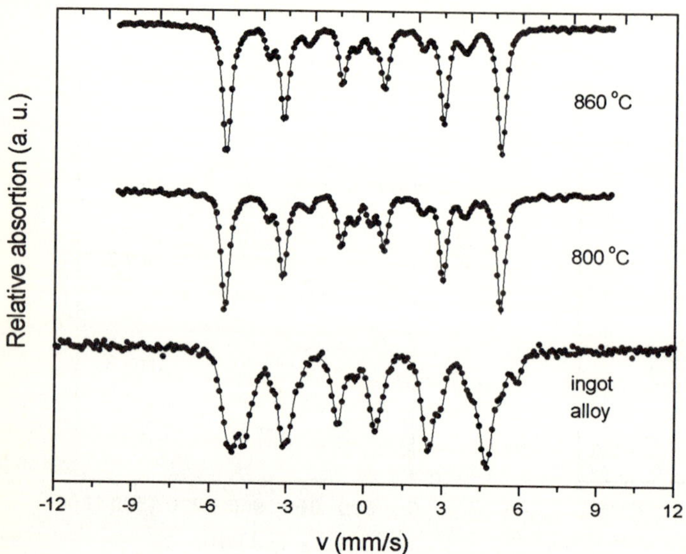

Fig. 3- ME spectra of the alloy 1: as made and after hydrogen treatments, during 30 minutes, at the indicated temperatures.

The curves indicate that about 2.4 H per Nd are absorbed for treatment times from 300 s on (in agreement with the known stoichiometry of the Nd hydride, NdH_{2+x}, $0.0 \leq x \leq 0.9$). Concerning the first step, the ratio H / Nd (at the grain border, estimated from the stoichiometry differences between the alloy and the ϕ-phase) is at least 6.6 indicating that not only the grain border Nd hydrides but also the matrix phase ϕ hydrogenates. However, for alloys 2 and 3 treated in hydrogen at 860 °C (Fig. 2) the estimated value of the H / Nd at the grain border is approximately 2.1, suggesting that only the Nd-rich phase has been hydrided.

Fig. 2 shows the ratio of absorbed H/total Nd for 30 min hydrogen treatments at 860 °C of alloys 1, 2 and 3, where the same two steps are observed. After 1800 s these ratios are about 2.4, 2.2, and 0.8 for alloys 1, 2 and 3, respectively. These results are consistent with the Mössbauer spectra shown in Fig. 4. where it can be seen that disproportionation is complete for alloy 1, almost complete for alloy 2 and partial for alloy 3. Furthermore, the presence of additives modifies the disproportionation rates (Fig. 3), being slowest for alloy 3.
In Fig. 5 the Mössbauer spectra obtained from partially recombined alloy 1 samples are displayed. It can be seen that they are quite complicated. Typically, these samples are obtained by interrupting

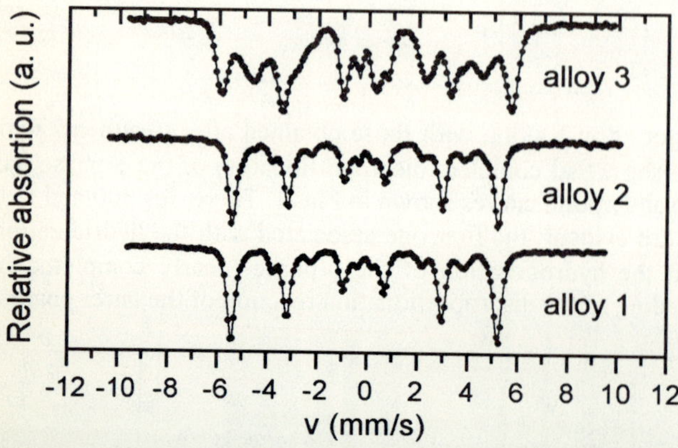

Fig. 4- ME spectra of alloys 1, 2 and 3 treated in hydrogen at 860 °C , during 30 minutes.

the DR reaction 12 Fe + Fe_2B +2 NdH_{2+x} $\rightarrow$ $Nd_2Fe_{14}B$ + (1+x/2)H_2. Then, besides the η-phase, they consist of a mixture of $Nd_2Fe_{14}B$, Fe_2B, α-Fe and NdH_{2+x}. From this previous knowledge, a fitting model should be proposed including: 6 sextets with relative intensities constrained to be in the same ratio as the 4c, 4e, $8j_1$,$8j_2$, $16k_1$ and $16k_2$ sites are in the $P4_2/mnm$ structure (ϕ phase), one sextet for the Fe_2B (which can not be resolved from the 4c interaction of the ϕ phase), another one from α-Fe and a doublet for the η-phase. Such a large number of fitting parameters being impracticable, we attempted to fit these spectra with a superposition of three components: a simulated spectrum of the ingot alloy, a simulated spectrum of the fully disproportionated alloy, and a doublet corresponding to the η-phase. The fitted parameters were: the relative fraction of the three components, the average isomer shift of the simulated components and hyperfine parameters of the doublet (δ_{IS}, quadrupole shift ε, linewidth Γ). With this procedure the spectra can be fairly well reproduced and the amount of ^{57}Fe in the transformed/ untransformed phases f determined (assuming equal recoil free fractions for all iron probes). The experimental quantity f follows an Avrami law,

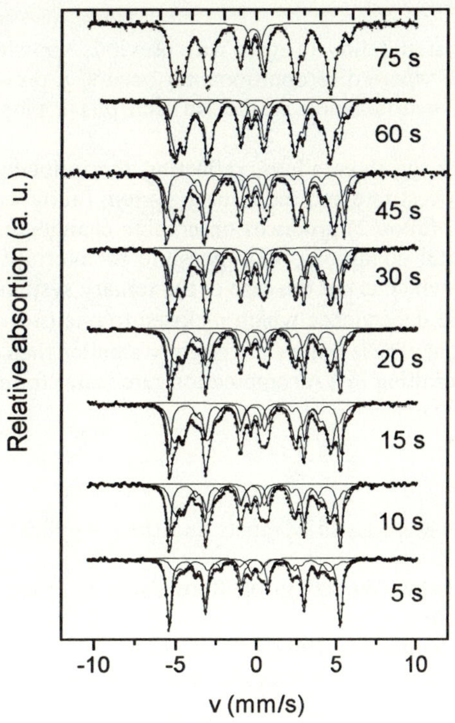

Fig. 5– ME spectra of alloy 1, partially recombined during the indicated time.

$$f = 1 - exp(-kt^n)$$

or

$$ln(ln(1/1-f)) = n\ lnt + lnk,$$

as shown in Fig. 6 (experimental values and fitted function). The parameters obtained from the fit are n ≈ 0.78 ± 0.08 and ln k ≈ -2.28 ± 0.16. This value of n is slightly smaller than that

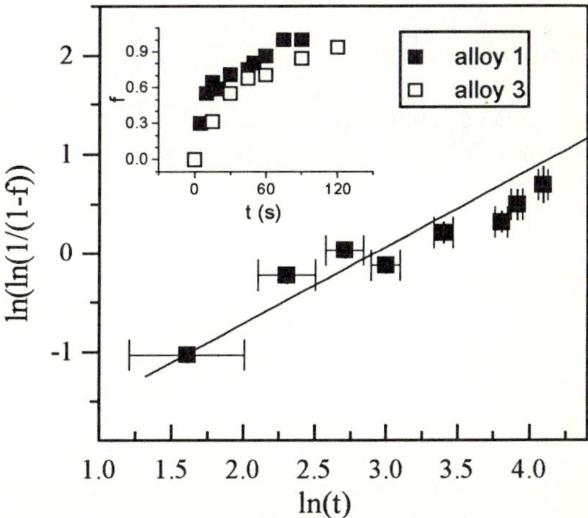

Figure 6 - Plot of ln (ln (1/(1/f))) vs. ln (t) for the recombination of the ingot alloy 1 at 800 °C. Also shown is the straight line fitted to the experimental data. In the inset the evolution of the recombined fraction f for experiments performed with alloys 1 and 3 is shown.

measured for alloy 3 [4], but still close to one which results as a limit case of the isothermal reaction curves for preferential growth at the grain boundaries from pre-existent nuclei. From the comparison of the value of ln k given above with the one measured for alloy 3 [4] an activation energy of 0.5 ± 0.15 eV/at is deduced for the recombination reaction. Again this value is slightly smaller than that corresponding to alloy 3. Although the variation of these two quantities has opposite consequences on the kinetics, the recombination of alloy 1 is faster (see inset of Fig. 6).

Summary and Conclusions

The results on the disproportionation of the ternary system (alloy 1) agree with previous knowledge on the $Nd_2Fe_{14}B$ + H phase diagram. In those cases where disproportionation occurs, a previous step in the absorption curve, corresponding to the hydridization of the Nd rich phase plus the hydrogenation of the ϕ-phase, is observed.

The disproportionation rate at 860 °C is different for the three alloys, reflecting the influence of additives. Alloy 3 (with almost 16 % Co) has the slowest rate and the ternary system (alloy 1) the fastest. The addition of small quantities of Ga and Zr (alloy 2) produces observable changes in the reaction rate. With this alloy, under our experimental conditions, it is possible to interrupt the disproportionation reaction within controllable times which is not the case of the ternary system.

The kinetics of recombination follows an Avrami-type dependence which n closed to one ($n \approx 0.78 \pm 0.08$) and an activation energy of 0.5 ± 0.15 eV/atom. These values are slightly smaller than that corresponding to the alloy with Co, Ga, Zr, and Al, resulting in a recombination rate faster for alloy 1.

References

[1] I.R. Harris, Proc. of the 12th Inter. Workshop on Rare-Earth Magnets and their Applications, Canberra, Australia 1992, pg.327.

[2] T.Takeshita and R. Nakayama, Proc. of the 12th Int. Workshop on Rare-Earth Magnets and their Applications, Canberra, Australia 1992, pg.670.

[3] R. Nakayama and T. Takeshita, J. of Alloys and Compounds, **193** (1993) 259-261

[4] C. E. Rodríguez Torres, M. B. Fernández van Raap, F. H. Sánchez, L. Mendoza Zélis, and F. Actis, Journal of Appl. Phys. (in press).

[5] M. B. Fernández van Raap, C. E. Rodríguez Torres, F. H. Sánchez, L. Mendoza Zélis, F. Actis, G. Cohen and H. Nagel. Proc. of the IV Latin American Conference on the Applications of the Mössbauer Effect, Peru (1996). To be published in Hyperfine Interacctions C.

Acknowledgements

The authors gratefully acknowledge the partial financial support from CONICET (through Programa TENAES) of Republica Argentina.

Materials Science Forum Vols. 302-303 (1999) pp. 319-323

The Effect of Ball Milling on the Magnetic Properties of HDDR Processed Nd-Fe-B Magnets

F.A. Actis[1], M.C. Saracco[1], J.C. Quinteros[1] and H. Nagel[2]

[1] CEMCOR - CIMM, Instituto Nacional de Tecnología Industrial, C.C. 884, 5000 Córdoba, Argentina

[2] Nagel & Krönert Eng. GmbH, Südfeld 2, DE-59174 Kamen, Germany

Keywords: HDDR, Milling, Magnetic Properties, Powders

Abstract

Nd-Fe-B powders produced through the HDDR (Hydrogenation, Disproportionation, Desortion, Recombination) process are coarse (mean particle size of 1000-2000 µm). They need to undergo a deagglomeration step prior to pressing. If they are to be used in the production of compression molded bonded magnets in order to improve die cavity filling conditions.

In the present work the effect of ball milling time on the particle size distribution, particle shape and tap density of isotropic powders is studied. Magnetic properties and density of compression molded magnets are also evaluated.

HDDR produced powders are very friable therefore a sharp decrease in the mean particle size is obtained even for short milling times. Powders milled for 15-20 minutes show improved flowability and reach higher densities during compaction. Angular shapes with sharp edges are characteristic of particles obtained through the HDDR process. Ball milling deagglomeration of such particles strongly changes the size distribution but without introducing dramatic changes in particle shape.

As a result of particle size reduction intrinsic coercivity is lowered from 1000 kA/m, in the as produced coarse powder, to 900 kA/m in the material milled for 60 minutes. However, because of its better compacting behaviour milled powders develop higher remanece.

Optimum density of 6.0 g/cm^3 was achieved for milling times of 15-20 minutes. For that conditions isotropic magnets with the following magnetic properties were produced: Hci=980.6 kA/m, Br=0.446 T and $(BH)_{max}$=32.2 kJ/m^3.

1. Introduction

The HDDR processing of NdFeB alloys is now a well established method to obtain powders suitable for the production of bonded magnets[1]. However particle agglomeration occurs because of liquid phase enhanced sintering process, that takes place during the recombination step. Therefore the powders that result from the final vacuum heat treatment are coarse. A deagglomeration step is needed in order to condition these powders for the subsequent forming operations. For this work ball milling has been chosen as the size reduction method due to scaling-up criteria.

2. Experimental procedure

Commercial $Nd_{14}Fe_{79}B$ alloy was disproportionated in a rotary kiln at 870°C, in 1 bar hydrogen pressure, for 60 minutes, and recombined at 780°C for 215 minutes at 100 ubar. The details of the heat treatment and kiln characteristics have been published elsewhere[2].

Lots of 400 g of HDDR processed powders were initially characterized. Afterwards they were deagglomerated in a steel ball mill[3], 107 mm in diameter and 130 mm in length, charged with 20 mm steel balls, for times varying between 2 and 60 minutes.

Partcile size distribution was determined through sieving analysis. Particle shape was observed by SEM. Tap density measurements were conducted in order to asses the powders flowability. Finally the as processed and milled powders were pressed[4] in a hard metal die, at pressures of 1-2 t/cm^2. After measuring their density the pressed pieces were epoxy impregnated and magnetized in a 35 kOe continuous field. Finally demagnetizing curves of the isotropic magnets were traced in a computerized hystersisgraph.

3. Results and discussion

3.1 Particle size analysis

Fig. 1 shows the particle size distribution of the as-processed material. It can be seen that the powder is rather coarse. Particles larger than 1000 µm stand for more than 50 wt % of the sample.

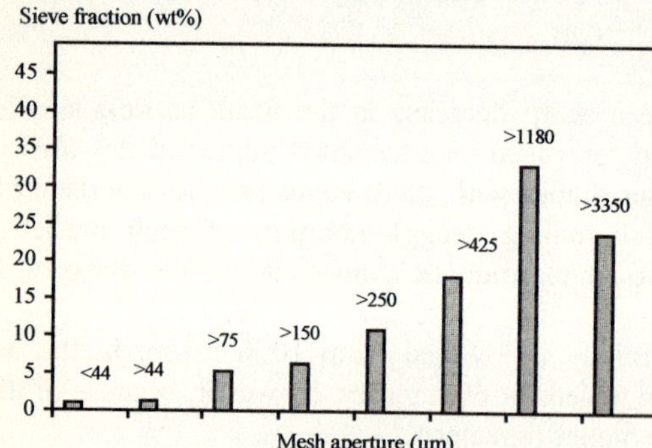

Fig. 1: Particle size distribution in the original powder.

Ball milling is a very effective way to condition these powders for compression molding forming. Fig. 2 displays the mean particle size evolution with milling time. The fast size reduction obtained indicates that HDDR processed NdFeB powders are very friable. Milling times of less than 20 minutes are enough to lower the mean particle size from 1600 to 200 µm. Times longer than 40 minutes introduce only very small additional changes. Fig. 3 to 6 exhibit the particle size distribution evolution with milling time.

It can be seen that milling times of up to 30 minutes are enough to break the large agglomerates. Then particle size changes become very slow indicating that milling of the individual particles has started.

In Fig. 7 the particle size distribution of commercial MQ-I powder, is displayed for comparison. The HDDR powder milled for 30 minutes exhibits a rather similar distribution.

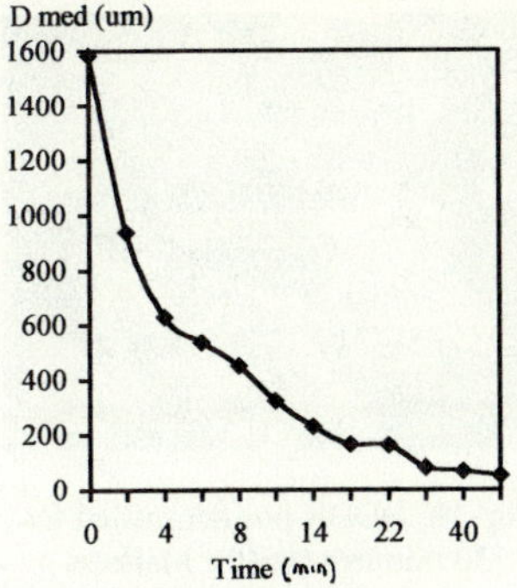

Fig. 2: Effect of milling time on mean particle size.

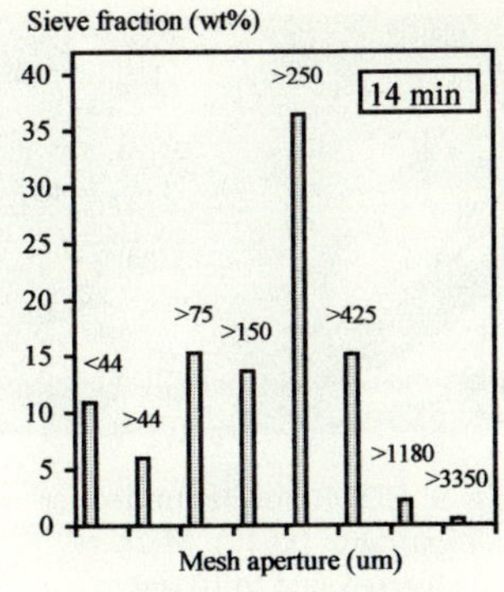

Fig. 3: Particle size distribution after milling for 14 minutes.

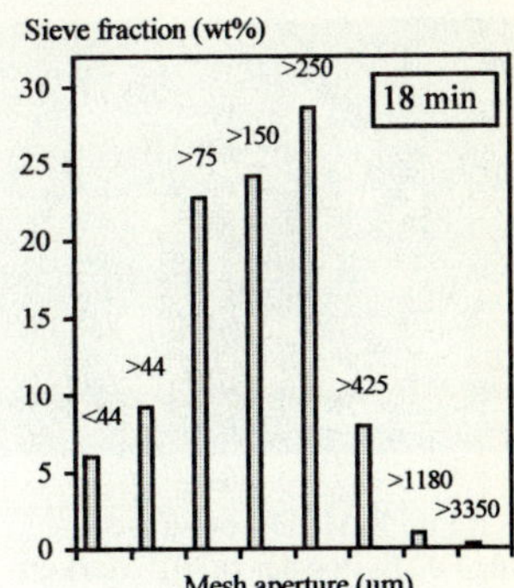

Fig. 4: Particle size distribution after milling for 18 minutes.

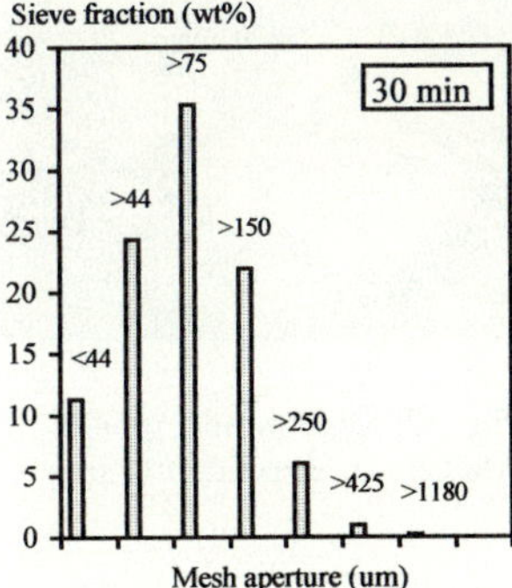

Fig. 5: Particle size distribution after milling for 30 minutes.

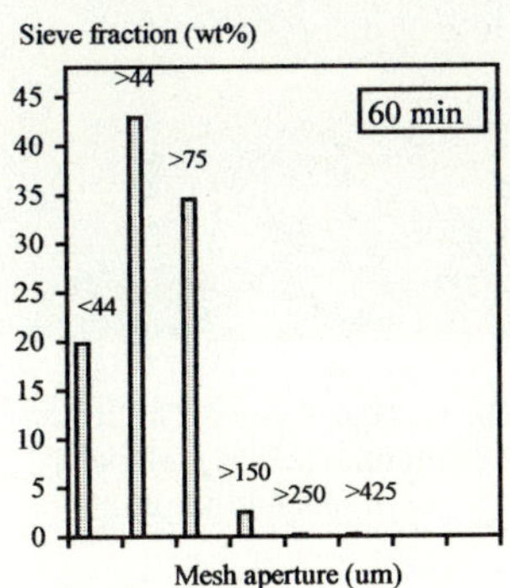

Fig. 6: Particle size distribution after milling for 60 minutes.

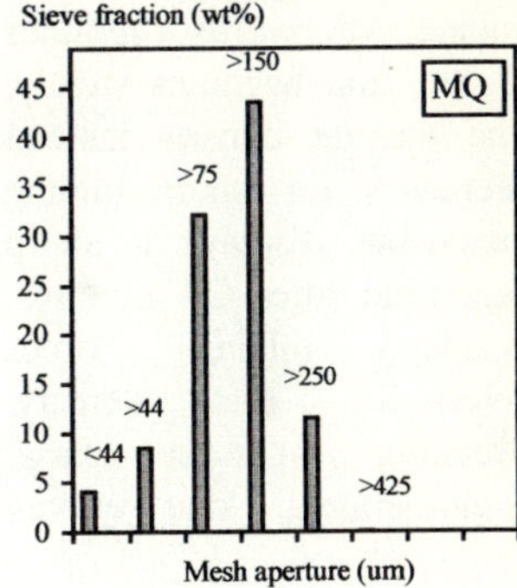

Fig. 7: Particle size distribution of commercial MQ-I powder.

3.2 Powder morphology

Fig. 8 shows the particle shape of the as heat treated powder and Fig. 9 to 11 correspond to powders subjected to different milling times. The MQ-I powder morphology is given in Fig. 12.

As ball milling proceeds the initially angular, sharp edged HDDR particles suffer, as already mentioned, a significant size reduction. The shape of the large particles does not change very much but an increasing number of small and rounded fine particles appear. The morphology of the MQ-I powder is characteristic of crushed flakes.

3.3 Compaction

A good flowability is essential in powders to be used in the production of compression molded magnets. Tap density is an accepted indication of how powders fill pressing cavities.

Remanence is directly correlated with density. Density in compression molded magnets will be determined by the applied pressing force and the powder behaviour. Fig. 13 represents the effect of milling time on both Tap and final density, in powders and compression molded magnets, respectively.

Fig. 8: HDDR processed $Nd_{14}Fe_{79}B_7$ powder (x80). Markers correspond to 100 μm.

Fig. 9: HDDR powder milled for 14 minutes (x640). Markers correspond to 10 μm.

Fig. 10: HDDR powder milled for 30 minutes (x640). Markers correspond to 10 μm.

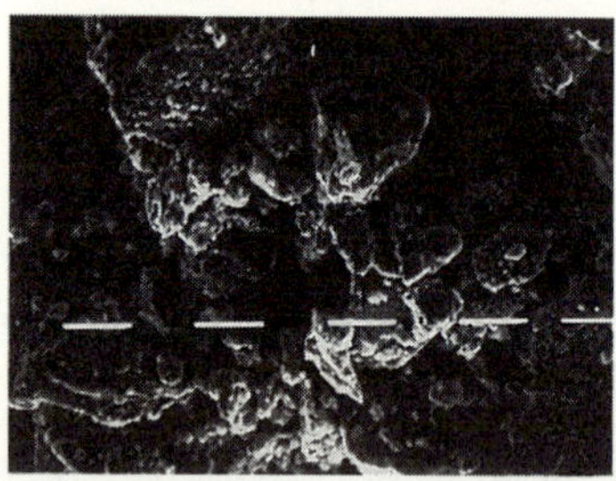

Fig. 11: HDDR powder milled for 60 minutes (x1250). Markers correspond to 10 μm.

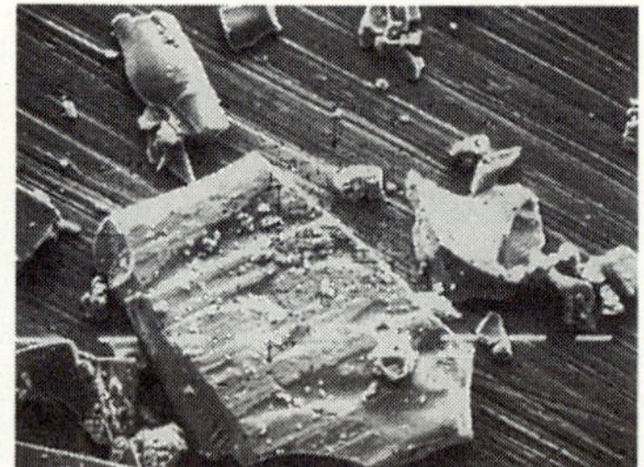

Fig. 12: MQ-I powder (x640). Markers correspond to 10 μm.

At the beginning Tap density grows and reaches a maximum for 18 minutes milling. Afterwards it reduces slightly and becomes steady. The magnet density instead decreases for short milling times but displays a sharp maximum after 20 minutes. Longer milling times deteriorate this density. Breaking of the large agglomerates, that takes places at the begining of the milling operation, greatly improves the powder flowability but with low influence on its compaction behaviour. Milling for times longer than 20 minutes increases the fine particles content of the powder deteriorating its flowing and compaction behaviour. The clear maximum in densitty obtained for 20 minutes shows that at that time the powder has got an adequate particle size distribution.

3.4 Magnetic properties

Fig. 14 is a plot of the magnetic properties, measured on compression molded magnets, vs. milling time. As expected intrinsic coercivity diminish as the sepecific surface of the powder rises. The remanence, by the contrary, shows a dependece quite similar to that observed for density.

The best energy product obtained was 32.2 kJ/m^3, corresponding to a milling time of 18 minutes, where both remanence and density maximums are located.

4. Conclusions

Ball milling for short times has proven to be a very effective way to obtain HDDR processed Nd-Fe-B powders with particle size distribution suitable for compression molding. As a result of milling densities of 6 g/cm^3, at 1-2 t/cm^2 compacting pressures, can be achieved. Regarding magnetic properties, moderate improvements in remanence and energy product are obtained, although a 5 %

reduction in coercivity is a non avoidable effect of size reduction.

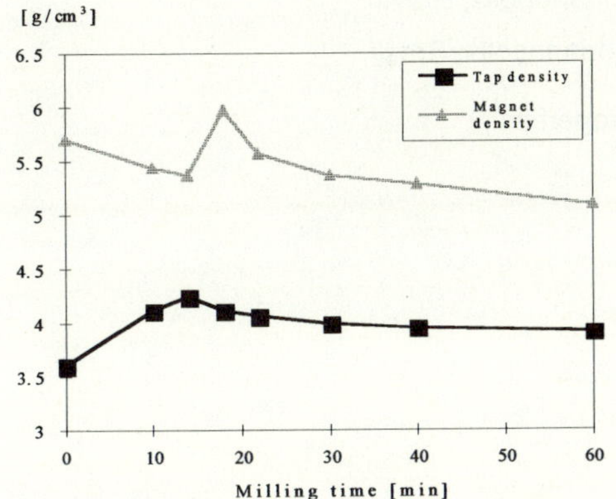

Fig. 13: Effect of milling time on tap and compression molded magnet density.

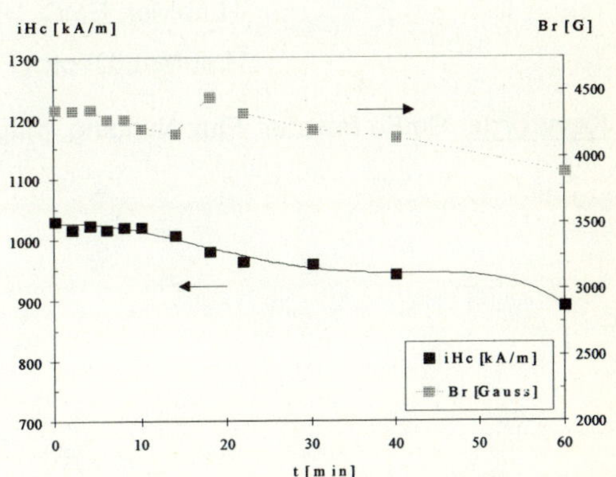

Fig. 14: Effect of milling time on intrinsic coercivity and remanence of compression molded magnets.

5. Acknowledgements

This work received financial support from CONICOR. Thanks are due to Mr. Niceto Esquenazi, from Artic company, for his assistance in pressing.

6. References

1. Hirosawa, S.; Uehara, S.; Mino, S.; Ishigaki, N.; Tomida, T. *"New aspects of Nd-Fe-B based hydrogenation-disproportion-desorption-recombination powders and anisotropic bonded magnets made from them: Microstructure and magnetic properties (invited)"* Symposium on the Technology and Applications of Bonded Magnets. J. Appl. Phys. 81 (8), 15 April 1997.

2. Actis, F.A.; Nagel, H.; Cohen, G.B.; Ragg, O.; Harris, I.R. *"Critical Parameters on the HDDR Processing of NdFeB Powders"*, Proceedings of the 14th Int. Workshop on Rare-Earth Magnets and Their Appl., Vol.1, 174, ed. F. P. Missell et al, World Scientific Publ. Co. (1996).

3. Kuhn, W.E.; Friedman, I.L.; Summers, W.; Szeguari, A. *"Milling of Brittle and Ductile Materials"*, (The ASM Commition on Milling) Production of Metal Powders. p 56-70.

4. Chin, T.; Wu, K.; Hung, M.; *"An Evaluation on the Orientation Alignment of a Nd-Fe-B Powder During Magnetic Compaction"*, 12th Workshop on Rare-Earth Magnets & Their Applications. Session 3: Posters, p. 158-190.

Materials Science Forum Vols. 302-303 (1999) pp. 324-328

Nitriding under N_2 Flux and Magnetic Properties of $Sm_2Fe_{17}N_x$ Powders

J.P. Leite[1], A.V.A. Pinto[2], T.R. Taylor[2] and P.A.P. Wendhausen[1]

[1] LabMat, EMC, UFSC, Florianópolis, Brazil

[2] LabMat, Depto. Física, Florianópolis, Brazil

Keywords: SmFe Powder, Flux Nitriding, Magnetic Properties

Abstract

The magnetic properties of $Sm_2Fe_{17}N_X$ powder, under flux processing,, as a function of the following parameters: flux (from 50 to 450 sccm, time and temperature constant), time (from 60 to 300 minutes, flux and temperature constant), temperature (from 670 to 870 K, time and flux constant), and mass of the sample in the reactor (from 50 to 200 mg, time, flux and temperature constant) were investigated. Also, the influence of the particle size under fixed nitriding conditions was investigated. The influence of variation of the above parameters on the magnetic properties were investigated using VSM measurements .

Introduction

It was demonstrated by Coey [1] that the interstitial compound of Sm_2Fe_{17} with nitrogen, has much better intrinsic magnetic properties than the alloy. The nitriding process improves the spontaneous magnetization of alloy about fifty per cent; the planar anisotropy changes to a strong uniaxial anisotropy, and the Curie temperature increases from 390 to 750 K. As first produced by Coey, the interstitial modification from Sm_2Fe_{17} to $Sm_2Fe_{17}N_X$ ($x \approx 3$) was obtained by gas-solid reaction in a closed reactor at a pressure of about 750 Torr and the nitrogen source may be molecular N_2 or another nitrogen containing gas such as ammonia [2]. At a typical nitrogenation temperature of 720 K, the nitrogenation reaction takes a few hours, depending on the particle size. Nitrogenation is not the only process used to produce interstitial modification. Carbon can equally well be introduced by gas-phase reaction using hydrocarbon gases to yield carbides $Sm_2Fe_{17}C_X$ ($x \approx 2.2$) and the intrinsic magnetic properties of the carbides (Curie temperature, magneto-crystalline anisotropy) are nearly as good as those of the nitrides[3]. Hydrogen has been long known as an interstitial in rare-earth intermetalics. At a temperature of about 520 K hydrogen gas easily reacts with Sm_2Fe_{17}, but the anisotropy remains easy-plane and with much less increase in Curie temperature [2]. In this work we investigate the flux nitriding efficiency at low pressure (from 1 to 6 Torr); moderate flux (from 50 to 450 sccm); variable processing time (from 0 to 240 minutes); variable temperature (from

670 to 870 K); and the sample mass in the reactor (from 50 to 200 mg), using as control parameters the behavior of magnetic properties (VSM measurements).

Experimental Procedure

Sm_2Fe_{17} ingots, from commercial grade material, were broken and submitted to HD process (p = 700 Torr, t = 120 minutes, T ≈ 300 K). The powder was sieved (mesh < 38 μm) and protected under toluene to avoid air contact. Before the nitriding process the residual hydrogen was removed, under vacuum at T = 620 K during 120 minutes.

The nitrides were produced by heating 100 mg of the powder in N_2 gas atmosphere under the following conditions:

I) Variable flux, from 50 to 450 sccm, time 60 minutes and temperature 820 K, both constant.
II) Variable time, from 0 to 240 minutes, flux 200 sccm and temperature 820 K, both constant.
III) Variable temperature, from 670 to 870 K, time 60 minutes and flux 200 sccm both constant.

The effect of sample mass was studied by varying the mass from 50 to 200 mg under the conditions: time 60 minutes, temperature 820 K and flux 200 sccm. Finally the influence of milling was measured (vibrating ball mill f = 30 Hz), by milling from 0 to 1200 minutes and nitriding under the conditions: time 60 minutes, flux 200 sccm and temperature 820 K.

For magnetic measurements, the powders were mixed with epoxy resin and spherical samples (diameter: 4.80±0,05 mm) were made and oriented in a magnetic field strength (μ_0H = 2,0 T). The measurements of the magnetic moment in the easy and perpendicular to the easy direction were performed using a vibrating sample magnetometer (VSM), with magnetic field strength up to μ_0H = 2,0 T.

Results and Discussion

In the figures, we use the following conventions and units:

Mf - Magnetic moment in the easy direction (μ_0M [T]);
Md - Magnetic moment perpendicular to Mf (μ_0M [T]);
Br - Remanence ([T]);
BHmax - Energy Product ([kJm^{-3}]);
jHc - Coercitivy (μ_0H [T]).

Figure 1 shows the influence of flux on the magnetic properties in the range studied. From this graph we can see that only the energy product is an increasing function of the flux. The deviation from smoothness from 300 to 400 sccm was confirmed and is real and we have no explanation for this irregular behavior. The easy and perpendicular direction moments and the remanence remain more or less constant.

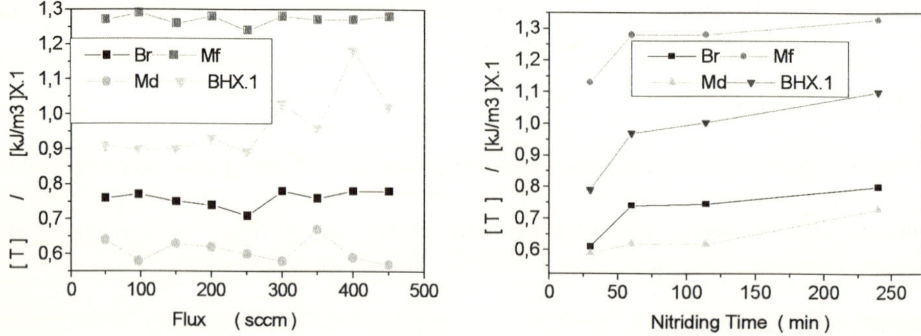

Figure 1 Influence of the flux on the magnetic parameters.

Figure 2 Variation of the magnetic parameters as function of nitriding time.

Figure 2 shows the variation of magnetic parameters as a function of the nitriding time. We can see that the initial sixty minutes are determinant to the process. After this interval of time the improvement of the magnetic properties is less significant and it is hard to justify a long processing time (with energy and N_2 gas loses) for such a modest gain in performance.

Figure 3 shows the influence of the nitriding temperature on the magnetic properties. From this graph we can see that there is a general tendency for improvement of the magnetic properties for all parameters between $T \geq 623$ K and $T \leq 823$ K. The remanence, easy direction moment and energy product increase in this temperature range, but the moment perpendicular to the easy direction decreases as required to increases the difference (Mf - Md) related to magnetic anisotropy. Beyond 823 K the magnetic properties begin to deteriorate.

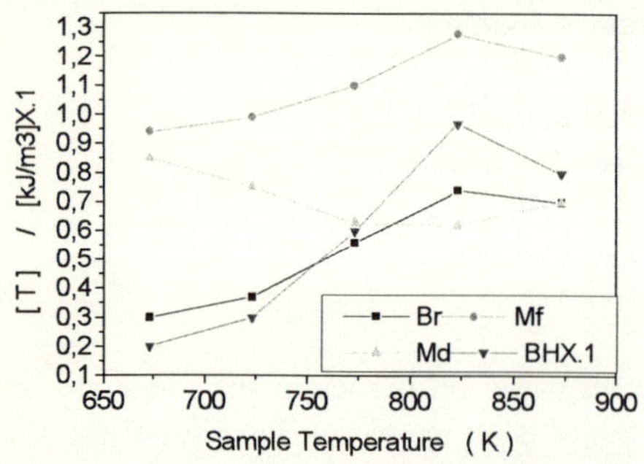

Figure 3 Magnetic properties as a function of the temperature of nitriding.

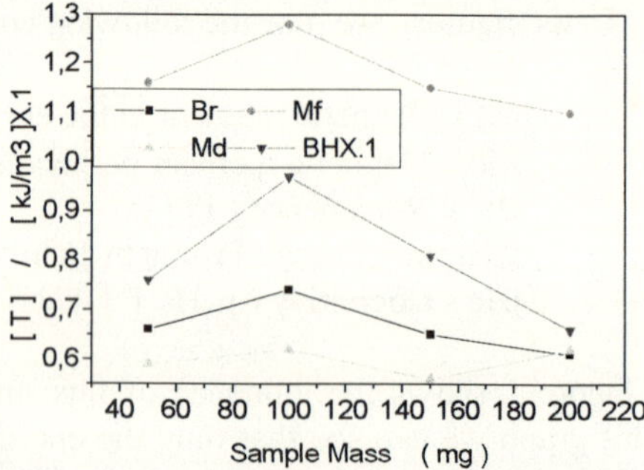

Figure 4 Influence of the mass on the magnetic properties.

Figure 4 shows that there is a relation between the sample mass and our particular experimental setup. Best results are for samples with mass around 100 mg. For other masses, the magnetic properties are less satisfactory.

Figure 5 shows the effect of milling on the magnetic properties (mass = 100 mg, time of nitriding = 60 min, flux = 200 sccm, temperature = 820 K). When the sample was milled for twenty hours, in vibrator mill, the particle size decreases by a factor of five or more (decrease of the mean particle area from ≈ 100 μm^2 to ≤ 20μm^2) and the result of nitriding is more complex. At the first milling stage (60 min), remanence and energy product have a important increase and above this time both decrease monotonically. In the range studied, the coercivity increases until around 800 min then remains constant. Finally, easy direction moment is a montonic decreasing function of the milling time.

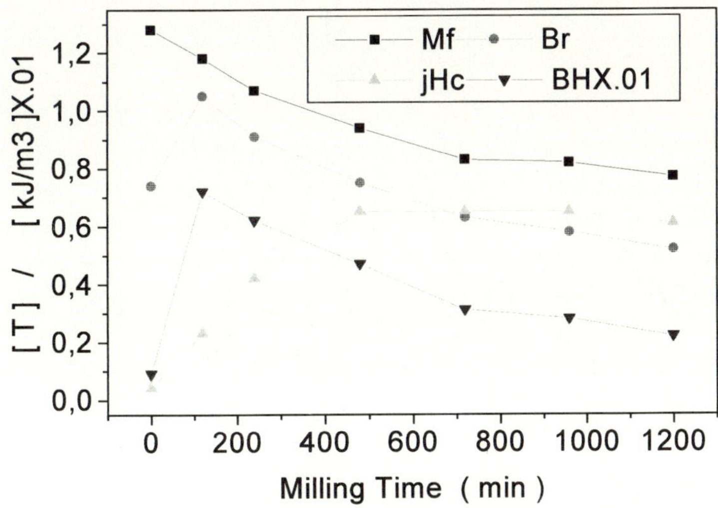

Figure 5 Magnetic properties as a function of milling time.

Conclusion

It is relevant to compare the flux process with the "traditional" nitriding process (closed reactor with fixed N_2 volume) [4].

a) It is more 'efficient' by a factor of five or more . To obtain the same magnetic properties, the time of nitriding under flux can be reduced from five to one hour.

b) For a nitriding time of sixty minutes, and temperatures between 720 and 820 K magnetic properties do not vary in traditional processing. However, in the same temperature interval under N_2 flux the nitriding process is more efficient more than doubling the remanence , the energy product and the difference between the easy and the hard moments.

References

1 - J. M. D. Coey, and L. Hong Sun, J.Magn.Magn. Mater., 87, (1990) L251

2 - R. Skomski - Rare-Earth Iron Permanent Magnets - Ed. J.M.Coey -Oxf. Sci.Publications, New York (1996), p. 178.

3 - J. M. D. Coey, H Sun, Y Otani and D. P. F. Hurley, J.Magn.Magn. Mater., 104 / 7, (1992), p. 1439

4 - Josinaldo P Leite, Paulo A P Wendhausen, Abio V A Pinto and Ted R Taylor - *First International Latin-American Conference on Powder Technology*, São Paulo -BR (1997)

Materials Science Forum Vols. 302-303 (1999) pp. 329-333

Magnetic Properties and Microstructure of the Alnico 8

J. Matutes Aquino, C. Domínguez Ríos, M. Miki Yoshida and O. Ayala Valenzuela

Centro de Investigación en Materiales Avanzados, S.C. Miguel de Cervantes 120, Complejo Industrial Chihuahua, Chihuahua, Chih. 31110, México

Keywords: Alnico, Permanent Magnets, Magnetic Properties, Microstructure

ABSTRACT

Alnico is a family of Fe-Co-Ni-Al materials with additions of Cu, Ti and Nb for permanent magnets. Alnico is heat-treated to produce two bcc phases: α_1 and α_2. The α_1 phase forms strongly ferromagnetic Fe-Co rich elongated particles in the non-ferromagnetic Ni-Al rich matrix. The shape anisotropy of the α_1 particles provides coercivity to the material. In this work we present a study of the microstructure and magnetic properties of sintered Alnico 8. It is very important to characterize the microstructure, because the magnetic properties are influenced by crystallographic orientation and proper <100> crystal texture is required to achieve the best properties in the Alnicos.

1. INTRODUCTION

Numerous experiments have shown that the effect of heat treatment (either during cooling or during isothermal treatment) in a magnetic field is to elongate the iron-cobalt rich particles in a direction which is parallel to, or close to, the applied field direction. The best magnetic properties are obtained when the applied field is parallel to <100> directions.

When the homogeneous alloy is cooled from the high-temperature single-phase (bcc) state to a critical temperature T_S of around 850°C, it undergoes a phase separation by spinodal decomposition to form α_1 and α_2 (b.c.c.) phases [1,2].

The Fe-Co rich particles (α_1 phase) are nearly spherical and regularly spaced. When alnico 8 is cooled below its Curie temperature T_C ($T_S - T_C \approx 20$-30°C) in the presence of a magnetic field, the particles grow preferentially in the field direction. Thermodynamically this elongation occurs because the magnetic free energy is lower when the easy magnetization axis is parallel to the applied field [3].

Although the elongation increases the surface area of the particles and, hence, the interface area between them and the matriz, the increase in the interfacial energy is small compared with the reduction in the magnetic energy which accompanies the elongation.

It has been established that the addition of Co to alnico reduces the interfacial energy. This is important because the thermomagnetic treatment is most effective in the alloys of high Co content. The elongation in the alloys is also favored by the addition of Co because it increases the

difference in the magnetic behavior of the Fe-Co rich particles and the matrix, reducing the magnetic energy when the particles are elongated in the field direction.
This work aims to determine the relationship of microstructure with the magnetic properties of a sintered alnico 8.

2. EXPERIMENTAL PROCEDURE

The samples are disks of 10.41 mm of diameter and 1.62 mm of thickness, supplied for Altec Electrónica. The nominal composition for the alnico 8 is shown in table No.1:

Table 1. Alnico 8 Composition

Element	Al	Ni	Co	Cu	Ti	Nb	Si	Fe
Wt. %	6.75	12.6	35.0	2.7	4.5	0.5	0.45	37.5

Cu K_{α}-radiation from a Phillips X Pert. MPD diffractometer was used for phases identification in samples of the alnico 8 and pure iron. The X-ray diffraction patterns from an alnico 8 disk sample and 99.9% pure iron powder were scanned for 2θ angles ranging from 30° to 120°.
SEM examination was performed in a JEOL JSM-5800 LV scanning electron microscope. The samples were polished with traditional procedures and etched with Marble's reagent to reveal the grain boundaries.
TEM observation was performed in a Phillips Model CM 200 transmission electron microscope. The samples for TEM was prepared with a Precision Ion Polishing System (Gatan, model 691 PIPs).
The magnetic field intensity in the perimetral surface of samples was measured with a Hall effect Gaussmeter (Walker Scientific Inc., model MG-3ADP).
Powder samples for Mössbauer spectroscopy were milled in a Vibratory Micromill (Fritsch model Pulverizette 0) during 60 hours. The Mössbauer spectra were taken in a Mössbauer Spectrometer from Austin Science Associates.
The hysteresis loops were obtained along different directions on the plane of the disk and for the powdered sample with vibrating-sample magnetometer model LDJ 9600 from LDJ Electronic Inc.

3. RESULTS AND DISCUSSION

X-ray diffraction patterns for alnico 8 and pure iron are shown in figure 1. The alnico lines are broader because the α_1 and α_2 lattice parameters are very close and because of the smallness of α_1 particles.
Figure 2 is a scanning electron micrograph of the surface of the disk showing a microstructure of equiaxed grains with a size of about 50 μm. Notice also the porosity accumulated in grain boundary.

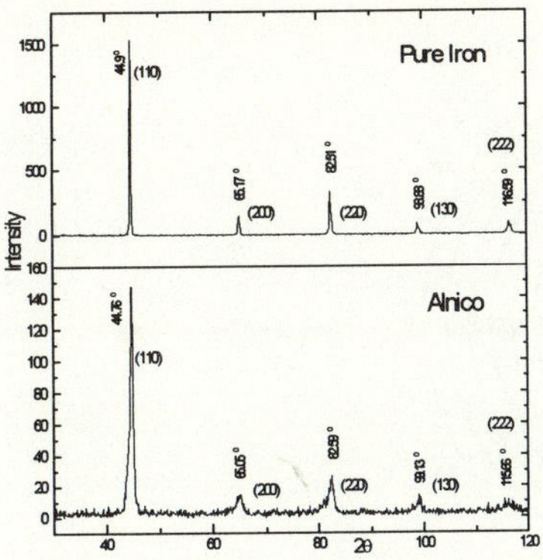

Figure 1. X-ray diffraction patterns of pure iron and alnico 8.

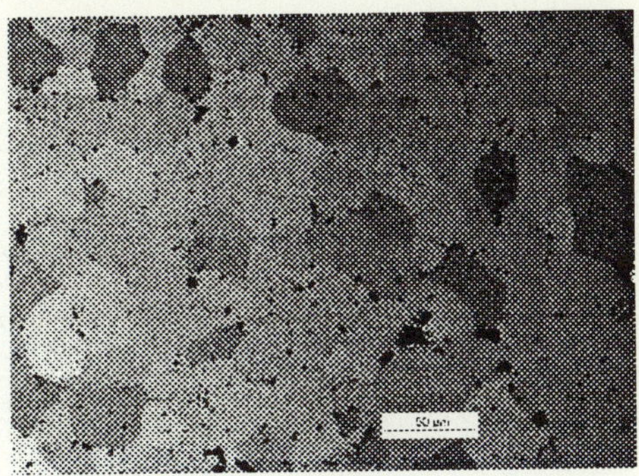

Figure 2. Scanning electron micrograph of the alnico 8.

Figure3a shows the dark field image from the microstructure of alnico 8. It consists of rodlike α_1 particles (light regions) with a average diameter of about 34 nm, and the α_2 matriz (dark regions). Note the regularity of the spacing between the α_1 particles (of about 45 nm), which is caracteristic of the phase separation by spinodal descomposition.
Figure 3b shows the indexed electron diffraction pattern corresponding to the [113] zone axis of the α_1 b.c.c. phase.

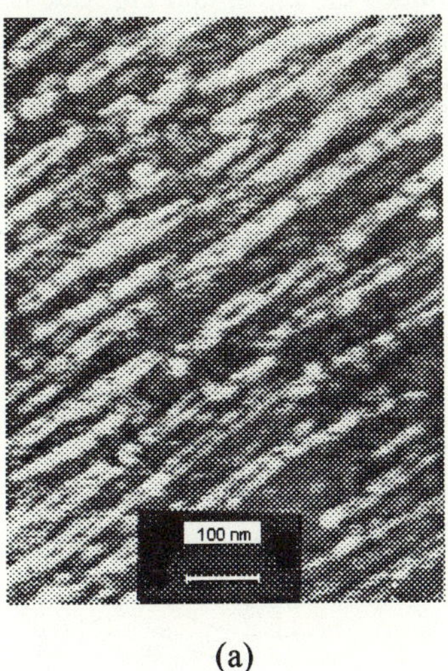

(a)

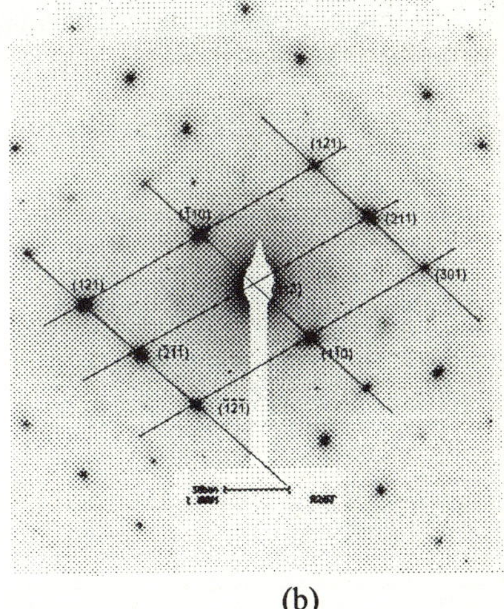

(b)

Figure 3. Transmission electron micrographs of alnico 8: (a) dark field image (b) electron diffraction pattern.

Figure 4 shows a polar plot of the magnetic field intensity as a function of the angle at the surface of the disk sample of alnico 8. The magnetic field intensity is rather symmetric around the 0° direction of magnetization with minima of magnetic field intensity perpendicular to this direction.

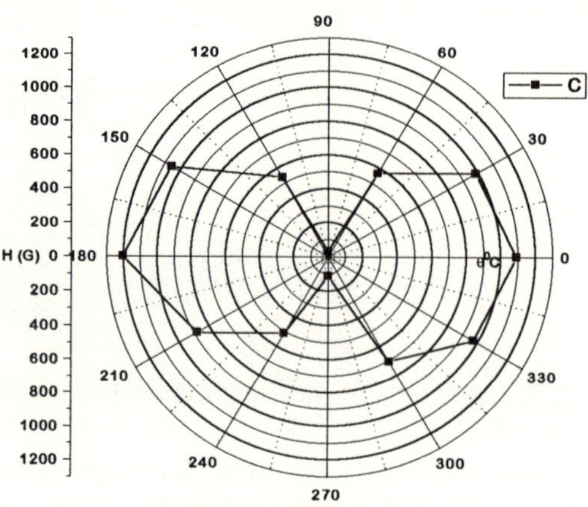

Figure 4. Magnetic field intensity for different angles at the surface.

Figure 5a shows the hysteresis loops of alnico 8 disk sample taken along different directions together with the hysteresis loop for the powdered sample. Figure 5b shows polar plots for the remanent specific magnetization σ_r , the saturation specific magnetization σ_s, and coercivity H_C obtained from the hysteresis loops in figure 5a. The highest values of σ_r, σ_s and H_C occur along the 90° direction of magnetization. The values obtained for σ_s and H_C are in the ranges reported by Mc. Currie for alnico 8 [4].

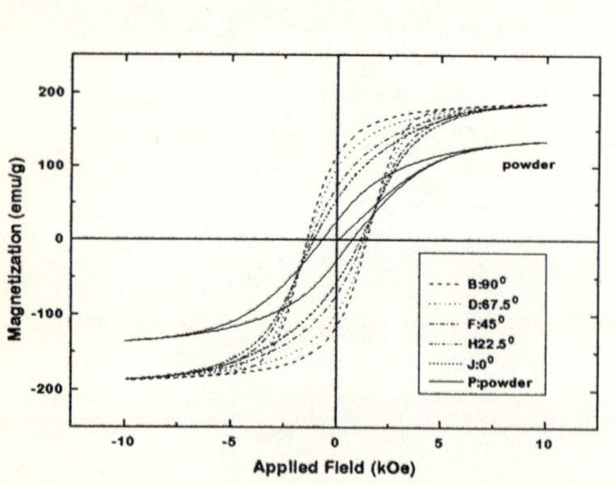

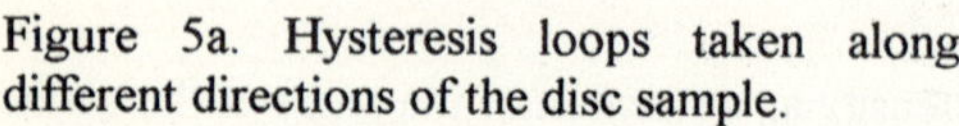

Figure 5a. Hysteresis loops taken along different directions of the disc sample.

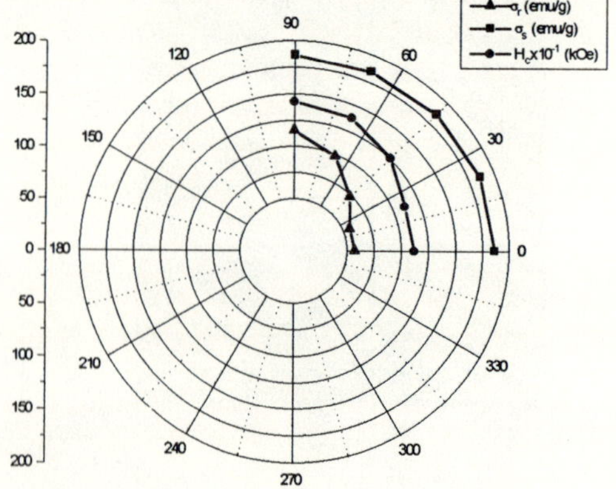

Figure 5b. Polar plot of the magnetic properties.

Figure 6a shows the Mössbauer spectrum of alnico 8 powdered sample taken at room temperature. It looks like the spectrum obtained by Albanese et al [5]. Figure 6b is an enlargement of the central zone of the spectrum. The wide central line of figure 6b is a non-resolved doublet due to the non-magnetic Al-Ni-rich matrix phase [6].

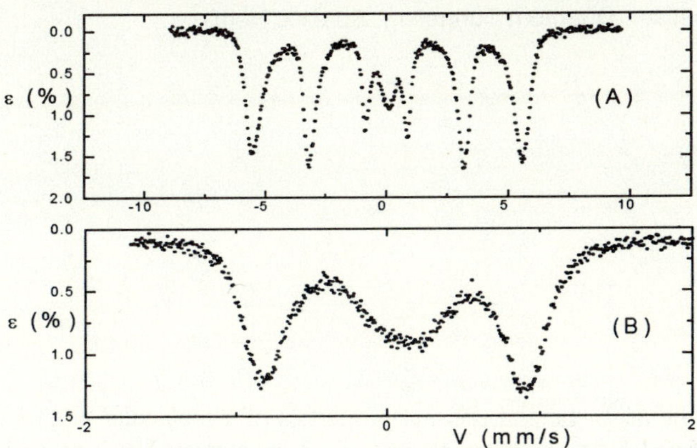

Figure 6. Room temperature Mössbauer spectra of the alnico powder. The wide central line of the figure 6b is a non-resolved doublet due to the non-magnetic Al-Ni matriz phase.

4. CONCLUSIONS

1.The alnico 8 have a microstructure composed of equiaxed grains with a diameter of about 50 μm inside which the α_1 and α_2 phases precipitate.
2.The α_1 elongated particles precipitate along the <100> directions of the matrix. These particles have a diameter of about 35 nm and an interparticle spacing of about 45 nm.
3. The x-ray diffraction pattern shows widening of the lines due to the smallness of the precipitates and to the superposition of matrix and precipitate diffraction lines.
4.Coercivity and remanence depend on the direction of magnetization due to the shape anisotropy of the elongated particles and to the parallelism of them.

REFERENCES

[1]. B.D. Cullity : Introduction To Magnetic Materials (Addison-Wesley Publishing Co. 1972).
[2]. Peter Campbell : Permanent Magnet Materials and Their Applications (Cambridge University Press 1994).
[3]. R.A. McCurrie:Ferromagnetic Materials, Structure and Properties (Academic Press, 1994).
[4]. R.A. McCurrie, Ferromagnetic Materials Vol. 3, Edited by E.P. Wohlfarth, North-Holland Publishing Company, 1982.
[5]. G. Albanese, G. Asti, and R. Criscouli, IEEE Trans. Magn. MAG-6, 161 (1970).
[6]. L.H. Bowen, E. De Grave, R.E. Vandenberghe in : Mössbauer Spectroscopy Applied to Magnetism and Materials Science, Volume 1 (Plenum Press, 1993).

Materials Science Forum Vols. 302-303 (1999) pp. 334-343

Quantum Molecular Nonomagnets

M.A. Novak, A.M. Gomes, W.S.D. Folly and R.E. Rapp

Instituto de Física, Universidade Federal do Rio de Janeiro,
C.P. 68528, Rio de Janeiro, RJ 21944, Brazil

Keywords: Nanomagnetism, Quantum Tunneling, Specific Heat

Abstract: We describe the magnetic and thermal properties of a molecular crystal of nanomagnets. Each nanomagnet is formed by a magnetic cluster of 12 interacting Mn ions and behaves at low temperature as macrospins. High field magnetization, complex susceptibility and specific heat results show that the clusters have a S=10 ground state with a high magnetocrystaline anisotropy and are magnetically well isolated from each other. The relaxation time is approximated by an Arrhenius law $\tau=\tau_0 e^{\Delta/kT}$ where Δ is the anisotropy energy barrier, with values being in fair agreement with high field magnetization and EPR measurements. Deviations from this ideal superparamagnetic behavior in a wide range of relaxation times and an unusual dependence with magnetic field are explained to be due to Quantum Tunneling of the Magnetization through the anisotropy energy barrier. Zero field specific heat results indicate that the usual uniaxial Hamiltonian is not enough to explain the results, being necessary to use the recently proposed fourth order corrections. Very low temperatures specific heat data indicate a ground state splitting much larger than expected. Finally, specific heat measurements done under magnetic field permitted to observe and measure the lattice spin relaxation by calorimetric methods with time constants that are in agreement with the magnetic relaxation measurements.

Introduction.

Since the pioneering work of Néel [1], ultra fine magnetic particles have been a very interesting and studied subject, never loosing its importance, and recently renewed by many evidences that quantum tunneling of the magnetization vector occurs in many systems [2]. The stability of information stored in high density magnetic media is rapidly approaching the superparamagnetic limit with the evolving miniaturization technology. The stability may be recovered by cooling the media, but would be eventually limited by quantum tunneling effects that are temperature independent. Quantum Tunneling of the Magnetization (QTM) in very small magnetic particles is also conceptually very interesting and a new area to test fundamental questions related to the quantum theory at a level between the atomic and the macroscopic scale. Those are some of the reasons why nanomagnetic organic clusters became a disputed matter in the last years [3].

In this scale, new theoretical approaches have to be developed for treating systems between single paramagnetic ions in a matrix and standard solid state theory based on translational symmetry. A giant spin model for the molecules, as compared to single ions, is used to explain the unusual magnetic properties observed at low temperatures. Histeresys and magnetic bistability associated to blocking of this macrospin in its effective anisotropy field, lead us previously to propose the possible use of them as magnetic memory elements at the molecular level [4]. This was later shown to be prevented by QTM, so it became important to study carefully how QTM can be hampered in highly anysotropic molecules.

In 1994, during the QTM-94 workshop [2] it became clear that it was important to study systems that avoid misinterpretations due to the existence a distribution of sizes and/or shapes, and the presence of interparticle interactions. At the same time two new systems were introduced. The study of the switching dynamics of single particles [6], and the study of a new system of strictly identical and non interacting magnetic clusters in metal-organic compounds [7]. Spin clusters became important in the last years as new quantum phenomena became evidenced at the nanoscopic level, specially in Mn_{12}ac ($Mn_{12}O_{12}(CH_3COO)_{16}(H_2O)_4)\cdot 2CH_3COOH\cdot 4H_2O$ where histeresys with steps were observed at fields where the relaxation time is accelerated due to thermally activated quantum tunneling [8]. In this paper we will present some of the developments reached by us in the last years. A better determination of the Spin Hamiltonian for the macrospin by high field EPR experiments [9] complemented with magnetic and specific heat measurements [10], as well as inelastic neutron scattering [11] started to give new insights and surely other techniques will be applied in this new fascinating subject.

Structural and Magnetic Properties

We will first describe shortly the main characteristics of Mn_{12}ac and its magnetic properties. Mn_{12}ac form a molecular crystal of tetragonal symmetry (space group I4) with a=17.319 Å, c= 12.388 Å, having elongated prism shape along the unique c axis. The two molecules per unit cell with a molecular weight of 2060 are surrounded disorderly by 4 waters and 2 acetic acids of crystallization. Each Mn_{12} cluster is formed by a crown of 8 Mn^{+3} (S=2) surrounding a tetrahedron of 4 Mn^{+4} (S=3/2) ions at the vertices (see Fig. 1), all of them linked by oxo and carboxilato bridges providing the antiferromagnetic superexchange paths among them. [12].

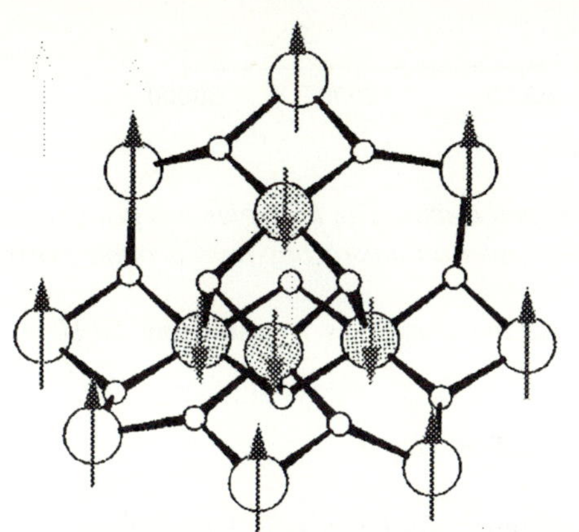

Fig. 1. Mn_{12}Ac cluster. Big circles are Mn IV and shaded ones Mn III. The small atoms are Oxygen. The other atoms are omitted for clarity.

The effective spin per cluster, determined from the powder χT value is 12μ_B at room temperature, a value lower than the expected value for 8 S=2 plus 4 S=3/2 Mn free spins (16.87μ_B), indicating the influence of antiferromagnetic correlations already at room temperature. The effective moment starts to increase steadily below ~150K down to ~20 K where it reaches a plateau with a value of 20.4μ_B. This points to a S=10 ground state that can be obtained with the 8 S=2 Mn^{4+} parallel to each other and antiparallel to the remaining 4 S=3/2 Mn^{3+}. From the susceptibility data between 4 and 10 K, we obtained a Curie Weiss temperature of ≈-0.05 K. This indicates that the average interaction between the clusters is very small and antiferromagnetic, presumably due to classical dipolar interactions. EPR experiments done with far infrared radiation [13] showed resonance lines with a linear field dependence indicating an uniaxial type zero field splitting of the S=10 multiplet, with a m_s=±10 ground state. Further confirmation of this ground state comes from magnetization data in Fig.2, showing a saturation value around 20 μ_B. From these data we determined an anisotropy constant

K=3.7*10^6erg/cm^3, in fair agreement with the cited EPR results. The easy axis magnetization between 3 and 10K are very well fitted to tanh($g\mu_B SB/k_BT$), in agreement with a $\pm$ m_S=10 ground state, like a perfect highly anisotropic superparamagnet. Above this temperature, the other m_S states become populated and the fit is not so simple. In addition, at still higher temperature (above 20K) the other multiplet (S=9,8...) with different spin configuration complicates the simple picture and the effective spin decreases [12].

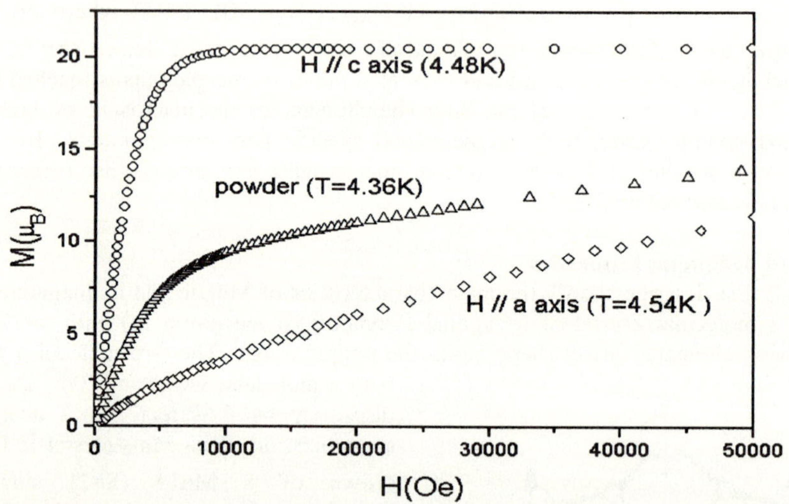

Fig.2. Magnetization isotherms along the c-axis and a-axis. The triangles are powder data. A rough extrapolation of powder and a-axis data gives an anisotropy field around 110kOe.

The above results are well explained by treating the cluster as a macrospin S=10 in an uniaxial crystal field with the following Hamiltonian:

$$H= -DS_z^2 - g\mu_B S\cdot B \qquad (1)$$

so the total anisotropy energy barrier between the up and down groundstates is Δ=100D.

Dynamical magnetic properties

After any magnetic field change, superparamagnetic particles are know to relax exponentially with time by exchange of energy with the lattice. The relaxation time for particles with uniaxial anisotropy barrier Δ=KV, where V is the volume of the particle and K the total anisotropy energy density, is expected to follow the well-known Arrhenius law [1] typical of a thermal activation process over the energy barrier:

$$\tau = \tau_0 \, e^{\Delta/kT} \qquad (2)$$

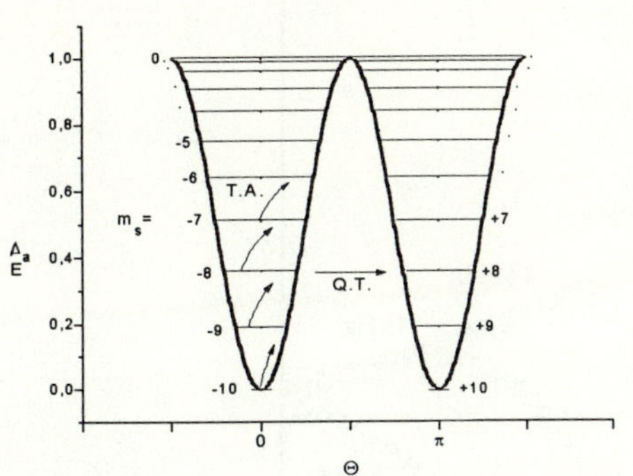

Figure 3. Anisotropy energy as function of the angle of the magnetization with the easy axis. The switching from one orientation to the opposite may be through Thermal Activation (TA) or underbarrier Quantum Tunneling (QT).

with τ_0 of the order of 10^{-10} s. We have reported [4] that the relaxation times of $Mn_{12}Ac$ determined by the imaginary maxima of the ac susceptibility and by the exponential relaxation of the remanent magnetization, followed the Arrhenius law (Eq. 2) with the anisotropy barrier $\Delta/k_B \cong 62$ K and $\tau_0 \sim 10^{-7}$ s, an extremely high value compared to typical superparamangets. This was explained [14] within a double well model (see Fig. 3) by a generalized Orbach relaxation process. The system is excited over several states stepwise until it reaches the top of the anisotropy barrier and decays to the bottom of the opposite direction well. Measurements done at lower temperature showed a marked deviation from Eq. 2 around 2K with the relaxation time becoming temperature independent and attributed to QTM. [5].

More evidence for QTM came from the field dependence of the relaxation time. The application of an external dc field is expected to decrease the relaxation time. Surprisingly it was observed that first the relaxation time increased with the field up to about 2 kOe and then it decreased with the field. The relaxation times continued to follow the Arrhenius law, with a field dependent energy barrier and field independent τ_0 [7]. Measurements of the field dependence of the relaxation time in single crystals along the c-axis are shown in Fig. 4, Where a structure of dips and maxima is clearly seen. This was explained for early powder results as due to thermally activated QTM [7]. In zero field the 10 states with $m_s = 10,9,...,1$ for one direction and $m_s=-10,-9,...,-1$ states for the opposite direction are degenerate (Fig.3). If no tunneling is present, the passage from one well to the opposite occur only by a classical thermal activation process over the barrier. If there is any quantum tunneling between some of those quasi-degenerated states (depending on some selection rule), the presence of a small magnetic field would inhibit this tunneling by energy mismatch until a new field is reached where tunneling will be favored between other matching states, with a decrease of the relaxation time (dips in Fig 4).

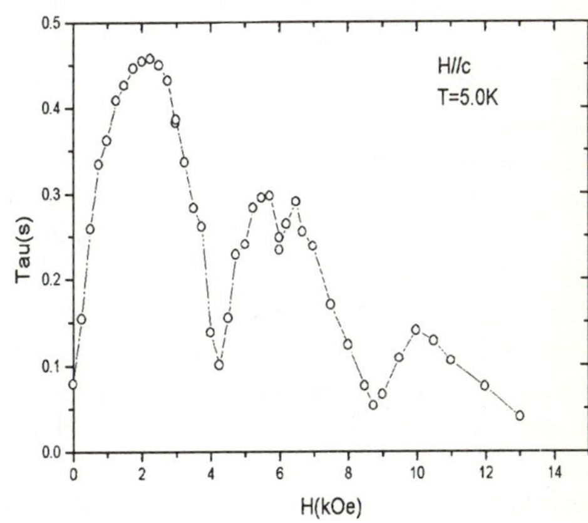

Fig 4. Relaxation time versus external field along the c axis.

From ac susceptibility measurements done in an extended range of frequency and temperature, another aspect of the quantum behavior of this system starts to emerge. Some selected runs in zero field are shown in Fig. 5. At lower temperatures another feature is observed: a second and faster relaxation seen as a decrease of χ' and another small maximum of χ''. This will be discussed further at the end of the paper.

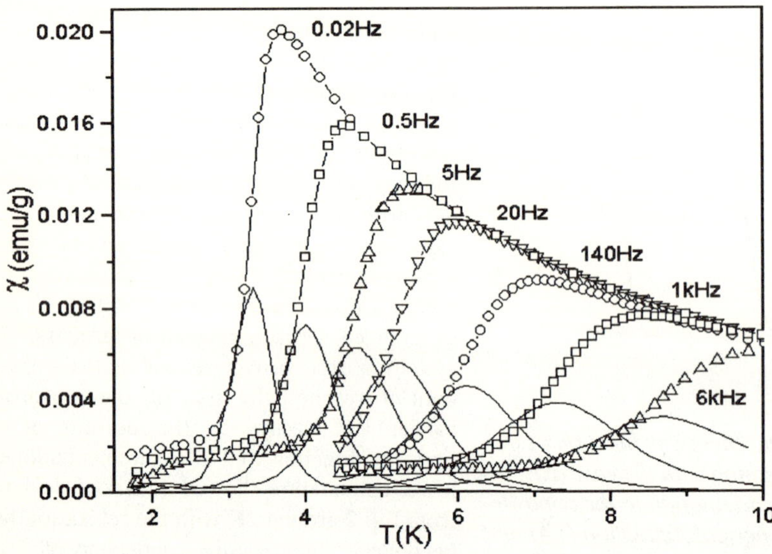

Figure 5. Real (line and symbols) and imaginary (full line) susceptibility as function of selected frequencies in zero dc field.

From the maxima of the imaginary component, the relaxation time is determined, being well fitted by an Arrhenius law (Eq.1) if a small range of frequencies is used. The single crystal values obtained around 5 K, Δ/k_b=68K and τ_0=1.1·10^{-7}s, are slightly different from the previously published results for powder sample, Δ/k_b = (64.1±0.7) K and τ_0 = (2.6±0.4)·10^{-7}s. In an extended range of frequency and temperature, the relaxation times shown in Fig. 6, present clearly a continuos curvature and one may thus define an effective (temperature dependent) energy barrier by the slope d(lnτ)/d(1/T) as well as an apparent τ_0. This last one believed to be unusually high depends on the range from where the extrapolation is taken. From the Ahrrenius plot in Fig. 6 we obtained the temperature dependence of this effective energy barrier shown in the inset. Our high frequency data indicated that τ_0 is of the order of 10^{-10} s, the usual number found for superparamagnets, and the effective energy barrier of the order of 100K, a value much higher than the one found by other techniques.

Further belief that the Hamiltonian Eq. 1 provides a good description of this system came from the steps in the magnetization histeresys in the blocked regime, reported to be equally spaced in the external H or B field [8,15]. At the same fields, maxima in the real part of the susceptibility versus field are observed near the blocking temperature at the measurement frequency [15]. We made extensive measurements in a wide temperature and field ranges. In Fig. 7 we show typical curves of Ac susceptibility data obtained under external field along the easy c-axis which permitted us to determine with good precision up to 4 resonant fields at H= 0.0, 4.3, 8.8, and 12.9 kOe, and independent of temperature (see inset). Those fields are close to the ones found at lower temperatures in magnetization measurements [8], but have a much narrower width. The real part of the susceptibility shows a maximum with a Lorentzian shape, characteristic of resonant phenomena.

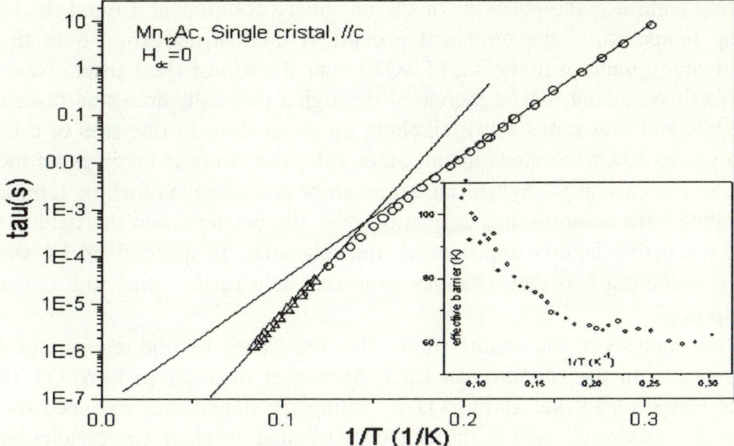

Fig 6. Temperature dependence of the relaxation time. The two lines show different Ahrrenius fits depending on the range used. Inset shows the temperature dependence the effective energy barrier.

The main resonance at zero field has a half width around 250 Oe. This is much larger than the expected linewidth for tunneling resonance [16], and indicates the presence of a distribution of local fields of different origins: hyperfine, dipolar and demagnetizing.

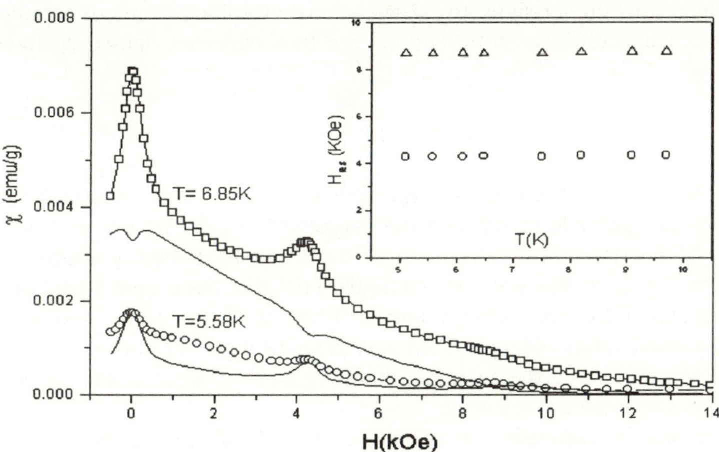

Fig. 7. Resonances at level crossing as seen by susceptibility under field. Full line are the imaginary susceptibility and symbols are the real part. Inset shows the temperature dependence of the position of the maxima.

An interesting feature is the behavior of the imaginary component around the level crossing. Below the blocking temperature, the dominant process is thermal activation over the barrier and some thermally activated quantum tunneling (TAQT) near the top of the barrier. Near a resonance field the back and forth switching will be increased through a thermally activated quantum tunneling path (involving middle and low states also), climbing up some steps in one side of the double well, tunneling and then going down the steps in the other side, this process involves an increase of the relaxation and thus an increase in χ''. When the temperature is above the blocking temperature, again the dominant switching path is by thermal activation over the barrier when the field is far from any resonant field, but when this field is approached, the relaxation is speeded up by some shortcuts through the barrier, involving less step changes from one side to the other and consequently less dissipation and a dip in χ'' .

A quantitative analysis of the results shows that the values for the resonances do not agree well with the expected from the Hamiltonian Eq.1. Moreover in order to have QT there must be another term in the Hamiltonian that allows QT by lifting the degeneracy between the + and - m_S sates. QTM theory [2] is usually based in the existence of small in plane (perpendicular) anisotropy term to allow this, but this anisotropy could not yet be detected neither is expected from the x-ray determined structure. Recently new EPR [9] as well as neutron [11] diffraction studies showed that is important to include fourth order terms in the Hamiltonian. We will now present specific heat measurements that can elucidate independently about the validity of the energy levels obtained from different Hamiltonians.

Specific Heat and the Ground State.

The specific heat measurements were done using the semi-adiabatic method from 1.4 K up to 25 K [10], and from 0.1 to 1.5K in another calorimeter attached to the mixing chamber of a dilution refrigerator [17]. The results do not show any sharp anomaly resulting from cooperative interactions between the clusters. Three different contributions are thus expected: lattice c_l, hyperfine c_h and magnetic Schottky c_S:

$$c_t = c_l + c_h + c_S \tag{3}$$

The hyperfine contribution is assumed to be negligible above 2K. As no isostructural diamagnetic compound exists, it is not possible to separate the magnetic from the lattice contribution. The data can be analyzed by comparison with calculation of the magnetic Schottky contribution from the energy levels given by Eq. 2. In this way we concluded [10] that the simple Hamiltonian Eq.2 only have a physical meaning if the total energy barrier 100D is about 100K, well above the value expected by using the usual values obtained from the Ahrrenius plot. The specific heat results depend on the energy level spacing, while the ac-susceptibility relaxation depends on the effective energy barrier which is affected by quantum tunneling.

A better agreement is found using the recently proposed Hamiltonian [8]:

$$H = DS_z^2 + AS_z^4 + B(S_+^4 + S_-^4) \tag{4}$$

with D=-0.56 K, A=1.11x10^{-3} K and B=2.9x10^{-5} K, used to simulate the EPR spectra. Even so a sizable linear term is found besides an excess contribution below 3K [10]. After taking care of the T^3 Debye and the calculated Schottky terms, a small rounded maximum at 1.8K is left with an upturn at lower temperature not having the expected a T^{-2} hyperfine behavior, as shown in Fig. 8. The recent high field EPR measurements show accurately the presence of fourth order terms, but the ground state splitting using Eq. 4 is too small to explain this contribution. Neutron scattering experiments [10] which agrees with this last Hamiltonian also reported the existence of a small and distinct energy level around 0.2 THz corresponding to a ground state splitting of 5 K. Our very low temperature data can be explained with a 4.5K splitting of the ground state, but having to assume that only 17 % of the total clusters contribute to this anomaly. With this we recover the expected T^{-2} contribution due to the Mn nuclear hyperfine interaction.

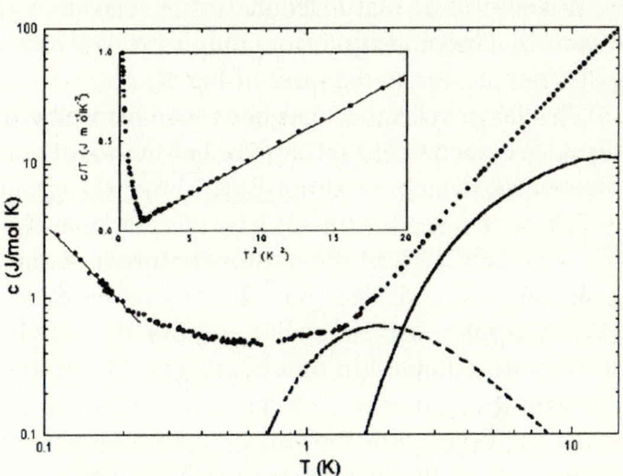

Fig.8. Specific heat in zero field.

The origin of this splitting, equivalent to 3.3 kOe, cannot be related to the combination of Hyperfine and dipolar fields, estimated [18] to give about 0.5 kOe, in agreement with the width of the Ac susceptibility resonance at zero field (Fig.7). Another possibility for the ground state splitting may be due to canting of individual Mn spins within each cluster occurring at low temperature, a process which is not cooperative, thus not giving rise to any sharp anomaly in the specific heat. In fact recent powder susceptibility measurement shown in Fig.9, presents a frequency independent small peak at 3.3K in the real part. Further investigation is needed in order to clarify this point.

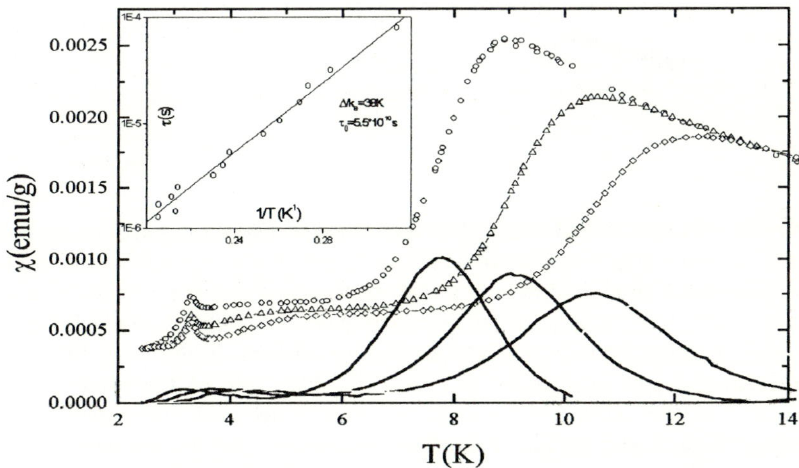

Fig 9. High frequency (2, 10 and 40 kHz) susceptibility in powder sample.

It is also clear that a second faster relaxation process with a smaller intensity appears at low temperatures. The relaxation time follows roughly an Ahrrenius behavior down to 2K, with a lower energy barrier as seen in the inset of Fig. 9.

This faster relaxation has been seen before by several authors but was never quantified neither explained. We speculate that is may be due to thermally activated quantum tunneling between the $|\pm m_S=8\rangle$ states, which lie about 36K above the ground state if we consider 100K the full energy barrier. There remains to explain why this fast relaxation is limited to a fraction of the total relaxation and also why only 17% of the cluster contribute to the low temperature specific heat anomaly.

Finally we consider the effect of an external magnetic field, leading to a field dependent Schottky anomaly. For simplicity we use the simple Hamiltonian (Eq.1). One may consider the system in quasi-equilibrium and count only the states that are accessible in the measuring time. It may be regarded as two sets of spins with the same energy level spacing, one with the $+m_S$ or down states and the other with the $-m_S$ or up states separated by the anisotropy barrier. At low enough temperature, when the time scale to switch sides is much longer than the typical measuring time, the two sets of energy levels are independent from each other, reducing the number of the accessible states to about one half. Fig.10 shows the specific heat data in a field of 3 kOe as

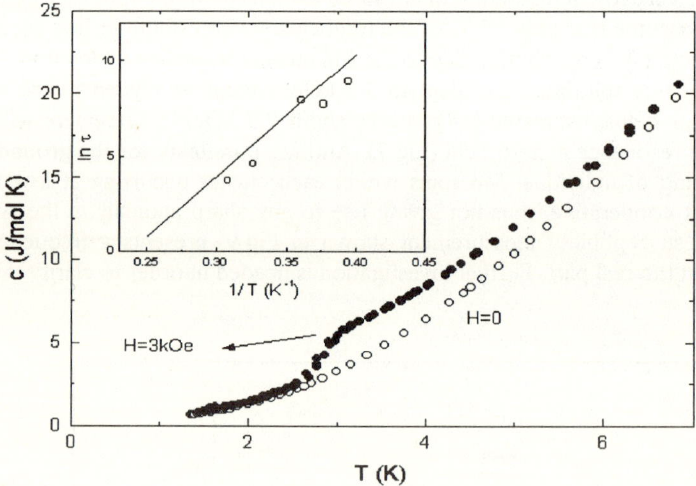

Fig 10. Specific heat measurements under field.

well as in zero field. A clear shoulder around 3 K is observed. This is because the excess contribution due to the field is mostly suppressed at temperatures below 3 K, which corresponds to the blocking temperature for the experimental time window. At the lower temperatures the relaxation time is orders of magnitude longer than the time scale of our measurements. Each cluster spin remains blocked in either side of the barrier, and the specific heat is almost the same as in the zero field case, the field changes very little the spacing between the levels in each side of the barrier. As the temperature is increased and the relaxation time (300s at 3K) becomes comparable to the experimental time the spins unblock and the state in both sides becomes progressively accessible and the specific heat increases. In fact in the temperature evolution after the heat pulse we could detect an extra exponential dependence and determine what we call the lattice-spin relaxation time, having a fair agreement with the magnetic relaxation time constant as can be seen in the inset of fig.10.

We point here that this specific heat behavior could only be observed in this sample due to the fact that it is a set of identical superparamagnetic particles having the same relaxation time. We have estimated the specific heat in a field by adding a Zeeman term to the Hamiltonian (Eq. 1), only to the lowest energy levels. The sum of this contribution and the zero field results agrees well with the data above the blocking temperature.

Conclusions

We have presented clear experimental evidence from specific heat measurements the need to use a Hamiltonian with fourth order terms, in agreement with recent EPR and neutron scattering experiments. The existence of a ground state splitting from very low temperature specific heat as well as a susceptibility peak at 3.3K suggests that there are still more to be done in this system. The relaxation time deviations from pure Ahrrenius law observed by ac-susceptibility is explained by use of an effective anisotropy energy barrier combining thermally activated quantum tunneling plus pure thermal activation over the barrier. Thermally activated quantum tunneling also explains the unusual field dependence of the relaxation time as well as maxima in the susceptibility versus field curves. Finally, we have seen superparamagnetic relaxation effects in specific heat measurements.

Acknowledgments: The authors would like to thank A. Caneschi for supplying us with samples, R.Sessoli , C. Paulsen and P. Stamp for fruitful discussions, F.A. Aguiar for the Squid magnetometer measurements and the financial support from FINEP, CNPq and FUJB (Brazil).

References:

[1] Néel, L (1949) *Ann. Geophys.* 5, 99.

[2] "Quantum Tunneling of Magnetization-QTM'94"- Nato ASI Series E, Vol 301, Edited by L. Gunther and B. Barbara, Kluwer (1995).

[3] P. C. Stamp, Nature 383, 125 (1996).

[4] R. Sessoli, D. Gatteschi, A. Caneschi and M. A. Novak, Nature 365,141 (1993).

[5] W. Wernsdorfer, E. Bonet Orozco, K. Hasselbach, A. Benoit, B. Barbara, N. Demoncy, A. Loiseau, and D. Mailly, Phys. Rev. Lett. 78, 1791 (1997).

[6] M. A. Novak and R. Sessoli in ref. [2], pp 171.

[7] J. R. Friedman, M. P. Sarachik, J. Tejada and R. Ziolo, Phys. Rev. Lett. 76, 20, 1996; see also L. Thomas, F. Lionti, R. Ballou, D. Gateschi, R. Sessoli and B. Barbara, Nature 145, 83,(1996).

[8] A.L.Barra, D. Gatteschi and R. Sessoli, Phys. Rev B**56**, 1 (1997).

[9] A.M. Gomes, M.A.Novak, R. Sessoli, A. Caneschi, D. Gatteschi, Phys. Rev. B **57**, 5021 (1998)

[10] M.Hennion, L.Pardi, I.Mirebeau, E.Suard, R.Sessoli and A.Caneschi, Phys. Rev.B **56**, 8819 (1997).

[11] T. Lis, Acta Cryst. B36, 2042,(1980).

[12] Caneschi A., Gateschi D. and Sessoli R. (1991), J. Am. Chem. Soc. 113, 5873.

[13] Villain J., Hartman-Boutron F., Sessoli R. and Rettori A.,Europhys. Lett, 27 (2), 159 (1994).

[14] C. Paulsen and J.G.Park in ref.[2], pp 189.

[15] J.M Hernandez, X.X.Zhang, F.Luis, J.Bartolomé, J.Tejada, and R.Ziolo, Europhys. Lett.35, 301 (1996).

[16] N. V. Prokof'ev and P. C. Stamp, J. Low Temp. Phys. 104, 143 (1996)

[17] M. A. Novak, A. M. Gomes and R. E. Rapp, Jour. Appl.Phys.83, 1 (1998)

[18] F. Hartman-Boutron, P Politi and J.Villain , Int. Jour. Mod. Phys. B 10, 2577 (1996).

Materials Science Forum Vols. 302-303 (1999) pp. 344-348

Spin-Glass Phase in Antiferromagnetic CrVMn Alloy

A.J.A. de Oliveira[1], W.A. Ortiz[1] and P.C. de Camargo[2]

[1] Grupo de Supercondutividade e Magnetismo, Departamento de Física, Universidade Federal de São Carlos, CP 676, 13565-905, São Carlos, SP, Brazil

[2] Laboratório de Materiais, DF, Universidade Federal do Paraná, CP 19081, 81531-990, Curitiba, PR, Brazil

Keywords: Spin-Glass, Spin-Density-Wave, Antiferromagnetism, Magnetic Susceptibility

Abstract

The DC magnetic susceptibility of $Cr_{0.98}V_{0.01}Mn_{0.01}$ is found to show a spin-glass behavior at low temperatures. The system also presents another phase above the Néel temperature, associated to a local spin-density-wave state. Effects of the magnetic field on both phases are studied, indicating a strong dependence on magnetic history.

1 - Introduction

Chromium and its alloys are well known to exhibit a spin-density-wave (SDW) itinerant magnetism [1], characterized by a wave-vector Q which is determined by the Fermi surface properties. The introduction of impurities, as transition metals, causes important changes in the density of states, energy gap and Fermi level, observed through magnetic, elastic, thermal and transport properties of the material. For alloys of the type CrV, CrMn, $(Cr_{1-x}Fe_x)_{1-y}(V,Mn)_y$ and $(Cr_{1-x}Si_x)_{1-y}(V,Mn)_y$, it has been found that the rigid band model can be successfully used to describe the Néel temperature (T_N) dependence on the ratio electron/atom. For those compounds, T_N varies at about 80 K/at%V (T_N decreases) and 50 K/at%Mn (T_N increases) [1].

Recently we have shown that for $Cr_{1-x}V_x$, with x = 0.001 and 0.0067, a Curie-Weiss (CW) behavior is observed above a characteristic temperature $T_{CW} > T_N$ [2]. This behavior is strongly dependent on applied magnetic field (H) and V concentration [2-5]. However, this feature is not observed either in pure Cr, because of the nature of its itinerant magnetism, or in bulk V, which exhibits electronic paramagnetism. The origin of local magnetic moments in CrV alloys is associated with the formation of local spin-density-waves around nonmagnetic impurities [2]. On the other hand, Galkin et al. [6] identified a new type of spin-glass (SG) in CrMn and CrSiMn alloys, with a characteristic freezing temperature almost independent on Mn concentration. This type of SG behavior seen in CrMn alloys differs fundamentally from a canonical spin-glass in that moments on the antiferromagnetic host lattice become frustrated at a constant characteristic pinning temperature, whereas in the latter case the moments on the impurity atoms are frustrated until they freeze into a spin-glass at a freezing temperature determined by their separation. However, Cr and CrV alloys do not exhibit this behavior, even at low magnetic fields [7,8].

In this work we present the DC magnetic susceptibility χ_{DC} data obtained for a sample of $Cr_{0.98}V_{0.01}Mn_{0.01}$. This sample exhibits not only a SG behavior at low temperatures but also another magnetic phase at higher temperatures, which is associated to a local spin-density-wave (LSDW).

2 - Experimental Procedure

Our sample with nominal composition of $Cr_{0.98}V_{0.01}Mn_{0.01}$ was prepared by arc melting in argon atmosphere. Measurements of magnetic susceptibility as a function of temperature, $\chi_{DC}(T)$ =

M(T)/H, magnetization as a function of magnetic field at a fixed temperature, M(H), and magnetization as a function of time, M(t), were taken with a SQUID magnetometer, MPMS-5S, supplied by Quantum Design. Each point of $\chi_{DC}(T)$ resulted from two scans over a length of 6 cm, with 40 point readings per scan. The standard deviation in general was very small, roughly of the same size of the plotted symbols in the figures. In each run of $\chi_{DC}(T)$ the sample temperature was lowered from 300 K to 5 K without magnetic field. After temperature stabilization the magnetic field was applied and the temperature increased at a rate of 2 K/min up to 300 K (ZFC). After that, the sample was measured on cooling down to 5 K (FC). The low-field AC-susceptibility of the samples was also measured using the SQUID magnetometer with an excitation field h_{AC} = 3.0 Oe, the imaginary part of which, being proportional to the electrical resistivity [10], was used to establish the Néel temperature of the sample.

3 - Results and Discussion

Figure 1 shows curves of $\chi_{DC}(T)$ with different magnetic fields: (a) H = 10 Oe; (b) H= 100 Oe; (c) H= 1 kOe; and (d) H= 5 kOe, respectively. We see two characteristic features of a SG behavior: the peak in the temperature dependence of the magnetic susceptibility in the ZFC run, and the hysteresis in relation to the FC measurement. In a typical metallic spin-glass, in which the RKKY interaction between magnetic impurity atoms is responsible for their frustration, the peak in the ZFC curve is identified as the freezing temperature (T_f). Above this temperature, $\chi_{DC}(T)$ obeys a Curie-Weiss law that describes the thermal fluctuations of the moments on the impurity atom [8]. Table 1 presents the values of T_f for different applied fields. This strong dependence is typical of spin-glass systems.

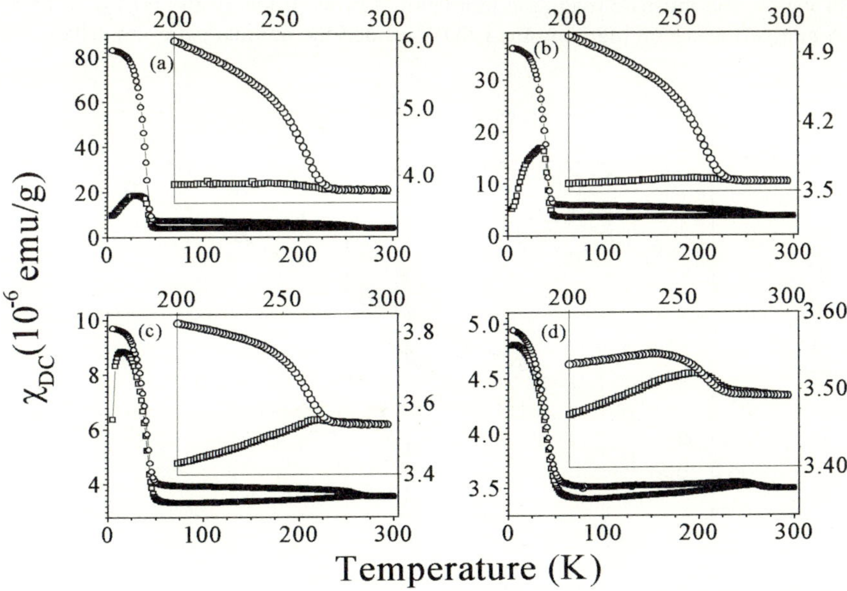

Figure 1. Temperature dependence of the magnetic susceptibility of the ternary alloy $Cr_{0.98}V_{0.01}Mn_{0.01}$, measured in different fields: (a) H = 10 Oe; (b) H = 100 Oe; (c) H = 1 kOe and (d) H = 5 kOe, after zero-field cooling (ZFC) and warming at the same field (FC). The freezing temperature (T_f) is shown in Table 1. The irreversibility temperature (T_i) of this sample is T = 275 K.

The Néel temperature has been determined as the inflection point of the resistivity curve that occurs just below its minimum, as demonstrated in a previous paper [2], using the imaginary component χ'' of the AC magnetic susceptibility. For the $Cr_{0.98}V_{0.01}Mn_{0.01}$ sample, we have determined $T_N = 235$ K. This temperature does not depend on the magnetic field, as is also the case for pure Cr[1]. We have observed that this temperature corresponds to an inflection point in $\chi_{DC}(T)$, in contrast to what is suggested by several authors, who associate T_N with the peak (or maximum) in the ZFC run [1].

$T_f \pm 1.0$ (K)	41.6	40.0	36.8	31.2	25.2	15.7	8.4
H (Oe)	10	50	100	500	1000	5000	10000

Table 1 – Values of the freezing temperature for different values of the of applied magnetic field, as determined from the maximum of ZFC curves.

The ZFC curve for this sample features a large thermal hysteresis when compared to the FC run. The characteristic temperature T_i at which the ZFC curve diverges from the FC is 275 K for this sample, being thus above T_N. We have also observed that this temperature does not depend on magnetic fields up to H = 10 kOe. Galkin and co-workers measured $\chi_{DC}(T)$ for $(Cr_{1-x}Co_x)_{1-y}V_y$, x = 0.032 and $0.001 \le y \le 0.01$ [11] and for $(Cr_{1-x}Fe_x)_{1-y}V_y$, x = 0.027 and $0.007 \le y \le 0.05$ [12], suggesting that T_i is associated with a transition to a new magnetic phase, identified as a local-spin-density-wave phase appearing above the Néel temperature. This phenomenon is analogous to the appearance of a Curie-Weiss behavior in CrV alloys above T_{CW} [2]. Nevertheless, in CrV samples the hysteresis is not observed, perhaps indicating that only V impurities are insufficient to produce this behavior. Probably, the intrinsic magnetic moments of other atoms of the host (Co, Fe or Mn) matrix are much more effective on inducing the LSDW around the nonmagnetic impurities of V.

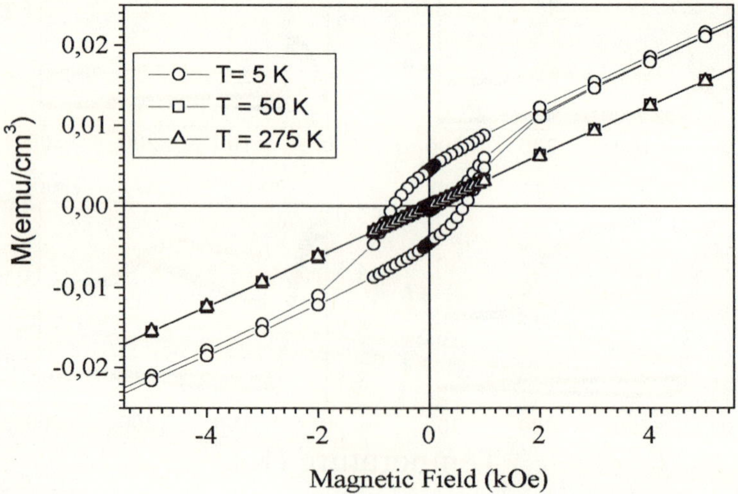

Figure 2. Field dependence of the magnetization of the ternary alloy $Cr_{0.98}V_{0.01}Mn_{0.01}$, measured at different temperatures: (o) T = 5 K; (□) T = 50 K; (Δ) T = 275 K. Notice that a hysteresis loop is present only for T = 5 K.

In the range 50 K ≤ T ≤ T_i we observed the evolution of $\chi_{DC}(T)$ with different magnetic fields. In low fields, the FC data show that the value of $\chi_{DC}(T)$ increases with decreasing temperatures. For higher fields (H= 5 kOe), $\chi_{DC}(T)$ diminishes with decreasing temperatures. A possible explanation for this behavior is that the magnetic moments associated with the local spin-density-wave reach a complete saturation, while the antiferromagnetic phase (SDW) does not, the magnetic field necessary to saturate it being of the order of tens of kOe. In this scenario, a FC measurement would reveal LSDW states brought into the antiferromagnetic state by action of the applied field, and they would then coexist with the SDW state of the host.

The field dependence of the magnetization, M(H), is illustrated in Figure 2 for T = 5, 50 and 275 K. It is strongly non-linear and largely hysteretic at lower temperatures, as expected in the SG phase. At T = 5 K, the coercive field was determined to be H_c = 635 Oe and the remnant magnetization in zero field, $M_r = 4.53 \times 10^{-3}$ emu/cm^3. Measurements made up to H = 50 kOe do not show saturation of the magnetization at any temperature of the experimental range, as in pure Cr, which has a linear magnetic response up to H = 120 kOe, due to its antiferromagnetic behavior [7]. In spite of the appreciable thermomagnetic hysteresis exhibited for measurements in the whole range of fields investigated, runs of M(H) at temperatures above T_f (50 and 275 K) are linear and magnetically reversible. We conclude thus that $T_f(H)$ determined in this work is, in fact, the upper boundary of a spin-glass phase.

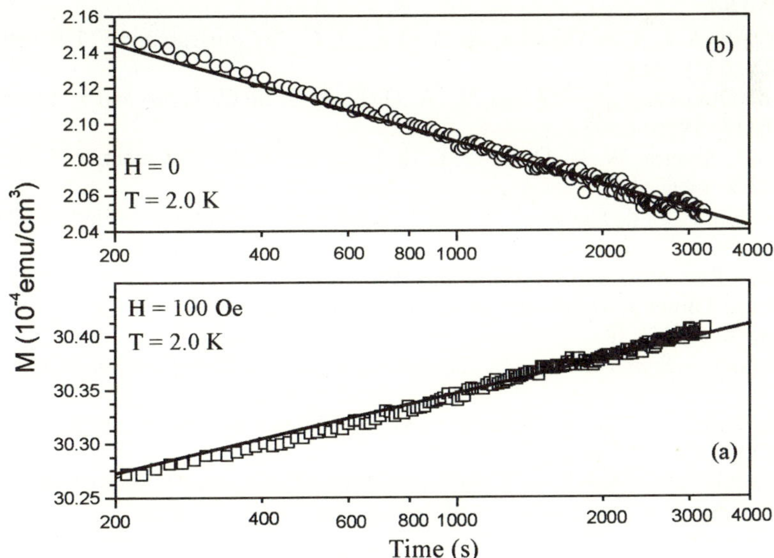

Figure 3. Time relaxation of the magnetization of the ternary alloy $Cr_{0.98}V_{0.01}Mn_{0.01}$, measured at T = 2 K: (a) H = 100 Oe after ZFC; (b) H = 0 after (a). The time axis is set to logarithmic scale; the straight line indicates a standard logarithmic relaxation, common in spin-glasses.

Figure 3 shows the time relaxation of the magnetization, measured at T = 2 K. Curve (a) is a plot of M(t) after ZFC the sample and application of H = 100 Oe. As seen, M(t) features a logarithmic increase, indicating that the system is in a metastable state and the magnetic domains progressively align with the field. Curve (b) is taken after curve (a) is over and the field turned to zero. The expected logarithmic decrease is clearly seen, as the system evolves from one metastable state to another. Equivalent runs taken at 50 and 275 K do not exhibit the metastable response.

In conclusion, we have shown experimental evidences that, as in CrMn alloys, $Cr_{0.98}V_{0.01}Mn_{0.01}$ present a spin-glass behavior below a characteristic temperature T_f, which is strongly field-dependent. Besides, a local spin-density wave phase is also observed, as previously reported in $(Cr_{1-x}Co_x)_{1-y}V_y$ and $(Cr_{1-x}Fe_x)_{1-y}V_y$. Further studies are in progress, in which we vary the relative concentration of V and Mn, to better understand the formation of both magnetic phases reported in this work.

Acknowledgements

This work was performed with partial support by brazilian agencies Fundação de Amparo à Pesquisa do Estado de São Paulo (FAPESP), proc. 96/12781-0, Conselho Nacional de Desenvolvimento Científico e Tecnológico (CNPq) proc. 520844/97-1, and Financiadora de Estudos e Projetos (FINEP) procs. 77.97.0074.00 and 41.96.0858.00.

References

[1] – E. Fawcett, H. L. Alberts, V.Yu. Galkin, D. R. Noakes and J. V. Yakhmi, Rev. Mod. Phys. 66 (1994), 25.

[2] – A. J. A. de Oliveira, W. A. Ortiz, P. C. de Camargo and V. Yu. Galkin, J. Magn. Magn. Mat. 152 (1996) 86;

[3] – P. Hill, N. Ali, A. J. A. de Oliveira, W. A. Ortiz, P. C. de Camargo and E. Fawcett, J. of Phys.: Cond. Mat. 6 (1994) 1761

[4] – A. J. A. de Oliveira, O. F. de Lima, W. A. Ortiz, P. C. de Camargo and E. Fawcett, J. Phys.: Cond. Mat. 8 (1996) L403;

[5] – A. J. A. de Oliveira, W. A. Ortiz, O. F. de Lima and P. C. de Camargo, J. Appl. Phys. 81 (1997) 4209

[6] – V. Yu. Galkin, P. C. de Camargo, N. Ali, J. Schaf and E. Fawcett – J. Phys.: Condens. Matter 7 (1995) L649

[7] – E. Fawcett – Rev. Mod. Phys. 60 (1988) p.209

[8] – A. J. A. de Oliveira, O. F. de Lima, W. A. Ortiz and P. C. de Camargo – Solid State Comumm. 96 (1995) 383

[9] – J. A. Mydosh – *Spin Glasses: An Experimental Introduction (London: Taylor & Francis)* (1993)

[10] – C. C. de Faria, A. J. A. de Oliveira, F. M. A. Moreira and W. A. Ortiz – IEEE Trans. on Magn.31 (1995) 3403

[11] – V. Yu. Galkin, W. A. Ortiz and E. Fawcett – J. Phys.: Condens. Matter 9 (1997) L577

[12] – V. Yu. Galkin, W. A. Ortiz and E. Fawcett – J. Appl. Phys. (in press)

*Corresponding author. e-mail: adilson@power.ufscar.br; fax: +55-16-2614835

Materials Science Forum Vols. 302-303 (1999) pp. 349-352

Magnetic Properties of Cu(L-aspartato)(H_2O)$_2$: An Alternating Linear Chain Antiferromagnet

R. Calvo[1], M.C.G. Passeggi[1], N.O. Moreno[2], D. Rao[2], G.E. Barberis[2], A.B. Cháves[3] and B.C.M. Torres[3]

[1] INTEC and Facultad de Bioquímica y Ciencias Biológicas (CONICET-UNL), Santa Fe, Argentina

[2] Instituto de Física "Gleb Wataghin", Universidade Estadual de Campinas, Campinas, SP, Brazil

[3] Instituto de Física, Universidade Federal do Rio de Janeiro, Rio de Janeiro, Brazil

Keywords: Mixed-Ligand Complexes, L-aspartic Acid, Cooper (II) Ion

Abstract

Specific heat [c(T)] and magnetization data [M(B,T)] were obtained in the copper complex of the amino acid L-aspartic acid [called Cu(L-asp)]. c(T) was measured between 3 and 15 K, where it is largest. M(B,T) was measured in the temperature range between 1.8 and 300 K for magnetic fields up to 9 Tesla. The c(T) and the susceptibility χ(T) curves show peaks at 3.95 K and 6.9 K, respectively, and approach zero at lower and higher temperatures. The data are well fitted proposing for Cu(L-asp) a behavior characteristic of an alternating linear chain antiferromagnet, where each S=1/2 copper ion is coupled with a magnitude J_o with one copper neighbor in the chain, and with αJ_o with the other. We evaluate an antiferromagnetic coupling parameter J_o/k = –5.3 K and α = 0.29, and g = 2.165±0.001 for the average g-value. A ferromagnetic coupling with a magnitude J'/k = 0.7 K between neighbor chains, is also estimated using a mean field approach. The alternating chain behavior indicates that, besides the exchange interactions transmitted through the σ-skeleton of the amino acid, interactions paths involving H-bonds are also important.

1. Introduction

Recently we reported the crystal structure, magnetic susceptibility and EPR measurements for the copper complex of L-aspartic acid,[1] Cu(L-aspartato)(H_2O)$_2$, [Cu($CO_2NH_2CHCH_2CO_2$)(H_2O)$_2$] (hereafter called Cu(L-asp)). The magnetic susceptibility [$\chi(T)$] versus temperature [T] data[1] displayed a broad maximum at 6.78 K, decreasing toward lower and higher temperatures, as expected for a 1D spin chain. The temperature of the maximum was an indication of short-range magnetic order of the spins in the chain. These values were compared to the predictions of Bonner and Fisher (BF)[2] for uniform antiferromagnetic chains. This model reproduced the general features of $\chi(T)$, but failed to agree with the magnitude and shape of the data around the maximum, and at low T. To account for the discrepancies we proposed[1] an additional interchain interaction J'. Thus, using the mean field approximation of Hatfield et *al.*,[3] we obtained J_o/k = –5.3 K and J'/k = 2.2 K. The value of J' was not much smaller than the intrachain antiferromagnetic interaction J_o and raises doubts about the interpretation of the results. Thus lower temperature measurements were required to progress in the understanding of the magnetic behavior of Cu(L-asp). An interesting result in Ref. 1 was the observation of a relatively large coupling J_o, in a system where the copper ions are connected by long σ-bonds which involve five diamagnetic atoms, and by hydrogen bonds. In this paper we report new and more detailed magnetic data for Cu(L-asp). Specific heat [c(T)] measurements have been performed in the temperature range $3 \leq T \leq 15$ K. Magnetization data have been obtained in the temperature range $1.8 \leq T \leq 300$ K with magnetic fields of up to 9 T. The temperature dependence of c(T) resembles that of a spin chain, but its magnitude at the peak is much higher than that expected for a uniform chain. The observed $\chi(T)$ tends to vanish at low temperatures, a result not seen in the previous measurements,[1] performed at higher T. We show here that the new c(T), $\chi(T)$ and M(B,T) data for Cu(L-asp) can be interpreted in terms of spin chains with alternating exchange interactions.

2. Experimental Results and Discussion

Cu(L-asp) was prepared as in Ref. 1. The specific heat data c(T)/R obtained in the temperature range $3 \leq T \leq 15$ K is shown in Fig 1. The observed maximum value is $c_{max}/R=0.473\pm0.004$ at 3.95 K, much larger than the value $c_{max}/R=0.35$ predicted for a uniform Heisenberg spin chain.[2] The molar magnetic susceptibility $\chi(T)$ of Cu(L-asp) was measured in an applied field of B = 0.1 T and the results for $1.8 \leq T \leq 100$ K are shown in Figure 2a. The magnetization measured at fixed temperatures (T = 2, 5, 7 and 10 K) and fields up to 9 T is displayed in Fig 2b.

In Cu(L-asp) the aspartic acid molecule binds one Cu(II) ion through the carboxylate oxygen and the amino nitrogen of the glycine ring, and to another copper through one of the oxygens of the other carboxylate group, displaying polymeric chains asp – Cu – asp - Cu – asp – Cu – asp along the ***c*** axis. Within each chain, every copper atom and L-aspartic molecule is obtained by simple lattice translations along ***c***. This may indicate that the superexchange pathways within the chains are provided by the σ-skeleton of the amino acid molecule, which involves five diamagnetic atoms and suggests that the magnetic behavior of Cu(L-asp) would be that of a uniform spin chain with a nearest neighbors coupling J_o. The results of BF[2] for the specific heat and the magnetic susceptibility are shown as dashed lines in Figs 1 and 2a. In both cases the maximum observed is much larger than the prediction. Also, in the case of the susceptibility the experiments show that it approach zero for low T, while the uniform chain model gives a minimum finite value.[2] These discrepancies between the experimental data for c(T) and $\chi(T)$ and the expectations are surprising in view of the crystallographic structure for Cu(L-asp) described above. In order to explain these experimental results we propose for Cu(L-asp) the behavior of an infinite alternating Heisenberg spin chain. The Hamiltonian corresponding to such system for n spins (n even) is given by:

$$\mathbf{H}=|J_o|\widetilde{\mathbf{H}}, \text{ with } \widetilde{\mathbf{H}}=\widetilde{\mathbf{H}}_{ex}+\widetilde{\mathbf{H}}_z=-2\frac{J_o}{|J_o|}\sum_i(S_{2i-1}.S_{2i}+\alpha S_{2i}.S_{2i+1})+\sum_i yS_{iz} \tag{1}$$

(i=1,.. n/2) where $\widetilde{\mathbf{H}}_{ex}$ and $\widetilde{\mathbf{H}}_z$ represent the exchange and Zeeman interactions, respectively. In Eq. 1 $y=g\mu_B B/|J_o|$, and B is the applied magnetic field. The thermodynamic properties of this system for $0 \leq \alpha \leq 1$, and $n\leq10$ were calculated by Duffy and Barr,[4] and extrapolated to chains with $n\to\infty$. Results for chains with n=12 and 14 spins have also been reported.[5] From $\widetilde{\mathbf{H}}$, the molar specific heat c(T) is obtained as:

$$c(T)=\frac{Nk}{x^2}\left[\left\langle\widetilde{\mathbf{H}}^2\right\rangle-\left\langle\widetilde{\mathbf{H}}\right\rangle^2\right] \quad . \tag{2}$$

Also, the molar magnetic susceptibility $\chi(T)$ and the molar magnetization M are obtained from:

$$\frac{\chi(T)\,|J_o|}{Ng^2\mu_B^2}=\frac{1}{x}\left\langle S_z^2\right\rangle_{y=0}, \quad \text{and} \quad M(B,T)=-\frac{Ng^2\mu_B^2 B}{y|J_o|}\left\langle S_z\right\rangle \tag{3}$$

where $x = kT/|J_o|$ and N is the Avogadro's number. The averages $\langle A \rangle$ of an operator A are calculated per ion in the n-spins chain after solving the Hamiltonian $\widetilde{\mathbf{H}}$. This was diagonalized for n=14 spins rings and used to calculate c(T), χ(T) and M(B,T). A least squares fit allowed to obtain g, J_o, α, and J' given in Figs 1-2. The agreement between the predictions of the model and the experimental results is very good. The small difference between J_o obtained from $c(T)$ and from $\chi(T)$ and $M(B,T)$ is attributed to the neglect of the interchain interactions in the calculation of $c(T)$. The lead us to conclude that the σ-bonds through the skeleton of the amino acid cannot be the only contribution to J_o, and that other important exchange paths involving H-bonds should also be considered.

References

1.- R. Calvo, C. A. Steren, O. E. Piro, T. Rojo, F. J. Zuñiga, and E. E. Castellano, *Inorg. Chem.* **32**, 6016 (1993).

2.- J. C. Bonner and M. E. Fisher, *Phys. Rev.* **135**, A640 (1964).

3.- W. E. Hatfield, R. R. Weller, J. W. Hall, *Inorg. Chem.* **19**, 3825 (1980).

4.- W. Duffy Jr. and K. P. Barr, *Phys. Rev.* **165**, 647 (1968).

5.- K. M. Diederix, J.P. Groen, L.S.J.M. Henkens, T.O. Klaassen and N.J. Poulis, *Physica* **93**B, 99 (1978); *Physica* **94**B, 9 (1978).

Figure Captions

Fig. 1. Molar specific heat data as a function of T for Cu(L-asp). The solid line gives the best fit to the data given by the alternating spin chain model. The values of α and J_o calculated from the data with this model are included. The dotted line corresponds to the prediction of the BF model[2] for a uniform chain.

Fig. 2. (a) Molar magnetic susceptibility data of Cu(L-asp). The solid line is the prediction of the alternating spin chain model including interchain couplings. The dashed line corresponds to the BF model.[2] (b) Molar magnetization observed at 2, 5, 7 and 10 K as a function of the magnetic field. The solid lines give the fitting of the data with the alternating chain model when a interchain interaction J' is allowed.

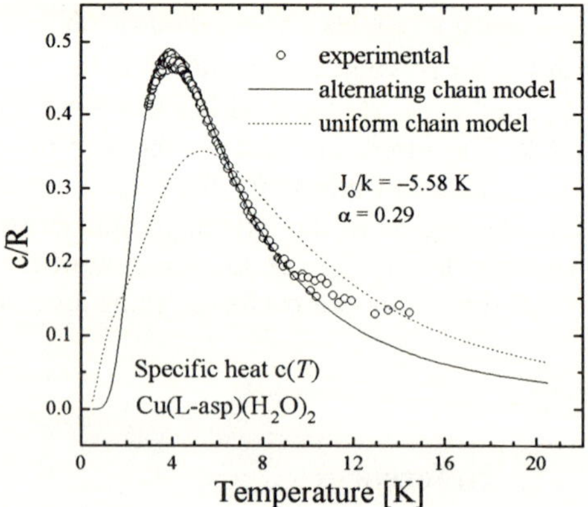

experimental
alternating chain model
uniform chain model
$J_o/k = -5.58$ K
$\alpha = 0.29$
c/R
Specific heat c(T)
Cu(L-asp)(H2O)2
Temperature [K]

Fig. 1

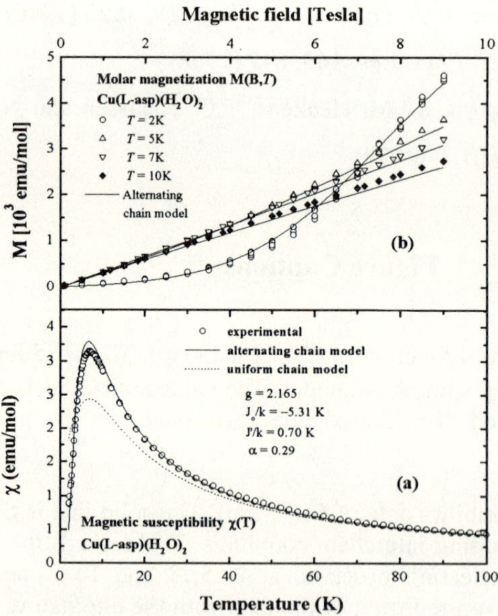

Magnetic field [Tesla]
Molar magnetization M(B,T)
Cu(L-asp)(H2O)2
T = 2K
T = 5K
T = 7K
T = 10K
Alternating chain model
M [10^3 emu/mol]
(b)
experimental
alternating chain model
uniform chain model
g = 2.165
$J_o/k = -5.31$ K
$J'/k = 0.70$ K
$\alpha = 0.29$
χ (emu/mol)
(a)
Magnetic susceptibility χ(T)
Cu(L-asp)(H2O)2
Temperature (K)

Fig. 2

Materials Science Forum Vols. 302-303 (1999) pp. 353-357

Frustrated Magnetization in $Pr_xLa_{1-x}BaCuO_5Fe$

W.A. Ortiz[1*], H. Pardo[2], F.M. Araújo-Moreira[1,3], E. Quagliata[4], K. Prassides[5], R. Mariezcurrena[2] and A.W. Mombrú[2]

[1] Departamento de Física, Universidade Federal de São Carlos, Caixa Postal 676, 13565-905 São Carlos, SP Brazil

[2] Laboratorio de Cristalografia, Cátedra de Física, Facultad de Química, Universidad de la República, P.O. Box 1157, Montevideo, Uruguay

[3] Center for Superconductivity Research, Dept. of Physics, University of Maryland, College Park, 20740-411 MD, USA

[4] Instituto de Física, Facultad de Ingeniería, Univ. de la República, Montevideo, Uruguay

[5] School of Chemistry, Physics and Environmental Sciences, University of Sussex, Falmer, Brighton, BN1 9QJ, United Kingdom

Keywords: Spin Glass, Antiferromagnetism

Abstract *The crystal structure of the system $Pr_xLa_{1-x}BaCuO_5Fe$ has been recently reviewed. The magnetic structure of samples with x = 0 and x = 1 is mainly due to effective local moments of iron and copper. In Pr-rich samples, Fe ions occupy two non-equivalent positions, making it substantially plausible that two or more magnetic subsets might coexist in the system. This contribution presents magnetization studies on five samples of the $Pr_xLa_{1-x}BaCuO_5Fe$ system (x = 0.0, 0.2, 0.5, 0.7 and 1.0). All samples exhibit a strong irreversible behavior between zero-field-cooled and field-cooled procedures below a certain irreversibility temperature T_i. Above T_i, both branches are coincident and well described by a Curie-Weiss fitting. Decreasing the temperature below T_i, the zero-field-cooled response increases less than the field-cooled curve, indicating some degree of frustrated antiferromagnetic couplings.*

In a recent paper [1], we have reviewed the crystal structure of the system $Pr_xLa_{1-x}BaCuO_5Fe$, using X-ray and neutron diffraction analysis and Mössbauer spectroscopy. The system was shown to have the same tetragonal structure exhibited by other members of the rare-earth series [2-8], $REBaCuFeO_{5+\delta}$, RE = Y, Nd, Sm, Gd, Dy, Tm, Lu. The magnetic structure of samples with x = 0 and x = 1, as revealed by neutron scattering at 10 K, is mainly due to effective local moments of iron and copper. The study has also revealed that, in Pr-rich samples, Fe ions occupy two non-equivalent positions. It is thus conceivable that two or more magnetic subsets might coexist in the system, encouraging a thorough study of its magnetic properties.

In this article we present magnetization studies on five samples of the $Pr_xLa_{1-x}BaCuO_5Fe$ family (x = 0.0, 0.2, 0.5, 0.7 and 1.0). Specimens were prepared annealing in air at 1000°C, as previously reported [4,9,10]. Quality and structural characteristics of the studied samples were confirmed by Rietveld analysis using FULLPROF [5] to x-ray powder diffraction data, obtained using CuKα radiation in a Seifert Scintag PAD-II diffractometer, with the scanning angle 2θ ranging from 10° to 110°. All compositions prepared exhibit a major $REBaCuFeO_5$-type phase and impurity phases identified as RE_2BaCuO_5 in an amount always less than 2%. The nonstoichiometric oxygen concentration was determined by back titration with $FeCl_2$ and $K_2Cr_2O_7$ [6,7] and yielded values that ranged from 0.03 to 0.04, the error being less than ± 0.01. The evolution of the lattice constants with x provides evidence that the system forms a solid solution. Magnetization measurements were carried out in a Quantum Design MPMS-5 SQUID magnetometer, in the temperature (T) range 1.8 - 400 K, with applied magnetic fields (H) up to 5 T. The DC-susceptibility (χ_{DC}), presented in emu/g units, is calculated as χ_{DC} = moment/(field*mass). Magnetic measurements were taken in two different manners:

(i) ZFC - warming after zero-field cooling the sample;

(ii) FCC - measuring while cooling in an applied magnetic field, H.

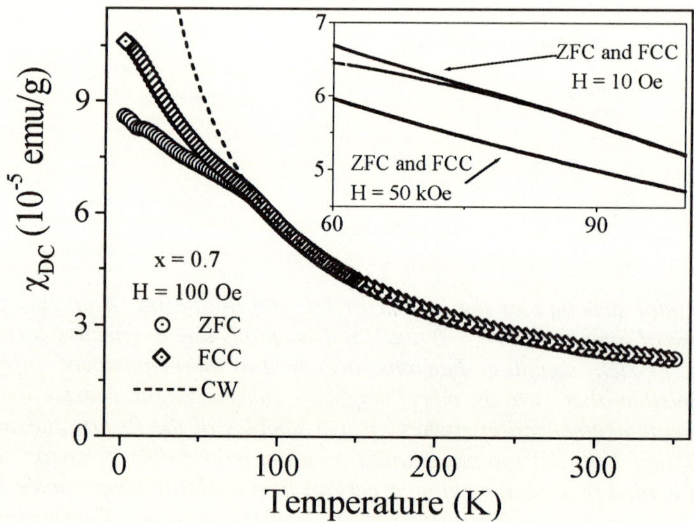

***Figure 1 -** ZFC, FCC and Curie-Weiss (CW) fitting of $\chi_{DC}(T)$ for x = 0.7 at 100 Oe. Inset: data taken at H = 10 Oe and H = 50 kOe; hysteresis is exhibited for low fields and disappears as H increases.*

Fig. 1 shows $\chi_{DC}(T)$ at H = 100 Oe for a sample with x = 0.7. The irreversible behavior among ZFC and FCC curves is clear, taking place at the irreversibility temperature T_i = 80 K for this concentration. Also shown in the figure is the Curie-Weiss (CW) fitting for the higher temperature portion of the magnetic response, revealing that the ordering subsystem coexists with a paramagnetic background. The inset is a close up of the irreversibility region: at small fields (e.g., 10 Oe, upper curve), T_i is present, whereas the response is completely reversible for higher fields as, for example, for the 50 kOe curve shown.

Fig. 2 presents ZFC and FCC curves, taken at H = 100 Oe, for samples with x = 0.0, 0.2, 0.5 and 1.0. As for x = 0.0 (Fig. 1), these samples also exhibit a strong irreversible behavior below T_i. Above it, both branches are coincident and well described by a Curie-Weiss fitting. In the hysteretic portion of the curve, the ZFC response is always below the FCC curve, indicating some degree of frustrated antiferromagnetic couplings.

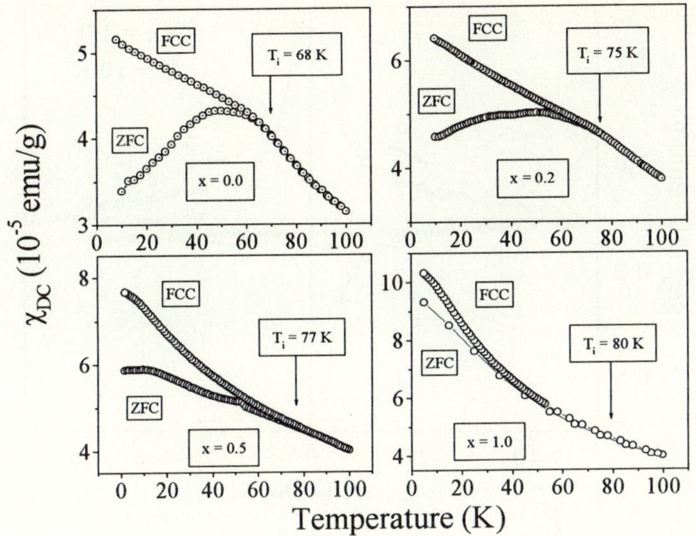

Figure 2 - *ZFC and FCC runs of $\chi_{DC}(T)$ for samples with x = 0.0, 0.2, 0.5 and 1.0, measured at H = 100 Oe. The irreversible temperature T_i is shown for each concentration.*

To further investigate the characteristics of this frustrated behavior, we have measured $\chi_{DC}(H)$ curves for all samples, above and below T_i. Without exceptions, no hysteresis is observed above T_i, while characteristic loops do appear for temperatures below T_i. The field dependence of the magnetization is very close to linear, being far from saturation even at 50 kOe. The hysteretic behavior of M(H) might be seem through a large magnification. Alternatively, one can plot the susceptibility $\chi_{DC}(H)$ to show the existence of a magnetic loop. Fig. 3 shows a typical hysteresis curve for the composition x = 1.0, at T = 10 K. Branch 1 is the virgin curve, starting from values close to $8X10^{-5}$ emu/g, decreasing and increasing again to approach that same level. On decreasing H (branch 2), $\chi_{DC}(H)$ increases toward infinity, a signature of remanence at zero magnetic field. The other two branches (3 and 4) close the cycle, completing a symmetric picture.

Relaxation curves at low temperatures have also been taken. Two experimental procedures were used:

(i) ZFC-relaxation – after cooling the sample, the field was turned on for five minutes; and then off again, measurements being taken at a fixed time interval;

(ii) FC-relaxation – the sample was cooled down in a certain field, which was then turned off for the measurements versus time.

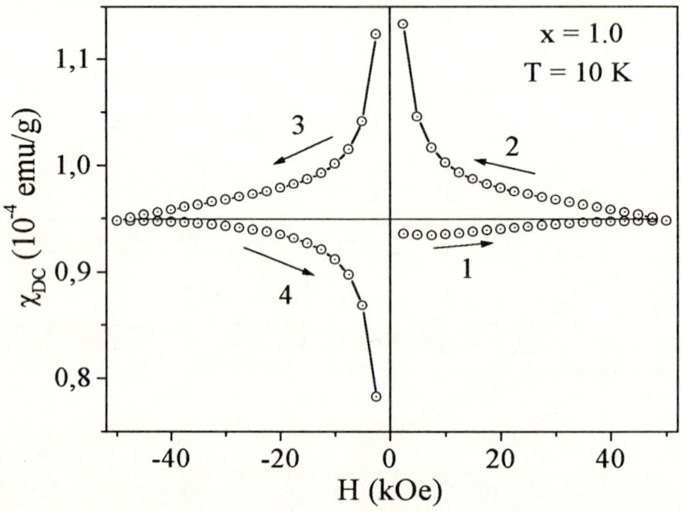

Figure 3 *- Hysteresis curve for the composition x = 1.0, at T = 10 K.*

Fig. 4 presents ZFC-relaxation and FC-relaxation curves for the composition x = 0.0, at T = 10 K, with H = 50 kOe. The time dependence of the remnant magnetization is well described by a stretched exponential form [11]. Similar curves have been taken for the other samples studied.

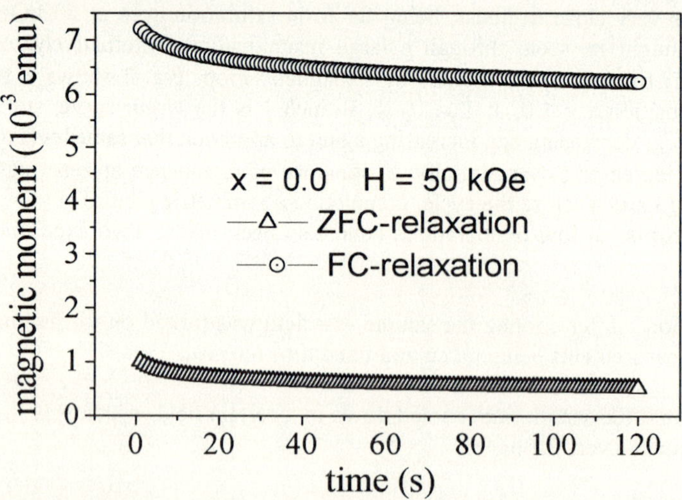

Figure 4 - *ZFC-relaxation and FC-relaxation curves for the composition x = 0.0, at T = 10 K, with H = 50 kOe.*

In summing up we stress that, from all the experimental evidence presented, every sample belonging to the system $Pr_xLa_{1-x}BaCuO_5Fe$ exhibits some characteristic signatures of spin-glasses. In addition to that, an ordinary Curie-Weiss paramagnetic background is always present. Further work is in progress to certify these and other spin-glass features, including the frequency behavior of the AC-susceptibility and some scaling properties.

Acknowledgments

This work was supported by CONICYT, PEDECIBA and CSIC (Uruguay), FAPESP, FINEP and CNPq (Brazil), SERC and the British Council (UK). Authors wish to thank NIST for beam time and Dr. Ross Erwin for his help in collecting neutron data. We are also indebted to Mr. Leopoldo Suescun for his help in data format conversion.

References

[1] H. Pardo, W. A. Ortiz, F. M. Araújo-Moreira, E. Quagliata, K. Prassides, R. Mariezcurrena and A. W. Mombrú, Phys. Rev. B (submitted)

[2] L. Er-Rakho, C. Michel, P. Lacorre and B. Raveau, J. Solid State Chem., **73** (1988), p. 531.

[3] C. Meyer, F. Hartmann-Boutron, Y. Gros and P. Strobel, Solid State Comm., **76** (1990), p. 163.

[4] A. W. Mombrú, C. Christides, A. Lappas, K. Prassides, M. Pissas, C. Mitros and D. Niarchos, Inorg. Chem., **33** (1994) p.1255.

[5] M. Pissas, C. Mitros, G. Kallias, V. Psycharis, D. Niarchos, A. Simopoulos, A. Kostikas, C. Christides and K. Prassides, Phys. C, **185** (1991) p.553.

[6] M. Pissas, C. Mitros, G. Kallias, V. Psycharis, A. Simopoulos, A. Kostikas and D. Niarchos, Phys. C, **192** (1992) p.35.

[7] M. Ruiz-Aragon, U. Amador, E. Moran and N.H. Andersen, Phys. C, **235-240** (1994) p.1609.

[8] M. Pissas, G. Kallias, V. Psycharis, H. Gamari-Seale, D. Niarchos and A. Simopoulos, Phys. Rev. B, **55** (1997) p.397.

[9] V. Caignaert, I. Mirebeau, F. Bouree, N. Nguyen, A. Ducouret, J. M. Greneche and B. Raveau, Journal of Solid State Chem., **114** (1995) p.24.

[10] A. W. Mombrú, K. Prassides, C. Christides, M. Pissas, C. Mitros, D. Niarchos and R. Erwin, submitted to Journal of Phys.: Cond. Matter.

[11] J. A. Mydosh, in *Spin glasses: an experimental introduction*; ed. by Taylor and Francis, (1993), p. 96.

*Corresponding author - e-mail: wortiz@power.ufscar.br - fax: +55.16.2614835.

Materials Science Forum Vols. 302-303 (1999) pp. 358-362

Compative Study of the Antiferromagnetic and Spin Glass Transitions of the $Fe_xMn_{0.6-x}Al_{0.4}$ Disordered Alloys, x = 0.20 and x = 0.375

C. Gonzáles, J.A. Tabares, L.E. Zamora, G.A. Pérez Alcázar and A. García

Departamento de Física, Universidad del Valle, Cali, Colombia, A.A. 25360
Fax: 57 2 3393237

Keywords: Antiferromagnetism, Spin Glass, FeMnAl Alloys

Abstract

In this work we present the results of an experimental study of the alloys $Fe_{0.20}Mn_{0.40}Al_{0.40}$ and $Fe_{0.375}Mn_{0.225}Al_{0.400}$ which were carried out by ac magnetic susceptibility (including frequency and dc field variations) and calorimetry, in the 12 - 300 K range. The ac magnetic susceptibility curves present one or more peaks that allow us to determine the transition temperatures. A discussion of the occurrence of these transitions and their relationship with frequency (ac susceptibility) and with dc field are presented. Ac calorimetry results are also included and confirm the former results although some shift in the corresponding transition temperature occurs.

1. Introduction.

The FeMnAl alloys system in the bcc disordered phase with competitive interaction shows an ordered phase (ferromagnetic or antiferromagnetic) and a disordered phase (spin glass phase). Also, depending on the system composition the occurrence of a re-entrant spin phase to a ferromagnetic phase or to an antiferromagnetic phase, has been reported [1-4], at very low temperatures.

Previous theoretical and experimental results of the $Fe_xMn_{0.7-x}Al_{0.3}$ [5] series proved the existence of pure and re-entrant spin glass phases as well as the paramagnetic, ferromagnetic, and antiferromagnetic phases. The competitive ferromagnetic and antiferromagnetic exchange interaction together with the disordered structure makes possible this behavior. The transition temperatures for the spin glass phases of the $Fe_xMn_{0.7-x}Al_{0.3}$ alloys are located below 70 K. A study on the $Fe_{0.89-x}Mn_{0.11}Al_x$ systems[6] showed a strong dependence of this transition temperature on the aluminum concentration. These series show reentrant spin behavior at $x = 0.4$ and $T = 50\ K$.

This work presents a study of the $Fe_xMn_{0.6-x}Al_{0.4}$ alloys series with $x = 0.20$ and $x = 0.375$, respectively, by using ac magnetic susceptibility (including frequency and dc field variations and ac calorimetry measurements, both in the *12 - 300 K* range.

2. Experimental procedure

The $Fe_{0.20}Mn_{0.40}Al_{0.40}$ and $Fe_{0.375}Mn_{0.225}Al_{0.400}$ samples were prepared by employing Fe, Mn, and Al powder with 99.9% purity. The samples were melted in an arc furnace in argon atmosphere, and then encapsulated in quartz tubes (previously evacuated) in order to be homogenized by annealing at 1000°C for a week. Finally, they were quenched in ice water. As it was proved in previous reports [1,2], after this heat treatment, the high temperature disordered phase is retained. Afterwards, there were made cylinders (3 mm diameter 4 mm height , weigh ≈ 80 mg) from powders of each sample in order to make ac susceptibility measurements. A closed cycle he susceptometer / magnetometer was used. In order to make the calorimetric measurements the sheets (1 mm^2 - 170 μm width) were made. A closed cycle He ac calorimeter was used.

3. Results and Discussion

Figure 1 shows the real (X') and the imaginary part (X") of the ac magnetic susceptibility measurements versus temperature in the *12 - 300 K* range, for the $x = 0.20$ and $x = 0.375$ alloys respectively. Figure 1(a) shows transition peaks at *219 K*, accordingly the X' curve, which can be associated to an antiferromagnetic to a paramagnetic phase transition, and at 49.9 K accordingly the X" curve, which is associated to a reentrant spin glass to an antiferromagnetic phase transition. Further, these conclusions will be discussed in more detail.

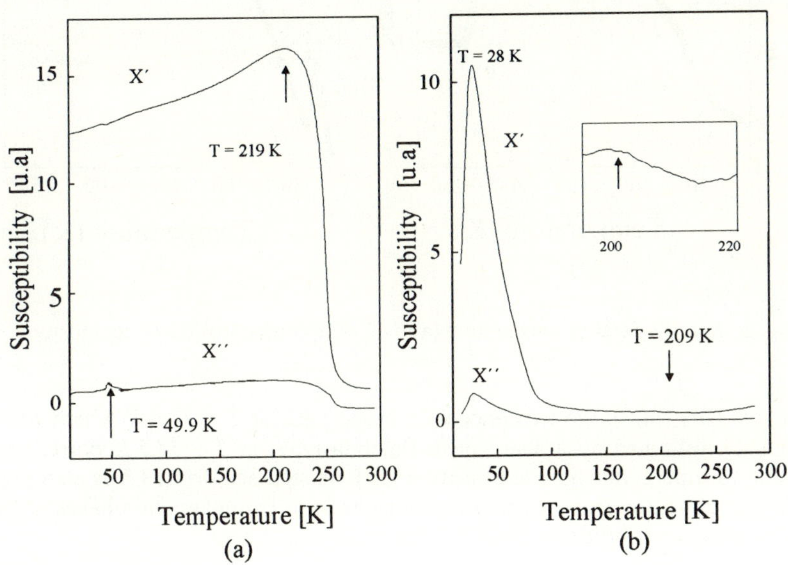

Fig. 1 Susceptibility vs temperature (a) 0.20 composition, (b) 0.375 composition.

From figure 1(b) it can be observed two peaks. One occurs at *209 K*, probably corresponding to a transition to a paramagnetic phase from an unknown phase and a transition at $T = 28$ *K* from a spin glass phase to the mentioned unknown phase. There are some evidences [10] that allow us to propose that this intermediate phase behaves as a superparamagnetic phase. This superparamagnetic is attributed to the formation of magnetic clusters of Fe and Mn separated by Al atoms as a consequence of the disordered character of the samples.

Figure 2 presents the specific heat versus temperature measurements, for the two samples, in the *12 - 300 K* interval. It corroborates the transition found by ac susceptibility measurements. In figure 2(a) it can be observed a singularity at $T = 209$ *K* for $x = 0.20$ which confirms the postulated transition from an antiferromagnetic to a paramagnetic phase. At low temperatures it was not observed any peaks close to *49.9 K* as was observed in the ac susceptibility measurement. The fact that this peak, which was attributed in the ac magnetic susceptibility measurements to a transition from re-entrant to antiferromagnetic phase was not obtained in the specific heat measurements confirms the re-entrant spin glass character of this transition. In some cases by specific heat measurements this transition appears as a broad line (pure spin glass) and in other cases it does not appear (re-entrant spin glass)[11].

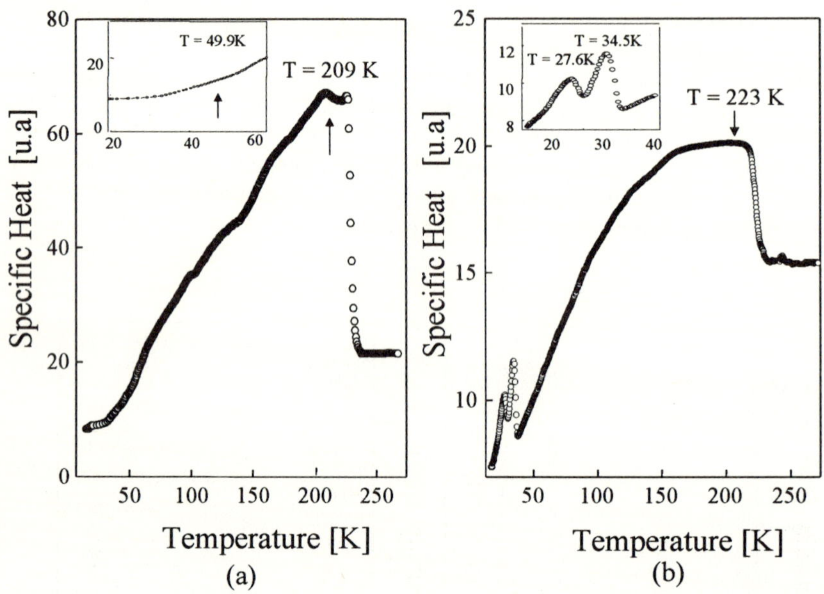

Fig. 2. Specific Heat vs temperature (a) 0.20 composition, (b) 0.375 composition.

Figure 2(b) ($x = 0.375$) shows the existence of a broad peak at $T = 27.6\ K$ which proves the spin glass behavior [11], followed by a sharp peak (lambda type) at $T = 34.5\ K$ corresponding to the antiferromagnetic to superparamagnetic transition [12]. A step near to *223 K* is also present, it may be produced by the transition from the superparamagnetic to paramagnetic phases. This transition was registered by ac susceptibility in *209 K*.

In order to prove in a more conclusive way the postulated transitions, ac susceptibility measurements were made for different frequencies and dc fields. Figure 3 shows the transition temperatures versus frequency for the two samples. Figure 3(a) corresponds to the variation of the temperature of the antiferromagnetic transition for the $x = 0.20$ sample. It can be seen that for a frequency range from *40 Hz* to *1500 Hz* an increase in the transition temperatures occurs. Above *1500 Hz* a decreasing in the transition temperature is observed. Figure 3(b) corresponds to the variation of the temperature of the spin glass transition for the $x = 0.375$ sample. As it can be seen, in the experimental frequency range, the transition temperature increases near monotonically. These two behaviors are characteristic of the form to the antiferromagnetic - paramagnetic transition, and the latter, to spin glass to paramagnetic transition [13,14].

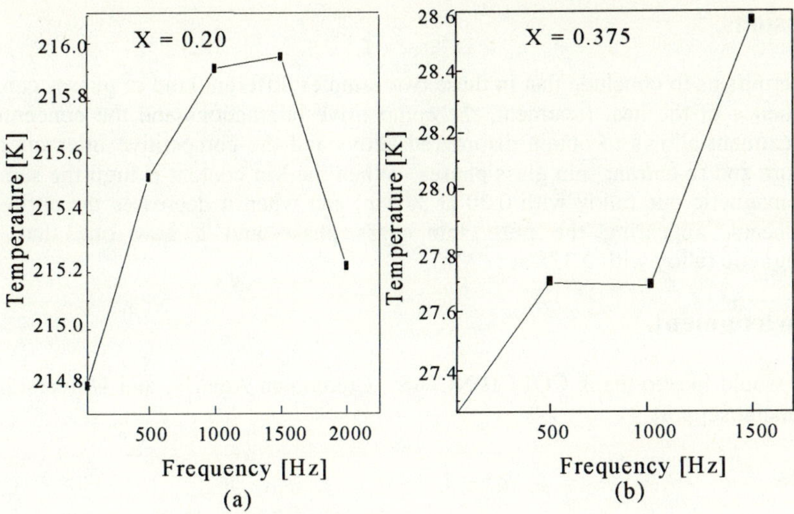

Fig. 3. Temperature vs frequency (a) 0.20 composition, (b) 0.375 composition.

Figure 4 represents the transition temperature versus the dc field. For the $x = 0.20$ sample there exist a random transition temperature behavior as a of function dc field. For the $x = 0.375$ sample, the transition temperature decreases with the dc field. These results are in accordance with those previously reported [13,14] in the literature.

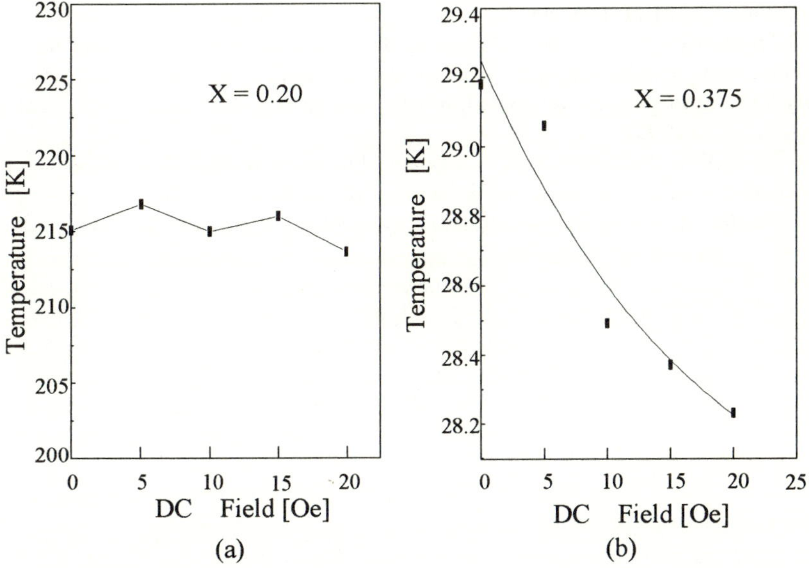

Fig. 4. Temperature vs dc field (a) 0.20 composition, (b) 0.375 composition.

4. Conclusions.

This work permits us to conclude that in these two samples different kind of phases can be obtained as a consequence of the heat treatment, the competitive interactions and the concentration. The used heat treatment allows to obtain disordered alloys and the competitive interactions permit to obtain the pure and re-entrant spin glass phases. When the Mn content is high the sample phase is the antiferromagnetic one (alloy with 0.20 at % Mn) and when it decreases the antiferromagnetic phase disappears, appearing the pure spin glass phase and a new one that behaves as superparamagnetic (alloy with 0.375 at % Mn).

5. Acknowledgment.

The authors would like to thank COLCIENCIAS , Colombian Agency, and Universidad del Valle for their financial support.

References.

1. G. A. Pérez Alcázar, E. Galvao da Silva, and J. A. Plascak, Phys Rev **B 38** (1988) 2816.
2. A. Rosales Rivera, G. A. Pérez, and J. A. Plascak, Phys Rev **B 41** (1990) 4774.
3. M. A. Kobeissi, Phys. C.. Condens Matter **3** (1991) 4983
4. Ligia E. Zamora, G. A. Pérez Alcázar, A. Bohórquez, and J. A. Tabares. JMMM. **137** (1994) 137
5. Ligia E. Zamora, G. A. Pérez Alcázar, A. Bohórquez, and J. F. Marco. Presented in ICAME'95 Rimini, Italy, September, (1995).
6. H. Bremers, Ch. Jarms, J. Hesse, S. Chajivasiliou, K. G. Efthimiadis, and Y. Tsoukalas. JMMM **140 - 144** (1995) 63.
7. J. Nogués, and K. V. Rao, JMMM **13** (1994) 11.
8. A. Wulfes, Ch. Botter, J. Hesse, J. Silver, and H. Ahlers. JMMM **104 - 107** (1992) 2069.
9. G. Grest, Phys Rev **21** (1980) 165.
10. Ligia E. Zamora. Doctorate Thesis "Estudio de las propiedades magnéticas de las aleaciones FeMnAl en la fase bcc desordenada". Universidad del Valle, Cali, Colombia (1996).
11. J. A. Mydosh, Spin glasses: an experimental introduction, Ed. Taylor and Francis, London, Washington, DC. (1993)
12. F. Borsa, M. G. Pini, A. Retteri, and V. Tognetti, Phys Rev **B 28** (1983) 5173.
13. Ch. Bötter, R. Starsch, A. Wulfes, and J. Hesse. JMMM **99** (1991) 280.
14. C. A. M. Mulder, A. J. Van Duynveldt, and J. A. Mydosh, Phys. Rev. **B**, (1981) 1384

Materials Science Forum Vols. 302-303 (1999) pp. 363-370

Antiferromagnetic Interactions in Diluted Systems from 20 mK Magnetometry

V. Bindilatti[1], N.F. Oliveira Jr.[1], E. ter Haar[1] and Y. Shapira[2]

[1] Instituto de Física, Universidade de São Paulo, C.P. 66318, CEP 05315-970, S. Paulo, SP, Brazil

[2] Department of Physics and Astronomy, Tufts University, Medford, MA 02155, USA

Keywords: Exchange Interaction, Magnetization Steps, Diluted Magnetic Semiconductors

Abstract. Magnetometry at 20 mK is being used to study week antiferromagnetic (AF) interactions in diluted magnetic systems. The key feature of these systems is the occurrence of magnetization steps due to small clusters, from which the AF antiferromagnetic constants, J_i, can be measured. A description of the experimental setup and results of our recent work on diluted magnetic semiconductors are presented. The results include a) the determination and identification of the Eu–Eu dominant AF exchange constant in $Pb_{1-x}Eu_xX$ (X = S, Se, Te), and b) an experimental study on the distant-neighbor Mn–Mn exchange constants (J_2, J_3, etc.) in the $Zn_{1-x}Mn_xX$ (X = S, Se, Te) series.

1. Introduction

Magnetization-steps (MST's) spectroscopy is a relatively new method of measuring both the interactions in magnetic clusters and the populations of such clusters [1]. Here, the word cluster refers either to a finite (usually small) number of magnetic ions (pairs, triplets, etc.) which are antiferromagneticaly coupled. Some magnetic materials contain only one type of cluster, but in dilute magnetic materials, such as the Diluted Magnetic Semiconductors (DMS's), many cluster types exist at the same time.

Basically, magnetization steps occur when the ground state of a cluster of spins changes due to a level crossing as a function of the magnetic field. Since the magnetic moment μ of a state is given by $-\partial E/\partial B$, the level crossing leads to an increase of μ of the ground state. At low temperatures, when the properties of an ensemble of such clusters are dominated by the ground state, the macroscopic magnetization M exhibits MST's at the fields where the level crossings occur.

For a single ion, the value of B at the steps depends on the crystal field parameters and the g-value. Two decades ago, Varret *et al.* [2] proposed the MST as a method of measuring crystal field parameters. A newer application is the measurement of the antiferromagnetic (AF) exchange constants, J_i, from MST's generated by larger clusters (mostly pairs) [1]. Here, the AF interactions lead to level crossings at fields which are directly related to the J_i.

The magnetization rise δM, associated with any MST, is directly related to the number of clusters that generated this MST. As a result, if the cluster type has been identified, the number of clusters of that type can be determined. This can be used to check, for instance, if the distribution of magnetic ions in the lattice is random. Conversely, if one assumes a random distribution, the measured δM can be of help in identifying the type of cluster responsible for the MST.

A rare example of the use of MST's to determine crystal field parameters is given in Ref. 3. More recent results include our work on the axial anisotropy of Co^{++} ions diluted in the wurtzite semiconductors CdS [4] and CdSe [5], and the Jahn-Teller distortion in the case of Cr^{++} diluted in ZnTe [6]. On the other hand, the MST's method has become the leading technique for the study of exchange interactions (mostly in DMS's) [1]. MST's from pairs in DMS's were first observed in 1984

[7, 8]. They were used to determine the dominant AF interaction, identified as J_1 between nearest neighbor (NN) spins. Summaries of results for MST's from NN pairs in II–VI DMS's can be found in Refs. 1 and 9. There has also been some experimental work on NN triplets [10] and quartets [11] in these materials. MST's arising from a much larger cluster, the so-called molecular ferric wheel (found in a magnetic molecule) have also been reported [12].

The observation of MST's is contingent on the availability of high magnetic fields and low temperatures. For Mn^{++} ions in II–VI DMS's, J_1 is of order -10 K and the observation of the related MST's required fields up to 20-30 T. For Co^{++} in II-VI DMS's, J_1 is even higher, and pulsed fields were used [13]. For interactions of order 1 K or less, fields of superconducting magnets are sufficient, but for such weak interactions the use of very low temperatures is imperative. Here, resolving the MST's is the main problem and cutting down the thermal broadening may require temperatures in the dilution refrigerator range (the thermal width of a step is $\Delta B \cong 3.5\, k_B T/g\mu_B$, while the spacing between successive steps is $\Delta B \cong 2|J|/g\mu_B$).

The present paper focuses on recent results in MST spectroscopy, from magnetometry performed at 20 mK. They became possible after the development of a sensitive force magnetometer

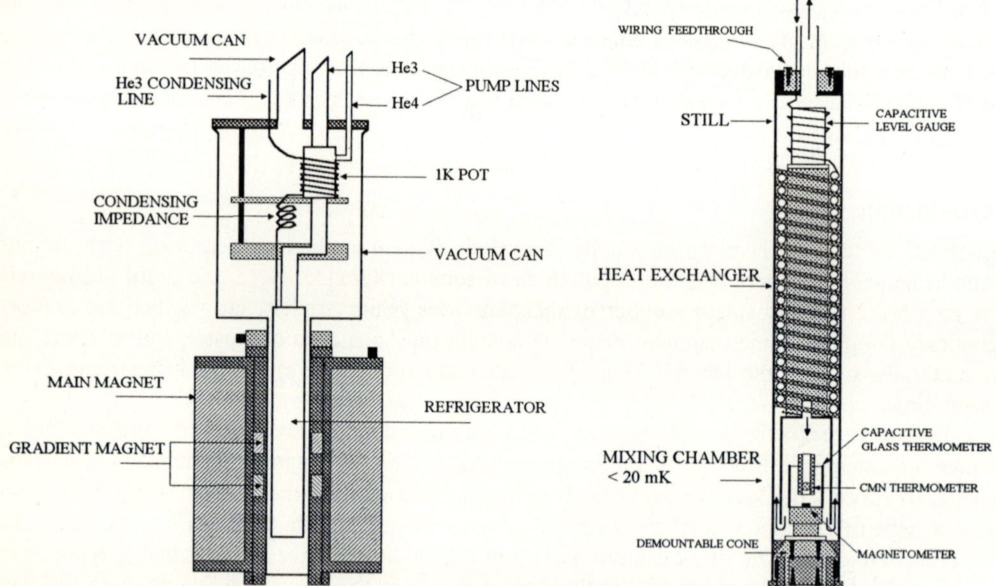

Figure 1. Left: schematic overview of the experimental setup. Right: the plastic dilution refrigerator.

operating inside a plastic dilution refrigerator. We begin by presenting a brief description of the experimental setup, followed by a discussion of the basic ideas involved the MST's method. Then, a brief account of recent experimental results will be presented. The topics covered include the weak AF interactions of Eu^{++} ions in PbS, PbSe and PbTe and the distant neighbor interactions of Mn^{++} in ZnS, ZnSe and ZnTe.

2. Experimental setup

Magnetization measurements in a dilution refrigerator may be done with the force method. The sample sits on a flexible medium (a membrane or a cantilever mechanism) and the displacement due to the magnetic force is measured capacitively. To provide the necessary field gradient, $\partial B/\partial z$, it is common to place the device out of the center of the magnet. Two main difficulties then have to be dealt with. One is that magnetically anisotropic samples will be subjected to torques, which will pro-

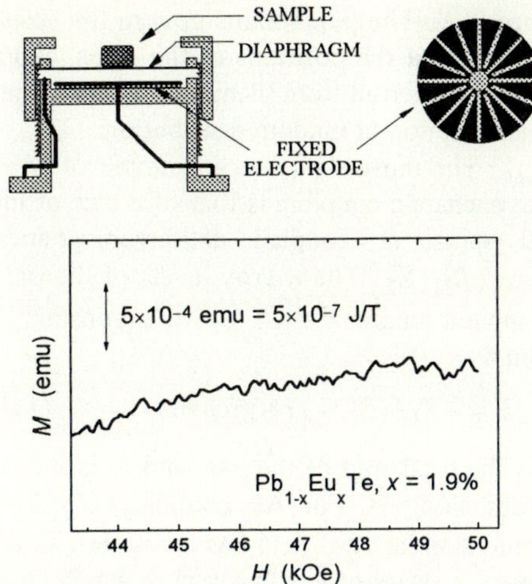

Figure 2. Upper: the force magnetometer. The diaphragm is a mylar sheet, 15 μm thick, aluminized in one side. The fixed electrode, built from printed circuit board, has radial grooves in the copper layer to cut down eddy currents. Lower: magnetic moment of a $Pb_{1-x}Eu_xTe$ (x=1.9%) sample, near saturation fields at T=20 mK. The trace was taken in *dc*-mode with $\partial H/\partial z$=1kOe/cm.

duce a spurious signal. This signal can easily be larger than the magnetization signal, making it difficult to subtract its effect. Another comes from the fact that the off center gradient is proportional to the field, cutting down the sensitivity at low fields.

The first experiments at $T\cong 0.1$ K were performed at the Francis Bitter National Magnet Laboratory, MIT. The apparatus consisted of a commercial dilution refrigerator adapted to Bitter magnets. The force magnetometer was operated out of the center of the magnet. The need for higher sensitivity at low fields led us to design a new system (at the University of S. Paulo), with the field gradient produced independently of the main field. For this, a pair of coils, wound with a 0.2 mm dia. NbTi superconducting wire, was fitted to the bore of a NbTi magnet. A simple plastic dilution refrigerator (PDR) produces the low temperatures. Fig. 1 shows the experimental arrangement. The left part is a sketch of the magnet, gradient coils, and refrigerator. The bottom part shows details of the PDR. The construction of PDR's has been discussed in detail in Ref. 14. In our case, for simplicity, we have built one with only a coiled Teflon tubing heat exchanger. This was enough to produce base temperatures under 20 mK. The diaphragm magnetometer is similar to the one described in Ref. 15. Fig. 2 shows a sketch of the device, and a portion of a typical trace. The sensitivity is near 10^{-4} emu.

The combination of independent gradient coils together with a PDR makes it possible to feed the gradient coils with an *ac* current, thus producing an oscillating $\partial B/\partial z$ (*ac*-mode). Because of the fully plastic construction of the fridge, practically no heat from eddy-currents is generated in the presence of fast varying fields. The advantage of operating in the *ac* mode is that synchronous detection gives directly the part of the signal due to the force, rejecting the part due to torque. This is because only the force is proportional to $\partial B/\partial z$. The *ac* operation of the gradient coils proved to be an essential feature in the investigation of anisotropic samples. For operation in *dc*-mode, a background trace is taken without field gradient, and is later subtracted from the trace with the steady gradient. The results obtained with the two modes of operation have always been very consistent, thus demonstrating the reliability of the technique. Typically we use gradients of 0.5 kOe/cm at about 5 Hz for the *ac*-mode and 1 kOe/cm for *dc* operation. A preliminary account of the performance of this system was given in Ref. 16.

3. MST's from AF exchange coupling

The value of the AF exchange constants between magnetic ions in a diluted magnetic system (such as the DMS's) can be determined from MST's. For low concentrations of the magnetic ions, these systems can be described in terms of collections of singles, pairs, triplets, etc. The types of clusters that are included in a particular "cluster model" depend on the exchange constants considered. Each cluster type will produce magnetization steps at fields characteristic of this cluster. The populations of the various clusters can be calculated if a random distribution of the magnetic ions is as-

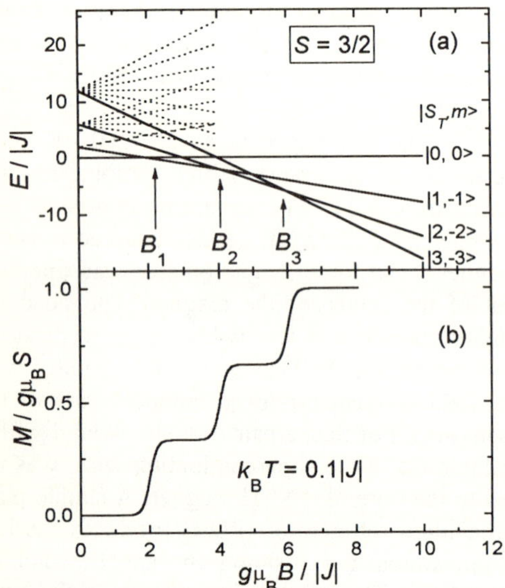

Figure 3. The energy levels of an AF-coupled pair (Eq. 1) as a function of field (a), and the resulting magnetization steps (b). The fields at which the steps occur determine the zero field splitting of the energy levels, and the exchange constant.

(k_BT=0.1|J|) is sketched in Fig. 3 (b). The values of B_n in this simple model, and the g-factor, yield J directly via Eq. (2) [1, 9]. Inclusion of smaller exchange couplings with distant neighbors [18] or other interactions may modify Eq. (2), but as a first approximation, the differences between successive MST's, will remain as a good measure of J.

Fig. 4 shows a simulation of the magnetization of a diluted magnetic system, for S=5/2 ions in a fcc-lattice and with AF interactions between NN only (J_1-model). The simulation, which assumes a random distribution, includes singles, pairs, triplets and quartets. The upper part of the figure shows the upper portion of the magnetization curve, when the singles are near complete saturation. The staircase-like behavior is built up mainly from the MST's of the pairs and triplets. Each MST gives a peak in the susceptibility dM/dB, as seen in the lower graph of Fig. 4. The five larger peaks, at fields which are even multiples of $|J|/g\mu_B$, are due mainly to the pairs. The sumed [17]. The populations govern the sizes of the steps, but the positions of the steps, and the value of J inferred from them, are independent of the assumption of random distribution.

The most elementary example of MST's from exchange coupling is that of a pair of identical spins, S, coupled antiferromagneticaly, $H = -2J\mathbf{S}_1 \cdot \mathbf{S}_2$. The energy levels of the pair in an applied magnetic field, up to a constant, are given by

$$E = -JS_T(S_T+1) + g\mu_B Bm. \qquad (1)$$

S_T is the total spin of the pair and m is its component along $\boldsymbol{B}$. For AF coupling, $J<0$, the ground state at zero field has S_T=0, but as B increases, a series of level crossings which change the ground state takes place. This is illustrated in Fig. 3 (a), for S=3/2. Each change of the ground state increases the value of m by one unit, generating a MST at low T. There are 2S MST's at fields B_n given by

$$g\mu_B B_n = 2n|J|, \qquad (2)$$

where n=1,2,...,2S. The magnetization curve of a collection of such pairs at a low temperature

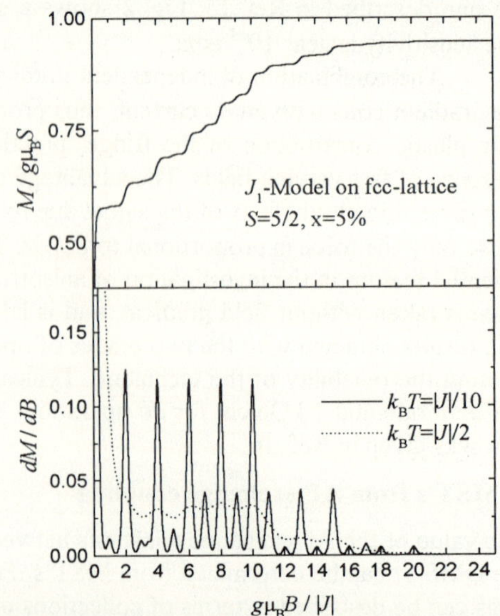

Figure 4. Calculated magnetization and susceptibility of a diluted magnetic system, including the contributions of singles, pairs, triplets, and quartets.

smaller peaks originate from triplets. For a half-integer S, the steps from the triplets occur at odd multiples of $|J|/g\mu_B$. There are two superposed series: one from closed triplets, starting at 3, and another from open triplets, starting at 7. Both series end at $B=15|J|/g\mu_B$. Still smaller peaks can be seen in the dM/dB curve which are due to the quartets. In the model considered there are six types of quartets [11] each with its own series of MST's. They are all saturated after $B=20|J|/g\mu_B$. The presented simulation was made with $T=|J|/k_B/10$. For comparison, the susceptibility for $T=|J|/k_B/2$ is also shown as a dotted line. At this higher T the structure of the susceptibility curve is already almost wiped out.

The model discussed above involved only one exchange interaction. The inclusion of additional AF exchange constants to the model will lead to several added features to the magnetization curve. Each new exchange constant, e.g. J_2 between next-nearest neighbors, will contribute new series of steps from the related pairs, triplets, etc. The field ranges of these steps are determined by the strength of the interaction. If the strength of the interactions in the model differ by an order of magnitude or so, the series of steps will be well separated in field. If they have comparable strengths, however, the structure of the final magnetization curve can become quite complicated.

4. Eu^{++} in lead chalcogenides

An important group of IV-VI DMS's are the lead chalcogenides, PbTe, PbSe and PbS, with a fraction of the lead ions replaced by Eu^{++} (S=7/2). The dominant AF exchange constant in these DMS's is very small. Only recently it has become possible to determine this exchange constant accurately by observing the MST's from pairs. The result for PbSe [19] is $J/k_B=-0.24\pm0.03$ K. All steps are below 30 kOe and their observation required not only the very low temperatures used, but also good sensitivity at low fields.

In Fig. 5 we show the data for three samples of Eu^{++} in PbTe [20]. The six large peaks correspond to the second to seventh MST's arising from pairs. From their separation ΔB_n, we obtained $J/k_B=-0.265\pm0.005$ K. The three less pronounced peaks above 30 kOe are due to triplets.

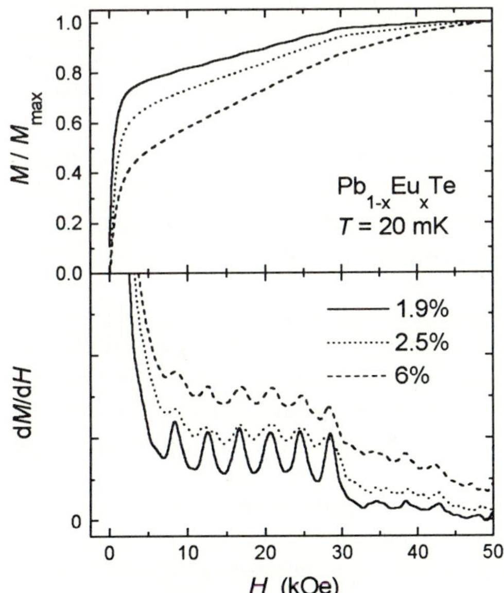

Figure 5. Magnetization and susceptibility (numerically derived from M) for three samples of $Pb_{1-x}Eu_xTe$. Six of the seven steps from pairs are resolved, giving $J/k_B=-0.265\pm0.005$ K.

For the third material in the series, $Pb_{1-x}Eu_xS$ [21], the dominant AF exchange constant is also weak, $J/k_B=-0.228\pm0.007$ K. In this case, however, the structure of the steps from pairs is significantly affected by the cubic crystal field (stronger than in the preceding materials). Besides, a splitting of the steps observed when the field was applied parallel to one of the [001] axes was interpreted as due to anisotropic exchange. The value quoted above for J is the isotropic part of the interaction.

In the pure materials, EuS, EuSe and EuTe, J_1 is ferromagnetic, J_2 being the stronger AF interaction [22]. To identify the measured J's, which are the dominant AF interaction in the diluted materials, quantitative analyses of the magnetization traces were made. The experimental data were compared with simulations based on random distribution, similar to the one shown in Fig. 4, in which the dominant J was taken to be either J_1 or J_2. For all the three diluted materials studied, the analyses indicated that the measured J, is actually J_1 between nearest neighbors.

5. Magnetization steps from distant neighbor pairs.

The distance dependence of the exchange interaction between Mn^{++} ions (S=5/2) in type II-VI DMS's has been at issue for about a decade. Although the NN-exchange constants J_1 in these materials have been measured by several techniques, available experimental information about the distant-neighbor J_i's is only indirect, the main reason for that being their small strength. A theory by Larson *et al.* [23] predicts a monotonic decay of J with distance with a fixed ratio between succeeding J_i's of about an order of magnitude (starting from the known J_1's of order −10 K).

Distant-neighbor J_i's have been directly measured only on the Co-based materials $Zn_{1-x}Co_xS$, $Zn_{1-x}Co_xSe$ [24] and $Zn_{1-x}Co_xTe$ [25]. For them two additional constants of the order or a few K, identified as J_2 and J_3, have been measured by the MST method.

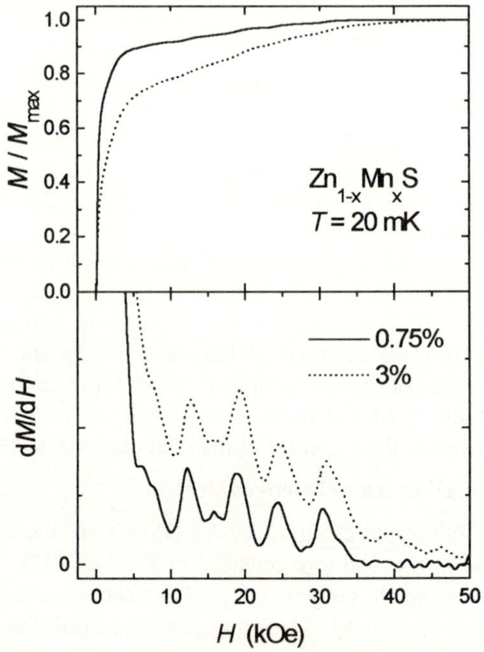

Figure 6. Magnetization and susceptibility (numerically derived from M) of $Zn_{1-x}Mn_xS$. The steps are due to distant neighbor exchange interactions.

Employing the apparatus at the University of S. Paulo, MST's from distant neighbors pairs of Mn^{++} (S=5/2) in ZnS, ZnSe and ZnTe were observed recently [26]. Fig. 6 shows data for two samples of $Zn_{1-x}Mn_xS$. The top half of the figure displays the magnetization traces (normalized at 50 kOe). The magnetic susceptibilities, obtained numerically from the experimental data, are shown in the lower half of the figure.

The last 4 steps of a series due to pairs are clearly visible in the susceptibility traces. Besides, the contribution of the triplets can also be observed. These steps correspond to J/k_B=−0.42±0.01 K. From additional experiments, which show no other series of steps up to about 20 T, it has been established that this is the second AF-coupling in strength for this material, to be compared with a J_1 of −16.1 K [9].

The exchange constants from two additional types of distant-neighbor pairs were measured from these traces. One of them, J/k_B=−0.27±0.01 K, gives rise to a series of MST's which ends a little under 20 kOe. It is responsible for the higher susceptibility at the two first peaks of the major series. The distinctive small peak at about 15 kOe is associated with its fourth MST. The magnetization ramp ending near 3 kOe is a series of unresolved MST's that gives another exchange constant $J/k_B \cong -0.04$ K.

The identification of these new three J's was made by comparison of the experimental data with simulations of the magnetization based on the cluster model and a random distribution of the magnetic impurities. The cluster statistics in this case is very involved because of the number of interactions to be considered. A computer program was developed to make the simulations. We started with the monotonic decay prediction and considered the model with the known value of J_1, and new values assigned to J_2, J_3 and J_4, in order of decreasing strength. The result was very poor. Instead of abandoning the assumption of random distribution, we tested the other possible assignments of the three measured values among J_2, J_3 and J_4. Somewhat surprisingly, a very good agreement between experiment and simulation is achieved when the assignment is such that $|J_4|>|J_2|>|J_3|$.

For the two other materials studied, $Zn_{1-x}Mn_xSe$ and $Zn_{1-x}Mn_xTe$, we also measured at least three new exchange constants. The statistical analyses in both cases resulted in the same non-monotonic decay, with J_4 and not J_2 being the second constant in strength, after J_1. The fact that this same result has emerged in all the three materials studied is a sound support to our conclusion. It also emerged from our results that the ratios between the J_i's are not material independent as predicted by existing theories [23,27].

Conclusion

The magnetization curve for a collection of magnetic ions diluted in a host lattice shows series of steps that are directly related to the AF couplings between the ions. Measuring these steps is an effective experimental method of studying these interactions. A sensitive magnetometer operating in a dilution refrigerator has been shown to be extremely useful in the study of weak interactions, of order 1 K and below.

FAPESP, CNPq, and FINEP supported the work at the University of São Paulo. The work at Tufts University was partially supported by NSF grants DMR-9219727 and INT-9216424.

REFERENCES

[1] Y. Shapira, J. Appl. Phys. **67**, 5090 (1990).
[2] F. Varret, Solid State Commun. **14**, 17 (1974).
[3] C.C. Becerra, A. Paduan, and N.F. Oliveira Jr., Phys. Status Solidi B **129**, K173 (1985).
[4] V. Bindilatti, A.N. Anisimov, N.F. Oliveira Jr., Y. Shapira, M. Goiran, F. Yang, S. Isber, M. Averous, and M. Demianiuk, Phys. Rev. B **50**, 16464 (1994).
[5] S. Isber, M. Averous, Y. Shapira, V. Bindilatti, A.N. Anisimov, N.F. Oliveira Jr., V.M. Orera, and M. Demianiuk, Phys. Rev. B **51**, 15211 (1995).
[6] G.H. McCabe , Y. Shapira, V. Bindilatti, N.F. Oliveira Jr., A. Twardowski, W. Mac, E.J. McNiff Jr., and M. Demianiuk, Solid State. Commun. **95**, 841 (1995).
[7] Y. Shapira, S. Foner, D.H. Ridgley, K. Dwight, and A. Wold, Phys. Rev. B **30**, 4021 (1984).
[8] R.L. Aggarwal *et al.*, in the Proc. 17th Intl. Conf. on the Phys. of Semicond., edited by J.D. Chadi and W.A. Harrison, (Springer, New York 1985), p. 419.
[9] S. Foner, Y. Shapira, D. Heiman, P. Becla, R. Kershaw, K. Dwight, A. Wold, Phys. Rev. B **39**, 11793 (1989).
[10] X. Wang, D. Heiman, S. Foner, and P. Becla, Phys. Rev. B **41**, 1135 (1990).
[11] M.T. Liu, Y. Shapira, E. ter Haar, V. Bindilatti, and E.J. McNiff Jr., Phys. Rev. B **54**, 6457 (1996).
[12] K.L. Taft, C.D. Delfs, G.C. Papaefthymiou, S. Foner, D. Gatteschi, and S.J. Lippard, J. Am. Chem. Soc. **116**, 823 (1994).
[13] S. Foner, D. Heiman, T.Q. Vu, R. Fries, G.H. McCabe, Y. Shapira, and M. Demianiuk, Solid State Commun. **95**, 455 (1995).
[14] E. ter Haar, R. Wagner, C.M.C.M. van Woerkens, S.C. Steel, G. Frossati, L. Skrbek, M.W. Meisel, V. Bindilatti, A.R. Rodrigues, R.V. Martin, and N.F. Oliveira Jr., J. Low Temp. Phys. **99**, 151 (1995).
[15] A.G. Swanson, Y.P. Ma, J.S. Brooks, R.M. Markiewicz, N. Miura, Rev. Sci. Instrum. **61**, 848 (1990).
[16] V. Bindilatti and N.F. Oliveira Jr., Physica B **194-196**, 63 (1994).
[17] U.W. Pohl and W. Busse, J. Chem. Phys. **90**, 6877 (1989).
[18] B.E. Larson, K.C. Hass, and R.L. Aggarwal, Phys. Rev. B **33**, 1789 (1986).
[19] V. Bindilatti, N.F. Oliveira Jr., Y. Shapira, G.H. McCabe, M.T. Liu, S. Isber, S. Charar, M. Averous, E.J. McNiff Jr., and Z. Golacki, Phys. Rev. B **53**, 5472 (1996).
[20] E. ter Haar, V. Bindilatti, N.F. Oliveira Jr., G.H. McCabe, Y. Shapira, Z. Golacki, S. Charar, M. Averous, and E.J. McNiff Jr., Phys. Rev. B **56**, 8912 (1997).
[21] V. Bindilatti, E. ter Haar, N.F. Oliveira Jr., M.T. Liu, Y. Shapira, X. Gratens, S. Charar, S. Isber, P. Masri, M. Averous, Z. Golacki, and E.J. McNiff Jr., Phys. Rev. B **57**, 7854 (1998).
[22] A. Mauger, and C. Godart, Phys. Repts. **141**, 51 (1986).
[23] B.E. Larson, K.C. Hass, H. Ehrenreich, and A.E. Carlsson, Phys. Rev. B **37**, 4137 (1988).
[24] Y. Shapira, T.Q. Vu, B.K. Lau, S. Foner, E.J. McNiff Jr., D. Heiman, C.L.H. Thieme, C-M. Niu, R. Kershaw, K. Dwight, and A. Wold, Solid State Commun. **75**, 201 (1990).

[25] T.Q. Vu, V. Bindilatti, Y. Shapira, E.J. McNiff Jr., C.C. Agosta, J. Papp, R. Kershaw, K. Dwight, and A. Wold, Phys. Rev. B **46**, 11617 (1992).
[26] V. Bindilatti, E. ter Haar, N.F. Oliveira Jr., Y. Shapira, M.T. Liu, Phys. Rev. Lett. **80**, 5425 (1988).
[27] A. Bruno and J. P. Lascaray, Phys Rev B **38**, 9168 (1988).

Materials Science Forum Vols. 302-303 (1999) pp. 371-374

Random-Field Effects and Glassy Behavior in $Fe_xZn_{1-x}F_2$

F.C. Montenegro[1*], K.A. Lima[1], M.S. Torikachvili[2] and A.H. Lacerda[3]

[1] Departamento de Física, UFPE, 50670-901 Recife, Pernambuco, Brazil

[2] Department of Physics, San Diego State University, San Diego CA 921182, USA

[3] National Magnetic Field Laboratory, Mail Stop E536, Los Alamos NM 8745, USA

Keywords: Random Fields, Spin-Glass Behavior, Magnetic Phase Diagrams

Abstract - The effects of strong random fields in the magnetic phases and critical behavior of the diluted antiferromagnet $Fe_xZn_{1-x}F_2$ are reported. We emphasize the appearance of a random-field induced glassy behavior in samples with concentrations far above the percolation threshold of this compound.

Introduction - A great deal of efforts has been required from theoreticians and experimentalists in the last three decades for understanding the effects of disorder in magnetic systems. As an attempt to simplify this hard task, a large amount of the research papers in this area is devoted to a few theoretical models, and to some magnetic compounds which can be recognized as experimental realizations of these models. Diluted antiferromagnets (DAFs) with uniaxial anisotropy are largely used as testing ground of theoretical models of random Ising magnets.[1] In three dimensions ($d = 3$), one of the best experimental realizations of the pure Ising model is the highly anisotropic insulating antiferromagnet FeF_2. The diluted system $Fe_xZn_{1-x}F_2$ represents experimentally the $d = 3$ random-exchange Ising model (REIM). In the REIM the exchange interaction is random in magnitude. This compound has a body centered tetragonal structure, with the Fe atoms in the corners and the body center of the tetrahedron, as shown in Fig. 1. The exchange constants, J_ℓ, have the experimental ratios $j_1 \equiv J_1/J_2 = -0.013$ and $j_3 \equiv J_3/J_2 = +0.053$.[2] The third-neighbor exchange constant J_3 introduces a weak frustration in the exchange interactions of this system.

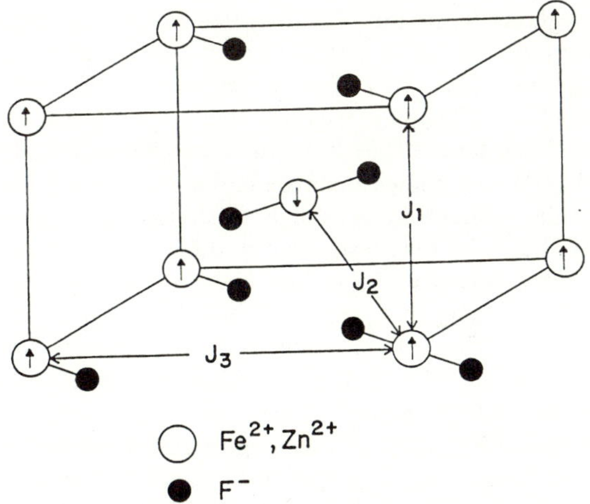

Fig. 1. Magnetic structure of $Fe_xZn_{1-x}F_2$. The first three nearest-neighbor interactions between Fe^{2+}ions are illustrated.

REIM and spin-glass behavior at zero field (H = 0) – Above the percolation threshold, x_p, and in the absence of exchange-frustration, DAF compounds at H = 0 are expected to present a zero-temperature ground state with AF long-range order (LRO). The phase transition which occurs at the Néel temperature, T_N, is predicted (and experimentally confirmed) to be governed by a characteristic REIM critical behavior. Recent results from local mean-field (LMF) numerical simulations have shown[3] that a weak frustration, possibly present in the exchange interactions of $Fe_xZn_{1-x}F_2$, has no influence on the magnetic configuration of this compound for moderate dilution ($0.6 \lesssim x < 1$). However, for more intense dilution ($x_p < x \lesssim 0.5$), even a tiny frustration reduces dramatically the magnitude of the AF order parameter, the staggered magnetization $M_S(x)$. Considering the degree of frustration in FeF_2 ($j_3 = +0.053$) in the simulations of the diluted system, M_S vanishes only at and below $x_p = 0.24 \pm 0.01$. This is in agreement with experimental results: at H = 0, AF LRO and REIM critical behavior have been reported in all measured samples of $Fe_xZn_{1-x}F_2$ with $x > 0.3$.[1] However, the LMF simulations indicate that the small frustration, combined with random dilution, lead the system to a spin-glass phase ($M_S = 0, Q \neq 0$) at H =0 in the vicinity of x_p. This feature explains the spin-glass behavior experimentally discovered[4] in a sample of $Fe_{0.25}Zn_{0.75}F_2$.

Weak random fields - The application of a uniform field parallel to the easy direction of a DAF compound produces an equivalence to the random-field Ising model (RFIM) model for the ferromagnet.[5] In the RFIM problem, the local field is random in direction (and probably magnitude). In ferromagnets a crossover from pure Ising to RFIM critical behavior occurs when the random field is "turned on". In DAFs the magnitude of the local random field can be tuned by the amplitude of the external field applied.[6] In d =3, weak random fields do not destroy the long-range order (LRO) present at H = 0. In $Fe_xZn_{1-x}F_2$, the transition from AF LRO to the paramagnetic (P) phase, through the critical temperature $T_c(H)$, is governed by a REIM to RFIM crossover scaling, i. e., $T_N - bH^2 - T_c(H) = CH^{2/\phi}$, with $\phi \cong 1.4$ being the REIM to RFIM crossover exponent. The equilibrium boundary, $T_{eq}(H)$, follows a similar scaling, i.e., $T_N - bH^2 - T_{eq}(H) = C_{eq}H^{2/\phi}$. $T_{eq}(H)$ was defined as the temperature above which different field cycling gives the same results in all measurements made in the same time scale. Under weak random fields $T_{eq}(H)$ lies just above $T_c(H)$, with the same concave curvature ($\phi \approx 1.4$). These curves join at the Néel temperature, $T_N(x)$ (see Fig. 3).

Strong random fields and glassy behavior - Random fields of *strong* magnitudes are predicted to make the AF LRO *unstable*,[7] giving rise to a random-field induced glassy phase[8] in the upper part of the (H,T) phase diagram. This was first observed[9] experimentally by magnetization measurements in a sample of $Fe_{0.31}Zn_{0.69}F_2$ (slightly above x_p). When H increases beyond the weak random-field regime, which for x = 0.31 lies within the interval 0 < H < 1.5 T, no AF LRO is observed below $T_{eq}(H)$.[10] $T_{eq}(H)$ now shows a convex curvature with $\phi \cong 3.4$, scaling with H as the de Almeida-Thouless boundary found in exchange frustrated spin glasses. The nature of the random-field-induced glassy behavior manifests itself in other characteristic field cycling and time dependent properties.[9,10] Very recently,[11] it has been shown by high field magnetization measurements that the $T_{eq}(H)$ boundary of a sample of $Fe_{0.60}Zn_{0.40}F_2$ presents a reversal of curvature relative to the small-H behavior. The critical boundary, $T_c(H)$, departs from the REIM to RFIM crossover scaling, with $\phi \cong 1.4$, for H > 5 T. In addition, the uniform magnetization increases monotonically with time in the high–H regime. These features are similar to the behavior found in the strong random-field regime of $Fe_{0.31}Zn_{0.69}F_2$, leading to the interpretation that the random-field induced glassy phase, found close to x_p, is extensive to samples with $x >> x_p$. To ascertain this scenario carefully, we present here high field magnetization measurements in a sample of $Fe_{0.56}Zn_{0.44}F_2$ in comparison with the $Fe_{0.60}Zn_{0.40}F_2$ results. The field dependence of the magnetization is illustrated in Fig. 2 for both samples. In these measurements, the sample was first

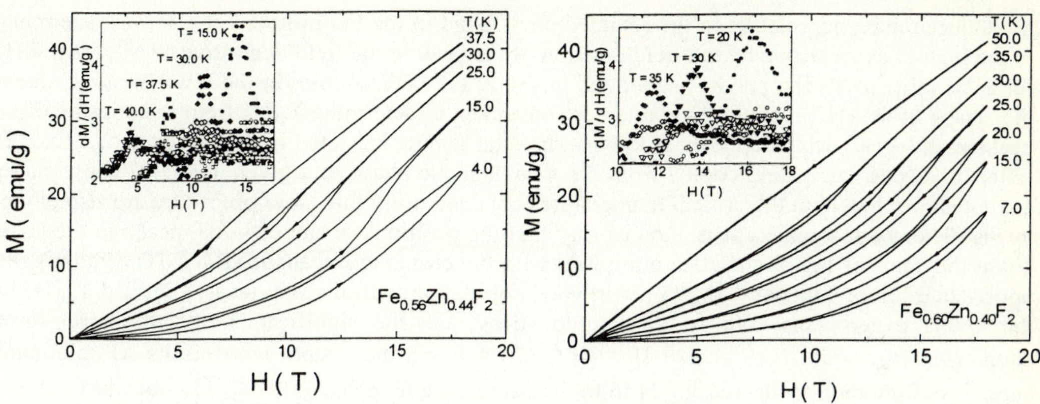

Fig. 2. M versus H plots for zero-field cooled $Fe_{0.56}Zn_{0.44}F_2$ and $Fe_{0.60}Zn_{0.40}F_2$, at several T. The arrows indicate field increasing FI, and field decreasing FD segments. The inset shows dM/dH versus H at fixed values of T, for each sample. Full (or open) symbols indicate the FI (or FD) procedure.

zero-field cooled (ZFC) from the paramagnetic phase to the temperature where the field increasing (FI) and field decreasing (FD) cycling takes place. An excess of magnetization ($\Delta M \equiv M_{FD} - M_{FI}$) appears in the FD branch (relative to FI), for all $T < T_N$, in an interval of H which depends on T and x. This defines upper, $H_{eq}^{u}(T)$, and lower, $H_{eq}^{\ell}(T)$, equilibrium boundaries in the (H,T) diagram (see Fig. 3). Hysteresis can only be observed within the field interval $H_{eq}^{\ell}(T) < H < H_{eq}^{u}(T)$. The field derivative of M is plotted against H (for fixed T) in the inset of Fig. 2. dM_{FI}/dH curves undergo

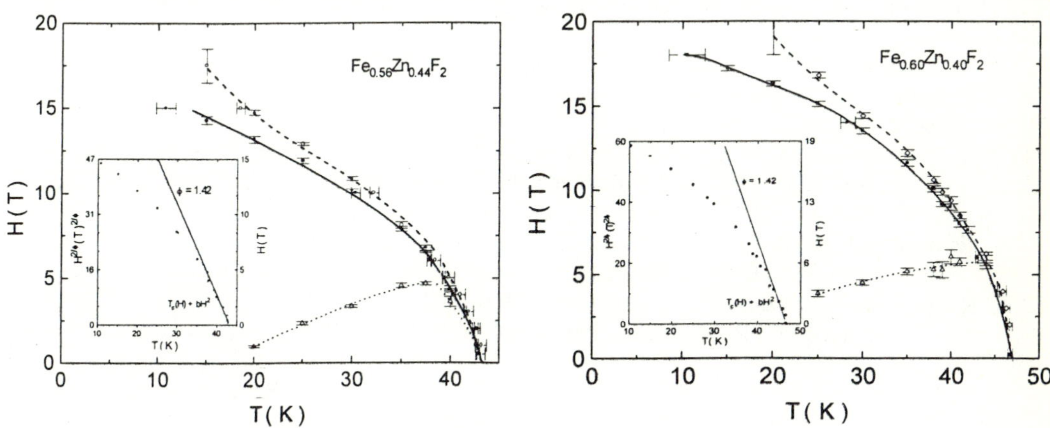

Fig. 3. Phase diagrams of $Fe_{0.56}Zn_{0.44}F_2$ and $Fe_{0.60}Zn_{0.40}F_2$. $T_c(H)$ is represented by full symbols. Equilibrium boundaries are represented by open symbols. Vertical and horizontal error bars are discussed in text. The insets show $T_c(H)$ data in $H^{2/\phi}$ versus T plots.

pronounced maxima, relative to the rounded ones found in the FD protocol. $T_c(H)$ was determined in the high-H experiments by the field position of the peak in the field derivative of M_{FI} (dM_{FI}/dH), for each value of T. The results are plotted in Fig. 3 for each sample. In the low-H regime. due to the shape of the (H,T) phase diagram, it is convenient to determine $T_c(H)$ from the position of the peaks in dM_{FH}/dT, which appears at low H when the sample is heated in the presence of a fixed H, (after it has been zero-field cooled from the paramagnetic phase to a given $T < T_N$). These curves are not shown here, but the critical temperatures obtained using this latter procedure are also plotted in Fig. 3 for both samples. It is noteworthy that the positions of the dM_{FI}/dT peaks in the (H,T) phase diagrams of Fig. 3 correlate quite well with the customary peaks in dM_{FH}/dT. Similarly, the upper equilibrium boundary, $H_{eq}^u(T)$ is in good correlation with the previously defined $T_{eq}(H)$ for the low-H experiments. We would like to stress here the significance of the novel lower equilibrium boundary, $H_{eq}^{\ell}(T)$. For $H < H_{eq}^{\ell}$, $\Delta M = 0$ and the system recovers its AF minimum-energy configuration by decreasing H from the paramagnetic phase. The $H_{eq}^{\ell}(T)$ boundary tends to reach the H = 0 line at a temperature T*(x) (this is more visible in the plot of $Fe_{0.56}Zn_{0.44}F_2$). For $T < T^*$, the H = 0 AF LRO is not recovered by FD in the time scale of the experiments. On the other hand by increasing T, $H_{eq}^{\ell}(T)$ line approaches a singular region in the (H,T) phase diagram. A multicritical (or possibly end-critical) point separates a low-H line of phase transitions, governed by the universal REIM-RFIM crossover scaling, from a pseudo-critical boundary at high-H, where strong random fields induce a glassy behavior.

In conclusion, the experiments in $Fe_{0.56}Zn_{0.44}F_2$ confirm the strong-random-field features previously reported[11] on a sample of $Fe_{0.60}Zn_{0.40}F_2$.

Acknowledgements – This work was supported by CAPES, CNPq and FINEP (Brazil). Work at Los Alamos was performed under auspices of the NSF and US DOE.

References

[1] For a review, see: D. P. Belanger and A. P. Young, J. Magn. Magn. Mater. 100 (1991) 272; D. P. Belanger, in: Spin Glasses and Random Fields, A. P. Young (Ed.), World Scientific, Singapore, 1997.

[2] M. T. Hutchings, B. D. Rainford and H. J. Guggenheim, J. Phys C 3 (1970) 307.

[3] E. P. Raposo, M. D. Coutinho-Filho and F. C. Montenegro, Europhys. Lett. 29 (1995. 507; J. Magn. Magn. Mater. 154, (1996) L155.

[4] F. C. Montenegro, M. D. Coutinho-Filho and S. M. Rezende, J. Appl. Phys. 63 (1988) 3755; Europhys. Lett. 8 (1989) 382; S. M. Rezende, F. C. Montenegro, M. D. Coutinho-Filho, C. C. Becerra and A. Paduan-Filho, J. Phys. (Paris) 49 (1989) 1267.

[5] S. Fishman and A. Aharony, J. Phys. C 12 (1979) L729.

[6] J. L. Cardy, Phys. Rev. B 29 (1984) 505.

[7] C. Ro, G. Grest, C. Soukoulis, K. Levin, Phys. Rev. B 31 (1985) 1682.

[8] J. R. L de Almeida and R. Bruinsma, Phys. Rev. B 35 (1987) 7267; U. Nowak and K. Usadel, Phys. Rev. B 44, (1991) 7426.

[9] F. C. Montenegro, U. A. Leitão, M. D. Coutinho-Filho and S. M. Rezende, J. Appl. Phys 67 (1990) 5243; F. C. Montenegro, A. R. King, and V. Jaccarino, Phys. Rev. B 44, (1991) 2155.

[10] D. P. Belanger, W. E. Murray, F. C. Montenegro, A. R. King, V. Jaccarino and R. W. Erwin, Phys. Rev. B 44 (1991) 2161.

[11] F. C. Montenegro, K. A. Lima, M. S. Torikachvili and A. H. Lacerda, J. Magn. Magn. Mater. 177-181 (1998) 145.

* Corresponding author, Tel.: + 55 81 271 8450; fax: + 55 81 271 0359; e-mail: fcm@df.ufpe.br.

Materials Science Forum Vols. 302-303 (1999) pp. 375-379

Evolution of Physical Properties in the $Ce_{1-x}La_xFe_2Ge_2$ Series

S.L. Bud'ko[1], M.B. Fontes[1], G.H. Walf[1,2], D. Mienert[2] and E.M. Baggio-Saitovitch[1]

[1] Centro Brasileiro de Pesquisas Fisicas, CBPF/CNPq, Rua Dr. Xavier Sigaud 150, Urca, Rio de Janeiro, RJ, 22290-180, Brazil

[2] Institut für Metallphysik und Nukleare Festkörperphysik, Technische Universität Braunschweig, Mendelssohnstraβe 3, DE-38106 Braunschweig, Germany

Keywords: Kondo Lattice, Single Ion Kondo Impurity, Magnetoresistance, $Ce_{1-x}La_xFe_2Ge_2$

Abstract

Evolution of structural, magnetic and transport properties of the $Ce_{1-x}La_xFe_2Ge_2$ series was studied by powder X - ray diffraction, temperature and field dependent magnetization and by resistivity in zero and applied magnetic field. The substitution of *Ce* by *La* in the $Ce_{1-x}La_xFe_2Ge_2$ series causes linear increase of lattice parameters in agreement with Vegard's law. Zero field resistance measurements together with the magnetization results suggest gradual change from the concentrated Kondo system - like behavior at low *x - La* to single Kondo impurity-like behavior for high *x - La* and finally to the normal metal for pure $LaFe_2Ge_2$ passing through the narrow region of the magnetically ordered ground state for the intermediate *La* concentrations. This evolution qualitatively resembles the behavior observed in the $Ce_{1-x}La_xRu_2Si_2$ compounds and can be qualitatively understood within the Doniach's diagram using the concept of negative chemical pressure.

Introduction

Recently a new moderate heavy fermion (HF) material $CeFe_2Ge_2$ ($\gamma \approx 210$ mJ/mol K^2), with integral valence of *Ce* (Ce^{3+}) at high temperatures, was reported [1,2]. This compound is one of the few rare earth heavy fermion compounds containing (nonmagnetic) *Fe*. Its *La* counterpart ($LaFe_2Ge_2$) is a normal metal with no anomalies in transport or magnetic properties [1]. Both pure materials crystallize in the tetragonal $ThCr_2Si_2$ -type structure in which a number of heavy fermions, as well as compounds with intermediate valence, HF superconductivity, exotic complex magnetic phase diagrams etc. were reported [3]. The isoelectronic substitution in the $CeFe_2Ge_2$ compound, performed in the transition metal site (the $Ce(Fe_{1-x}Ru_x)_2Ge_2$ series) [2] leads to the magnetically ordered ground state passing the quantum critical point [4]. It was shown [2,5] that for the $Ce(Fe_{1-x}Ru_x)_2Ge_2$ family the effect of chemical (substitution) and physical pressure are equivalent and both lead to the evolution of the physical properties which can be understood in general terms within the Doniach's phase diagram [6] and in greater detail in terms of the phase diagram of Continentino et al. [4] for concentrated Kondo systems (CKS). The goal of the present work is to study in detail the evolution of the properties of the $Ce_{1-x}La_xFe_2Ge_2$ series, where the situation is more complex: in addition to chemical pressure, *La* substitution causes a decrease of average magnetic moment at the rare earth site.

Experimental

The polycrystalline samples of the $Ce_{1-x}La_xFe_2Ge_2$ series, with $x = 0$, 0.1, 0.25, 0.3, 0.4, 0.45, 0.5, 0.55, 0.6, 0.75, 0.9 and 1, were synthesized by conventional arc-melting of the stoichiometric amounts of high purity elements in partial (~200 Torr) Ar atmosphere, with subsequent annealing at ~950°C for 4-5 days. The room temperature X-ray diffraction (XRD) patterns do not contain any impurity lines. The lattice parameters were determined using Rietveld refinement program [7]. Temperature and field dependent resistivity was measured by conventional four - probe technique using LR-700 AC resistance bridge (frequency f = 16 Hz, current 1-3 mA) in an Oxford cryostat with superconducting magnet (up to 80 kG) and VTI (1.6 - 300 K). DC magnetization for a subset of the samples was measured using Foner - type magnetometer in TU Braunschweig (T=2-300 K, H=0-12 kG).

It has to be noted that there are some intrinsic problems for this series, which have to be taken into account in the measurements and their interpretation: a)polycrystalline samples always have voids and micro-cracks and (uncontrolled) preferential orientation of the crystallites due to the thermal gradient in the time of their formation; b)furthermore, samples of this particular family are very brittle, so it is extremely difficult to cut them to regular rectangular bars, therefore the normalized rather than absolute values of resistivity will be presented; c)in addition, despite no spurious peaks were observed on XRD patterns, the M(H) curves reveal very small (< 1%) amount of magnetically ordered impurities, which saturate in field of less than 10 kG, these impurities have to be taken into account in the interpretation of the M(T) measurements (we guess that similar problem was responsible for the misinterpretation of the ground state of $CeFe_2Ge_2$ in [8]).

Results

Room temperature X-ray diffraction patterns reveal that the single phase compounds are formed in the whole range of concentrations. The lattice parameters and the unit cell volume increase linearly with the *La* concentration (Fig. 1) in agreement with Vegard's law. Their overall changes from $CeFe_2Ge_2$ to $LaFe_2Ge_2$ are about 1%, 0.8% and 2.8% for *a*, *c* and the unit cell volume, respectively. The *a/c* ratio remains almost constant through the series.

Zero field resistance measurements show residual resistance ratio RRR = ρ(300K)/ρ(2K) around 25, quite high for polycrystalline intermetallics, for the end compounds of the series, while it is approximately following the Nordheim's rule for the intermediate substitutions (Fig. 2).

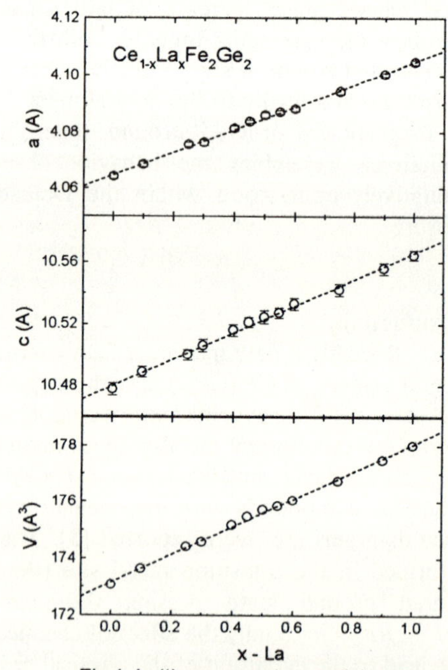

Fig.1. Lattice parameters and the unit cell volume for the $Ce_{1-x}La_xFe_2Ge_2$ series.

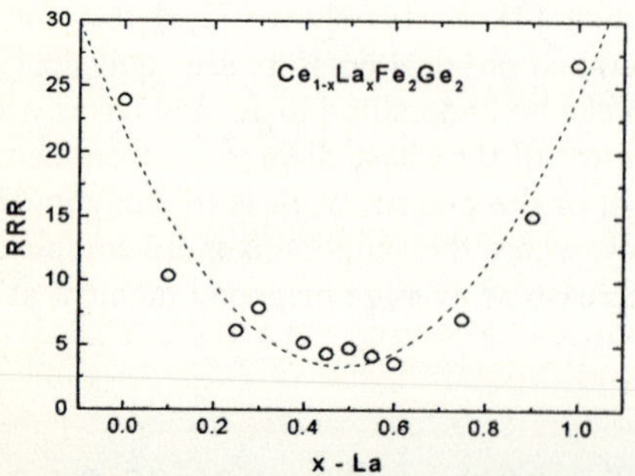

Fig. 2. Residual resistance ratios for the $Ce_{1-x}La_xFe_2Ge_2$ series

The representative curves of the temperature dependent resistance for the series are shown in Fig. 3a. For the pure $CeFe_2Ge_2$ the temperature dependence is typical for condensed Kondo system with high Kondo (and coherence) temperature (about 100 K) [1,2,5]. The overall similar behavior persists

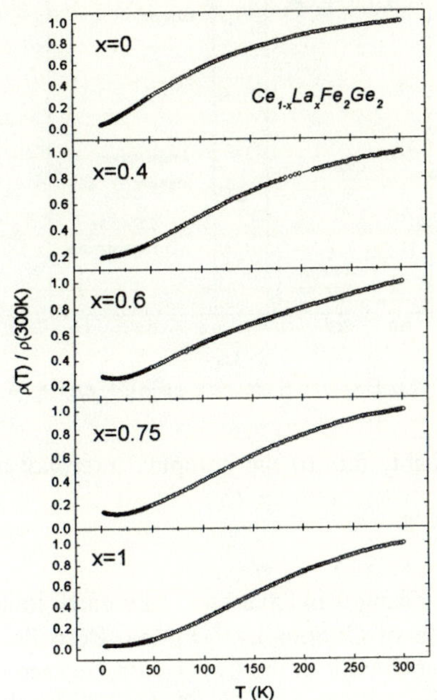

Fig. 3a. Temperature dependent resistance for representative $Ce_{1-x}La_xFe_2Ge_2$ compounds.

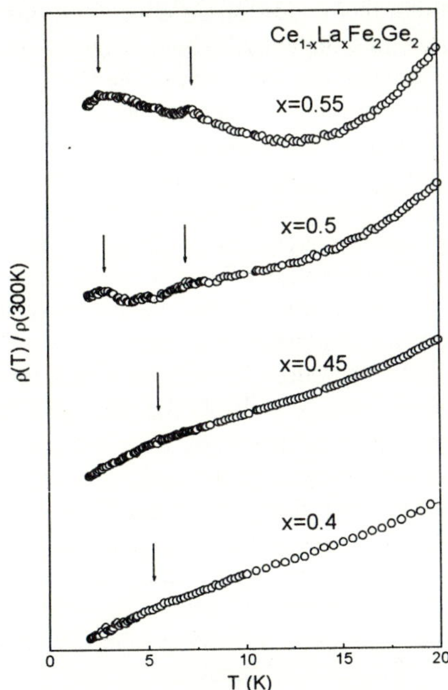

Fig. 3b. Low temperature part of normalized resistivity curves for intermediate *La* concentrations. Arrows mark anomalies.

up to *La* concentration of about $x = 0.6$. The other end compound, $LaFe_2Ge_2$, is a normal metal with no anomalies in temperature dependent resistivity. When *La* is partly substituted for *Ce*, a low temperature minimum, typical for the single impurity Kondo systems appears in ρ(T) curves. This behavior persists up to about 40% of *Ce* ($x = 0.6$), a value rather high for "conventional" single ion Kondo impurity systems. Finally, detailed examination of the low temperature part of the ρ(T) curves for *x* values around 0.5 (Fig. 3b) reveals that, for limited concentration range, the slope of the ρ(T) changes and/or bump - like anomalies appear. This fact, together with the preliminary magnetization data, favors the interpretation of the anomalies as the appearance of some kind of magnetic order. Examples of the field dependent magnetoresistance curves at T = 4.2K are shown in Fig. 4. It is positive and close to linear in field for pure $LaFe_2Ge_2$, negative, approximately following H^2 behavior, for single impurity Kondo region, and positive again for lower *La* concentrations in concentrated Kondo system region (Fig. 5).

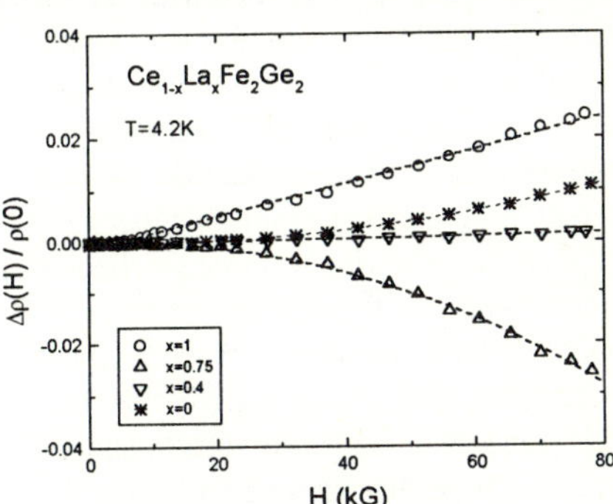

Fig. 4. Examples of field dependent magnetoresistance at T=4.2K

Based on the results of transport and preliminary magnetization measurements, which will be reported elsewhere [9], a tentative phase diagram for the evolution of the ground state of the $Ce_{1-x}La_xFe_2Ge_2$ system can be constructed (Fig. 6). The temperature of the Kondo minimum in the single Kondo

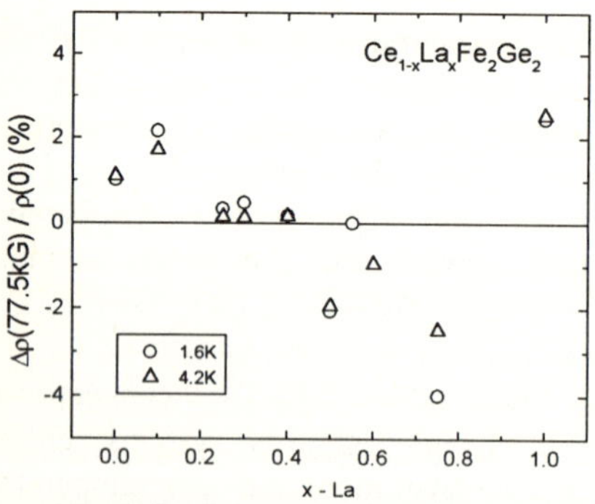

Fig. 5. Relative change of resistivity in applied field of 77.5kG

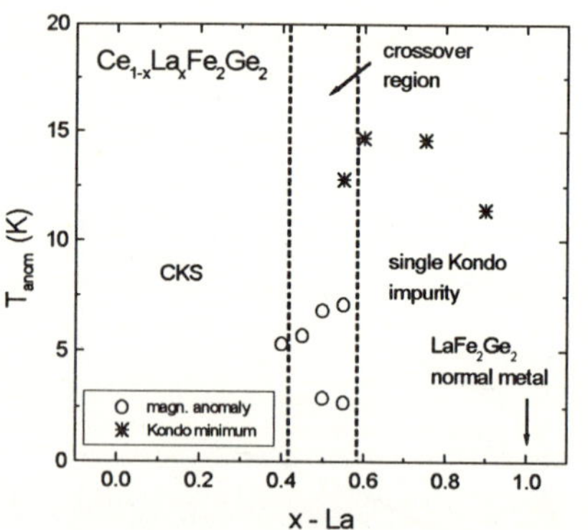

Fig. 6. Evolution of ground state in $Ce_{1-x}La_xFe_2Ge_2$

impurity region is not monotonic with *Ce* concentration, probably due to the complex interplay of the interactions close to the crossover region.

Discussion

The variation of the lattice parameters is following the changes in "average" rare earth ionic radius and is consistent with the expected random substitution of *Ce* ions by *La* ones. Zero field resistivity measurements reveal that single Kondo impurity-like behavior in the $Ce_{1-x}La_xFe_2Ge_2$ series holds down to $x \leq 0.55$, a rather high (≥45%) concentrations of *Ce*, i.e. contribution from the independent Kondo centers is dominant in a wide range of concentrations. Similar behavior was observed, for example, in $Ce_xLa_{1-x}Cu_6$ [10] and $Ce_xLa_{1-x}Pb_3$ [11] systems where single Kondo impurity-like behavior dominates up to as high as 80% of Ce concentration, as well as in the series with closely related tetragonal structure, $Ce_xLa_{1-x}FeGe_3$, where this region of *Ce* concentrations is about 50% [12,13]. The behavior is different for *Uranium* intermetallics, where single Kondo impurity regime occurs only for very small *U* concentrations (usually, less than 10%). The theoretical treatment of this difference between intermetallic Kondo systems containing *Ce* and *U*, using Nozieres' approach in Fermi liquid theory for magnetic impurities as scattering centers at low ($T < T_K$) temperatures, was presented in works of Gor'kov and Kim [14]. Low temperature field dependent magnetoresistance is positive for pure $LaFe_2Ge_2$, as expected for normal metal. In the single Kondo impurity regime it becomes negative and approximately proportional to H^2 in agreement with the theoretical views [15] for this region. In the concentrated Kondo system regime the low temperature magnetoresistance is positive again, consistent in sign with the one observed for pure $CeFe_2Ge_2$ [1]. For the CKS the theory [16] predicts positive magnetoresistance for $T \ll T_K$, which is consistent with our results since the estimation of T_K for the pure $CeFe_2Ge_2$ is about 100K [1]. It is predicted to turn negative for $T \geq T_K$, however magnetic fields available for this study are not high enough ($H \leq$ 80kG) to observe this change of sign at high temperatures with confidence. Finally, the anomalies observed in a narrow region of intermediate *La* concentrations suggest, that as a result of competition between the negative chemical pressure (which favors the appearance of the magnetic order according to the phase diagrams in [4,6]) and the loss of the "average" magnetic moment in the rare earth site (which opposes the formation of the ordered state), the system enters into a region where some type of magnetic order exists at low temperatures, similar to what was observed, for example, in $Ce_{1-x}La_xRu_2Si_2$ [17]. Additional studies, involving specific heat measurements, ^{57}Fe Mössbauer spectroscopy and neutron diffraction are necessary to clarify the nature of the observed low temperature anomalies.

Conclusions

The *La* substitution in the $Ce_{1-x}La_xFe_2Ge_2$ series causes the evolution of its ground state from concentrated Kondo system to single Kondo impurity regime and then to normal metal with a possible small intermediate region of magnetic order. The change of the magnetoresistance throughout the series is consistent with the theoretical views for each regime. The constructed tentative phase diagram can be viewed as a result of the interplay of the opposite effects of the negative chemical pressure and the loss of "average" magnetic moment on the rare earth site and is in agreement with predictions of the the Doniach's (or Continentino's) phase diagram for heavy fermion systems.

Acknowledgments

We appreciate experimental help of R.Pereira and W.Vanoni in the XRD measurements. This work was partially supported by CNPq (Brazil) (SLB, MBF), FAPERJ (Brazil) (SLB) and DAAD/CAPES (PROBRAL 193-97) Germany - Brazil exchange program (MBF, GHW, EMBS).

References

[1] T.Ebihara, K.Motoki, H.Toshima, M.Takashita, N.Kimura, H.Sugawara, K.Ichihashi, K.Settai, Y.Onuki, Y.Aoki, and H.Sato, Physica B, **206-207** (1995) 219.
[2] M.B.Fontes, M.A.Continentino, S.L.Bud'ko, M.El-Massalami, L.C.Sampaio, A.P.Guimarães, E.Baggio-Saitovitch, M.F.Hundley and A.H.Lacerda, Phys.Rev. B, **53** (1996) 11678.
[3] see for example: "Handbook of crystal structures and magnetic properties of rare earth intermetallics", ed. A.Szytula and J.Leciejewicz, CRC Press, Boca Raton, 1994.
[4] M.A.Continentino, Phys.Rev. B, **47** (1993) 11587; M.A.Continentino, G.M.Japiassu and A.Troper, Phys.Rev. B, **39** (1989) 9734.
[5] S.L.Bud'ko, M.B.Fontes, M.A.Continentino and E.Baggio-Saitovitch, Physica B, **217** (1996) 111.
[6] S.Doniach, in "Valence instabilities and related narrow band phenomena", ed. R.D.Parks, Plenum, New York, 1997, p.169.
[7] A.Sathkivel and R.A.Young, "Users guide to programs DBWS-9006 and DBWS9006PC", Georgia Inst. Techn., 1993.
[8] I.Felner, I.Mayer, A.Grill and M.Schieber, Solid State Comm., **16** (1975) 1005.
[9] S.L.Bud'ko, M.B.Fontes, G.H.Walf, D.Mienert and E.Baggio-Saitovitch, unpublished.
[10] Y.Onuki and T.Komatsubara, J. Magn. Magn. Mat., **63-64** (1987) 281.
[11] C.L.Lin, A.Wallash, J.E.Crow, T.Mihalishin and P.Schlottman, Phys. Rev. Lett., **58** (1987) 1232.
[12] H.Yamamoto and M.Ishikawa, J. Phys. Soc. Japan, **65** (1996) 50.
[13] S.L.Bud'ko, M.B.Fontes and E.M.Baggio-Saitovitch, J.Phys.: Cond. Matter, *accepted* (1998).
[14] L.P.Gor'kov and Ju H. Kim, Physica C, **235-240** (1994) 2451; Phys. Rev. B, **51** (1995) 3970.
[15] see for example: P. Schlottmann, Phys. Rep. **181** (1989) 1.
[16] C.Chen, Z.-Z. Li and W.Hu, J.Phys.: Cond. Matter, **5** (1993) 95 and refs. therein.
[17] M.J.Besnus, P.Lehman and A.Meyer, J.Magn.Magn.Mat., **63-64** (1987) 323.

Materials Science Forum Vols. 302-303 (1999) pp. 380-383

Local Spin Waves in Heisenberg Ferromagnets with Dipolar Field Coupling

E.H. Vivas C.[1,2] and J.C. Granada E.[1]

[1] Departamento de Física, Universidad del Valle, Santiago de Cali, Colombia, A.A. 25360
[2] E-Mail: hvivas@galois.univalle.edu.co

Keywords: Local Magnetic Excitations, Heisenberg Ferromagnet

The Heisenberg exchange interaction model is used to study the bulk and local magnetic excitations of a ferromagnet with a planar defect. A formalism is established to obtain the Green magnetic tensor components, which describe the dynamical response and the dispersion relations in the long wavelength approximation, taking into account the local magnetic field proposed by Phillips-Rosenberg [Rep. Prog. Phys. **29**, 285, (1966)] and the semiclassical magnetic torque relation.

PACS numbers: 75.25.+z, 76.50.+g, 71.45.+G

I. INTRODUCTION

There has been remarkable interest in the study of surface excitations in various magnetically ordered systems. It has been shown recently by many authors, that within the framework of the Heisenberg exchange model in addition to the bulk spin wave excitations there may be surface spin waves whose amplitudes decay exponentially away from the surface of the material toward the interior. Surface magnetic modes have also been treated extensively by macroscopic methods involving Maxwell's equations in cases where the effects of long range dipole-dipole interaction prevails over the short range exchange interactions, for ferromagnetic slab [1], and the influence of the dipolar interaction on the classical ground state and on quantum fluctuations in one dimensional antiferromagnets systems has been calculated [2]. Gopalan and Cottan [3] present a microscopic calculation of the surface and bulk excitation in degenerate ferromagnetic semiconductors using the s-d (or s-f) model which consists of a Heisenberg ferromagnet interacting with itinerant spins of the conduction electrons via a contact type s-d (or s-f) interaction. Wei G. [4], considered a layered ferrimagnetic Heisenberg model in which the coupling strengths between two nearest neighbor spins are J within a layer and $J_\perp$ between two layers. The effect of interlayer coupling on the sublattice magnetization is studied using spin-wave theory. In this paper, we will consider a broken symmetry ferromagnetic system and we use the spin wave theory in order to study the effect of dipolar coupling on the dispersion laws and the spectral density function of the local modes arising in such system.

II. FORMALISM FOR GREEN TENSOR COMPONETS

We assume a simple cubic ferromagnet (with lattice spacing a and the crystal axes orientation parallel to coordinate axes) and take exchange interactions only between nearest neighbors [5]. For our model, we consider breakdown in translational symmetry at $x_3 = 0$, where $J_{jk}(x, x+\delta)$ take the value J_o. In analytical form:

$$J_{jk}(\boldsymbol{x}, \boldsymbol{x}+\delta) = [J + (J_0 - J)\delta(x_3)]\delta(\boldsymbol{x} - \boldsymbol{a})\delta_{jk} \; ; \tag{1}$$

J is the interaction in places where $x_3 \neq 0$. The (Bloch equation) motion equation which relates the total magnetization $\boldsymbol{M}(\boldsymbol{x},t)$ and total field $\boldsymbol{H}(\boldsymbol{x},t)$ is a convenient macroscopic description of the long wavelength behavior of a ferromagnet and it is given by:

$$\left(\frac{d\,\boldsymbol{M}(\boldsymbol{x},t)}{dt}\right)_k = -\gamma(\boldsymbol{M}(\boldsymbol{x},t) \times \boldsymbol{H}(\boldsymbol{x},t))_k - \Gamma(\boldsymbol{M}(\boldsymbol{x},t) - M_0\,\boldsymbol{x}_3)_k \; ; \tag{2}$$

with:

$$M_k(\boldsymbol{x},t) = (M_0 - m_k(\boldsymbol{x},t))\delta_{3k} + m_k(\boldsymbol{x},t), \tag{3}$$

$$H_k(\boldsymbol{x},t) = (H_0 - m_k(\boldsymbol{x},t))\delta_{3k} + h_k(\boldsymbol{x},t), \tag{4}$$

where M_0 is the saturation magnetization, γ the gyromagnetic ratio ($g\mu_B/\hbar$ in S.I. units), and H_0 is the applied magnetic field on x_3 direction. In general, the single phenomenological Γ gives an inadequate representation of the complicated relaxation mechanism [6].

It does however provide a particularly compact and not too incorrect formulation of a number of dynamic processes. For a Heisenberg ferromagnet the fluctuating field $h_j^e(x,t)$ arises due to the exchange interactions. Following Phillips and Rosenberg (1966), this magnetic field can be expressed as:

$$h_j^e(x,t)=\frac{2}{\gamma^2}\sum_k\sum_\delta J_{jk}(x,x+\delta)M_k(x+\delta,t) \tag{5}$$

Where the coupling tensor $J_{jk}(x,x+\delta)$ between sites x and $x+\delta$ is given by (1). The dipolar field effect is considered through a *magnetostatic scalar potential* $\varphi(x,t)$:

$$h_j^d(x,t)=-\nabla_j\varphi(x,t). \tag{6}$$

The complete set of equations of motion at the medium can be written in compact form by introducing the relation

$$\hat{u}_j(x,\omega)=\sum_k\int d^3x'\,g_{jk}(x,x';\omega)\hat{f}_k(x',\omega)\ ; \tag{7}$$

where:

$$\hat{u}_j(x,\omega)\rightarrow(m_1(x,\omega),m_2(x,\omega),\varphi(x,\omega)) \tag{8}$$

$$\hat{f}_k(x,\omega)\rightarrow(\omega_m H_1(x,\omega),\omega_m H_2(x,\omega),0), \tag{9}$$

with

$$\omega_m=-\gamma M_0 \tag{10}$$

and $g_{jk}(x,x';\omega)$ are the components of the magnetic Green tensor. In this section we present calculations for obtain the magnetic Green tensor components $g_{jk}(x,x';\omega)$ for a broken symmetry ferromagnet. Let $x_\perp$ be the projection of x on the plane perpendicular to x_3 direction, although the presence of the planar defect removes infinitesimal translational invariance from our physical system on this direction, the system retains it in directions parallel to $x_\perp$. We can therefore simplify the determination of the Green's tensor $g_{jk}(x,x';\omega)$ by Fourier analyzing it according to:

$$\tilde{g}_{jk}(x,x';\omega)=\frac{1}{4\pi^2}\int d^2k_\perp\,\tilde{g}_{jk}(x_3,x_3';\omega,k_\perp)exp[ik_\perp\cdot(x_\perp-x_\perp')] \tag{11}$$

The general solutions for the components of the Green tensor must satisfy the following system of equations: (for $0<x_3, x'_3<0.$)

$$\hat{\Omega}(x_3;\omega,k_\perp)\tilde{g}_{ij}(x_3,x_3';\omega,k_\perp)=\hat{N}_{ik}(x_3;\omega,k_\perp)\delta_{jk}\delta(x_3-x_3'), \tag{12}$$

where $\hat{\Omega}(x_3;\omega,k_\perp)$ is a differential operator defined by

$$\hat{\Omega}(x_3;\omega,k_\perp)=-D^2\prod_{j=1}^{3}\left(\frac{d^2}{dx_3^2}+\zeta_j^2(\omega,k_\perp)\right), \tag{13}$$

where $D=-2J\omega_m a^2/\gamma^2$ and $\zeta_j(\omega,k_\perp)$ is any root of the condition $det[\tilde{N}_{ik}(\omega,k_\perp)]=0$, where $\tilde{N}_{ik}(\cdots)$ result of the replacement of d/dx_3 by $i\zeta_j(\omega,k_\perp)$ in $\hat{N}_{ik}(x_3,\omega,k_\perp)$. The non-homogeneous solutions of eqs. (12) can be written in compact form as:

$$\tilde{g}_{jk}^{(p)}(x_3,x_3';\omega,k_\perp)=-\frac{1}{D^2}\sum_{i=1}^{n}\frac{[G_{jk}(\omega,k_\perp)_i]}{\Delta_i(\omega,k_\perp)}\frac{exp(i\zeta_i|x_3-x_3'|)}{2i\zeta_i}; \tag{14}$$

$$\Delta_i(\omega,\boldsymbol{k}_\perp)=\prod_{j=1}^{n}\left[\zeta_j^2(\omega,\boldsymbol{k}_\perp)-\zeta_i^2(\omega,\boldsymbol{k}_\perp)\right];\quad i\neq j \tag{15}$$

where $\left[G_{jk}(\omega,\boldsymbol{k}_\perp)_i\right]$ are cumbersome expressions of ω and $\boldsymbol{k}_\perp$, and n is the number of roots $\zeta_j(\omega,\boldsymbol{k}_\perp)$ subject to the constraint: $Im[\zeta_j(\omega,\boldsymbol{k}_\perp)]>0$. The general solution of (12) is given by:

$$\tilde{g}_{jk}(x_3,x_3';\omega,k_\perp)=\tilde{g}_{jk}^{(p)}(x_3,x_3';\omega,k_\perp)+\sum_{l=1}^{n}K_l^j(x_3';\omega,k_\perp)exp(i\zeta_l\, x_3); \tag{16}$$

where the terms $K_l^j(\cdots)$ are functions determinated by the boundary condition of continuity of $m_1(\omega,\boldsymbol{x})$, $m_2(\omega,\boldsymbol{x})$ and $d\,\varphi(\omega,\boldsymbol{x})/dx_3$ at $x_3=0$ [7].

III. RESULTS

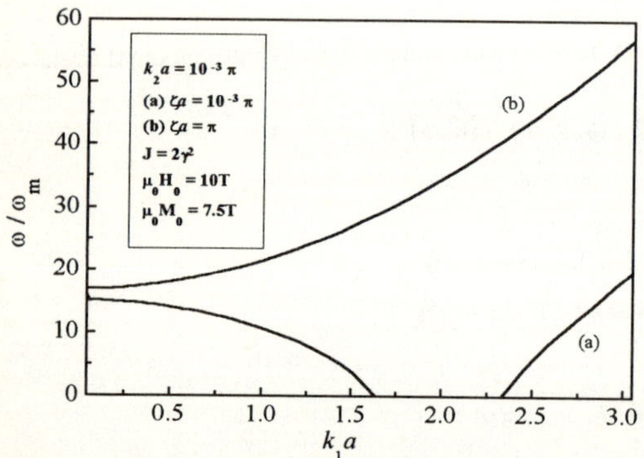

Figure 1. Spin wave spectrum of the bulk excitations (in units of ω_m) against k_1a in a Heisenberg ferromagnet for different propagation wavevectors. (a) (k_1, $10^{-3}\pi/a$, $10^{-3}\pi/a$), (b) (k_1, $10^{-3}\pi/a$, π/a). The applied magnetic field is such that μ_0H_0=10 T.

The spin wave excitations can be deduced by examining the poles of (14) and (16). In the figure 1 we show the frequency of the spin waves for bulk excitations as function of the wavevector k_1 for (a) $\zeta a = 10^{-3}\pi$ and (b) $\zeta a = \pi$. We observe that the dipolar coupling produces a decay in the frequency for values of $J/\gamma^2 \cong 2$ close the centre of the first Brillouin zone. A similar behavior of the low frequency modes in antiferromagnetic chains with dipolar interaction was reported recently by Hummed et. al.[2]. In figure 2 we plotted several local modes

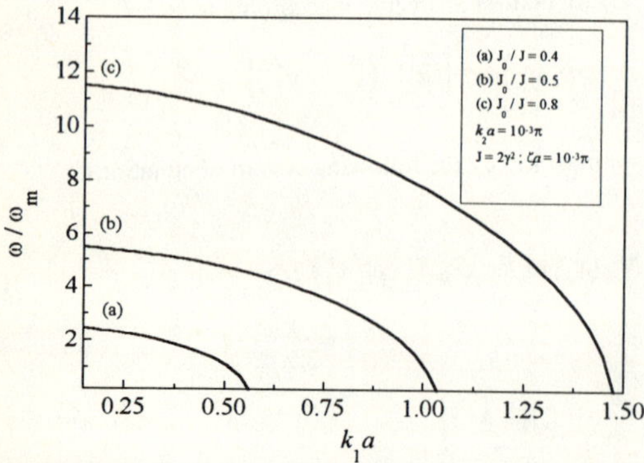

Figure 2: Dispersion curves of the local spin waves for different values of J_0/J taking into account the dipolar field coupling, where $\zeta a = 10^{-3}\pi$.

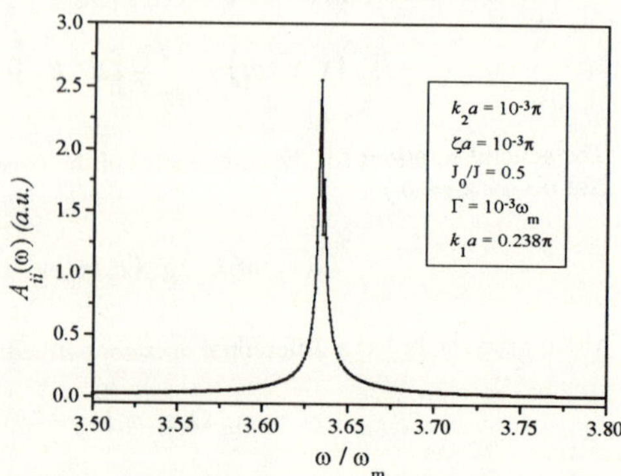

Figure 3: Spectral density function $A_{ii}(\omega)$ (in arbitrary units) associated to planar anisotropy at $x_3 = 0$, as function of ω/ω_m, in the particular case where: J_0/J =0.5, $\Gamma = 10^{-3}\omega_m$ and $k_1a = 0.238\pi$.

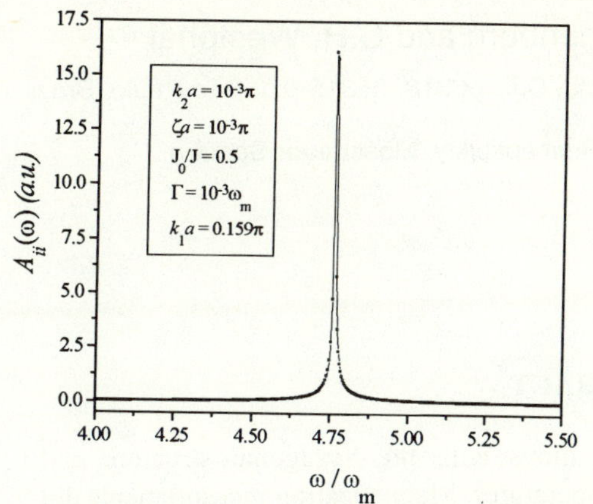

Figure 4: Spectral density function $A_{ii}(\omega)$ (in arbitrary units) associated to planar anisotropy at $x_3 = 0$, as function of ω/ω_m, in the particular case where: $J_0/J = 0.5$, $\Gamma = 10^{-3}\omega_m$ and $k_1a = 0.159\pi$.

associated with the existence of a layer breaking the translational symmetry at $x_3 = 0$.
The curves were plotted for the same bulk parameters as in fig. [1] and for different relations $J_0 / J < 1$. We see that the frequencies of the local modes lie in the forbidden region of the bulk excitations below the frequencies corresponding to the centre of the Brillouin zone. For a given value of J_0 / J there exists an interval of values of $k_1a < (k_1a)_{crit}$ in which the frequency of the local mode decreases until it becomes zero. For $k_1a > (k_1a)_{crit}$ the squared frequency becomes negative, which indicate that the ferromagnet will spontaneously create such spin wave and become demagnetized [6]. We see additionally that for a given value $k_1a < (k_1a)_{crit}$ the frequency of the local mode increases with J_0/J, and for $J_0 / J = 1$ it enters the bulk continuum. In figure [3] we have plotted the spectral density function $A_{ii}(x_3;\omega,\boldsymbol{k}_\perp)$ which is defined by:

$$A_{ii}(x_3;\omega,\boldsymbol{k}_\perp) = Im\{Sp[g(x_3,x_3,\omega,\boldsymbol{k}_\perp)]\}; \quad (17)$$

and provides a measure of the amplitude of the Fourier components with wavevector $\boldsymbol{k}_\perp$ and frequency ω of the thermal fluctuation in $m_i(x,t)$ with the system [8]. In our case, we study the contributions of anisotropy at $x_3 = 0$, and get different peaks associate to reduce frequency distribution for localized spin wave excitations, where $J/J_0 = 0.5$ and $k_2a=10^{-3}\pi$. We see that the peak corresponding to the local mode shifts to left with the increasing of the in-plane wavevector. This fact is in accordance with the behavior of the modes described in fig. [2].

IV. CONCLUSION

We have shown that in the regime when the exchange and dipole-dipole effects take place in a Heisenberg ferromagnets with broken translational symmetry, is possible the appearance of local modes which exists for a certain intervals of the in-plane wavevector.

ACKNOWLEDGMENTS

This work was supported by **COLCIENCIAS** under contract N^0 1106-05-025-96.

V. REFERENCES

[1] M. G. Cottam, D. R. Tilley in *Introduction to Surface and Superlattice Excitations*. Chapter **3**, 333 (Cambridge University Press, London, 1989).
[2] M. Hummel, C. Pich and F. Schwald. Preprint cond-mat/9703218, (1997).
[3] S. Gopalan and M.G. Cottam, Phys. Rev. B. **42** 10331, (1990).
[4] G. Wei, R. Qiu and A. Du, Phys. Lett. A **205**. 335, North Holland, (1995).
[5] H. S. Bennett and P. C. Martin, Phys. Rev., **138**, N^0 **2A**, 608, (1965).
[6] E. H. Vivas C. in *Study of magnons spectrum in Heisenberg ferromagnets with breakdown in the translational symmetry*. Chapter **4**. Thesis (in Spanish). Universidad del Valle, (1997).
[7] R. E. Camley and D. L. Mills, Phys. Rev. B. **18**, N^0 9, 4821, (1978).

Materials Science Forum Vols. 302-303 (1999) pp. 384-388

Magnetic Properties of $Zr(Fe_xCr_{1-x})_2$ Laves Phase Alloys

J.A.H. Coaquira, H.R. Rechenberg and C.H. Westphal

Instituto de Física, Universidade de São Paulo, C.P. 66318, 05315-970 São Paulo, Brazil

Keywords: Laves Phase, Spin Glass, Magnetic Susceptibility, Mössbauer Spectra

ABSTRACT

The magnetic properties of $Zr(Fe_xCr_{1-x})_2$ alloys with the hexagonal structure and $0.3 \leq x \leq 0.7$ have been investigated at low temperatures. Magnetization measurements did not reveal ferromagnetic ordering at any temperature. Low-field AC susceptibility *vs* temperature showed a sharp peak at $T = T_p$; T_p had a strong concentration dependence, vanishing at $x=0.26$. The T_p dependence on measuring frequency was measured in the 17-6000 Hz range. Mössbauer spectra were measured as a function of temperature and fitted with hyperfine field distributions. The average hyperfine field vanished at a temperature T_M ($T_M > T_p$) indicating spin freezing at the Mössbauer time scale. The T_p *vs* frequency and T_M data could be described by means of the Vogel-Fulcher equation with fitted parameters for each sample.

1. INTRODUCTION

Pseudobinary Laves-phase compounds of the general formula $A(B'_xB''_{1-x})_2$, where A is a rare earth or a nonmagnetic metal (e.g. Y, Zr, or Hf) and at least one of (B', B'') is an iron-group 3d element, provide a convenient framework for investigating the onset of magnetism in transition metal alloys [1,2]. Considerable attention has been paid to the $Zr(Fe_xM_{1-x})_2$ system with M another 3d metal or Al. While $ZrFe_2$ is a ferromagnet (T_C=630 K, M_s=1.60 μ_B/Fe), all other end compounds are nonmagnetic, leading to various kinds of magnetic order in the corresponding solid solutions [2-5]. For M=Cr, magnetic studies are still fragmentary [6,7]. The end compounds $ZrFe_2$ and $ZrCr_2$ (a Pauli paramagnet) are cubic (C15 structure), while their solid solutions crystallize in the hexagonal C14 structure in the $x \approx 0.2$-0.8 range [8]. Kanematsu and Fujita investigated the $Zr(Fe,Cr)_2$ system mostly near the Fe-rich end ($x \geq 0.85$, C15 structure) and proposed a nearest-neighbors model to interpret their measured ferromagnetic moments [3]. Jacob *et al.* reported ferromagnetic behavior for a few alloys in the C14 range, however noting that their magnetization curves were very far from saturation [9]. No reliable Curie temperature or magnetization data are as yet available for this system.

We have undertaken a systematic investigation of the magnetic properties of the C14-phase $Zr(Fe_xCr_{1-x})_2$ system for $0.3 \leq x \leq 0.7$. This composition interval was chosen in order to avoid possible complications arising from structural phase transformations. We here report on low-field AC susceptibility measurements which have revealed a well-characterized spin glass behavior for all samples studied.

2. EXPERIMENTAL DETAILS

$Zr(Fe_xCr_{1-x})_2$ samples with x=0.3, 0.4, 0.5, 0.6 and 0.7 were prepared by arc melting high-purity elements under argon. Powder x-ray diffractograms were analyzed with the Rietveld method. All samples were found to be single phase with the C14 structure. Magnetization measurements were made with a Foner-type vibrating-sample magnetometer, in magnetic fields up to 9 and at temperatures between 4.2 and 300 K. AC susceptibility was measured on powdered samples in the 1.5-200 K temperature range, using a Hartshorn bridge coupled to a lock-in amplifier. Mössbauer spectra were measured in a variable-temperature helium cryostat using a ^{57}Co(Rh) source kept at room temperature.

3. RESULTS AND DISCUSSION

Magnetization measurements indicated that none of the studied alloys could be characterized as ferromagnetic at T=4.2 K. The magnetization increased smoothly with applied field, showing no saturation tendency. Though some hysteresis was observed for x=0.7 and 0.6, Arrott plots did not extrapolate to a positive M^2 for any sample. M vs. $1/H$ extrapolation yielded a saturation magnetic moment of 0.74 μ_B/Fe at x=0.7, and smaller values for smaller x.

Low-field AC susceptibility measurements as a function of temperature did not exhibit a well-defined Curie-Weiss behavior, but a sharp peak was evident at low temperatures for all samples. Figure 1 shows χ vs. (T/T_p) data around the peak temperature T_p. The concentration dependence of T_p (measured at 155 Hz) is given in Table I, and it can be described by an x^2 law with $T_p \to 0$ for x=0.26.

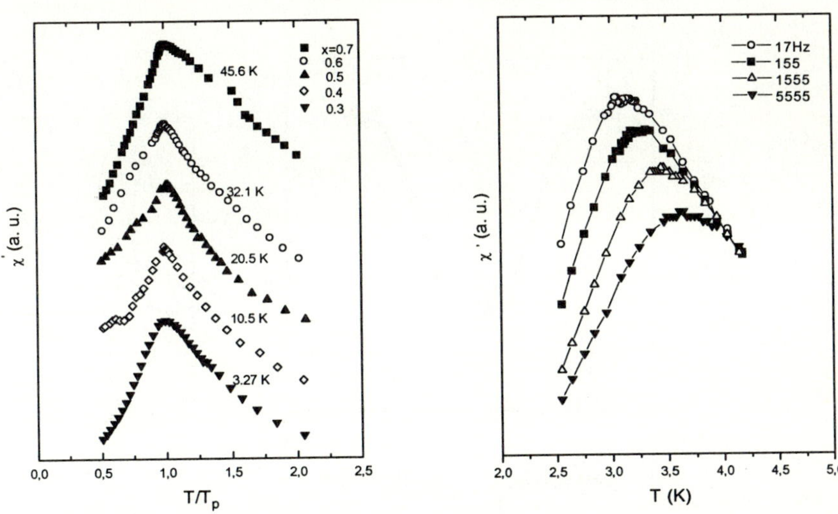

Fig. 1. AC susceptibility *vs* temperature. Peak temperatures are indicated.

Fig.2. $Zr(Fe_{0.3}Cr_{0.7})_2$ susceptibility *vs* T at various frequencies.

Figure 2 illustrates the dependence of the χ vs. T curve on measuring frequency. This behavior is easily understood if the spin-glass transition is pictured as the blocking process of effective magnetic clusters. The frequency effect is usually described by the condition $\omega_{meas}=\tau(T_p)^{-1}$, where $\tau(T)$ is a relaxation time given by the Vogel-Fulcher law [10]

$$\tau(T) = \tau_0 \exp\left(\frac{E_a}{k(T - T_0)}\right) \quad (1)$$

T_0 has been interpreted as a measure of intercluster interactions [11]. We have measured T_p at four frequencies in the 17-6000 Hz range and fitted the data with Eq. (1), using $\tau_0 = 10^{-13}$ s. Fitted E_a and T_0 values are given in Table I. E_a varies nearly linearly with x, and T_0 tends to zero at x=0.26.

Fig. 3 shows some Mössbauer spectra of $Zr(Fe_{0.7}Cr_{0.3})_2$ at various temperatures. Good fits could be obtained with a $P(B_{hf})$ given by two Gaussian functions. The T=60 K spectrum could be fitted with two quadrupolar doublets without any magnetic broadening.

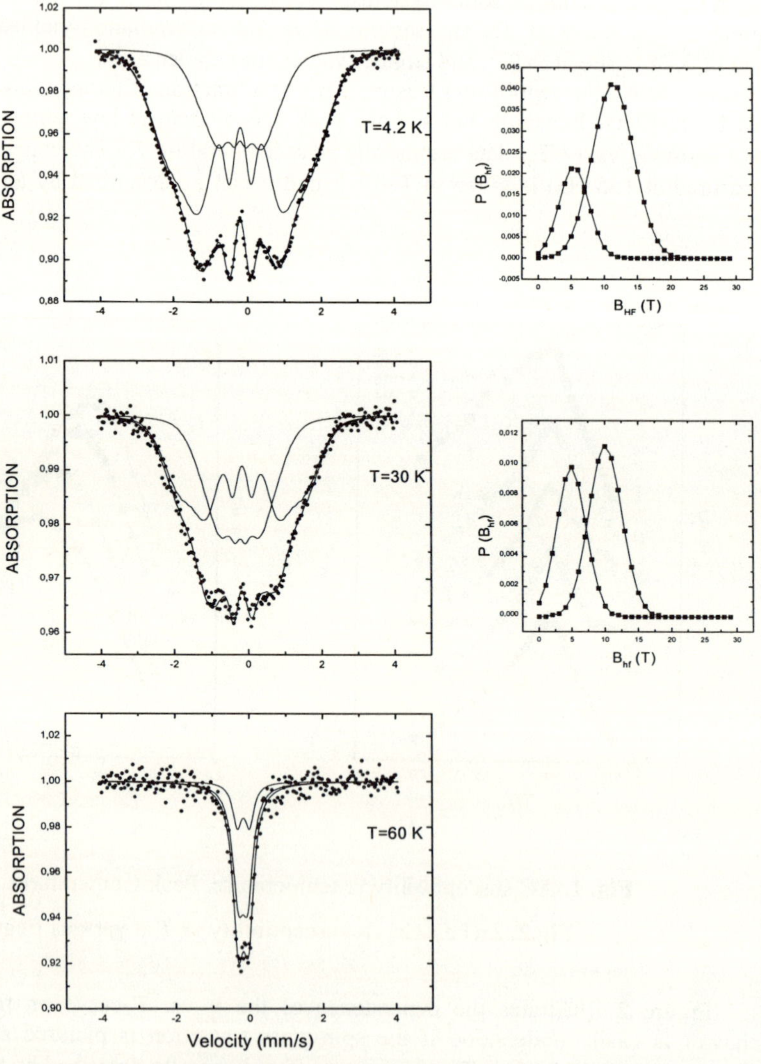

Fig. 3. Mössbauer spectra of $Zr(Fe_{0.7}Cr_{0.3})_2$ at various temperatures. The spectra at T=4.2 K and 30 K were fitted with a double-Gaussian hyperfine field distribution, also shown.

TABLE I. Concentration dependence of freezing temperature from AC susceptibility data at 155 Hz and parameters obtained from fitting their measured frequency dependence with Vogel-Fulcher's law.

x	0.3	0.4	0.5	0.6	0.7
T_p (K)	3.27	10.5	20.5	32.1	45.6
T_0 (K)	1.33	7.6	15.6	25.0	36.9
E_a/k (K)	49	67	123	173	210

The fitted hyperfine field average $\langle B_{hf} \rangle$ is plotted as a function of temperature in Figure 4, and can be seen to vanish at T=58±3 K. This value is significantly higher than the freezing temperature obtained from χ_{ac} (T_p=46 K), but it is in excellent agreement with Eq. (1) for $\tau=10^{-9}$ s, *i.e.* the characteristic time for Mössbauer hyperfine measurements.

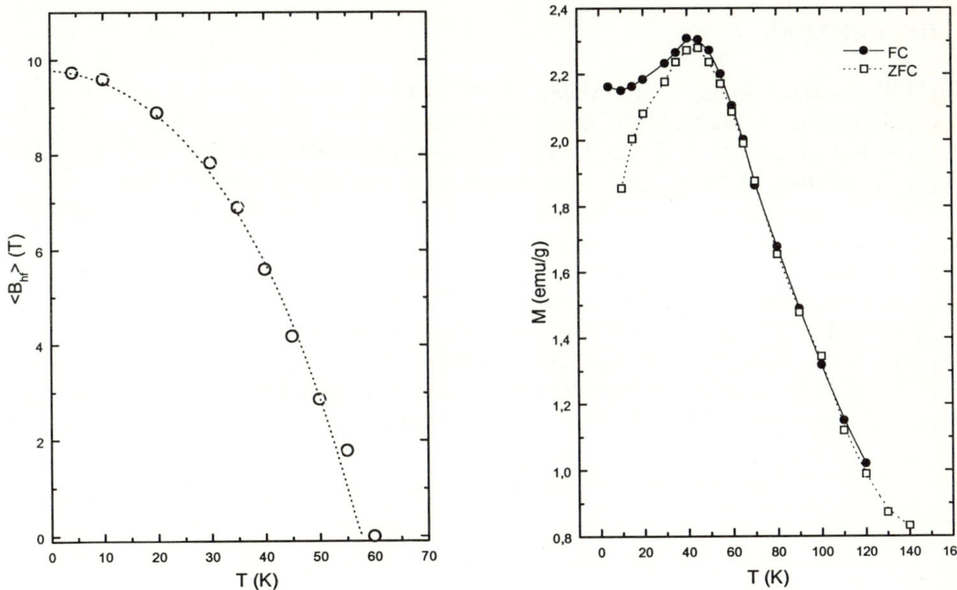

Fig. 4. Average hyperfine field for $Zr(Fe_{0.7}Cr_{0.3})_2$ as a function of temperature.

Fig. 5. Zero field-cooled and field-cooled magnetization ($\mu_0 H$=0.21 T) of $Zr(Fe_{0.7}Cr_{0.3})_2$.

The $Zr(Fe_{0.7}Cr_{0.3})_2$ alloy was investigated in more detail for other signatures of spin-glass behavior. Zero field-cooled and field-cooled magnetization data, taken at 0.21 T applied field and shown in Figure 5, revealed characteristic spin freezing behavior. In addition, a small magnetization remaining after turning off the field at T=4.2 K exhibited a slow logarithmic decay.

4. CONCLUSIONS

In summary, we have found that the $Zr(Fe_xCr_{1-x})_2$ system exhibits spin freezing behavior without long-range magnetic order for Fe concentrations ranging from x=0.26 up to, at least, x=0.7. The ferromagnetic phase boundary at the Fe-rich side still has to be determined. It would be interesting to see to what extent the magnetic properties are sensitive to the C14-C15 structural change in this system. Finally, it is of interest to compare our results to the extensively studied $Zr(Fe_xCo_{1-x})_2$ system [5,12]. In the latter, ferromagnetism extends from $ZrFe_2$ to $x \approx 0.5$ and cluster-glass behavior [12] is limited to the narrow $0.25<x<0.5$ interval. The larger stability range of the ferromagnetic phase in $Zr(Fe,Co)_2$ is related to the fact that the Fe moment decreases more slowly upon alloying than for $Zr(Fe,Cr)_2$ (cf. [3,5]). In addition, the magnetic properties of $Zr(Fe,Co)_2$ have been reported to be very sensitive to preparation conditions: *e. g.*, the Curie temperature is lowered by thermal annealing. This fact has been ascribed [12] to the occurrence of Fe atoms at Zr sites in as-cast samples; such anti-structure atoms are believed to play an essential role in the building up of an infinite magnetic cluster. We are not aware of such effects for $Zr(Fe,Cr)_2$ but, as far as the Rietveld analysis of our diffractograms indicates, no significant site inversion appears to occur for the as-cast samples used in the present study.

ACKNOWLEDGEMENTS: This work was supported by CNPq and FAPESP.

REFERENCES

[1] W. Steiner, J. Magn. Magn. Mater. 14 (1979) 47
[2] G. Hilscher, J. Magn. Magn. Mater. 27 (1982) 1
[3] K. Kanematsu and Y. Fujita, J. Phys. Soc. Japan 29 (1970) 864
[4] Y. Muraoka, M. Shiga, and Y. Nakamura, phys. stat. sol.(a) 42 (1977) 369
[5] Y. Muraoka, M. Shiga, and Y. Nakamura, J. Phys. F: Metal Phys. 9 (1979) 1889
[6] D. Shaltiel, I. Jacob, and D.Davidov, J. Less-Common Met. 53 (1977) 117
[7] H. Fujii, F. Pourarian, and W. E. Wallace, J. Less-Common Met. 88 (1982) 187
[8] K. Kanematsu, J. Phys. Soc. Japan 31 (1971) 1355
[9] I. Jacob, D. Davidov, and D. Shaltiel, J. Magn. Magn. Mater. 20 (1980) 226
[10] J. L. Tholence, Solid St. Commun. 35 (1980) 113
[11] S. Shtrikman and E. P. Wohlfarth, Phys. Lett. 85A (1981) 467
[12] G. Wiesinger and G. Hilscher, J. Phys. F: Metal Phys. 12 (1982) 497

Materials Science Forum Vols. 302-303 (1999) pp. 389-392

Reentrant Transition and Spin-Glass State in $CoIn_{2-2x}Cr_{2x}S_4$ Single Crystals

V. Sagredo[1], J. Mantilla[1] and R. Zysler[2]

[1] Physics Department, Faculty of Sciences, Universidad de los Andes, Mérida, Venezuela

[2] Centro Atómico Bariloche, Bariloche, Argentina

Keywords: Spin Glass, Reentrant Ferrimagnet

Abstract. The magnetic properties of system $CoIn_{2-2x}Cr_{2x}S_4$ ($0 \leq x \leq 0.5$) with spinel structure were investigated by means of dc magnetization measurements in magnetic fields 1- 50 kOe.
The results of magnetic measurements show that the compounds with $x \leq 0.15$ exhibits features typical for spin-glass-like behavior and one with $x = 0.5$ presents a reentrant ferrimagnetic behaviour.

Introduction

Sulfur spinels exhibit a variety of physical properties which make them interesting materials. Sulfospinels presents metallic conduction and superconductivity as well as semiconductivity[1]. Spinels with the general formula $CuCr_2X_4$ (X : S,Se,Te) have been found to be metallic ferromagnet with high Curie temperature. These properties are in contrast with those of the other chromium chalcogenide spinels which are semiconductors with magnetic critical temperature lower than room temperature, e.g., $MnCr_2S_4$ with T_c= 77 K, $FeCr_2S_4$ with T_c = 177 K and $CoCr_2S_4$ with T_c = 240 K [2,3].
The space group of the spinel (AB_2X_4) is $Fd\overline{3}m$. The chalcogenide X anions approximate to a close packed cubic lattice.The cations occupy twice as many octahedral sites B as tetrahedral sites A. The local symmetry of sites A is tetrahedral with point group T_d, whereas for the B sites the point group is D_{3d}. The crystal structure is completely described by the cubic edge a. The angles and distances between the cations are important in determining the magnetic properties of the spinels.The exchange interactions in ferrites with this structure are of the indirect type.The strongest interaction is between the A cations in tetrahedral sites and the B cations in the octahedral sites. It is possible to realize either a selective magnetic dilution in one of the two sublattices or a distribution, at different extent of magnetic ions between the sublattices. This leads to different distribution and ratios between the strenght of the exchange interactions
The aim of this work is to study the influence of the substitution of indium ions by chromium ions in the magnetic properties of the spinel series $CoIn_{2-2x}Cr_{2x}S_4$ ($0 \leq x \leq 0.5$) The compound with x = 0.0 is an spin-glass semiconductor with a transition temperature of 12.4 K and that with x = 1.0 is a ferrimagnetic semiconductor [3,4].
Gashimov et.al [5] reported the magnetic and electrical properties of that system for ($0.5 \leq x \leq 1$) on sinthetized samples. With increasing x the following observations were made: a decrease of the Curie temperature down to 187 K for $CoInCrS_4$ as well as a decrease of the saturation intensity of magnetization per molecule.

Experimental details

The samples were prepared by reaction of the stoichiometric amounts of the high purity elements. The component elements were sealed under vacuum in a quartz capsule and held at 1100 °C for

three days, the resulting ingot was cooled to room temperature and then was carefully ground. About 2 g of the sinthetized material were introduced into a quartz ampoule and a suitable amount of $CrCl_3$ transport agent added.
The crystals were grown by placing the ampoule in a two zone temperature furnace keeping the source temperature and the deposition temperature at 1000 and 950°C respectively.
X-ray powder diffraction analysis of all samples has been performed to identify the unit cell of $CoIn_{2-2x}Cr_{2x}S_4$.From these x-ray measurements it is concluded that all the compounds under study are spinels. However more precise structural data are required to know the distribution of Co,In and Cr atoms among the tetrahedral and octahedral sites in the lattice. DC magnetic susceptibility measurements have been performed in the temperature range $2 \leq x \leq 300$ K by using a Squid magnetometer.

Results and discussion

Pure $CoIn_2S_4$ and $CoCr_2S_4$ are referred to as nearly inverse spinels.In the former compound the tetrahedral sites are occupied to the extent of 37 % by Co^{2+} and 63% by In^{3+} ions [6,7]. The structure of $CoIn_{2-2x}Cr_{2x}S_4$ samples was verified by x-ray measurements. The results of the x-ray studies show lattice parameters values which agree with previous reported values [5]. At x = 0.15 and x = 0.5 the lattice parameters are equal to 10.605 Å and 10.050 Å respectively. Figures 1 and 2 shows the temperature dependence of the magnetization for $CoIn_{2-2x}Cr_{2x}S_4$ with x = 0.15 and x = 0.5.

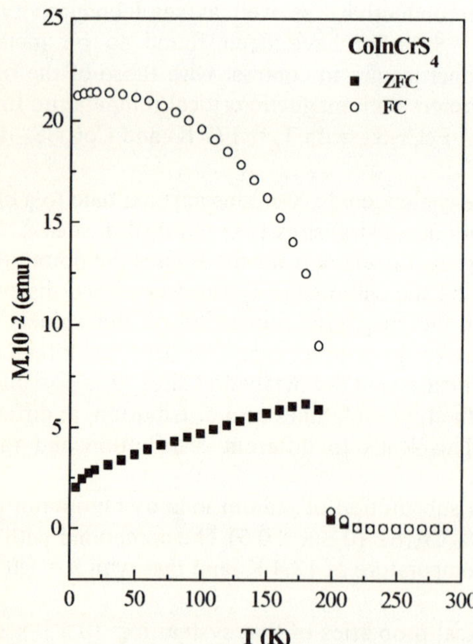

Fig. 1. Temperature dependence of the FC and ZFC magnetization for x = 0.5.

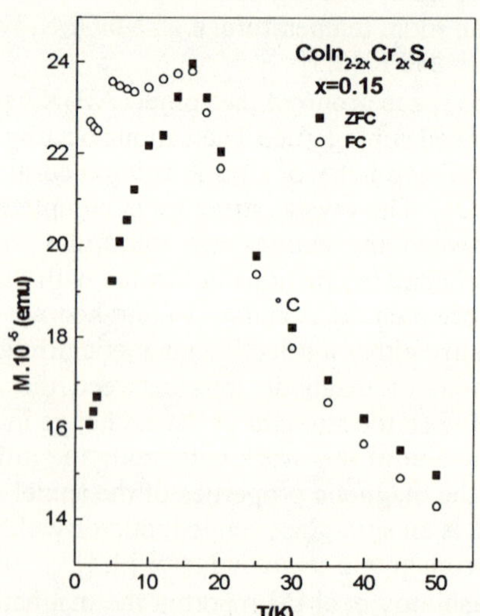

Fig. 2. Temperature dependence of the FC and ZFC magnetization for x = 0.15.

The behaviour of these samples indicates a quite different magnetic behavour. For the low chromium concentration sample in Fig.2 it is possible to observe a peak in the ZFC data at T = 16 K, this value is very close to the one reported for E.Agostinelli et.al [4] for $CoIn_2S_4$ (T = 12.2 K)

A different behaviour is observed from the FC magnetization data,it remain almost constant with the decreasing temperature.
This differences should reflect a different extent of magnetic correlations.The appearance of irreversibility just below the freezing temperature is indicative of spin-glass like behaviour and suggest a cooperative freezing between the Co and Cr ions.
Because of the random positions of the Co^{2+} and Cr^{3+} ions in the sublattices the ferro and antiferromagnetic interactions are the main cause of the spin-glass like behaviour in these samples.
However the strong negative interaction between Co^{2+} and Cr^{3+} ions gives the ferrimagnetic order observed in Fig.1.The undiluted $CoCr_2S_4$ is ferrimagnetic (the A site moments are oriented antiparallel to the B site moments) with B-B and A-A interactions remaining unsatisfied. We could expect that when non-magnetic ions are introduced, the frustration produces local canted states [8]. On dilution, well below the percolation threshold for the A-B bonds only, the collinear ferrimagnetic phase breaks down to a so called reentrant spin-glass state, as is shown in Fig.1 . Where the longitudinal component orders ferrimagnetically and the transverse one exhibits a spin-glass order. Two transitions are then expected: one at T_n for the longitudinal component and a second one at T_f =197 K for the transverse component. A similar low temperature canted state has been predicted by Gabay and Toulouse on the basis of the Mean Field model in the ferromagnetic order [9].
The reentrant transition is expected if a spin concentration is as required for the onset of the ferrimagnetic ordering. The system could be considered as consisting of a ferrimagnetic network

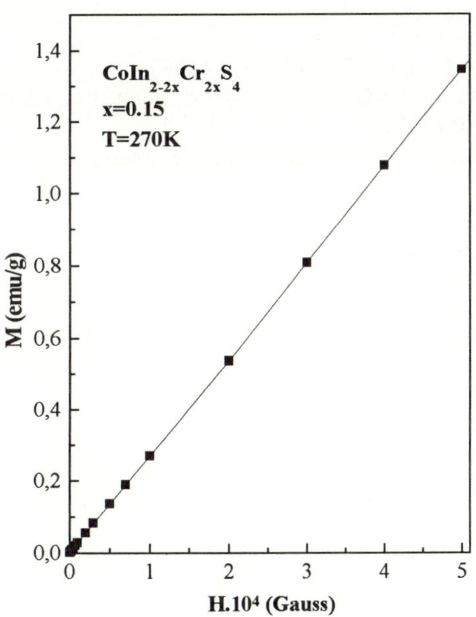

Fig. 3. Field dependence of the magnetization of $CoIn_{2-2x}Cr_{2x}S_4$ with x = 0.15.

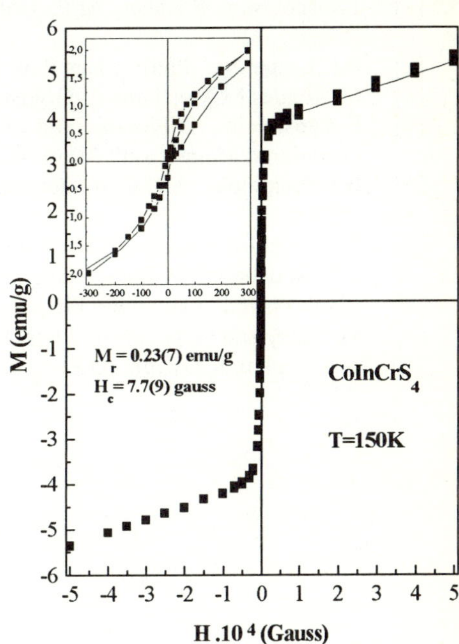

Fig. 4. Magnetization curve of $CoInCrS_4$ at 150 K.

and finite spin-clusters. If the intra-clusters coupling exist, then below the freezing temperature the ferrimagnetic state becomes unstable and evolves into a spin glass like state.
Fig.3 and Fig 4 presents the magnetization isotherms for samples with x = 0.15 and x = 0.5. As can be noticed the isotherms of the magnetization were not saturated in the fields in which

measurements were carried out: up to 5 T at 270 and 150 K respectively. However the application of the high magnetic field (5T) at 150 K causes a ferrimagnetic ordering in sample with x = 0.5.
As we have suggested the magnetic structure of the $CoIn_{2-2x}Cr_{2x}S_4$ compounds is a result of the competition of the exchange interaction between the magnetic moments. As it is known in the spinel structure direct negative BB interactions are possible, however the 90° Cr-S-Cr superexchange integral is positive [10]. These appears to be competition leading to a very small value for the Cr-Cr-interaction, but it seems that the negative interaction between Co^{2+} in A sites and Cr^{3+} in B sites is strong enough to gives a ferrimagnetic order.

Conclusions.

It is shown that substitution of In^{3+} by Cr^{3+} in the system $CoIn_{2-2x}Cr_{2x}S_4$ gives rise to spin-glass like state and in high chromiun concentration samples the reentrant ferrimagnetic state appears. The examination of the magnetic behaviour of the studied samples allow us to suggest that as the ratio of the interactions increases the frustration of the antiferromagnetic bonds increases favoring the appearance of the reentrant ferrimagnetic state.

References.

[1] I.Sosnowska, P.Fisher, M.F. Collins, B.D. Gaulin and M.Gogolowicz, J.M.M.M. **80** 311 (1989).
[2] F.K.Lotgering, Phillips Res.Rep. **11**, 190 (1956).
[3] M.Nogues,M.Mejai and L. Goldstein, J.Phys.Chem.Solids **40**, 375 (1979).
[4] E.Agostinelli, D.Fiorani and A.M.Testa "Fundamental and Applicative Aspects of Disordered Magnetism" World Scientific ,30 (1979).
[5] G.I.Gashimov, A.G.Rustamov and M.B.Gadzhlev. Neorganicheskie Materialy, **28** 2053 (1992).
[6] D.Fiorani,L.Gastaldi and S.Viticoli, Solid state Comm. **48**, 869 (1983).
[7] Y.R.Bouchard,P.A.Russo and A.Wold. Inorg.Chem., **4**, 685 (1965).
[8] J.Hubsch and G.Gavoille, Phys.Rev.B **26,** 3815 (1982).
[9] M.Gabay and G.Toulouse. Phys. Rev. Lett., **47**, 201 (1981).
[10] T.A.Kaplan, K.Dwight, D.Lyon and N. Menyuk. J.Appl.Phys.Supp.**32**,125 (1961).

Materials Science Forum Vols. 302-303 (1999) pp. 393-396

Magnetic Ordering in the $TbNi_5$ Hexagonal Compound

V.M.T.S. Barthem[1], H.S. Amorim[1], D. Schmitt[2] and D. Gignoux[2]

[1] Instituto de Física-UFRJ, Cx. P. 68528, CEP 21945-970, Rio de Janeiro, RJ, Brazil

[2] Laboratoire Louis Néel, CNRS, 166X, FR-38042 Grenoble, France

Keywords: Intermetallic Compounds, Magnetic Properties, Helimagnetism

Abstract. The magnetic properties of the hexagonal compound $TbNi_5$ have been studied. The AC susceptibility measurements on $TbNi_5$ single crystal show that, at zero DC magnetic field, the compound is ferromagnetic for temperatures smaller than 16.5 K. At this temperature the compound becomes helimagnetic. For fields greater than 450 Oe the compound becomes ferromagnetic at any temperature in the ordering phase. The thermal variation of the magnetization, as its field dependence, indicates the presence of domains in the helical phase.

1-Introduction

The intermetallic $TbNi_5$ compound crystallizes in the $CaCu_5$ - type hexagonal structure (p6/mmm space group)[1]. In this compound, only the terbium ions are magnetic. At temperatures smaller than 23.5 K, and at high magnetic fields, the compound is ferromagnetic. It has a strong magnetocrystalline anisotropy where there is an easy plan and a hard c-axis [2]. In a previous work [3], magnetic measurements in a polycrystalline sample of $TbNi_5$ have indicated that, at low field (H<400 Oe), the compound is helimagnetic in the temperature range $17.5\ K < T < 23.5\ K$. The present paper is an extension of this work. We present new evidences of the helical phase with some additional magnetic measurements in a single crystal of $TbNi_5$ that allow us to determine more precisely the transition ferro-helimagnetic temperature and the critical field that make the heli-ferromagnetic transition.

2-Experimental Procedure

The single crystal has been prepared by Czochralski method in a crucible induction furnace at the Laboratories Louis Néel (CNRS) in Grenoble/France. The sample was oriented, within 0.5^0, by x-ray diffraction using the conventional back-reflection Laue method. It also follows from it that the compound crystallizes in the $CaCu_5$ structure but there is also a spurious phase. This phase was also observed in the magnetic measurements and isolated. The measurements were performed in a Lake Shore 7100 Magnetometer and AC Susceptometer.

3-Results and discussion

Figure 1 shows the magnetization in the easy a-axis of $TbNi_5$, in a field of 50 Oe, as function of temperature. The lower curve (zero field cold, ZFC) was obtained by first cooling the system in a zero magnetic field from 50 K to 4.5 K. The magnetic field is then applied and the magnetization is measured while the temperature is increased. The upper curve (field cold, FC) is obtained in a similar manner except that the sample is cooled in a measuring field. The different behaviour, as the sample follows a FC or a ZFC process, can be explained as due to the existence of magnetic domains. This explanation agreed with reference [3] in which it was observed a dependence of the AC susceptibility with the H_{AC} amplitude, characterizing the presence of domains in the order phase. The ZFC curve shows the progressive unblocking of the domains until T_{fh}=16.0 K, when the compound becomes helimagnetic and this curve merges with the FC curve up to 21.0 K. In a narrow temperature interval, just below T_N (the peak temperature 23.5K) there is a shift between the two curves. So, from these measurements we identify, like in ref. [3] (except by the T_{fh} value), two different magnetic behaviors in the two ranges of temperature; for $T<T_{fh}$ the compound is ferromagnetic and for $T_{fh} \leq T \leq 23.5K$ the compound is helimagnetic. Above T_N, in the FC curve, one can see the presence of the impurity phase observed in the X-ray analysis.

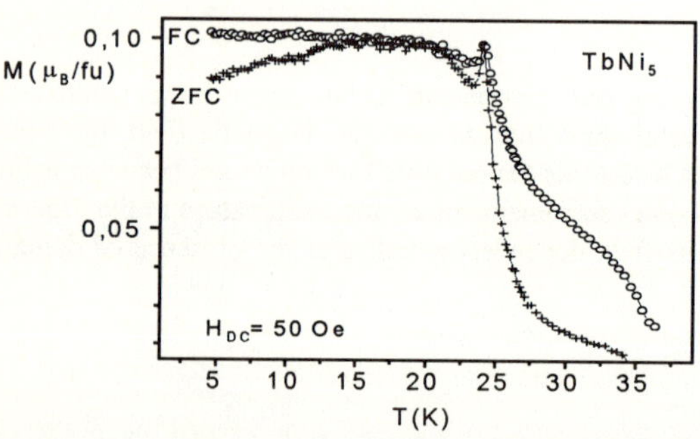

Figure-1. Magnetization temperature dependencies of $TbNi_5$ along the easy *a*-axis magnetization ZFC and FC refer to zero field cooling and field cooling of the sample, respectively.

The thermal variation of the real part of the AC susceptibility, at different DC field along the $TbNi_5$ easy *a* axis, is presented in figure 2(a) and, along the hard *c* axis, in figure 2(b). It can be seen that the peak associated to the helimagnetic transition goes to higher temperatures, that is with the field the ferromagnetic phase exists at a higher temperature. The peak disappears, that is there is no more the helical phase, when the field reaches a critical value of about 450 Oe. So, the compound becomes ferromagnetic at all temperatures.

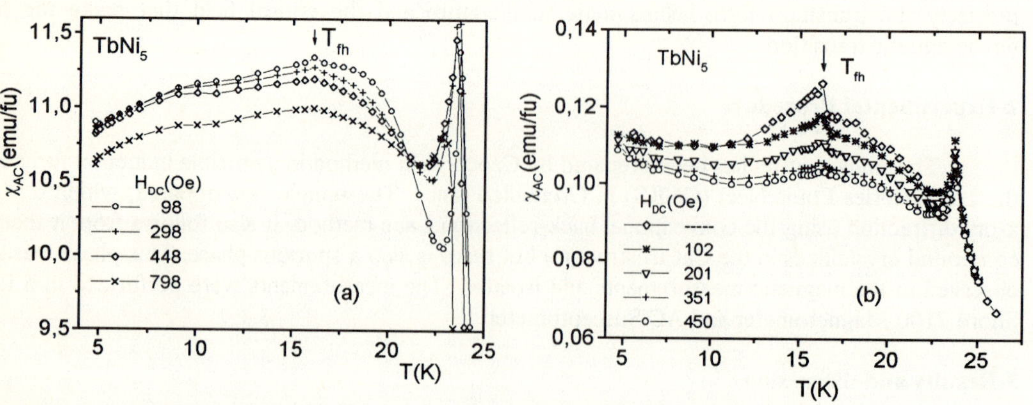

Figure-2. The real part of the AC susceptibility at different DC field, (a) in the easy and (b) in the hard axis. (H_{AC}= 10 Oe and f= 95 Hz).

Another way to observe the presence of the helimagnetic phase in this compound is through the analysis of the magnetization curves at different temperatures. The change in the thermal variation of the coercive field can reflect a magnetic transition. However, in figure 3, we can see that no magnetic hysteresis is observed in the magnetization curves along *a*-axis. This is not true when the magnetization is measured along *c* axis (hard direction) as it can be seen in figure 4.

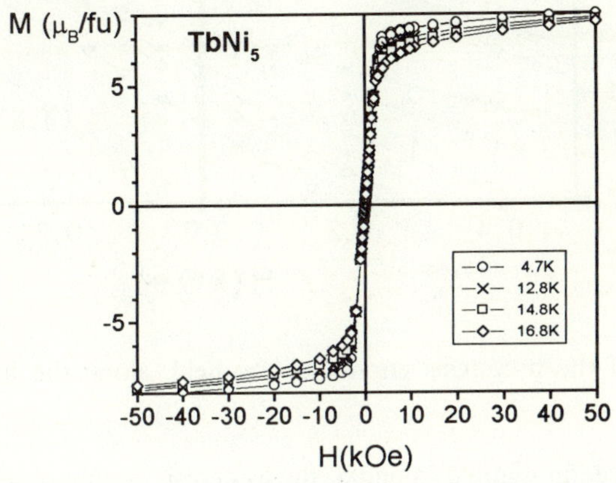

Figure-3. Field dependence of the magnetization along the easy magnetization *a* axes at different temperatures.

The magnetization curve of $TbNi_5$, at 4.9 K along c-axis, is shown in figure 4. A small hysteresis is observed at low field. Figure 5 details the magnetization curves at low fields, along the same direction, at different temperatures. The rate at which the coercive field decreases in the 4.5K-16.5K range of temperature is about 9 Oe/K, and the rate in the range 16.5K-23.5 K is about 2.5 Oe /K. The fact that H_c does not go to zero at the transition temperature T_N is due to magnetic impurity phases. So from figure 5 one can see that there are two magnetic phases. The first one, with the higher H_c decreasing rate, is the ferromagnetic phase. The other, with the very small rate, is a helimagnetic phase with ferromagnetic domains. The domain contribution in the helical phase came, as discussed in reference [3], from the ferromagnetic layer that involves helical structures spiraling in opposite senses.

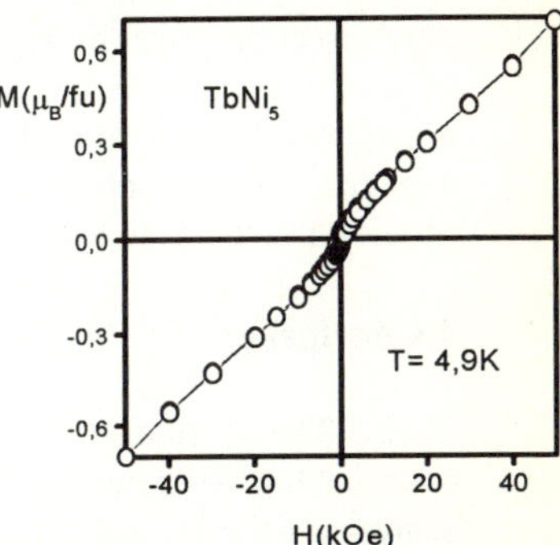

Figure-4. The hysteresis curve of $TbNi_5$ in the hard magnetic direction at 4.9K.

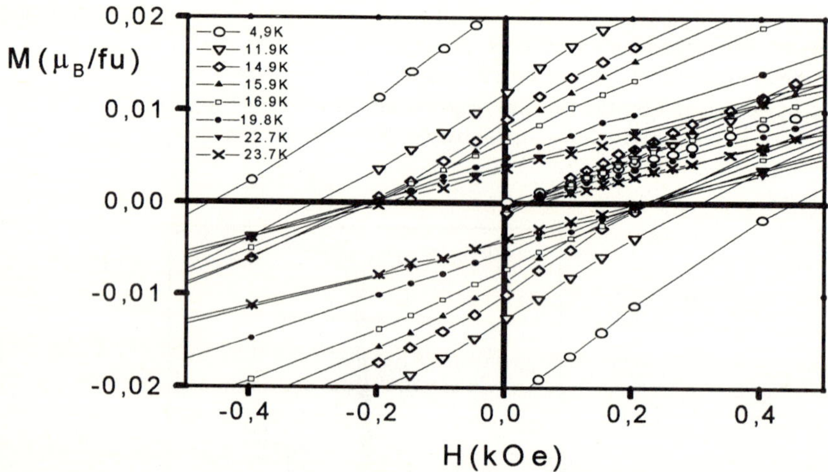

Figure-5. Detail of the hysteresis curves at low field, along the hard magnetization for different temperatures.

As we can see in figure 3, 4 and 5, the hysteresis is only observed in the hard axis. When the coercive field is smaller than the applied field sensitivity, or of the same order of magnitude, we do not see any remanence in the easy direction. In other words, the smallest field that we can produce is greater than the coercive field. So, the hysteresis can not be seen. On the measurement along the hard direction we intent to stay with the magnetic field direction perpendicular to the easy one ($\theta_{MH}=\pi/2$). A small disorientation results in a very small component of the magnetic field in the easy direction ($\alpha=(\pi/2-\theta_{MH})<<1$). As this disorientation α is small, we have the field component in the easy direction given by $H_{appl}\cos(\theta_{MH}) = H_{appl}\sin(\alpha) \cong H_{appl}\,\alpha$. For $\alpha \cong 10^{-2}$ rad, and an applied field of about 100 Oe, there is a magnetic field of about 1 Oe in the easy direction. Thus, we can observe a magnetic hysteresis with a coercive field smaller than the smallest field that we can produce. This explains why we can observe a magnetic hysteresis in the hard magnetic direction while it is not observed in the easy one.

4-Conclusion

The thermal and field variation of the magnetization in $TbNi_5$ can be explained when we consider the compound is ferromagnetic for $T<T_{fh}$ and helimagnetic between T_{fh} and 23.5 K. From analysis of the FC and ZFC curves, the AC susceptibility thermal variation at zero DC field and the hysteresis curves in the hard direction we have T_{fh} =16.5K. The thermal variations of the AC susceptibility, at different applied fields, show that, for a critical field of about 450 Oe, the compound becomes ferromagnetic at all temperatures.

5-References

[1] Wernick J.H. and Geller S., Acta Crystallographica 12 (1959) 662.
[2] Blanco J. A., Gignoux D., Schmitt D, Tari A. and Zhang F.Y. 6 (1994) 4335.
[3] Barthem V. M T.S. and Moreira da Gama E. A., J. Phys. : Condens. Matter 9 (1997) 7609.

Materials Science Forum Vols. 302-303 (1999) pp. 397-400

Temperature Calibration Devices Based on First Order Magnetic Phase Transition of $(R_{1-x}R'_x)Co_2$ Pseudo-Binaries Compounds

F. Garcia, G.R.C. Cernicchiaro and A.Y. Takeuchi

Centro Brasileiro de Pesquisas Físicas,
Rua Dr. Xavier Sigaud, 150, Urca, Rio de Janeiro 22.290/180, Brazil
E-Mail: geraldo@cat.cbpf.br

Keywords: Temperature Sensor, First Order Magnetic Phase Transition, Calibration, Pseudo-Binary Compounds

Abstract

The present paper is a contribution to low temperature sensor calibration techniques. It is known that first order phase transitions are useful tools for standard calibration of several temperature scales. Normally, the temperature scale is defined by using repeatable temperature points referred to as fixed points. We propose a method of manufacturing temperature calibration devices based on the magnetic features of the pseudo-binaries compounds of formula $(R_{1-x}R'_x)Co_2$, where R and R' are heavy rare earth elements. Previous studies revealed that a first order magnetic phase transition from ferrimagnetism to paramagnetism is observed in compounds as $TmCo_2$ (3.7K) [1], $ErCo_2$ (33K) [2], $HoCo_2$ (75K) [3] and $DyCo_2$ [4] (140K). These transitions correspond to a large anomaly in the characteristic properties as a function of temperature. For instance, the electrical resistivity curve of $ErCo_2$ presents a discontinuity of 500% in an interval smaller than 0.1 K. In this work, we present the electrical resistivity data for $(Er,Ho)Co_2$ compounds showing the viability of this proposed devices from 33K up to 75K. This range can be extended below and above by substitution of the chemical components. The choice of the compound and its concentration depends of the operational device characteristics. The number of transitions can be fixed by a convenient arrangement of several serial units. The main contribution of this technique is to be able to set well defined first order transition temperatures by a stoichiometric composition. The magnetic phase transition was determined in this system by electrical resistivity, magnetization and AC susceptibility measurements.

1. Introduction

Temperature is an extremely important parameter in science and industry. The accuracy with which temperature is measured often affects the outcome of an experiment. The accuracy of a temperature sensor is associated to how close the parameter measured (resistance or voltage) can be converted to some recognized temperature scale [5,6]. The signal sensitivity/resolution is the change in the measured parameter per change in temperature. It must be calibrated to determine its response curve. Temperature scales are defined by fixed points. These points are based on physical premises that occur reproducibly at the same temperature. There are numerous examples of fixed points: boiling points, freezing points, triple points, superconducting transitions, superfluid transitions, etc. The temperature scale is transferred from primary thermometers to a secondary thermometer maintained by governmental and commercial calibration labs. The most common of these reference-standard thermometers is the resistance thermometer (a high purity platinum or a high purity rhodium-iron alloy). These materials are highly stable. However, it is very expensive and time consuming to maintain these standard devices and systems for calibration process of temperature sensors. On the other hand, usual temperature sensors are complex assemblies of wires, electrical connections, packages, seals, etc., and consequently they have the potential for drift in calibration. Thermal cycling, improper use, aging, thermal / vibration / mechanical shock, environmental effects (magnetic field / electrical stress / electromagnetic interference / electrostatic discharge.), etc. affect the reproducibility (repeatability, precision, stability) and the response curve of the device. Even when this degradation is not very apparent, it is desirable to minimize errors in recalibration and errors associated with installation and operation of the thermometer. We propose a method of manufacturing temperature working standard calibration devices based on the magnetic features of the pseudo-binaries $(R_{1-x}R'_x)Co_2$ compounds, where R and R' are heavy rare earth elements.

2. Experimental Results

The samples were prepared by arc-melting the constituent elements in an argon atmosphere. To avoid the appearance of spurious phases, 6% wt excess of rare earth over the stoichiometric composition was necessary. The ingots were then annealed in evacuated sealed quartz tubes for a period of one week at T = 800°C. The crystal structure and the lattice parameters were determined, at room temperature, by X-ray powder diffraction. All the samples are single phase cubic $MgCu_2$ type. Electrical resistivity measurements were performed using a standard dc four points methods.

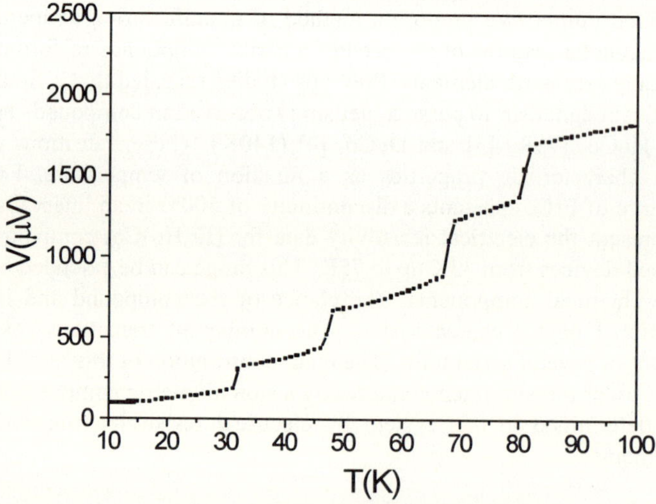

Fig. 1 – Voltage drop of four serial connected elements of $(Er_{1-x}Ho_x)Co_2$ (geometrical parameters of each element: cross section 1x1mm^2, length 3mm; current 100mA).

A magnetic phase transition temperature calibration device is a solid sensor which provides easily detectable first order magnetization transition, at fixed temperature points. Usually the temperature characteristics of commercial temperature sensors are very predictable and exhibit good uniformity from device to device. Fabrication processes are likely to become controllable to the point where interchangeable sensors will become commonplace, within a given tolerance. Actually, it is possible to purchase uncalibrated sensors, which have voltage or resistance readings at helium, nitrogen and room temperature provided by the manufacturers. They must be calibrated by the user. With the application of sensor specific interpolation/extrapolation techniques this feature allows an abbreviated calibration, based on two or three calibration points, to generate a resistance-versus-temperature or a voltage-to-temperature curve over the useful range of selected sensors with remarkable accuracy. However, a two or three calibration point calibration is as good as the accuracy of the calibration points. Helium boiling temperature is very well defined, but depends on atmospheric pressure. Boiling nitrogen and ice point temperatures can vary ± 0.5 K. Therefore, the interest in fabricating a low cost device with a controllable number and range of temperature calibration points.

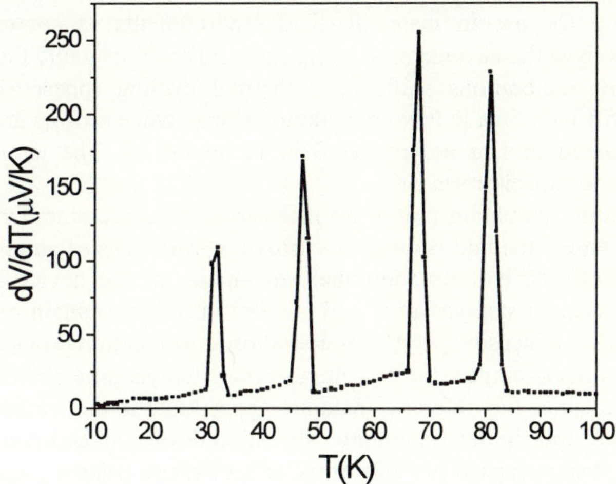

Fig. 2. A differencial voltage dV/dT versus T of four serial connected elements of $(Er_{1-x}Ho_x)Co_2$ with x = 0.00, 0.40, 0.80 and 1.00 respectivelly, data refered to Fig. 1.

Calibration costs reflect time and equipment needed for the calibration. The main advantages of our proposed method are to allow calibration in situ, limited instrumentation requirements, and improvements in the accuracy without the cost of a full calibration. First order magnetic phase transition from ferrimagnetism to paramagnetism are observed in compounds such as $TmCo_2$ (3.7K) [7], $ErCo_2$ (32K) [8], $HoCo_2$ (75K) [9] and $DyCo_2$ [10] (140K). These transitions correspond to a large anomaly in the characteristic properties as function of temperature. One limitation on the use of these compounds to base standard calibration devices is the restriction of the useful temperature range of the sensor.

In fact the physical phenomena must occur at the appropriate temperature range, and the temperature dependence of the property being measured must occur at a measurable level. In this work (Fig. 1), we present the electrical resistivity data for $(Er_{1-x}Ho_x)Co_2$ compounds showing the viability of this proposed devices from 32K up to 75K. The step-like behavior is associated to the first order phase transition, from ferrimagnetic to paramagnetic state, of $ErCo_2$ (32.07 K), $(Er_{0.60}Ho_{0.40})Co_2$ (47.08 K), $(Er_{0.20}Ho_{0.80})Co_2$ (68.02 K) and $HoCo_2$ (81.02 K). This range can be extended below and above for substitution of the chemical components. The choice of the compound and its concentration depends of the operational device characteristics. The number of transitions can be fixed by a convenient arrangement of several serial units. The main contribution of this technique is to be able to set well-defined first order transition temperatures by a stoichiometric composition. The transition is associated with high signal size variation (Fig. 2), and it can be easily determined by resistivity or magnetization measurements, via standard read-out electronics.

3. Discussion and Conclusion

The temperatures Tc of the magnetic phase transitions of $(Er_{1-x}Ho_x)Co_2$ compounds are weakly affected by applied magnetic field at least up to 10 kOe, less than 1% [11]. Therefore, a suitable application of magnetic phase transition devices are the temperature control in magnetic environments in conjunction with the capacitance sensors. Usually, magnetic fields cause calibrations shifts, which produce false temperature measurements in conventional sensors. The shift is not permanent and sensors return to their zero-field calibration when the field is turned off. Capacitors are good for use in magnetic field environments as control sensors. Capacitance temperature sensors have the advantage of being insensitive to magnetic fields. However, they have a tendency to show calibrations shifts after thermal cycling (phase shifts in the ferroeletric materials) and to drift with time at low temperature. Capacitance sensors are only sold uncalibrated. They can be calibrated in situ before the field is turned on. The temperature control is then accomplished with the capacitor sensor.

The temperature transition points are insensitive to temperature cycling, and the transition temperature measurement method is quite insensitive to variations of amplitude, background noise, electronic read-out offsets, etc. As additional advantages of the device it allows a sequence of transitions with convenient step variation and number in order to obtain more accuracy in selected ranges. The application of sensor specific interpolation/extrapolation provides the means to generate accurate resistance-versus-temperature or voltage-versus-temperature curves using a limited number of calibration points, since this reference data are appropriate to the temperature range of use. The use of magnetic phase transition temperature calibration devices in addition with these interpolation techniques allows inexpensive calibrations of the entire system (sensors, wires, connections, etc) for selected temperature sensors to use with standard readout electronics.

Acknowledgments: M.R.Soares for useful discussions and CNPq by financial support.

References

[1] E. Gratz, R. Hauzer, A. Lindbaum, M. Maikis, R. Resel, G. Schaud, R.Z. Levitin, A.S. Marcosian, I.S. Dubenko, A. Yu Sokolov and S.W. Zochowski, J. Phys: Condensed Mater 7 (1995) 597.

[2] G. Petrich and L.R. Mossbauer, Phys. Lett. 26A (1968) 403.

[3] R. Lemaire, Cobalt 33 (1966) 201.

[4] F. Givord and J.S. Sahah, C.R. Hebd. Séan, Acad. Sci. Paris B 274 (1972) 923.

[5] Lounasmaa, (Academic Press Inc, London) (1974): "Experimental principles and methods below 1K"

[6] Lake Shore – Technical Reports – Temperature Measurement and Control (1995).

[7] E. Gratz, R. Hauzer, A. Lindbaum, M. Maikis, R. Resel, G. Schaud, R.Z. Levitin, A.S. Marcosian, I.S. Dubenko, A. Yu Sokolov and S.W. Zochowski, J. Phys: Condensed Mater 7 (1995) 597.

[8] G. Petrich and L.R. Mossbauer, Phys. Lett. 26A (1968) 403.

[9] R. Lemaire, Cobalt 33 (1966) 201.

[10] F. Givord and J.S. Sahah, C.R. Hebd. Séan, Acad. Sci. Paris B 274 (1972) 923.

[11] F.garcia, M.R.Soares, A.Y.Takeuchi and S.F. da Cunha, J. Alloys and Compounds 000 (1998) 000 (to be publish).

Materials Science Forum Vols. 302-303 (1999) pp. 401-405

Commercially Fabricated Silicon Levers for Torque Magnetometry

S. Spagna, A. Wilson, M.B. Simmonds, J. Diederichs and R.E. Sager

Quantum Design Inc., San Diego CA 91212, USA

Keywords: Torquemeter, Magnetism, Superconductivity

Abstract. We describe a fully automated torque-magnetometer system featuring microfabricated silicon levers. The system uses a piezoresistive technique to measure the torsion of the lever about its symmetry axis and achieves a typical torque resolution of 2 x 10^{-9} Nm with 40 seconds averaging time. An integrated loop on the torquelever is used as a calibrating standard with an accuracy of 1%. We also present specific measurements demonstrating the capabilities of this system.

1 Introduction

The widespread use of silicon micromachined technology in recent years has made possible the fabrication of new torquemeter sensors that can be used as a powerful tool in investigating the magnetic properties of small anisotropic samples [1-3]. In general, torque-magnetometers can be used as a complimentary tool to superconducting quantum interference device (SQUID) based magnetometers to study the magnetic properties of materials. The use of a torque-magnetometer is particularly useful at high fields [3] where the sensitivity of the device increases linearly with increasing field and shielding problems make the employment of SQUID based devices more difficult.

A novel torque-magnetometer based on a microfabricated silicon p^+-doped piezoresistive cantilever has been successfully demonstrated by Rossel and coworkers [1-2]. Designed for sensitively measuring the magnetic properties of microgram particles, this miniature torquemeter employs a microscopic cantilever on a length scale of one hundred microns as a sample platform. By measuring the deflection of the cantilever in an external applied magnetic field, a careful determination of the magnetic torque, τ = **m** x **B**, generated by a small particle deposited at the extremity of the lever is possible. The detection of the lever deflection is achieved via a linear change in resistance ΔR of one or more piezoresistors lithographically patterned on the lever surface. In this paper, we present a highly integrated system that extends this new technique to larger samples and can be easily and advantageously used to measure the magnetic properties of anisotropic crystals weighing few milligrams. The torque-magnetometer (Tq-MAG) readily incorporates in a multi-purpose top-loading commercial temperature and magnetic field platform [4], which enables measurements in applied magnetic fields as high as 14 T, and at temperatures from 1.8 K to 400 K.

2 Torquemeter design

A schematic diagram illustrating the active component of the piezoresistive silicon sensor employed in the Tq-MAG is shown in Fig. 1. The device is microfabricated from a polished Si (100) wafer using a standard anisotropic etching technique. The process is applied to both sides of the wafer resulting in a lever thickness of approximately 25 μm. The torquelever is suspended in a trench and protected by a thick silicon perimeter (not shown) which facilitates safe handling of the sensor by the edges of the silicon chip. The sensor comprises a 2 x 2 mm^2 sample stage upon which a crystalline sample can be attached using small amounts of grease paste. A torque applied to the sample stage produces a mechanical stress in each of the two legs connecting the stage to the sensor perimeter. Two costantan piezoresistor grids patterned in the region of high stress of each leg are used to sense the component of the stress along the <110> direction. The high symmetry of the design allows for a sensitive differential measurement of the imbalance $\Delta(R_1\text{-}R_2)$ that develops between the two piezoresistors. In order to minimize the contribution of the gravitational force component to torque measurements a Wheatstone bridge circuit is used to detect preferentially the torsion, or twisting of the lever about its symmetry axis, rather than the flexion or bending of the lever.

A removable rotator platform provides the sensor housing and an interface to horizontal and vertical rotator probes. These rotator probes allows for a continuous 380 degrees excursion employing a computer controlled stepper motor with a step resolution up to 0.0005 degrees. The rotator platform features gold plated copper-beryllium spring contacts, which mate to the torquelever sensor circuit pads for convenient electrical interface. The modular design of the rotator platform allows the user to easily mount the sample on the torquelever with the help of a microscope and avoids the difficulty of manipulating the entire rotator probe.

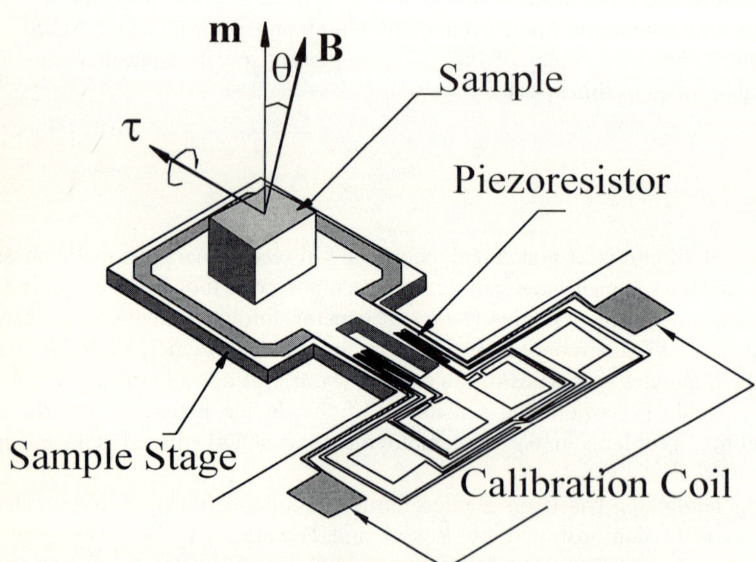

Fig. 1 Diagram of the piezoresistive lever.

A convenient in-situ calibration of the torquemeter can be made using a metallic coil integrated directly on the lever. The coil consists of a 50 μm thick copper trace to maintain its overall resistance below 10 Ω and thus minimizing self heating effects. A typical excitation current of I_{coil} = 1 mA produces a maximum dissipation power of 0.1 mW, thus allowing useful low temperature operation of the calibration coil. A thin palladium overlayer on the copper traces prevents the electrical circuit from tarnishing upon thermal cycling. In general, the magnitude of the torque generated by a dc excitation current in the coil is given by: $\tau_{coil} = I_{coil} A B_{app} \sin\theta$. In this equation, B_{app} is the external applied magnetic field, $A \approx 2.9 \times 10^{-6}$ m^2 is the effective area enclosed by the coil, and θ is the angle between the normal direction to the coil plane and the external applied magnetic field. During routine calibrations performed as function of temperature we use $\theta = 90°$, $I_{coil} = 1$ mA, and $B_{app} = 9$ T. Fig. 2 shows the results of a magnetic field linearity test employing a large excitation current of ± 5 mA in the calibrating coil producing a measurable magnetic torque of $\tau_{coil} = \pm 1.5 \times 10^{-8}$ Nm at 1 T. The data indicates that the resistance imbalance between the two piezoresistors produced by reversing the current in the calibration coil increases linearly with increasing applied magnetic field up to 14 T. Experimentally, it is found that the linear response of the torquelever persists well up to torque signal magnitudes of 5×10^{-5} Nm thus allowing a typical overall operating dynamic range of approximately $\pm 10^{-5}$ Nm.

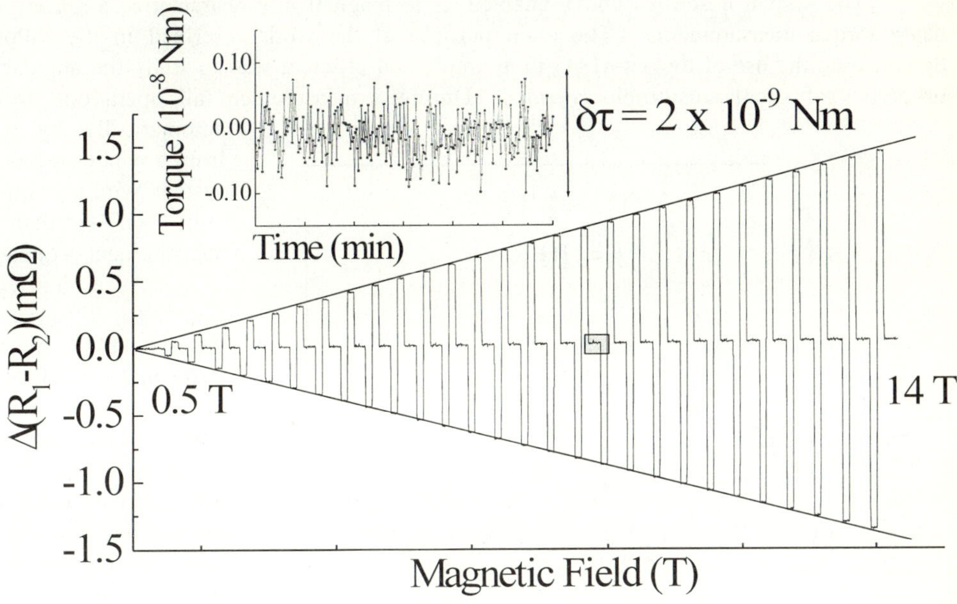

Fig. 2 Magnetic field calibration.

The inset of Fig. 2 depicts the typical noise level in the measurement at zero magnetic field for the shaded region in the graph. The dispersion of this signal can be extracted to be 2×10^{-9} Nm for a 40 seconds averaging time per point at T = 250 K. Encouraging results were also obtained from preliminary noise measurements in a cryocooled (liquid helium free) version of the cryostat [5], which employs a conventional 4.2 K closed cycled refrigerator (Gifford-McMahnon type) to provide the cooling necessary for the superconducting solenoid and refrigeration of the sample space. These tests have shown that owing to the relatively high resonant frequency of the torquelever, the ultimate noise level, for the torsion mode, is unaffected by the low frequency vibrations generated by the displacer strokes in the cooler head. As the use of cryocoolers in measurement systems grows, such stability of signal to the environment becomes important.

During routine laboratory operation, the intrinsic temperature dependence of the residual resistance imbalance between the two piezoresistors of a bare lever (with $I_{coil} = B_{app} = 0$) is calibrated before an experiment. The temperature dependence of the residual imbalance is typically measured to vary less than 10% over the temperature range between 300 K and 1.8 K. For a careful determination of the absolute torque, the magnetic field dependence of the residual resistance imbalance is also determined and, while negligible at room temperature, it is measured to vary about 1 to 2 % at 1.8 K over the magnetic field range from 0 T to 14 T.

3 Magnetic torque measurements

The system described above enabled us to magnetically characterize a variety of samples using torque measurements. The main purpose of the work described in the following is to demonstrate the use of the Tq-MAG in a simple and efficient way to study the angular dependent properties of small anisotropic crystals. During a measurement all operations are controlled automatically by a computer from a windows based software system, with user input required only on initiation of a measurement sequence.

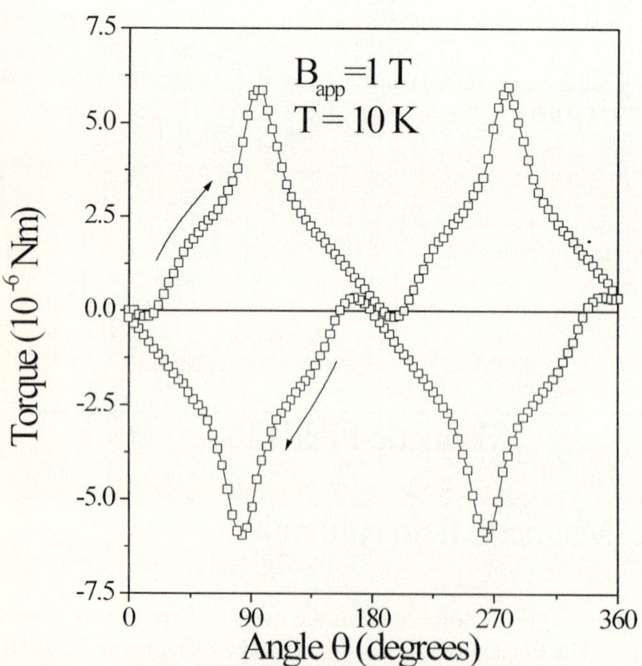

Fig. 3 Rotational hysteresis loops, τ(θ), for a $Bi_2Sr_2CaCuO_{8+\delta}$ single crystal.

As a first example we describe a dynamical mode of the Tq-MAG in which the torquemeter itself rotates slowly in a constant field and the torque is recorded as function of angle. Using this technique the internal field component responsible for the magnetic torque in the sample varies as $B_{\perp} = B_{app} \sin\theta$. This method is well suited for studying the irreversible magnetization of a 0.13 mg superconducting $Bi_2Sr_2CaCu_2O_{8+\delta}$ single crystal sample with critical temperature of Tc ≈ 80 K. Fig. 3 depicts the isothermal rotational magnetization hysteresis loops, τ(θ), at T = 10 K with an external applied magnetic field of B_{app} = 1T. From the torque curves, large irreversibility can be observed when crossing the ab-planes (i.e. CuO_2 planes) at 90° and 270° due to the flux pinning of the magnetic field lines along the c-axis for this extremely anisotropic layered superconductor.

As a second example, we show that isothermal magnetic torque measurements as function of magnetic field can be used to observe the de Haas-van Alphen (dHvA) effect in samples possessing an anisotropic Fermi surface. Displayed in Fig. 4 are the data recorded for the magnetic field applied ≈ 15° from the <001> crystal direction in a Lanthanum hexaboride, LaB_6, single crystal doped with 1% Cerium. In this compound the dHvA oscillations manifest themselves at pumped liquid He^4 temperatures. The data were recorded by continuously sweeping the magnetic field at a rate of 5 Oe/sec. A more detailed study of this effect will be presented in a forthcoming report [6].

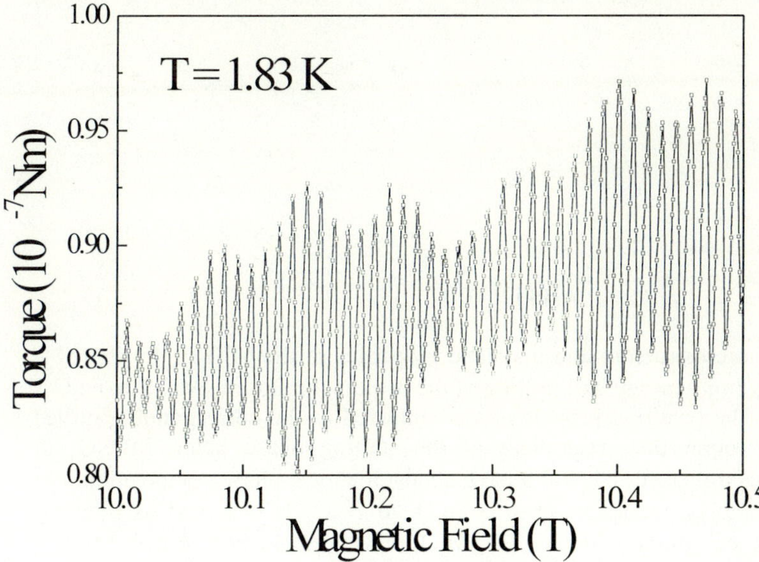

Fig. 4 Large torque amplitude oscillations measured on a LaB_6 - 1%Ce single crystal.

The torquelever sensors were developed in collaboration with C. Rossel and M. Willemin at the IBM Research Division, Zürich Research Laboratory, and the Physik-Institut der Universität Zürich, Switzerland. One of the authors, SS, would like to thank D. Hall and B. L. Brandt at the National High Magnetic Field Laboratory, Tallahassee Florida, USA for helpful discussions.

References

[1] C. Rossel, P. Bauer, D. Zech, J. Hofer, M. Willemin, and H. Keller, J. Appl. Phys. **79** (1996), p. 8166.

[2] M. Willemin, C. Rossel, J. Brugger, M. Despont, H. Rothuizen, P. Vettiger, J. Hofer, and H. Keller, J. Appl. Phys. **83** (1997), p. 1163.

[3] P. A. Crowell, A. Madouri, M. Specht, G. Chaboussant, D. Mailly, L. P. Levy, Rev. Sci. Instrum. **67** (1996), p. 4161; and M. J. Naughton, J. P. Ulmet, A. Narjis, S. Askenazy, M. V. Chaparala, A. P. Hope, Rev. Sci. Instrum. **68** (1997), p. 4061.

[4] Physical Property Measuring System (PPMS) manufactured by Quantum Design.

[5] Dynacool® manufactured by Quantum Design.

[6] D. Hall, S. Spagna, S. Tripp, and B. L. Brandt (to be published).

Materials Science Forum Vols. 302-303 (1999) pp. 406-410

Microstructural and Magnetic Properties of Mechanosynthesized Ferrites

G.F. Goya and H.R. Rechenberg

Instituto de Física, Universidade de São Paulo, 05389-970 SP São Paulo, Brazil

Keywords: Spinel Ferrites, Ball Milling, Nanoparticles

ABSTRACT

We present a study on microstructure and magnetic properties of the MFe_2O_4 ferrite (M = Cu and Zn) produced by high-energy ball milling of the precursor mixture $\{MO + \alpha\text{-}Fe_2O_3\}$ for times up to 600 h. The resulting properties are compared to the corresponding milled MFe_2O_4 ferrites. It is found that, regardless of the starting phase being MFe_2O_4 or $\{MO+\alpha\text{-}Fe_2O_3\}$, samples milled for $t > 400$ h have almost identical phase composition. For milled $\{MO+Fe_2O_3\}$ samples, reduction from $\alpha\text{-}Fe_2O_3$ to Fe_3O_4 is observed as previous stage to the formation of spinel structure $Fe_{3-x}M_xO_4$ having $x\approx1$. For milled MFe_2O_4 samples, segregation of $\alpha\text{-}Fe_2O_3$ and subsequent reduction lead to the same $Fe_{3-x}M_xO_4$ phase after $t\approx400$ h. X-ray diffraction (XRD), Mössbauer and magnetization data show that this phase has microstructural and magnetic properties corresponding to $x\approx1$ and substantial short-range disorder. An effective magnetic moment $\mu_{eff} = 2.23\mu_B$ is obtained for the resulting $ZnFe_2O_4$ spinel phase, which implies an inversion degree of $\delta=0.2$. Annealing of milled sample gradually restores the bulk hyperfine parameters and μ_{eff} values. However, the inverted structure is noticed even after annealing at temperatures $T\approx900$ K.

INTRODUCTION

A wide variety of techniques are being used to synthesize nanostructural materials, such as electrodeposition, aerogel procedures, chemical processing and precipitation under hydrothermal conditions [1,2]. High energy ball milling has proved to be a suitable method for obtaining nanometer-size ceramic particles and, more recently, the synthesis of metastable alloys and ceramic compounds with improved properties for permanent magnets and recording media applications [2-6]. Although theoretical and empirical models of collision frequency and impact energy during the milling process have been proposed [7], the fundamental mechanisms involved are not yet fully understood. Ball milling is usually considered as a means to obtain solid-state chemical reactions, which occur across the welded interfaces of the powder particles when impacted by the grinding media. As solid reactions are governed by the geometry and diffusion rates of the reactants and products, they require high temperatures (i.e. several hundred degrees) to overcome the energy barrier. Mechanical milling minimizes the energy barrier of the products by increasing the reaction interface, through continuously welding and fracturing the powder particles at the nanometer scale. Recently, the use of ball milling as a synthesis route for ferrites has called the attention on their unusual magnetic properties arising from high microstrain, vacancy densities and/or altered site populations, originated in the milling process.

The distributions of M^{+2} and Fe^{+3} ions in the spinel structure can be represented by

$$[M_{\delta}Fe_{1-\delta}]^{A}[M_{1-\delta}Fe_{1+\delta}]^{B}O_{4}$$

where δ=0 and δ=1 stands for the inverse and normal cases respectively [8]. For $ZnFe_2O_4$ and $CuFe_2O_4$ the final δ values depend on the synthesis conditions, and may vary from 0 to ~0.2-0.3 [9]. Copper ferrite is ferrimagnetic at room temperature, with Néel temperature T_N=780 K. By replacing Cu^{+2} with non-magnetic Zn^{+2} at A sites, the ordering temperature shifts down, and for $ZnFe_2O_4$ is T_N=9 K [10]. At this temperature, Fe^{+3} ions order antiferromagnetically through B-B exchange interactions. The ordering temperature as well as saturation magnetization and coercive forces are strongly influenced by the final site population [9]. Ball milling is therefore a valuable tool to attain new metastable phases with improved magnetic properties. In this work, high-energy ball milled samples of a) {α-Fe_2O_3+MO} and b) MFe_2O_4 are studied with the main purpose of comparing their evolution, regarding the resulting microstructure and magnetic properties.

EXPERIMENTAL

To compare the properties of ball-milled MFe_2O_4 ferrites with the milled mixture of its reagents {Fe_2O_3+MO}, two kinds of starting samples were used in our ball milling experiments:

a) Mixtures of 1:1 parts of the MO (M=Zn and Cu) and α-Fe_2O_3 oxides (99.999% purity, ≈200μm particle size), as precursors of the corresponding spinel phases. These will be labeled ZFM and CFM samples, for Zn and Cu respectively.
b) Polycrystalline spinel powders. These were prepared by mixing stoichiometric amounts of the same α-Fe_2O_3 and MO precursors used in (a), and heating in air at 1200 K for 20 hours, repeating the procedure three times. The final cooling rate was set to 2 K/min to avoid quenching of metastable phases. These will be labeled hereafter as ZFS and CFS for Zn and Cu, respectively.

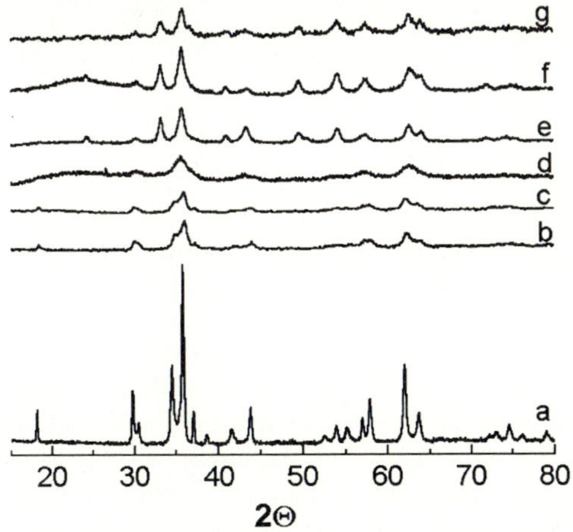

Figure 1. XRD data of $CuFe_2O_4$ sample a) pure, and milled for b) 15, c) 45, d) 180, e) 2280, f) 2.5 10^4, g) 3.8 10^5 min.

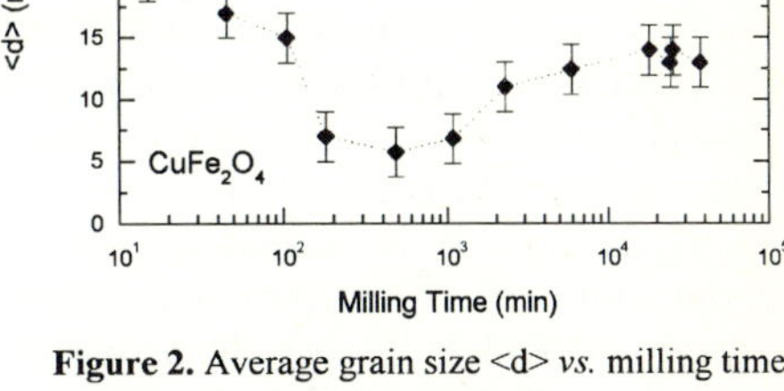

Figure 2. Average grain size <d> *vs.* milling time for $CuFe_2O_4$. Note the logarithmic x-scale. The initial grain size <d_0> is also shown.

All samples were milled in a planetary ball mill (Fritsch Pulverisette 7) in closed containers of hardened steel, during different intervals from 15 to ~38000 min. Acetone was added to higher homogeneity during milling. Afterwards, samples were dried in air at 323 K. X-ray diffraction measurements were performed using a Philips PW-1140 diffractometer with Cu K_α radiation in the 2θ range of 10°-80°. Average grain sizes <d> were estimated using the Scherrer equation. Mössbauer measurements were performed with a conventional constant-acceleration spectrometer in transmission geometry between T=296 and 4.2K. Isomer shifts are given relative to that of α-Fe at room temperature. Lorentzian lineshapes were used to fit the spectra. Magnetization measurements were performed in a vibrating sample magnetometer at 4.2 and 300 K, in fields up to 9 T.

RESULTS AND DISCUSSION

A. $CuFe_2O_4$

Typical XRD data for *as-prepared* and milled $CuFe_2O_4$ samples are shown in figure 1. The major features along the series are the broadening of the peaks, the smaller intensity and the disappearing of the <103> hkl reflection ($2\theta = 34.5°$) of tetragonal $CuFe_2O_4$ after the first minutes of milling. Figure 2 shows the evolution of the average grain size <d> *vs.* milling time for $CuFe_2O_4$. It is observed that <d> drops abruptly after the first minutes of milling (note the logarithmic scale of time axis in fig. 2) and, for times longer than $\sim 10^3$ min, <d> attains a constant value ≈10-15 nm. The same qualitative behavior was found in all other M-Fe-O powders studied in this work.

The phase evolution of CFS sample, deduced from the XRD data of fig. 1, has three major stages: 1) decomposition of $CuFe_2O_4$ with α-Fe_2O_3 segregation after $\approx 10^3$ min of milling, 2) reduction of α-Fe_2O_3 to Fe_3O_4 at $\approx 2x10^3$ min, and 3) formation of a spinel-like structure $Fe_{3-x}Cu_xO_4$. The α-Fe_2O_3 to Fe_3O_4 reduction has been previously observed in milled $CuFe_2O_4$ samples [11]. It is related to the milling conditions of our experiment, since the closed containers do not allow oxygen exchange with the environment. This Mössbauer data for samples milled more than $6x10^3$ min show that this reduction continues as long as hematite is present. To illustrate this, reduced hyperfine fields H/H_0 (H_0 = hyperfine field of bulk sample) *vs.* milling time ($t > 6x10^3$ min) for CFM and ZFM samples are displayed in figure 3. Since the mean particle size <d> remains nearly constant in this time range, the observed decreasing of H_{hyp} cannot be related to particle size effects, such as collective magnetic excitations. A more likely source of this may be the inclusion of Cu^+ and the reduction of Fe^{3+} to Fe^{2+} in the $M_xFe_{3-x}O_4$ structure produced by milling. This corresponds to stage (3) described above. A similar evolution is shown for ZFM samples milled for $t > 6x10^3$ min (see fig. 3).

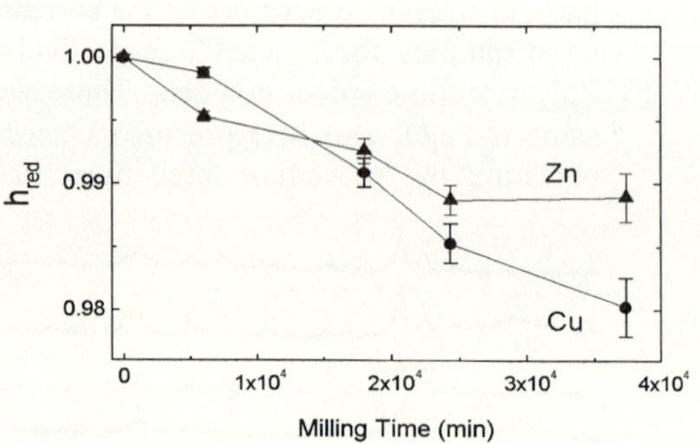

Figure 3. Reduced hyperfine fields of Fe_2O_3 in CuO and ZnO after milling for different times.

To exclude the influence of milling parameters (sample-to-ball mass ratio, rotation velocity, etc) on the final results, an additional milling experiment was performed using $CuFe_2O_4$ and $\{CuO+Fe_2O_3\}$ as starting powders (labeled CFS and CFM respectively) which were milled in separated containers, but simultaneously, during t=2.5x10^4 min. The corresponding XRD data of these samples are shown in figure 4, where a remarkable similitude between both results can be noticed. The XRD peaks are broad as result of the small particle size, and can be assigned to the same spinel like phase mentioned above, plus α-Fe_2O_3 phase. In agreement with this view, magnetization curves at T=4.2 K (not shown) do not reach full saturation even at H=90 kOe. This is likely to be originated in the weak ferromagnetism of the Fe_2O_3 present. Assuming that at our highest fields, samples are fully oriented by coherent rotation, the relation

$$\sigma = \sigma_s + \chi H \qquad (1)$$

gives σ_s = 18.5emu/g and 17.1emu/g for $CuFe_2O_4$ and $\{CuO+Fe_2O_3\}$ respectively. These values are lower than the σ_s = 33.4emu/g found for pure $CuFe_2O_4$, as expected from the presence of unreacted/segregated hematite in both samples. These results demonstrate that long milling times lead to the same metastable phase composition, regardless of the initial structure of the Cu-Fe-O sample. This phase has a spinel-like structure, and can be regarded as the resulting solid solution of Cu and Fe oxides with composition $Fe_{3-x}Cu_xO_y$.

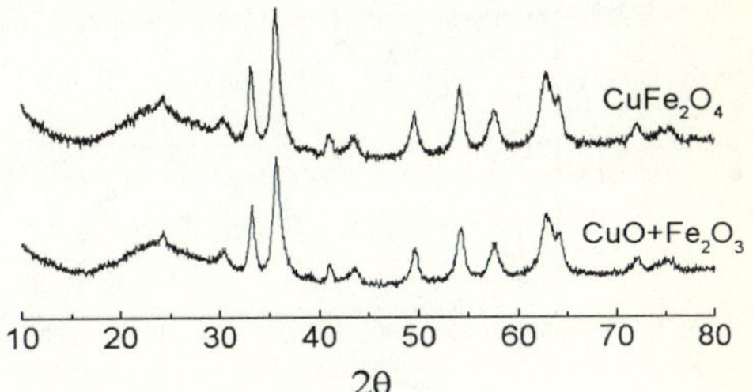

Figure 4: XRD data for $CuFe_2O_4$ and $\{CuO+Fe_2O_3\}$ samples milled for 2.5x10^4 min.

B. $ZnFe_2O_4$

The evolution of the grain size and microstructure of milled Fe-Zn-O system showed essentially the same main features as those already discussed for Fe-Cu-O powders. However, in this case larger amounts of the spinel phase result from the extended milling of ZFM samples, and only a small fraction of segregated α-Fe_2O_3 is observed in milled ZFS sample. After 3.8x10^4 min of milling, the relative amounts of phases are 95 % for spinel phase and 5 % for hematite (for *both* ZFM and ZFS samples), thus mechanochemical reaction is nearly complete. Figure 5 shows the room temperature hyperfine parameters for sample milled 623 h, and subsequently annealed during 1 hour at different temperatures. It can be seen that the values of ΔQ and Γ decrease continuously, approaching the bulk Zn-ferrite values even for annealing temperatures T≈ 900 K, indicating that partial local disorder persists even at this temperature. As previously reported, Zn-ferrite nanoparticles show larger saturation magnetization σ_s [12], although it is not yet clear whether this enhancement is due to particle size effects, ionic disorder, or both. For our ZFS and ZFM milled samples, σ_s decreases with increasing annealing temperatures (see figure 6) until a value somewhat larger than the bulk. This higher σ_S values are related to the increased Fe population at A sites, which makes possible the FM ordering of B-site ions, and also favors the AFM A-B interactions. A higher limit for the inversion degree can be roughly estimated assuming a full collinear alignment of Fe ions in both sublattices, which gives the simple relation $\mu_{eff} = 2\delta\mu(Fe)$. From sample ZFM milled 3.8x10^4 min, the obtained $\mu_{eff} = 2.23\mu_B$ value correspond to δ=0.2. From figures 5 and 6 it turns out that a measurable amount of inversion remains even after annealing at T ≈ 1000 K. This can be regarded as a bulk property, since at this temperature, XRD data gives <d> = 0.15(1) μm. This relatively stable configuration involves locally distorted Fe coordination, as seen from the higher ΔQ values, and non-compensated Fe ions at A sites leading to higher saturation moments. A more detailed study of magnetic properties is needed to establish whether this corresponds to an actual stable phase.

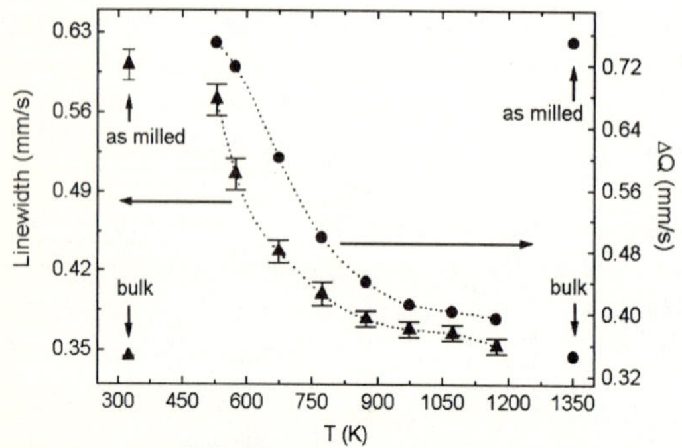

Figure 5. Quadrupole splitting ΔQ (circles) and linewidth Γ (triangles) versus annealing temperature for milled $ZnFe_2O_4$. As-milled and bulk values are shown for comparison.

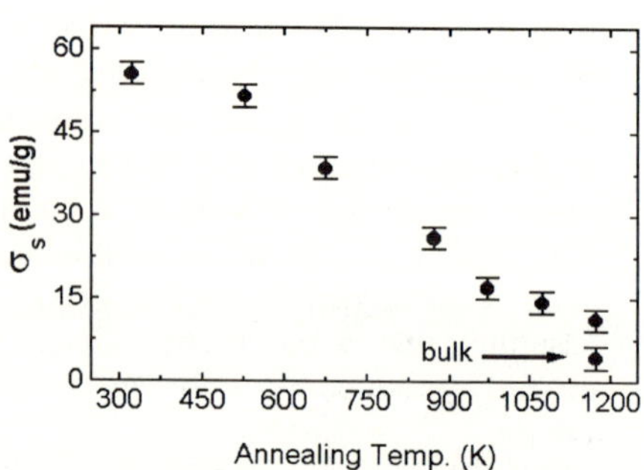

Figure 6. Saturation magnetization σ_s of sample ZFM *vs.* annealing temperature. The value of bulk $ZnFe_2O_4$ is also shown.

In summary, Fe-M-O powders processed by high-energy ball milling can produce chemical reactions leading to a $M_xFe_{3-x}O_4$ spinel phase, indistinguishable of the corresponding MFe_2O_4 phase milled under identical conditions. The α-Fe_2O_3 to Fe_3O_4 reduction seems to be necessary as a previous step to obtain the $M_xFe_{3-x}O_4$ spinel solution. The resulting ferrite has substantial structural disorder, observed as a distribution of Fe local environments. This local disorder is preserved after annealing at rather high temperatures, since even for T = 1100 K the bulk properties are not fully recovered.

ACKNOWLEDGEMENTS: One of us (G.F.G.) thanks financial support from FAPESP through a post-doctoral fellowship.

REFERENCES

[1] C.C. Koch, Nanostructured Materials, 9 (1997) 13, and references therein.
[2] "Third International Conference and Business Forum on Polymer Bonded Magnets" Atlanta, Georgia, April 1996.
[3] G. Nicoara, D. Fratiloiu, M. Nogues and J.L. Dormann, Mat. Sci. Forum 235 (1997) 145.
[4] S. Linderoth, J.Z. Jiang and S. Mørup, Mat. Sci. Forum, 235-238 (1997) 205.
[5] J. Ding, P.G. McCormick and R. Street, JMMM 171 (1997) 309.
[6] G. Concas, F. Congiu and M. Bionducci, J. Phys. Chem. Solids 58 (1997) 1341.
[7] U. Köster, Mat. Sci. Forum 235 (1997) p. 377. L. Battezzati, Ibid. p. 317, M. Magini and A. Iasonna, Mat. Sci. Forum 225 (1996) p.229, M. Rustici, G. Mulas and G. Cocco, Ibid. p. 243, R.M. Davis, B. McDermott and C.C. Koch, Metall. Trans. A 19A (1988) 2867,
[8] J. Smit and H.P.J. Wijn, *"Ferrites"* (Wiley, New York, 1959).
[9] S. Krûpica and P. Novak, "Oxide Spinels" in *Ferromagnetic Materials*, vol. 3, Ed. by E.P. Wolfarth (North-Holland Publishing Company) 1982.
[10] J.M. Hastings and L.M. Corliss, Phys. Rev. 102, (1956) 1460.
[11] G.F. Goya, H.R. Rechenberg and J. Jiang, J. App. Phys. In press.
[12] J. Battle, T. Clark and B.J.Evans, J. de Physique IV, 7 (1997) C1-257.

Materials Science Forum Vols. 302-303 (1999) pp. 411-415

Frequency Dependence of Soft Ferrite Magnetic Properties

M.I. Rosales[1] and R. Valenzuela[2]

[1] Centro de Investigación en Materiales Avanzados, Miguel de Cervantes 120, Complejo Industrial Chihuahua, Chihuahua, Chih. 31110 México

[2] Instituto de Investigaciones en Materiales, UNAM, Apdo. P. 70-360, México D.F., 04510 México

Keywords: Soft Ferrites, Magnetic Properties, Ni-Zn Ferrites, Inductance Spectroscopy, Microstructure

ABSTRACT.
This research work presents an exhaustive study on the magnetic properties of soft magnetic materials of Ni-Zn ferrites by AC techniques to establish some relations with microstructure they present, so final magnetic properties and frequency performance can be also established by microstructure considerations. Three commercial compositions of nickel-zinc ferrites were prepared using the traditional ceramic process. Different microstructures were obtained by using different dwell time during sintering at high temperature. Characterization of samples included physical and chemical properties as density changes during sintering, grain size and grain size distribution, composition by X-ray diffraction, point to point chemical composition and Curie temperature. Inductance spectroscopy technique was used to analyze frequency response in the 5Hz to 13 MHz range. This performance was evaluated at room temperature and at different temperatures from -160 °C to 20-30 °C above Curie point for each composition. Spectroscopy results were analyzed by complex permeability formalism to establish an electric circuit to simulate frequency response of this polycrystalline magnetic materials. Electric circuit elements were directly associated to physical parameters of the material. Finally, correlations among macroscopic magnetic properties that can be easily evaluated, as the relaxation frequency (*fr*), and intrinsic magnetic properties as saturation magnetization (M_s) were established.

INTRODUCTION.
Soft magnetic materials are very important in almost all aspects of the modern electric and electronic technology, due to their ability to effectively shape and concentrate magnetic flux. The desired characteristics for most applications of these materials are: high induction, high permeability, low coercivity and low core losses. Specifically, at low frequencies, as core transformers, high permeability materials are required, while at high frequencies low losses are the most important parameter to control. Because of their high electric resistance Ni-Zn ferrites present low eddy current losses, which made these ferrites appropriated for high frequency applications.

Soft ferrites are being continuously improved to accomplish the growing quantitative and qualitative requirements from consumers. However, development of ferrites is a difficult task where economic aspects play an important role. As a consequence of the impossibility of improving at the same time all required properties for a specific application, there exist a wide variety of soft ferrites.

The increasing demand for switching power supplies is one of the most important factors for recent development of better ferrites, whose market is actually growing at 8-9 % annual rate [1].

Understanding of the microstructure effects on the DC properties of magnetic materials has reached a mature state, while AC properties are still in constant and intensive investigation. Determination of the frequency behavior in soft ferrites by inductance spectroscopy is of great importance for basic investigation of the magnetization processes, and of course, for applications. An equivalent electric circuit that represents materials behavior can be established from the analysis of the complex permeability data, whose elements can be correlated to physical parameters of the material.

EXPERIMENTAL PROCEDURE.

Samples were prepared by the traditional ceramic technique, using the corresponding oxides (reagent grade) as raw materials. Three different composition of mixed ferrites, having commercial application, were synthesized. These compositions are (Ni_x Zn_{1-x}) Fe_2O_4 with x = 0.3, 0.35 and 0.4. A small excess of divalent oxides was used to compensate the Zn volatilization during sintering. Oxides were mixed in a ball mill for 8 hr in an ethanol suspension and dried. Mixture was pressed into a toroidal shape, applying 1.5 ton/cm^2 for a few minutes. PVA was added to the oxide mixture to easily hand compacted parts. Sintering was carried out in a tubular laboratory furnace at a temperature of 1150 °C for several dwell times from 6 to 96 hr. Air forced atmosphere with a flux of 2.4 lt/min was used for all sintering processes.

Composition of samples was established by x-ray diffraction using a Siemens D-5000 with Cu Kα radiation. Average grain size for each composition and sintering time was evaluated in polished samples by scanning electron microscopy (SEM) in a Zeiss DSM-960. Point composition was determined in the grains and in the grain boundary by energy dispersive analytical x-ray (EDAX). Curie temperature (T_C) was established from the initial permeability thermal behavior, using an experimental setup described elsewhere [2]. This evaluation is also indicative of the chemical homogeneity of the sample.

Frequency measurements were carried out in samples wound with a low capacitance coil in a HP-4192 impedance analyzer at temperatures in the 110-450 °K range. This evaluation was made at 94 discrete frequencies in the 5 Hz to 13 MHz frequency range. Frequency results were interpreted according to the complex inductance formalism, to establish the equivalent electric circuit to model materials performance.

RESULTS AND DISCUSSION.

During sintering, samples showed a density increase of the order of 60%, reaching densities of 98% of the theoretical value. The samples quality was verified by the Zn content variation and by the X-ray diffraction pattern. Since the volatilization of Zn reached less than 2% of total Zn, and X-ray diffraction showed a single spinel phase, samples were considered suitable for further analysis.

Grain size determination showed an average value of 2-2.5 microns for samples sintered at 6 hr and 4.5-5 microns for samples sintered at 96 hr. In Figure 1 a direct relation can be observed between grain size and sintering time.

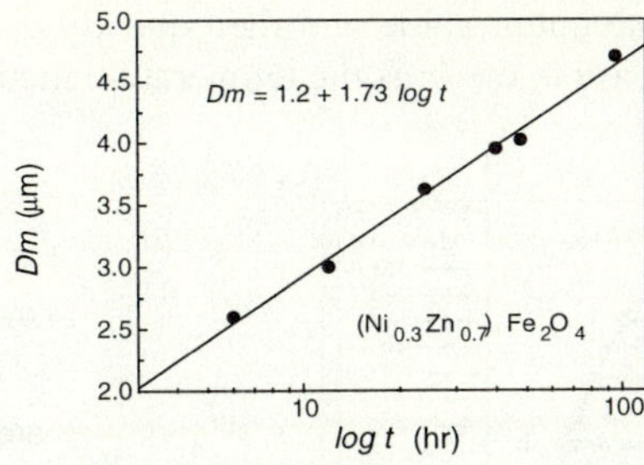

Fig. 1.- Grain size as a function of sintering time.

Variation of initial permeability as a function of temperature shows first a smooth increasing slope and then at the T_C a sharp drop. It has been shown that the sharpness of the drop is an indicative of the chemical homogeneity of the sample [3].

It has been also demonstrated that inductance spectroscopy is a useful technique to identify magnetization-polarization mechanisms occurring in magnetic materials according to their velocity response to an applied AC field [4]. As this study was limited to reversible mechanisms, i.e. domain wall bowing mechanisms, very low fields, of the order of 16 mOe, were used.

In order to establish an equivalent electric circuit to simulate sample response, was necessary to eliminate coil effect from the overall experimental response. To do this each sample was evaluated at a temperature quite above the T_C. These results were subtracted point to point from the overall response of the same sample evaluated at any other temperature, below Curie point.

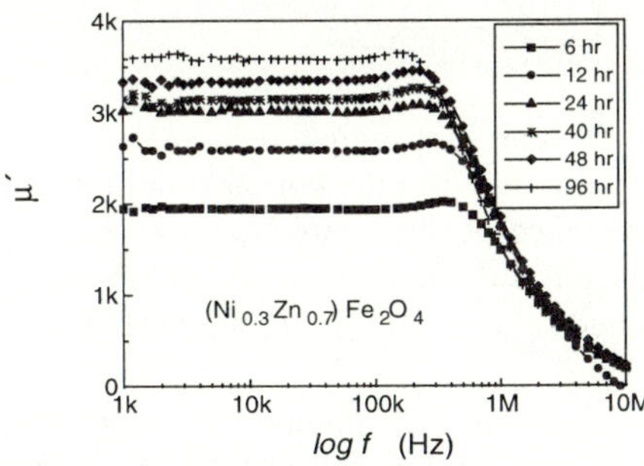

Fig. 2.- Real permeability vs. frequency for each sintering time.

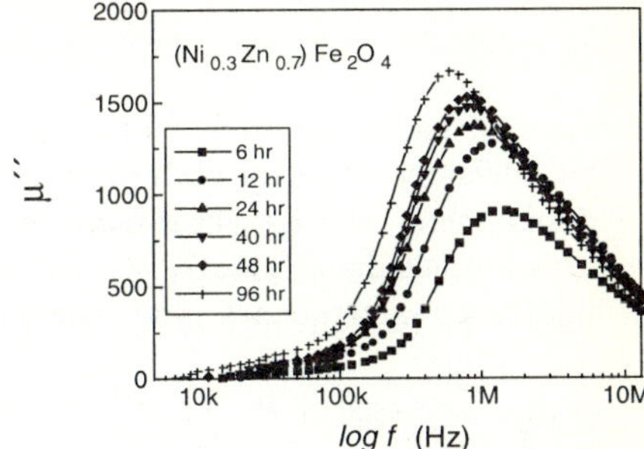

Fig. 3.- Imaginary permeability vs. frequency for each sintering time.

Permeability was determined from the inductance values through the toroidal geometry equation. Figures 2 and 3 show "cleaned" real (μ') and imaginary (μ'') permeability respectively, at room temperature, as a function of frequency for each sintering time. In these figures it can be observed that the μ increases as the sintering time increases, i.e. as grain size increases; on the other hand relaxation frequency *(fr),* maximum in the μ'' vs. f curve, decreases as sintering time increases. That is, μ is directly related to the grain size while *fr* is inverse to it [5]. For the composition range studied, permeability increases and relaxation frequency *(fr)* decreases as Zn content increases.

Samples were also evaluated by inductance spectroscopy at different temperature. Results are shown in the Figures 4 and 5. In Figure 4, as the temperature increases it is observed an increase in real permeability until a maximum value and then quickly decreases to a zero value. In Figure 5 we can see a constant decrease in the *fr* as the temperature increases.

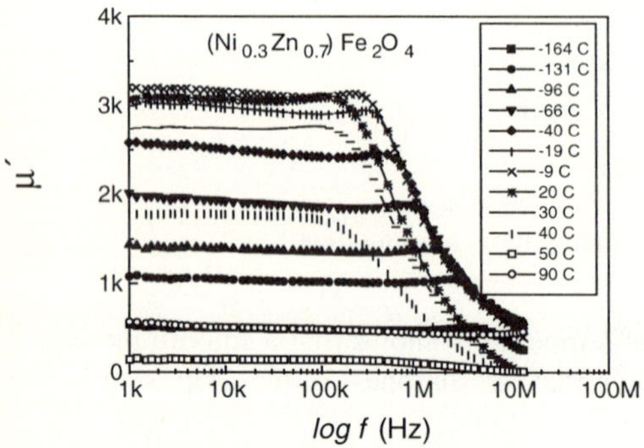

Fig. 4.- Real permeability vs. frequency at different temperatures.

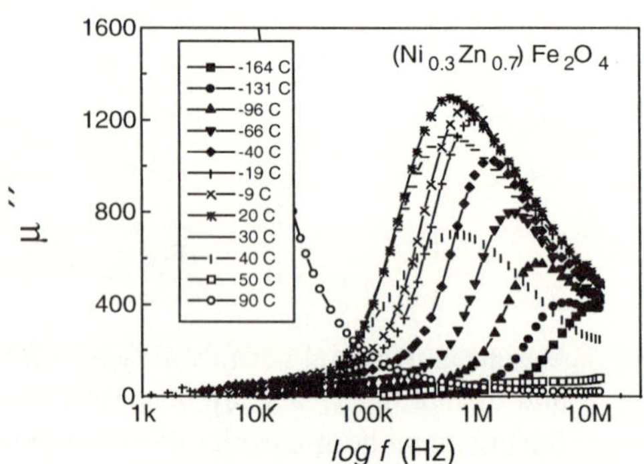

Fig. 5.- Imaginary permeability vs. frequency at different temperatures.

To simulate these results an electric circuit having elements representing coil response in series with elements for sample response is required, so coil signal can be subtracted from that of the sample itself. The circuit proposed is as follow:

Rp

Rs Ls

Lp

where *Rs* and *Ls* represent coil response and *Rp* and *Lp* represent sample response. *Rs* and *Ls* were considered constant for each sample. *Rs* was directly measured in the coil at room temperature by an ohmmeter, and *Ls* was established from the inductance of the sample having same number of turns evaluated above de T_C. *Lp* values were established as the inductance at 10 kHz from "clean" inductance vs. frequency plots, while *Rp* was calculated from *fr* and *Lp*, through relation:

$$Rp = 2\pi \; fr \; Lp \qquad \text{(Eq. 1)}$$

In this way sets of *Rs*, *Ls*, *Rp* and *Lp* were established for each composition and temperature. Validation of this circuit and the calculated real and imaginary permeability as a function of frequency for Ni-Zn composition prepared have been reported elsewhere [6,7].

The equivalent inductance element (*Lp*) represents the domain wall permeability μ and the resistor element (*Rp*) is associated with the domain wall damping term β of the domain wall motion equation [8]. In Figure 6 we observe a linear relation between resistance element and temperature. In order to find some dependence of this equivalent circuit element, we compared *Rp* with basic magnetic parameters such as saturation magnetization (M_s) reported in literature [9,10]. This comparison between different samples can have a full physical meaning, as far as saturation magnetization is an intrinsic property, which depends essentially on the composition and crystal structure, and is insensitive to the granular microstructure. A linear dependence of $(4\pi M_s)^2$ with *Rp* was found for all compositions, as can be seen in Figure 7. Fitting parameters showed a dependence not only on composition, but on average grain size as well.

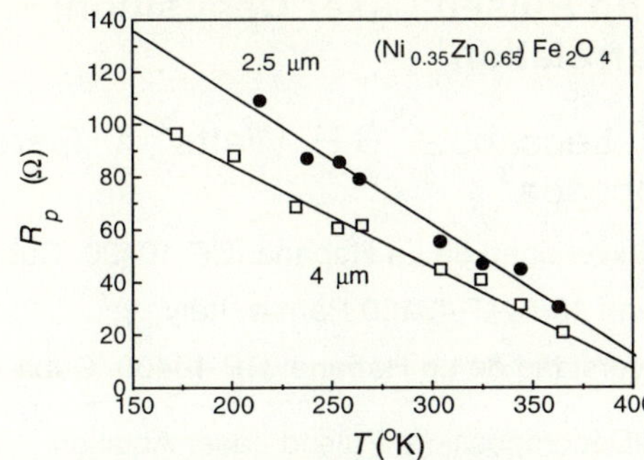

Fig. 6.- Resistance element (Rp) vs. T

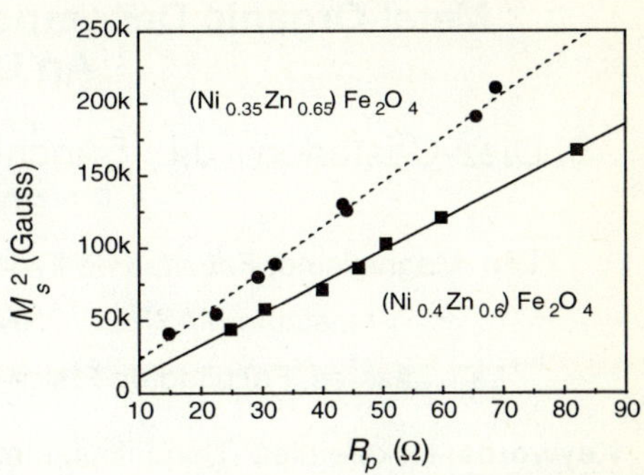

Fig. 7.- Saturation magnetization vs. Rp

Resistor element is related to wall damping term by: [8,11]

$$R = cte\left[\frac{1}{Dm\beta}\right]M_s^{\,2} \qquad \text{(Eq. 2)}$$

where Dm is the average grain size. According to our experimental results, $(Dm\beta)$ product must be a constant and equivalent to the slope in the $(4\pi M_s)^2$ vs. Rp plot, i.e. β is constant and independent of temperature for samples of same composition and average grain size. Similar results have been obtained in amorphous magnetic alloys using different experimental techniques [12].

SUMMARY.

Frequency evaluation of soft ferrite through inductance spectroscopy permits to analyze magnetic mechanisms and to establish an electric circuit to model material behavior.

Complex plane plots of the real and imaginary parts of permeability showed relaxation behavior, which can be related to an equivalent circuit having an $RpLp$ parallel arrangement.

A linear dependence of M_s^2 with Rp was found for all compositions, whose fitting parameters showed a dependence not only on composition, but on average grain size as well. This linear dependence confirm the validity of Eq. 2 and more important shows the possibility of evaluating microscopic parameters from macroscopic measurements.

REFERENCES.

1.- T. Abraham, *ACerS Bulletin* **75**(2), (1996), p. 47

2.- E. Cedillo, J. Ocampo, V. Rivera and R. Valenzuela, *J. Phys. E: Sci. Instrum.* **13**, (1980), p. 383

3.- J.M. Yellup and B.A. Parker, *J. Phys. Sta. Sol.* **A 55**, (1979), p. 137

4.- R. Valenzuela, Adv. Topics in Mat. Sci. & Eng. Ed. Morán-López, Sánchez, Plenum Press, (1993).

5.- I. Betancourt and R. Valenzuela, *IEEE Trans. Magn* **33**, (1997), p. 3973

6.- M.I. Rosales, *Ph.D Thesis*, **UNAM**, (1997).

7.- M.I. Rosales, E.Amano, M.P.Cuautle and R.Valenzuela, *J. Mat.Sci. and Eng.* **B49** (1997), p.221

8.- G. Aguilar-Sahagún, P. Quintana, E. Amano, J.T.S. Irvine and R. Valenzuela, *J. Appl. Phys.* **75**, (1994).

9.- J. Smit and H.P. Wijn, "Ferrites", *Philips*, (1961), p. 158

10.- W.H. Aulock, "Handbook of Microwave Ferrite Materials", *Academic Press*, (1965), p. 381

11.- M. Guyot and V. Cagan, *J. Appl. Phys.* **76**, (1994), p. 6209

12.- C. Aroca, E. López and P. Sánchez, *Phys. Review B*, **30**(7), (1984), p. 4024

Materials Science Forum Vols. 302-303 (1999) pp. 416-420

Magnetic Properties of Polycrystalline Hexaferrites Thin Films Grown by Metal-Organic Decomposition and Pulsed Laser Deposition: An Useful Correlation

S. Díaz-Castañón[1], J.L. Sánchez Ll.[1], F. Leccabue[2], B.E. Watts[2], T. García[3] and E. Deposada[3]

[1] Lab. Magnetismo, Facultad de Física-IMRE, Universidad de La Habana, CP 10400, Cuba

[2] Instituto MASPEC/CNR, Via Chiavari 18/A, IT-43100 Parma, Italy

[3] Lab. Láseres, Facultad de Física-IMRE, Universidad de La Habana, CP 10400, Cuba

Keywords: Hexaferrites, Thin Films, Metal-Organic Decomposition, Pulsed Laser Ablation

Abstract

The evaluation of different routes to prepare polycrystalline thin films based on hexagonal ferrites is of considerable interest due to the excellent properties of these materials as high density magnetic recording media. In particular, the metal-organic decomposition and the pulsed laser ablation deposition have been used for us to grow high coercivity polycrystalline hexaferrite thin films and the results obtained are compared. Using MOD technique $BaFe_{12}O_{19}$ thin films were deposited onto ZrO_2/Si substrates, and the influence of different parameters such as the annealing temperature (500-850°C) and the atmosphere (oxygen and air) during thermal treatments on the magnetic properties are discussed. The formation of barium hexaferrite starts at 600 °C, the beneficial role of an oxygen atmosphere in the increasing of coercivity was established and the higher coercivity was 4.0 kOe. Using PLD technique $BaFe_{12}O_{19}$ and $SrFe_{12}O_{19}$ thin films were deposited on Si substrates using a Nd-YAG laser apparatus. As-deposited films were annealed in air at temperatures between 600 and 900 °C in order to improve the crystallinity. The samples annealed in the temperature range 700-850 °C show a higher coercivity (5.75-6.60 kOe) which are among the best values reported in thin films, powders and sintered samples. In all cases the saturation magnetisation of the films is in agreement with values reported in the literature.

Introduction

The renewed interest on the different ways for the preparation of ultrafine powders and polycrystalline thin films based on hexagonal ferrites are closely connected with the wide expansion of the magnetic recording media applications in the 90's. For this it is very important the development of techniques to produce powders and thin films with submicronic particles, homogeneous in size/shape, with high saturation magnetisation, moderate coercivity (iHc≤ 2.0 kOe), narrow switching field distribution, and orientation coefficient (Mr/Ms) near to 1. In this sense, the M-type hexagonal ferrites ($MFe_{12}O_{19}$, M= Ba, Sr or Pb) are one the most promising materials due to their uniaxial anisotropy [1]. In fact, perpendicular and longitudinal particulate recording media, based in M-type hexaferrites, with high recording densities with respect to the commercially available tapes or films have been reported. [2-5].

In order to obtain thin films several techniques such as RF sputtering [3,4,6], metal-organic chemical vapour deposition [7], pulsed laser ablation deposition (PLD)[5,8,9], metal-organic decomposition (MOD) [10-12], and sol-gel [13] have been evaluated. MOD route is an attractive possibility for industrial applications, because it is possible to prepare films with large areas at low cost, great homogeneity in grain distribution, and submicronic particles. The PLD is a new deposition technique with many advantages over other physical techniques for growing complex and multicomponent materials. In previous works the different conditions for the preparation of Ba-M hexaferrite thin films by MOD and PLD have been examined [9, 12]. These results are correlated with those obtained in the growth of Sr-M hexaferrite thin films by PLD, in which the coercivity was clearly higher (5.75-

6.60 kOe). In particular, the attention was devoted to analyse the effect of temperature and atmosphere used during thermal treatments on the magnetic properties.

Experimental procedure

The MOD process used to grow Ba-M films was previous discussed in ref. [12]. Ferric hexanoate in mineral oil and barium methoxyethoxide, as raw materials, were mixed in stoichiometric cationic ratio Ba^{+2}/Fe^{+3}= 1/12. The solution obtained was stirred for 30 mins and syringed through a 0.2 μm teflon filter on 10-20 mm^2 ZrO_2-coated Si substrates and, finally, was spun at speeds above 2500 rpm for 30 s. In order to increase the thickness all the procedure was repeated for a total of four depositions. After each spinning the films were pyrolyzed at 400 °C for 10 mins. A subsequent heat treatment between 500-850 °C for 1 h was performed in order to crystallise the ferrite. Samples were obtained in both oxygen and air atmosphere.

To growth Ba-M and Sr-M films by means of PLD technique a home-made Nd-YAG laser apparatus, working in the Q-switch regime, was employed [9] using ceramic disks of $BaFe_{12}O_{19}$ and $SrFe_{12}O_{19}$ as targets. The target-substrate distance was fixed at 40 and 25 mm for Ba-M and Sr-M growths, respectively. During the deposition process, the substrates were kept at room temperature under air atmosphere. As-deposited films were thus annealed at temperatures between 600 and 900 °C in air to improve the crystallinity.

Magnetic properties were measured at room temperature by applying the magnetic field parallel and perpendicular to the plane of the films; the measurements were carried out using an Oxford model 3001 vibrating sample magnetometer (with H_{max}= 16 kOe and a maximum sensibility of 10^{-5} emu). The system is equipped with low and high temperature facilities. In order to calculate the saturation magnetisation (Ms), the area of the films were determined. In the case of MOD films the thickness was determined using a JEOL JSM-530 scanning electron microscope [11].

Results and Discussion

Due to the extremely low magnetic moment of the films it is a critical problem to determine their magnetic properties. In Figure 1 the heating thermomagnetic curve of hexaferrite thin film is showed. This curve was recorded at 8.0 kOe in order to minimise the unfavourable signal-to-noise ratio. A well defined drop at 448 °C, corresponding to the Curie temperature of $BaFe_{12}O_{19}$, is observed, as a clear signal of the well reproduction of the target composition. This behaviour was the same for all films obtained. As an example, Figure 2 shows the correct hysteresis loop after the subtraction of the offset background signals as a function of the field corresponding to sample holder + substrate.

MOD films

Figure 3 shows the effect of the atmosphere and annealing temperatures on the coercivity of the Ba-M films. As can be seen, samples heated at 600 °C present a low coercivity in comparison with the other samples, corresponding to poorly crystallised particle state. Above this temperature, the samples present high coercivities, and their values are comparable to those measured for bulk isotropic sintered polycrystals [14]. Moreover, the coercivity of samples heated in oxygen is clearly superior. The microstructural study shows that the average grain size of samples heated in oxygen is significantly lower that those annealed in air; these data are summarised in Table 1. The smaller grains obtained during an oxygen annealing can be explain the increasing of the coercivity. The Mr/Ms coefficient, measured in the parallel/perpendicular directions in both series, does not show significant differences and the corresponding values near to 0.5 denotes the isotropic character of the films.

In Figure 4, the typical hysteresis loop for the best film is displayed. Note that the minor loops were measured in order to appreciate the magnetisation mechanism. The virgin magnetisation curve, characterised by a low initial susceptibility, suggests that the grains tend to behave as single domains in the thermally demagnetised state. Moreover, the values of Ms (276 emu/cm^3) and iHc (4.0 kOe) are of the same order of those obtained for Ba-M sintered magnets.

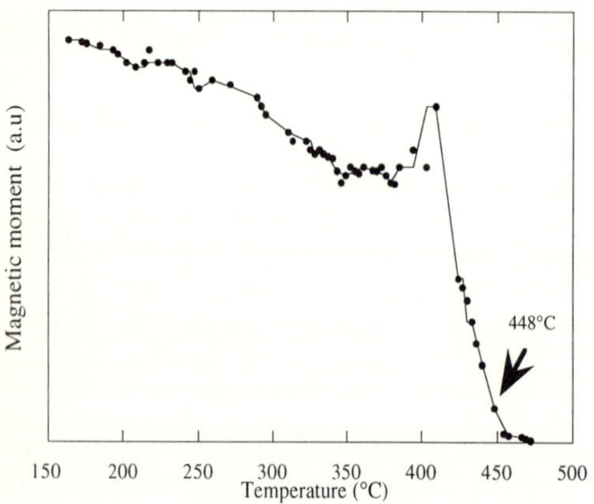

Figure 1. Thermomagnetic curve for Ba-M hexaferrite thin film, recorded at 8.0 kOe.

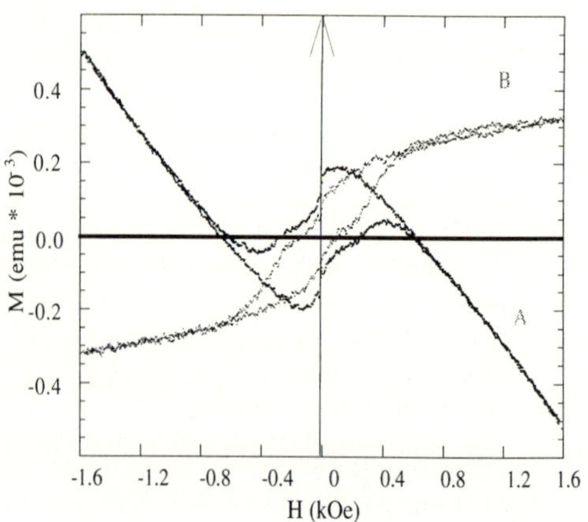

Figure 2. Hysteresis loop for a hexaferrite thin film. A) signal directly recorded; B) signal obtained after subtracting the off-set background.

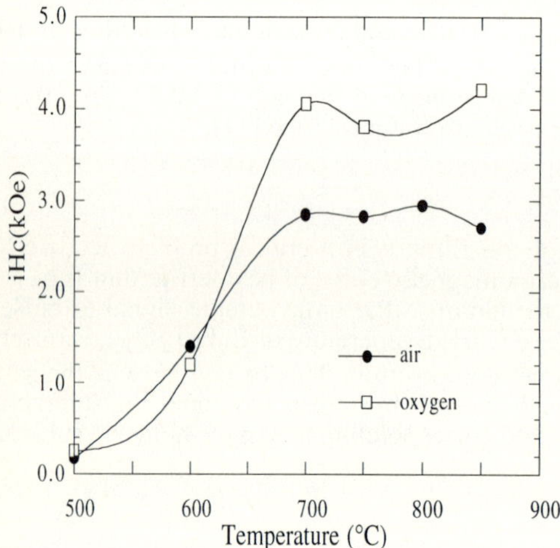

Figure 3. Coercivity as a function of annealing temperatures for different atmosphere.

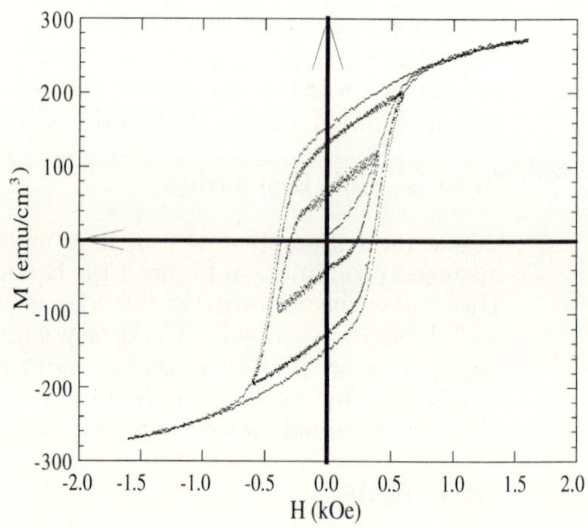

Figure 4. Hysteresis loop for a MOD Ba-M thin film annealed at 750 °C in oxygen.

PLD films

In Figure 5 the dependence of the intrinsic coercivity on the annealing temperatures for Ba-M films, grown on Si substrates, is reported. In the same line with the results obtained for MOD films, the coercivity is low for temperatures lower than 600 °C and increases rapidly exhibiting values comparable to sintered polycrystals (3.5 kOe) for temperatures between 700-800 °C. The submicronic particles in this temperature range are in agreement with the high values of the coercivity obtained (see Table 1). The decreasing tendency observed above 750°C can be attributed to the grain growth. The Mr/Ms values measured (not show here) confirm that the films obtained were magnetically isotropic.

Table 1. Growth parameters and grain size characteristics for the hexaferrite thin films. (see ref. [9,11,12]).

Films	Preparation methods	Temperature (°C)/time (h), atmosphere	Grain size (μm)
Ba-M	MOD	750/1, air	0.1-0.2
-	-	750/1, oxygen	<< 0.1
-	-	850/1, air	0.4-0.5
-	-	850/1, oxygen	0.2-0.3
-	PLD	750/1, air	0.2
Sr-M	PLD	750/1, air	≈ 0.3
-	-	800/1, air	≈ 0.35
-	-	840/1, air	≈ 0.4

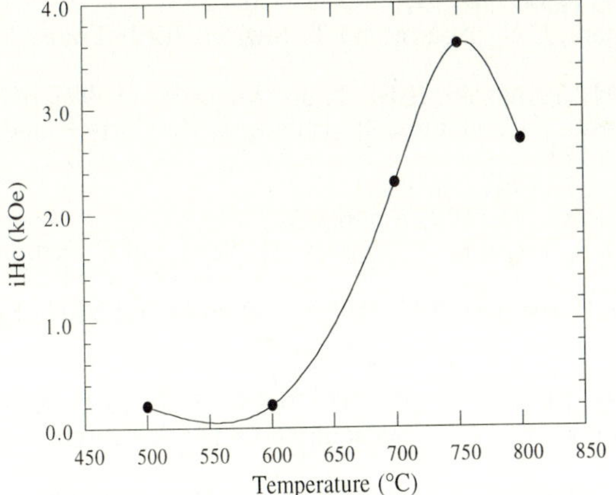

Figure 5. Coercivity as a function of annealing temperatures for Ba-M PLD thin films.

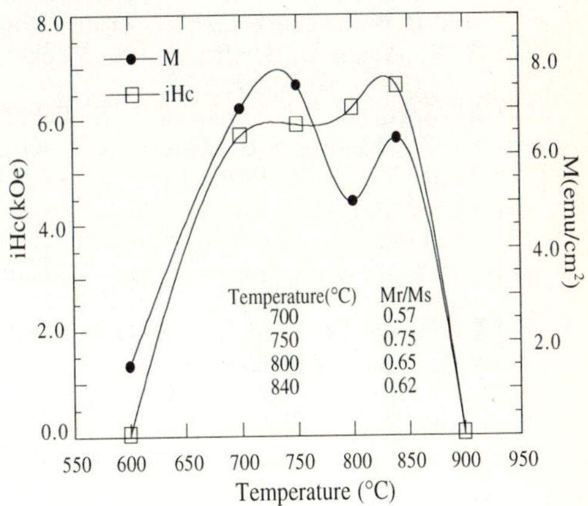

Figure 6. Magnetisation and coercivity as a function of annealing temperature for Sr-M PLD thin films.

Figure 6 shows the magnetic properties of the Sr-M films. In correspondence with Ba-M films examined previously, iHc is lower for temperatures inferior of 600 °C and for temperatures greater than 900°C. As can be noted, the coercivity in the range 700-850 °C is higher (5.7-6.65 kOe, respectively) and it is comparable with the values obtained for powders with submicronic and nanocrystalline particles prepared by chemical coprecipitation or glass crystallisation methods [14]. The preliminary grain size observations of these films support the existence of submicronic particles and further investigations are in progress. In addition, the Mr/Ms values show the tendency to the preferential orientation in the parallel direction for the films grown at 760 °C.

Conclusions

High coercivity polycrystalline $BaFe_{12}O_{19}$ and $SrFe_{12}O_{19}$ thin films were successfully grown by MOD and PLD procedures. In all cases, M-type hexaferrite films are formed at annealing temperatures above 600 °C for short times. The beneficial effect of an oxygen thermal treatment on iHc for Ba-M MOD thin films was encountered; the values of Ms measured in these samples are in agreement with those obtained for Ba-M sintered magnets. The Sr-M PLD thin films exhibit the best value of coercivity (6.65 kOe). Ba-M films obtained by MOD and PLD were magnetically isotropic;

on the contrary, for Sr-M PLD films a certain degree of magnetic anisotropy was found in the sample treated at 750 °C.

Acknowledgements

This work was carried out in the framework of the research project 96-192 RG/PHYS/LA of the Third World Academy of Sciences (TWAS/ICTP, Trieste). S. Díaz-Castañón thanks the ICTP for making possible his stay at MASPEC/CNR Institute of Parma.

References

[1] D. Speliotis, M. Dugas, J. Humenansky, K.L. Babcok and K. Peter, J. Appl. Phys. 81 (1997), p. 3830.
[2] P.M. Sollis, P.R. Bissell, S.M. McCann, T. Onions, M. Van Boven, P.T. Por, M.Th. Rekveldt and R.W. Chantrell, IEEE Trans. Magn. 33 (1997), p. 3073.
[3] B. R. Acharya, S. Prasad, N. Venkataramani, E.N. Abarra and T. Suzuki, IEEE Trans. Magn. 33 (1997), p. 3640.
[4] A. Morisako, M. Matsumoto, S. Takei and T. Yamazaki, IEEE Trans. Magn. 33 (1997), p. 3100.
[5] P. C. Dorsey, S.B. Qadri, K.S. Grabowski, D.L. Knies, P. Lubitz, D.B. Chrisey and J.S. Hortwitz, Appl. Phys. Lett. 70 (1997), p. 1173.
[6] X. Sui and M.H. Kryder, Appl. Phys. Lett. 63 (1993), p. 1582.
[7] E.J. Donahue and D.M. Schleich, J. Appl. Phys. 71 (1992), p. 6013.
[8] C.A. Carosella, D.B. Chrisey, P. Lubitz, J.S. Horwitz, P. Dorsey, R. Seed and C. Vittoria, J. Appl. Phys. 71 (1992), p. 5107.
[9] T. García, J.L. Sánchez, S. Díaz, L. Ponce, E. Estevez, J.L. Peña, P. Bartolo and M.H. Farias, Mat. Lett. 28 (1996), p. 65.
[10] D.F. Ryder Jr. and D. Bonnenberg, J. Amer. Ceram. Soc. 73 (1990), p. 282.
[11] B.E. Watts. F. Leccabue, R. Panizzieri, E. Melioli, S. Díaz, H. Rodríguez and J.L. Sanchez, Proc. 8th CIMTEC-World Ceramics Congress, Ed. P. Vincenzini (Techna srl, Florence,1995), p. 2677.
[12] S. Díaz, J.L. Sánchez B.E. Watts, F. Leccabue and R. Panizzieri, J. Magn. Magn. Mater. 151 (1995), p. 173.
[13] C. Surig, K.A. Hempel and D. Bonnenberg, Appl. Phys. Lett. 63 (1993), p. 2836.
[14] H. Kojima, in: Ferromagnetic Materials, vol. 3, Ed. E. P. Wohlfarth (North-Holland Publ. Co., Amsterdam, 1982), p. 305.

Materials Science Forum Vols. 302-303 (1999) pp. 421-424

Magnetic Viscosity in Pb-M Hexaferrites

J.C. Faloh Gandarilla, N. Suárez Almodovar and S. Díaz Castañón
Laboratorio de Magnetismo, IMRE - Facultad de Física, Universidad de La Habana, C.P. 10400, La Habana, Cuba

Keywords: Magnetic Viscosity, Hexaferrites, Coercivity

Abstract Magnetic viscosity measurements in $PbFe_{12}O_{19}$ powders obtained by metallorganic decomposition and chemical coprecipitation methods are reported in the temperature range 75-300 K. The measurements were carried out using an Oxford model 3001vibrating sample magnetometer with a maximum applied field of 16 kOe. The time dependence of magnetization was measured after saturation at a given steady reverse field. The ln(t) behavior of the magnetization was confirmed. The fluctuation field S_v was calculated by $S_v=S/\chi_{irr}$, where $\chi_{irr}=\chi_t-\chi_{rev}$ and χ_t is derived from a M(H, M'=cte) curve. S_v is found to be independent on the applied field, for applied fields around the coercive field H_c. The S_v values determined at room temperature are related to H_c within the limits of a Barbier plot $\log(S_v)= m\cdot\log (H_c)$, with $m \approx 1$. S_v and the activation volume, V_a, increase monotonically with temperature. V_a values ($\sim 10^{-17}$ cm^3) are in agreement with previous reports on Ba-M and Sr-M hexaferrites, corresponding to a critical diameter about the Bloch wall thickness δ.

Introduction.

Magnetic viscosity, the change in magnetization over time of a magnet at constant applied field, has been intensively studied in recent years for its importance from the fundamental standpoint and in various applications. At constant field thermal activation alone may lead to significant variation in magnetization. For a flat distribution of energy barriers [1,2] the magnetization decays with time as:

$$M=\textit{constant} + S \ln(t) \qquad \text{Eq 1}$$

where S is called the viscosity coefficient. The magnetic viscosity coefficient S is related to the irreversible susceptibility and the fluctuation field S_v by

$$S=S_v\chi_{irr} \qquad \text{Eq 2}$$

The concept of the fluctuation field was introduced by Néel [3] who assumed that thermal agitation could be represented by a fictitious field. The fluctuation field is linked with the micromagnetics of magnetization reversal. In principle independent measurements of S and χ_{irr} can be used to determine S_v.

The activation volume, it is the volume associated with the initiation of magnetization reversal [4], can be defined as $\Delta m/M_s$, where M_s is the spontaneous magnetization and Δm is the moment variation between the minimum energy state and the maximum energy state, reached at the instant that the energy barrier is overcome [5]. The activation volume can be estimated from S_v by the relation

$$V_a=kT/S_vM_s \qquad \text{Eq 3}$$

where k is the Boltzman constant.

In this work the magnetic viscosity phenomenon in $PbFe_{12}O_{19}$ powders obtained by the metallorganic decomposition (sample MOD) and chemical coprecipitation (sample cop) methods is studied. The temperature dependence of S_v and the activation volume is reported in the temperature range from 75 to 300 K.

Experimental procedure.

Measurements were made on small disc samples, prepared by pressing and sintering the $PbFe_{12}O_{19}$ powders for one hour at 920°C. The magnetoestructural characterization of the powders is in good agreement with the reported data, magnetic properties and grain size are shown in Table 1. As can be seen, the coercive field values are notably different, due to the particle size. Then the samples are considered to have a different coercive force mechanism. The initial magnetization curves suggest respectively a nucleation mechanism for the sample cop and a nearly coherent rotation mechanism for the sample MOD.

Table 1. Magnetic properties and grain size of $PbFe_{12}O_{19}$ samples.

Sample	T_c (°C)	H_a (kOe)	σ_s (emu/g)	iH_c (kOe)	Particle diameter (µm)
MOD	448	------	53	4.1	0.4
cop	448	13.4	58	1.5	1.5

Viscosity measurements were performed between 75 and 300 K using a vibrating sample magnetometer. M(t) was recorded for different steady fields on the demagnetization curve for 1200 s. Before each measurement the sample was saturated. Viscosity data were fitted to an equation of the form of Eq 1 to calculate S.

S_v was calculated directly by $S_v=S/\chi_{irr}$, where S_v and χ_{irr} values before correcting for the demagnetizing field were used. χ_{irr} can be expressed as $\chi_{irr}=\chi_t-\chi_{rev}$. After the measurements of viscosity the field was reversed by a small amount and returned to the measurement field. Then χ_{rev} was estimated as the slope of this minor loop assuming that all the processes which occur are reversible. In order to obtain χ_t from the slope of the demagnetization curve the (M, H) values were determined from the M'=dM/dt=constant condition. First dM/dt of the measured magnetization was plotted versus M and then the intersection of a horizontal line, dM/dt= constant, with the curves corresponding to the different fields applied gives the (M, H) values. This procedure was proposed by Givord et al. [6] and is compatible with the magnetic equation of state proposed by Estrin et al. [7].

Results and discussion.

Similar curvatures of the magnetic field dependencies of S and χ_{irr} were observed and these had maximum values around the coercive field for each temperature. Fig 1 shows S_v as a function of the specific magnetization. Over a wide range of specific magnetization, from +20 to -20 emu/g for the demagnetizing curve, S_v stays independent of σ. The relationship between log (S_v) and log (H_c) fits the Barbier linear plot [8] at room temperature fairly well, with a slope of about 1.0.

The temperature dependence of S_v and V_a is shown in Fig 2 and Fig 3. A clear increase of V_a with temperature is observed for the two samples. The V_a values are in agreement with previous reports on Ba-M and Sr-M hexaferrites, giving linear dimensions about the domain wall thickness δ.

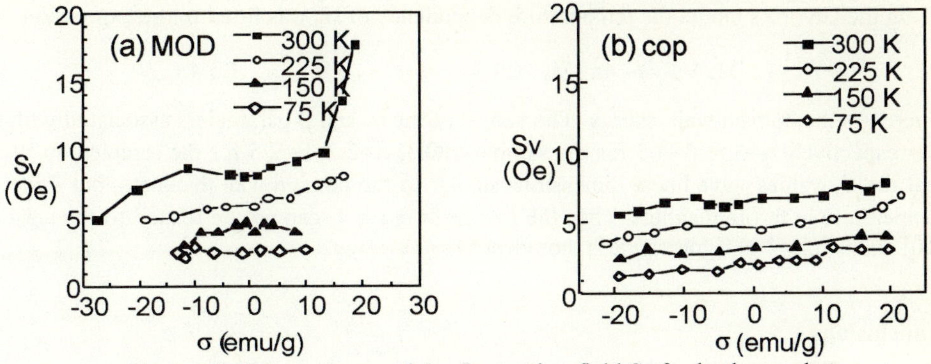

Fig 1. Field dependence of the fluctuation field S_v for both samples.

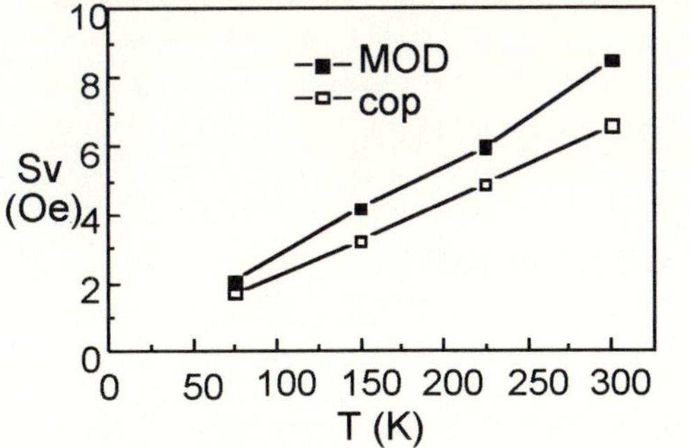

Fig 2. Temperature dependence of the fluctuation field S_V.

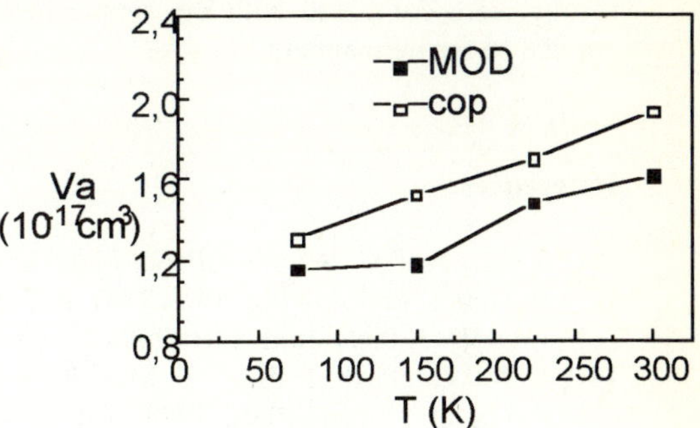

Fig 3. Temperature dependence of the activation volume.

Fig 4 shows the temperature dependence of the coercive field of the samples. The figure also shows the calculated H_c values applying a model due to Givord and co-workers in which the magnetic reversal occurs in a volume equal to the activation volume and is determined by the formation of a domain wall [9].

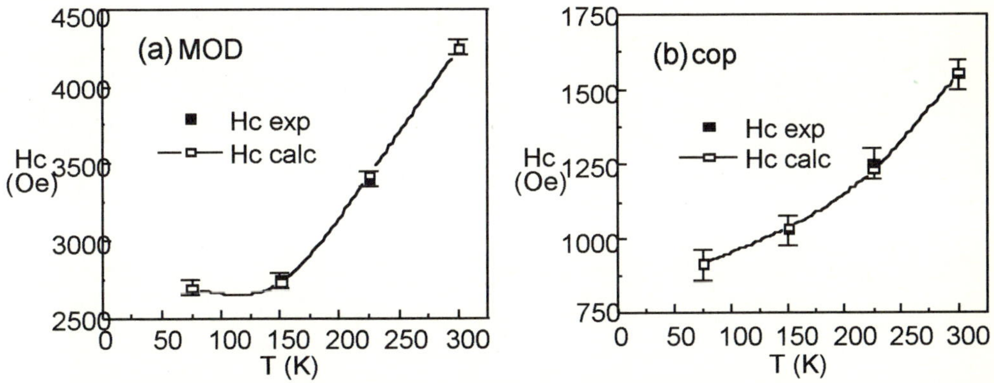

Fig 4. Temperature dependence of the experimental coercive field (Hc exp) and the calculated coercive field (Hc calc) .

Using the Givord's model the temperature dependence of H_c was fitted to the expression

$$H_c = (\alpha\,\gamma) / (M_s\, V_a^{1/3}) - 4\pi\beta M_s - 25\, S_v \qquad \text{Eq 4}$$

where γ is the domain wall energy. The values of the α and β parameters associated with the best fit are respectively α=3.6, β=4.5 for the sample MOD, α=2.2, β=2.5 for the sample cop. It was found that the V_a values give linear dimensions similar to the domain wall thickness, but δ increases with temperature in ferrite magnets. Then the first term in Eq 4 can not be related to the anisotropy field [10] and its physical significance is not clear.

Conclusions.

The fluctuation field S_v and the activation volume were measured in two $PbFe_{12}O_{19}$ samples with different coercive fields. The values of the V_a are in agreement with previous reports on ferrite samples, and increase with temperature in the studied range, from 75 to 300 K. The experimental results were compared with the simple model due to Givord and co-workers, which was already applied to ferrite magnets.

References.

[1] R. Street, J. C. Woolley. Proc. Phys. Soc. A62 (1949), p. 562.
[2] P. Gaunt. Philos. Mag. 34 (1976), p. 775.
[3] L. Néel. J. Phys. Rad. 11 (1959), p. 40.
[4] E. P. Wohlfarth. J. Phys. F: Met. Phys. 14 (1984), p. L155.
[5] P. Gaunt. J. Appl. Phys. 59 (1986), p. 4129.
[6] D. Givord, P. Tenaud, T. Viadieu. J. Magn. Magn. Mater. 72 (1988), p. 247.
[7] Y. Estrin, P. G. McCormick, R. Street. J. Phys.: Condens. Matter. 1 (1989), p. 4845.
[8] J. C. Barbier. Ann. Phys. 9 (1954), p. 84.
[9] D. Givord, A. Lienard, P. Tenaud, T. Viadieu. J. Magn. Magn. Mater. 67 (1987), p. L281.
[10] D. Givord, P. Tenaud, T. Viadieu. IEEE Trans. Magn. 24 (1988), p. 1921.

Materials Science Forum Vols. 302-303 (1999) pp. 425-429

Magnetoelectrical Characterization of Mixed Ba-Sr Ferrites

S.A. Palomares-Sánchez[1], S. Ponce-Castañeda[1], M. Mirabal-García[1], L. Narváez-Hernández[2], R. de Lira-Hueso[3], O. Pérez-Orta[3], A. Huanosta-Tera[4] and S. Diaz-Castañón[5]

[1] Instituto de Física, Universidad Autónoma de San Luis Potosí, 78000 San Luis Potosí, S.L.P., México

[2] Instituto de Metalurgia, Universidad Autónoma de San Luis Potosí, 78000 San Luis Potosí, S.L.P., México

[3] Fac. De Ciencias, Universidad Autónoma de San Luis Potosí, 78000 San Luis Potosí, S.L.P., México

[4] Instituto de Investigaciones en Materiales, UNAM, 04510 México, D.F., México

[5] Lab. de Magnetismo, Fac. Física-IMRE, Universidad de la Habana, La Habana, Cuba

Keywords: Ba-Sr Ferrites, Magnetization, Coercivity, Electrical Conductivity

Abstract

Results on structural, magnetic and electrical characterization of ferrimagnetic compounds of $Ba_xSr_{1-x}Fe_{12}O_{19}$ (x = 0.0, 0.25, 0.50, 0.75 and 1.0) prepared by the conventional ceramic method are reported. The composition dependence of saturation magnetization and coercitivity shows a maximum values at x=0.5 and x=0.25, respectively. The conductivity independent of frequency, obtained from direct calculation of impedance plots, exhibits an acceptable Arrhenius behavior. Temperature dependence of the total *ac* conductivity shows, at high and low temperatures, the *dc* and *ac* components. It was observed an increase in conductivity as the Ba concentration increases. All associated activation energies and dielectric constant are also calculated.

Introduction

In recent years a renewed interest in M type ferrites with different substitutions has arised due to possible uses in various novel devices. Ba and Sr hexaferrites find application as permanent magnets in motors and loudspeakers [1]. They are also applied in magnetic devices where dielectric media are necessary. Recently ultrafine particles of Ba ferrite have been considered as a promising medium for high-density perpendicular magnetic storage [2]. The properties of this kind of materials are well known; however, diverse magnetic and electrical properties arise when they are doped with other materials or when different techniques of preparation are used [3,4]. Studies on mixed Ba-Sr ferrites as single crystal [5] and as polycrystalline samples with admixtures of the group III oxides have been performed [6]. Although many studies have been performed on the magnetic properties of M-type ferrites, very few have been done on their electrical properties [7]. The most immediate interest in ferrites relies on its magnetic properties, but technological applications require a wider knowledge of general physical properties as well, especially when the materials are studied as function of composition or addition of different modifiers.

In this work results on *ac* and *dc* electrical properties, surface characterization and magnetic properties of mixed (Ba,Sr)-M type ferrites with different concentration of Ba and Sr are reported. Using the impedance spectroscopy technique, electrical conductivity of the samples was determined. This kind of measurements were made over a wide range of frequencies and temperatures. Magnetic characterization was performed in order to evaluate the coercivity (*iHc*), saturation (σ_s) and remanent (σ_r) magnetization, and orientation ratio (σ_r/σ_s).

Experimental procedure

The ceramic method was used in order to prepare mixed (Ba,Sr)-M type ferrites. Powders of $BaCO_3$, $SrCO_3$ and Fe_2O_3 with a purity higher than 99.99% were mixed stoichiometrically to obtain samples of general composition $Ba_xSr_{1-x}Fe_{12}O_{19}$, with x=0.0, 0.25, 0.50, 0.75 and 1.0. All the constituent powders were mixed homogeneously by ball-milling in ethanol for two hours. The resulting powders were calcinated during 2:45 hours at 1273K and samples with cylindrical shape were compacted by applying a pressure of 3.5 Tons/cm². For the sintering process a temperature ramp of 300K/h and a dwell of 873K during one hour and a ramp of 300K/h up to 1473K for one hour was applied. The topographic surface images were obtained using a JEOL model JSM-T330A scanning electron microscope (SEM). X-ray powder diffraction patterns were recorded with a

Rigaku 2200 diffractometer. Magnetic characterization was carried out in spherical samples, at room temperature, using an Oxford vibrating sample magnetometer with H_{max}= 16 kOe. Corrections due to the demagnetizing field were also done. Measurements of the electrical conductivity were recorded in the frequency range from 5 Hz to 13 MHz using an impedance analyzer HP4192A. An electric vertical furnace was used to carry out the frequency measurements in temperature range room temperature to about 500°C. The applied voltage was 1.0V, in all cases.

Results and discussion

X-ray diffractograms recorded for all the samples showed the same structure as that previously reported for the M type ferrite [8,9]. From SEM micrographs of samples with x=0.25, 0.50 and 0.75, it is observed that the average grain size for x=0.25 was larger than that for x=0.75 and x =0.50, but for the later case a more irregular grain shape was observed. Figure 1 shows the variation of the saturation magnetization and coercivity plotted as a function of the composition. The highest value of coercivity is obtained for x=0.25, whereas the highest saturation magnetization value is achieved for x=0.50. In the cases x=0.0 and x=1.0, low values of *iHc* are obtained. It is evident the beneficial role of the mixture Ba-Sr on the magnetic properties.

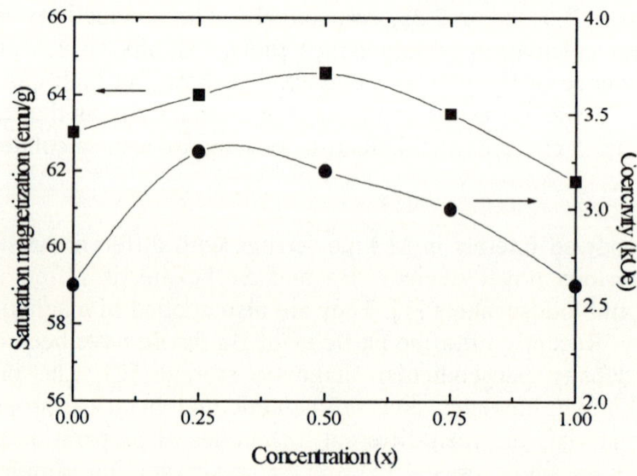

Fig. 1. Saturation magnetization and coercivity for mixed (Ba,Sr)-ferrites.

Electrical data were acquired in terms of the real (Z') and imaginary (Z'') components of the complex impedance. A simple equivalent circuit proposed to simulate the electrical behavior is constructed by two parallel RC loops connected in series corresponding, the first one (R_1C_1), to high frequency data and the second one (R_2C_2) to the low frequency region. Resistivity calculation was assisted by using the NLLS fitting routine [10] for each experimental temperature. Values of resistivity, calculated from the impedance curves, were used to plot Arrhenius curves of $\log\sigma$ against $1000/T$, as shown in figure 2. Conductivity data fit closely to an equation of the type $\sigma=\sigma_o \exp[-E_a/kT]$, where σ_o is the pre-exponential of conductivity, T is the absolute temperature, E_a is the activation energy and k the Boltzmann constant. The results associated with the region close to the origin will be referred to as the bulk properties σ_g. Capacitance values (C_1) on the high frequency regime are of the order of pF, which allows to assign that response as coming from the grains of the sample, whereas at low frequencies C_2 values are of the order of nF, an accepted [11] value for grain boundary phenomena, σ_{gb}. In figure 2 the total conductivity, σ_t, is also plotted.
The *ac* conductivity, $\sigma_t(w)$, is frequently reported in terms of the real part of admittance plotted as function of frequency; additional information can also be gained by plotting it against $1000/T$ at fixed frequencies. Using this fact we present *ac* results. Figure 3 shows the temperature dependence

of conductivity at selected fixed frequencies. In this figure each set of curves from (a) to (e) indicates decreasing content of Sr, in the understanding that the (a) set corresponds to x=0, (b) curve to x=0.25, and so on.

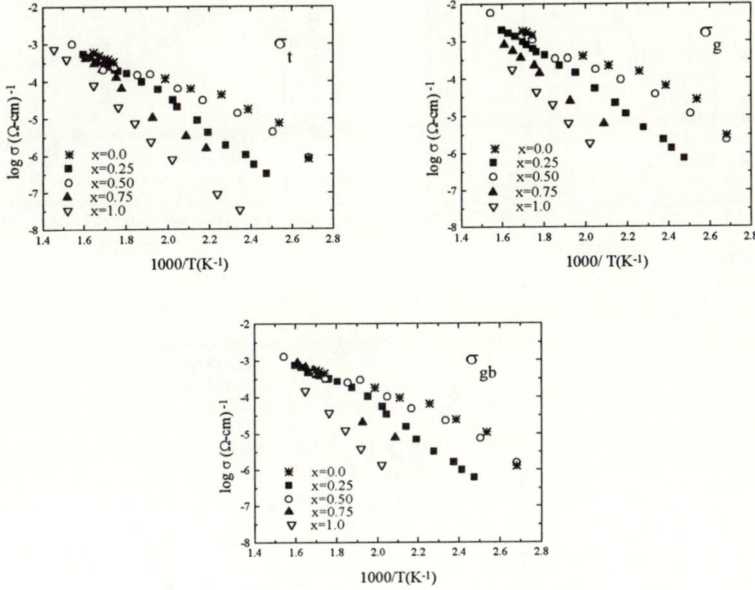

Fig. 2. Arrhenius curves of *log*σ versus *1000/T* for σ_t, σ_g and σ_{gb}.

It is noticeable to observe an Arrhenius behavior in two temperature intervals with different activation energies. At high temperatures all conductivity curves converge to frequency independent values. Such conductive behavior corresponds to the σ_{dc} component. Table I shows activation energies, E_{dc}, associated with the *dc* conductivity. The region of temperatures above the point where the slope changes, must be associated with that phenomenology in which long range charge carriers dominate the conductive process, whereas at low temperatures the pure σ_{ac} component is manifested by separating the low frequency conductivity from that of the high frequency one. As it can be appreciated in Table II for all studied compounds, E_{ac} decreases gradually at increasing frequencies in the low temperature range, accompanied by an increase in the magnitude of the *ac* conductivity. The largest value of the dielectric constant, ϵ', corresponds to x=0.0, decreasing as the barium content increases.

Conclusions

M-type ferrites of general composition $Ba_xSr_{1-x}Fe_{12}O_{19}$ were synthesized and their magnetic parameters, at room temperature, were found particularly interesting. The values of saturation magnetization and coercitive force are the highest for x=0.50 and x=0.25, respectively. Concerning the electrical measurements, at temperatures above the critical temperature of all samples investigated, the conductivity becomes frequency independent. Such regime corresponds to the σ_{dc} component; however, at low temperatures the pure σ_{ac} component manifests. It was also observed that a variation of the ratio Ba/Sr disturbs the electrical conductivity mechanism, leading to a continuous increase in conductivity as the barium content increases. Finally, it was also possible to calculate the activation energies for the *dc* and bulk, grain and grain boundary conductivities, as well as the activation energies for the pure *ac* behavior of conductivity.

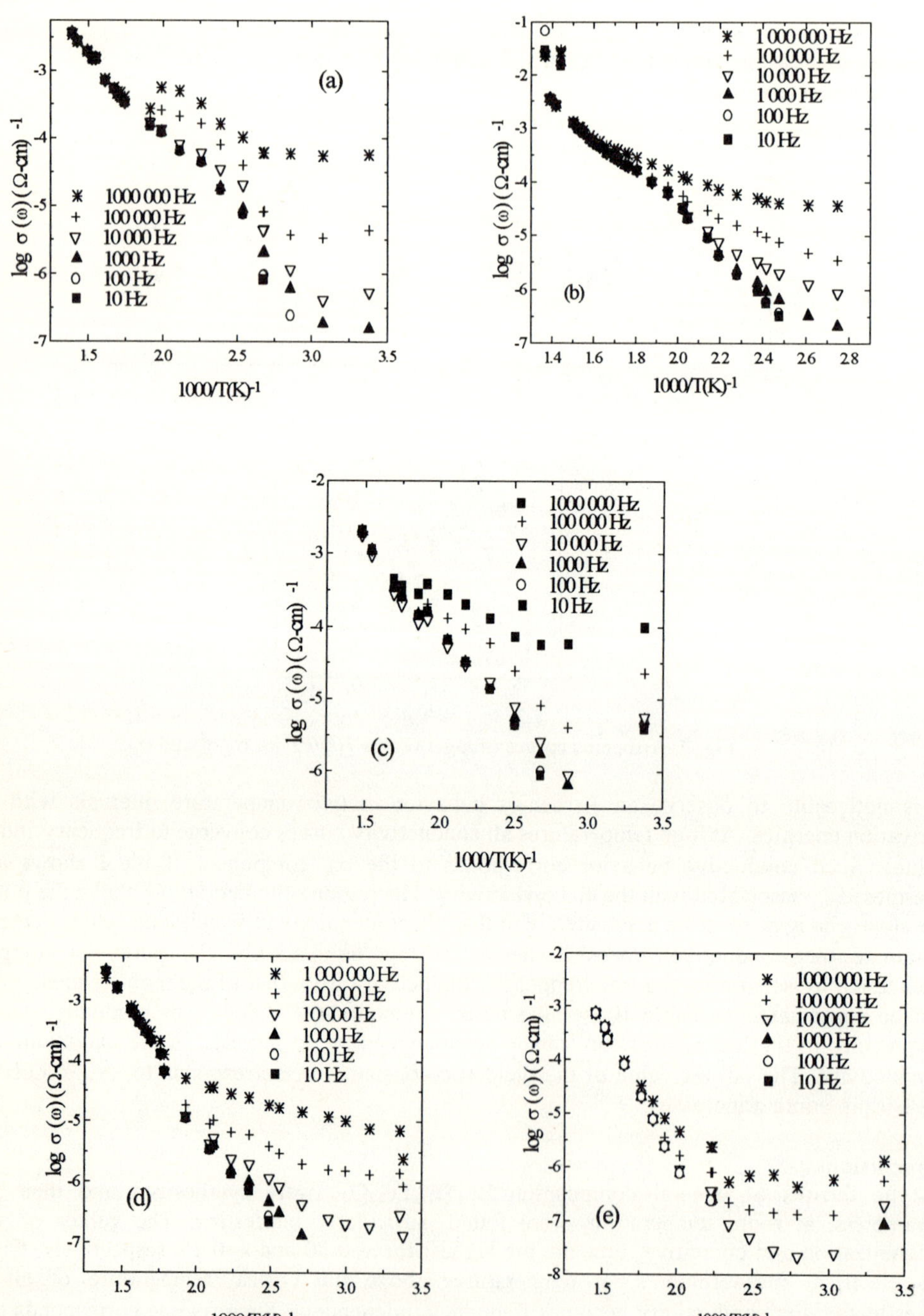

Fig. 3. Plots of log$\sigma(\omega)$ versus *1000/T* at fixed frequencies for (a) x=0.0, (b) x=0.25, (c) x=0.50, (d) x=0.75 and (e) x=1.0.

Table I. Activation energies for the *dc* (E_{dc}) and bulk conductivity (E_g). All activation energies were calculated within an error of ±0.003 eV.

x	E_{dc}	E_b	E_{gb}	E_t	Dielectric constant
0.0	0.60	0.42	0.38	0.39	1344
0.25	0.68	0.77	0.71	0.73	613
0.50	0.53	0.54	0.47	0.50	920
0.75	0.79	0.90	0.92	0.87	348
1.00	0.86	1.05	1.11	0.98	90

Table II. Activation energies for pure *ac* behavior.

Frecuency (KHz)	E_{ac} x=0.0	E_{ac} x=0.25	E_{ac} x=0.50	E_{ac} x=0.75	E_{ac} x=1.0
10^{-2}	0.57	0.85	0.56	0.59	
10^{-1}	0.63	0.84	0.54	0.56	
1	0.53	0.56	0.49	0.47	
10	0.45	0.40	0.40	0.29	0.30
10^2	0.37	0.31	0.36	0.18	0.18
10^3	0.16	0.17	0.18	0.11	0.13

Acknowledgments

This project was supported by the Consejo Nacional de Ciencia y Tecnología (CONACyT) under grants Nrs. F666-E9407 and 485100-5-2232P, Fondo de Apoyo a la Docencia and Fondo de Apoyo a la Investigación, UASLP, México. Thanks also to DGIA-UNAM for partial support, and to J. Guzmán for technical support.

References

[1] Kojima, H. *Ferromagnetic materials*, Vol. 3, E. P. Wohlfarth ed., North Holland, Amsterdam, 1982.

[2] Chantrell, R. W. and O'Grady, K., J. Phys. D: Appl. Phys. **25** (1992) 1.

[3] K. Haneda, C. Miyakawa, and H. Kojima, J. Am. Ceram. Soc. **57** (1974) 354.

[4] A. Chatterjee, D. Has, S. K. Pradhan and D. Chakravorty, J. Magn. Magn. Mater **127** (1993) 214.

[5] L. Jahn and H. G. Müller, phys. stat. sol. **35** (1969) 723.

[6] W. Wolski and J. Kowalewska, Jap. J. Appl. Phys. **9** (1970) 711.

[7] Pal, M., Brahma, P., Chakravorty, D. and Agrawal, D. C., J. Mag. Mag. Mater. **147** (1995) 208.

[8] V. Adelsköld, Ark. Kemi. Mineral. Geol. 12A (1938) 1.

[9] V. D. Townes, J. H. Fang and S. Perrota, Z. Kristallogr. **125** (1967) 1139.

[10] B. A. Boukamp, *Equivalent Circuit*, University of Twente Dept. Chemical Technology, Netherlands.

[11] A. K. Jonscher, *Dielectric Relaxation in Solids*, Chelsea Dielectric Press, London, (1983). J. R. Macdonald ed., *Impedance Spectroscopy*, John Wyley & Sons (1987). J. T. S. Irvine, D. C. Sinclair and A. R. West, Adv. Mater. **2** (1990) 132.

Materials Science Forum Vols. 302-303 (1999) pp. 430-434

Electrical and Magnetic Properties of Coprecipited Ba-Hexaferrites with Addition of SiO_2

S.A. Palomares-Sánchez[1], S. Ponce-Castañeda[1], M. Mirabal-García[1], L. Narváez-Hernández[2], R. de Lira-Hueso[3], O. Pérez-Orta[3], A. Huanosta-Tera[4] and S. Diaz-Castañón[5]

[1] Instituto de Física, Universidad Autónoma de San Luis Potosí, 78000 San Luis Potosí, S.L.P., México

[2] Instituto de Metalurgia, Universidad Autónoma de San Luis Potosí, 78000 San Luis Potosí, S.L.P., México

[3] Fac. De Ciencias, Universidad Autónoma de San Luis Potosí, 78000 San Luis Potosí, S.L.P., México

[4] Instituto de Investigaciones en Materiales, UNAM, 04510 México, D.F., México

[5] Lab. de Magntismo, Fac. Física-IMRE, Universidad de la Habana, La Habana, Cuba

Keywords: Ba Ferrite, SiO_2, Chemical Coprecipitation, Electrical Conductivity

Abstract

Preliminary results on the effects of magnetic, electrical and structural characteristics of $BaFe_{12}O_{19}$, with addition of SiO_2, prepared by the chemical coprecipitation method are reported. Several firing temperatures were used to investigate the admixture of SiO_2 in the samples. Experimental data of electrical conductivity were obtained by ac measurements. Results are presented in terms of thermal and frequency dependent conductivity, $\sigma(\omega)$. Universal $\sigma(0)+A\omega s$-type behavior is observed for $log\sigma(\omega)$ *vs* $logf$, while $log\sigma(\omega)$ against $1000/T$, at fixed frequencies, exhibit deviation of an Arrhenius behavior as frequency increases. Generally speaking, curves of ac measurements follow the jump relaxation model. The addition of SiO_2 modifies slightly the observed conductivity. Saturation magnetization and coercitivity as a function of the thermal treatment of the samples, in the interval 950°C-1220°C, are also reported.

Introduction

M-type hexaferrites with the general formula $MeO{\cdot}nFe_2O_3$, where n=6 and Me=Ba, Sr and Pb are widely used as permanent magnets. Although the ceramic method has been the most popular technique to produce hard and soft ferrites [1], other alternative methods have been explored to synthesize high quality magnetic hexaferrites. Among the most common preparation methods are the sol-gel technique [2], the chemical coprecipitation [3] and the self-propagating high temperature synthesis [4]. Several studies have also been performed in order to study the influence of the addition of other substances and the substitution of some Fe atoms by other ions. Increase in density and improvement of rate of synthesis of Pb, Sr and Ba ferrites is achieved when Bi_2O_3 is added to the ferrites [5]; for further details on modified hexaferrites a summary of their magnetic properties can be consulted in [6]. On the other hand, it has been found that an admixture of SiO_2 reduces the grain size and at the same time the coercivity is increased [7]. Small grain size (d<0.1 μm) and large values of coercivity ($iHc \approx$ 6,000 Oe) are also obtained when the chemical coprecipitation method is used to prepare $BaFe_{12}O_{19}$ [3]. In this work the effect of the admixture of SiO_2 on the structural, electrical and magnetic properties of coprecipited M-type Ba hexaferrites is investigated. The impedance spectroscopy method has been used in order to characterize the samples electrically [8], and it was observed that the dispersion of the electrical conductivity follows the jump relaxation model [9]. Magnetic measurements were performed in order to evaluate the coercivity (iHc), saturation (σ_s) and remanent magnetization (σ_r), and orientation ratio (σ_r/σ_s). These results are complemented with studies of scanning electron microscopy and X-ray diffraction.

Experimental procedure

Ba-hexaferrite powders were prepared using the chemical coprecipitation method. A solution of $FeCl_3{\cdot}6H2O$ and $BaCl_2{\cdot}2H_2O$ in water was poured into a $NaOH/Na_2Co_3$ alkaline solution. The coprecipitated powder was dried in air at 40°C. The resultant powder was separated into five parts and 0.2%, 0.4%, 0.6%, 0.8% and 1% of SiO_2 in weight was added to each one. These mixtures were intimately mixed and polyvinyl alcohol was used as binding medium. Samples of different sizes were pressed as discs by applying a pressure of 3.5 Ton/cm^2. Thermal treatment of the obtained samples was performed at 950 °C, 1040 °C, 1130 °C and 1220 °C, respectively to each set of samples during 1 hour.

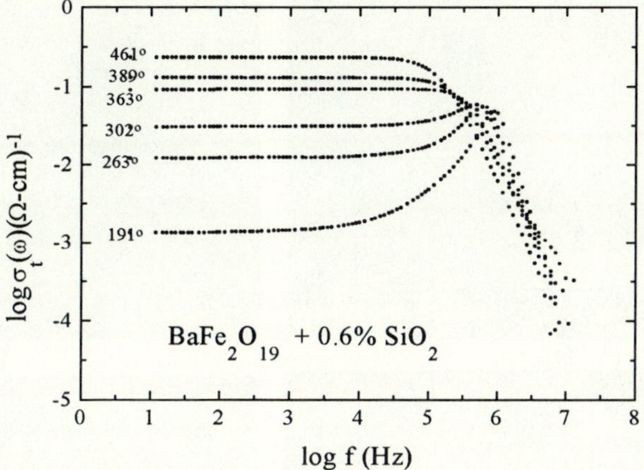

Figure 1. Conductivity as a function of the frequency for the Ba-ferrite with addition of 0.6% of SiO_2.

This procedure was carried out in normal atmosphere. Surface images were obtained using a JEOL model JSM-T330A scanning electron microscope. X-ray diffraction patterns were recorded with a Rigaku 2200 diffractometer and magnetic properties were measured using an Oxford vibrating sample magnetometer (H_{max}=16 kOe), in spherical samples. Corrections due to the demagnetizing field were done. Measurements of the electrical conductivity were carried out in the frequency range from 5 Hz to 13 MHz using an impedance analyzer HP4192A. Gold electrodes were previously fabricated on the opposite faces of the coin-shaped samples. Electrical characterization was performed in ferrite samples with 0.6% of SiO_2 in weight.

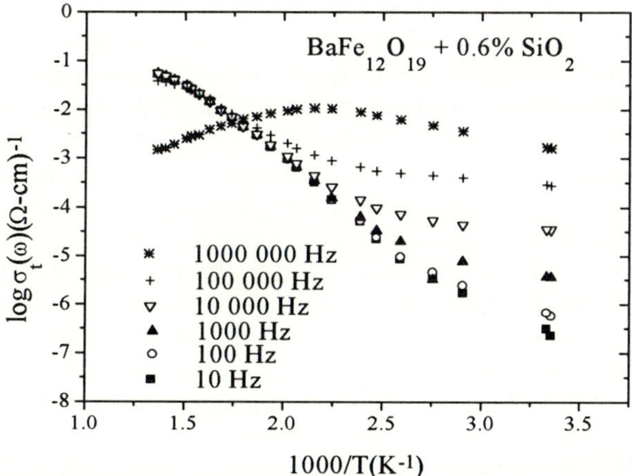

Figure 2. Conductivity as a function of the temperature for several frequencies.

The behavior of the total conductivity, $\sigma_t(\omega)$, of this modified Ba-hexaferrite will be described in terms of the ac electrical response. Magnetic characterization was performed in samples with 1.0%

of SiO_2 in weight, for different annealing temperatures. Observation of the surface topography of samples was also carried out using a scanning electron microscopy (SEM).

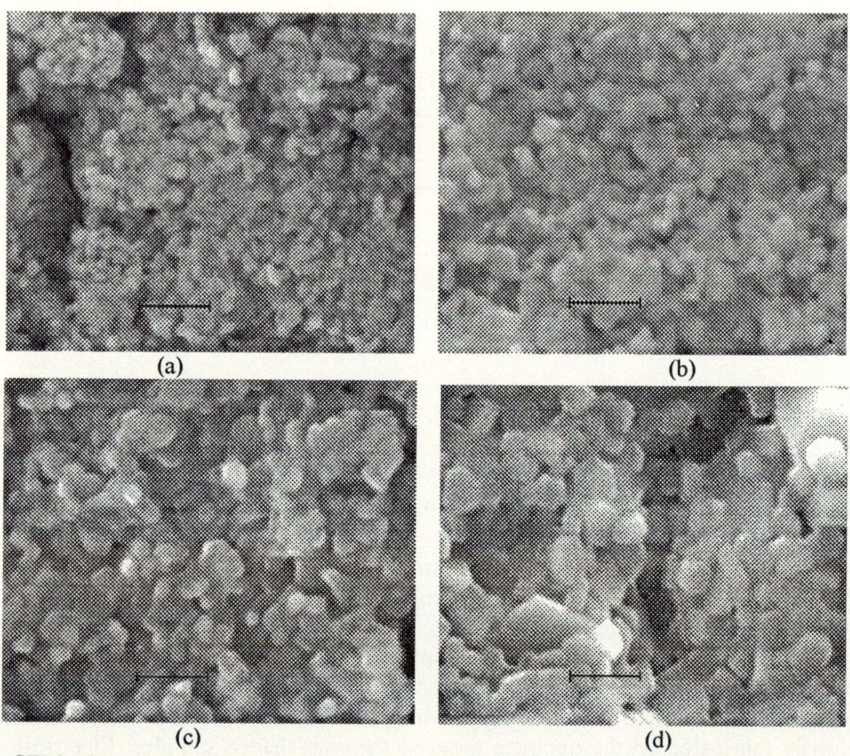

Figure 3. SEM micrography of sample with 1% SiO_2 fired at (a) 950 °C, (b) 1040 °C (c) 1130 °C and (d) 1220 °C. The scale corresponds to 1.0 μm.

Results and discussion

X-ray diffractograms indicate that all the samples have the same structure as that previously reported for the M type ferrite [6]. The behavior of the total conductivity, $\sigma_t(\omega)$, of this modified Ba-hexaferrite will be described in terms of the ac electrical response using the real part of the admittance, $Y^*=Y'+jY''$, where $j=\sqrt{-1}$,$Y'=Z'/[(Z')^2+(Z'')^2]$. Z' and Z'' are the real and imaginary parts of the impedance. Y' is a convenient parameter to describe the total conductivity, due to $\sigma_t(\omega)=gfY'$, where gf is a geometrical factor. When $\log\sigma_t(\omega)$ is plotted against $\log f$, all samples exhibit dispersion at high frequencies, whereas a frequency independent region is observed at low frequencies. In figure 1, a typical conductivity plot corresponding to the barium ferrite plus 0.6% of SiO_2, fired at 1220°C, is presented. The point where the conductivity changes from the frequency independent regime to the frequency dependent one, shifts towards higher frequencies as the temperature increases. The observed dispersion of conductivity as a function of the frequency follows the jump relaxation model [9]. This model predicts that the charge carriers can jump from one site to its neighboring vacant site at low frequencies, originating the dc conductivity. At high frequencies the charge carriers have not enough time to realize successful jumps, increasing the probability for the charges to go back to their initial site. This is the principal cause for the high frequency dispersion present in the curves of figure 1. $\sigma_t(\omega)$ plotted as a function of frequency does not display clearly its behavior with temperature. This can be emphasized plotting $\log\sigma_t(\omega)$ versus $1000/T$ for different fixed frequencies. Figure 2 shows plots of $\sigma_t(\omega)$ vs T for the same sample of figure 1. Curves show two almost linear stages in the temperature range studied; each one can be properly fitted by an exponential function $\sigma_t(\omega)=Aexp(-E_a/kT)$. At the fixed frequencies 10Hz, 100Hz, 1KHz and 10KHz, and high temperatures, the calculated individual conductivities take approximately equal values for the same temperatures.

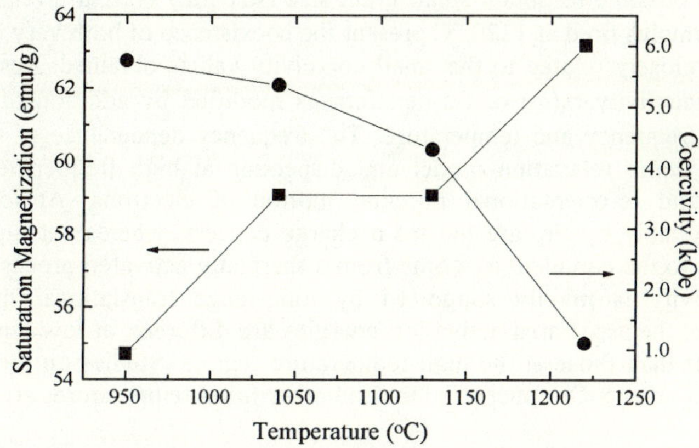

Figure 4. Saturation magnetization and coercivity of samples with 1% of SiO_2 with thermal treatment at different temperatures.

These values keep together at the high temperature tail, defining the dc conductivity, whose activation energy is 0.48 eV, for the case in figure 2. This behavior describes the process when diffusive hopping of electrons is the dominant mechanism of the conductivity. As the frequency increases the activation energy takes smaller values. That difference takes into account the fact that higher frequencies involve much more correlated forward-backward hopping of electrons and that in this case only a fraction of the required energy for long-range conduction will be spent. Departure from the above-described behavior is exhibited for the conductivity at 100kHz and 1MHz. This deviation, observed in figure 2, will be described somewhere. In table 1 values of the activation energies involved are presented.

Table 1. Activation energy for samples fired at 1220°C with 0.6% SiO_2 for different frequencies.

f (kHz)	E_{ac} (eV)
10^{-2}	0.60
10^{-1}	0.56
10^{0}	0.46
10^{1}	0.29
10^{2}	0.14
10^{3}	0.13

Surface morphology of samples with 1% SiO_2, fired at different temperatures is presented in figure 3. From SEM micrographs it is possible to measure the average grain sizes. For higher firing temperatures a larger grain size is obtained. It is noticeable to observe that samples sintered at 1220°C exhibit two different grain sizes, figure 3. One size ranges from 8 μm to 22 μm and the other one from 0.40 μm to 0.50 μm. Figure 4 shows the saturation magnetization and coercitive field of the samples with different thermal treatment. As can be seen, the higher the firing temperature the lower the coercivity (*iHc*). Low values of coercivity have been already observed in samples with large grain size. For T=1220 °C the small value of coercivity is due to the coexistence of very different grain sizes.

Conclusions

Barium M-type ferrites were synthesized using the chemical coprecipitation method. Using this method it is possible to obtain small grain size (< 1 μm) even at a relative low firing temperature (920 °C). Samples fired at 1220 °C present the coexistence of both very large and very small grains. This fact is closely related to the small coercivity values obtained. Results on the behavior of the total ac conductivity $\sigma_t(\omega)$ of Ba-hexaferrites modified by addition of SiO_2 were presented as a function of frequency and temperature. The frequency dependence of the total conductivity $\sigma_t(\omega)$ satisfies the jump relaxation model and dispersion at high frequencies can be attributed to the translation and re-orientational hopping motion of electrons. At low temperatures, electrons provided probably by Fe, are the main charge carriers, whereas at high temperatures, additional contribution to the conductivity come from a thermally activated process. At high temperatures the dc conductivity is mainly supported by long-range translational motion of electrons. As a consequence, the associated activation energies are different; at low temperatures they were found to be smaller than those at the high temperature regime. Studies on the physical characteristics of samples with other SiO_2 concentrations and other firing temperatures are in progress.

Acknowledgments

This project was supported by the Consejo Nacional de Ciencia y Tecnología (CONACyT) under grants Nrs. F666-E9407 and 485100-5-2232P, Fondo de Apoyo a la Docencia and Fondo de Apoyo a la Investigación, UASLP, México. Thanks also to DGIA-UNAM for partial support.

References

1. Kojima, H. *Ferromagnetic materials*, Vol. 3, E. P. Wohlfarth ed., North Holland, Amsterdam, 1982.
2. A. Chatterjee, D. Has, S. K. Pradhan and D. Chakravorty, J. Magn. Magn. Mater **127** (1993) 214.
3. K. Haneda, C. Miyakawa, and H. Kojima, J. Am. Ceram. Soc. **57** (1974) 354. V. Pankov, M. Pernet, P. Germi and P. Mollard, J. Magn. Magn. Mater. **120** (1993) 69.
4. I. P. Parkin, G. Elwin, L. Fernández, Q. T. Bui, Q. A. Pankhurst, A. V. Komarov, and Y. G. Morozov, Adv. Mater. **9** (1997) 643.
5. Okamura, T., Kojima, H and Watanawe, S., Sci. Rep. RITU **A7** (1955) 411, 418.
6. Kojima, H., Sci. Rep. RITU, **A7** (1955) 502. Kojima, H., Sci. Rep. RITU, **A7** (1955) 507. Van Uiter, L. G., J. Appl. Phys. **28** (1957) 317.
7. J. Esper, *High-Tech Ceramics, Viewpoints and Perspectives*, Edited by G. Kostorz, Academic Press, London (1989) 119-136.
8. A. K. Jonscher, *Dielectric Relaxation in Solids*, Chelsea Dielectric Press, London, (1983). J. R. Macdonald ed., *Impedance Spectroscopy*, John Wyley & Sons (1987). J. T. S. Irvine, D. C. Sinclair and A. R. West, Adv. Mater. **2** (1990) 132.
9. K. Funke, Prog. Solid State Chem. **22** (1993) 111.

Materials Science Forum Vols. 302-303 (1999) pp. 435-439

Synthesis of Ba Hexaferrite/Fe Magnetic Composites

P.G. Bercoff[1] and H.R. Bertorello[1,2]

[1] Facultad de Matemática, Astronomía y Física, Universidad Nacional de Córdoba H. de la Torre y M. Allende, Ciudad Universitaria, 5000 Córdoba, Argentina
Phone: (+54) 33 4051, Fax: (+54) 33 4054, E-Mail: berto@famaf.fis.uncor.edu

[2] Member of CONICET, Argentina

Keywords: Magnetic Composites, Nanostructured Materials, Ball Milling, Ba-Hexaferrite, Phase Transformations

Abstract

Magnetic composite materials (MCM) had been synthesized from commercially pure Ba hexaferrite and pure iron, by means of high energy ball milling under air or argon atmosphere. Thermal transformations of the ball-milled powder were studied in air and argon atmosphere from room temperature to 1300 K with the aid of DTA (Differential Thermal Analysis).

Milling and heat treating in Ar produces Fe_3O_4 + Fe+ BaO, with low remanence and coercivity. Milling in air or Ar and heat treating in air produces the formation of phase M + Fe_2O_3 with coercivity Hc= 4800 Oe, close to the larger values obtained by other methods. When milling is produced in air and the heat treatment in Ar, the final product is phase M + Fe_3O_4 with a nanostructure such that partial exchange coupling occurs among the particles of each phase.

Introduction

It has been shown that the synthesis of Ba- or Sr-hexaferrite by means of high-energy ball milling (HEBM) leads to improved magnetic properties [1]. On the other hand, prolonged milling of Ba-hexaferrite powders produce a decrease of both coercivity and magnetic saturation, either in dry-milled (in air [2]) or wet-milled [3] conditions. Controlled annealing of Ba-hexaferrite powders milled in air or low pressure vacuum resulted in improved coercivity values [4], as a result of fine crystal grain size and a defect and stress-free spin arrangement [5].

However, the magnetic properties of Ba-hexaferrite (such as the maximum energy product for permanent magnets) are far below those obtained in metallic compounds because of their low saturation magnetization, e.g. the magnetization of α-Fe is more than twice that of $BaFe_{12}O_{19}$. One way to improve magnetic properties is to combine two magnetic phases, one hard and the other soft but with high magnetization, such that the composite will have a combination of the properties of the components and will depend on the nature and strength of particle interactions.

Ceramic/metal nanocomposites, such as Al_2O_3/Fe, ZnO/Fe, Fe_3O_4/Cu [6] and Fe_3O_4/Fe [7], have successfully been prepared using HEBM. Also a nanocomposite Fe_3O_4/ $BaFe_{12}O_{19}$ was successfully synthesized [8] by milling Fe and $BaFe_{12}O_{19}$. in air and transforming by heat treating in Ar up to 1300 K.

In this work we use HEBM in air and in Ar atmospheres to synthesize nanocomposites starting with Fe and Ba-hexaferrite. The results on the structure and magnetic properties after HEBM and heat treatment are reported.

1. Experimental

The starting materials for the milling were commercially available pure (99.5wt%) barium dodecairon nonadecoxide $BaFe_{12}O_{19}$ ferrite (Ba-hexaferrite) and pure (99.95wt%) iron powder. The mean size of iron particles was ~40 μm and that of the ferrite ~50 μm. The composition of the mixture was 20 wt% Fe and 80 wt% Ba-hexaferrite.

HEBM was carried out for 60 hours in a high-energy ball mill using a hardened steel vial and balls of the same material. The ball-to-powder ratio was 10:1 and the frequency chosen for the process was 200 rpm. The milling processes were made in air and in Ar atmosphere. Thermal transformations of the ball-milled powders were induced by heating in air and in Ar atmospheres from room temperature to 1300K and were studied with the aid of DTA (Differential Thermal Analysis). The evolution of the structure was monitored by x-ray diffraction using CuK_{α} radiation. Maximum magnetization, M_{max}, remanence magnetization, M_r, and coercivity, H_c, of the powders were measured with a vibrating sample magnetometer with a maximum applied magnetic field of 1.4T.

2. Results

2.1. Effect of HEBM:

When milling in air or Ar atmosphere the chosen powder mixture, it is observed [8] that the structure of phase M (Ba-hexaferrite) degrade due to the milling process, and the peaks in the X-ray diffraction spectra become less intense and broad. The α-Fe peaks become also broadened.

Milling up to 60 hours produce a transformation of the powder structure, phase M desappears and the resulting phases are: when milling in Ar atmosphere, Fe_3O_4 and small amounts of α-Fe, as shown in Fig. 1a; (sample mAr). When milling in air, Fe_3O_4 and $BaCO_3$ coexists, with no free Fe present, as shown in fig. 2a (sample mAir). When milling in air, carbon can be absorbed from CO_2

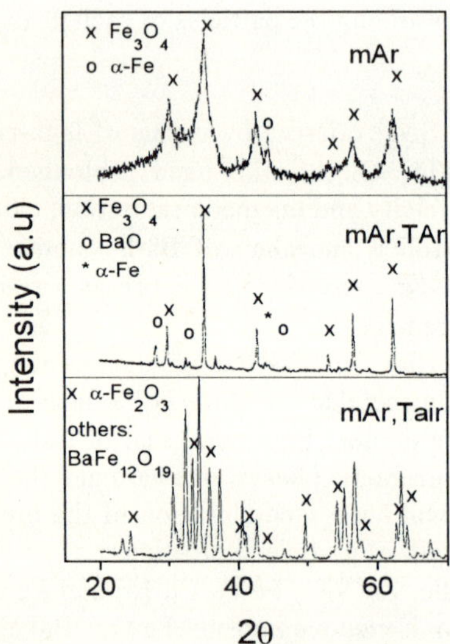

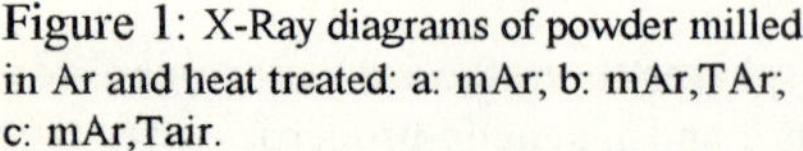
Figure 1: X-Ray diagrams of powder milled in Ar and heat treated: a: mAr; b: mAr,TAr; c: mAr,Tair.

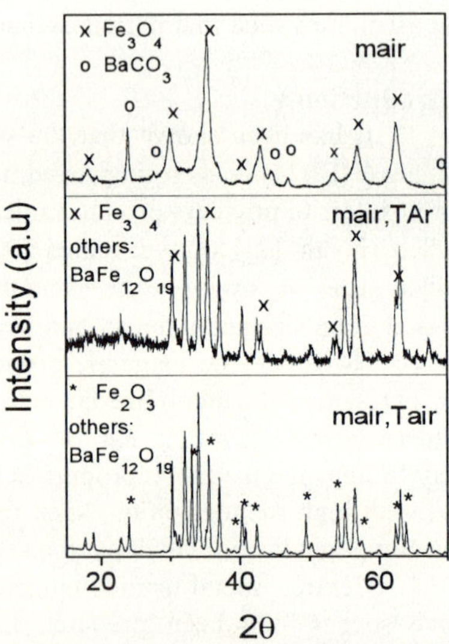

Figure 2: X-Ray diagrams of powder milled in air and heat treated: a: mair; b: mair,TAr; c: mair,Tair.

present in the air and combined with Ba to form $BaCO_3$. When milling in Ar, carbon is not available and $BaCO_3$ is not formed. In this case it is probable that Ba remain in solid solution inside Fe_3O_4.

It was observed [2,3,8] that coercivity, H_c, and saturation magnetization, M_s, progressively diminishes as a consequence of the damage produced in the ferrite structure, but after 60 hours of HEBM the sample shows an increase in saturation magnetization, due to the change in the crystalline structure where, as revealed by x-ray diffraction, the new magnetic phase is Fe_3O_4 (magnetite).

2.2 Effect of heat treatment:

Samples of powder milled in the above mentioned conditions where studied with the aid of DTA in Ar and air, by heating up to 1300 K at a rate of 10 K/min. These are identified as: (mAr,TAr) (milled in Ar, heat treated in Ar); (mAr,Tair) (milled in Ar, heat treated in air); (mair,TAr) (milled in air, heat treated in Ar) and (mair,Tair) (milled in air, heat treated in air).

Figure 3 show the DTA curves obtained in each case.

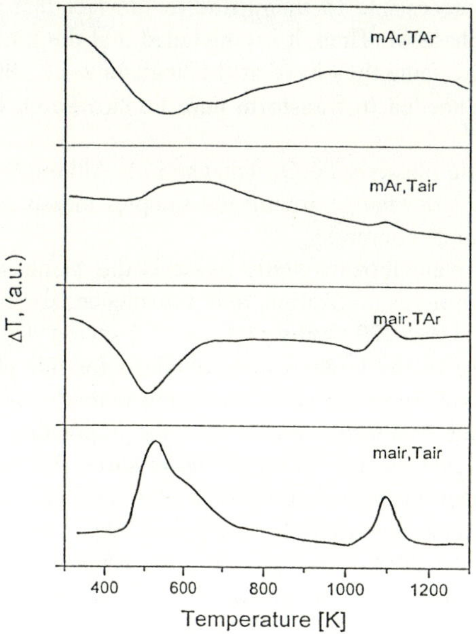

Figure 3: DTA curves obtained from powders milled in air or Ar atmospheres. See text for identification of samples.

In all cases studied exist broad exo or endothermic peaks at low temperatures, extending from 400 K up to 800 K. It was observed that this corresponds to a consolidation of the structural changes originated during the milling process and a recovery of defects produced during milling [8]. The endothermic peak at low temperatures appears when the heat treating is made in Ar atmosphere while the exothermic peak when it is made in air atmosphere.

Sample (mAr,TAr) show only an endothermic peak in the mentioned range of temperatures. Samples (mAr,Tair) and (mair,TAir) show an exothermic peak at low temperatures and an exothermic peak is observed at 1090 K, extended from 1000 K up to 1170 K. This high temperature peak is more important in sample (mair,Tair), while in sample (mAr,Tair) it is an small peak. Finally, sample (mair,TAr) shows an endothermic peak at low temperatures and an exothermic peak at high temperatures.

The X-ray diffraction analysis on the sample (mAr,TAr) shows that the main phase after heat treating is Fe_3O_4 with BaO and a small amount of Fe. Other small peaks were not assigned yet to another phase. The diffraction diagram is shown in Fig. 1b.

Samples (mAr,Tair) and (mair,Tair) contain phase M and α-Fe_2O_3 (hematite), as can be seen in Fig. 1c and 2c, respectively, while the other sample, (mair,TAr), contains phase M and Fe_3O_4 (ferrite) in minor proportion, see Fig. 2b. Table I show the results of magnetic measurements made.

Table I: Magnetic properties of milled and heat treated samples.

	mAr,TAr	mAr,Tair	mair,TAr	mair,Tair
Mmax [emu/g]	78	38.6	62.7	39
Mr [emu/g]	14.4	21.6	29.6	22.7
Hc [Oe]	93	4762	1906	4875

3. Discussion:

When pure $BaFe_{12}O_{19}$ (phase M) is exposed to HEBM the structure changes very slowly and for very long milling times (of the order of 1000 hours) $BaFe_{12}O_{19}$ decomposes into α-Fe_2O_3 (hematite) [10]. In our case, with 20 wt% Fe, the structural changes occur sooner and phase M decomposes into Fe_3O_4., when milling in Ar, or to Fe_2O_3 + $BaCO_3$ when milling in air.

When phase M is the main component after milling and heat treatment, it is supposed it is the result of the transformations that took place in the original M phase particles, transformed into Fe_3O_4 by the milling process and afterwards transformed back into phase M during the heat treatment. The pure Fe particles also transformed into Fe_3O_4 by the milling process (completely in samples milled in air and partially in samples milled in Ar) but they do not became M phase particles during the heat treatment because Ba was not accessible to them, due to the fact that Ba atoms remains inside the particles that originally were phase M. Then, it is concluded that the fraction of M phase present in each sample is approximately the same they have at the beginning, i.e., 80% in wt, not taking into account the extra oxygen atoms needed to transform pure Fe into Fe_3O_4 or into α-Fe_2O_3.

In sample (mAr,TAr) the main ferromagnetic phase is Fe_3O_4 (magnetite). Although it has the highest maximum magnetization at 1.4 T (M_{max}=78 emu/g) among the samples measured, it has a remanence of only M_r= 14.4 emu/g, and a negligible coercivity.

In samples (mAr,Tair) and (mair,Tair) the main ferromagnetic phase is the M phase and the other is weakly ferromagnetic. Hence, maximum magnetization and remanence are mainly the contributions of the M phase alone: in average, M_{max}=39 emu/g and M_r= 22 emu/g. Coercivity is also high, H_c= 4800 Oe in average, and is close to the larger values obtained for this phase [1,9] when prepared by mechanical alloying $BaCO_3$ and Fe_2O_3. It is worth noting that M_r/M_{max}= 0.56, above M_r/M_s=0.5 predicted by a Stoner and Wolfarth model of non interacting particles. Measurements of X rays diffraction peak broadening in the M phase peaks show that the particle size is around 30 nm, well below the single domain particle size. At such particle size it is expected that some particle interaction takes place, rising M_r at values above M_r= 0.5 M_s (enhanced remanence effect) and lowers the coercivity by means of the exchange interactions among hard particles.

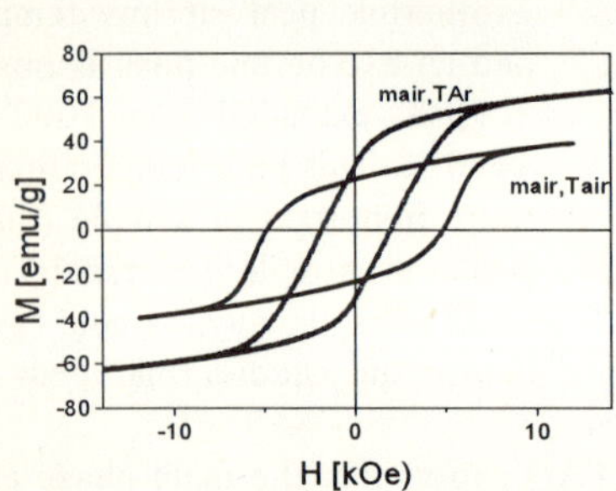

Figure 4: Histeresis cycles of powder milled in air and heat treated in air or Ar atmospheres.

The other sample, (mair,TAr) contains phase M and Fe_3O_4 (magnetite). The maximum magnetization is the sum of their partial contributions, and the remanence and coercivity are the result of the combined effect of dipolar and exchange interactions among the hard phase (M phase) and the softer phase (magnetite). Figure 4 show the histeresis cycle of samples (mair,Tair) and (mair,TAr). It can be seen that the behavior of sample (mair,Tair) corresponds to a good hard magnetic material, although remanence is low because only 80% of the sample is phase M. The other one, sample (mair,TAr), has coercivity higher than the coercivity of phase M alone in powder form with grain size of the order of several microns. Remanence is also higher than the sum of the partial contributions of each phase, indicating that some exchange interaction occurs in this composite. dM/dH is large near H=0, and is due to the large component of reversible inverse domain nucleation promoted by the dipolar interactions among the phase M and the magnetite particles [8]. This behavior of sample (mair,TAr) is due to the grain size, of the order of 30 nm, that promotes only a partial exchange coupling among the particles of each phase.

We define the apparent irreversible susceptibility as

$$\chi_{irr}^{app} = \chi_{Tot} - \chi_{rev}^{app} \quad (1)$$

where $\chi_{Tot} = dM/dH$ is the total susceptibility and χ_{rev}^{app} is taken as the total susceptibility evaluated in the first quadrant. Figure 5 show the apparent irreversible susceptibility of sample (mair,TAr) and of phase M, taken from sample (mair,Tair) or (mAr,Tair).

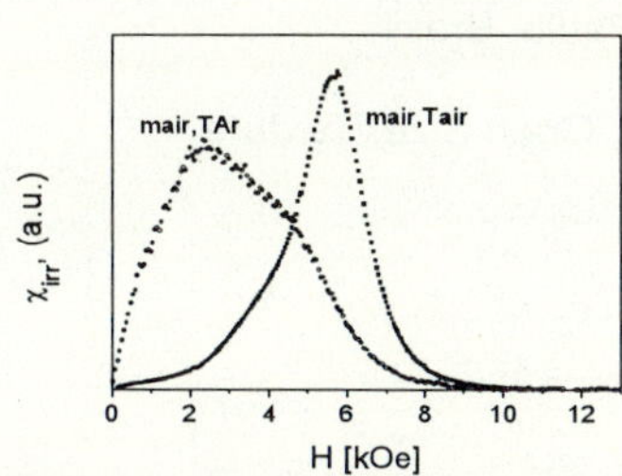

Figure 5: Apparent irreversible susceptibility of phase M alone (sample mairT,air) and of the composite (sample mair,TAr).

Phase M shows only one narrow peak at H=6000 Oe, near the coercive field. This is the field at which most of the inverse domains are nucleated. The sample (mair,TAr) shows a very broad peak that is composed of two distinct peaks: one centered at H= 1800 Oe and the other at H= 3900 Oe. We identify the peak at low fields as due to inverse domain nucleation of magnetite particles, promoted by the exchange coupling with M phase particles. The other peak is identify with the inverse domain nucleation of M phase particles, shifted from 6000 Oe to 3900 Oe by the effect of exchange coupling between M phase particles.

4.Conclusions

Using HEBM in a powder mixture made of 20wt% Fe + 80 wt% $BaFe_{12}O_{19}$ produces a fast degradation of the crystal structure of $BaFe_{12}O_{19}$, producing a transformation to Fe_3O_4 + Fe if the milling is done in Ar or to Fe_3O_4 + $BaCO_3$ if it is done in air. Subsequent heat treatment up to 1300 K produces a recovery of the structure if the milling and the heat treatment are done in Ar. Heat treating in air the powders milled in air or Ar produces the formation of phase M + Fe_2O_3 while if the milling is produced in air and the heat treatment in Ar the final products are phase M + magnetite. This last transformation produce a nanostructure such that a partial exchange coupling exists among the particles, rising the inverse domain nucleation field of the soft phase particles (magnetite) and lowering the nucleation field of the hard phase particles.

References

[1] J. Ding, D. Maurice, W. F. Miao, P. G. McCormick and R. Street, *J.of Mag.and Mag. Mat.* Vol 150, 1995,p. 417

[2] K. Haneda and H. Kojima, *J. Am. Ceram. Soc.* Vol. 57, 1974, p. 68,.

[3] W.A.Kaczmarek and B.W.Ninham, *IEEE Trans. Magn.*, **MAG-30**, 1994, p. 717,

[4] W.A.Kaczmarek and B.W.Ninham, *J.App.Phys.*,Vol.76, 1994, p. 6065,

[5] O.Kubo, T.Ido and H.Yokoyama, *IEEE Trans. Magn.*, **MAG-18,** 1982, p.1122)

[6] M. Pardavi-Horvath and L. Takacs, *Scripta Metall. et Mat.* Vol 33, 1995, p. 1731

[7] J. Ding, W. F. Miao, R. Street and P. G. McCormick, *Scripta Mat.*, Vol. 35, 1996, p. 1307,

[8] P.G.Bercoff and H.R.Bertorello, accepted for publication in *J.of Mag. and Mag.Mat.*

[9] J.Ding, R.Street and H.Nishio, *J.of Mag. and Mag.Mat.*, vol.164, 1996, p.385

[10] W. A. Kaczmarek, *Journal of Mat. Sci.* Vol. 34, pg. 5271, 1996.

This work was partially supported by CONICOR, Córdoba, Argentina, PID N° 4155 and by Secretaría de Ciencia y Tecnología, Universidad Nacional de Córdoba, PID N° 177/97.

Materials Science Forum Vols. 302-303 (1999) pp. 440-445

Separating Components of the Hysteresis Loss of Non-Oriented Electrical Steels

F.J.G. Landgraf[1], J.C. Teixeira[1], M. Emura[1], M.F. de Campos[2] and C.S. Muranaka[2]

[1] IPT, Instituto de Pesquisas Tecnológicas do Estado de São Paulo, Av. Prof. Almeida Prado 532, 05508-901, Sao Paulo, Brazil
E-Mail: landgraf@ipt.br

[2] USP-Universidade de São Paulo, São Paulo, Brazil

Keywords: Electrical Steels, Magnetic Losses, Hysteresis Loss, Grain Size, Texture

Abstract

This paper shows the separation of quasi-static hysteresis losses into two components, one related to Low Induction and other to High Induction levels. The separation line between those two regions has been arbitrarily chosen as the induction at maximum permeability. Results show that the grain size has stronger effect on the low induction component, while the high induction component is more sensitive to sample measuring direction. This may provide basis for better materials-properties correlation modeling and product development.

Introduction

In 1995, the world market for electrical steels has been estimated in 1.8 billion dollars [1]. Non-oriented electrical steels are widely used for the stators and rotors of rotating electrical machines. For those applications, where there is an ever-increasing demand for higher energy efficiency, new grades of steels are developed every year, offering lower magnetic losses and higher magnetic permeability [2].

There is much to be profited by the use of magnetic losses components separation for the development of electrical steels. It is now common practice to separate the magnetic losses into hysteresis, eddy-current and anomalous (or excess) components [3-9]. Based on that approach, it should be possible to identify the effects of chemical composition and microstructure in each of those components, providing a basis for better materials-properties correlation modeling and product development.

Two of the most investigated microstructural variables for electrical steels are grain size [4, 6, 10-13] and texture [11, 14-16]. Increasing the grain size (G.S.) decreases hysteresis loss and increases anomalous loss, not affecting the classical eddy-current loss. The anisotropy of magnetic losses and permeability is usually attributed to texture but, as texture description is complex, it is not easy to correlate it with hysteresis and anomalous losses.

Various papers have described the ideal texture for non-oriented steels as the fibre texture (100)[0kl], based on its effect on permeability. This texture has not been achieved in commercial practice. The theoretical limit for the isotropic B_{50} value (induction at 50 A/cm) of a steel with no silicon addition with that texture has been predicted as 1.85T. On the other hand, no predictions were made about the loss value of those materials, due to a lack of suitable models. We are not yet able to predict how far can the Total Magnetic Loss be lowered by texture control, for example, in a 0.5%Si steel, 0.5mm thick, that normally shows 6.0W/kg at 1.5T, 60Hz (4.7W/kg at 50Hz).

This paper proposes a methodology for the separation of the hysteresis loss component in two parts, one related to low induction and other to high induction, and discusses the results obtained in samples with different grain sizes, measured in the rolling and in the transverse direction. It is hoped that this methodology may lead to a better materials-properties correlation modeling.

Materials and methods

A low-carbon 0.5%Si steel was cast into a 50x350x300mm ingot, hot rolled to 2.5mm thickness, 350mm wide, cold rolled to about 0.55mm thickness, annealed to average grain size of 13μm, and subjected to several temper rolling elongations. 30x300mm Epstein samples were cut in the rolling and transverse directions and all samples were submitted to a typical industrial decarburizing annealing at 760°C for 2h. Due to different temper rolling elongation, six samples with different grain sizes and crystallographic textures were produced.

Core Loss measurements at 60Hz were performed separately in the rolling direction (RD) and transverse direction (TD) samples, with an instrument built in our lab, capable of measuring losses at frequencies from 100 to 0.01Hz, under sinusoidal induction. Total magnetic losses were measured by the wattmeter method; classical parasitic losses were calculated using electrical resistivity measured in each sample (average 19μΩcm). Hysteresis loss at 1.5T was measured by the quasi-static method, integrating the 0.01Hz hysteresis loop area. Anomalous losses were calculated from total, hysteresis and parasitic losses. Coercive force was also measured from the hysteresis loop. B_{50} values were determined at 60Hz.

The method for separation of hysteresis loss component into a low induction and high induction component will be described below.

Results and discussion

The annealing of samples with temper rolling in the range from 0 to 17% elongation resulted in grain sizes from 13 to 360μm. The sample with 4% elongation showed some inhomogeneities in the grain size distribution. No texture measurements were attempted in the samples with grain size larger than 50μm. An ODF- Orientation Distribution Function analysis of the 13μm sample texture showed a {111}<uvw> fiber component (average 3.5 times random) and a (110)[001] component reaching 1.8 times random. Assuming that the B_{50} value is strongly dependent on the texture, its measurement in the rolling and transverse direction should offer an idea of the texture evolution with grain size. From the B_{50} values of Table 1 we may assume that a larger Goss component could be present in the 48 to 163μm samples, and a larger {111}<uvw> fraction at the 360μm sample.

Table 1. Effect of prior temper rolling elongation on grain size, Induction at 5000 A/m (B_{50}) in the Rolling (RD) and Transverse (TD) direction of annealed samples.

elongation	%	0	17	13	11	7	4
G.S.	(μm)	13	48	104	125	163	360
B_{50} (RD)	(T)	1.75	1.75	1.75	1.75	1.77	1.71
B_{50} (TD)	(T)	1.75	1.70	1.68	1.68	1.68	1.67

The effect of grain size on the losses components can be seen in Figure 1. Hysteresis loss decreases and anomalous loss increases when grain size is increased, as expected from literature, except in the 13 μm sample, where anomalous loss is high, for unknown reason. It is also possible to see that the anisotropy of the Total Magnetic Losses can be attributed to the anisotropy of hysteresis loss, as the anomalous component shows very little anisotropy. Hysteresis losses at the transverse direction are from 12 to 36% higher than in the rolling direction.

The anisotropy of the hysteresis loss can be seen when the shapes of hysteresis loops in the two different directions of the same sample are compared, as in Figure 2. There we may see that, although the coercive force is slightly larger in the transverse direction (10% larger, on the average), loop area is 20% larger (on the average) in the transverse direction. We will see below that most of the loss difference is concentrated in region above maximum permeability.

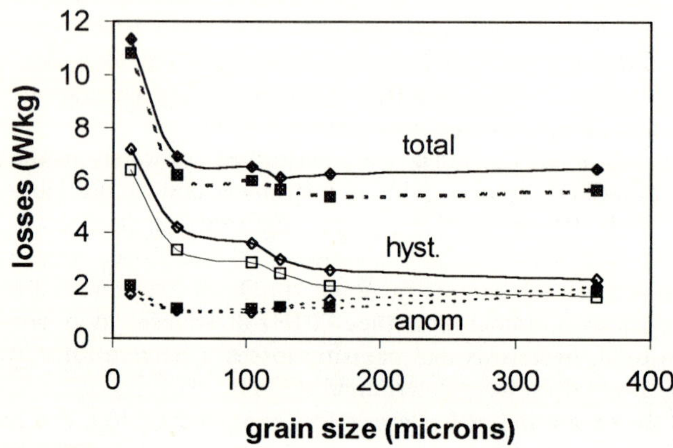

Figure 1. Effect of grain size on Total Magnetic Losses, Hysteresis Loss and Anomalous Loss at 1.5T 60Hz, measured in the Rolling direction (□) and Transverse Direction (◊).

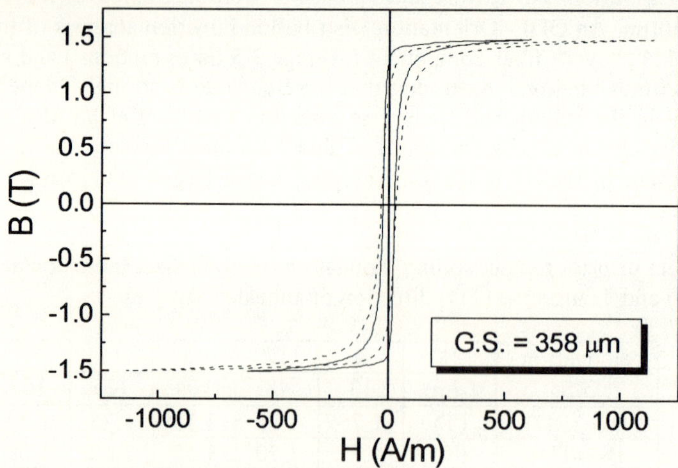

Figure 2. Hysteresis loops of the sample with grain size of 358μm, measured on the rolling (full line) and transverse direction (dotted line)

We have also observed that, in most cases, the initial magnetization curve joins the lower branch of the hysteresis loop just below the point of maximum permeability, which is about the position of the "knee" of the hysteresis loop. This "knee" of the curves is usually associated with a change in the magnetization mechanism, suggesting that the hysteresis loss can be separated in two components: a low induction (LI) and a high induction (HI) component.

We found no better indicator for a line to separate magnetization mechanisms regions than the maximum permeability point. Figure 3 gives the impression that the knee of the curve lies above the maximum permeability point, but the visual impression of the knee position changes with the x-scale used. The values of the Low Induction and High Induction components discussed below were calculated from the integration of the areas inside and outside the maximum permeability induction lines, as exemplified in Figure 3.

Figure 4 shows the effect of grain size on the energy per cycle values of the two components, in the rolling and transverse direction. It can be seen that, as grain size is increased, the Low Induction (LI) component is reduced from 600 to 100J/m^3 , and the High Induction (HI) component falls from 300 to 150J/m^3 . Besides that, it can also be seen that there is not much anisotropy in the LI component and a large anisotropy in the HI component. In all samples, more than 50% of the Total Loss anisotropy is due to the HI component anisotropy. As an example, the 54J/m^3 difference in the HI component of the 358μm sample, seen in figure 4, accounts for 50% of its total Magnetic Loss anisotropy of 0.8W/kg, seen in figure 1.

It is normally assumed that the anisotropy of Losses and B_{50} are due to texture. Nevertheless, it was not possible to correlate the High Induction component values with B_{50} values. For example, the large decrease in B_{50} value found in the 358μm rolling direction sample is not matched by a related increase in the HI component value.

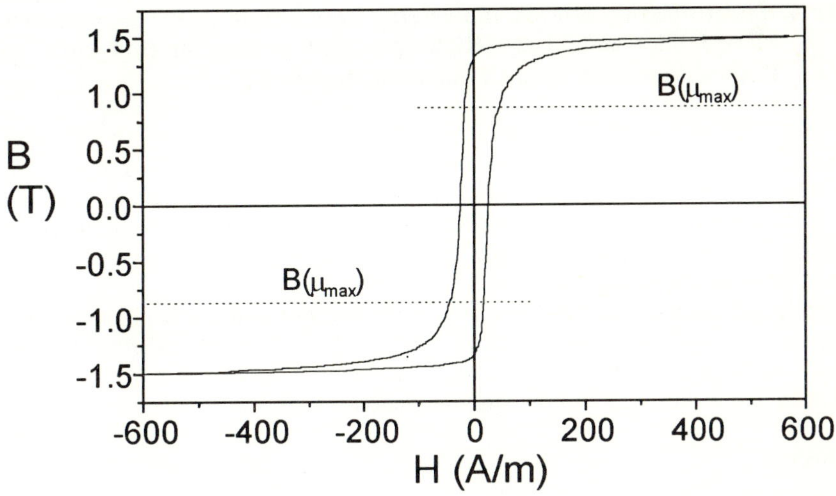

Figure 3. Hysteresis loop of the 358μm sample, rolling direction, showing the position of the maximum permeability induction.

Classical textbooks [17,18] state that the induction at the knee of the magnetization curve divides regions of different predominant magnetization behavior: below the knee, domain wall movement is the main magnetization process, whether domain rotation dominates above it. Although it must be true that domain wall movement is the main mechanism for losses in the Low Induction region, we have refrained from supposing Domain Rotation as the main mechanism for the High Induction region losses. The text books assume the reversible domain rotation as the main mechanism for magnetization above the knee, but reversible changes do not generate losses.

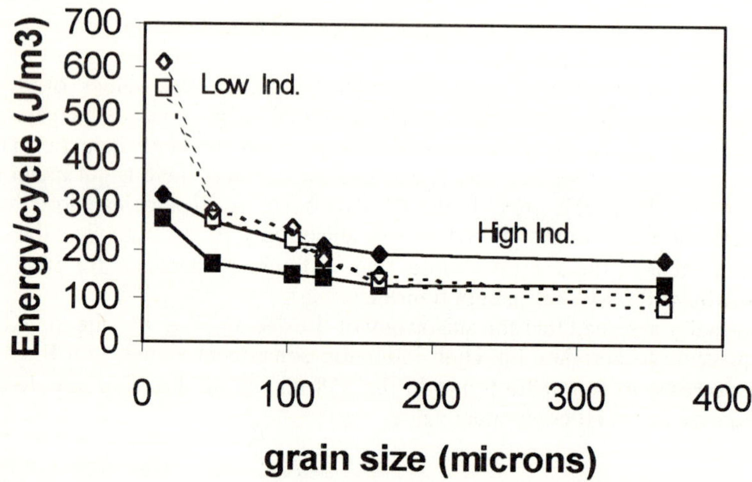

Figure 4. Effect of grain size on the Energy Loss / cycle in the Low Induction component (dotted line) and in the High Induction component (full line), measured in the Rolling direction (□) and Transverse Direction (◊).

Chikazumi mentions the existence of large Barkhausen jumps above the knee of the curve of the lower branch [18]. He attributes those jumps to domain wall displacement and/or irreversible domain rotation, but states that "irreversible rotation is expected in fine particles or in an extremely heterogeneous materials which contains a lot of inclusions and precipitates", which is not the case of the samples here investigated.

So, it is still open for discussion which is the predominant mechanism of energy dissipation in the high induction range of the hysteresis loss of electrical steels, whether irreversible rotation or domain wall movement. Whichever is the prevalent mechanism, it must be sensitive to texture to explain the strong anisotropy of the High Induction Losses.

Conclusions

A methodology for separating the hysteresis loss component into a High Induction and a Low Induction component is proposed, using the maximum permeability induction line to separate the two regions. This procedure may be useful, based on the observation that a significant amount of the hysteresis loss is dissipated in the High Induction region, and its anisotropy suggests it is sensitive to texture.

Based on that methodology, it is possible to show that grain size increase lowers much more the Low Induction component, and that most of the Total Magnetic Loss anisotropy can be attributed to the High Induction component.

Acknowledgements

Thanks to FAPESP (proc. 97/4877-0) and CSN for the research grant, Flavio Beneduce Neto for ingot casting, Saturnino Ltd for the hot and cold rolling, Quantech for the temper rolling, Tempel Ltd for the industrial heat treatment.

References

[1] T. Abraham, T. - Soft Magnetic Materials'96, Gorham/Intertech Consulting, 1996

[2] A. Hiura; Y. Oda; Y. Tanaka; S. Kazama; Y. Okami, NKK Tech. Review 77 (1997) p. 27-33

[3] K. Eloot, J. Penning, J. Dilewijns, Journal De Physique, to be published

[4] F. Fiorillo, E. Ferrara, L. Ferrando, C. Appino, C. Lebourg, J. Degauque, M. Baricco, IEEE Trans. Mag. 33 (1997) p. 3802-3804.

[5] A.D. Paepe, K. Eloot, J. Dilewijns, C. Standaert, JMMM 160 (1996) P. 129-130.

[6] C.K. Hou, IEEE Trans. Mag., 32 (1996) p. 471-477.

[7] B. Cornut, A. Kedous-Lebouc, T. Waeckerlé JMMM 160 (1996) p. 102-108.

[8] J.C.Bavay, J. Verdun, J. Materials Eng. Performance 2 (1993) p. 169-172.

[9] G. Bertotti, J. Appl. Phys. 57 (1985) p. 2110-2117.

[10] M. Shiozaki, Y. Kurosaki, J. Mat.Eng. 11 (1989) p. 37-43.

[11] E.T. Stephenson, M.R. Amann, Energy Efficient Electrical Steels, 1980.

[12] J. Degauque, B. Astie, J.L. Porteseil, R.Vergne, JMMM 26 (1982) p. 261-263.

[13] G. Bertotti, G.D. Schino, A.F. Milone, F. Fiorillo, J. de Physique 9 (1985) p. 385-388.

[14] H. Shimanaka, K. Matsumura, T. Irie, H. Nakamura, JMMM 19 (1980) p. 63-64.

[15] G.H. Shirkoohi, A.J. Moses, JMMM 133 (1994) p. 440-442, 1994.

[16] G. Bertotti, G. Ban, G. Abbruzzese, J. Appl. Phys. 73 (1993) p. 6612-6614.

[17] R.M. Bozorth, Ferromagnetism, Van Nostrand, 1951, p. 480

[18] S. Chikazumi, S. Physics of Magnetism, J.Wiley, 1964, P. 245

Materials Science Forum Vols. 302-303 (1999) pp. 446-449

Grain Growth in Low-Carbon Lamination Steels with Various Al Contents

C. Oldani[1], F. Actis[1], R. Mancini[1] and H. Rissone[2]

[1] CEMCOR-CIMM, INTI, CC 884 (5000) Córdoba, Argentina

[2] SIDERAR, (2900) San Nicolás, Buenos Aires, Argentina

Keywords: Electrical Steels, Lamination Steels, Grain Growth, Aluminum Nitride

ABSTRACT

In this paper, low-carbon lamination steels with Al contents ranging from 0% to 0,16% were subjected to decarburizing annealing thermal treatments. Carbon content and microstructure were evaluated at different annealing temperatures and times.

Al content strongly affects the grain growth behaviour of these steels intended for electrical use. Intermediate Al contents is deleterious for grain growth. Microstructures able to develop good magnetic properties (low losses and high permeability) with mean grain sizes of 120-180 μm were feasible even at relatively low annealing temperatures for steels with Al contents higher than 0,07%.

INTRODUCTION

Steels known as lamination or low carbon steels are the ones with non-oriented grain, partially processed and with low silicon content. They reach their maximum magnetic potential during the final steps of manufacture at the user's plant. In these operations a decarburization annealing heat treatment produces important effects such as removing punching residual tensions, decarburization to very low carbon content, ferritic grain growth and a favorable magnetic crystallographic texture. The aim of the producer of non-oriented steel is to produce materials with low core losses and high permeability, uniformly in the whole plane of the sheet.

The main factor affecting magnetic properties is the grain size [1] and its control in low-carbon steels is critical for their electromagnetic uses. The optimum grain size for a core material must be in the 120-180 μm range [2] but inclusion forming elements (i.e., C, Mn, and Al) present in the steel may affect the final grain size inhibiting grain growth. In these context, AlN precipitates strongly affects the grain growth behaviour of these steels intended for electrical use. Precipitates larger than a critical size will no longer pin grain boundaries [3], and small inclusions do not permit the grain boundary motion even at high temperatures.

EXPERIMENTAL PROCEDURE

For the experimental work, a vertical tubular furnace with controlled temperature and atmosphere was used. The samples were obtained from commercial steel laminations, with initial chemical composition as shown in Table 1.

Table 1: Chemical composition of the samples

Steel	%C	%S	%Mn	%P	%Si	%Al	%N
E	0,062	0,007	0,32	0,014	0,12	----	s/d
M	0,068	0,007	0,25	0,012	< 0,016	0,052	0,0047
H	0,050	0,004	0,61	0,045	0,40	0,075	0,0051
A	0,050	0,004	0,58	0,045	0,40	0,160	0,0050

In the heat treatments it was included a silicon killed steel, **E**, without Al, to compare with the aluminum killed steels with different Al content. Although the Al content may not be the only variable affecting final grain size because it must also be consider the previous deformation, all the steels used in this study had the same temper rolling deformation (about 7%) so it was not consider as a variable.

The samples were introduced in the furnace suspended from a stainless steel wire, that allows picking up the samples to a cold zone of the furnace. The treatments were carried out under a mixture of N_2+15%H_2 atmosphere, saturated in water at 30°C.

Some samples were cooled during the heating of the furnace, and the others during the isothermal treatment. The isothermal heat treatments were conducted at temperatures of 700° and 750°C with treatment times up to 120 minutes. Only one steel, **M**, was also treated at 800°C and times up to 120 minutes.

The processed samples were analyzed for carbon content and the ferritic grain size was measured using the intersection method in at least 25 fields in each sample.

RESULS AND DISCUSSION

The results of the carbon evolution during the heating up stage (at a heating rate of 6 °C/min.) are presented in Table 2. As it is shown, the decarburization kinetics is very fast above 650°C. When samples reach the lower heat treatment temperature (700°C) all of them are in the ferritic phase.

Table 2: Decarburization during the heating stage [%]

Steel	As received	600°C	650°C	700°C	750°C
E	0.062	0.062	0.060	0.0142	0.0026
M	0.068	0.068	0.068	0.0104	0.0057
H	0.050	0.049	0.042	0.0084	0.0060
A	0.050	0.041	0.031	0.0043	0.0031

In Table 3, the experimental grain growth results during heating are shown, and in Table 4 the grain size evolution during the isothermal treatments at different temperatures is presented. Microstructural examination of the samples indicates that the steel E shows a gradual grain growth during the heating of the sample and also while it is at temperature, both at 700 and at 750 °C. This behavior is characteristic of the normal grain growth mechanism and it can be explained because there was no aluminum to form precipitates that could pin the grain boundaries.

Table 3: Grain growth evolution during the heating stage [μm]

Steel	As received	500°C	600°C	650°C	700°C	750°C	800°C
E	13,8	17,8	19,3	42,5	77,8	85,0	- - -
M	10,5	13,5	15,1	17,3	17,9	19,4	Duplex 13 – 346
H	11,0	8,6	7,6	8,9	Duplex 7,6 – 87,0	120,8	- - -
A	12,1	8,8	9,2	9,5	Duplex 9,3 – 81,9	80,0	- - -

Table 4: Grain growth evolution during the isothermal treatments [μm]

Steel	700°C 0 min	700°C 120 min	750°C 0 min	750°C 120 min	800°C 0 min	800°C 120 min
E	77,8	96,9	85,0	109,9	- - -	- - -
M	17,9	33,8	19,4	Duplex 19,6 - 450	Duplex 13 – 346	533
H	Duplex 7,6 – 87,0	132,1	120,8	135,3	- - -	- - -
A	Duplex 9,3 – 81,9	80,2	80,0	100,2	- - -	- - -

On the other side, the samples M, H and A exhibit an explosive grain growth (abnormal grain growth). In steel M this mechanism happens at 750°C where it can be observed the first signs of abnormal growth. This kind of growth was also observed at the end of the heat treatment (120 minutes).

A 800°C annealing treatment was done only to the steel M to see the grain size evolution at this higher temperature. The results indicated that when the steel reached this temperature, it had a "duplex" structure with small grains around very big ones. After 120 minutes the sample had an homogeneous structure with a grain size of 530 μm. This behavior can be explained because very fine precipitates of AlN present in the steel effectively anchor the grain boundary till the temperature is sufficiently high. In that moment, only a little number of grains with some advantage over the others will grow to very big sizes.

On the contrary, steels A and H experienced the beginning of this abnormal growth at a relatively lower temperature. During the heating stage, recrystallization occurred in these two steels at temperatures between 500° and 600°C. At 700°C the secondary recrystallization promoted the growth of some grains in a duplex structure. At 750°C the sample had a homogeneous microstructure, with an average grain size of 80 μm in steel A and 120 μm in steel H. During the isothermal stage it was observed only small variations in the grain size till the 120 minutes of treatment. This is characteristic of a normal growth behavior. In the same way, it was obtained bigger grain sizes in sample H (135 μm) than in sample A (95 μm) at the finish of the isothermal heat treatments.

This different behavior must be attributed to different Al content in the steels H and A. In sample H the Al content is higher than the characteristic content in aluminum killed steels and it is sufficient for the growth of some grains, because of the AlN precipitates coalescence. This is also true for the sample A, but in this case, the aluminum content must be in excess because the grain grows to a smaller limiting grain size.

CONCLUSION

It could be concluded that:

- Decarburization is feasible at temperatures above 650°C and is completed (< 0,01%C) when the samples reach 700°C.
- The aluminum free steel (0%Al) exhibits a normal grain growth behavior during the heating stage and during the isothermal heat treatments.
- Secondary recrystallization with abnormal grain growth was observed in all the aluminum killed steels.
- A very big grain structure was developed in the sample with low aluminum content (0,052%Al): 533 μm at 800°C and 120 minutes at temperature.

- The steel with intermediate aluminum content (0,075%Al) arrived at an average grain size of 135 µm in two hours at 750°C. This microstructure is able to develop good magnetic properties.
- A smaller limiting grain size is characteristic of the steel with higher aluminum. An increase of Al content to 0,16% may be deleterious for the microstructure of the steel.

REFERENCES

[1] F. Actis ; C. Oldani; 1er Taller Argentino de Materiales Magnéticos y sus Aplicaciones, Córdoba Argentina, 1995

[2] M. Shiozaki ; Y. Kurosaki ; J.Mat.Engineering, 11, 1989, p.37-43

[3] G. Lyudkovsky ; P.D. Southwick ; Met.Trans.A, 17A, 8, 1986, p.1267-1275

ACKNOWLEGMENTS

The authors thank SIDERAR for the provision of the steel samples and the collaboration of K.Rafael in the metallographic analyses.

Materials Science Forum Vols. 302-303 (1999) pp. 450-454

Application of Heat Treated Low-Carbon Electrical Steels in Small Motors

A.C. Agüero[1], F.A. Actis[2], H.R. Moyano[1] and O. Casais[3]

[1] CEMCOR-CIMM, Instituto Nacional de Tecnologia Industrial,
Av. Vélez Sársfield 1561, C.C. 884-(5000) Córdoba, Argentina
E-Mail: aguero@com.uncor.edu

[2] SIDERAR S.A.I.C., Centro Siderúrgico Gral. Savio, C.C. 801-(2900) San Nicolás, Argentina
E-Mail: apaafe@siderar.com

[3] Tamyr S.A., Mendoza 553-(1621) Benavídez, Buenos Aires, Argentina
E-Mail: tamyr@netben.com.ar

Keywords: Heat Treatment, Electrical Steel, Synchronous Permanent Magnet Motor, Finite Element Method

Abstract

In this work the influence of the magnetic properties of steels, at different processing conditions, on the performance of a synchronous permanent magnet motor (SPMM) with 2.7 watt power is studied.
A commercial low-carbon steel was subjected to 2 different thermal treatments and characterized for microstructure, carbon content and magnetic properties.
A computer model using finite element method was developed to calculate the influence of the heat treatment properties on the ampere-turns reduction.
Several motor prototypes were constructed using steels in the as-received and heat-treated conditions. Torque measurements were used as indication of the motors performance.

1. Introduction

Laminations or low-carbon steels are used since the '50s as a low cost alternative to the silicon steels, to be employed in low and medium output machines like fractional power motors, ballasts and small transformers. These steels are usually produced in a semi processed condition and they reach their maximum magnetic potential during the final steps of manufacture at the user's plant. **[1-2].**

The application of heat treated laminations to small electric motors is unusual because magnetic losses are not of concern when low powers are involved. However an additional effect of the heat treatment is a significant increase in permeability (about 3 to 4 times) which shifts the magnetization curve to lower magnetizing field strengths. The consequence is that a given working flux can now be achieved with lower ampere-turns at the magnetizing coil.
In this work the application of this kind of steels is studied in a small SPMM. The SPMM is an inexpensive device employed as a time reference, which provides rotating movement with a low torque. Typical applications are chart recorders, programming devices, valve drives, etc. Fig. 1 presents a schematic drawing.

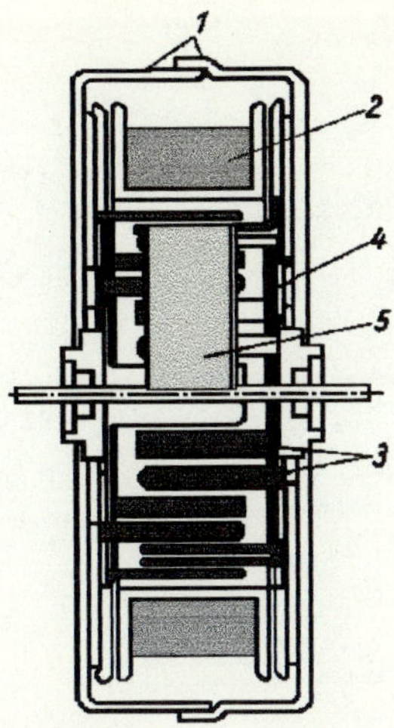

Fig. 1: Topology of a SPMM: 1- Stator housing, 2- coil, 3-claw poles,
4- short-circuiting ring (copper), 5- ring permanent magnet rotor

Typical characteristics, such as the developed torque can be obtained with the aid of 2D FE analysis. Although 2D axisymmetric representation of SPMMs provides less accurate results, when compared to 3D modelling, it allows faster, lower-cost simulations **[3]**.

2. Experimental procedures

A commercial low-carbon (0,07 %), 1 mm thickness, steel was subjected to 2 different heat treatments : 680 °C, 2.5 hours, in nitrogen atmosphere and 780 °C, 2.5 hours, in wet nitrogen (85%)/hydrogen (15%) atmosphere, in a laboratory furnace (dew point 35°C). Microstructure was analyzed in Nital 2% etched samples, on a Normanski effect optical microscope. Carbon content was determined in a Leco analyser and magnetic properties in a 28 cm Epstein frame.

Finite element analysis of the motor was performed using a 2D package (FLUX2D) **[4]**. Motor prototypes were constructed and evaluated in a laboratory work-bench where torque, exciting current and temperature could be measured.

3. Results and discussion

Fig. 2-a shows the microstructure of the steel in the as-received condition. Fig.2-b and 2-c display the microstructures of the samples treated at 680°C (in nitrogen) and 780°C (in wet nitrogen/hydrogen). Table 1 presents the carbon contents of the steels in the as-received and heat treated conditions.

a b c

Fig. 2: a- Microstructure of low-carbon cold-rolled steel, x200
b- Microstructure of the steel treated at 680°C in nitrogen, x200
c- Microstructure of the steel treated at 780°C in wet nitrogen/hydrogen, x200

Table 1: Carbon content of the steels.

Steel condition	% C
As-received	0.07
Heat treated at 680°C (in nitrogen)	0.0044
Heat treated at 780°C (in nitrogen/hydrogen)	0.0008

Comparison between microstructures Fig. 2-b and Fig. 2-c shown duplex (30-250 um) and homogeneous grains (180 um) respectively.

The steady state operation of the SPMM has been simulated using a 2D FE package and Fig. 3 presents a schematic layout of the motor geometry in two views: axisymmetric and transverse, displayed by the geometric modeler of FLUX2D.

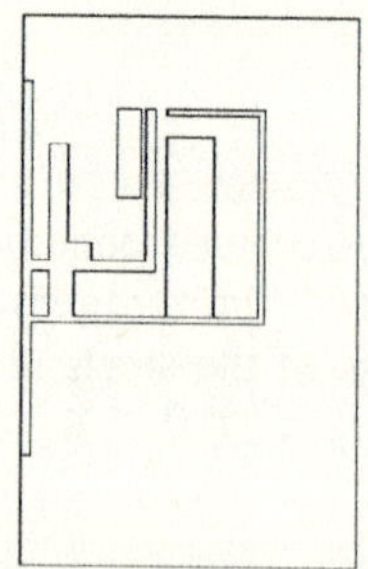

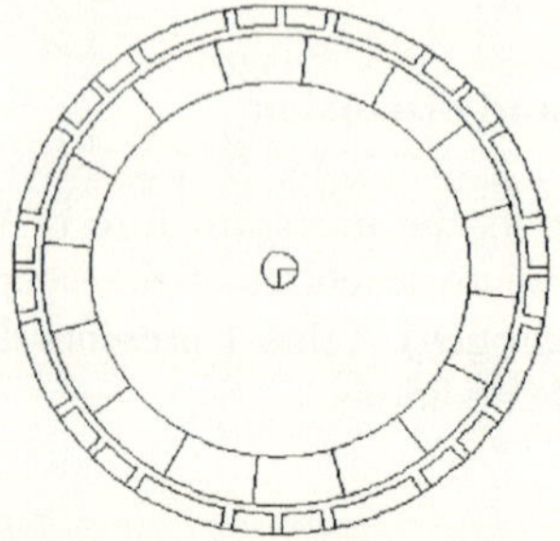

Fig. 3: Schematic of the motor using axisymmetrical and plane geometry

Fig. 4 shows the flux density distribution through contour plots of the transverse plane model where torque can be calculated. The prototype motor is displayed in Fig. 5.

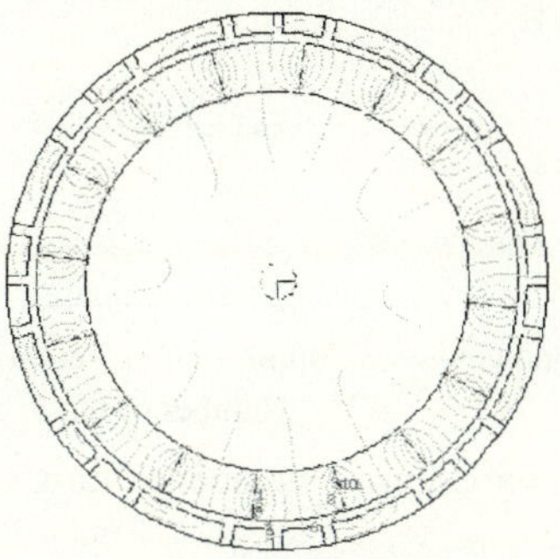

Fig. 4: Magnetic flux distribution

Fig. 5: Experimental model.

Prototype torques, turns, temperatures and currents measurements are indicated in Table 2. In Fig. 6 torque variation in function of number of turns of N_2/H_2 heat treated steel is shown. It can be appreciated that this material reaches a higher torque for a smaller number of turns, although some temperature increase results from the higher currents involved.

Table 2: Motors prototypes performance
TS: standard; TTNH: heat treatment in nitrogen/hydrogen; TTN: heat treatment in nitrogen

Motor Prototypes	Number of turns	Torque [kgcm]	Temperature [°C]	Current [mA]
TS	11000	11	33	12,8
TTNH1	9000	13,18	45	18,5
TTNH2	8000	9,89	50	24,5
TTN1	9000	10,7	54	21

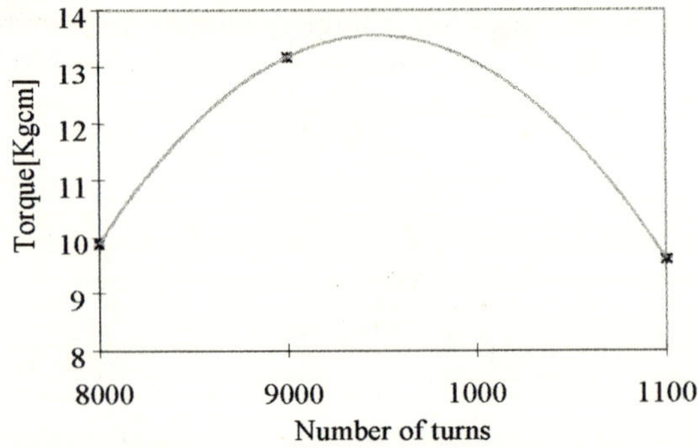

Fig. 6: Torque variation vs. number of turn (heat treatment: N_2 / H_2)

4. Conclusions

This paper shows that the substitution of the ordinary steel by a heat-treated one that permits a 20% increase in torque. Heat treatment in wet nitrogen/hydrogen atmosphere is more effective than with nitrogen atmosphere. At the same time an 18 % reduction in the number of turns of the magnetizing coil is obtained.
Results of the prototypes were in good agreement with numerical the solutions employed.

Acknowledgments

To Miss Karim Rafael for the metalographic work and Prof. José Roberto Cardoso of the LMAG, Escola Politécnica da Universidade de Sào Paulo, to have provided the helpful discussions about the problem solution and technical support.

References

[1] Actis, F.A.; Oldani, C.R.; Moyano, H.R.; Cohen, G.B. *"Composición de Pérdidas en Aceros Eléctricos de Bajo Carbono"*. Anales del I Taller Argentino Materiales Magnéticos y sus Aplicaciones, Córdoba 17-20 de Mayo de 1995, 35-41.

[2] Oldani, C.R.; Actis, F.A.; Mancini, R. *"Descarburación y Crecimiento de Grano en Laminaciones de Uso Eléctrico"* Primer Simposio Bilateral Argentino-Norteamericano Sobre Ciencia e Ingeniería de Materiales, Anales de la Asociación Química Argentina, 1996.

[3] Agüero, A.C.; Actis, F.A.; Cardoso, J.R.; Nabeta, S.; Silva, V.C.; *"Finite Element Analysis of a Synchronous Permanent Magnet Micromotor through Axisymmetric and Transverse Plane Simulations"*, IEEE Trans. on Magn., September 1998 (in press).

[4] FLUX2D - CAD package for 2D Electromagnetic Field Computation, CEDRAT.

Materials Science Forum Vols. 302-303 (1999) pp. 455-459

Synthesis and Study of a Magnetic Fluid

J. Matutes Aquino, M.P. González Sandoval, M. Miki Yoshida
and O. Ayala Valenzuela

Centro de Investigación en Materiales Avanzados, S.C. Miguel de Cervantes 120, Complejo Industrial Chihuahua, Chihuahua, Chih. 31110 México

Keywords: Magnetic Fluids, Magnetic Particles, Magnetite, Superparamagnetism

ABSTRACT

A magnetic fluid was obtained by a peptization method with kerosene as liquid carrier, oleic acid as stabilizer and magnetite as magnetic particles.
Nanometric magnetite powder was synthesized in aqueous medium by coprecipitation of ferric and ferrous salts.
The magnetic properties and structure of the particles and the fluid itself were determined with a vibrating sample magnetometer and a Mössbauer spectrometer. The rheological properties were determined with a cone and plate rheometer, the particles morphology was studied by transmission electron microscopy.

1. INTRODUCTION

A magnetic fluid is a colloidal system with an unique combination of fluidity and capability of interaction with a magnetic field. Its components are a liquid base, chosen to conform with its field of application; a dispersed phase, magnetic particles of colloidal size and a stabilizer to prevent colloid particles from aggregating.

Applications of magnetic fluids : lubrification and sealing systems with mineral oils as a liquid base, electroswitching devices and thermal energy conversors using liquid metals as a base and medical applications with an aqueous base.

The saturation magnetization is limited by the magnetization of the colloidal particles and by the volumetric fraction of particles that can be stabilized in the system without affecting its fluidity [1].

The dispersed phase determines the real magnetic properties of the fluid. For that reason is important an exhaustive study of the formation mechanism of the colloidal particles and their stabilization in the liquid phase.

2. EXPERIMENTAL PROCEDURE

The magnetic fluid was prepared in two steps. Firstly, highly dispersed magnetite was obtained by chemical coprecipitation. Second, a stable magnetic fluid was obtained by peptization.

For the preparation of the magnetite particles and the magnetic fluid the followings Baker Analyzed Reagents were used :Ferrous Chloride 4-Hydrate Crystal (>99.99% pure), Ferric Chloride 4-Hydrate Crystal (>99.99% pure), Ammonium Hydroxide (>28.4% NH_3), Kerosene KO8483, Oleic Acid G50610.

Preparation of the dispersed phase :

During the preparation of the highly dispersed magnetite by chemical coprecipitation the following chemical reaction occurred [2]:

$$FeCl_2\ 4H_2O + 2FeCl_3\ 6H_2O + 8NH_4OH \rightarrow Fe_3O_4 + 8NH_4Cl + 20H_2O$$

0.1M $FeCl_2\ 4H_2O$ and 0.1M $FeCl_3\ 6H_2O$ solutions were mixed at 70°C. The precipitation of Fe_3O_4 requires of the rapid neutralization of iron salts, for which ammonium solution in excess is rapidly poured to the iron salts solution with constant agitation. In the present case a 50% in excess of a 10% ammonium solution (NH_4OH) was rapidly poured with a constant agitation of 2500 R.P.M. There is an optimum value of pH to succeed in the precipitation of particles (approx. pH=6). The precipitate was washed several times with destilled water to remove the adsorbed ions from the surface of the formed magnetite particles and to allow the further adsorption of stabilizer molecules. Magnetite particles with a size distribution of about 7-20 nm were obtained.

Stabilization of the dispersed phase :

The particles of magnetite were stabilized in the fluid using a peptization method. Kerosene, which is a viable option for research purpose was used as the liquid base. The molecules of the stabilizer interact with the material of the magnetic particles forming a monomolecular cover to avoid the agglomeration of the particles. The long chain end of the molecule of the stabilizer should has affinity with the liquid carrier, should participate in the thermal motion and should avoid the approximation of the particles. The problem of selection of the stabilizer is solved experimentally. In the present case the stabilizer was oleic acid.

The peptization was carried out at 70°C during several hours with constant agitation, using a mixture of kerosene, oleic acid and magnetite in the proportions 20 :13 :67 [2]. During this process the residual water is removed and a thick paste is obtained. To obtain the stabilized magnetic fluid the paste is further diluted in a solution of kerosene and oleic acid at the same temperature and under constant agitation.

In order to obtain a stabilized magnetic fluid, it should be centrifuged at a separation factor of 6000g during 1 hour to remove the particles of diameter larger than 30 nm.

3. RESULTS AND DISCUSSION

The room temperature Mössbauer spectrum of the magnetic powder in figure 1a shows the superparamagnetic behavior of the particles due to the smallness of their sizes. This spectrum also shows that a small amount of larger particles are contributing to a broad and weak six line subspectrum [3]. The Mössbauer spectrum taken at -21°C (figure 1b) shows a decrease of the intensity of the superparamagnetic doublet. The Mössbauer spectrum taken at -196°C has a complex structure and the superparamagnetic behavior has apparently disappeared [4].

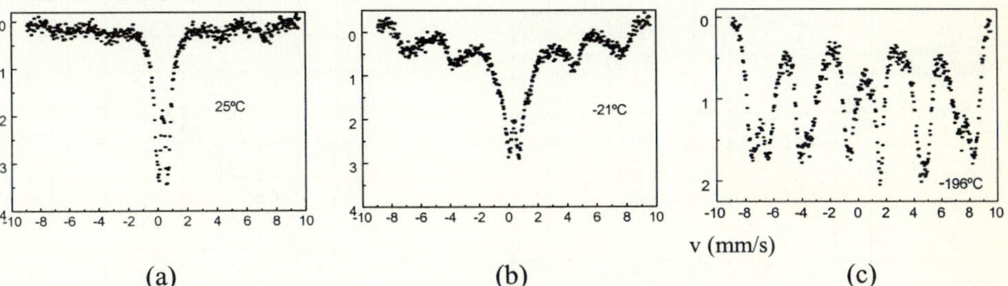

Figure 1 . Mössbauer spectra of the magnetic powder for temperatures from 25°C to -196°C.

The identification of the X-ray diffraction pattern of the powder (figure 2) corresponds to a mixture of magnetite and hematite. During the chemical coprecipitation, with stoichiometric proportions of ferrous and ferric salts, occurred the oxidation of some iron cations from ferrous to ferric. The excess of ferric cations favors the formation of hematite [5]. We try to solve this problem by modifying the initial amounts of $FeCl_2$ and $FeCl_3$ in the reaction; however, it also caused an increase of the particles size which produces a smaller fraction of magnetic particles in the fluid.

The magnetization curves of the powder and the magnetic fluid (figure 3) have the typical behavior of superparamagnetic particles (without hysteresis). The presence of hematite decreases the saturation magnetization of the magnetic powder and fluid [6].

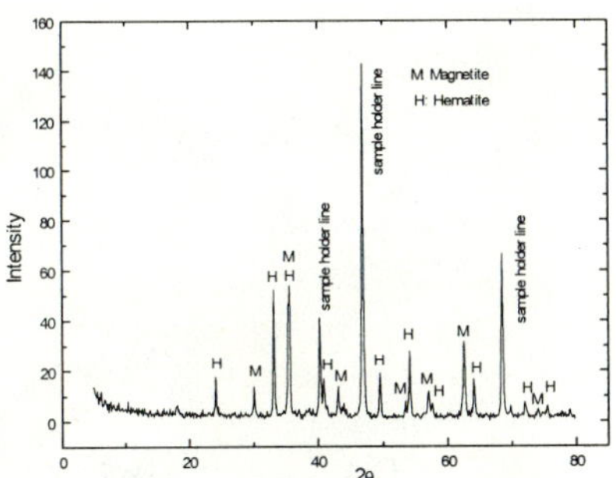

Figure 2. X-ray diffraction pattern of powder.

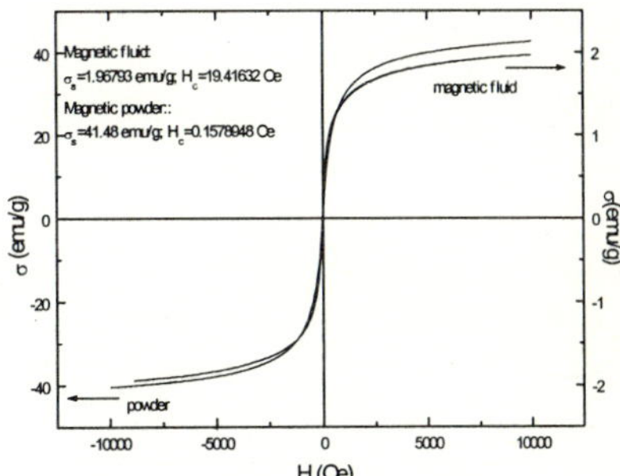

Figure 3. Hysteresis loops of the magnetic powder and the magnetic fluid.

Figure 4a shows the dependence of the magnetic fluid viscosity with the shear strain rate for different temperatures. For 30ºC, 40ºC, 50ºC, and 60ºC the viscosity changes for shear strain rate values lower than about 100 s^{-1} and remains essentially constant for shear strain rate values higher than 100 s^{-1}. The variation of viscosity with low values of the shear strain rate is due to the structured character of the fluid, which is destroyed for values of this rate higher than about 100 s^{-1}. Figure 4b shows the dependence of the shear stress with the shear strain rate for different temperatures. The magnetic fluid approaches to a Newtonian behavior for temperatures between 30ºC and 60ºC and for high values of the shear strain rate.

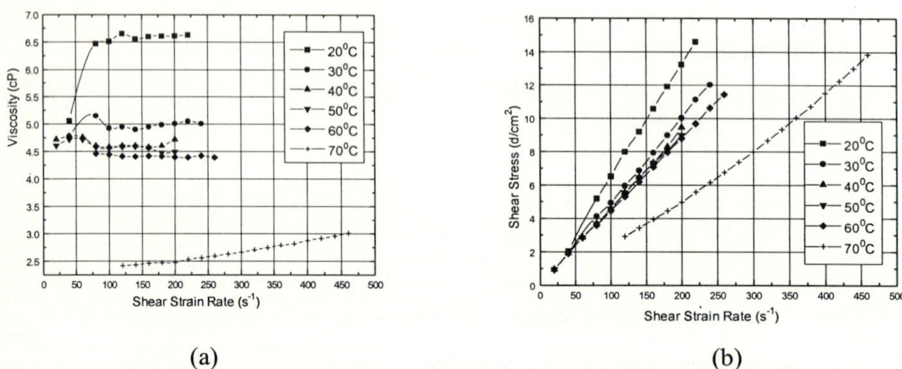

Figure 4. Dependence for different temperatures of: (a) viscosity with shear strain rate, (b) shear stress with shear strain rate.

Figure 5 shows the thermogravimetric analysis and differential thermal analysis curves of the magnetic fluid from room temperature up to 250ºC. A weight loss of about 60% occurs between 90ºC and 150ºC. The weight loss slows down at higher temperatures.
The transmission electron micrography (figure 6) indicates that most of the particles are smaller than about 20 nm [4,7,8].

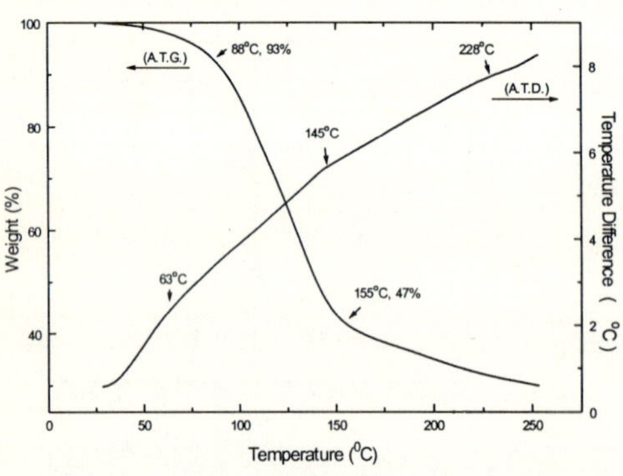

Figure 5. Thermogravimetric analysis and differential thermal analysis of the magnetic fluid.

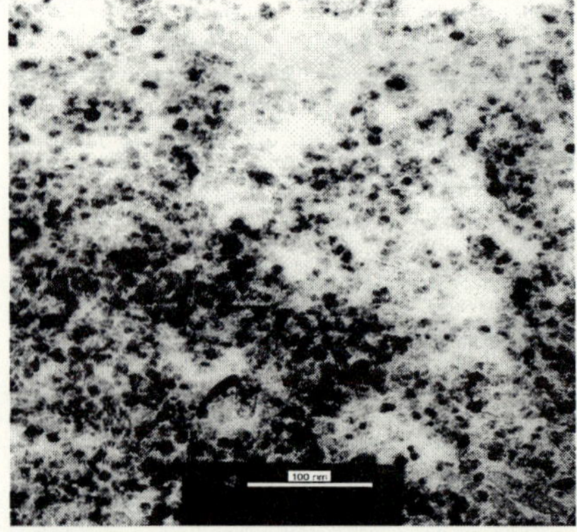

Figure 6. Transmission electron micrography of magnetic fluid particles.

4. CONCLUSIONS

1. Superparamagnetic particles were obtained by coprecipitation of ferrous and ferric salts. The particles are a mixture of magnetite and hematite. The hematite portion lowers the saturation magnetization.
2. The rheological curves indicate a different behavior at low and high shear strain rates. Below $100s^{-1}$, the structure of the magnetic fluid determine the dependence of viscosity with the shear strain rates. Above $100s^{-1}$ the magnetic fluid has a Newtonian behavior.
3. Between 90°C and 150°C there is a high weight loss in the magnetic fluid, because of the volatility of the liquid carrier.
4. The obtained magnetic fluid is a colloid with magnetic particles smaller than about 20 nm.

REFERENCES

1. Scholten, J of Magnetism and Magnetic Materials **39** (1983) 99.
2. B.M. Berkovsky, "Magnetic Fluids Engineering and Applications",(Oxford University Press, 1993), p. 13.
3. R. Keller, E. Schmidbauer, J. of Magnetism and Magnetic Materials **162** (1996) 86.
4. T. Pal, S. M. Mitra, Transactions of the Indian Ceramic Society, **48** (989) 115..
5. S.E. Khalafalla, G.W. Reimers IEEE Transactions on Magnetics, . **MAG-16** (1980) 178.
6. A.T. Skjeltorp, Phys. Rev. **51**(1983) 2306.
7. S. Morup, J. of Magnetism and Magnetic Materials **39** (1983) 45.
8. K. O`Grady, A. Bradbury, J. of Magnetism and Magnetic Materials **39** (1983) 91.

Materials Science Forum Vols. 302-303 (1999) pp. 460-463

Magnetic 3d Impurities in HCP Iron Host

C. Paduani[1], F. França[2] and J.C. Krause[3]

[1] Departamento de Física - UFSC, Florianópolis, CEP 88040-900, SC, Brazil

[2] CCT - FEJ - UDESC, Joenville, CEP 89223-100, SC, Brazil

[3] Instituto de Física - UFRGS, Porto Alegre, CEP 91501-970, RGS, Brazil

Keywords: HCP Iron, 3d Impurities, Magnetic Properties, DV Method

Abstract: The formation of localized moments in hcp iron host is investigated by electronic structure calculations of 3d impurities with the first-principles discrete variational (DV) molecular cluster method. The spin polarised case is considered within the framework of the local-spin-density (LSD) approximation. For the central atom of each cluster is obtained the magnetic moment, the hyperfine magnetic field and the charge transfer. The cluster is embedded in the long ranged Coulombian potential of the crystal field, consisting of few hundred of atoms. The Mn atom exhibits the largest values for both the magnetic moment (0.32 μ_B) and the contact field (-90 kG), whereas the Sc atom has the largest ionization (+0.41e). The magnetic moment increases in magnitude from Sc up to Mn.

1. Introduction.

The formation of localized magnetic moments in impurity ions within a metallic host system has long been the subject of intensive experimental and theoretical research [1-5]. Many theories developed to describe the formation of the local moment agree that, the formation will depend upon the width of the state as well as on its position with respect to the Fermi level. Thus, a complete description of this process requires a full investigation of the band structure of the host. Recently, the problem of impurities in systems with structural disorder was treated by the self-consistent tight-binding linear-muffin-tin-orbital recursion method (TB-LMTO-REC) [6], with good results.

Since the earliest treatments of the problem of dilute impurities in metallic hosts, several theoretical schemes of one-center character have been attempted. Delley [7] have investigated the electronic structure of iron impurities in metallic hosts by means of molecular cluster calculations [7]. Their results are in qualitative agreement with the Green function KKR method and could even reproduce fairly well experimental features. According to the authors, no discrepancy exists between cluster and continuum models if we choose clusters of a suitable size.

Cluster calculations have been performed with the DV method to study the local magnetic properties of hcp iron [8]. The results are in good agreement with experimental predictions for this phase and confirm an antiferromagnetic ordering at low temperatures. Considering the early difficulties to stabilize this phase it has not attracted attention of experimental workers. However, recent developments in experimental techniques have led to great advances in material science engineering. This iron phase was inittialy studied in substitutional alloys in terminal solid solutions where the properties are extrapolated to the pure case. Later on, high pressures martensitic-type mechanism were developed to produce pure hcp iron. In the last years, thin films of hcp iron have been grown on ruthenium substract [9-11]. Since then there is an increasing interest in the study of the properties of these new materials.

In this work the local magnetic properties of localized 3d impurities in hcp iron host are investigated by means of cluster calculations with the DV method [12]. These calculations rely on the local density approximation of density functional theory, where is used the exchange-correlation term of von Barth and Hedin [13]. The DV method is quite well descrite in the literature listed abovre. Therefore, we present here only the results of our calculations.

2. Computational method.

The molecular clusters were built up with 27 atoms as shown in fig. 1. The lattice parameters, $a=4.6273$ *a.u.* and $c = 7.4268$ *a.u.* were adopted for hcp iron. As is known, usually this is obtained under high pressures [14-16].The dependence of the total electronic energy of this cluster with the lattice spacing was investigated in our previous work [8]. A somewhat higher stability was achieved exactly at the value initially adopted for the lattice spacing in this phase, which were obtained from a plot of the ratio c/a against applied pressure, for that point corresponding to the onset of the irreversibility of the curves for loading and unloading process, at a pressure of 17 Gpa [16].

Except for zinc, the numerically generated atomic basis functions include 1s-4p orbitals, in zinc case, we included 5s orbital alloying for some extra ionization. The crystal potential was set up by using the converged configuration of the central atom for the cluster representing pure hcp iron. The atomic basis for each transition element was initially converged after few iterations an further kept constante through the iteration procedure.

3. Results and discussion.

The complete coordination of the central atom in the cluster makes it the adequate probe with bulk-like characteristics. The reduced coordination numbers of the surface atoms are responsible for their enhanced magnetic moments. This is a finite cluster ef egin{tabular}{10he truncation at the surface does not allow sufficient wavefunction delocalization to takes place. It is worthwhile to mention that lattice relaxations are not included in these calculations.

Table 1 summarizes calculated results for each impurity atom, located at the center of the cluster, besides the results for pure hcp iron, for comparison (from [8]). Note how the magnetic moment μ increases steadly from Sc up to Mn. The magnetic moment depends on the occupancy and should be larger for Mn, as observed, but it depends also on electron correlation, which increases for heavier impurities. As one observes, the Co atom exhibits a small magnetic moment.

The largest value for the magnetic moment (0.32 μ_B) belongs to Mn (as normally occurs in this group), ferromagnetically coupled with the surrounding iron atoms in that plane. This somewhat parabolic dependence for μ in this series is also expected from calculations based on the Anderson model, as was discussed by Moriya.

Table 1: Calculated results for the ionization (Q), local magnetic moment μ the contribution from the conduction band $4s+4p$ (μ_C) for the total magnetic moment, and the hyperfine magnetic field H_C (in Tesla) for the central atom of each cluster.

Cluster	Sc	Ti	V	Cr	Mn	Fe	Co	Ni	Cu	Zn
μ	0.04	0.05	0.17	0.22	0.32	0.29	-0.06	-0.08	-0.04	0.02
μ_C	0.02	0.01	0.02	0.02	0.03	0.05	-0.02	-0.01	0.01	0.01
Q	0.41	0.29	0.19	0.10	0.03	0.07	-0.05	-0.08	-0.15	0.10
H_C	-4	-6	-75	-81	-90	-95	-61	-61	-52	-76

The behavior of the partial $4s+4p$ contribution from the conduction band (μ_C) to the total magnetic moment gives information about the effective exchange interaction between localized and conduction electrons, typically an antiferromagnetic coupling for ferromagnetic materials [17], as well as about the behavior of the magnetic ordering temperature. Besides, the μ_C value gives also a quantitative description of the effect of replacing the impurities atoms for Fe atoms ($\mu_C=0.05$), on the magnetic polarization of the conduction-band. For instance, there is a competitive effect on this polarization arising from the contribution of μ_C from iron host atoms against that contribution from the addition of more impurity atoms, an interesting problem to be investigated in alloys [18]. Note in the table how the magnetic polarization μ_C is practically insensitive to the occupancy of the impurity site for early impurities (Sc-Mn). It can be observed also how μ_C is ferromagnetically coupled with μ for Sc-Mn. The rather small value for μ_C gives an indication of the weakness of the exchange coupling between localized and conductions electrons therein, and it is quite sensitive to the occupancy at the central site. This trend is well described in the present calculations. For the Zn atom the impurity may be considered simply as an extra positive charge, which is to be screened out by a deformation and modification of the conduction electron band itself. In this case one observes a collapsing value for the local moment, as expected.

As one can observe in table 1 the charge transfer Q has a strong dependence on the occupancy at the impurity site. The Sc atom has the largest (positive) ionization Q, suffering an expressive outflow of electrons (0.41). Starting from Ti, Q decreases steadly throughout this series down to a minimum for Mn. At the Co atom Q changes sign, indicating an acceptor character for electrons at the impurity site thereafter.

The calculated results for the magnetic hyperfine field correspond to the Fermi contact term H_C, as is typical for transition elements. In table 1, the largest H_C value belongs to Mn (-90 kG), followed by Cr (-81 kG), Zn (-76 kG), V (-75 kG) and Cu (-52 kG). The Sc atom has the smallest value (-4 kG). In all these impurities H_C has negative values. Note in the table that, starting from Sc, *0* H_C increases (in absolute value) up to the Mn atom, decreasing thereafter. This behavior reflects the contribution of the conduction electrons, which is dominant, and has, except for Ti, the same sign as H_C. In fact, the contribution for the electronic density at the Ti nucleus is almost entirely due to the core electrons; this is negative, and has opposite sign to that arising from conduction electrons. Furthermore, for late impurities (Co-Zn) the fast reduction observed in H_C for Cu comes essentially from the lower contribution in this case from both conduction and core electrons.

Magnetic susceptibility measurements indicate that hcp Fe-Mn alloys do not obey the Curie-Weiss law and increase markedly as decreasing Mn concentration [19]. According to the authors the results indicate also that a high density at the Fermi energy is expected in this case, which is confirmed in the present calculations. The hcp Fe-Mn alloys were obtained with 17.8, 25.9 and 28.6 at. % Mn. In all these alloys the average magnetic moment was about 0.25 μ_B at room temperature. The calculated value of μ_B in the present work for the Mn atom in this host agrees fairly well with that result, mainly if one considers the dependence of the local moment with the occupation at the neighboring sites besides the lattice spacing variation with composition and thermal effects.

In spite of the finite cluster size, the local symmetry of the central atom is well representative of the local environment in a hcp matrix. Thus one should expect that the calculated properties for the atom at the origin of coordinates can reproduce fairly well the results for bulk atoms or at least their features. In this case all electrons are included in the self-consistent procedure. One effect of the cluster size is enhanced values for the local moment. However, the behavior of the magnetization throughout this series is expected to follow these results, which are in good agreement with experimental predictions for pure hcp iron. Furthermore, it is known that when 3d impurities are introduced in amorphous iron, the average effective exchange interaction term on the neighboring sites of the impurity tends to grow, which should in turn increase the stability of the magnetic moment thereby. Such local mechanism could be considered in such a case of a compacted structure in order to explain the larger values obtained for the magnetic moment for early impurities (Sc-Mn) which exhibits a moment aligned parallel to the host moment. The increase of the electron correlation leads to the reversion of μ for late impurities (Co-Cu) through the increase of the exchange parameter favoring a ferromagnetic ordering in this matrix. This could explain the reversion of the μ value for leter impurities.

4. Conclusions.

Self-consistent cluster calculations were performed with the first-principles DV method in the LSD approximation to investigate the formation of local magnetic states at *3d* impurities in hcp iron host. The results indicate that, except for Sc, Cu and Zn, all these atoms are expected to exhibit magnetic moments in this matrix. The largest values belongs to Mn, Cr, V and Ni impurities, respectively. The Mn magnetic moment is ferromagnetically coupled with the remaining iron atoms in the first neighborhood, whereas the Co, Ni and Cu atoms have their magnetization aligned antiparallel to the surrounding iron atoms in that plane. In fact, the Co impurity tends to establish a ferromagnetic ordering in this host. Finally, these calculations indicate that the potential perturbation induced by 3d impurities in this host is expected to extend to the neighboring shells.

This work was supported by CAPES, CNPq and FINEP, Brazilian agencies.

References

[1] P. W. Anderson and A. M. Clogston, Bull. Am. Phys. Soc. **6** (1961) 124; P. W. Anderson, Phys. Rev. **124,** 1 (1961), p. 41.
[2] B. T. Matthias, M. Peter, H. J. Williams, A. M. Clogston, E. Corenzwit and R. C. Sherwood, Phys. Rev. Letters **5** (1960), p. 542.
[3] P. de Faget de Casteljau and J. Friedel, J. Phys. Radium **17** (1956), p. 27; A. Blandin and J. Friedel, J. Phys. Radium **19** (1958), p. 573.
[4] L. Wilk and S. H. Vosko, Phys. Rev. A **15**, 5 (1977), p. 1839.
[5] R. Zeller : Modelling Simul. Mater. Sci. Eng. **1**, (1993), p. 553.
[6] O. Yu. Kontsevoi and V. A. Gubanov, Phys. Rev. B **51**, (1995), p. 15125; O. Yu. Kontsevoi, O. N. Mryasov, R. F. Sabiryanov and V. A. Gubanov, J. Phys. Condens. Matter **6** (1994), p. 997.
[7] B. Delley, D. E. Ellis and A. J. Freeman, J. Magn. Magn. Mater. **30** (1982), p. 71.
[8] C. Paduani and F. Fran\c ca, J. Magn. Magn. Mat. **145** (1995), p. 147.
[9] J. P. Sanchez, M. F. Ravet, M. Piecuch and M. Maurer Hyp. Int. **57** (1990), p. 2077.
[10] T. Kraft, M. Methfessel, M. van Schilfgaarde and M. Scheffler, Phys. Rev. B **47**, 15 (1993), p. 9862.
[11] M. C. Saint-Lager, D. Raoux, M. Brunel, M. Piecuch, E. Elkaïm and J. P. Lauriat, Phys. Rev. B **51**, 4(1995), p. 2446.
[12] D. E. Ellis and G. S. Painter, Phys. Rev. **B2** (1970), p. 2887; A. Rosen, D. E. Ellis, H. Adachi and F. W. Averill, J. Chem. Phys. **65**, 9(1976), p. 3629.
[13] U. von Barth and L. hedin : J. Phys. **C5** (1972), p. 1629.
[14] D. N. Pipkorn, C. K. Edge, P. Debrunner, G. De Pasquari, H. G. Drickermer and H. Frauenfelder, Phys. Rev. **135** (1964), p. 1604.
[15] D. Bancroft, E. L. Peterson, S. Minshall, J. Appl. Phys. **27** (1956), p. 291.
[16] W. A. Basset and E. Huang : Science, **238** (1987), p. 780.
[17] A. Blandin, in Magnetism vol. V of *Magnetic Properties of Metallic Alloys* (Academic, New York, 1973).
[18] H. Chacham, E. G. da Silva, D. Guenzburger and D. E. Ellis, Phys. Rev. B **35**, 4 (1987), p.1602.
[19] H. Ohno and M. Mekata : J. Phys. Soc. Japan **31**,1 (1971), p. 102.

Materials Science Forum Vols. 302-303 (1999) pp. 464-468

Electrochemical Behaviour of Magnetic Materials in Various Solutions

J. Matutes Aquino, F. Almaraya Calderón, C. Gaona Tiburcio and A. Martínez-Villafañe

Centro de Investigación en Materiales Avanzados, S.C. Miguel de Cervantes 120, Complejo Industrial Chihuahua, Chihuahua, Chih. 31110 México

Keywords: Magnetic Materials, Electrochemical Techniques, Corrosion

ABSTRACT

This research work is devoted to analyse the electrochemical behaviour of magnetic materials like NdFeB and Alnico, which are used in automotive applications. The experimentation was carried out using the potentiokinetic polarisation technique according to the ASTM G5-94 standard . The solutions used were distilled water, sulphuric acid 1N and Ringer's solution. All the solutions were deaerated. In all the experiments a constant potential sweep rate of 0.6 V/hr was used, and the potential of the calomel saturated reference electrode was measured. The potentiokinetic sweeps for cathodic potentials were well defined, except for the transpassivation values. The experimental arrangement consists of a system of three electrodes, using a potentiostat/galvanostat instrument.

1. INTRODUCTION

Materials for permanent magnets like barium ferrites and alnicos are highly resistant to corrosion. However, corrosion is a serious problem for rare earth magnets like Nd-Fe-B. In this kind of materials corrosion occurs superficially if no coating is provided for protection, then oxygen diffuses inside the magnet, causing metallurgical changes on their surface layer [1]. Once the magnet is oxidised, the surface layer will show a lower intrinsic coercivity and this region is easily demagnetised.

Due to the coercivity of the Nd-Fe-B magnets, they have to be designed very thin; this increases the relation area to volume and then corrosion problems increases too.

The dependence of the depth of the oxidised layer d_0 with temperature and time have the form $d_0 = f(T)\sqrt{t}$. Because the function $f(T)$ is highly non-linear there is a critical temperature above which the uncoated magnets will experience degradation in their magnetic performance. Below this temperature they operate safely.

Nd_2O_3 is one of the oxidation products for the Nd-Fe-B system at high temperatures. At high humidity levels, the Nd-Fe-B system absorbs the hydrogen in the atmosphere into the surface and disintegrates [2].

The corrosion have serious consequences on the performance of the magnetic properties of the Nd-Fe-B system. The resistance against degradation of the magnetic properties can only be improved by the control of the composition and coating protection. The coating will be effective if it is free of porosity. The $Nd_{15}(Fe_{1-x} Co_x)_{77} B_7Al_1$ magnet can be protected against flux loss in 95% humidity at 80°C, with a nickel-plated coating of 10-15 μm thickness [2].

Another method to avoid corrosion consists in a dual layer system: one is a epoxy-type coating of 20-30 μm thickness and the other one an aluminium layer of about 10 μm of thickness applied by ion vapour deposition to ensure good homogeneity [2].

The inherent porosity of bonded magnets makes the process of blending the compound very important because in this stage the binder may be mixed in such a way as to coat the individual particles [2].

2. EXPERIMENTAL PROCEDURE

Electrochemical measurements of the magnetic materials NbFeB and Alnico specimens were carried out utilising the potentiokinetic polarisation technique [3]. An potentiostat/galvanostat instrument was used. The potenctiokinetic polarisation scans were carried out according to the ASTM G5-94 standard. Three solutions were studied: 1N sulphuric acid, Ringer's solution (NaCl 9g, KCl 0.425 g, $CaCl_2$ 0.119 g, $NaHCO_3$ 01.g in 1 litre aqueous solution), and distilled water. All solutions were deaerated. A constant potential sweep rate of 0.6 V/hr was used in all experiments. All potentials were measured with respect to a saturated SCE (0.244 V versus SHE a 25°C) as reference electrode. Two parallel graphite rods served as the counter electrode for the current measurements.

The potenciodynamic polarisation curves for the magnetic materials are shown in figures 2 and 3. There was a shift of the open circuit corrosion potential in the noble direction as the solution tested was changed. A positive shift in corrosion potential was expected, since, as shown in equation 1, the potential is directly proportional to the temperature [4].

$$E = \frac{RT \ln K}{nF} \qquad (1)$$

Where: E: corrosion potential, T: absolute temperature, K: solubility constant, n: number of electrons and F: Faraday constant

The corrosion rates were calculated using the Stern-Geary [5] relationship as shown in equation 2.

$$i_{corr} = \frac{\beta}{\Delta E / \Delta I} = \frac{\beta}{Rp} \qquad (2)$$

Where: β: Stern-Geary constant and ΔE/ΔI, Rp: Polarisation resistance

The Stern-Geary constant is calculated from the anodic and cathodic Tafel Slopes, as shown in equation 3.

$$\beta = \frac{b_a b_c}{2.3(b_a + b_c)} \qquad (3)$$

The anodic and cathodic Tafel slopes were taken from the polarisation curves with the lineal region in the interval of ± 100 mV, but the Rp were taken in the interval of ± 20 mV after the E_{corr} to open circuit; the curve of polarisation exhibited a lineal limited region in that interval with the loops of lineal Tafel of 0.3 to 0.5 decades, the Tafel slopes were not clearly appreciable in all the cases.

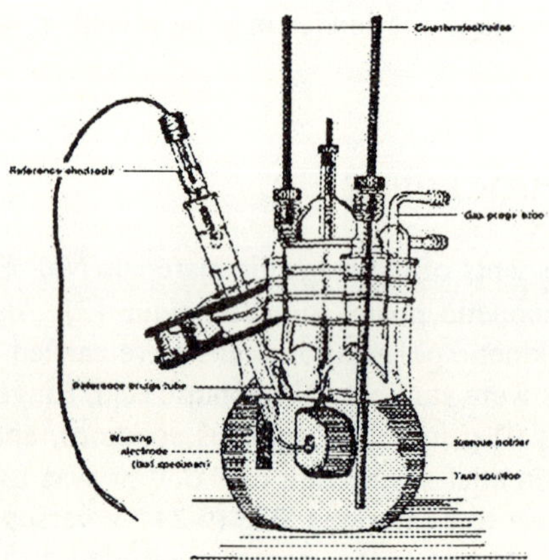

Figure 1. Experimental arrangement consists of a system of three electrodes [6].

3. RESULTS AND DISCUSSION

The results of the chemical composition of the magnetic samples are the following

Materials	Fe	Co	Ni	Al	Ti	Cu	Nb	B	Nd	Si
NdFeB	**71.62**	**0.034**	-	-	-	-	-	**1.016**	**26.79**	**0.059**
Alnico	**37.5**	**35.0**	**12.6**	**6.75**	**4.5**	**2.7**	**0.5**	-	-	**0.45**

Table 1. Chemical Composition (% wt.)

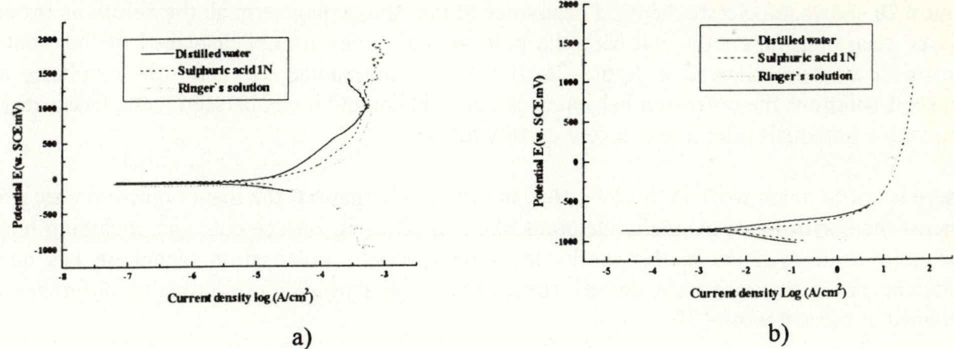

Figure 2. Potentiodynamic polarisation curves a) for the Supermagnet of NdFeB and b) for the magnet of Alnico

Materials	**Electrolyte**								
	Destilled water			**Sulphuric acid 1N**			**Ringer's solution**		
	E_{corr} mV	i_{corr} mA/cm^2	Corrosion Rate mm/yr	E_{corr} mV	i_{corr} mA/cm^2	Corrosion Rate mm/yr	E_{corr} mV	i_{corr} mA/cm^2	Corrosion Rate mm/yr
Supermagnet of NdFeB	-70	1.13×10^{-5}	1.31×10^{-5}	-47	1.26×10^{-5}	1.46×10^{-4}	-105	1.53×10^{-5}	1.77×10^{-4}
Magnet of Alnico	-859	5.07×10^{-1}	5.86	-857	6.14×10^{-1}	7.1	-876	6.71×10^{-1}	7.76

Table 2. Results of potentiodynamic polarisation scans and calculated corrosion rates from the Supermagnet of NdFeB and the magnet Alnico..

Table 2. shows the results of potentiodynamic polarisation scans and calculated corrosion rates of the Supermagnet of NdFeB and the Alnico magnet in various solutions. Figure 2a shows the electrochemical behaviour of the Supermagnet of NdFeB in all the solutions tested, and it can be seen that the anodic and cathodic polarisation curves are slight displaced of their potential of corrosion. If we look the anodic curve for the sulphuric acid and Ringer's solutions the corrosion behaviour is quite different from that destilled water solution, which means that corrosion increases in presence of aggressive solutions.

It could be observed that the values of E_{corr} in distilled water are less negative than those of the other solutions, similar to the values of the rate of corrosion. The Ringer's solution presents a greatest rate of corrosion than the others.

Figure 2b shows the electrochemical behaviour of the Alnico magnet in all the solutions tested, and it can be seen that the anodic and cathodic polarisation curves are not displaced of their potential of corrosion as those showed in figure 2a. If we look the anodic curve for the sulphuric acid and Ringer's solutions the corrosion behaviour is quite similar and it can be seen some passivation region at positive potentials after a big current density demand.

There is not to much work in this line. But, in other investigations the main objective was protection against the corrosion of magnetic materials like NdFeB using surface coatings, including nickel, zinc and gold coatings. Also in this works the potentiokinetic polarisation technique has been used, obtaining potential vs. current density curves that exhibit passivation zones, to difference of them obtained in present work [7].

4. CONCLUSIONS

- The electrochemical techniques used in this work: Polarisation curves offers a suitable alternative in order to evaluate the corrosion rate of those materials that are exposed to the presence of various solutions.

- From the results obtained of the potenciodynamic polarisation curves, it could be observed that the increase of the current density causes an increment in the corrosion rate, for the Alnico magnet.

- The results obtained in this investigation represent an advance on the study of the electrochemical behaviour of the magnetic materials under corrosion conditions.

REFERENCES

1. Adler E. and Marik H.J. 5th International Workshop on Rare Earth-Cobalt Permanent Magnets and their applications, pp. 335-56. Dayton : University of Dayton.
2. Campbell P., Permanent Magnet Materials and their Applications. Cambridge University Press, 1994, pp. 76-80.
3. *ASTM* G5-94 *Standard* Vol. 3.02 Wear and Erosion; Metal corrosion. Philadelphia, PA., USA.
4. Sykes J. M. "25 Years of Progress in Electrochemical Methods" En: Br. Corrs J. Vol. 25. No. 3, 1990. pp. 175-183.
5. Stern M. and Geary A.L. J. Electrochem, Soc. 104 (1) 1957. Pp. 56-61
6. EG&G Princeton Applied Research. Application note CORR1. *Basics of corrosion measurements.*
7. Man, H. H., Man H. C. and Leung L. K. Journal of Magnetism and Magnetic Materials. 152 (1996) pp. 47 - 53.

Materials Science Forum Vols. 302-303 (1999) pp. 469-473

Synthesis and Magnetic Properties of Submicronic Particles of Maghemite

J. Matutes Aquino[1], P. García Casillas[1], O. Ayala Valenzuela[1] and S. García García[2]

[1] Centro de Investigación en Materiales Avanzados, S.C. Miguel de Cervantes 120, Complejo Industrial Chihuahua, Chihuahua, Chih. 31110 México

[2] Facultad de Física - IMRE, Laboratorio de Magnetismo, Universidad de La Habana, Cuba

Keywords: Magnetic Particles, Thermal Decomposition, Maghemite, Organic Precursors

ABSTRACT

Iron oxide powders were obtained by the isothermal decomposition of ferric citrate in an oxygen atmosphere at different temperatures. Throughout the experiments, the product of decomposition was a mixture of maghemite (γ-Fe_2O_3) and hematite (α-Fe_2O_3) with different relative proportions and particle size distributions. After TGA and DTA measurements, a decomposition temperature of 300°C was chosen, and a maximum specific saturation magnetization of about 40 emu/g was obtained. Most of the product particles were less than 1μm in size, including a superparamagnetic fraction, as evidenced by Mössbauer spectroscopy.

1. INTRODUCTION

Nowadays, magnetic particles for recording of analogic and digital information is one of the most important applications of magnetic materials. The γ-Fe_2O_3, Fe_3O_4 and CrO_2 oxides comply with the required properties, but so far only γ-Fe_2O_3 and Co-doped modifications are the most widely used materials [1].

Several preparation methods have been used for synthesizing γ-Fe_2O_3 powders which aimed at producing uniform-sized particles for dispersed magnetic media [2]. To obtain elongated maghemite particles for magnetic recording, the traditional procedure is to start from α-FeOOH (ferric oxyhydroxide) in the form of small acicular crystals [1,3,4]. The process can be summarized by the equations:

$$6\ \alpha\text{-FeOOH}\ (400°C) \rightarrow 3\ \alpha\text{-}Fe_2O_3 + 3H_2O \quad (1)$$

$$3\ \alpha\text{-}Fe_2O_3 + H_2 \rightarrow 2\ Fe_3O_4 + H_2O \quad (2)$$

Careful oxidation of the resulting acicular magnetite at about 250°C produces acicular particles of γ-Fe_2O_3 according to:

$$2\ Fe_3O_4 + 0.5\ O_2 \rightarrow 3\ \gamma\text{-}Fe_2O_3 \quad (3)$$

Maghemite has a defect inverse spinel structure being the completely oxidized form of magnetite. It contains Fe^{3+} ions in tetrahedral A sites (fully occupied) and in octahedral B positions, shared with vacancies "( )", to compensate for the increase in the positive charge as a consequence of oxidation.

Thus, in analogy to magnetite, the stoichiometric formula can be expressed as $(Fe^{3+})[Fe^{3+}_{5/3}(\ )_{1/3}]O_4$; having a specific saturation magnetization of 74 emu/g due to the ferrimagnetic order of A and B sublattices. Since maghemite particles have a high degree of acicularity, shape anisotropy is a relevant contribution. However, incoherent mechanisms of magnetization reversal, as curling and fanning [1,5], can reduce noticeably the coercitivity values. Different microstructures reported in previous studies are due to differences in the conditions of preparation [6,7].
Organic precursors are an interesting approach to achieve new morphologies and to explore potential applications. In this work, maghemite powders obtained from the isothermal decomposition of ferric citrate were characterized by X-ray diffraction, TGA-DTA, magnetization and Mössbauer measurements.

2. EXPERIMENTAL PROCEDURE

High purity n-hydrated ferric citrate (16% Fe minimum) from J.T. Baker was milled during 8 hours in a vibratory micro-mill. The thermal decomposition of the obtained powders was followed by TGA-DTA measurements in a 2960 TA-Instruments analyzer, heating at a constant rate of 2°C/min from room temperature up to 400°C. Based on these results, the citrate powder was heat-treated during 1, 4 and 8 hours at 300°C in a oxygen USP grade (Praxair) flow in a programmable chamber furnace, looking for a high conversion of the ferric citrate particles into iron oxides and to dispose the various grain size distributions and relative proportions of the products. Any possible rest of ferric citrate was eliminated by filtration using a Whatman polycarbonate membrane filter with a pore size of 1μm and tridestilled water, followed by drying at 40°C.
Magnetic properties were measured in a vibrating sample magnetometer LDJ Electronics 9600 magnetometer, with a maximum applied field of 1.6 T. Transmission Mössbauer spectra were collected at liquid nitrogen and room temperatures in an ASA spectrometer in constant acceleration mode. Phase analysis of the products was performed by X-ray diffraction in a Siemens D5000 powder diffractometer at room temperature, using Cu-K_α radiation.

3. RESULTS AND DISCUSSION

Figure 1(a) shows the TGA-DTA curves for the decomposition of ferric citrate. The thermogravimetric analysis reflects a water weight loss of about 10% in the 90°C-150°C interval, followed by a rapid decrease, which matches with the emergence of two overlapped exothermic peaks; no further changes beyond 300°C were observed. After deconvolution, by fitting two Gaussian curves, the maxima of the peaks were determined to be at 192.7°C and 227.8°C, respectively, as shown in the Figure 1(b), reflecting two stages of the decomposition process.
Figure 2 shows the X-ray diffraction pattern corresponding to the isothermal decomposition of ferric citrate during 8 hours at (a) 192°C and (b) 300°C. The phases identification reveals a mixture of maghemite (γ-Fe_2O_3) and hematite (α-Fe_2O_3) in both cases.
The ferric citrate decomposed during 8 hours at 300°C was separated by filtration with a polycarbonate membrane into a fine fraction (<1μm) and a coarse fraction (>1μm); the former represents about 95 wt. % of the product. Fig. 3 shows the hysteresis loops for both fractions and for the original sample, previous to separation. The values of the specific saturation magnetization were 39.46 (fine), 35.77 (coarse) and 36.24 (sample) emu/g, respectively.

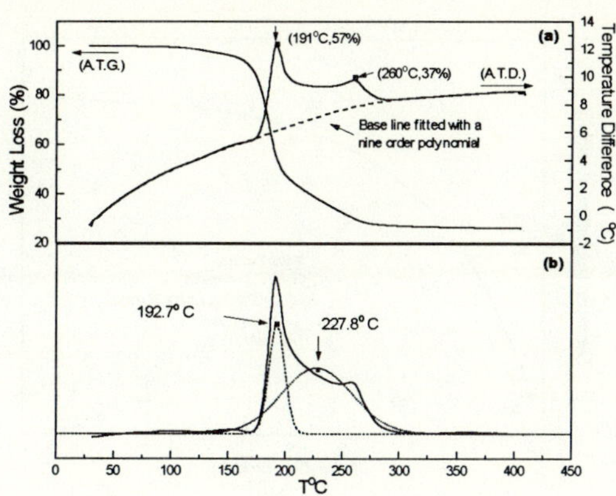

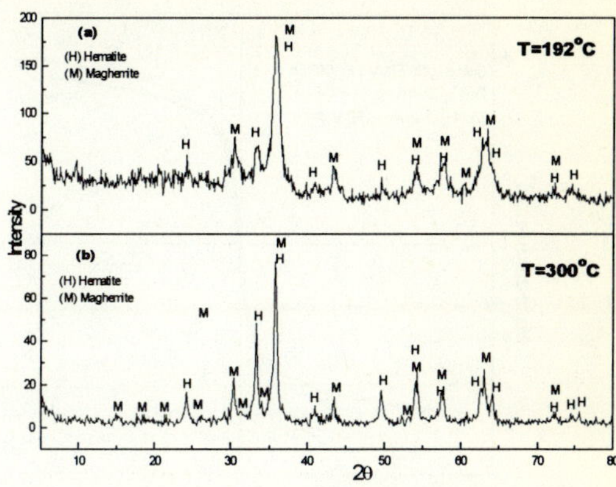

Figure 2. X-ray diffraction pattern of ferric citrate annealed during 8 h at (a) 192°C and (b) 300°C.

Figure 1. (a)Thermogram of ferric citrate in air flow. Heating rate: 2°C/min, (b) deconvolution by fitting two Gaussian curves.

Figure 4 shows the evolution of the specific saturation magnetization with the decomposition time at 300°C. The sample reaches a maximum of about 39 emu/g for an annealing time of about 8 hours; no further increment was observed because the ferric citrate is completely decomposed into a mixture of maghemite and hematite, as reflected by the X-ray diffraction pattern. This is also confirmed by the gradual reduction, with the decomposition time, of the paramagnetic contribution at the center of the Mössbauer spectra associated to residual ferric citrate, as shown in the Figure 5.

Figure 6 shows the Mössbauer spectra collected at room temperature and at liquid nitrogen temperature for the sample annealed 8 hours at 300°C. In figure 6(a), the broad lines of the magnetic sextets and their smooth inner slopes, along with the central paramagnetic component, suggest the existence of a wide size distribution of the obtained iron oxides, reaching the superparamagnetic dimensions and leading to relaxation effects. This is confirmed by an almost elimination of the central doublet at 77 K, and a better definition of the magnetic contribution.

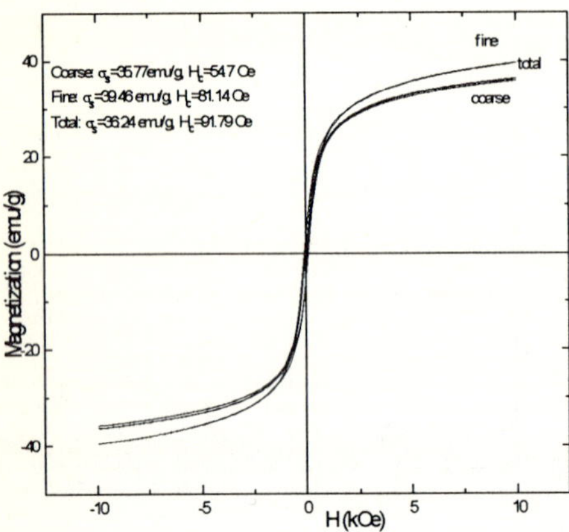

Figure 3. Hysteresis loops of total of ferric citrate decomposed during 8 hours at 300°C (total), and their fractions greater than 1μm (coarse) and smaller than 1 μm (fine).

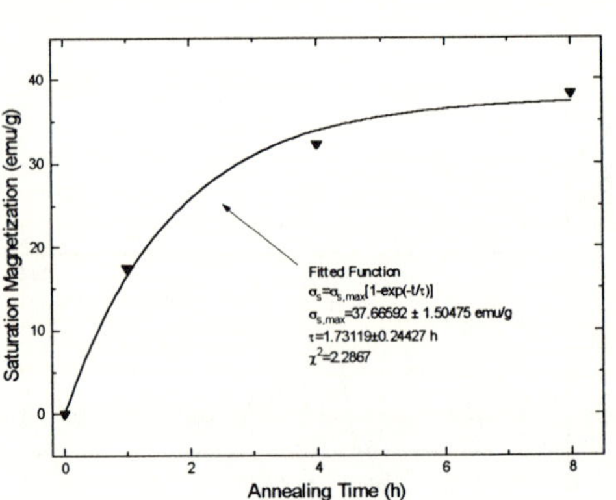

Figure 4. Evolution of ferric citrate with annealing time at 300°C. in oxygen atmosphere.

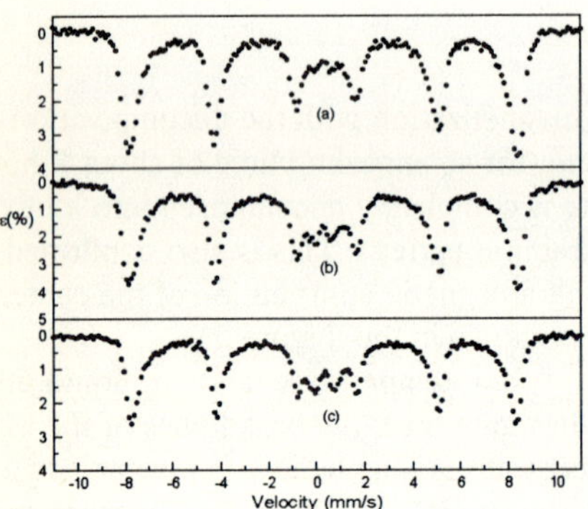

Figure 5. Evolution Mössbauer absorption spectra measured at 300°C during (a) 8 hours, (b) 4 hours and (c) 1 hours.

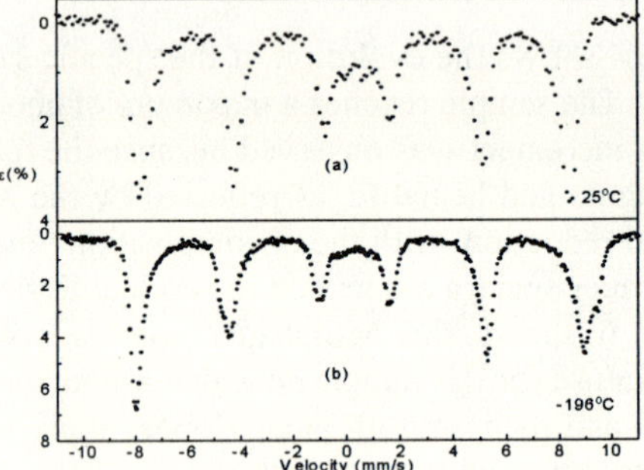

Figure 6. Mössbauer absorption spectra measured at : (a) 25°C and (b)T=-196°C of ferric citrate decomposed during 8 hours at 300°C.

4. CONCLUSIONS

n-hydrated ferric citrate isothermally decomposes at 300°C into a mixture of maghemite and hematite. The decomposition is complete for an annealing time of 8 hours with about 90 wt. % of the product particles having a size smaller than 1μm; the specific saturation magnetization was 39 emu/g. A certain fraction of the particles reach a nanometric scale and show a superparamagnetic behavior at room temperature. The low temperatures involved in the heat treatments of this organic precursor is a promising perspective to attain small sized magnetic powders for a wide scope of applications. An optimization of the sample procedure is in course.

REFERENCES

1. R.A. McCurrie in : Ferromagnetic Materials, Structure and Properties (Academic Press, 1994) p. 171.
2. V.K. Pillai in : Synthesis of Magnetic Oxide Nanoparticles Using Microemulsions (Thesis for the degree of PhD, University of Florida, U.S.A. 1995) p. 51.
3. G. Bate in : Magnetic Oxides, Part 2, de. D.J. Craik (Wiley Interscience, New York, 1975) p. 698.
4. M.P.Sharrock, IEEE Trans. on Magnetics 25 (1989) 4374.
[5]. L.H. Bowen, E. De Grave, R.E. Vandenberghe in : Mössbauer Spectroscopy Applied to Magnetism and Materials Science, Volume 1 de. G.J. Long and F. Grandjean (Plenum Press, 1993) p. 139.
6. A.E. Berkowitz, R.P. Goehner, E.L. Hall and P.J. Flanders, J. Appl. Phys. 57 (1985) 3928
7. G. Bate in : Ferromagnetic Materials, Vol. 2 de. E.P. Wohlfarth (North-Holland Pub. Co., 1980) p. 381.

Materials Science Forum Vols. 302-303 (1999) pp. 474-478

Nature and Origin of the Magnetic Mineralogy of Some South American Sedimentary Rocks: First Results

T.S. Berquó[1], M. Ernesto[1] and C.S.M. Partiti[2]

[1] Instituto Astronômico e Geofísico, Universidade do São Paulo, Rua de Matão 1226, Cidade Universitária, 05508-900, São Paulo, SP, Brasil

[2] Instituto de Física, Universidade de São Paulo, C.P. 66318, 05315-970, São Paulo,SP, Brasil

Keywords: Rock Magnetism, Magnetic Minerals, Sandstones, Natural Magnetization

Abstract: Magnetic properties of South American sediments formed during the Early Paleozoic Era were investigated. Magnetic tests detected the presence of hematite, goethite and possibly magnetite as the main carriers of the magnetization. The main purpose of this study is to investigate the origin of those minerals as the paleomagnetic results indicate that they carry a non-primary magnetization.

1. Introduction

In the last years great efforts have been made in the identification and characterization of the magnetic carriers of the natural remanent magnetization (NRM) of rocks, as well as for understanding the processes of the NRM acquisition by the magnetic minerals present in the various rock types. Those studies are fundamental for the correct interpretation of the paleomagnetic (the record of the ancient geomagnetic field in rocks) results, since different components of magnetization may be carried by different magnetic minerals. Magnetic minerals in sediments can be either of depositional or diagenetic (physical and chemical processes that affects the sediments during their history) origin. If these minerals are non-depositional, and did not crystallized during the initial stages of diagenesis, they are not primary, and probably acquired a secondary magnetization. Igneous and metamorphic rocks are also subject to alterations of their initial mineralogical composition, allowing the formation of secondary magnetic minerals. Therefore, for paleomagnetic purposes it is always necessary to investigate the nature and origin of the magnetic carriers in order to identify the primary and secondary magnetizations [1,2].

One of the main goals in paleomagnetism is to calculate paleomagnetic poles for the various geological periods and for each continent in order to infer apparent polar wander paths (APWP) from which the drifting of the continents can be inferred [3]. For most of the Paleozoic times the South American APWP is based on data from sedimentary rocks, some of which are very weakly magnetized. The depositional or chemical character of the primary magnetic mineralogy in these rocks, and the prevailing chemical origin of the secondary minerals, make it difficult to distinguish the two generations of magnetic minerals. Thorough investigations of the magnetic mineralogy and the remanence acquisition processes are fundamental to establish whether the stable magnetization component in those rocks corresponds to the sediment formation age.

The subject of this study are sandstones from the Caacupé Group, Eastern Paraguay, and from the Parnaíba Basin in Northern Brazil, both of Early Paleozoic ages. The magnetic record found in these sandstones are generally associated with hematite and goethite, although some magnetite is also present; goethite is as typical weathering product whereas hematite can be also detritic. In this way the direction of magnetization preserved by those minerals may not represent the direction of the local geomagnetic field at the time of deposition.

Samples were analysed in the Laboratory of Paleomagnetism, IAG/USP, using standard procedures. The NRM components were identified by means of thermal and/or alternating magnetic field (a.f.) demagnetizations. A preliminary identification of the magnetic carriers was possible by investigating their coercivity and temperature spectra during the demagnetization processes.

However, for a better identification and characterization of these carriers other techniques were applied.

2. Techniques and results

X-ray diffraction: is a traditional technique used for identification of the magnetic mineralogy in sediments. Diffractograms were obtained using a Phlips PW3710 diffractometer with Cu Kα radiation. In figure 1 a typical X-ray diffractogram is presented, for samples of the Caacupé Group where it is possible to notice the presence of quartz and other minerals like hematite and goethite. The peaks corresponding to those last two minerals were substantially broadened, indicating that those minerals probably grew chemically and are not well crystallized.

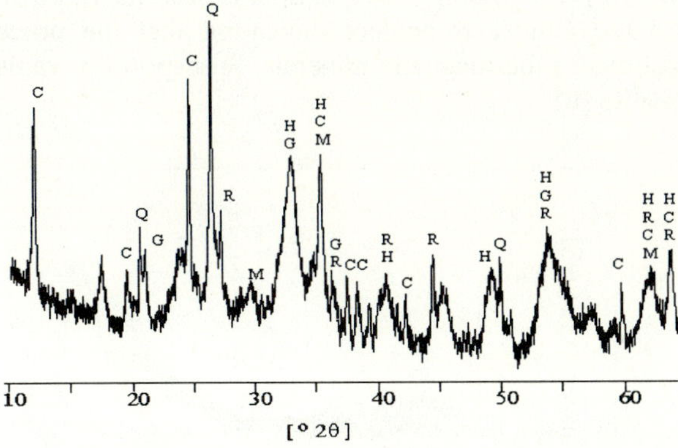

Figure 1. Typical diffractogram of Caacupé Group samples. The symbols indicate the minerals identified: C - Kaolinite, Q - quartz, G - goethite, R - rutile, M - magnetite, H - hematite.

Magnetic susceptibility: is a measure of the ease with which a material can be magnetized. It depends on the concentration, size, and composition of the magnetic minerals [4]. In sedimentary rocks strongly magnetic minerals like magnetite are generally not abundant, and susceptibility is in average weaker than in other types of rocks. Hematite and other high coercivity minerals impart low susceptibilities, which is influenced by paramagnetic (or diamagnetic, like quartz) minerals. During the thermal treatment, the variation of the initial magnetic susceptibility can give indications of physical and chemical alterations of the magnetic minerals.

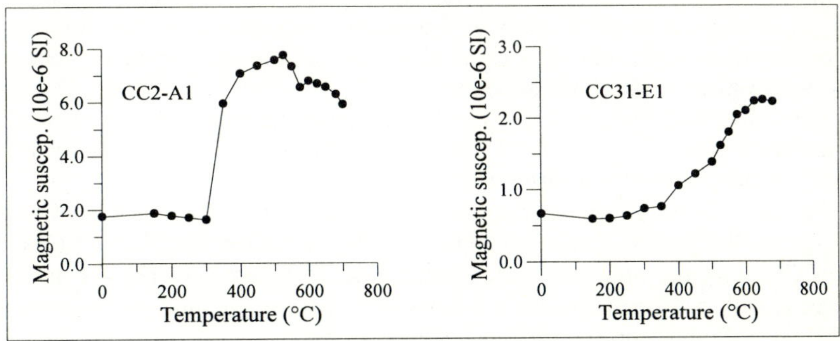

Figure 2. Variations of magnetic susceptibility during thermal treatment. Measurements at room temperature.

Susceptibility of the studied samples were measured in an bridge SI-1 from *Sapphire Instruments*. Figure 2 shows variations of magnetic susceptibility obtained during the thermal treatment of the Caacupé samples. Measurements were made at room temperature. It is possible to notice the low values of the initial magnetic susceptibility indicating low content of ferrimagnetic minerals, and presence of goethite, characterized by the increase of susceptibility above 350°C due to the transformation of this mineral into hematite [5].

Curie Temperature: Magnetic measurements at high or low temperature are generally related to the identification of magnetic minerals through the determination of their Curie temperatures or though some other transition dependent of temperature.

In this study Curie temperatures were determined by measuring variations in susceptibility, when samples were heated in a furnace CS-3 of the KLY-3S *Kappabridge* system. A thermomagnetic curve for a Parnaíba Basin sample is seen in figure 3. In this example the heating and cooling curves are almost coincident indicating that the process was reversible with no significant modifications of the magnetic minerals. Susceptibility vanishes close to 680°C attesting the presence of hematite [6].

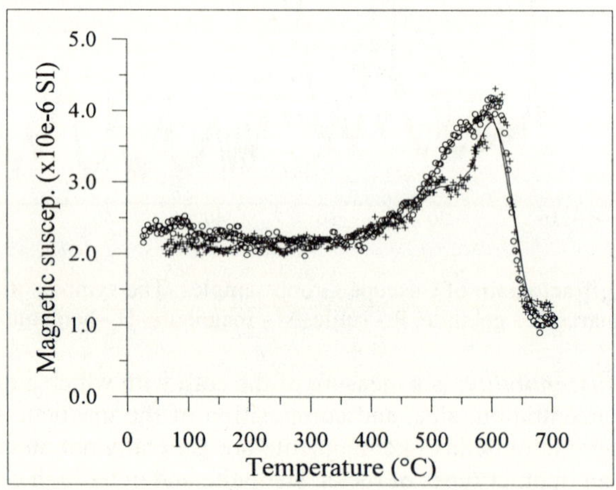

Figure 3. Thermomagnetic curve showing variations of the magnetic susceptibility against temperature. Circles represent the heating stages and crosses represent the cooling stages.

Mössbauer spectroscopy: is a technique of great mineralogical interest because it differentiates the various oxidation states, spin states and structural settings. It also presents advantages such as: *i*) quantification of the relative proportions of the minerals in a sample; *ii*) identification of minerals not well crystallized, and *iii*) identification of very small grains, that techniques like X-ray diffraction cannot resolve [7].

Mössbauer absorbers were prepared by pressing the powder between two lucite plates, with a sample thickness of 10 $mgcm^{-2}$ Fe. A conventional Mössbauer spectrometer was used with a liquid nitrogen cryostat to keep the absorber at 78 K. The source was ^{57}Co in Rh.

Figure 4 shows Mössbauer spectra for samples from the Caacupé Group. These spectra consist of a six-line pattern that could be fitted with two different hyperfine fields, with exception of sample CC53 that presents only one sextet. The fit parameters [7] are given in table 1. By comparasion with published hyperfine fields values for pure goethite [8] and pure hematite [9] at 77 K, these two different spectra can be assigned to goethite and hematite. The ratio between the amounts of goethite and hematite could be calculated from the relative spectral areas. The relative area for goethite was larger than that for hematite for samples CC30 and CC50. And for samples CC21 and CC47 the relative area for hematite was larger than that for goethite

Hysteresis loops: This analysis gives an indication of the presence of diamagnetic, paramagnetic and ferromagnetic minerals. For the ferrimagnetic minerals it is possible to calculate coercivities, and infer about grain sizes [10]. The shape of the hysteresis loops depends on the mineralogical composition of a sample, as well as on grain sizes.

Hysteresis loops for the Caacupé samples are shown in Figure 5. The loops are strongly influenced by the presence diamagnetic minerals (sample CC-1), as expected in samples dominated by quartz, and with a small content of ferrimagnetic minerals. The curves do not saturate indicating the presence of minerals with high coercivity like hematite and goethite. The hysteresis loops in this study were built in a vibrating sample magnetometer (VSM) from *Molspin Ltd.,* that allows fields up to 1 T.

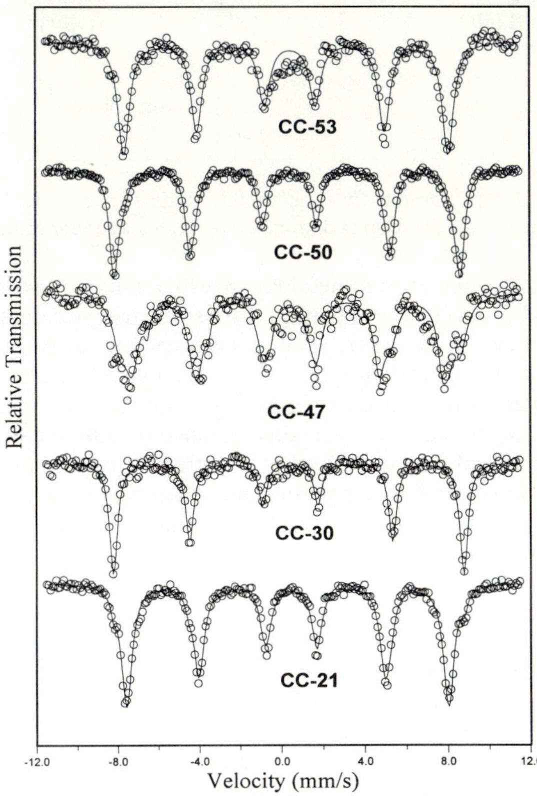

Figure 4. Mössbauer spectra for samples from the Caacupé Group. Circles represent experimental data, lines represent the best fit.

Table 1. Mössbauer parameters at 78K- Hyperfine field (Bhf), quadrupole splitting (DQ), isomer shift with respect to metallic iron (IS) and relative spectral area (AR). Numbers in parentheses indicate standard deviation on last significant figure.

Goethite

Sample	Bhf (kG)	DQ (mm/s)	IS(mm/s)	AR
CC21	483(2)	-0.25(5)	0,48(3)	6
CC30	492(2)	-0.26(2)	0.48(2)	99
CC47	473(1)	-0.27(2)	0.47(1)	12
CC50	494(2)	-0.19(2)	0.48(1)	78
CC53	487(3)	-0.25(9)	0.47(4)	100

Hematite

Sample	Bhf (kG)	DQ (mm/s)	IS (mm/s)	AR
CC21	527(1)	-0.19(2)	0.46(1)	94
CC30	526(3)	-0.16(1)	0.48(5)	1
CC47	522(1)	-0.18(3)	0.48(1)	88
CC50	520(3)	-0.18(2)	0.48(2)	22

3. Discussion and conclusions

The routine procedures used in paleomagnetic studies for demagnetizing samples (thermal and a.f. demagnetizations) were able to detect the presence of hematite and goethite, and possibly magnetite. More refined techniques used to investigate the magnetic mineralogy, like X-ray diffraction, Mössbauer spectroscopy and hysteresis loops confirmed those observations, and gave additional information on the mineralogical compositon of the studied samples. Particularly, the hysteresis loops showed a wasp-waisted shape (Fig. 5) due to the presence of a fast-saturating component (possibly magnetite) superimposed to an unsaturated one (hematite and goethite) [9]. In the Mössbauer spectra we see a low hyperfine field if compared to synthetic samples, suggesting isomorfic substitution (Al, Ti) or defects on lattice [12].

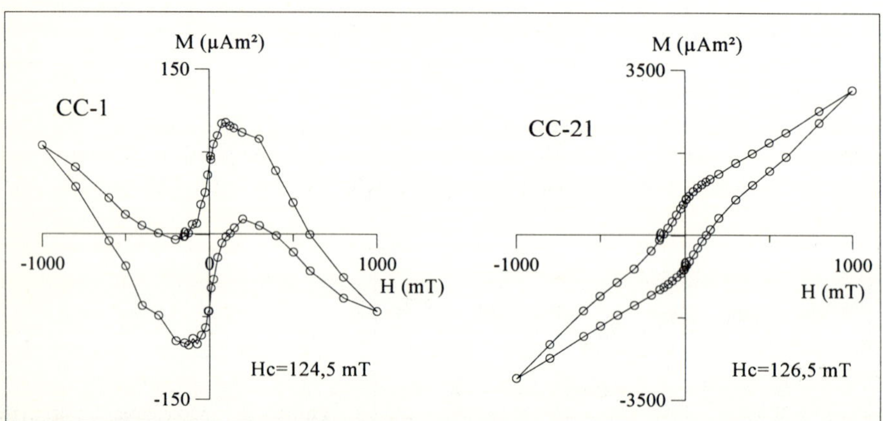

Figure 5. Hysteresis loops of the Caacupé Group. Sample CC-1 is weakly magnetic and the loop is influenced by diamagnetic minerals; for sample CC-21 the loop is dominated by high coercivity minerals.

From the paleomagnetic results and the magnetic mineralogy investigations it is possible to conclude that *i)* goethite, which is a typical product of weathering, carries a remanence component correspondent to the present geomagnetic field; *ii)* hematite, is always present and, in the Caacupé samples, is responsible by two components of magnetization in the samples, one of them related to the present field. The other component was identified also in samples with traces of magnetite, indicating that the age of this remanent component can be contemporaneous with the sediment formation (Early Paleozoic). However, this hypothesis needs further investigation relying mainly on the identification of the magetization direction carried by magnetite, and whether at least part of the hematite can be of depositional origin. In this way, scanning electronic microscopy (SEM) observations are already in progress.

4. References

[1] L. Tauxe, D. V. Kent, N. D. Opdyke, Earth Planet. Sci. Letters **47**, (1980) 279.
[2] T. R. Walker, E. E. Larson, R. P. Hoblitt, J. Geophys. Res. **86**, (1981) 317.
[3] C. M. Conti, A. E. Rapalini, J. F. Vilas, Geophys. J. Int. **121**, (1995) 848.
[4] D. W. Collinson, Methods in rock magnetism and palaeomagnetism: techniques and instrumentation. Chapman and Hall Ltd., New York. 503 pp, (1983).
[5] M. J. Dekkers, Geophys. J. Int. **97**, (1989) 323.
[6] W. O'Reilly, Rock and mineral magnetism. Blackie, Glasgow. 220pp, (1984).
[7] E. Murad, Properties and behavior of iron oxides as determined by Mössbauer spectroscopy. p. 309-350. *In* W. Stuckt et al (eds.), Iron in soil and clay minerals. D. Reidel publishing company, (1988).
[8] J. B. Forsyth, J. G. Hedley, C. E. Jonhson, J. Phys. C, 1, (1968) 179.
[9] W. Künding, H. Bömmel, G. Constalaris, R. H. Lindquist, Phys. Rev. **142**, (1966) 327.
[10] R. Day, M., Fuller, V. A. Schimidt, Phys. Earth Planet. Inter. **13,** (1977) 260.
[11] W. Williams, D. J. Dunlop, J. Geophys. Res. **100**, (1995) 3859.
[12] R. E. Vandenberghe, E. De Grave, C. Landuydt, L. H. Bowen, Hyp. Int. **53**, (1990) 175.

Materials Science Forum Vols. 302-303 (1999) pp. 479-483

Magnetic Relaxation Studies in Ferronematics

S.M. Shibli[1] and D. Walton[2]

[1] Instituto de Física, Universidade de São Paulo,
Caixa Postal 66318, 05315-970, São Paulo, SP, Brazil

[2] Physics Dept., McMaster University Hamilton, Ontario, Canada L8S 4M1

Keywords: Liquid Crystals, Ferrofluids, Ferronematics, Magnetic Susceptibility

ABSTRACT

The change in magnetization with time of a variety of ferronematics (which are lyotropic nematic liquid crystals with containing a small amount of ferrofluid) is reported. The relaxation of the magnetization follows a logarithmic law and the magnetic viscosity, S, depends on the ferrofluid concentration.

1. INTRODUCTION

Magnetic relaxation phenomena have been studied in several ferrofluids [1]. In this work we are interested in performing a similar study in another complex fluid, called ferronematics. Lyotropic nematic liquid crystals in the presence of a magnetic field tend to become oriented. The degree of orientation depends on the temperature and composition conditions. We report the results of such measurements on plain liquid crystals and ferronematics (FN's),[2] which are lyotropic nematic liquid crystals doped with small amounts of ferrofluids. This material is an anisotropic liquid crystalline phase described in details elsewhere, [3] with the addition of a conventional water-based magnetic ferrofluid. A strong orientational coupling exists between the disperse phase (ferroparticles) and the liquid crystalline matrix. Thus, doping a liquid crystal with dispersed ferromagnetic particles enables us to easily control, through an application of rather weak magnetic field, the orientational state of FN's. It has already been established [4] that the molecules in these anisotropic materials tend to align either parallel or perpendicular to the applied magnetic field. A magnetic field applied parallel to the director induces a certain amount of magnetization, while a magnetic field applied perpendicular to the director induces a different amount of magnetization.[4] In this work we are studying the case where the magnetic field is applied parallel to the director due to experimental apparatus limitations.

The lyotropic liquid crystals employed are a mixture of potassium laurate /1−decanol

/water ($KL/DeOH/H_2O$), with the concentration in weight percent of 28.50/7.20/64.30, respectively. The ferrofluid added to this material consists of nanoparticles of magnetite (Fe_3O_4) suspended in water, acquired at Ferrofluidcs Co. The concentration, in weight percent, of ferrofluid added to the liquid crystal varied from 0 to 10.0. The samples were encapsulated in a pyrex sample holder with diameter of 1.0 cm thick. An either variable or fixed magnetic field induced the molecules main axis to be parallel to them. The measurements were performed at several different temperatures, between 276 and 306 K, so the phase analysed varied from discotic to calamitic nematic ones, which was verified and confirmed by direct observation with a polarizing microscope. The measurements were performed in a superconducting quantum-interference device (SQUID) magnetometer utilizing the SHE Corp. (San Diego) model 30 electronic system.

The magnetic susceptibility measurements were analysed as a function of time. The evolution of both magnetic susceptibility and relaxation was observed for a plain lyotropic liquid crystal (Fig.1) and a ferronematic sample with ferrofluid concentration of 1.0 % by weight (Fig.2).

Comparing these two figures we observed that it was necessary to apply a magnetic field 30 times higher, in the case of the plain material, for obtaining values of magnetization similar, in magnitude, to those of the doped ones. Varying the magnetic field we have observed that the magnetic effect over these samples increases with the increase of ferrofluid concentrations. Furthermore, at lower fields ($H < 200$ G) the magnetic susceptibility shows higher variation with time, which denotes that the ferrofluid is not saturating in this specific field range. At higher magnetic fields ($H > 10{,}000$G) the magnetic saturation is observed. The highest doped sample (10 wt %) shows the highest magnetic susceptibility and also the most stable data throughout the time range.

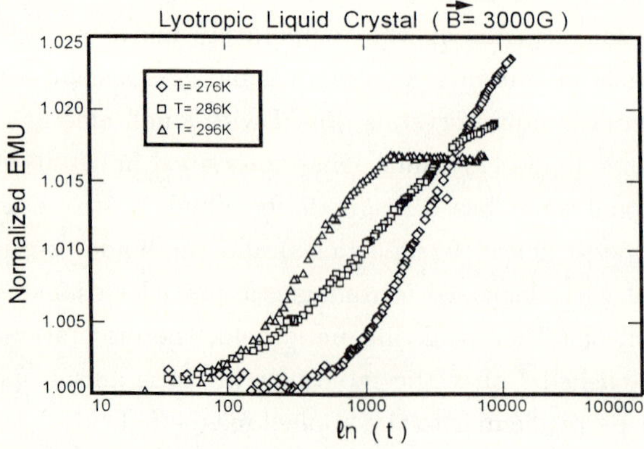

Figure 1: Normalized magnetic susceptibility measurements for plain lyotropic liquid crystal. The same

sample was measured at three different temperatures.

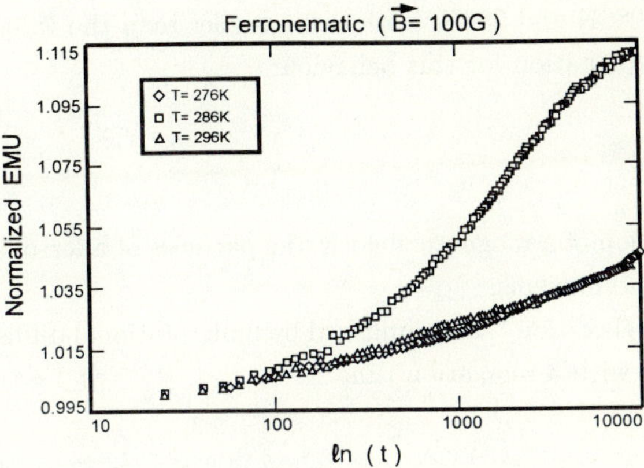

Figure 2: Normalized magnetic susceptibility measurements for ferronematic sample. The same sample was measured at three different temperatures.

Measurements of magnetization with different temperatures were performed and are shown in Figure 3. Observing these results we have verified a maximum magnetic susceptibility value in T= 286 K, which is the region where a biaxial nematic phase is measured.

FERRONEMATIC

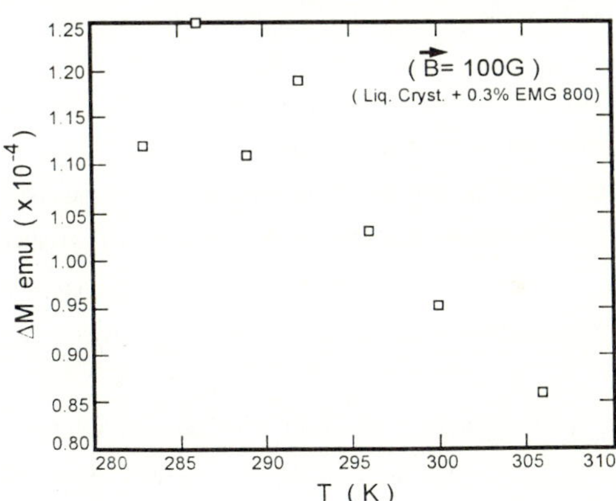

Figure 3: Magnetic susceptibility data for a ferronematic sample measured at different temperatures.

We have also verified a striking change at the material viscosity, which increased as the magnetic field and ferrofluid concentration increased.

The most interesting feature of the data is the fact that the relaxation rate, $S = \frac{\partial M}{\partial lnt}$ is lower at both 276K and 296K than it is at 286K. However the initial values of M are roughly equal at 286K and 296K, both being higher than the value at 276K. The following is a qualitative explanation for this behaviour.

2. THEORY

After the application of a magnetic field H the particles of a ferrofluid approach equilibrium by two different mechanisms:

(1) Brownian relaxation [5], accomplished by bulk rotational diffusion of the particle within the carrier liquid, with a relaxation time

$$\tau_B = 3V'\eta/kT \qquad (1)$$

where V' is the hydrodynamic volume of the particle including the coat of surfactant, η the dynamic viscosity of the carrier liquid, k is the Boltzmann's constant and T is the absolute temperature, and

(2) Neèl relaxation [6], i.e. rotation of the internal magnetisation vector inside the particle, with a time constant of

$$\tau_N = \tau_0 exp(\delta E/kT) \qquad (2)$$

where τ_0 is usually approximated as 10^{-9} s, and

$$\delta E = KV(1-h)^2 \qquad (3)$$

is the energy barrier assuming uniaxial noninteracting particles. $h = H/H_k$ is the reduced field and H_k is the internal field due to anisotropy.

In real ferrofluids there is always a distribution of energy barriers due to a distribution of the magnetic anisotropy constants K (due to crystalline and predominantly shape anisotropy) as well as of the volumes of the magnetic particle cores V. Particles with relaxation times faster than the measurement time (a common value is 100s) are called superparamagnetic. [7-9]

Particles that do not relax by the Neèl mechanism within the measurement period are called blocked particles and do not contribute to the magnetisation change within the laboratory time. These particles relax by the Brownian process, with a time constant τ_B

$$M(t) = M(t_0)\exp(-t/\tau_B) \qquad (4)$$

However, if the relaxation by the Neèl mechanism is complete in a time shorter than a laboratory time, the only relaxation left to take place is due to Brownian motion, the relaxation time is once again τ_0, and the above equation applies except that now $M(t_0)$ will be larger due to the fact that the moments of the particles have reached their equilibrium distribution following the Neèl relaxation process.

At an intermediate temperature both processes are taking place, and the relaxation rate represents some combination of τ_N and τ_B.

In summary, for the above scenario to be correct, the blocking temperature of the magnetic grains must be in the vicinity of 286 K. The anisotropy field is mainly a shape anisotropy, and from the shapes of the grains we deduce a demagnetizing factor, from which we can deduce a blocking temperature.

REFERENCES

1. X.X. Zhang, R. Ziolo, E.C. Kroll, X. Bohigas and J. Tejada, J.Mag.Mag.Mater. 140(1995),p. 1853.

2. S.V. Burylov and Y.L. Raikher, Mat.Sci.and Engineering C2(1995), p. 235.

3. S.M. Shibli, A.L.L. Dantas and D. Walton, Appl.Phys.Lett. 72(1998),p. 674.

4. P.J. Collings, "Liquid Crystals", Oxford (1947).

5. P. Debye, Polar Molecules (Chemical Catalog Company, NY, 1929).

6. L. Nel, Ann. Geophys. 5(1949),p. 99.

7. M.I. Shliomis and Y.L. Raiker, IEEE Trans.Magn. 16(1980),p. 237.

8. J. Tejada, X.X. Zhang and E.M. Chudnovsky, Phys.Rev. B 47(1993),p. 14977, and references therein.

9. B. Barbara et al., J.Appl.Phys. 73(1993),p. 6703.

Materials Science Forum Vols. 302-303 (1999) pp. 484-488

Magnetic Properties, Hyperfine Interactions and Thermal Stability of Perovskite-Type Fe-Ni Nitrides γ-$(Fe_{1-x}Ni_x)_4N$ ($0 \le x \le 0.9$)

X.G. Diao[1], A.Y. Takeuchi[1], F. Garcia[1], I.S. Oliveira[1], A.P. Guimarães[1], H.R. Rechenberg[2] and R.B. Scorzelli[1]

[1] Centro Brasileiro de Pesquisas Físicas, 22290-970, Rio de Janeiro, Brazil

[2] Instituto de Física, Universidade de São Paulo, C.P. 66318, 05315-970, São Paulo, Brazil

Keywords: Fe-Ni Nitrides, Magnetic Properties, Hyperfine Field, Thermal Stability

Abstract:

Magnetic properties, hyperfine interactions and thermal stability of the perovskite-type Fe-Ni nitrides γ-$(Fe_{1-x}Ni_x)_4N$ have been investigated by vibrating sample magnetometer, nuclear magnetic resonance and thermogravimetric analysis combined with X-ray diffraction and Mössbauer spectroscopy. The composition dependence of lattice parameter, saturation magnetization and Curie temperature has been found to decrease monotonically with increasing Ni concentration. This behavior contrasts with that of the corresponding (fcc) γ-$Fe_{1-x}Ni_x$ alloys, for which each of these properties exhibits a maximum at some intermediate composition. The thermal decomposition temperature also decreases with x; the Fe-Ni nitrides, however, show a higher resistance against oxidation on heating the pure Fe nitride. Good agreement was observed between the hyperfine fields, which are greatly influenced by Ni substitution, by NMR and by Mössbauer spectroscopy.

1. Introduction

The cubic iron nitride γ-Fe_4N is a potentially interesting material for high-density magnetic recording [1-4]. The average magnetic moment per iron atom in this perovskite iron nitride is close to that observed in pure α-iron [2] but γ-Fe_4N has a larger corrosion and wear resistance than pure iron [3,4]. On the other hand, the crystal structure of γ-Fe_4N is the face-centered cubic (fcc) γ-iron lattice with a nitrogen atom placed at the body center of the unit cell [5]. Its simple crystallographic structure makes it attractive for fundamental investigations, both theoretical and experimental.

Keeping the original crystal structure, element substitution is an effective way to improve properties of materials. Substitution for Fe in γ-Fe_4N by other transition metals such as Co, Mn, Pt, Pt, Ni, etc. [6-9] has been tried. Only the γ-$(Fe_{1-x}Ni_x)_4N$ nitrides, however, have been synthesized with a relatively high nickel concentration [10]. Although some investigations performed on γ-$(Fe_{1-x}Ni_x)_4N$ showed some interesting physical phenomena in the series of the nitrides [9-12], the understanding of some properties still remains uncomplete, especially for the compounds with higher nickel concentrations. In the present work, crystalline structure, magnetic and thermal stability properties and hyperfine interactions of perovskite-type Fe-Ni nitrides γ-$(Fe_{1-x}Ni_x)_4N$ were investigated using vibrating sample magnetometer (VSM), X-ray diffraction, Mössbauer spectroscopy, nuclear magnetic resonance (NMR) and thermogravimetric analysis.

2. Preparation of samples

Samples of single phase nitrides γ-$(Fe_{1-x}Ni_x)_4N$ ($0 \le x \le 0.75$) have been synthesized by annealing $(Fe_{1-x}Ni_x)C_2O_4 \cdot nH_2O$ oxalates, obtained by the conventional chemical coprecipitation method, in NH_3/H_2 mixture gas flow at different temperatures. The absolute gas flow, ratio of NH_3/H_2 and annealing temperature have to be adjusted according to the different nickel concentrations. The single phase samples with higher nickel concentrations turned out to be more difficult to prepare than the lower ones. For very high nickel concentrations, e. g., x=0.8 and x=0.9, γ-$(Fe_{1-x}Ni_x)_4N$ nitrides are also obtained, but with a mixture of small amounts of Fe-Ni alloys and ε-phase.

3. Results and discussion

All the samples were identified by X-ray diffraction exhibiting a peroskite-type crystalline structure similar to γ-Fe_4N. The lattice parameter of each nitride γ-$(Fe_{1-x}Ni_x)_4N$, including γ-Ni_4N [13] decreases monotonically with increasing nickel concentration.

Magnetic hysteresis loops for γ-$(Fe_{1-x}Ni_x)_4N$ nitrides measured with VSM at room temperature are shown in Fig.1. The results of saturation magnetization versus nickel

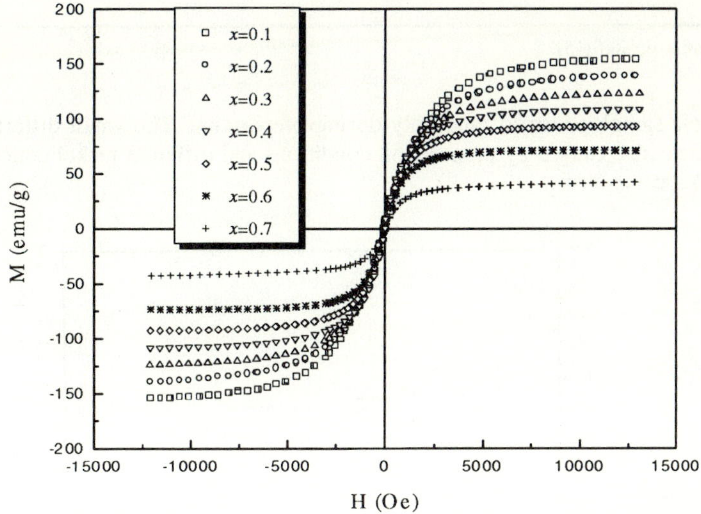

Fig.1. Magnetic hysteresis loops for γ-$(Fe_{1-x}Ni_x)_4N$ nitrides at room temperature.

concentration derived from the loops, as well as the results measured at 80K are listed in Table 1 and plotted in Fig. 2. The saturation magnetization at room temperature decreases almost linearly with increasing nickel concentration. This can be mainly attributed to the fact that nickel atoms have smaller magnetic moment than iron. At 80K, the saturation magnetization also decreases monotonically with increasing nickel concentration. Nevertheless, it is reasonable that the saturation magnetization does not decrease linearly since the local magnetic moment of Fe and Ni in the nitrides cannot be kept unchanged with the different nickel concentrations, as can be seen in the NMR results. The average magnetic moment per unit cell of γ-$(Fe_{1-x}Ni_x)_4N$ is calculated according to the equation

$$\mu = 1.79\times10^{-4}P_m{*}\sigma_s \tag{1}$$

where P_m is the formula weight of γ-$(Fe_{1-x}Ni_x)_4N$ (Table 1). The average magnetic moment results from the contribution of iron and nickel atoms. Both of them have different values at different nonequivalent crystalline sites and therefore we have not been able to determine the local magnetic moment of Fe and Ni atoms from the above results. We did not try to control the shape

Table 1. Saturation magnetization, average magnetic moment per formula, hyperfine field, and thermal decomposition temperature of perovskite-type γ-$(Fe_{1-x}Ni_x)_4N$ nitrides.

x	σ (emu/g)		μ	B_{hf}(T)		T_{decom}
	80K	300K	(μ_B)	FeI	FeII	(K)
0	205[a]	193[b]	8.71	36.7	23.5	973
0.1	200	168	8.54	39.6	24.5	
0.2	162	148	6.95	42.1	24.8	953
0.3	150	130	6.47			
0.4	146	114	6.32			898
0.5	114	98	4.96			
0.6	110	75	4.81			873
0.7	92	45	4.04			
0.75	77		3.39			803
0.8	68		3.00			
0.9						788

[a]Ref.[14]; [b]Ref.[15]

of the particle samples formed randomly during preparation. The small difference of the shapes and the particle size caused by preparation conditions and different nickel concentration results in almost no shape anisotropy.

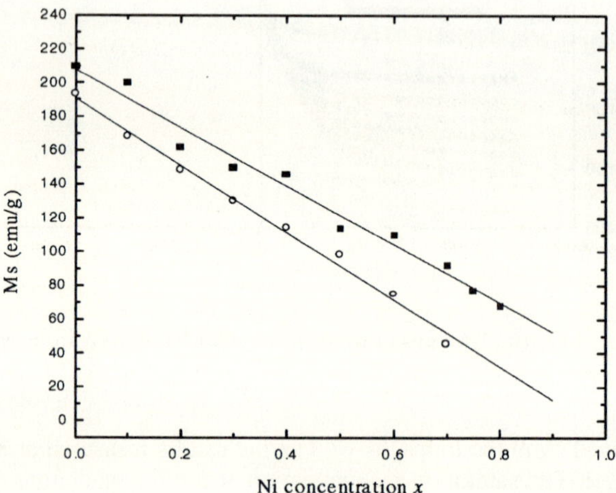

Fig. 2. Saturation magnetization of γ-$(Fe_{1-x}Ni_x)_4N$ nitrides versus nickel concentration at 80K and 300K.

The Curie temperatures of the nitrides were also measured with VSM. The results indicate that the Curie temperature also decreases monotonically with the increase of nickel concentration reaching values below the room temperature for nickel concentrations larger than 70at.% Ni. Combining the above results, one can see that both saturation magnetization and Curie temperature decrease with increasing Ni concentration in the perovskite-type Fe-Ni nitrides

γ'-$(Fe_{1-x}Ni_x)_4N$. This behavior contrasts with that of the corresponding (fcc) γ-$Fe_{1-x}Ni_x$ alloys, for which each of these properties exhibits a maximum at some intermediate composition.

NMR were performed on all the γ'-$(Fe_{1-x}Ni_x)_4N$ samples at 4.2K but only samples with x=0, 0.1 and 0.2 were investigated, due to severe loss of signal intensity for increasing Ni concentration. The ^{57}Fe NMR spectra of these three nitrides are shown in Fig.3. A strong line broadening is

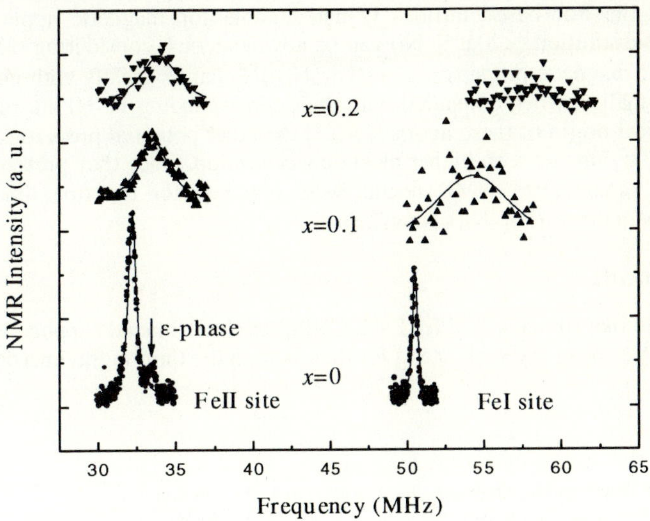

Fig. 3. ^{57}Fe NMR spectra of γ'-$(Fe_{1-x}Ni_x)_4N$ at 4.2K.

apparent, which is probably attributed to disorder due to Ni substitution. The frequency shift for increasing x shows a change of hyperfine field with Ni substitution. The results of hyperfine field obtained from the NMR spectra fits are listed in Table 1. Except for the scattered experimental points for FeI site for x=0.2, all the other peaks corresponding to the individual crystalline sites are well-defined enough to determine the hyperfine field. The NMR results showed for FeI and FeII sites an increase in hyperfine field per nickel concentration in the range 0-20%. Good agreement was observed between the hyperfine fields measured by NMR and Mössbauer spectroscopy. A little amount of ε-$Fe_{3\pm y}N$ impurity in γ'-Fe_4N sample was obsearved by NMR and its hyperfine field is a little larger than FeII site of γ'-Fe_4N. The rapid decrease in NMR signal with Ni concentration at the FeI site arises mainly from a rapid decrease of iron atom number, caused by the priority of Ni occupation at this site.

By thermogravimetric analysis, the thermal decomposition temperatures, T_{decom}, of γ'-$(Fe_{1-x}Ni_x)_4N$ nitrides have been determined and are also listed in Table 1. The measurement was performed in argon gas flow when heating the nitride samples. Even so, γ'-Fe_4N has been oxidized when heated above 250°C. In the Ni-substituted samples, however, the resistance to oxidation is enhanced. The results show that the decomposition temperature of γ'-$(Fe_{1-x}Ni_x)_4N$ decreases with increasing nickel concentration, especially for the higher nickel concentration, above 75at.% Ni, for which T_{decom} goes down to ~500°C, which is the typical temperature used for nitrogenation of iron-based alloys. Formation and decomposition of the nitrides occur simultaneously during the preparation, which explains why the single phase nitrides γ'-$(Fe_{1-x}Ni_x)_4N$ with higher nickel concentration are usually difficult to prepare.

4. Conclusion

In summary, we have studied in this work the magnetic properties, hyperfine interactions and thermal stability of perovskite-type Fe-Ni nitrides γ-$(Fe_{1-x}Ni_x)_4N$ in a quite large region of nickel concentration. Although the stability against oxidation has been enhanced by substitution of nickel for iron in the nitrides, their macro-magnetic properties and thermal stability have deteriorated with the increasing nickel concentration. Therefore, a large proportion of nickel substitution in the iron-based nitrides is not suitable for magnetic applications, but a small percentage of substitution (<20at.% Ni) can be advantageous, considering the ensemble of useful parameters. The magnetic properties of γ-$(Fe_{1-x}Ni_x)_4N$ change greatly with nickel substitution and have some anomalies, when compared with their corresponding Fe-Ni alloys. It is worthwhile to study the physical origin of these anomalies and the other potential properties versus temperature, especially for the samples with higher nickel concentration, since they present Curie temperatures close to room temperature. NMR results were obtained for the first time to investigate the hyperfine fields in the γ-$(Fe_{1-x}Ni_x)_4N$ nitrides.

Acknowledgements

We would like to thank FAPERJ and CNPq for their finacial support. We are also indebted to F. C. Rizzo Assunção and C. L. Zeng for their help in the thermogravimetric measurements.

References:

[1] C. Gao, W. D. Doyle and M. Shamsuzzoha, J. Appl. Phys., 73 (1993) p.6579.
[2] S. Matar, G. Demazeau and B. Siberhicot, IEEE Trans. Magn., MAG-26 (1990) p.60.
[3] K. Umeda, E. Kita and A. Tasaki and A. Tasaki, IEEE Trans. Magn., MAG-22 (1986) p.591.
[4] M. Kume, T. Tsujioka, M. Matsuura and Y. Abe, IEEE Trans. Magn., Mag-23 (1987) p.3633.
[5] K. H. Jack, Proc. Roy. Soc., London, A. 208 (1951) p.216.
[6] S. Matar, I. Fournas, S. Cherubin-Jeannette and G. Demazeau, Eur. J. Solid State Inorg. Chem., 30 (1993) p.871.
[7] B. Siberchicot, S. Matar, L. Fournes, G. Demazeau and P. Hagenmuller, J. Solid State Chem., 84 (1990) p.10.
[8] C. Cordier-Robert and J. Foct, Eur. J. Solid State Inorg. Chem., 29 (1992) p.39.
[9] G. W. Wiener and J. A. Berger, J. Metal, 7 (1955) p.360.
[10] X. G. Diao, F. S. Li, Z. J. Zhao and S. Q. Zhou, J. Mat. Sci. Lett., 15 (1996) p.1590.
[11] G. Shirane, W. J. Takei and S. L. Ruby, Phys. Rev., 126 (1962) p.49.
[12] P. Mohn, K. Schwarz, S. Matar and G. Demazeau, Phys. Rev., B45 (1992) p.4000.
[13] R. Baranova, Yu. Khodyrev and S. Semiletov, Sov. Phys. Crystallogr., 27 (1982) p.554.
[14] Qinian Qi, K. O'Donnell, E. Touchais and J. M. D. Coey, Hyperfine Interactions, 94 (1994) p.2067.
[15] K. Tagawa, E. Kita, A. Tasaki, Jpn. J. Appl. Phys., 21 (1982) p.1596.

Materials Science Forum Vols. 302-303 (1999) pp. 489-492

Study of the Mean Hyperfine field of Fe-Ni Alloys by CEMS

B. Cruz, J.A. Tabares and G.A. Pérez Alcázar

Departamento de Física, Universidad del Valle, Cali, Colombia, A.A. 25360
Fax: (57)339339

Keywords: Fe-Ni Alloys, CEMS, Hyperfine Fields

Abstract. A study by Conversion Electron Mössbauer Spectroscopy (CEMS) carried out by using a Parallel Plate Avalanche Counter with samples of Fe-Ni alloys (25 to 50 at.% Ni, steps 2.5), including the invar composition (35 at.% Ni) is reported. All CEMS measurements were carried out at room temperature. In all samples (25 to 50 at.% Ni), the Mössbauer spectra showed a six line magnetic spectrum according to their ferromagnetic character, with a broad Hyperfine Field Distribution (HFD), according to the disordered character of the alloys. The obtained Mean Hyperfine Fields (MHF) agree with values previously reported by Transmission Mössbauer Spectroscopy (TMS), specially the minimum in the MHF value exhibited at the invar composition (35 at.% Ni). Results for treated samples (with oxidation at 200 °C), close to the invar composition are also included. The spectra presented a high relative area of ferromagnetic oxides (α-Fe_2O_3 and Fe_3O_4). The formation of the magnetic oxides at this relatively low temperature is not absolutely clear, but apparently is related to the singular behaviour of the invar alloy.

1. Introduction

A great deal of experimental work has been done on Fe-Ni alloys. In particular, much of the work reported in the literature is devoted to Invar alloys concerning the abnormally large magnetovolume and magnetoelastic effect [1,2]. Mössbauer spectroscopy is one of the useful techniques in determining magnetic properties of the Fe-Ni alloys. The first systematic study of the Ni content dependence of the Mean Hyperfine Field (MHF) was carried out by Johnson et al. [3], that reveals the abrupt deviation of the mean magnetic moment per atom in the Slater-Pauling curve at C_{Ni} 35 % at. Ni, where the Invar alloy occurs. Experimental results, by using Mössbauer spectroscopy in the transmision mode (TMS), of disordered $Fe_{x-1}Ni_x$, $0.05<x<0.95$ alloys, were reported by Restrepo et al. [4], that are in complete accordance with Johnson's results. The appearence of a non-magnetic phase, that corresponds to a fcc γ-phase with Ni $\leq$ 30 at.%, proposed as a low spin γ-Fe-Ni phase, was reported by Scorzelli [5]. Other reported results [6] allow to ascribe the mentioned phase to surface effects induced by mechanical work.

In this work we report a study by Conversion Electron Mössbauer Spectroscopy of disordered Fe-Ni alloys in the composition range 25 to 50 % at. Ni, in 2.5 steps, that includes the Invar composition. The Hyperfine Field Distribution (HFD) of each sample and the products for the Invar sample oxidized at 200 °C in oxygen atmosphere are determined and the Slater-Pauling curve presented.

2. Experimental procedure

Alloys were prepared in the composition range 25 to 50 at.% Ni, in 2.5 steps, by using pure (99.9%) Fe and Ni powders. After weighing the nominal compositions, the samples were melted in an arc furnace in Ar atmosphere and then sealed in quartz tubes previously evacuated at high vacuum. A thermal treatment at 1000°C for a week was made and then rapidly quenched in ice water in order to retain the high temperature disordered phase. Disks of 14 mm diameter were machined, and polished using silicon carbide papers and fine abrasive powders. Finally, they were cleaned with acetone in an ultrasonic bath. For Invar composition oxidation treatment an oven consisting of a cylindrical heater with a long quartz tube in its axis, and inner diameter 25 mm, was used. Two rubber bottle caps, each one with a hole for the oxygen inlet and outlet, were put at the ends of the quartz tube. A flexible hose (5 mm of diameter) introduced into a glass containing water, 40 mm below the water level and counting the number of bubbles per minute, was used to measure and control the oxygen flux. The Invar sample was exposed to the oxygen atmosphere with a flux of ~100 bubbles per minute at 200 °C during 8 hours. After this, the oxygen flux was interrupted and the oven turned off. The Conversion Electron Mössbauer Spectra (CEMS) were carried out using a Parallel Plate Avalanche Counter[7] and a Mössbauer spectrometer in the constant acceleration mode. All spectra were fitted using the Normos program [8]. In each case, and according to the disordered character of the samples, two hyperfine distributions and some additional sites were necessary. All Mössbauer parameters were referred to α-Fe.

3. Results and Discussion

Figure 1 shows the recorded Mössbauer spectra for 25 to 32.5 at.% Ni. The corresponding HFD are

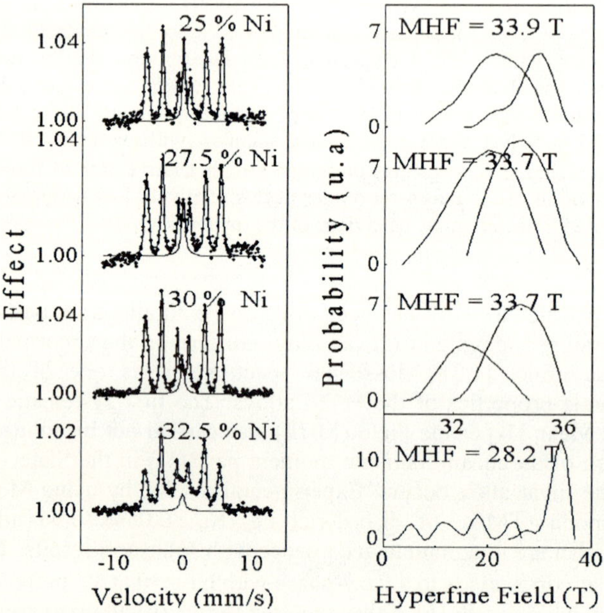

Fig. 1. Mössbauer spectra for 25 to 32.5 at.% Ni and hyperfine field distribution (HFD).

also included. It shows six line magnetic spectra according to their ferromagnetic character, with a narrow hyperfine field distribution (HFD), around 33.5 T, corresponding to a Fe-rich concentration. The fitted Mean Hyperfine Fields are in good accordance with Restrepo et al.[4] TMS results. Although, and contrary to the mentioned previous results, a single line, with relative area between 16 % for 25 at.% Ni to 6 % for 32.5 at.% Ni, corresponding to a fcc paramagnetic phase, is present. Remembering that this composition range presents a bcc structure phase [2] the presence of this fcc phase is not expected. The fitting parameters and some previous results on surface analysis [9] allow us to consider that this corresponds to austenite, maybe due to the metastable character of the alloy in the neighborhood of the Invar composition or to surface structural changes induced by mechanical work. There seems to be agreement that the mean reason of the presence of this fcc phase at compositions corresponding to bcc phase is the sample preparation and the heat treatment [5,6].

Figure 2 corresponds to the Mössbauer spectra for the 35 to 50 at.% Ni and the respective HFD's. Like Figure 1 the spectra consist on six line ferromagnetic spectra, but a broad hyperfine field distribution (HFD) is present in this composition range, related to the disordered character of the samples. Also, the fitted Mean Hyperfine Fields are in good accordance with Restrepo et al.[4] TMS results, specially the singular low value for the MHF at Invar composition. It is important to note that the previously discussed single line is absent in this range of compositions, which is expected because of their fcc structure [2].

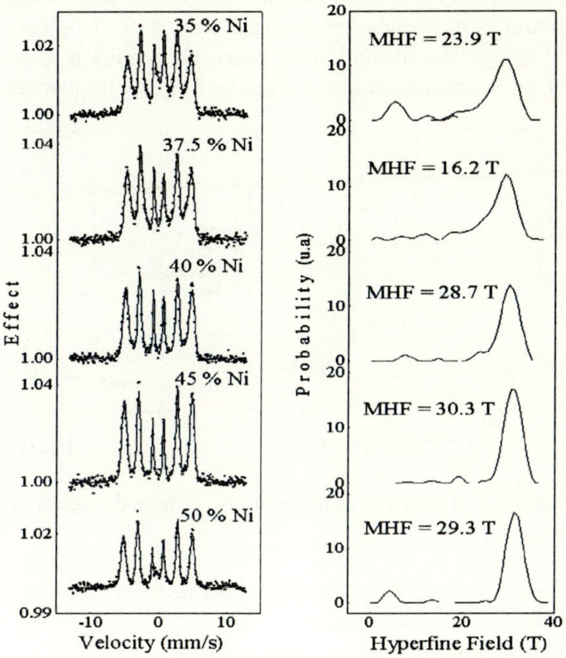

Fig.2. Mössbauer spectra for the 35 to 50 at.% Ni and the respective HFD's.

Figure 3 shows the experimental Mean Hyperfine Field dependence on the Ni concentration. Below Invar composition, in the Fe-rich region, H presents a curve with near zero variation, then suddenly decreases from 33.7 T to 23.9 T for Invar concentration (35 at.% Ni) and then it increases up to a maximun value of 30.3 T for the concentration 45 at.% Ni. For the concentration 50 at.% Ni, a lower value (29.3 T) is determined according with a gradual decrease down to 26.1 T for the pure Ni value reported by Restrepo's et al.[4]. Thus, our experimental results of the Ni content dependence of the MHF are in good agreement with Johnson's et al. [2] indicating that CEMS is a good technique in the determination of the magnetic properties of the Fe-Ni alloys.

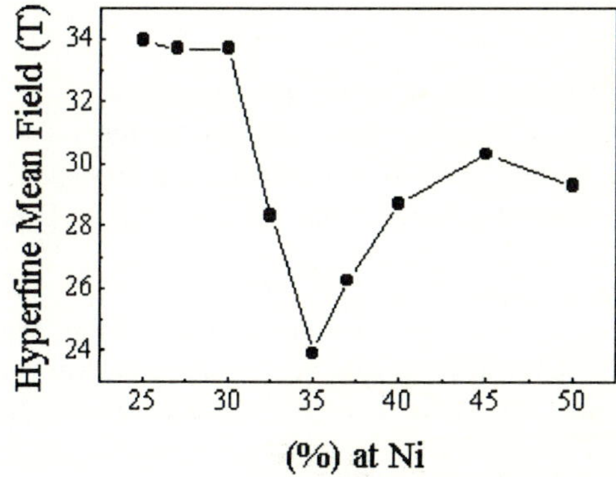

Fig. 3. Experimental mean hyperfine field dependence on the Ni concentration.

Figure 4 shows the spectra of the Invar concentration before the oxidizing treatment (a) and after the oxidizing treatment (b). A clear difference is noticed between these spectra. The latter presents a well defined sextet over a wide and low component with a MHF of 24.0 T maybe corresponding to the substrate. The sextet corresponds to ferromagnetic oxides (α-Fe_2O_3 and Fe_3O_4) whose relative area is ~ 85.0%. Although the formation of magnetic oxides is a consequence of the oxidizing treatment their occurrence is not clear at the employed relatively low temperature.

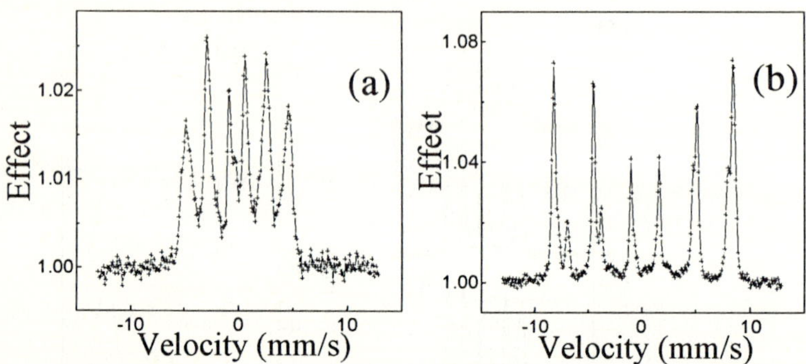

Fig. 4. Mössbauer spectra of the Invar concentration before the oxidizing treatment (a) and after the oxidizing treatment (b).

But in advance, this seems to be a technical process to form Hematite and Magnetite on surfaces, that could be used in magnetic store media.
It follows from the present work that the results obtained by CEMS agree with TMS ones but the former seems to be better in determining the fcc paramagnetic phase present when the sample had been subjected to a surface treatment by mechanical work or irradiated with particles or some class of radiation [5,10]. Likewise, the structural unstability at the Invar composition may be the main reason for the production of magnetic oxides at a relative low temperature.

4. Acknowledgement

This work was supported by COLCIENCIAS, Colombian Agency for Science Development , and the ALAMED Program of the European Community. One of us (JAT) express his acknowledgement to Mössbauer Spectroscopy Group at the IQFR (CSIC) Madrid Spain.

References

[1] Y. Nakamura. IEEE Trans. Magn. MAG-12. 278 (1976).
[2] A.Z. Menshikov. J. Magn. Magn. Mater. 10, 205 (1979).
[3] C.E. Johnson, M.S. Ridout and T.F. Cranshaw. Proc. Phy. Soc. 81, 1079 (1963).
[4] J. Restrepo, L. E. Zamora, G.A. Pérez-Alcázar and A. Bohórquez to be published in Hyp. Interact.
[5] R.B. Scorzelli, Hyp. Interact. 110, 143 (1997)
[6] J.Y. Ping, D.G. Rancourt and R.A. Dunlap, J. Magn. Magn. Mater. 103, 286 (1992).
[7] J.R. Gancedo and M. Gracia, Hyp. Interact. 29, 1097 (1986)
[8] R.A. Brand, Nucl. Instrum. Methods Phys. Res. B 28, (1987), and Nucl. Instrum. Methods Phys. Res. B 28, 417 (1987)
[9] G. Principi, in Mössbauer Spectroscopy Applied to Magnetism and Materials Science, G.J. Long and F. Grandjean, eds., p. 33, Plenum Press, New York (1993)
[10] R.H. Tendler, C. Rodriguez, L.J. Gallego and J.A. Alonso, J. Mater. Sci. 31, 6395 (1996).

AUTHOR INDEX

KEYWORD INDEX

Advanced Reading

II-VI Compounds and Semimagnetic Semiconductors

Eds. H. Heinrich and J.B. Mullin

Joint Proc. of the 3rd European Workshop on II-VI Compounds and the 4th Intl. Workshop on Semimagnetic (Diluted Magnetic) Semiconductors, both held in Linz, Austria 1994

Materials Science Forum Vols. 182-184

ISBN 0-87849-689-0
1995, 830 pp, CHF 320.00 / US$ 266.00

The main emphasis of this volume is on the fundamental physics of the II-VI compounds focusing on the properties of epitaxial layers, heterostructures, multilayers and superlattices. Main topics are doping of wide gap II-VI compounds, quantum wires and quantum dots, exitonic gain as well as basic processes in lasers including the microgun laser. New aspects covered are dephasing phenomena.

As coherent femtosecond spectroscopy has revealed that semimagnetic (diluted magnetic) semiconductors, DMS, are materials of highest scientific as well as technological interest, a comprehensive treatment of the scientific and technological „state of the art" has been felt to be very timely. With 88 papers on DMS, the present publication excellently fills that gap.

Contents Sections:

- Doping of II-VI Compounds
- Structured Materials, Quantum Wires and Dots
- Fundamental Properties and Characterization
- Basic Processes in Devices
- Diode Lasers
- Developments in DMS
- Growth
- Spin Orientation Studies
- Mn-Based II-VI, IV-VI and III-V DMS
- Excitonic Effects
- Magnetic Polarons and Polaron Dynamics
- Interfaces and Superlattices
- Tunneling, New Exchange Effects
- Quasilocalization and Mesoscopic Effects
- Characterization and Related Studies

Detailed information on this title – including the complete table of contents – is available on the internet at http://www.ttp.net or through TTP's E-Mail Preview Service. For further information, please send an e-mail to 'preview@ttp.net' with the word 'help' as the body of the message.

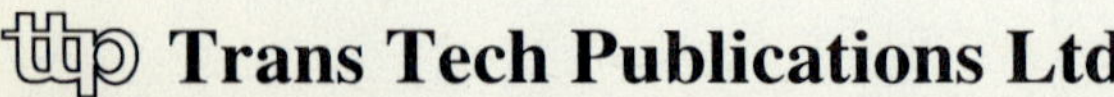

Brandrain 6
CH-8707 Uetikon-Zuerich
Switzerland

Fax: +41 (1) 922 10 33
e-mail: ttp@ttp.net
Web: http://www.ttp.net